Multiplication Patterns

$(a + b)^2 = a^2 + 2ab + b^2$

$(a - b)^2 = a^2 - 2ab + b^2$

$(a + b)(a - b) = a^2 - b^2$

$(a + b)^3 = a^3 + 3a^2b + 3ab^2 + b^3$

$(a - b)^3 = a^3 - 3a^2b + 3ab^2 - b^3$

$(a + b)^n = \binom{n}{0}a^n + \binom{n}{1}a^{n-1}b + \binom{n}{2}a^{n-2}b^2 + \cdots + \binom{n}{n}b^n$

Properties of Exponents and Radicals

$b^n \cdot b^m = b^{n+m}$ $\qquad \left(\dfrac{b^n}{b^m}\right) = b^{n-m}$

$(b^n)^m = b^{mn}$

$(ab)^n = a^n b^n$ $\qquad \sqrt[n]{ab} = \sqrt[n]{a}\sqrt[n]{b}$

$\left(\dfrac{a}{b}\right)^n = \dfrac{a^n}{b^n}$ $\qquad \sqrt[n]{\dfrac{a}{b}} = \dfrac{\sqrt[n]{a}}{\sqrt[n]{b}}$

Equations Determining Functions

Linear function: $\qquad f(x) = ax + b$

Quadratic function: $\qquad f(x) = ax^2 + bx + c$

Polynomial function: $\qquad f(x) = a_n x^n + a_{n-1}x^{n-1} + \cdots + a_1 x + a_0$

Rational function: $\qquad f(x) = \dfrac{g(x)}{h(x)}$, where g and h are polynomials

Exponential function: $\qquad f(x) = b^x$, where $b > 0$ and $b \neq 1$

Logarithmic function: $\qquad f(x) = \log_b x$, where $b > 0$ and $b \neq 1$

Properties of Absolute Value

$|a| \geq 0$

$|a| = |-a|$

$|a - b| = |b - a|$

$|a^2| = |a|^2 = a^2$

Properties of Logarithms

$\log_b b = 1$

$\log_b 1 = 0$

$\log_b rs = \log_b r + \log_b s$

$\log_b \left(\dfrac{r}{s}\right) = \log_b r - \log_b s$

$\log_b r^P = p(\log_b r)$

Factoring Patterns

$a^2 - b^2 = (a + b)(a - b)$

$a^3 - b^3 = (a - b)(a^2 + ab + b^2)$

$a^3 + b^3 = (a + b)(a^2 - ab + b^2)$

Set Notation

$\{x \mid x > a\}$

$\{x \mid x < b\}$

$\{x \mid a < x < b\}$

$\{x \mid x \geq a\}$

$\{x \mid x \leq b\}$

$\{x \mid a < x \leq b\}$

$\{x \mid a \leq x < b\}$

$\{x \mid a \leq x \leq b\}$

Interval Notation

(a, ∞)

$(-\infty, b)$

(a, b)

$[a, \infty)$

$(-\infty, b]$

$(a, b]$

$[a, b)$

$[a, b]$

Algebra for College Students

Algebra for College Students

FIFTH EDITION

Jerome E. Kaufmann

PWS Publishing Company

I(T)P An International Thomson Publishing Company

Boston • Albany • Bonn • Cincinnati • Detroit • London • Madrid • Melbourne • Mexico City
New York • Paris • San Francisco • Singapore • Tokyo • Toronto • Washington

PWS PUBLISHING COMPANY
20 Park Plaza, Boston, Massachusetts 02116-4324

International Thomson Publishing
The trademark ITP is used under license.

For more information, contact:

PWS Publishing Co.
20 Park Plaza
Boston, MA 02116

International Thomson Publishing Europe
Berkshire House 168-173
High Holborn
London WC1V 7AA
England

Thomas Nelson Australia
102 Dodds Street
South Melbourne, 3205
Victoria, Australia

Nelson Canada
1120 Birchmount Road
Scarborough, Ontario
Canada M1K 5G4

International Thomson Editores
Campos Eliseos 385, Piso 7
Col. Polanco
11560 México D.F., México

International Thomson Publishing GmbH
Königswinterer Strasse 418
53227 Bonn, Germany

International Thomson Publishing Asia
221 Henderson Road
#05-10 Henderson Building
Singapore 0315

International Thomson Publishing Japan
Hirakawacho Kyowa Building, 31
2-2-1 Hirakawacho
Chiyoda-ku, Tokyo 102
Japan

Library of Congress Cataloging-in-Publication Data

Kaufmann, Jerome E.
    Algebra for college students / Jerome E. Kaufmann. -- 5th ed.
       p.   cm.
    Includes index.
    ISBN 0-534-94896-0
    1. Algebra.   I. Title.
QA154.2.K36   1995
512.9--dc20
                                      95-33000
                                       CIP

Sponsoring Editor: David Dietz
Production Coordinator: Robine Andrau
Market Development Manager:
Marianne C. P. Rutter
Manufacturing Coordinator: Marcia A. Locke
Production: Susan Graham

Interior/Cover Designer: Julia Gecha
Interior Illustrator: Network Graphics
Typesetter: York Graphic Services, Inc.
Cover Photo: ©Peter Miller/The Image Bank
Cover Printer: Henry N. Sawyer Co., Inc.
Text Printer: Quebecor/Hawkins

Printed and bound in the United States of America
95 96 97 98 99—10 9 8 7 6 5 4 3 2 1

CONTENTS

4 Rational Expressions 170

5 Exponents and Radicals 226

6 **Quadratic Equations and Inequalities   282**

7 **Coordinate Geometry and Graphing Techniques   334**

14 Counting Techniques and Probability   704

PREFACE

Algebra for College Students, Fifth Edition, was written for those college students who need an algebra course to bridge the gap between elementary algebra and the more advanced courses in precalculus mathematics. The first six chapters contain intermediate algebra topics, and the last eight chapters contain a blend of intermediate and college algebra topics. All of the material is at an intermediate level.

The basic concepts of algebra are presented in a simple, straightforward manner. Algebraic ideas are developed in a logical sequence, but in an easy-to-read manner without excessive formalism. Concepts are frequently developed through examples, continuously reinforced through additional examples, and then applied in a variety of problem-solving situations.

The examples demonstrate a large variety of situations; other situations are left for students to think about in the problem sets. In the examples students are guided to organize their work and to decide when a meaningful shortcut might be used.

In the preparation of this edition, a special effort was made to incorporate improvements suggested by reviewers and by users of the earlier editions, while at the same time preserving the book's many successful features.

New in This Edition

- Problems called **Thoughts into Words** are now included in every problem set except the review exercises. These problems are designed to encourage students to express in written form their thoughts about various mathematical ideas. See, for example, Problem Sets 2.1, 3.5, 4.7, 5.5, and 10.2.

- Miscellaneous problem sections, now called **Further Investigations**, have been enhanced by the addition of more problems that lend themselves to small group work. These problems remain as "extras" but add flexibility for the instructor. See, for example, Problem Sets 1.2, 2.7, 5.6, 6.5, and 9.5.

- A **Chapter Test** has been included at the end of each chapter. Along with the **Chapter Review Problem Sets**, these practice tests should provide students with ample opportunity to prepare for the "real" tests. **Cumulative Review Problem Sets** appear at the ends of Chapters 3, 5, 7, and 12.

- The **chapter introductions** have been rewritten in an effort to provide more motivation for students to study algebra. Each introduction begins with at least one application that leads into the material of the chapter.

- **Applications** have been added in several sections, including the following:

  Sections 3.1, 3.2, and 3.3: Examples and problems that connect geometry and the study of polynomials
  Section 5.2: Applications involving radicals
  Section 5.5: Applications involving radical equations
  Section 6.2: Applications of the Pythagorean theorem
  Section 7.1: Applications of slope

- The use of a **graphing utility** is introduced in Section 7.2. Graphics calculator examples (designated by an icon) are then incorporated, as appropriate, throughout Chapters 7 through 14. These examples are written so that students without a graphing utility can read and benefit from them. For example, see Sections 7.2, 7.4, 8.3, 9.4, 9.5, and 10.5.

- A new section of problems called **Graphics Calculator Activities** has been added to many of the problem sets in Chapters 7 through 14. These activities, which are good for either individual or small group work, have been designed to reinforce concepts already presented and lay the groundwork for concepts about to be discussed. They also help students to predict shapes and locations of graphs based on earlier graphing experiences. Through working these problems, students should become more familiar with the capabilities and limitations of a graphics calculator. For example, see Problem Sets 7.2, 7.4, 8.2, 8.3, 9.4, 9.5, and 10.5.

- Parts of **Chapter 8** have been reorganized and a new Section 8.3 has been added. Section 8.2 now discusses linear and quadratic functions and applications of quadratic functions. New Section 8.3 presents transformations of some basic curves. These transformations are then used as appropriate in later sections. Section 8.4 contains the composition of functions in preparation for inverse functions presented in Section 8.5.

- A focal point of every revision is the **problem sets.** Users of the previous editions were very helpful in suggesting problems to be added, deleted, or changed in some way.

Other Special Features

- A common thread runs throughout the book: namely, *learn a skill,* next *use the skill to help solve equations and inequalities,* and then *use equations and inequalities to solve word problems.* This thread influenced some other decisions.

  1. Numerous word problems are scattered throughout the text. These problems deal with a large variety of applications and constantly show the connections between mathematics and the real world.

2. Many problem-solving suggestions are offered throughout, with special discussions in several sections. The problem-solving suggestions are demonstrated in more than 80 worked-out examples.

3. Newly acquired skills are used as soon as possible to solve equations and inequalities, which are, in turn, used to solve word problems. Therefore, the concept of solving equations and inequalities is introduced early and developed throughout the text. The concepts of factoring, solving equations, and solving word problems are tied together in Chapter 3.

- As recommended by the American Mathematical Association of Two-Year Colleges, many basic geometric concepts are integrated in a problem-solving setting. Contained in this text are approximately 20 worked-out examples and 100 problems that connect algebra, geometry, and the real world. Specific discussions of geometric concepts are contained in the following sections:

  Section 2.2: Complementary and supplementary angles; the sum of the angles of a triangle equals $180°$

  Section 2.4: Area and volume formulas

  Section 3.4: More on area and volume formulas, perimeter, and circumference formulas

  Section 3.7: Pythagorean theorem

  Section 6.2: More on the Pythagorean theorem, including work with isosceles right triangles and $30°–60°$ right triangles

- Specific graphing ideas (intercepts, symmetry, restrictions, asymptotes, and transformations) are introduced and used throughout Chapters 7 and 8. In Section 8.3 the work with parabolas from Chapter 7 is used to develop definitions for translations, reflections, stretchings, and shrinkings. These transformations are then applied to the graphs of $f(x) = x^3$, $f(x) = \frac{1}{x}$, $f(x) = \sqrt{x}$, and $f(x) = |x|$. Transformations are also used later with polynomial, exponential, and logarithmic curves.

- All answers for Chapter Review Problem Sets, Chapter Tests, and Cumulative Review Problem Sets appear in the back of the text.

Additional Comments About Some of the Chapters

- Chapter 1 is written so that it can be covered quickly, and on an individual basis if so desired, by those needing only a brief review of some basic algebraic concepts.

- Chapter 2 presents an early introduction to the heart of an intermediate algebra course. Problem solving and the solving of equations and inequalities are introduced early so they can be used as unifying themes throughout the text.

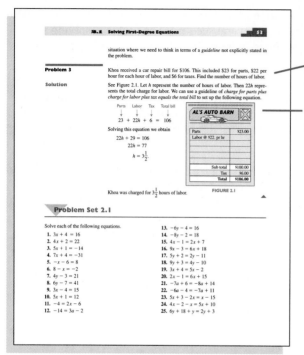

Many sample word problems are fully solved in sections specifically emphasizing problem solving.

Clearly rendered representational art lends interest and helps students visualize the problem.

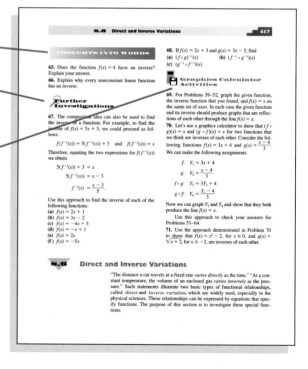

"Thoughts into Words" problems encourage students to express their mathematical understanding verbally.

"Further Investigations" problems, which require skills learned in the section, are especially appropriate for group work.

"Graphics Calculator Activities," which reinforce concepts and lay the groundwork for new material, ask students to predict the shape and locations of graphs and to draw conclusions from what they see.

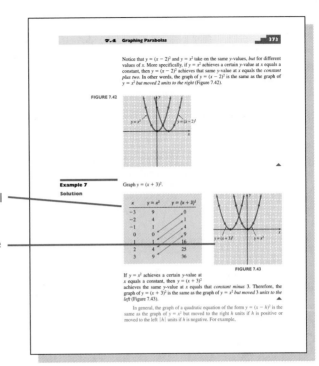

In sections on graphing, tables of values show numerical approaches to problems.

Large, clear graphs depict curves accurately and restate the equation being graphed.

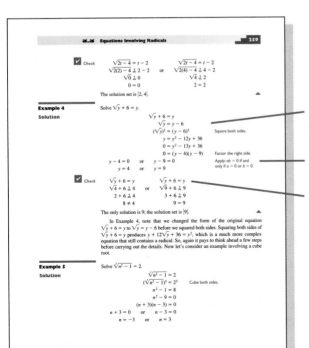

Many worked examples show careful, step-by-step problem solving.

Annotations make clear each step of the problem.

The "Check" feature in worked examples and problems reminds students to complete this important problem-solving step.

- Chapter 6 is organized to give students the opportunity to learn, on a day-to-day basis, different techniques for solving quadratic equations. The process of completing the square is treated as a viable equation-solving process for certain types of quadratic equations. The emphasis on completing the square in this setting pays dividends in Chapter 7 when we graph parabolas and circles. Section 6.5 offers some guidance as to when to use a particular technique for solving quadratic equations. In addition the often overlooked relationships involving the sum and product of roots are discussed and used as an effective checking procedure.

- Chapter 7 is written on the premise that students at this level need more work with coordinate geometry concepts—specifically graphing techniques—*before* functions are defined. My experience indicates a need for students in this course to become proficient at graphing straight lines, parabolas, and circles. In addition at least a little work with graphing ellipses and hyperbolas seems appropriate. Graphing suggestions are offered throughout the chapter.

- Chapter 8 is devoted entirely to functions and the issue is not clouded by the jumping back and forth between functions and relations that are not functions. It includes some work on the composition of functions and the use of quadratic functions in problem-solving situations.

- Chapter 10 presents a modern-day version of the concepts of exponents and logarithms. The emphasis is on making the concepts and their applications understood. The calculator is used as a tool to help with the complicated computational aspects.

- Chapters 11 and 12 contain the various techniques for solving systems of linear equations. This material is organized so that instructors can use as much of the two chapters as needed for a particular course. The work with the elimination-by-addition method in Sections 11.2 and 11.3 emphasizes equivalent systems and sets the stage for the use of matrices in Chapter 12.

- Problem solving is the unifying theme of Chapters 13 and 14. In contrast to most texts of this type, Chapter 14 contains a significant amount of probability. These two chapters lend themselves to individual or small group work.

Ancillaries for Instructors

The following useful ancillaries are available to adopters of this text:

- **Annotated Instructor's Edition** includes answers to all problems in the text—most printed adjacent to the problem.

- **Instructor's Solutions Manual** contains solutions for even-numbered problems and answers for all odd-numbered problems.

- **Test Bank with Chapter Tests** contains all questions and answers from the computerized test bank and three sample tests (two multiple choice, one open ended) for each chapter. These tests may be duplicated for student testing by instructors using the text.

- **Computerized testing software** is available for the IBM and compatibles and for the Macintosh. The computerized testing programs contain multiple-choice and open-ended questions that allow users to edit, rearrange, and add to the question bank.
- **Videotape Series** follows the organization and style of the textbook. Video lectures include basic instruction and worked examples.

Ancillaries for Students

- **Student's Solutions Manual** contains complete worked-out solutions for all odd-numbered problems.
- **Worksheets and Study Guide** is a text-specific study resource in worktext format. It includes examples and exercises for topics keyed to sections in the text so that students have the opportunity for additional practice and study assistance. The manual is designed to be integrated as an interactive component to lectures or for instructional use outside the classroom.
- **MathQuest Tutorial Software** is an interactive, text-specific intuitive tutorial that runs on both Windows and Macintosh platforms. The program provides fill-in, multiple-choice, and true/false questions. If a student answers a question incorrectly, the program will first respond with hints; if the student answers incorrectly a second time, the program will supply a step-by-step solution. Record-keeping capabilities enable students to monitor their progress.

Acknowledgments

I would like to take this opportunity to thank the following people who served as reviewers for this edition:

Janice Burgan
Chaffey Community College

Paul A. Dirks
*Miami-Dade Community College
—Wolfson*

R.A. Evans
Santa Fe Community College

Mary Leeseberg
Manatee Community College

Richard J. Lesnak
Robert Morris College

Allan Riveland
Washburn University

I am very grateful to the staff of PWS, especially David Dietz and Mary Beckwith, for their continuous cooperation and assistance throughout this project. I would also like to express my sincere gratitude to Robine Andrau and to Susan Graham. They continue to make my life as an author so much easier by carrying out the details of production in a dedicated and caring way.

In addition, I would like to thank Karen Schwitters for her work on the *Student's Solutions Manual* and the *Instructor's Solutions Manual*; Kay Haralson and Jennie Preston-Sabin for creating the *Worksheets and Study Guide*; and Karen Sharp for developing the videos.

Again, very special thanks are due to my wife, Arlene, who spends numerous hours typing and proofreading manuscripts.

Jerome E. Kaufmann
Marble Falls, Texas

Basic Concepts and Properties

The temperature at 6 P.M. was $-3°$F. By 11 P.M. the temperature had dropped another 5°F. We can use the **numerical expression** $-3 - 5$ to determine the temperature at 11 P.M.

Justin has p pennies, n nickels, and d dimes in his pocket. The algebraic expression $p + 5n + 10d$ represents that amount of money in cents.

Algebra is often described as a generalized arithmetic. That description may not tell the whole story, but it does indicate an important idea: A good understanding of arithmetic provides a sound basis for the study of algebra. In this chapter we use the concepts of numerical expression and algebraic expression to review some ideas from arithmetic and to begin the transition to algebra. Be sure that you thoroughly understand the basic concepts we review in this first chapter.

1.1 Sets, Real Numbers, and Numerical Expressions

In arithmetic, we use symbols such as 6, $\frac{2}{3}$, $.27$, and π to represent numbers. The symbols $+$, $-$, $\cdot$, and $\div$ commonly indicate the basic operations of addition, subtraction, multiplication, and division, respectively. Thus, we can form specific **numerical expressions**. For example, we can write the indicated sum of six and eight as $6 + 8$.

In algebra, the concept of a variable provides the basis for generalizing arithmetic ideas. For example, by using x and y to represent *any* numbers, we can use the expression $x + y$ to represent the indicated sum of *any* two numbers. The x and y in such an expression are called **variables** and the phrase $x + y$ is called an **algebraic expression**.

We can extend to algebra many of the notational agreements we make in arithmetic, with a few modifications. The following chart summarizes the notational agreements that pertain to the four basic operations.

Operation	Arithmetic	Algebra	Vocabulary
Addition	$4 + 6$	$x + y$	The **sum** of x and y
Subtraction	$14 - 10$	$a - b$	The **difference** of a and b
Multiplication	$7 \cdot 5$ or 7×5	$a \cdot b$, $a(b)$, $(a)b$, $(a)(b)$, or ab	The **product** of a and b
Division	$8 \div 4$, $\frac{8}{4}$, or $4\overline{)8}$	$x \div y$, $\frac{x}{y}$, or $y\overline{)x}$	The **quotient** of x and y

Note the different ways to indicate a product, including the use of parentheses. The *ab form* is the simplest and probably the most widely used form. Expressions such as abc, $6xy$, and $14xyz$ all indicate multiplication. We also call your attention to the various forms that indicate division. In algebra, we usually use the fractional form $\frac{x}{y}$, although the other forms do serve a purpose at times.

Use of Sets

We can use some of the basic vocabulary and symbolism associated with the concept of sets in the study of algebra. A **set** is a collection of objects and the objects are called **elements** or **members** of the set. In arithmetic and algebra the elements of a set are usually numbers.

The use of set braces, $\{\ \}$, to enclose the elements (or a description of the elements), and the use of capital letters to name sets provide a convenient way to communicate about sets. For example, we can represent a set A, which consists of the vowels of the alphabet, as

$A = \{\text{vowels of the alphabet}\}$ Word description,

$A = \{a, e, i, o, u\}$ List or roster description, or

$A = \{x\,|\,x \text{ is a vowel}\}$ Set builder notation.

We can modify the listing approach if the number of elements is quite large. For example, all of the letters of the alphabet can be listed as

$\{a, b, c, \ldots, z\}.$

We simply begin by writing enough elements to establish a pattern, then the three dots indicate that the set continues in that pattern. The final entry indicates the last element of the pattern. If we write

$\{1, 2, 3, \ldots\}$

the set begins with the counting numbers 1, 2, and 3. The three dots indicate that it continues in a like manner forever; there is no last element. A set that consists of no elements is called the **null set** (written $\varnothing$).

The **set builder notation** combines the use of braces and the concept of a variable. For example, $\{x\,|\,x \text{ is a vowel}\}$ is read "the set of all x such that x is a vowel." Note that the vertical line is read "such that." We can use set builder notation to describe the set $\{1, 2, 3, \ldots\}$ as $\{x\,|\,x > 0 \text{ and } x \text{ is a whole number}\}$.

We use the symbol $\in$ to denote set membership. Thus, if $A = \{a, e, i, o, u\}$, we can write $e \in A$, which we read as "e is an element of A." The slash symbol, $/$, is commonly used in mathematics as a negation symbol. Therefore, $m \notin A$ is read as "m is *not* an element of A."

Two sets are said to be *equal* if they contain exactly the same elements. For example,

$\{1, 2, 3\} = \{2, 1, 3\},$

because both sets contain the same elements; the order in which the elements are written doesn't matter. The slash mark through the equality symbol denotes *not equal to*. Thus, if $A = \{1, 2, 3\}$ and $B = \{1, 2, 3, 4\}$, we can write $A \neq B$, which we read as "set A is not equal to set B."

Real Numbers

We refer to most of the algebra that we will study in this text as the **algebra of real numbers**. This simply means that the variables represent real numbers. Therefore, it is necessary for us to be familiar with the various terminology that classifies different types of real numbers.

$\{1, 2, 3, 4, \dots\}$	Natural numbers, counting numbers, positive integers,
$\{0, 1, 2, 3, \dots\}$	Whole numbers, nonnegative integers,
$\{\dots -3, -2, -1\}$	Negative integers.
$\{\dots -3, -2, -1, 0\}$	Nonpositive integers.
$\{\dots -3, -2, -1, 0, 1, 2, 3, \dots\}$	Integers

We define a **rational number** as any number that can be expressed in the form $\frac{a}{b}$, where a and b are integers and b is not zero. The following are examples of rational numbers.

$-\frac{3}{4}, \quad \frac{2}{3}, \qquad\qquad 6 \quad$ Because $6 = \frac{6}{1}$,

$-4 \quad$ Because $-4 = \frac{-4}{1} = \frac{4}{-1}, \qquad 0 \quad$ Because $0 = \frac{0}{1} = \frac{0}{2} = \frac{0}{3} \dots$,

$.3 \quad$ Because $.3 = \frac{3}{10}, \qquad\qquad 6\frac{1}{2} \quad$ Because $6\frac{1}{2} = \frac{13}{2}$

We can also define a rational number in terms of a decimal representation. Before doing so, let's briefly review the different possibilities for decimal representations. We can classify decimals as **terminating**, **repeating**, or **nonrepeating**. Some examples follow.

$\begin{bmatrix} .3 \\ .46 \\ .789 \\ .6234 \end{bmatrix}$ Terminating decimals $\qquad \begin{bmatrix} .6666 \dots \\ .141414 \dots \\ .694694694 \dots \\ .2317171717 \dots \\ .5417283283283 \dots \end{bmatrix}$ Repeating decimals

$\begin{bmatrix} .276314583 \dots \\ .21411811161111 \dots \\ .673183329333 \dots \end{bmatrix}$ Nonrepeating decimals

A repeating decimal has a block of digits that repeats indefinitely. This repeating block of digits may be of any number of digits and may or may not begin immediately after the decimal point. A small horizontal bar is commonly used to indicate the repeat block. Thus, .6666 . . . is written as $.\overline{6}$ and .2317171717 . . . as $.23\overline{17}$.

In terms of decimals, we define a **rational number** as a number that has either a terminating or repeating decimal representation. The following examples illustrate some rational numbers written in $\frac{a}{b}$ form and in decimal form.

$$\frac{3}{4} = .75, \qquad \frac{3}{11} = .\overline{27}, \qquad \frac{1}{8} = .125, \qquad \frac{1}{7} = .\overline{142857}, \qquad \frac{1}{3} = .\overline{3}$$

We define an **irrational number** as a number that *cannot* be expressed in $\frac{a}{b}$ form, where a and b are integers and b is not zero. Furthermore, an irrational number has a nonrepeating and nonterminating decimal representation. Some examples of irrational numbers and a partial decimal representation for each are as follows.

$$\sqrt{2} = 1.414213562373095\ldots, \qquad \sqrt{3} = 1.73205080756887\ldots,$$
$$\pi = 3.14159265358979\ldots$$

The entire set of **real numbers** is composed of the rational numbers along with the irrationals. The following tree diagram summarizes the various classifications of the real number system.

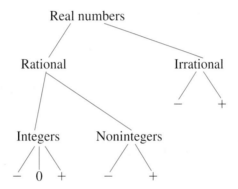

We can trace any real number down through the diagram as follows.

7 is real, rational, an integer, and positive;

$-\frac{2}{3}$ is real, rational, noninteger, and negative;

$\sqrt{7}$ is real, irrational, and positive;

.38 is real, rational, noninteger, and positive.

REMARK   We usually refer to the set of nonnegative integers, $\{0, 1, 2, 3, \ldots\}$, as the set of **whole numbers**, and the set of positive integers, $\{1, 2, 3, \ldots\}$, as the set of **natural numbers**. △

The concept of subset is convenient to use at this time. A set A is a **subset** of a set B if and only if every element of A is also an element of B. This is written as $A \subseteq B$ and read as "A is a subset of B." For example, if $A = \{1, 2, 3\}$ and $B = \{1, 2, 3, 5, 9\}$, then $A \subseteq B$ because every element of A is also an element of B. The slash mark again denotes negation, so if $A = \{1, 2, 5\}$ and $B = \{2, 4, 7\}$, we can say that A *is not a subset of* B by writing $A \nsubseteq B$.

We can use the subset vocabulary to make the following kinds of statements about the real number system.

1. The set of rational numbers is a subset of the set of real numbers.

2. The set of irrational numbers is a subset of the set of real numbers.

3. The set of integers is a subset of the set of rational numbers. The set of integers is also a subset of the set of real numbers.

Equality

The relation **equality** plays an important role in mathematics — especially when manipulating real numbers and algebraic expressions that represent real numbers. An equality is a statement in which two symbols, or groups of symbols, are names for the same number. The symbol $=$ is used to express an equality. Thus, we can write

$$6 + 1 = 7, \qquad 18 - 2 = 16, \qquad 36 \div 4 = 9.$$

(The symbol $\neq$ means *is not equal to*.) The following four basic properties of equality are self-evident, but we do need to keep them in mind. (We will expand this list in Chapter 2 when we work with solutions of equations.)

Properties of Equality

Reflexive Property

For any real number a,

$$a = a.$$

Examples $14 = 14$; $x = x$; $a + b = a + b$.

Symmetric Property

For any real numbers a and b,

if $a = b$, then $b = a$.

Examples   If $13 + 1 = 14$, then $14 = 13 + 1$.
If $3 = x + 2$, then $x + 2 = 3$.

Transitive Property

For any real numbers a, b, and c,

if $a = b$ and $b = c$, then $a = c$.

Examples If $3 + 4 = 7$ and $7 = 5 + 2$, then $3 + 4 = 5 + 2$.
If $x + 1 = y$ and $y = 5$, then $x + 1 = 5$.

Substitution Property

For any real numbers a and b: If $a = b$ then a may be replaced by b, or b may be replaced by a, in any statement without changing the meaning of the statement.

Examples If $x + y = 4$ and $x = 2$, then $2 + y = 4$.
If $a - b = 9$ and $b = 4$, then $a - 4 = 9$.

Numerical Expressions

Let's conclude this section by **simplifying some numerical expressions** that involve whole numbers. The following examples illustrate several important ideas that pertain to this process. Study them carefully and be sure that you agree with each result.

Example 1

Simplify $18 + 16 - 9 + 14 - 12 - 10$.

Solution

The additions and subtractions are to be done from left to right in the order in which they appear. Thus, $18 + 16 - 9 + 14 - 12 - 10$ simplifies to 17. ▲

Example 2

Simplify $7 \cdot 4 \div 2 \cdot 3 \cdot 2 \div 4$.

Solution

The multiplications and divisions are to be done from left to right in the order in which they appear. Thus, $7 \cdot 4 \div 2 \cdot 3 \cdot 2 \div 4$ simplifies to 21. ▲

Example 3

Simplify $5 \cdot 3 + 4 \div 2 - 2 \cdot 6 - 28 \div 7$.

Solution

First, we do the multiplications and divisions in the order in which they appear. Then we do the additions and subtractions in the order in which they appear. Our work may take on the following format.

$$5 \cdot 3 + 4 \div 2 - 2 \cdot 6 - 28 \div 7 = 15 + 2 - 12 - 4$$
$$= 1$$

▲

Example 4

Simplify $(4 + 6)(7 + 8)$.

Solution

We use the parentheses to indicate the *product* of the quantities $4 + 6$ and $7 + 8$. Perform the additions inside the parentheses first and then multiply.

$$(4 + 6)(7 + 8) = (10)(15) = 150$$ ▲

Example 5

Simplify $(3 \cdot 2 + 4 \cdot 5)(6 \cdot 8 - 5 \cdot 7)$.

Solution

First, we do the multiplications inside the parentheses.

$$(3 \cdot 2 + 4 \cdot 5)(6 \cdot 8 - 5 \cdot 7) = (6 + 20)(48 - 35)$$

Then, we do the addition and subtraction inside the parentheses.

$$(6 + 20)(48 - 35) = (26)(13)$$

Then, we find the final product.

$$(26)(13) = 338$$ ▲

Example 6

Simplify $6 + 7[3(4 + 6)]$.

Solution

We use brackets for the same purposes as parentheses. In such a problem we need to simplify *from the inside out*; that is, we perform the operations in the innermost parentheses first. We would thus obtain

$$6 + 7[3(4 + 6)] = 6 + 7[3(10)]$$
$$= 6 + 7[30]$$
$$= 6 + 210$$
$$= 216.$$ ▲

Example 7

Simplify $\dfrac{6 \cdot 8 \div 4 - 2}{5 \cdot 4 - 9 \cdot 2}$.

Solution

First, we perform the operations above and below the fraction bar. Then, we find the final quotient.

$$\frac{6 \cdot 8 \div 4 - 2}{5 \cdot 4 - 9 \cdot 2} = \frac{12 - 2}{20 - 18}$$
$$= \frac{10}{2}$$
$$= 5$$ ▲

REMARK   With parentheses we could write the problem in Example 7 as $(6 \cdot 8 \div 4 - 2) \div (5 \cdot 4 - 9 \cdot 2)$. △

The following list summarizes how to simplify numerical expressions. When evaluating a numerical expression perform the operations in the following order.

1. Perform the operations inside the symbols of inclusion (parentheses, brackets, and braces) and above and below each fraction bar. Start with the innermost inclusion symbol.
2. Perform all multiplications and divisions in the order in which they appear from left to right.
3. Perform all additions and subtractions in the order in which they appear from left to right.

▼ Problem Set 1.1

For Problems 1–10, identify each statement as true or false.

1. Every irrational number is a real number.
2. Every rational number is a real number.
3. If a number is real, then it is irrational.
4. Every real number is a rational number.
5. All integers are rational numbers.
6. Some irrational numbers are also rational numbers.
7. Zero is a positive integer.
8. Zero is a rational number.
9. All whole numbers are integers.
10. Zero is a negative integer.

From the list $0, 14, \frac{2}{3}, \pi, \sqrt{7}, -\frac{11}{14}, 2.34, 3.2\bar{1}, 6\frac{7}{8},$ $-\sqrt{17}, -19,$ and -2.6 identify each of the following. (Problems 11–18)

11. The whole numbers
12. The natural numbers
13. The rational numbers
14. The integers
15. The nonnegative integers
16. The irrational numbers
17. The real numbers
18. The nonpositive integers

For Problems 19–32, use the following set designations.

$N = \{x \,|\, x$ is a natural number$\}$.
$Q = \{x \,|\, x$ is a rational number$\}$.
$W = \{x \,|\, x$ is a whole number$\}$.
$H = \{x \,|\, x$ is an irrational number$\}$.
$I = \{x \,|\, x$ is an integer$\}$.
$R = \{x \,|\, x$ is a real number$\}$.

Place $\subseteq$ or $\nsubseteq$ in each blank to make a true statement.

19. R _____ N 20. N _____ R
21. I _____ Q 22. N _____ I
23. Q _____ H 24. H _____ Q
25. N _____ W 26. W _____ I
27. I _____ N 28. I _____ W
29. $\{1, 3, 5, 7, \ldots\}$ _____ I
30. $\{0, 2, 4, \ldots\}$ _____ W
31. $\{0, 3, 6, 9, \ldots\}$ _____ N
32. $\{-2, -1, 0, 1, 2\}$ _____ W

For Problems 33–42, list the elements of each set. For example, the elements of $\{x \,|\, x$ is a natural number less than 4$\}$ can be listed as $\{1, 2, 3\}$.

33. $\{x \,|\, x$ is a natural number less than 3$\}$
34. $\{x \,|\, x$ is a natural number greater than 3$\}$

35. $\{n \mid n$ is a whole number less than $6\}$

36. $\{y \mid y$ is an integer greater than $-4\}$

37. $\{y \mid y$ is an integer less than $3\}$
38. $\{n \mid n$ is a positive integer greater than $-7\}$

39. $\{x \mid x$ is a whole number less than $0\}$
40. $\{x \mid x$ is a negative integer greater than $-3\}$

41. $\{n \mid n$ is a nonnegative integer less than $5\}$

42. $\{n \mid n$ is a nonpositive integer greater than $3\}$

For Problems 43–50, replace each question mark to make the given statement an application of the indicated property of equality. For example, $16 = ?$ becomes $16 = 16$ because of the reflexive property of equality.

43. If $y = x$ and $x = -6$, then $y = ?$ (Transitive property of equality)
44. $5x + 7 = ?$ (Reflexive property of equality)

45. If $n = 2$ and $3n + 4 = 10$, then $3(?) + 4 = 10$ (Substitution property of equality)
46. If $y = x$ and $x = z + 2$, then $y = ?$ (Transitive property of equality)
47. If $4 = 3x + 1$, then $? = 4$ (Symmetric property of equality)
48. If $t = 4$ and $s + t = 9$, then $s + ? = 9$ (Substitution property of equality)
49. $5x = ?$ (Reflexive property of equality)
50. If $5 = n + 3$, then $n + 3 = ?$ (Symmetric property of equality)

For Problems 51–74, simplify each of the numerical expressions.

51. $16 + 9 - 4 - 2 + 8 - 1$
52. $18 + 17 - 9 - 2 + 14 - 11$
53. $9 \div 3 \cdot 4 \div 2 \cdot 14$
54. $21 \div 7 \cdot 5 \cdot 2 \div 6$

55. $7 + 8 \cdot 2$
56. $21 - 4 \cdot 3 + 2$
57. $9 \cdot 7 - 4 \cdot 5 - 3 \cdot 2 + 4 \cdot 7$
58. $6 \cdot 3 + 5 \cdot 4 - 2 \cdot 8 + 3 \cdot 2$
59. $(17 - 12)(13 - 9)(7 - 4)$
60. $(14 - 12)(13 - 8)(9 - 6)$
61. $13 + (7 - 2)(5 - 1)$
62. $48 - (14 - 11)(10 - 6)$
63. $(5 \cdot 9 - 3 \cdot 4)(6 \cdot 9 - 2 \cdot 7)$
64. $(3 \cdot 4 + 2 \cdot 1)(5 \cdot 2 + 6 \cdot 7)$
65. $7[3(6 - 2)] - 64$
66. $12 + 5[3(7 - 4)]$
67. $[3 + 2(4 \cdot 1 - 2)][18 - (2 \cdot 4 - 7 \cdot 1)]$
68. $3[4(6 + 7)] + 2[3(4 - 2)]$
69. $14 + 4\left(\dfrac{8 - 2}{12 - 9}\right) - 2\left(\dfrac{9 - 1}{19 - 15}\right)$
70. $12 + 2\left(\dfrac{12 - 2}{7 - 2}\right) - 3\left(\dfrac{12 - 9}{17 - 14}\right)$
71. $[7 + 2 \cdot 3 \cdot 5 - 5] \div 8$
72. $[27 - (4 \cdot 2 + 5 \cdot 2)][(5 \cdot 6 - 4) - 20]$
73. $\dfrac{3 \cdot 8 - 4 \cdot 3}{5 \cdot 7 - 34} + 19$
74. $\dfrac{4 \cdot 9 - 3 \cdot 5 - 3}{18 - 12}$

75. You should be able to do calculations like those in Problems 51–74 *with* and *without* a calculator. Different types of calculators handle the *priority-of-operations* issue in different ways. Be sure you can do Problems 51–74 with *your* calculator.

THOUGHTS INTO WORDS

76. Explain in your own words the difference between the reflexive property of equality and the symmetric property of equality.

77. Your friend keeps getting an answer of 30 when simplifying $7 + 8(2)$. What mistake is he making and how would you help him?

78. Do you think $3\sqrt{2}$ is a rational or irrational number? Defend your answer.

1.2 Operations with Real Numbers

Before we review the four basic operations with integers, let's briefly discuss some concepts and terminology we commonly use with this material.

The symbol -1 can be read as "negative one, opposite of one, or additive inverse of one." The opposite-of and additive-inverse-of terminology is especially meaningful when we work with variables. For example, the symbol $-x$, read as "opposite of x" or "additive inverse of x," emphasizes an important issue. Since x can be any real number, $-x$ (opposite of x) can be zero, positive, or negative. If x is a positive number, then $-x$ is negative. If x is a negative number, then $-x$ is positive. If x is zero, then $-x$ is zero.

$-(6) = -6$ The opposite of six (additive inverse of six) is negative six.

$-(-4) = 4$ The opposite of negative four is four.

$-(0) = 0$ The opposite of zero is zero.

In general, it can be stated that **the opposite of the opposite of any real number is the real number itself.** This is symbolically expressed as $-(-a) = a$; we sometimes refer to it as the **double negative** property.

Just as we use the symbol $=$ to represent *is equal to*, we also use the symbols $<$ and $>$ to represent *is less than* and *is greater than*, respectively. Thus, we can make statements such as $3 < 4$, $5 > 2$, $-4 < -1$, and $-2 > -6$. On a number line the statement $-4 < -1$ means that -4 is to the left of -1. Likewise, the statement $-2 > -6$ means that -2 is to the right of -6. The symbol $\leq$ means *is less than or equal to* and the symbol $\geq$ means *is greater than or equal to*.

Absolute Value

We can use the concept of **absolute value** to precisely describe how to operate with positive and negative numbers. Geometrically, the absolute value of any number is the distance between the number and zero on the number line. For example, the absolute value of 2 is 2. The absolute value of -3 is 3. The absolute value of 0 is 0 (see Figure 1.1).

FIGURE 1.1

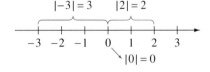

Symbolically, absolute value is denoted with vertical bars. Thus, we write

$$|2| = 2 \qquad |-3| = 3 \qquad |0| = 0.$$

More formally, we define the concept of absolute value as follows.

DEFINITION 1.1

For all real numbers a,
1. If $a \geq 0$, then $|a| = a$.
2. If $a < 0$, then $|a| = -a$.

According to Definition 1.1 we obtain

$|6| = 6$ by applying part 1,

$|0| = 0$ by applying part 1,

$|-7| = -(-7) = 7$ by applying part 2.

Notice that the absolute value of a positive number is the number itself, but the absolute value of a negative number is its opposite. Thus, the absolute value of any number, except 0, is positive and the absolute value of 0 is 0.

Adding Integers

We can use various physical models to describe addition of integers. For example, profits and losses pertaining to investments provide a meaningful model. A loss of \$25 (written as -25) on one investment along with a profit of \$60 (written as 60) on a second investment produces an overall profit of \$35. We can write this as $(-25) + 60 = 35$. Think in terms of profits and losses for each of the following examples.

$$50 + 75 = 125, \qquad\qquad 20 + (-30) = -10,$$
$$(-10) + (-40) = -50, \qquad (-50) + 75 = 25,$$
$$100 + (-50) = 50, \qquad\quad (-50) + (-50) = -100$$

Although all problems that involve addition of integers could be solved using the profit-and-loss interpretation, it is sometimes convenient to give a more precise description of the addition process. For this purpose we can use the concept of absolute value. Suppose that we want to describe the process of adding -50 and 75. We could say "subtract the absolute value of -50 from the absolute value of 75." Thus,

$$(-50) + 75 = |75| - |-50|$$
$$= 75 - 50$$
$$= 25.$$

We could describe the addition problem $20 + (-30)$ as, "subtract the absolute value of 20 from the absolute value of -30 and then take the opposite." Thus,

$$20 + (-30) = -(|-30| - |20|)$$
$$= -(30 - 20)$$

$$= -(10)$$
$$= -10.$$

We could describe the addition problem $(-10) + (-40)$ as, "add the absolute value of -10 and the absolute value of -40 and then take the opposite." We would write

$$(-10) + (-40) = -(|-10| + |-40|)$$
$$= -(10 + 40)$$
$$= -(50)$$
$$= -50.$$

In general, we can describe **addition of integers** as follows.

Two Positive Integers The sum of two positive integers is the sum of their absolute values.

Two Negative Integers The sum of two negative integers is the opposite of the sum of their absolute values.

One Positive and One Negative Integer We can find the sum of a positive integer and a negative integer by subtracting the smaller absolute value from the larger absolute value and giving the result the sign of the original number that has the larger absolute value. If the integers have the same absolute value, then their sum is 0.

Zero and Another Integer The sum of 0 and any integer is the integer itself.

Now consider the following examples in terms of the previous list, which concerns the addition of integers.

$$(-6) + (-8) = -(|-6| + |-8|) = -(6 + 8) = -14,$$
$$(-19) + (-11) = -(|-19| + |-11|) = -(19 + 11) = -30,$$
$$18 + (-14) = (|18| - |-14|) = 18 - 14 = 4,$$
$$(-17) + 25 = (|25| - |-17|) = 25 - 17 = 8,$$
$$14 + (-21) = -(|-21| - |14|) = -(21 - 14) = -7,$$
$$(-32) + 17 = -(|-32| - |17|) = -(32 - 17) = -15,$$
$$19 + (-19) = 0, \qquad -49 + 0 = -49,$$
$$-21 + 21 = 0, \qquad 0 + (-72) = -72$$

It is true that this *absolute value approach* precisely describes the process of adding integers, but don't forget about the profit-and-loss interpretation. The next problem set also includes other physical models that describe the addition of integers. Some people find such models very helpful.

Subtracting Integers

We describe subtraction of integers in terms of addition.

> **Subtraction of Integers** If a and b are integers, then $a - b = a + (-b)$.

It may be helpful for you to read $a - b = a + (-b)$ as "a minus b is equal to a plus the opposite of b." In other words, we can change every subtraction problem to an equivalent addition problem. Consider the following examples.

$$7 - 9 = 7 + (-9) = -2, \qquad -5 - (-13) = -5 + 13 = 8,$$
$$8 - (-12) = 8 + 12 = 20, \qquad -16 - (-11) = -16 + 11 = -5,$$
$$-9 - 6 = -9 + (-6) = -15$$

It should be apparent that addition of integers is a key operation. The ability to add integers effectively is a necessary skill for future algebraic work. To simplify numerical expressions that involve addition and subtraction of integers, first change all subtractions to additions and then perform the additions.

Example 1

Solution

Simplify $7 - 9 - 14 + 12 - 6 + 4$.

$$7 - 9 - 14 + 12 - 6 + 4 = 7 + (-9) + (-14) + 12 + (-6) + 4$$
$$= -6 \qquad\qquad ▲$$

Example 2

Solution

Simplify $-12 + 17 - (-10) + 9 - 3$.

$$-12 + 17 - (-10) + 9 - 3 = -12 + 17 + 10 + 9 + (-3)$$
$$= 21 \qquad\qquad ▲$$

It is helpful if you can *mentally* convert subtractions to additions. In the next two examples the work shown in the dashed boxes could be done mentally.

Example 3

Solution

Simplify $4 - 9 - 18 + 13 - 10$.

$$4 - 9 - 18 + 13 - 10 = \boxed{4 + (-9) + (-18) + 13 + (-10)}$$
$$= -20 \qquad\qquad ▲$$

Example 4

Simplify $(3 - 7) - (4 - 9)$.

Solution

$$(3 - 7) - (4 - 9) = [3 + (-7)] - [4 + (-9)]$$
$$= (-4) - (-5)$$
$$= -4 + 5$$
$$= 1$$

Multiplying Integers

We can interpret multiplication of whole numbers as repeated addition. For example, $3 \cdot 2$ means three 2s; thus, $3 \cdot 2 = 2 + 2 + 2 = 6$. We can use this same repeated-addition interpretation of multiplication to find the product of a positive integer and a negative integer as illustrated by the following examples.

$$2(-3) = -3 + (-3) = -6, \qquad 3(-2) = -2 + (-2) + (-2) = -6,$$
$$4(-5) = -5 + (-5) + (-5) + (-5) = -20$$

Note the use of parentheses to indicate multiplication. Sometimes both numbers are enclosed in parentheses, as with $(3)(-4)$.

When multiplying whole numbers we realize that the order in which we multiply two factors does not change the product. For example, $2(3) = 6$ and $3(2) = 6$. Using this idea, we can handle a negative integer times a positive integer as follows.

$$(-2)(3) = (3)(-2) = (-2) + (-2) + (-2) = -6,$$

$$(-3)(4) = (4)(-3) = (-3) + (-3) + (-3) + (-3) = -12,$$

$$(-4)(3) = (3)(-4) = (-4) + (-4) + (-4) = -12$$

Finally, let's consider the product of two negative integers. The following pattern helps with the reasoning.

$$4(-2) = -8,$$
$$3(-2) = -6,$$
$$2(-2) = -4,$$
$$1(-2) = -2,$$
$$0(-2) = \quad 0, \qquad \text{The product of zero and any number is zero.}$$
$$(-1)(-2) = \quad ?$$

Certainly, to continue this pattern the product of -1 and -2 has to be 2. In general, this type of reasoning would help us realize that the product of any two negative integers is a positive integer.

We can use the concept of absolute value to describe the **multiplication of integers** as follows.

1. The product of two positive integers or two negative integers is the product of their absolute values.
2. The product of a positive and a negative integer (either order) is the opposite of the product of their absolute values.
3. The product of zero and any integer is zero.

The following examples illustrate this description of multiplication.

$$(-6)(-7) = |-6| \cdot |-7| = 6 \cdot 7 = 42,$$

$$(8)(-9) = -(|8| \cdot |-9|) = -(8 \cdot 9) = -72,$$

$$(-5)(12) = -(|-5| \cdot |12|) = -(5 \cdot 12) = -60,$$

$$(-19)(0) = 0, \qquad (0)(-31) = 0$$

The previous examples illustrated a step-by-step process for multiplying integers. In practice, however, the key issue is to remember whether the product is positive or negative. In other words, we need to remember that the product of two positive or two negative integers is positive and the product of a positive and a negative integer (either order) is negative.

Dividing Integers

The relationship between multiplication and division provides the basis for dividing integers. For example, we know that $8 \div 2 = 4$ because $2 \cdot 4 = 8$. In other words, the quotient of two numbers can be found by looking at a related multiplication problem. In the following examples we used this same type of reasoning to determine some quotients that involve integers.

$$\frac{6}{-2} = -3 \quad \text{because } (-2)(-3) = 6;$$

$$\frac{-12}{3} = -4 \quad \text{because } (3)(-4) = -12;$$

$$\frac{-18}{-2} = 9 \quad \text{because } (-2)(9) = -18;$$

$$\frac{0}{-5} = 0 \quad \text{because } (-5)(0) = 0;$$

$$\frac{-8}{0} \quad \text{is undefined.} \qquad \text{Remember that division by zero is undefined!}$$

A precise description for **division of integers** follows.

1. The quotient of two positive or two negative integers is the quotient of their absolute values.

2. The quotient of a positive integer and a negative integer or a negative integer and a positive integer is the opposite of the quotient of their absolute values.

3. The quotient of zero and any nonzero integer (zero divided by any nonzero integer) is zero.

The following examples illustrate this description of division. Again, for practical purposes the key idea is to remember whether the quotient is positive or negative.

$$\frac{-16}{-4} = \frac{|-16|}{|-4|} = \frac{16}{4} = 4, \qquad\qquad \frac{28}{-7} = -\left(\frac{|28|}{|-7|}\right) = -\left(\frac{28}{7}\right) = -4,$$

$$\frac{-21}{3} = -\left(\frac{|-21|}{|3|}\right) = -\left(\frac{21}{3}\right) = -7, \qquad \frac{0}{-9} = 0$$

Now let's simplify some numerical expressions that involve the four basic operations with integers.

Example 5

Simplify $-6 + 8(-3) - (-4)(-3)$.

Solution

$$-6 + 8(-3) - (-4)(-3)$$
$$= -6 + (-24) - 12 \qquad \text{Do the multiplications first.}$$
$$= -42$$
▲

Example 6

Simplify $-24 \div 4 + 8(-5) - (-5)(3)$.

Solution

$$-24 \div 4 + 8(-5) - (-5)(3)$$
$$= -6 + (-40) - (-15) \qquad \text{Do the multiplications and divisions first.}$$
$$= -31$$
▲

Example 7

Simplify $-7 - 5[-3(6 - 8)]$.

Solution

$$-7 - 5[-3(6 - 8)]$$
$$= -7 - 5[-3(-2)] \qquad \text{Start with the innermost parentheses.}$$
$$= -7 - 5[6]$$
$$= -7 - 30$$
$$= -37$$
▲

Example 8

Solution

Simplify $[3(-7) - 2(9)][5(-7) + 3(9)]$.

$$[3(-7) - 2(9)][5(-7) + 3(9)] = [-21 - 18][-35 + 27]$$
$$= [-39][-8]$$
$$= 312 \qquad \blacktriangle$$

We have just reviewed the basic operations with *integers*; however, the descriptions we gave for manipulating integers apply to *real numbers* in general. For example, the sum of two negative real numbers is the opposite of the sum of their absolute values.

$$(-3.21) + (-4.47) = -(|-3.21| + |-4.47|)$$
$$= -(3.21 + 4.47)$$
$$= -7.68$$

Likewise, the product of two negative real numbers is the product of their absolute values.

$$(-1.2)(-2.3) = |-1.2| \cdot |-2.3|$$
$$= (1.2)(2.3)$$
$$= 2.76$$

▼ Problem Set 1.2

Perform the following operations with integers.

1. $8 + (-15)$ **2.** $9 + (-18)$

3. $(-12) + (-7)$

4. $(-7) + (-14)$

5. $-8 - 14$ **6.** $-17 - 9$

7. $9 - 16$ **8.** $8 - 22$

9. $(-9)(-12)$ **10.** $(-6)(-13)$

11. $(5)(-14)$ **12.** $(-17)(4)$

13. $(-56) \div (-4)$

14. $(-81) \div (-3)$

15. $\dfrac{-112}{16}$ **16.** $\dfrac{-75}{5}$

17. $-24 + 52$ **18.** $-19 + 38$

19. $19 - (-14)$ **20.** $11 - (-25)$

21. $(-17)(12)$ **22.** $(-15)(9)$

23. $78 \div (-13)$ **24.** $72 \div (-6)$

25. $0 \div (-14)$

26. $(-19) \div 0$

27. $(-21) \div 0$

28. $0 \div (-11)$ **29.** $-21 - 39$

30. $-23 - 38$ **31.** $-42 - (-25)$

32. $-53 - (-26)$

33. $(-4)(-3)(6)$ **34.** $(-5)(7)(-2)$

35. $9(-8)(5)$ **36.** $6(-8)(5)$

37. $(-3)(-7)(-6)$

38. $(-4)(-7)(-9)$

39. $-\left(\dfrac{-56}{-7}\right)$ **40.** $-\left(\dfrac{-48}{-6}\right)$

Simplify each of the following numerical expressions.

41. $9 - 12 - 8 + 5 - 6$

42. $6 - 9 + 11 - 8 - 7 + 14$

43. $-21 + (-17) - 11 + 15 - (-10)$

44. $-16 - (-14) + 16 + 17 - 19$

45. $7 - (9 - 17)$

46. $-7 - (8 - 15)$

47. $16 - 18 + 19 - [14 - 22 - (31 - 41)]$

48. $-19 - [15 - 13 - (-12 + 8)]$

49. $[14 - (16 - 18)] - [32 - (8 - 9)]$

50. $[-17 - (14 - 18)] - [21 - (-6 - 5)]$

51. $6 - 5(-3)$ **52.** $-12 - 2(-6)$

53. $-5 + (-2)(7) - (-3)(8)$

54. $-9 - 4(-2) + (-7)(6)$

55. $5(-6) - (-4)(11)$

56. $-7(9) + (-8)(12)$

57. $(-6)(-9) + (-7)(4)$

58. $(-7)(-7) - (-6)(4)$

59. $3(5 - 9) - 3(-6)$

60. $7(8 - 9) + (-6)(4)$

61. $(6 - 11)(4 - 9)$

62. $(7 - 12)(-3 - 2)$

63. $-6(-3 - 9 - 1)$

64. $-8(-3 - 4 - 6)$

65. $56 \div (-8) - (-6) \div (-2)$

66. $-65 \div 5 - (-13)(-2) + (-36) \div 12$

67. $-3[5 - (-2)] - 2(-4 - 9)$

68. $-2(-7 + 13) + 6(-3 - 2)$

69. $\dfrac{-6 + 24}{-3} + \dfrac{-7}{-6 - 1}$

70. $\dfrac{-12 + 20}{-4} + \dfrac{-7 - 11}{-9}$

71. $[(-3)(4) - (-2)(1)][(-2)(-7) - (-8)(6)]$

72. $[4(-6) - 3(-9)][(-1)(-6) - 7(-11)]$

73. $(7 - 11)[-3 - (6 - 7)] - 12$

74. $(-4 - 3)[-2 - (-1 + 6)] - 16$

Perform the following operations with decimals.

75. $(-5.4) + (-7.2)$

76. $(-1.8) + (-2.7)$

77. $(12.2) + (-5.6)$

78. $(13.7) + (-7.9)$

79. $-17.3 + 12.5$

80. $-16.3 + 9.6$

81. $(-2.1)(-3.5)$

82. $(-8.5)(-3.3)$

83. $(5.4)(-7.2)$

84. $(2.9)(-3.6)$

85. $\dfrac{-1.2}{-6}$ **86.** $\dfrac{-6.3}{.7}$

87. $(16.8) \div (-1.2)$

88. $(16.5) \div (-1.5)$

89. $-21.4 - (-14.9)$

90. $-32.6 - (-9.8)$

91. $1.42 - 7.29$

92. $2.73 - 8.14$

93. $-2.11 - 4.67$

94. $-3.27 - 4.15$

☐C **95.** Do Problems 41–94 with your calculator.

THOUGHTS INTO WORDS

96. Explain why $\dfrac{0}{8} = 0$ but $\dfrac{8}{0}$ is undefined.

97. The following simplification problem is incorrect. (The answer should be -11.) Find and correct the error.

$$8 \div (-4)(2) - 3(4) \div 2 + (-1) = (-2)(2) - 12 \div 1$$
$$= -4 - 12$$
$$= -16$$

Further Investigations

98. To find the sum of 65, 68, 71, 73, and 69 we might consider the deviation of each of these numbers from some other number, say 70. That is, consider $-5 + (-2) + 1 + 3 + (-1) = -4$. Thus, the sum of the original numbers is $(5 \cdot 70) + (-4) = 346$. Use this procedure to find the following sums.

(a) $84 + 78 + 82 + 75$
(b) $136 + 145 + 142 + 137 + 132$
(c) $56 + 63 + 61 + 58 + 52 + 67$
(d) $119 + 123 + 121 + 116 + 114 + 128$

99. The temperature at 4 A.M. was $-19°F$. By noon the temperature had increased by $12°F$. Use addition of integers to model this situation and determine the temperature at noon (see Figure 1.2).

100. The temperature at 5 P.M. was $-2°F$ and by 11 P.M. the temperature had dropped another $7°F$. Use subtraction of integers to describe this situation and determine the temperature at 11 P.M. (see Figure 1.2).

101. In one week a small company showed a profit of $565 for one day and a loss of $85 for each of the other four days. Use multiplication and addition of integers to model this situation and determine the company's profit or loss for the week.

FIGURE 1.2

1.3 Properties of Real Numbers and the Use of Exponents

At the beginning of this section we will list and briefly discuss some of the basic properties of real numbers. Be sure that you understand these properties, for not only do they facilitate manipulations with real numbers, they also serve as the basis for many algebraic computations.

Closure Property for Addition

If a and b are real numbers, then $a + b$ is a unique real number.

Closure Property for Multiplication

If a and b are real numbers, then ab is a unique real number.

We say that the set of real numbers is *closed* with respect to addition and also with respect to multiplication. That is, the sum of two real numbers is a unique real number and the product of two real numbers is a unique real number. We use the word *unique* to indicate *exactly one*.

Commutative Property of Addition

If a and b are real numbers, then

$$a + b = b + a.$$

Commutative Property of Multiplication

If a and b are real numbers, then

$$ab = ba.$$

We say that addition and multiplication are commutative operations. This means that the order in which you add or multiply two numbers does not affect the result. For example, $6 + (-8) = (-8) + 6$ and $(-4)(-3) = (-3)(-4)$. It is also important to realize that subtraction and division *are not* commutative operations; order does make a difference. For example, $3 - 4 = -1$ but $4 - 3 = 1$. Likewise, $2 \div 1 = 2$ but $1 \div 2 = \dfrac{1}{2}$.

Associative Property of Addition

If a, b, and c are real numbers, then

$$(a + b) + c = a + (b + c).$$

Associative Property of Multiplication

If a, b, and c are real numbers, then

$$(ab)c = a(bc).$$

Addition and multiplication are **binary operations**. That is, we add (or multiply) *two* numbers at a time. The associative properties apply if more than two numbers are to be added or multiplied; they are grouping properties. For example, $(-8 + 9) + 6 = -8 + (9 + 6)$; changing the grouping of the numbers does not affect the final sum. This is also true for multiplication, which is illustrated by $[(-4)(-3)](2) = (-4)[(-3)(2)]$. Subtraction and division *are not* associative operations. For example, $(8 - 6) - 10 = -8$, but $8 - (6 - 10) = 12$. An example showing that division is not associative is $(8 \div 4) \div 2 = 1$, but $8 \div (4 \div 2) = 4$.

Identity Property of Addition

If a is any real number, then

$$a + 0 = 0 + a = a.$$

Zero is called the identity element for addition. This merely means that the sum of any real number and zero is identically the same real number. For example, $-87 + 0 = 0 + (-87) = -87$.

Identity Property of Multiplication

If a is any real number, then

$$a(1) = 1(a) = a.$$

We call 1 the identity element for multiplication. The product of any real number and 1 is identically the same real number. For example, $(-119)(1) = (1)(-119) = -119$.

Additive Inverse Property

For every real number a, there exists a unique real number $-a$ such that

$$a + (-a) = -a + a = 0.$$

The real number, $-a$ is called the **additive inverse of** a or the **opposite of** a. For example, 16 and -16 are additive inverses and their sum is 0. The additive inverse of 0 is 0.

Multiplication Property of Zero

If a is any real number, then

$$(a)(0) = (0)(a) = 0.$$

The product of any real number and zero is zero. For example, $(-17)(0) = 0(-17) = 0$.

Multiplication Property of Negative One

If a is any real number, then

$$(a)(-1) = (-1)(a) = -a.$$

The product of any real number and -1 is the opposite of the real number. For example, $(-1)(52) = (52)(-1) = -52$.

Multiplicative Inverse Property

For every nonzero real number a, there exists a unique real number $\dfrac{1}{a}$, such that

$$a\left(\frac{1}{a}\right) = \frac{1}{a}(a) = 1.$$

The number $\dfrac{1}{a}$ is called the **multiplicative inverse** or the **reciprocal of a**. For example, the reciprocal of 2 is $\dfrac{1}{2}$ and $2\left(\dfrac{1}{2}\right) = \dfrac{1}{2}(2) = 1$. Likewise, the reciprocal of $\dfrac{1}{2}$ is $\dfrac{1}{\frac{1}{2}} = 2$. Therefore, 2 and $\dfrac{1}{2}$ are said to be reciprocals (or multiplicative inverses) of each other. Since division by zero is undefined, zero does not have a reciprocal.

Distributive Property

If a, b, and c are real numbers, then

$$a(b + c) = ab + ac.$$

The distributive property ties together the operations of addition and multiplication. We say that **multiplication distributes over addition**. For example, $7(3 + 8) = 7(3) + 7(8)$. Since $b - c = b + (-c)$, it follows that **multiplication also distributes over subtraction**. This can be symbolically expressed as $a(b - c) = ab - ac$. For example, $6(8 - 10) = 6(8) - 6(10)$.

The following examples illustrate the use of the properties of real numbers to facilitate certain types of manipulations.

Example 1

Solution

Simplify $(74 + (-36)) + 36$.

In such a problem it is much more advantageous to group -36 and 36.

$$(74 + (-36)) + 36 = 74 + ((-36) + 36)$$
$$= 74 + 0 = 74$$

By using the associative property for addition ▲

Example 2

Solution

Simplify $[(-19)(25)](-4)$.

It is much easier to group 25 and -4. Thus,

$$[(-19)(25)](-4) = (-19)[(25)(-4)]$$
$$= (-19)(-100)$$
$$= 1900$$

By using the associative property for multiplication ▲

Example 3

Solution

Simplify $17 + (-14) + (-18) + 13 + (-21) + 15 + (-33)$.

We could add in the order in which the numbers appear. However, since addition is commutative and associative we could change the order and group in any convenient way. For example, we could add all of the positive integers and add all of the negative integers and then find the sum of these two results. It might be convenient to use the vertical format as follows.

$$
\begin{array}{rrr}
 & -14 & \\
17 & -18 & \\
13 & -21 & -86 \\
\underline{15} & \underline{-33} & \underline{45} \\
45 & -86 & -41
\end{array}
$$

▲

Example 4

Solution

Simplify $-25(-2 + 100)$.

For this problem it might be easiest to first apply the distributive property and then to simplify as follows.

$$-25(-2 + 100) = (-25)(-2) + (-25)(100)$$
$$= 50 + (-2500)$$
$$= -2450$$

▲

Example 5

Solution

Simplify $(-87)(-26 + 25)$.

For this problem it would be better not to apply the distributive property, but instead to add the numbers inside the parentheses first and then find the indicated product.

$$(-87)(-26 + 25) = (-87)(-1)$$
$$= 87 \qquad \blacktriangle$$

Example 6

Simplify $37(104) + 37(-4)$.

Solution

Remember that the distributive property allows us to change from the form $a(b + c)$ to $ab + ac$ or from $ab + ac$ to $a(b + c)$. In this problem we want to use the latter change. Thus,

$$37(104) + 37(-4) = 37(104 + (-4))$$
$$= 37(100)$$
$$= 3700. \qquad \blacktriangle$$

Examples 4, 5, and 6 illustrate an important issue. Sometimes the form $a(b + c)$ is more convenient, but at other times the form $ab + ac$ is better. In these cases, as well as in the cases of other properties, you should *think first* and decide whether or not the properties can be used to make the manipulations easier.

Exponents

Exponents are used to indicate repeated multiplication. For example, we can write $4 \cdot 4 \cdot 4$ as 4^3 where the "raised 3" indicates that 4 is to be used as a factor 3 times. The following general definition is helpful.

DEFINITION 1.2

If n is a positive integer and b is any real number, then

$$b^n = \underbrace{bbb \cdots b}_{n \text{ factors of } b}.$$

We refer to the b as the **base** and n as the **exponent**. The expression b^n can be read as "b to the nth power." We commonly associate the terms *squared* and *cubed* with exponents of 2 and 3, respectively. For example, b^2 is read "b squared" and b^3 as "b cubed." An exponent of 1 is usually not written, so b^1 is written as b. The following examples illustrate Definition 1.2.

$$2^3 = 2 \cdot 2 \cdot 2 = 8, \qquad \left(\frac{1}{2}\right)^5 = \frac{1}{2} \cdot \frac{1}{2} \cdot \frac{1}{2} \cdot \frac{1}{2} \cdot \frac{1}{2} = \frac{1}{32},$$

$$3^4 = 3 \cdot 3 \cdot 3 \cdot 3 = 81, \qquad (.7)^2 = (.7)(.7) = .49,$$

$$-5^2 = -(5 \cdot 5) = -25, \qquad (-5)^2 = (-5)(-5) = 25$$

Please take special note of the last two examples. Note that $(-5)^2$ means -5 is the base and is to be used as a factor twice. However, -5^2 means that 5 is the base and after it is squared then we take the opposite of that result.

Simplifying numerical expressions that contain exponents creates no trouble if we keep in mind that exponents are used to indicate repeated multiplication. Let's consider some examples.

Example 7

Solution

Simplify $3(-4)^2 + 5(-3)^2$.

$$3(-4)^2 + 5(-3)^2 = 3(16) + 5(9) \quad \text{Find the powers.}$$
$$= 48 + 45$$
$$= 93 \qquad \blacktriangle$$

Example 8

Solution

Simplify $[3(-1) - 2(1)]^3$.

$$[3(-1) - 2(1)]^3 = [-3 - 2]^3$$
$$= [-5]^3$$
$$= -125 \qquad \blacktriangle$$

Example 9

Solution

Simplify $2(2)^2(-3) - 3(2)(-3)^2 - (-3)^3$.

$$2(2)^2(-3) - 3(2)(-3)^2 - (-3)^3 = 2(4)(-3) - 3(2)(9) - (-27)$$
$$= -24 - 54 + 27$$
$$= -51 \qquad \blacktriangle$$

Problem Set 1.3

For Problems 1–14, state the property that justifies each of the statements. For example, $3 + (-4) = (-4) + 3$ because of the commutative property of addition.

1. $[6 + (-2)] + 4 = 6 + [(-2) + 4]$

2. $x(3) = 3(x)$

3. $42 + (-17) = -17 + 42$

4. $1(x) = x$

5. $-114 + 114 = 0$

6. $(-1)(48) = -48$

7. $-1(x + y) = -(x + y)$

8. $-3(2 + 4) = -3(2) + (-3)(4)$

9. $12yx = 12xy$

10. $[(-7)(4)](-25) = (-7)[4(-25)]$

11. $7(4) + 9(4) = (7 + 9)4$

12. $(x + 3) + (-3) = x + (3 + (-3))$

13. $[(-14)(8)](25) = (-14)[8(25)]$

14. $\left(\dfrac{3}{4}\right)\left(\dfrac{4}{3}\right) = 1$

For Problems 15–26, simplify each numerical expression. Be sure to take advantage of the properties whenever they can be used to make the computations easier.

15. $36 + (-14) + (-12) + 21 + (-9) - 4$
16. $-37 + 42 + 18 + 37 + (-42) - 6$
17. $[83 + (-99)] + 18$
18. $[63 + (-87)] + (-64)$
19. $(25)(-13)(4)$
20. $(14)(25)(-13)(4)$
21. $17(97) + 17(3)$
22. $-86(49 + (-48))$
23. $14 - 12 - 21 - 14 + 17 - 18 + 19 - 32$
24. $16 - 14 - 13 - 18 + 19 + 14 - 17 + 21$
25. $(-50)(15)(-2) - (-4)(17)(25)$
26. $(2)(17)(-5) - (4)(13)(-25)$

For Problems 27–54, simplify each of the numerical expressions.

27. $2^3 - 3^3$
28. $3^2 - 2^4$
29. $-5^2 - 4^2$
30. $-7^2 + 5^2$
31. $(-2)^3 - 3^2$
32. $(-3)^3 + 3^2$
33. $3(-1)^3 - 4(3)^2$
34. $4(-2)^3 - 3(-1)^4$
35. $7(2)^3 + 4(-2)^3$
36. $-4(-1)^2 - 3(2)^3$
37. $-3(-2)^3 + 4(-1)^5$
38. $5(-1)^3 - (-3)^3$
39. $(-3)^2 - 3(-2)(5) + 4^2$
40. $(-2)^2 - 3(-2)(6) - (-5)^2$
41. $2^3 + 3(-1)^3(-2)^2 - 5(-1)(2)^2$
42. $-2(3)^2 - 2(-2)^3 - 6(-1)^5$
43. $[4(-3) + 5(-2)^2]^2$
44. $[-3(2) - 2(-1)^3]^3$
45. $[3(-2)^2 - 2(-3)^2]^3$
46. $[-3(-1)^3 - 4(-2)^2]^2$
47. $2(-1)^3 - 3(-1)^2 + 4(-1) - 5$
48. $(-2)^3 + 2(-2)^2 - 3(-2) - 1$

49. $2^4 - 2(2)^3 - 3(2)^2 + 7(2) - 10$
50. $3(-3)^3 + 4(-3)^2 - 5(-3) + 7$
51. $5(-1)^4 - 4(-1)^3 - 3(-1)^2 + 8(-1) - 14$
52. $-3(-1)^5 - 2(-1)^4 + 4(-1)^3 - 7(-1)^2 + 9 \cdot (-1) + 12$
53. $-3^2 - 2^3 + 4(-1)^4 - (-2)^3 + 11$
54. $-2^4 + 2^3 - 2^2 + 2 - 15$

C **55.** Do Problems 41–54 with your calculator and be sure that your calculator can handle the use of exponents.

For Problems 56–64, use your calculator to evaluate each numerical expression.

C **56.** 2^{10} **57.** 3^7 **58.** $(-2)^8$
59. $(-2)^{11}$ **60.** -4^9
61. -5^6 **62.** $(3.14)^3$
63. $(1.41)^4$
64. $(1.73)^5$

THOUGHTS INTO WORDS

65. State, in your own words, the multiplication property of negative one.

66. Explain how the associative and commutative properties can help simplify $[(25)(97)](-4)$.

67. Your friend keeps getting an answer of 64 when simplifying -2^6. What mistake is she making and how would you help her?

1.4 Algebraic Expressions

Algebraic expressions such as

$$2x, \qquad 8xy, \qquad 3xy^2, \qquad -4a^2b^3c, \qquad \text{and} \qquad z$$

are called **terms**. A term is an indicated product that may have any number of factors. The variables involved in a term are called **literal factors** and the numerical factor is called the **numerical coefficient**. Thus, in $8xy$, the x and y are literal factors and 8 is the numerical coefficient. The numerical coefficient of the term $-4a^2bc$ is -4. Since $1(z) = z$, the numerical coefficient of the term z is understood to be 1. Terms that have the same literal factors are called **similar terms** or **like terms**. Some examples of similar terms are

$$3x \quad \text{and} \quad 14x, \qquad\qquad\qquad 5x^2 \quad \text{and} \quad 18x^2,$$

$$7xy \quad \text{and} \quad -9xy, \qquad\qquad\qquad 9x^2y \quad \text{and} \quad -14x^2y,$$

$$2x^3y^2, \quad 3x^3y^2, \quad \text{and} \quad -7x^3y^2.$$

By the symmetric property of equality we can write the distributive property as

$$ab + ac = a(b + c).$$

Then the commutative property of multiplication can be applied to change the form to

$$ba + ca = (b + c)a.$$

This latter form provides the basis for simplifying algebraic expressions by **combining similar terms**. Consider the following examples.

$$3x + 5x = (3 + 5)x \qquad\qquad -6xy + 4xy = (-6 + 4)xy$$
$$= 8x, \qquad\qquad\qquad\qquad = -2xy,$$

$$5x^2 + 7x^2 + 9x^2 = (5 + 7 + 9)x^2 \qquad 4x - x = 4x - 1x$$
$$= 21x^2, \qquad\qquad\qquad = (4 - 1)x = 3x$$

More complicated expressions might first require some rearranging of the terms by applying the commutative property for addition.

$$7x + 2y + 9x + 6y = 7x + 9x + 2y + 6y$$
$$= (7 + 9)x + (2 + 6)y$$
$$= 16x + 8y,$$

$$6a - 5 - 11a + 9 = 6a + (-5) + (-11a) + 9$$
$$= 6a + (-11a) + (-5) + 9$$
$$= (6 + (-11))a + 4$$
$$= -5a + 4$$

As soon as you feel that you understand the various simplifying steps you may want to do the steps mentally. Then you could go directly from the given expression to the simplified form as follows.

$$14x + 13y - 9x + 2y = 5x + 15y,$$

$$3x^2y - 2y + 5x^2y + 8y = 8x^2y + 6y,$$

$$-4x^2 + 5y^2 - x^2 - 7y^2 = -5x^2 - 2y^2$$

Applying the distributive property to remove parentheses and then to combine similar terms sometimes simplifies an algebraic expression (as the next examples illustrate.)

$$
\begin{aligned}
4(x + 2) + 3(x + 6) &= 4(x) + 4(2) + 3(x) + 3(6) \\
&= 4x + 8 + 3x + 18 \\
&= 4x + 3x + 8 + 18 \\
&= (4 + 3)x + 26 \\
&= 7x + 26,
\end{aligned}
$$

$$
\begin{aligned}
-5(y + 3) - 2(y - 8) &= -5(y) - 5(3) - 2(y) - 2(-8) \\
&= -5y - 15 - 2y + 16 \\
&= -5y - 2y - 15 + 16 \\
&= -7y + 1,
\end{aligned}
$$

$$
\begin{aligned}
5(x - y) - (x + y) &= 5(x - y) - 1(x + y) \qquad \text{Remember } -a = -1(a). \\
&= 5(x) - 5(y) - 1(x) - 1(y) \\
&= 5x - 5y - 1x - 1y \\
&= 4x - 6y
\end{aligned}
$$

When multiplying two terms such as 3 and $2x$, the associative property for multiplication provides the basis for simplifying the product.

$$3(2x) = (3 \cdot 2)x = 6x$$

This idea can be put to use in the following example.

$$
\begin{aligned}
3(2x + 5y) + 4(3x + 2y) &= 3(2x) + 3(5y) + 4(3x) + 4(2y) \\
&= 6x + 15y + 12x + 8y \\
&= 6x + 12x + 15y + 8y \\
&= 18x + 23y
\end{aligned}
$$

After you are sure of each step, a more simplified format may be used, as the following examples illustrate.

$$
\begin{aligned}
5(a + 4) - 7(a + 3) &= 5a + 20 - 7a - 21 \qquad \text{Be careful with this sign.} \\
&= -2a - 1,
\end{aligned}
$$

$$3(x^2 + 2) + 4(x^2 - 6) = 3x^2 + 6 + 4x^2 - 24$$
$$= 7x^2 - 18,$$

$$2(3x - 4y) - 5(2x - 6y) = 6x - 8y - 10x + 30y$$
$$= -4x + 22y$$

Evaluating Algebraic Expressions

An algebraic expression takes on a numerical value whenever each variable in the expression is replaced by a real number. For example, if x is replaced by 5 and y by 9, the algebraic expression $x + y$ becomes the numerical expression $5 + 9$, which simplifies to 14. We say that $x + y$ has a value of 14 when x equals 5 and y equals 9. If $x = -3$ and $y = 7$, then $x + y$ has a value of $-3 + 7 = 4$. The following examples illustrate the process of finding a value of an algebraic expression. We commonly refer to the process as **evaluating algebraic expressions**.

Example 1

Find the value of $3x - 4y$ when $x = 2$ and $y = -3$.

Solution

$$3x - 4y = 3(2) - 4(-3), \quad \text{when } x = 2 \text{ and } y = -3$$
$$= 6 + 12$$
$$= 18 \qquad\qquad \blacktriangle$$

Example 2

If $a = -1$ and $b = 4$, find the value of $2a^2 - 5b^2$.

Solution

$$2a^2 - 5b^2 = 2(-1)^2 - 5(4)^2, \quad \text{when } a = -1 \text{ and } b = 4$$
$$= 2(1) - 5(16)$$
$$= 2 - 80$$
$$= -78 \qquad\qquad \blacktriangle$$

Example 3

Evaluate $x^2 - 2xy + y^2$ for $x = -2$ and $y = -5$.

Solution

$$x^2 - 2xy + y^2 = (-2)^2 - 2(-2)(-5) + (-5)^2, \quad \text{when } x = -2 \text{ and } y = -5$$
$$= 4 - 20 + 25$$
$$= 9 \qquad\qquad \blacktriangle$$

Example 4

Evaluate $(3x + 2y)(2x - y)$ for $x = 4$ and $y = -1$.

Solution

$$(3x + 2y)(2x - y) = [3(4) + 2(-1)][2(4) - (-1)], \quad \text{when } x = 4$$
$$\text{and } y = -1$$
$$= (12 - 2)(8 + 1)$$
$$= (10)(9)$$
$$= 90 \qquad\qquad \blacktriangle$$

Simplifying by combining similar terms will help you evaluate some algebraic expressions. The following examples illustrate this idea.

Example 5

Evaluate $7x - 2y + 4x - 3y$ for $x = -3$ and $y = 6$.

Solution

Let's first simplify the given expression.

$$7x - 2y + 4x - 3y = 11x - 5y$$

Now we can evaluate for $x = -3$ and $y = 6$.

$$11x - 5y = 11(-3) - 5(6), \quad \text{when } x = -3 \text{ and } y = 6$$
$$= -33 - 30$$
$$= -63 \qquad \blacktriangle$$

Example 6

Evaluate $2(3x + 1) - 3(4x - 3)$ for $x = -6$.

Solution

$$2(3x + 1) - 3(4x - 3) = 6x + 2 - 12x + 9$$
$$= -6x + 11$$

Substituting $x = -6$, we obtain

$$-6x + 11 = -6(-6) + 11$$
$$= 36 + 11$$
$$= 47. \qquad \blacktriangle$$

Example 7

Evaluate $2(a^2 + 1) - 3(a^2 + 5) + 4(a^2 - 1)$ for $a = 10$.

Solution

$$2(a^2 + 1) - 3(a^2 + 5) + 4(a^2 - 1) = 2a^2 + 2 - 3a^2 - 15 + 4a^2 - 4$$
$$= 3a^2 - 17$$

Substituting $a = 10$, we obtain

$$3a^2 - 17 = 3(10)^2 - 17$$
$$= 3(100) - 17$$
$$= 300 - 17$$
$$= 283. \qquad \blacktriangle$$

Translating from English to Algebra

To use the tools of algebra to solve problems we must be able to translate from English to algebra. This translation process requires that we recognize key phrases in the English language that translate into algebraic expressions (which involve the operations of addition, subtraction, multiplication, and division). Some of these key phrases and their algebraic counterparts are listed in the following table. The variable n represents the number being referred to in each phrase.

English phrase	Algebraic expression
Addition	
The sum of a number and 4	$n + 4$
7 more than a number	$n + 7$
A number plus 10	$n + 10$
A number increased by 6	$n + 6$
8 added to a number	$n + 8$
Subtraction	
14 minus a number	$14 - n$
12 less than a number	$n - 12$
A number decreased by 10	$n - 10$
The difference between a number and 2	$n - 2$
A number subtracted from 13	$13 - n$
Multiplication	
14 times a number	$14n$
The product of 4 and a number	$4n$
$\frac{3}{4}$ of a number	$\frac{3}{4}n$
Twice a number	$2n$
Multiply a number by 12	$12n$
Division	
The quotient of 6 and a number	$\frac{6}{n}$
The quotient of a number and 6	$\frac{n}{6}$
A number divided by 9	$\frac{n}{9}$
The ratio of a number and 4	$\frac{n}{4}$
Mixture of operations	
4 more than three times a number	$3n + 4$
5 less than twice a number	$2n - 5$
3 times the sum of a number and 2	$3(n + 2)$
2 more than the quotient of a number and 12	$\frac{n}{12} + 2$
7 times the difference of 6 and a number	$7(6 - n)$

An English statement may not always contain the key words *sum, difference, product,* or *quotient.* Instead, the statement may describe a physical situation and from this description we must deduce the operations involved. We make some suggestions for handling such situations in the following examples.

Example 8

Sonya can type 65 words per minute. How many words will she type in m minutes?

Solution

The total number of words typed equals the product of the rate per minute and the number of minutes. Therefore, Sonya should be able to type $65m$ words in m minutes.   ▲

Example 9

Russ has n nickels and d dimes. Express this amount of money in cents.

Solution

Each nickel is worth 5 cents and each dime is worth 10 cents. We represent the amount in cents by $5n + 10d$.   ▲

Example 10

The cost of a 50-pound sack of fertilizer is d dollars. How much is the cost per pound for the fertilizer?

Solution

We calculate the price per pound by dividing the total cost by the number of pounds. We represent the price per pound by $\dfrac{d}{50}$.   ▲

The English statement we want to translate to algebra may contain some geometric ideas. Tables 1.1 and 1.2 contain some of the basic relationships that pertain to linear measurement in the English and metric systems, respectively.

TABLE 1.1

English System

12 inches	= 1 foot
3 feet	= 1 yard
1760 yards	= 1 mile
5280 feet	= 1 mile

TABLE 1.2

Metric System

1 kilometer	=	1000 meters
1 hectometer	=	100 meters
1 dekameter	=	10 meters
1 decimeter	=	.1 meter
1 centimeter	=	.01 meter
1 millimeter	=	.001 meter

Example 11

The distance between two cities is k kilometers. Express this distance in meters.

Solution

Since 1 kilometer equals 1000 meters, the distance in meters is represented by $1000k$.   ▲

Example 12

The length of a rope is y yards and f feet. Express this length in inches.

Solution

Since 1 foot equals 12 inches and 1 yard equals 36 inches, the length of the rope in inches can be represented by $36y + 12f$.   ▲

Example 13

The length of a rectangle is l centimeters and the width is w centimeters. Express the perimeter of the rectangle in meters.

Solution

A sketch of the rectangle may be helpful (Figure 1.3).

FIGURE 1.3

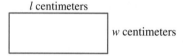

l centimeters

w centimeters

The perimeter of a rectangle is the sum of the lengths of the four sides. Thus, the perimeter in centimeters is $l + w + l + w$, which simplifies to $2l + 2w$. Now, since 1 centimeter equals .01 of a meter, the perimeter, in meters, is $.01(2l + 2w)$. This could also be written as $\dfrac{2l + 2w}{100} = \dfrac{2(l + w)}{100} = \dfrac{l + w}{50}$.   ▲

▸ Problem Set 1.4

Simplify each of the following algebraic expressions by combining similar terms.

1. $-7x + 11x$
2. $5x - 8x + x$
3. $5a^2 - 6a^2$
4. $12b^3 - 17b^3$
5. $4n - 9n - n$
6. $6n + 13n - 15n$
7. $4x - 9x + 2y$
8. $7x - 9y - 10x - 13y$
9. $-3a^2 + 7b^2 + 9a^2 - 2b^2$
10. $-xy + z - 8xy - 7z$
11. $15x - 4 + 6x - 9$
12. $5x - 2 - 7x + 4 - x - 1$
13. $5a^2b - ab^2 - 7a^2b$
14. $8xy^2 - 5x^2y + 2xy^2 + 7x^2y$

Simplify each of the following algebraic expressions by removing parentheses and combining similar terms.

15. $3(x + 2) + 5(x + 3)$
16. $5(x - 1) + 7(x + 4)$
17. $-2(a - 4) - 3(a + 2)$
18. $-7(a + 1) - 9(a + 4)$
19. $3(n^2 + 1) - 8(n^2 - 1)$
20. $4(n^2 + 3) + (n^2 - 7)$
21. $-6(x^2 - 5) - (x^2 - 2)$
22. $3(x + y) - 2(x - y)$
23. $5(2x + 1) + 4(3x - 2)$
24. $5(3x - 1) + 6(2x + 3)$
25. $3(2x - 5) - 4(5x - 2)$

26. $3(2x - 3) - 7(3x - 1)$

27. $-2(n^2 - 4) - 4(2n^2 + 1)$

28. $-4(n^2 + 3) - (2n^2 - 7)$

29. $3(2x - 4y) - 2(x + 9y)$

30. $-7(2x - 3y) + 9(3x + y)$

31. $3(2x - 1) - 4(x + 2) - 5(3x + 4)$

32. $-2(x - 1) - 5(2x + 1) + 4(2x - 7)$

33. $-(3x - 1) - 2(5x - 1) + 4(-2x - 3)$

34. $4(-x - 1) + 3(-2x - 5) - 2(x + 1)$

Evaluate each of the following algebraic expressions for the given values of the variables.

35. $3x + 7y, \quad x = -1$ and $y = -2$

36. $5x - 9y, \quad x = -2$ and $y = 5$

37. $4x^2 - y^2, \quad x = 2$ and $y = -2$

38. $3a^2 + 2b^2, \quad a = 2$ and $b = 5$

39. $2a^2 - ab + b^2, \quad a = -1$ and $b = -2$

40. $-x^2 + 2xy + 3y^2, \quad x = -3$ and $y = 3$

41. $2x^2 - 4xy - 3y^2, \quad x = 1$ and $y = -1$

42. $4x^2 + xy - y^2, \quad x = 3$ and $y = -2$

43. $3xy - x^2y^2 + 2y^2, \quad x = 5$ and $y = -1$

44. $x^2y^3 - 2xy + x^2y^2, \quad x = -1$ and $y = -3$

45. $7a - 2b - 9a + 3b, \quad a = 4$ and $b = -6$

46. $-4x + 9y - 3x - y, \quad x = -4$ and $y = 7$

47. $-9a^2 + 6 + 7a^2 - 14, \quad a = -5$

48. $9x - y - 3y + 8y, \quad x = 6$ and $y = -7$

49. $-2a - 3a + 7b - b, \quad a = -10$ and $b = 9$

50. $3(x - 2) - 4(x + 3), \quad x = -2$

51. $-2(x + 4) - (2x - 1), \quad x = -3$

52. $-4(2x - 1) + 7(3x + 4), \quad x = 4$

53. $2(x - 1) - (x + 2) - 3(2x - 1), \quad x = -1$

54. $-3(x + 1) + 4(-x - 2) - 3(-x + 4), \quad x = 2$

55. $3(x^2 - 1) - 4(x^2 + 1) - (2x^2 - 1), \quad x = -4$

56. $2(n^2 + 1) - 3(2n^2 - 3) + 3(5n^2 - 2),$
$n = -3$

57. $5(x - 2y) - 3(2x + y) - 2(x - y), \quad x = -1$
and $y = -2$

C For Problems 58–63, use your calculator and evaluate each of the algebraic expressions for the indicated values. Express the final answers to the nearest tenth.

58. $\pi r^2, \quad \pi = 3.14$ and $r = 2.1$

59. $\pi r^2, \quad \pi = 3.14$ and $r = 8.4$

60. $\pi r^2 h, \quad \pi = 3.14, r = 1.6,$ and $h = 11.2$

61. $\pi r^2 h, \quad \pi = 3.14, r = 4.8,$ and $h = 15.1$

62. $2\pi r^2 + 2\pi rh, \quad \pi = 3.14, r = 3.9,$ and $h = 17.6$

63. $2\pi r^2 + 2\pi rh, \quad \pi = 3.14, r = 7.8,$ and $h = 21.2$

For Problems 64–78, translate each English phrase into an algebraic expression and use n to represent the unknown number.

64. The sum of a number and 4

65. A number increased by 12

66. A number decreased by 7

67. Five less than a number

68. A number subtracted from 75

69. The product of a number and 50

70. One-third of a number

71. Four less than one-half of a number

72. Seven more than three times a number

73. The quotient of a number and 8

74. The quotient of 50 and a number

75. Nine less than twice a number

76. Six more than one-third of a number

77. Ten times the difference of a number and 6

78. Twelve times the sum of a number and 7

For Problems 79–99, answer the question with an algebraic expression.

79. Brian is *n* years old. How old will he be in 20 years?

80. Crystal is *n* years old. How old was she 5 years ago?

81. Pam is *t* years old and her mother is 3 less than twice as old as Pam. What is the age of Pam's mother?

82. The sum of two numbers is 65 and one of the numbers is *x*. What is the other number?

83. The difference of two numbers is 47 and the smaller number is *n*. What is the other number?

84. The product of two numbers is 98 and one of the numbers is *n*. What is the other number?

85. The quotient of two numbers is 8 and the smaller number is *y*. What is the other number?

86. The perimeter of a square is *c* centimeters. How long is each side of the square?

87. The perimeter of a square is *m* meters. How long, in centimeters, is each side of the square?

88. Jesse has *n* nickels, *d* dimes, and *q* quarters in his bank. How much money, in cents, does he have in his bank?

89. Tina has *c* cents, which is all in quarters. How many quarters does she have?

90. If *n* represents a whole number, what represents the next larger whole number?

91. If *n* represents an odd integer, what represents the next larger odd integer?

92. If *n* represents an even integer, what represents the next larger even integer?

93. The cost of a 5-pound box of candy is *c* cents. What is the price per pound?

94. Larry's annual salary is *d* dollars. What is his monthly salary?

95. Mila's monthly salary is *d* dollars. What is her annual salary?

96. The perimeter of a square is *i* inches. What is the perimeter expressed in feet?

97. The perimeter of a rectangle is *y* yards and *f* feet. What is the perimeter expressed in feet?

98. The length of a line segment is *d* decimeters. How long is the line segment expressed in meters?

99. The distance between two cities is *m* miles. How far is this, expressed in feet?

THOUGHTS INTO WORDS

100. Explain the difference between *simplifying a numerical expression* and *evaluating an algebraic expression*.

101. How would you help someone who is having difficulty expressing *n* nickels and *d* dimes in terms of cents?

SUMMARY

(1.1) A **set** is a collection of objects; the objects are called **elements** or **members** of the set. Set *A* is a subset of set *B* if and only if every member of *A* is also a member of *B*. The sets of **natural numbers, whole numbers, integers, rational numbers,** and **irrational numbers** are all subsets of the set of **real numbers.**

We can evaluate **numerical expressions** by performing the operations in the following order.

1. Perform the operations inside the parentheses and above and below fraction bars.
2. Find all powers or convert them to indicated multiplication.
3. Perform all multiplications and divisions in the order in which they appear from left to right.
4. Perform all additions and subtractions in the order that they appear from left to right.

(1.2) The **absolute value** of a real number a is defined by (1) if $a \geq 0$, then $|a| = a$, and (2) if $a < 0$, then $|a| = -a$.

The sum of two positive integers is the sum of their absolute values. The sum of two negative integers is the opposite of the sum of their absolute values. The sum of a positive and a negative integer is given by: (a) If the positive integer has the larger absolute value, then their sum is the difference of their absolute values when you subtract the smaller absolute value from the larger; (b) If the negative integer has the larger absolute value, then their sum is the opposite of the difference of their absolute values when you subtract the smaller absolute value from the larger.

The statement $a - b = a + (-b)$ changes every subtraction problem to an equivalent addition problem.

The **product** of two positive or two negative integers is the product of their absolute values. The product of one positive and one negative integer is the opposite of the product of their absolute values.

The **quotient** of two positive or two negative integers is the quotient of their absolute values. The quotient of one positive and one negative integer is the opposite of the quotient of their absolute values.

(1.3) The following basic properties of real numbers help with numerical manipulations and serve as a basis for algebraic computations.

Closure Properties $a + b$ is a real number
ab is a real number

Commutative Properties $a + b = b + a$
$ab = ba$

Associative Properties $(a + b) + c = a + (b + c)$
$(ab)c = a(bc)$

Identity Properties $a + 0 = 0 + a = a$
$$a(1) = 1(a) = a$$

Additive Inverse Property $a + (-a) = (-a) + a = 0$

Multiplication Property of Zero $a(0) = 0(a) = 0$

Multiplication Property of Negative One $-1(a) = a(-1) = -a$

Multiplicative Inverse Property $a\left(\dfrac{1}{a}\right) = \left(\dfrac{1}{a}\right)a = 1$

Distributive Properties $a(b + c) = ab + ac$
$$a(b - c) = ab - ac$$

(1.4) Algebraic expressions such as $2x$, $8xy$, $3xy^2$, $-4a^2b^3c$, and z are called **terms**. A term is an indicated product and may have any number of factors. We call the variables in a term the **literal factors** and we call the numerical factor the **numerical coefficient**. Terms that have the same literal factors are called **similar** or **like terms**.

The distributive property in the form $ba + ca = (b + c)a$ serves as the basis for **combining similar terms**. For example,

$$3x^2y + 7x^2y = (3 + 7)x^2y = 10x^2y.$$

To translate English phrases into algebraic expressions, we must be familiar with the standard vocabulary of *sum*, *difference*, *product*, and *quotient*, as well as other terms that express the same ideas.

Chapter 1 Review Problem Set

1. From the list 0, $\sqrt{2}$, $\dfrac{3}{4}$, $-\dfrac{5}{6}$, $8\dfrac{1}{3}$, $-\sqrt{3}$, -8, $.34$, $.2\overline{3}$, 67, and $\dfrac{9}{7}$ identify each of the following.

(a) The natural numbers
(b) The integers
(c) The nonnegative integers
(d) The rational numbers

(e) The irrational numbers

For Problems 2–10, state the property of equality or the property of real numbers that justifies each of the statements. For example, $6(-7) = -7(6)$ because of the *commutative property for multiplication*; and if $2 = x + 3$, then $x + 3 = 2$ is true because of the *symmetric property of equality*.

2. $7 + (3 + (-8)) = (7 + 3) + (-8)$

3. If $x = 2$ and $x + y = 9$, then $2 + y = 9$

4. $-1(x + 2) = -(x + 2)$

5. $3(x + 4) = 3(x) + 3(4)$
6. $[(17)(4)](25) = (17)[(4)(25)]$

7. $x + 3 = 3 + x$
8. $3(98) + 3(2) = 3(98 + 2)$
9. $\left(\frac{3}{4}\right)\left(\frac{4}{3}\right) = 1$
10. If $4 = 3x - 1$, then $3x - 1 = 4$.

For Problems 11–22, simplify each of the numerical expressions.

11. $-8 + (-4) - (-6)$
12. $9 - 12 + (-4) - (-1)$
13. $-8 - 7 + 4 + 6 - 10 + 8 - 3$
14. $4(-3) - 12 \div (-4) + (-2)(-1) - 8$
15. $-3(2 - 4) - 4(7 - 9) + 6$
16. $(48 + (-73)) + 74$
17. $[5(-2) - 3(-1)][-2(-1) + 3(2)]$
18. $-4^2 - 2^3$
19. $(-2)^4 + (-1)^3 - 3^2$
20. $2(-1)^2 - 3(-1)(2) - 2^2$
21. $[4(-1) - 2(3)]^3$
22. $3 - [-2(3 - 4)] + 7$

For Problems 23–32, simplify each of the algebraic expressions by combining similar terms.

23. $3a^2 - 2b^2 - 7a^2 - 3b^2$
24. $4x - 6 - 2x - 8 + x + 12$
25. $ab^2 - 3a^2b + 4ab^2 + 6a^2b$
26. $3(x + 1) - 2(x - 4) + 5(x + 4)$
27. $3(2n^2 + 1) + 4(n^2 - 5)$
28. $-2(3a - 1) + 4(2a + 3) - 5(3a + 2)$

29. $-(n - 1) - (n + 2) + 3$
30. $3(2x - 3y) - 4(3x + 5y) - x$

31. $4(a - 6) - (3a - 1) - 2(4a - 7)$
32. $-5(x^2 - 4) - 2(3x^2 + 6) + (2x^2 - 1)$

For Problems 33–42, evaluate each of the algebraic expressions for the given values of the variables.

33. $-5x + 4y$ for $x = 3$ and $y = -4$
34. $3x^2 - 2y^2$ for $x = -1$ and $y = -2$
35. $-5(2x - 3y)$ for $x = 1$ and $y = -3$
36. $(3a - 2b)^2$ for $a = -2$ and $b = 3$
37. $a^2 + 3ab - 2b^2$ for $a = 2$ and $b = -2$
38. $3n^2 - 4 - 4n^2 + 9$ for $n = 7$
39. $3(2x - 1) + 2(3x + 4)$ for $x = 4$
40. $-4(3x - 1) - 5(2x - 1)$ for $x = -2$
41. $2(n^2 + 3) - 3(n^2 + 1) + 4(n^2 - 6)$ for $n = 3$

42. $5(3n - 1) - 7(-2n + 1) + 4(3n - 1)$ for $n = -5$

For Problems 43–50, translate each English phrase into an algebraic expression and use n to represent the unknown number.

43. Four increased by twice a number
44. Fifty subtracted from three times a number
45. Six less than two-thirds of a number
46. Ten times the difference of a number and 14
47. Eight more than five times a number
48. The quotient of a number and three less than the number
49. Three less than five times the sum of a number and 2
50. Three-fourths of the sum of a number and 12

For Problems 51–60, answer the question with an algebraic expression.

51. The sum of two numbers is 37 and one of the numbers is n. What is the other number?

52. Yuriko can type w words in an hour. What is her typing rate per minute?

53. Harry is y years old. His brother is 7 years less than twice as old as Harry. How old is Harry's brother?

54. If n represents a multiple of three, what represents the next largest multiple of three?

55. Celia has p pennies, n nickels, and q quarters. How much, in cents, does Celia have?

56. The perimeter of a square is i inches. How long, in feet, is each side of the square?

57. The length of a rectangle is y yards and the width is f feet. What is the perimeter of the rectangle expressed in inches?

58. The length of a piece of wire is d decimeters. What is the length expressed in centimeters?

59. Joan is f feet and i inches tall. How tall is she, in inches?

60. The perimeter of a rectangle is 50 centimeters. If the rectangle is c centimeters long, how wide is it?

CHAPTER I TEST

1. State the property of equality that justifies writing $x + 4 = 6$ for $6 = x + 4$.

2. State the property of real numbers that justifies writing $5(10 + 2)$ as $5(10) + 5(2)$.

For Problems 3–11, simplify each numerical expression.

3. $-4 - (-3) + (-5) - 7 + 10$

4. $7 - 8 - 3 + 4 - 9 - 4 + 2 - 12$

5. $5(-2) - 3(-4) + 6(-3) + 1$

6. $(-6) \cdot 3 \div (-2) - 8 \div (-4)$

7. $-4(3 - 6) - 5(2 - 9)$ **8.** $[48 + (-93)] + (-49)$

9. $3(-2)^3 + 4(-2)^2 - 9(-2) - 14$

10. $[2(-6) + 5(-4)][-3(-4) - 7(6)]$

11. $[-2(-3) - 4(2)]^5$

12. Simplify $6x^2 - 3x - 7x^2 - 5x - 2$ by combining similar terms.

13. Simplify $3(3n - 1) - 4(2n + 3) + 5(-4n - 1)$ by removing parentheses and combining similar terms.

For Problems 14–20, evaluate each algebraic expression for the given values of the variables.

14. $-7x - 3y$ for $x = -6$ and $y = 5$

15. $3a^2 - 4b^2$ for $a = -1$ and $b = 3$

16. $6x - 9y - 8x + 4y$ for $x = 4$ and $y = -8$

17. $-5n^2 - 6n + 7n^2 + 5n - 1$ for $n = -6$

18. $-7(x - 2) + 6(x - 1) - 4(x + 3)$ for $x = -2$

19. $-2xy - x + 4y$ for $x = 3$ and $y = 9$

20. $4(n^2 + 1) - (2n^2 + 3) - 2(n^2 + 3)$ for $n = -4$

For Problems 21 and 22, translate the English phrase into an algebraic expression using n to represent the unknown number.

21. Thirty subtracted from six times a number

22. Four more than three times a number

(continued on next page)

CHAPTER 1 TEST (*continued*)

For Problems 23–25, answer each question with an algebraic expression.

23. The product of two numbers is 72 and one of the numbers is n. What is the other number?

24. Tao has n nickels, d dimes, and q quarters. How much money, in cents, does she have?

25. The length of a rectangle is x yards and the width is y feet. What is the perimeter of the rectangle expressed in feet?

Equations, Inequalities, and Problem Solving

A retailer of sporting goods bought a putter for $18. He wants to price the putter to make a profit of 40% of the selling price. What price should he mark on the putter? **The equation $s = 18 + .4s$ can be used to determine that the putter should be sold for $30.**

Throughout this text we develop algebraic skills, use the skills to help solve equations and inequalities, and then use equations and inequalities to solve applied problems. In this chapter we review and expand concepts that are important for our problem solving development.

2.1 Solving First-Degree Equations

In Section 1.1 we stated that *an equality (equation) is a statement where two symbols, or groups of symbols, are names for the same number.* It should be further stated that an equation may be true or false. For example, the equation $3 + (-8) = -5$ is true, but the equation $-7 + 4 = 2$ is false.

Algebraic equations contain one or more variables. The following are examples of algebraic equations.

$$3x + 5 = 8, \qquad 4y - 6 = -7y + 9, \qquad x^2 - 5x - 8 = 0,$$
$$3x + 5y = 4, \qquad x^3 + 6x^2 - 7x - 2 = 0$$

An algebraic equation such as $3x + 5 = 8$ is neither true nor false as it stands and we often refer to it as an *open sentence*. Each time that a number is substituted for x, the algebraic equation $3x + 5 = 8$ becomes a numerical statement that is true or false. For example, if $x = 0$, then $3x + 5 = 8$ becomes $3(0) + 5 = 8$, which is a false statement. If $x = 1$, then $3x + 5 = 8$ becomes $3(1) + 5 = 8$, which is a true statement. **Solving an equation** refers to the process of finding the number (or numbers) that makes an algebraic equation a true numerical statement. We call such numbers the **solutions** or **roots** of the equation that **satisfy** the equation. We call the set of all solutions of an equation its **solution set.** Thus, $\{1\}$ is the solution set of $3x + 5 = 8$.

In this chapter we will consider techniques for solving **first-degree equations in one variable.** This means that the equations contain only one variable and that this variable has an exponent of one. The following are examples of first-degree equations in one variable.

$$3x + 5 = 8, \qquad \frac{2}{3}y + 7 = 9,$$
$$7a - 6 = 3a + 4, \qquad \frac{x - 2}{4} = \frac{x - 3}{5}$$

Equivalent equations are equations that have the same solution. For example,

1. $3x + 5 = 8$
2. $3x = 3$
3. $x = 1$

are all equivalent equations since $\{1\}$ is the solution set of each.

The general procedure for solving an equation is to continue replacing the given equation with equivalent *but simpler* equations until we obtain an equation of the form *variable = constant* or *constant = variable*. Thus, in the example above, $3x + 5 = 8$ was simplified to $3x = 3$, which was further simplified to $x = 1$, from which the solution set $\{1\}$ is obvious.

To solve equations we need to use the various properties of equality. In addition to the *reflexive, symmetric, transitive,* and *substitution properties* we listed in Section 1.1, the following properties of equality play an important role.

Addition Property of Equality

For all real numbers a, b, and c,

$$a = b \quad \text{if and only if } a + c = b + c.$$

Multiplication Property of Equality

For all real numbers a, b, and c, where $c \neq 0$,

$$a = b \quad \text{if and only if } ac = bc.$$

The addition property of equality states that *any number can be added to both sides of an equation and an equivalent equation is produced*. The multiplication property of equality states that *we obtain an equivalent equation whenever we multiply both sides of an equation by the same nonzero real number*. The following examples demonstrate the use of these properties to solve equations.

Example 1

Solution

Solve $2x - 1 = 13$.

$$2x - 1 = 13$$
$$2x - 1 + 1 = 13 + 1 \qquad \text{Add 1 to both sides.}$$
$$2x = 14$$
$$\frac{1}{2}(2x) = \frac{1}{2}(14) \qquad \text{Multiply both sides by } \frac{1}{2}.$$
$$x = 7$$

The solution set is $\{7\}$.   ▲

To check an apparent solution we can substitute it into the original equation and see if we obtain a true numerical statement.

 Check

$$2x - 1 = 13$$
$$2(7) - 1 \overset{?}{=} 13$$
$$14 - 1 \overset{?}{=} 13$$
$$13 = 13$$

Now we know that $\{7\}$ is the solution set of $2x - 1 = 13$. We will not show our checks for every example in this text, but do remember that checking is a way to detect arithmetic errors.

Example 2

Solution

Solve $-7 = -5a + 9$.

$$-7 = -5a + 9$$

$$-7 + (-9) = -5a + 9 + (-9) \qquad \text{Add } -9 \text{ to both sides.}$$

$$-16 = -5a$$

$$-\frac{1}{5}(-16) = -\frac{1}{5}(-5a) \qquad \text{Multiply both sides by } -\frac{1}{5}.$$

$$\frac{16}{5} = a$$

The solution set is $\left\{\dfrac{16}{5}\right\}$. ▲

Notice that in Example 2 the final equation is $\dfrac{16}{5} = a$ instead of $a = \dfrac{16}{5}$. Technically, the symmetric property of equality (if $a = b$, then $b = a$) would permit us to change from $\dfrac{16}{5} = a$ to $a = \dfrac{16}{5}$, but such a change is not necessary to determine that the solution is $\dfrac{16}{5}$. Notice that we could use the symmetric property at the very beginning to change $-7 = -5a + 9$ to $-5a + 9 = -7$; some people prefer working with the variable on the left side of the equation.

Let's clarify another point. We stated the properties of equality in terms of only two operations, addition and multiplication. We could also include the operations of subtraction and division in the statements of the properties. That is to say, we could think in terms of *subtracting the same number from both sides of an equation* and also in terms of *dividing both sides of an equation by the same nonzero number*. For example, in the solution of Example 2, we could subtract 9 from both sides rather than add -9 to both sides. Likewise, we could divide both sides by -5 instead of multiplying both sides by $-\dfrac{1}{5}$.

Example 3

Solution

Solve $7x - 3 = 5x + 9$.

$$7x - 3 = 5x + 9$$

$$7x - 3 + (-5x) = 5x + 9 + (-5x) \qquad \text{Add } -5x \text{ to both sides.}$$

$$2x - 3 = 9$$

$$2x - 3 + 3 = 9 + 3 \qquad \text{Add 3 to both sides.}$$

$$2x = 12$$

$$\frac{1}{2}(2x) = \frac{1}{2}(12) \qquad \text{Multiply both sides by } \frac{1}{2}.$$

$$x = 6$$

The solution set is $\{6\}$. ▲

Example 4

Solve $4(y - 1) + 5(y + 2) = 3(y - 8)$.

Solution

$$4(y - 1) + 5(y + 2) = 3(y - 8)$$
$$4y - 4 + 5y + 10 = 3y - 24$$

Remove parentheses by applying distributive property.

$$9y + 6 = 3y - 24$$

Simplify left side by combining similar terms.

$$9y + 6 + (-3y) = 3y - 24 + (-3y)$$

Add $-3y$ to both sides.

$$6y + 6 = -24$$
$$6y + 6 + (-6) = -24 + (-6)$$

Add -6 to both sides.

$$6y = -30$$
$$\frac{1}{6}(6y) = \frac{1}{6}(-30)$$

Multiply both sides by $\frac{1}{6}$.

$$y = -5$$

The solution set is $\{-5\}$. ▲

We can summarize the process of solving first-degree equations in one variable as follows.

STEP 1 Simplify both sides of the equation as much as possible.

STEP 2 Use the addition property of equality to isolate a term that contains the variable on one side and a constant on the other side of the equation.

STEP 3 Use the multiplication property of equality to make the coefficient of the variable 1. That is, multiply both sides of the equation by the reciprocal of the numerical coefficient of the variable. The solution set should now be obvious.

STEP 4 Check each solution by substituting it in the original equation and verify that the resulting numerical statement is true.

Use of Equations to Solve Problems

In order to use the tools of algebra to solve problems we must be able to translate back and forth between the English language and the language of algebra. More specifically, we need to translate *English sentences* into *algebraic equations*. Such translations allow us to use our knowledge of *equation solving* to solve word problems. Let's consider an example.

Problem 1

If we subtract 27 from three times a certain number, the result is 18. Find the number.

Solution

Let n represent the number to be found. The sentence *if we subtract 27 from three times a certain number, the result is 18* translates into the equation $3n - 27 = 18$. Solving this equation, we obtain

$$3n - 27 = 18$$
$$3n = 45 \qquad \text{Add 27 to both sides.}$$
$$n = 15. \qquad \text{Multiply both sides by } \tfrac{1}{3}.$$

The number to be found is 15.   ▲

We often refer to the statement *let n represent the number to be found* as **declaring the variable.** We need to choose a letter to use as a variable and indicate what it represents for a specific problem. This may seem like an insignificant idea, but as the problems become more complex the process of declaring the variable becomes even more important. Furthermore, it is true that you could probably solve a problem such as Problem 1 without setting up an algebraic equation. However, as problems increase in difficulty the translation from English to algebra becomes a key issue. Therefore, even with these relatively easy problems we suggest that you concentrate on the translation process.

The next example involves the use of integers. Remember that the set of integers consists of $\{ \ldots -2, -1, 0, 1, 2, \ldots \}$. Furthermore, the integers can be classified as *even*, $\{ \ldots -4, -2, 0, 2, 4, \ldots \}$ or *odd*, $\{ \ldots -3, -1, 1, 3, \ldots \}$.

Problem 2

The sum of three consecutive integers is 13 greater than twice the smallest of the three integers. Find the integers.

Solution

Since consecutive integers differ by 1, we will represent them as follows: let n represent the smallest of the three consecutive integers; then $n + 1$ represents the second largest, and $n + 2$ represents the largest.

The sum of the three consecutive integers   13 greater than twice the smallest

$$n + (n + 1) + (n + 2) = 2n + 13$$
$$3n + 3 = 2n + 13$$
$$n = 10$$

The three consecutive integers are 10, 11, and 12.   ▲

To check our answers for Problem 2 we must determine whether or not they satisfy the conditions stated in the original problem. Since 10, 11, and 12 are consecutive integers whose sum is 33 and since twice the smallest plus 13 is also 33 $(2(10) + 13 = 33)$, we know that our answers are correct. (Remember, when checking a result for a word problem it is *not* sufficient to check the result in the equation set up to solve the problem; the equation itself may be in error!)

In the two previous problems, the equation formed was almost a direct translation of a sentence in the statement of the problem. Now let's consider a situation where we need to think in terms of a *guideline* not explicitly stated in the problem.

Problem 3

Khoa received a car repair bill for $106. This included $23 for parts, $22 per hour for each hour of labor, and $6 for taxes. Find the number of hours of labor.

Solution

See Figure 2.1. Let h represent the number of hours of labor. Then $22h$ represents the total charge for labor. We can use a guideline of *charge for parts plus charge for labor plus tax equals the total bill* to set up the following equation.

$$\underset{\underset{23}{\downarrow}}{\text{Parts}} + \underset{\underset{22h}{\downarrow}}{\text{Labor}} + \underset{\underset{6}{\downarrow}}{\text{Tax}} = \underset{\underset{106}{\downarrow}}{\text{Total bill}}$$

Solving this equation we obtain

$$22h + 29 = 106$$
$$22h = 77$$
$$h = 3\frac{1}{2}.$$

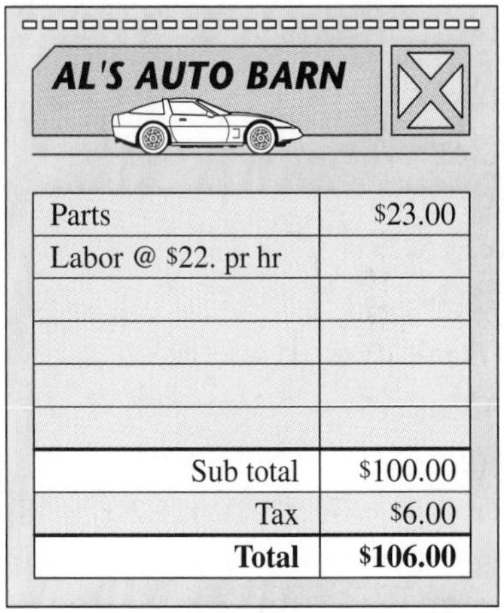

Khoa was charged for $3\frac{1}{2}$ hours of labor.

AL'S AUTO BARN	
Parts	$23.00
Labor @ $22. pr hr	
Sub total	$100.00
Tax	$6.00
Total	**$106.00**

FIGURE 2.1

Problem Set 2.1

Solve each of the following equations.

1. $3x + 4 = 16$
2. $4x + 2 = 22$
3. $5x + 1 = -14$
4. $7x + 4 = -31$
5. $-x - 6 = 8$
6. $8 - x = -2$
7. $4y - 3 = 21$
8. $6y - 7 = 41$
9. $3x - 4 = 15$
10. $5x + 1 = 12$
11. $-4 = 2x - 6$
12. $-14 = 3a - 2$

13. $-6y - 4 = 16$
14. $-8y - 2 = 18$
15. $4x - 1 = 2x + 7$
16. $9x - 3 = 6x + 18$
17. $5y + 2 = 2y - 11$
18. $9y + 3 = 4y - 10$
19. $3x + 4 = 5x - 2$
20. $2x - 1 = 6x + 15$
21. $-7a + 6 = -8a + 14$
22. $-6a - 4 = -7a + 11$
23. $5x + 3 - 2x = x - 15$
24. $4x - 2 - x = 5x + 10$
25. $6y + 18 + y = 2y + 3$

26. $5y + 14 + y = 3y - 7$

27. $4x - 3 + 2x = 8x - 3 - x$

28. $x - 4 - 4x = 6x + 9 - 8x$

29. $6n - 4 - 3n = 3n + 10 + 4n$

30. $2n - 1 - 3n = 5n - 7 - 3n$

31. $4(x - 3) = -20$

32. $3(x + 2) = -15$

33. $-3(x - 2) = 11$

34. $-5(x - 1) = 12$

35. $5(2x + 1) = 4(3x - 7)$

36. $3(2x - 1) = 2(4x + 7)$

37. $5x - 4(x - 6) = -11$

38. $3x - 5(2x + 1) = 13$

39. $-2(3x - 1) - 3 = -4$

40. $-6(x - 4) - 10 = -12$

41. $-2(3x + 5) = -3(4x + 3)$

42. $-(2x - 1) = -5(2x + 9)$

43. $3(x - 4) - 7(x + 2) = -2(x + 18)$

44. $4(x - 2) - 3(x - 1) = 2(x + 6)$

45. $-2(3n - 1) + 3(n + 5) = -4(n - 4)$

46. $-3(4n + 2) + 2(n - 6) = -2(n + 1)$

47. $3(2a - 1) - 2(5a + 1) = 4(3a + 4)$

48. $4(2a + 3) - 3(4a - 2) = 5(4a - 7)$

49. $-2(n - 4) - (3n - 1) = -2 + (2n - 1)$

50. $-(2n - 1) + 6(n + 3) = -4 - (7n - 11)$

Solve each of the following problems by setting up and solving an algebraic equation.

51. If 15 is subtracted from three times a certain number the result is 27. Find the number.

52. If 1 is subtracted from seven times a certain number the result is the same as if 31 is added to three times the number. Find the number.

53. Find three consecutive integers whose sum is 42.

54. Find four consecutive integers whose sum is −118.

55. Find three consecutive odd integers such that three times the second minus the third is 11 more than the first.

56. Find three consecutive even integers such that four times the first minus the third is 6 more than twice the second.

57. The difference of two numbers is 67. The larger number is 3 less than six times the smaller number. Find the numbers.

58. The sum of two numbers is 103. The larger number is 1 more than five times the smaller number. Find the numbers.

59. Angelo is paid double time for each hour he works over 40 hours in a week. Last week he worked 46 hours and earned $572. What is his normal hourly rate?

60. Suppose that a plumbing repair bill, not including tax, was $130. This included $25 for parts and an amount for 5 hours of labor. Find the hourly rate that was charged for labor.

61. Suppose that Maria has 150 coins consisting of pennies, nickels, and dimes. The number of nickels she has is 10 less than twice the number of pennies; the number of dimes she has is 20 less than three times the number of pennies. How many coins of each kind does she have?

62. Hector has a collection of nickels, dimes, and quarters totaling 122 coins. The number of dimes he has is 3 more than four times the number of nickels, and the number of quarters he has is 19 less than the number of dimes. How many coins of each kind does he have?

63. The selling price of a ring is $750. This represents $150 less than three times the cost of the ring. Find the cost of the ring in Figure 2.2.

FIGURE 2.2

64. In a class of 62 students, the number of females is one less than twice the number of males. How many females and how many males are there in the class?

65. An apartment complex contains 230 apartments each having one, two, or three bedrooms. The number of two-bedroom apartments is 10 more than three times the number of three-bedroom apartments. The number of one-bedroom apartments is twice the number of two-bedroom apartments. How many apartments of each kind are in the complex?

66. Barry sells bicycles on a salary-plus-commission basis. He receives a monthly salary of $300 and a commission of $15 for each bicycle that he sells. How many bicycles must he sell in a month to have a total monthly income of $750?

THOUGHTS INTO WORDS

67. Explain the difference between a numerical statement and an algebraic equation.

68. Are the equations $7 = 9x - 4$ and $9x - 4 = 7$ equivalent equations? Defend your answer.

69. Suppose that your friend shows you the following solution to an equation.

$$17 = 4 - 2x$$
$$17 + 2x = 4 - 2x + 2x$$

$$17 + 2x = 4$$
$$17 + 2x - 17 = 4 - 17$$
$$2x = -13$$
$$x = \frac{-13}{2}$$

Is this a correct solution? What suggestions would you have in terms of the method used to solve the equation?

70. Explain in your own words what it means *to declare a variable* when solving a word problem.

Further Investigations

71. Solve each of the following equations.
(a) $5x + 7 = 5x - 4$
(b) $4(x - 1) = 4x - 4$
(c) $3(x - 4) = 2(x - 6)$
(d) $7x - 2 = -7x + 4$
(e) $2(x - 1) + 3(x + 2) = 5(x - 7)$
(f) $-4(x - 7) = -2(2x + 1)$

72. Verify that for any three consecutive integers, the sum of the smallest and largest is equal to twice the middle integer. [*Hint*: Use n, $n + 1$, and $n + 2$ to represent the three consecutive integers.]

73. Verify that no four consecutive integers can be found such that the product of the smallest and largest is equal to the product of the other two integers.

2.2 Equations Involving Fractional Forms

To solve equations that involve fractions it is usually easiest to begin by **clearing the equation of all fractions**. This can be accomplished by multiplying both sides of the equation by the least common multiple of all the denominators in the equation. Remember that the least common multiple of a set of whole numbers is the smallest nonzero whole number that is divisible by each of the numbers. For example, the least common multiple of 2, 3, and 6 is 12. When working with fractions, we refer to the least common multiple of a set of denominators as the **least common denominator** (LCD). Let's consider some equations involving fractions.

Example 1

Solve $\frac{1}{2}x + \frac{2}{3} = \frac{3}{4}$.

Solution

$$\frac{1}{2}x + \frac{2}{3} = \frac{3}{4}$$

$$12\left(\frac{1}{2}x + \frac{2}{3}\right) = 12\left(\frac{3}{4}\right) \qquad \text{Multiply both sides by 12, which is the LCD of 2, 3, and 4.}$$

$$12\left(\frac{1}{2}x\right) + 12\left(\frac{2}{3}\right) = 12\left(\frac{3}{4}\right) \qquad \text{Apply distributive property on left side.}$$

$$6x + 8 = 9$$

$$6x = 1$$

$$x = \frac{1}{6}$$

The solution set is $\left\{\frac{1}{6}\right\}$.

✔ **Check**

$$\frac{1}{2}x + \frac{2}{3} = \frac{3}{4}$$

$$\frac{1}{2}\left(\frac{1}{6}\right) + \frac{2}{3} \overset{?}{=} \frac{3}{4}$$

$$\frac{1}{12} + \frac{2}{3} \overset{?}{=} \frac{3}{4}$$

$$\frac{1}{12} + \frac{8}{12} \overset{?}{=} \frac{3}{4}$$

$$\frac{9}{12} \overset{?}{=} \frac{3}{4}$$

$$\frac{3}{4} = \frac{3}{4}$$

Example 2

Solve $\frac{x}{2} + \frac{x}{3} = 10$.

Solution

$$\frac{x}{2} + \frac{x}{3} = 10 \qquad \text{Recall that } \frac{x}{2} = \frac{1}{2}x.$$

$$6\left(\frac{x}{2} + \frac{x}{3}\right) = 6(10) \qquad \text{Multiply both sides by the LCD.}$$

$$6\left(\frac{x}{2}\right) + 6\left(\frac{x}{3}\right) = 6(10) \qquad \text{Apply distributive property on left side.}$$

$$3x + 2x = 60$$

$$5x = 60$$

$$x = 12$$

The solution set is $\{12\}$.

As you study the examples of this section, pay special attention to the steps shown in the solutions. Certainly, there are no rules as to which steps should be performed mentally; this is an individual decision. When you solve problems, show enough steps to allow the flow of the process to be understood and to minimize the chances of making careless computational errors.

Example 3

Solve $\dfrac{x-2}{3} + \dfrac{x+1}{8} = \dfrac{5}{6}$.

Solution

$$\frac{x-2}{3} + \frac{x+1}{8} = \frac{5}{6}$$

$$24\left(\frac{x-2}{3} + \frac{x+1}{8}\right) = 24\left(\frac{5}{6}\right) \qquad \text{Multiply both sides by the LCD.}$$

$$24\left(\frac{x-2}{3}\right) + 24\left(\frac{x+1}{8}\right) = 24\left(\frac{5}{6}\right) \qquad \text{Apply distributive property to left side.}$$

$$8(x-2) + 3(x+1) = 20$$

$$8x - 16 + 3x + 3 = 20$$

$$11x - 13 = 20$$

$$11x = 33$$

$$x = 3$$

The solution set is $\{3\}$. ▲

Example 4

Solve $\dfrac{3t-1}{5} - \dfrac{t-4}{3} = 1$.

Solution

$$\frac{3t-1}{5} - \frac{t-4}{3} = 1$$

$$15\left(\frac{3t-1}{5} - \frac{t-4}{3}\right) = 15(1) \qquad \text{Multiply both sides by LCD.}$$

$$15\left(\frac{3t-1}{5}\right) - 15\left(\frac{t-4}{3}\right) = 15(1) \qquad \text{Apply distributive property to left side.}$$

$$3(3t-1) - 5(t-4) = 15$$

$$9t - 3 - 5t + 20 = 15 \qquad \text{Be careful with this sign!}$$

$$4t + 17 = 15$$

$$4t = -2$$

$$t = -\frac{2}{4} = -\frac{1}{2} \qquad \text{Reduce!}$$

The solution set is $\left\{-\dfrac{1}{2}\right\}$. ▲

Problem Solving

As we expand our skills for solving equations, we also expand our capabilities for solving word problems. There is no one definite procedure that will ensure success at solving word problems, but the following suggestions can be helpful.

Suggestions for Solving Word Problems

1. Read the problem carefully and make certain that you understand the meanings of all of the words. Be especially alert for any technical terms used in the statement of the problem.

2. Read the problem a second time (perhaps even a third time) to get an overview of the situation being described. Determine the known facts as well as what is to be found.

3. Sketch any figure, diagram, or chart that might be helpful in analyzing the problem.

4. Choose a meaningful variable to represent an unknown quantity in the problem (perhaps t, if time is an unknown quantity) and represent any other unknowns in terms of that variable.

5. Look for a *guideline* that you can use to set up an equation. A guideline might be a formula such as *distance equals rate times time*, or a statement of a relationship such as *the sum of the two numbers is 28*.

6. Form an equation containing the variable that translates the conditions of the guideline from English to algebra.

7. Solve the equation and use the solution to determine all facts requested in the problem.

8. Check all answers back into the **original statement of the problem.**

Keep these suggestions in mind as we continue to solve problems. We will elaborate on some of these suggestions at different times throughout the text. Now let's consider some problems.

Problem 1

Find a number such that three-eighths of the number minus one-half of it is 14 less than three-fourths of the number.

Solution

Let n represent the number to be found.

$$\frac{3}{8}n - \frac{1}{2}n = \frac{3}{4}n - 14$$

$$8\left(\frac{3}{8}n - \frac{1}{2}n\right) = 8\left(\frac{3}{4}n - 14\right)$$

$$8\left(\frac{3}{8}n\right) - 8\left(\frac{1}{2}n\right) = 8\left(\frac{3}{4}n\right) - 8(14)$$

$$3n - 4n = 6n - 112$$

$$-n = 6n - 112$$

$$-7n = -112$$

$$n = 16$$

The number is 16. Check it!   ▲

Problem 2

The width of a rectangular parking lot is 8 feet less than three-fifths of the length. The perimeter of the lot is 400 feet. Find the length and width of the lot.

Solution

Let l represent the length of the lot. Then $\frac{3}{5}l - 8$ represents the width (Figure 2.3).

FIGURE 2.3

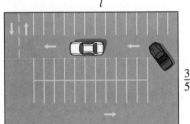

A *guideline* for this problem is the formula *perimeter of a rectangle equals twice the length plus twice the width* $(P = 2l + 2w)$. Use this formula to form the following equation.

$$
\begin{array}{ccc}
P & = 2l + & 2w \\
\downarrow & \downarrow & \downarrow \\
400 & = 2l + & 2\left(\frac{3}{5}l - 8\right)
\end{array}
$$

Solving this equation we obtain

$$400 = 2l + \frac{6l}{5} - 16$$

$$5(400) = 5\left(2l + \frac{6l}{5} - 16\right)$$

$$2000 = 10l + 6l - 80$$

$$2000 = 16l - 80$$

$$2080 = 16l$$

$$130 = l.$$

The length of the lot is 130 feet and the width is $\frac{3}{5}(130) - 8 = 70$ feet.   ▲

In Problems 1 and 2 notice the use of different letters as variables. It is helpful to choose a variable that has significance for the problem you are working on. For example, in Problem 2 the choice of l to represent the length seems natural and meaningful. (Certainly this is another matter of personal preference, but you might consider it.)

In Problem 2 a geometric relationship ($P = 2l + 2w$) serves as a guideline for setting up the equation. The following geometric relationships pertaining to angle measure may also serve as guidelines.

1. **Two angles** whose sum of measures is 90° are called **complementary angles**.

2. **Two angles** whose sum of measures is 180° are called **supplementary angles**.

3. The sum of the measures of the three angles of a triangle is 180°.

Problem 3

One of two complementary angles is 6° larger than one-half of the other angle. Find the measure of each of the angles.

Solution

Let a represent the measure of one of the angles. Then $\frac{1}{2}a + 6$ represents the measure of the other angle. Since they are complementary angles, the sum of their measures is 90°.

$$a + \left(\frac{1}{2}a + 6\right) = 90$$

$$2a + a + 12 = 180$$

$$3a + 12 = 180$$

$$3a = 168$$

$$a = 56$$

If $a = 56$, then $\frac{1}{2}a + 6$ becomes $\frac{1}{2}(56) + 6 = 34$. The angles have measures of 34° and 56°. ▲

Problem 4

Find the measures of the three angles of a triangle if the smallest angle is one-sixth of the largest and the other angle is twice the smallest angle.

Solution

Let a represent the measure of the largest angle. Then $\frac{1}{6}a$ represents the measure of the smallest angle and $2\left(\frac{1}{6}a\right)$, which simplifies to $\frac{1}{3}a$, represents the other angle. Since the sum of the measures of the angles of a triangle is 180°, we can set up and solve the following equation.

$$a + \frac{1}{6}a + \frac{1}{3}a = 180$$

$$6a + a + 2a = 1080$$

$$9a = 1080$$

$$a = 120$$

If $a = 120$, then $\frac{1}{6}a$ becomes $\frac{1}{6}(120) = 20$ and $\frac{1}{3}a$ becomes $\frac{1}{3}(120) = 40$. The angles have measures of $120°$, $20°$, and $40°$.

Problem Set 2.2

Solve each of the following equations.

1. $\frac{3}{4}x = 9$

2. $\frac{2}{3}x = -14$

3. $\frac{-2x}{3} = \frac{2}{5}$

4. $\frac{-5x}{4} = \frac{7}{2}$

5. $\frac{n}{2} - \frac{2}{3} = \frac{5}{6}$

6. $\frac{n}{4} - \frac{5}{6} = \frac{5}{12}$

7. $\frac{5n}{6} - \frac{n}{8} = \frac{-17}{12}$

8. $\frac{2n}{5} - \frac{n}{6} = \frac{-7}{10}$

9. $\frac{a}{4} - 1 = \frac{a}{3} + 2$

10. $\frac{3a}{7} - 1 = \frac{a}{3}$

11. $\frac{h}{4} + \frac{h}{5} = 1$

12. $\frac{h}{6} + \frac{3h}{8} = 1$

13. $\frac{h}{2} - \frac{h}{3} + \frac{h}{6} = 1$

14. $\frac{3h}{4} + \frac{2h}{5} = 1$

15. $\frac{x - 2}{3} + \frac{x + 3}{4} = \frac{11}{6}$

16. $\frac{x + 4}{5} + \frac{x - 1}{4} = \frac{37}{10}$

17. $\frac{x + 2}{2} - \frac{x - 1}{5} = \frac{3}{5}$

18. $\frac{2x + 1}{3} - \frac{x + 1}{7} = -\frac{1}{3}$

19. $\frac{n + 2}{4} - \frac{2n - 1}{3} = \frac{1}{6}$

20. $\frac{n - 1}{9} - \frac{n + 2}{6} = \frac{3}{4}$

21. $\frac{y}{3} + \frac{y - 5}{10} = \frac{4y + 3}{5}$

22. $\frac{y}{3} + \frac{y - 2}{8} = \frac{6y - 1}{12}$

23. $\frac{4x - 1}{10} - \frac{5x + 2}{4} = -3$

24. $\frac{2x - 1}{2} - \frac{3x + 1}{4} = \frac{3}{10}$

25. $\frac{2x - 1}{8} - 1 = \frac{x + 5}{7}$

26. $\frac{3x + 1}{9} + 2 = \frac{x - 1}{4}$

27. $\frac{2a - 3}{6} + \frac{3a - 2}{4} + \frac{5a + 6}{12} = 4$

28. $\frac{3a - 1}{4} + \frac{a - 2}{3} - \frac{a - 1}{5} = \frac{21}{20}$

29. $x + \frac{3x - 1}{9} - 4 = \frac{3x + 1}{3}$

30. $\frac{2x + 7}{8} + x - 2 = \frac{x - 1}{2}$

31. $\frac{x + 3}{2} + \frac{x + 4}{5} = \frac{3}{10}$

32. $\frac{x - 2}{5} - \frac{x - 3}{4} = -\frac{1}{20}$

33. $n + \frac{2n - 3}{9} - 2 = \frac{2n + 1}{3}$

34. $n - \frac{3n + 1}{6} - 1 = \frac{2n + 4}{12}$

35. $\frac{3}{4}(t - 2) - \frac{2}{5}(2t - 3) = \frac{1}{5}$

36. $\frac{2}{3}(2t + 1) - \frac{1}{2}(3t - 2) = 2$

37. $\frac{1}{2}(2x - 1) - \frac{1}{3}(5x + 2) = 3$

38. $\frac{2}{5}(4x - 1) + \frac{1}{4}(5x + 2) = -1$

39. $3x - 1 + \frac{2}{7}(7x - 2) = -\frac{11}{7}$

40. $2x + 5 + \frac{1}{2}(6x - 1) = -\frac{1}{2}$

Solve each of the following problems by setting up and solving an algebraic equation.

41. Find a number such that one-half of the number is 3 less than two-thirds of the number.

42. One-half of a number plus three-fourths of the number is 2 more than four-thirds of the number. Find the number.

43. Suppose that the width of a certain rectangle is 1 inch more than one-fourth of its length. The perimeter of the rectangle is 42 inches. Find the length and width of the rectangle.

44. Suppose that the width of a rectangle is 3 centimeters less than two-thirds of its length. The perimeter of the rectangle is 114 centimeters. Find the length and width of the rectangle.

45. Find three consecutive integers such that the sum of the first plus one-third of the second plus three-eighths of the third is 25.

46. Lou is paid $1\frac{1}{2}$ times his normal hourly rate for each hour he works over 40 hours in a week. Last week he worked 44 hours and earned $276. What is his normal hourly rate?

47. A board 20 feet long is cut into two pieces such that the length of one piece is two-thirds of the length of the other piece. Find the length of the shorter piece of board.

48. Jody has a collection of 116 coins consisting of dimes, quarters, and silver dollars. The number of quarters is 5 less than three-fourths of the number of dimes. The number of silver dollars is seven more than five-eighths of the number of dimes. How many coins of each kind are in her collection?

49. The sum of the present ages of Angie and her mother is 64 years. In eight years Angie will be three-fifths as old as her mother at that time. Find the present ages of Angie and her mother.

50. Annilee's present age is two-thirds of Jessie's present age. In 12 years the sum of their ages will be 54 years. Find their present ages.

51. Aura took three biology exams and has an average score of 88. Her second exam was ten points better than her first and her third exam was four points better than her second exam. What were her three exam scores?

52. The average of the salaries of Tim, Maida, and Aaron is $24,000 per year. If Maida earns $10,000 more than Tim, and Aaron's salary is $2000 more than twice Tim's salary, find the salary of each person.

53. One of two supplementary angles is 4° more than one-third of the other angle. Find the measure of each of the angles.

54. If one-half of the complement of an angle plus three-fourths of the supplement of the angle equals 110°, find the measure of the angle.

55. If the complement of an angle is 5° less than one-sixth of its supplement, find the measure of the angle.

56. In $\triangle ABC$, angle B is 8° less than one-half of angle A and angle C is 28° larger than angle A. Find the measures of the three angles of the triangle.

THOUGHTS INTO WORDS

57. Explain why the solution set of the equation $x + 3 = x + 4$ is the null set.

58. Explain why the solution set of the equation $\frac{x}{3} + \frac{x}{2} = \frac{5x}{6}$ is the entire set of real numbers.

59. Why must potential answers to word problems be checked back into the original statement of the problem?

▼ 2.3   Equations Involving Decimals

Certainly we can solve the equation $x - .17 = .18$ by adding .17 to both sides. However, as equations that contain decimals become more complex, it is often easier to first **clear the equation of all decimals** by multiplying both sides by an appropriate power of 10. Let's consider some examples.

Example 1

Solution

Solve $.6x = 1.2$.

$$.6x = 1.2$$
$$10(.6x) = 10(1.2) \qquad \text{Multiply both sides by 10.}$$
$$6x = 12$$
$$x = 2$$

✔ Check

$$.6x = 1.2$$
$$.6(2) \stackrel{?}{=} 1.2$$
$$1.2 = 1.2$$

The solution set is $\{2\}$.   ▲

Example 2

Solution

Solve $.07x + .11x = 3.6$.

$$.07x + .11x = 3.6$$
$$100(.07x + .11x) = 100(3.6) \qquad \text{Multiply both sides by 100.}$$
$$7x + 11x = 360$$
$$18x = 360$$
$$x = 20$$

✔ Check

$$.07x + .11x = 3.6$$
$$.07(20) + .11(20) \stackrel{?}{=} 3.6$$
$$1.4 + 2.2 \stackrel{?}{=} 3.6$$
$$3.6 = 3.6$$

The solution set is $\{20\}$.   ▲

Example 3

Solution

Solve $s = 1.95 + .35s$.

$$s = 1.95 + .35s$$
$$100(s) = 100(1.95 + .35s) \qquad \text{Multiply both sides by 100.}$$
$$100s = 195 + 35s$$
$$65s = 195$$
$$s = 3$$

The solution set is $\{3\}$. Check it!   ▲

Example 4

Solution

Solve $.12x + .11(7000 - x) = 790$.

$$.12x + .11(7000 - x) = 790$$
$$100[.12x + .11(7000 - x)] = 100(790) \qquad \text{Multiply both sides by 100.}$$
$$12x + 11(7000 - x) = 79{,}000$$
$$12x + 77{,}000 - 11x = 79{,}000$$
$$x + 77{,}000 = 79{,}000$$
$$x = 2000$$

The solution set is $\{2000\}$. ▲

Back to Problem Solving

We can solve many consumer problems with an algebraic approach. For example, let's consider some discount sale problems involving the relationship *original selling price minus discount equals discount sale price.*

Problem 1

Solution

Karyl bought a dress at a 35% discount sale for $32.50. What was the original price of the dress?

Let p represent the original price of the dress. Using the discount sale relationship as a guideline, the problem translates into an equation as follows.

Original selling price	Minus	Discount	Equals	Discount sale price
↓		↓		↓
$(100\%)(p)$	$-$	$(35\%)(p)$	$=$	$\$32.50$

Switching this equation to decimal form and solving, we obtain,

$$(100\%)(p) - (35\%)(p) = 32.50$$
$$(65\%)(p) = 32.50$$
$$.65p = 32.50$$
$$65p = 3250$$
$$p = 50.$$

The original price of the dress was $50. ▲

Problem 2

Solution

A pair of jogging shoes that was originally priced at $25 is on sale for 20% off. Find the discount sale price of the shoes.

Let s represent the discount sale price.

Original price	Minus	Discount	Equals	Sale price
↓		↓		↓
$\$25$	$-$	$(20\%)(\$25)$	$=$	s

Solving this equation we obtain

$$25 - (20\%)(25) = s$$
$$25 - (.2)(25) = s$$
$$25 - 5 = s$$
$$20 = s.$$

The shoes are on sale for $20.   ▲

REMARK   Keep in mind that if an item is on sale for 35% off, then you are going to pay
100% − 35% = 65% of the original price. Thus, in Problem 1 you could begin
with the equation $.65p = 32.50$. Likewise, in Problem 2, you could start with the
equation $s = .8(25)$.   △

Another basic relationship that pertains to consumer problems is *selling
price equals cost plus profit*. We can state profit (also called markup, markon,
and margin of profit) in different ways. It may be stated as a percent of the selling price, a percent of the cost, or simply in terms of dollars and cents. We shall
consider some problems for which the profit is calculated either as a percent of
the cost or as a percent of the selling price.

Problem 3   A retailer has some shirts that cost $20 each. She wants to sell them at a profit of
60% of the cost. What selling price should be marked on the shirts?

Solution   Let s represent the selling price. Use the *selling price equals cost plus profit*
relationship as a guideline.

Selling price	Equals	Cost	Plus	Profit
↓		↓		↓
s	=	$20	+	(60%)($20)

Solving this equation yields

$$s = 20 + (60\%)(20)$$
$$s = 20 + (.6)(20)$$
$$s = 20 + 12$$
$$s = 32.$$

The selling price should be $32.   ▲

REMARK   A profit of 60% of the cost means that the selling price is 100% of the cost plus
60% of the cost, or 160% of the cost. Thus, in Problem 3 we could solve the
equation $s = 1.6(20)$.   △

Problem 4   A retailer of sporting goods bought a putter for $18. He wants to price the putter
to make a profit of 40% of the selling price. What price should he mark on the
putter?

Solution

Let s represent the selling price.

Selling price	Equals	Cost	Plus	Profit
↓	↓	↓		↓
s	$=$	$\$18$	$+$	$(40\%)(s)$

Solving this equation yields

$$s = 18 + (40\%)(s)$$
$$s = 18 + .4s$$
$$10s = 180 + 4s \qquad \text{Multiply both sides by 10.}$$
$$6s = 180$$
$$s = 30.$$

The selling price should be $30. ▲

We can solve certain types of investment problems by using an algebraic approach. Consider the following examples.

Problem 5

A man invests $8000; part of it at 11% and the remainder at 12%. His total yearly interest from the two investments is $930. How much did he invest at each rate?

Solution

Let x represent the amount he invested at 11%. Then $8000 - x$ represents the amount he invested at 12%. Use the following guideline.

Interest earned from 11% investment	+	Interest earned from 12% investment	=	Total amount of interest earned
↓		↓		↓
$(11\%)(x)$	$+$	$(12\%)(8000 - x)$	$=$	$\$930$

Solving this equation yields

$$(11\%)(x) + (12\%)(8000 - x) = 930$$
$$.11x + .12(8000 - x) = 930$$
$$11x + 12(8000 - x) = 93{,}000 \qquad \text{Multiply both sides by 100.}$$
$$11x + 96{,}000 - 12x = 93{,}000$$
$$-x + 96{,}000 = 93{,}000$$
$$-x = -3000$$
$$x = 3000.$$

Therefore, $8000 - x = 5000$. $3000 was invested at 11% and $5000 at 12%.

▲

Problem 6

A certain amount of money is invested at 8% and $1500 more than that amount is invested at 9%. The annual interest from the 9% investment exceeds the annual interest from the 8% investment by $160. How much is invested at each rate?

Solution

Let x represent the amount invested at 8%. Then $x + 1500$ represents the amount invested at 9%. Use the following guideline.

$$\begin{array}{ccccc} \text{Interest earned} & = & \text{Interest earned} & + & \$160 \\ \text{from 9\% investment} & & \text{from 8\% investment} & & \end{array}$$

$$(9\%)(x + 1500) \quad = \quad (8\%)(x) \quad + \quad \$160$$

Solving this equation, we obtain

$$(9\%)(x + 1500) = (8\%)(x) + 160$$
$$.09(x + 1500) = .08(x) + 160$$
$$9(x + 1500) = 8x + 16{,}000 \qquad \text{Multiply both sides by 100.}$$
$$9x + 13{,}500 = 8x + 16{,}000$$
$$x = 2500.$$

Therefore, $x + 1500 = 4000$. $2500 was invested at 8% and $4000 at 9%.

Don't forget to check word problems; determine whether the answers satisfy the conditions stated in the original problem. A check for Problem 6 is as follows.

 Check

We claim that $2500 was invested at 8% and $4000 at 9% and this satisfies the condition, $1500 more was invested at 9% than at 8%. The $2500 at 8% produces $200 of interest and the $4000 at 9% produces $360. Therefore, the interest from the 9% investment exceeds the interest from the 8% investment by $160. The conditions of the problem are satisfied and our answers are correct.

Problem Set 2.3

Solve each of the following equations.

1. $.14x = 2.8$

2. $1.6x = 8$

3. $.09y = 4.5$

4. $.07y = .42$

5. $n + .4n = 56$

6. $n - .5n = 12$

7. $s = 9 + .25s$

8. $s = 15 + .4s$

9. $s = 3.3 + .45s$

10. $s = 2.1 + .6s$

11. $.11x + .12(900 - x) = 104$

12. $.09x + .11(500 - x) = 51$

13. $.08(x + 200) = .07x + 20$

14. $.07x = 152 - .08(2000 - x)$

15. $.12t - 2.1 = .07t - .2$

16. $.13t - 3.4 = .08t - .4$

17. $.92 + .9(x - .3) = 2x - 5.95$

18. $.3(2n - 5) = 11 - .65n$

19. $.1d + .11(d + 1500) = 795$

20. $.8x + .9(850 - x) = 715$

21. $.12x + .1(5000 - x) = 560$

22. $.10t + .12(t + 1000) = 560$

23. $.09(x + 200) = .08x + 22$

24. $.09x = 1650 - .12(x + 5000)$

25. $.3(2t + .1) = 8.43$

26. $.5(3t + .7) = 20.6$

27. $.1(x - .1) - .4(x + 2) = -5.31$

28. $.2(x + .2) + .5(x - .4) = 5.44$

Solve each of the following problems by setting up and solving an algebraic equation.

29. Judy bought a coat at a 20% discount sale for $72. What was the original price of the coat?

30. Jim bought a pair of slacks at a 25% discount sale for $24. What was the original price of the slacks?

31. Find the discount sale price of a $64 item that is on sale for 15% off.

32. Find the discount sale price of a $72 item that is on sale for 35% off.

33. A retailer has some skirts that cost $30 each. She wants to sell them at a profit of 60% of the cost. What price should she charge for the skirts?

34. The owner of a pizza parlor wants to make a profit of 70% of the cost for each pizza sold. If it costs $2.50 to make a pizza, at what price should each pizza be sold?

35. If a ring costs a jeweler $200, at what price should it be sold to make a profit of 50% on the selling price?

36. If a head of lettuce costs a retailer $.32, at what price should it be sold to make a profit of 60% on the selling price?

37. If a pair of shoes costs a retailer $24 and he sells them for $39.60, what is his rate of profit based on the cost?

38. A retailer has some skirts that cost her $45 each. If she sells them for $83.25 per skirt, find her rate of profit based on the cost.

39. Mitsuko's salary for next year is $34,775. This represents a 7% increase over this year's salary. Find Mitsuko's present salary.

40. Don bought a car with 6% tax included for $15,794. What was the price of the car without the tax?

41. Eva invested a certain amount of money at 10% interest and $1500 more than that amount at 11%. Her total yearly interest was $795. How much did she invest at each rate?

42. A total of $4000 was invested, part of it at 8% interest and the remainder at 9%. If the total yearly interest amounted to $350, how much was invested at each rate?

43. If $500 is invested at 6% interest, how much additional money must be invested at 9% so that the total return for both investments averages 8%?

44. A sum of $2000 is split between two investments, one paying 7% interest and the other 8%. If the return on the 8% investment exceeds that on the 7% investment by $40 per year, how much is invested at each rate?

45. Suppose that Javier has a handful of coins consisting of pennies, nickels, and dimes worth $2.63. The number of nickels is one less than twice the number of pennies and the number of dimes is 3 more than the number of nickels. How many coins of each kind does he have?

46. Sarah has a collection of nickels, dimes, and quarters worth $15.75. She has 10 more dimes than nickels and twice as many quarters as dimes. How many coins of each kind does she have?

47. A collection of 70 coins consisting of dimes, quarters, and half-dollars has a value of $17.75. There are three times as many quarters as dimes. Find the number of each kind of coin.

48. Abby has 37 coins, consisting only of dimes and quarters, worth $7.45. How many dimes and how many quarters does she have?

49. From a consumer's viewpoint, would you prefer that a retailer figure his profit based on the cost or on the selling price of an item? Explain your answer.

50. Is a 10% discount followed by a 30% discount the same as a 30% discount followed by a 10% discount? Justify your answer.

51. What is wrong with the following solution and how should it be done?

$$1.2x + 2 = 3.8$$
$$10(1.2x) + 2 = 10(3.8)$$
$$12x + 2 = 38$$
$$12x = 36$$
$$x = 3$$

Further Investigations

Solve each of the following equations and express the solutions in decimal form. Check all of your solutions. Use your calculator whenever it seems helpful.

52. $1.2x + 3.4 = 5.2$

53. $.12x - .24 = .66$

54. $.12x + .14(550 - x) = 72.5$

55. $.14t + .13(890 - t) = 67.95$

56. $.7n + 1.4 = 3.92$

57. $.14n - .26 = .958$

58. $.3(d + 1.8) = 4.86$

59. $.6(d - 4.8) = 7.38$

60. $.8(2x - 1.4) = 19.52$

61. $.5(3x + .7) = 20.6$

62. The following formula can be used to determine the selling price of an item when the profit is based on a percent of the selling price.

$$\text{Selling price} = \frac{\text{Cost}}{100\% - \text{Percent of profit}}$$

Show how this formula is developed.

63. A retailer buys an item for $90, resells it for $100, and claims that he is only making a 10% profit. Is his claim correct?

64. Is a 10% discount followed by a 20% discount equal to a 30% discount? Defend your answer.

2.4 Formulas

To find the distance traveled in four hours at a rate of 55 miles per hour we multiply the rate times the time; thus, the distance is $55(4) = 220$ miles. We can state the rule *distance equals rate times time* as a formula: $d = rt$. **Formulas** are rules we state in symbolic form, usually as equations.

The techniques we have considered for solving equations can be used to solve a formula for a specified variable if we are given numerical values for the other variables in the formula. Let's consider some examples.

Example 1

If we invest P dollars at r percent for t years, the amount of simple interest i is given by the formula $i = Prt$. Find the amount of interest earned by $500 at 7% for 2 years.

Solution

By substituting $500 for P, 7% for r, and 2 for t, we obtain

$$i = Prt$$
$$i = (500)(7\%)(2)$$

$$i = (500)(.07)(2)$$
$$i = 70.$$

Thus, we earn $70 in interest. ▲

Example 2

If we invest P dollars at a simple rate of r percent, then the amount A accumulated after t years is given by the formula $A = P + Prt$. If we invest $500 at 8%, how many years would it take to accumulate $600?

Solution

Substituting $500 for P, 8% for r, and $600 for A, we obtain

$$A = P + Prt$$
$$600 = 500 + 500(8\%)(t).$$

Solving this equation for t yields

$$600 = 500 + 500(.08)(t)$$
$$600 = 500 + 40t$$
$$100 = 40t$$
$$2\frac{1}{2} = t.$$

It will take $2\frac{1}{2}$ years to accumulate $600. ▲

When using a formula it is sometimes convenient to first change its form. For example, suppose we are to use the *perimeter* formula for a rectangle ($P = 2l + 2w$) to complete the following chart.

Perimeter (P)	32	24	36	18	56	80	
Length (l)	10	7	14	5	15	22	All in centimeters
Width (w)	?	?	?	?	?	?	

Since w is the unknown quantity, it would simplify the computational work if we first solved the formula for w in terms of the other variables as follows.

$$P = 2l + 2w$$
$$P - 2l = 2w \qquad \text{Add } -2l \text{ to both sides.}$$
$$\frac{P - 2l}{2} = w \qquad \text{Multiply both sides by } \frac{1}{2}.$$
$$w = \frac{P - 2l}{2} \qquad \text{Apply the symmetric property of equality.}$$

Now for each value for P and l, we can easily determine the corresponding value for w. Be sure you agree with the following values for w: 6, 5, 4, 4, 13, and 18. We can also solve the formula $P = 2l + 2w$ for l in terms of P and w as follows.

$$P = 2l + 2w$$

$$P - 2w = 2l \qquad \text{Add } -2w \text{ to both sides.}$$

$$\frac{P - 2w}{2} = l \qquad \text{Multiply both sides by } \frac{1}{2}.$$

$$l = \frac{P - 2w}{2} \qquad \text{Apply the symmetric property of equality.}$$

Let's consider some other often used formulas and see how we can use the properties of equality to alter their forms. Throughout this section, we will identify formulas when we first use them. (Some geometric formulas are also given on the endsheets.)

Example 3

Solve $A = \frac{1}{2}bh$ for h (Area of a triangle).

Solution

$$A = \frac{1}{2}bh$$

$$2A = bh \qquad \text{Multiply both sides by 2.}$$

$$\frac{2A}{b} = h \qquad \text{Multiply both sides by } \frac{1}{b}.$$

$$h = \frac{2A}{b} \qquad \text{Apply the symmetric property of equality.}$$ ▲

Example 4

Solve $A = P + Prt$ for t.

Solution

$$A = P + Prt$$

$$A - P = Prt \qquad \text{Add } -P \text{ to both sides.}$$

$$\frac{A - P}{Pr} = t \qquad \text{Multiply both sides by } \frac{1}{Pr}.$$

$$t = \frac{A - P}{Pr} \qquad \text{Apply the symmetric property of equality.}$$ ▲

Example 5

Solve $A = P + Prt$ for P.

Solution

$$A = P + Prt$$

$$A = P(1 + rt) \qquad \text{Apply the distributive property to the right side.}$$

$$\frac{A}{1 + rt} = P \qquad \text{Multiply both sides by } \frac{1}{1 + rt}.$$

$$P = \frac{A}{1 + rt} \qquad \text{Apply the symmetric property of equality.}$$ ▲

Example 6

Solve $A = \frac{1}{2}h(b_1 + b_2)$ for b_1 (Area of a trapezoid).

Solution

$$A = \frac{1}{2}h(b_1 + b_2)$$

$2A = h(b_1 + b_2)$ Multiply both sides by 2.

$2A = hb_1 + hb_2$ Apply the distributive property to right side.

$2A - hb_2 = hb_1$ Add $-hb_2$ to both sides.

$\dfrac{2A - hb_2}{h} = b_1$ Multiply both sides by $\frac{1}{h}$.

$b_1 = \dfrac{2A - hb_2}{h}$ Apply the symmetric property of equality.   ▲

In Example 5, notice that we used the distributive property to change from a form of $P + Prt$ to $P(1 + rt)$. However, in Example 6 we used the distributive property to change $h(b_1 + b_2)$ to $hb_1 + hb_2$. In both problems the key issue is to *isolate the term* that contains the variable being solved for so that an appropriate application of the multiplication property of equality will produce the desired result. Also note the use of *subscripts* to identify the two bases of a trapezoid. Subscripts allow us to use the same letter b to identify the bases, but b_1 represents one base and b_2 the other.

Sometimes we are faced with equations such as $ax + b = c$, where x is the variable and a, b, and c are referred to as *arbitrary constants*. Again we can use the properties of equality to solve the equation for x as follows.

$$ax + b = c$$

$ax = c - b$ Add $-b$ to both sides.

$x = \dfrac{c - b}{a}$ Multiply both sides by $\frac{1}{a}$.

In Chapter 7 we will be working with equations such as $2x - 5y = 7$, which are called equations of *two* variables in x and y. Often we need to change the form of such equations by *solving for one variable in terms of the other variable*. The properties of equality provide the basis for doing this.

Example 7

Solve $2x - 5y = 7$ for y in terms of x.

Solution

$$2x - 5y = 7$$

$-5y = 7 - 2x$ Add $-2x$ to both sides.

$y = \dfrac{7 - 2x}{-5}$ Multiply both sides by $-\frac{1}{5}$.

$y = \dfrac{2x - 7}{5}$ Multiply the numerator and denominator of the fraction on the right by -1. (The final step would not be absolutely necessary, but usually we prefer to have a positive number as a denominator.)   ▲

Equations of two variables may also contain arbitrary constants. For example, the equation $\frac{x}{a} + \frac{y}{b} = 1$ contains the variables x and y, and the arbitrary constants a and b.

Example 8

Solve the equation $\frac{x}{a} + \frac{y}{b} = 1$ for x.

Solution

$$\frac{x}{a} + \frac{y}{b} = 1$$

$$ab\left(\frac{x}{a} + \frac{y}{b}\right) = ab(1) \qquad \text{Multiply both sides by } ab.$$

$$bx + ay = ab$$

$$bx = ab - ay \qquad \text{Add } -ay \text{ to both sides.}$$

$$x = \frac{ab - ay}{b} \qquad \text{Multiply both sides by } \frac{1}{b}.$$ ▲

REMARK Traditionally, equations that contain more than one variable, such as those in Examples 3-8, are called **literal equations**. As illustrated, it is sometimes necessary to solve a literal equation for one variable in terms of the other variables. △

Formulas and Problem Solving

We often use formulas as *guidelines* for setting up an appropriate algebraic equation when solving a word problem. Let's consider an example to illustrate this point.

Problem 1

How long will it take $500 to double itself if we invest it at 8% simple interest?

Solution

For $500 to grow into $1000 (double itself) it must earn $500 in interest. Thus, let t represent the number of years it will take $500 to earn $500 in interest. Now we can use the formula $i = Prt$ as a guideline.

$$i = Prt$$

$$500 = 500(8\%)(t)$$

Solving this equation we obtain

$$500 = 500(.08)(t)$$

$$1 = .08t$$

$$100 = 8t$$

$$12\frac{1}{2} = t.$$

It will take $12\frac{1}{2}$ years. ▲

Sometimes we use formulas in the analysis of a problem but not as the main guideline for setting up the equation. For example, uniform motion problems involve the formula $d = rt$, but the main guideline for setting up an equation for such problems is usually a statement about either *times*, *rates*, or *distances*. Let's consider an example to demonstrate.

Problem 2

Mercedes starts jogging at 5 miles per hour. One-half hour later, Karen starts jogging on the same route at 7 miles per hour. How long will it take Karen to catch Mercedes?

Solution

First, let's sketch a diagram and record some information (Figure 2.4).

FIGURE 2.4

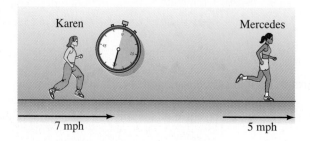

If we let t represent Karen's time, then $t + \dfrac{1}{2}$ represent Mercedes' time. We can use the statement, Karen's distance equals Mercedes' distance, as a guideline.

$$
\underset{\displaystyle 7t}{\underset{\uparrow}{\text{Karen's distance}}} \qquad = \qquad \underset{\displaystyle 5\left(t + \tfrac{1}{2}\right)}{\underset{\uparrow}{\text{Mercedes' distance}}}
$$

Solving this equation we obtain

$$7t = 5t + \frac{5}{2}$$

$$2t = \frac{5}{2}$$

$$t = \frac{5}{4}.$$

Karen should catch Mercedes in $1\dfrac{1}{4}$ hours. ▲

Note that in Problem 2 we used a simple arrow diagram to record and organize the pertinent information of the problem. Some people find it helpful to use a chart with uniform motion problems. We shall use a chart in Problem 3 but keep in mind that we are not trying to dictate a particular approach; you decide what works best for you.

Problem 3

Two trains leave a city at the same time, one traveling east and the other traveling west. At the end of $9\frac{1}{2}$ hours, they are 1292 miles apart. If the rate of the train traveling east is 8 miles per hour faster than the other train, find their rates.

Solution

If we let r represent the rate of the westbound train, then $r + 8$ represents the rate of the eastbound train. Now we can record the *times* and *rates* in a chart and then use the distance formula ($d = rt$) to represent the *distances*.

	Rate	Time	Distance (d = rt)
Westbound train	r	$9\frac{1}{2}$	$\frac{19}{2}r$
Eastbound train	$r + 8$	$9\frac{1}{2}$	$\frac{19}{2}(r + 8)$

Since the distance that the westbound train travels plus the distance that the eastbound train travels equals 1292 miles, we can set up and solve the following equation.

$$\frac{19r}{2} + \frac{19(r + 8)}{2} = 1292$$
$$19r + 19(r + 8) = 2584$$
$$19r + 19r + 152 = 2584$$
$$38r = 2432$$
$$r = 64$$

The westbound train travels at a rate of 64 miles per hour and the eastbound train at a rate of $64 + 8 = 72$ miles per hour. ▲

Now let's consider a problem that is often referred to as a mixture-type problem. There is no basic formula that applies to all of these problems but we suggest that you think in terms of a pure substance, which is often helpful in setting up a guideline. Also keep in mind that a statement such as *a 40% solution of some substance* means that the solution contains 40% of that particular substance and 60% of something else mixed with it. For example, a 40% salt solution contains 40% salt and the other 60% is something else, probably water. Now let's illustrate what we mean by the suggestion *think in terms of a pure substance*.

Problem 4

How many liters of pure alcohol must we add to 20 liters of a 40% solution to obtain a 60% solution?

Solution

The key idea to solving such a problem is to recognize the following guideline.

$$\begin{pmatrix} \text{Amount of pure} \\ \text{alcohol in the} \\ \text{original solution} \end{pmatrix} + \begin{pmatrix} \text{Amount of pure} \\ \text{alcohol to be} \\ \text{added} \end{pmatrix} = \begin{pmatrix} \text{Amount of pure} \\ \text{alcohol in the} \\ \text{final solution} \end{pmatrix}$$

Let l represent the number of liters of pure alcohol to be added and the guideline translates into the following equation.

$$(40\%)(20) + l = 60\%(20 + l)$$

Solving this equation yields

$$.4(20) + l = .6(20 + l)$$
$$8 + l = 12 + .6l$$
$$.4l = 4$$
$$l = 10.$$

We need to add 10 liters of pure alcohol. (Perhaps you should check this answer back into the original statement of the problem!)   ▲

▼ Problem Set 2.4

1. Solve $i = Prt$ for i, given that $P = \$300$, $r = 8\%$, and $t = 5$ years.

2. Solve $i = Prt$ for i, given that $P = \$500$, $r = 9\%$, and $t = 3\frac{1}{2}$ years.

3. Solve $i = Prt$ for t, given that $P = \$400$, $r = 11\%$, and $i = \$132$.

4. Solve $i = Prt$ for t, given that $P = \$250$, $r = 12\%$, and $i = \$120$.

5. Solve $i = Prt$ for r, given that $P = \$600$, $t = 2\frac{1}{2}$ years, and $i = \$90$. Express r as a percent.

6. Solve $i = Prt$ for r, given that $P = \$700$, $t = 2$ years, and $i = \$126$. Express r as a percent.

7. Solve $i = Prt$ for P, given that $r = 9\%$, $t = 3$ years, and $i = \$216$.

8. Solve $i = Prt$ for P, given that $r = 8\frac{1}{2}\%$, $t = 2$ years, and $i = \$204$.

9. Solve $A = P + Prt$ for A, given that $P = \$1000$, $r = 12\%$, and $t = 5$ years.

10. Solve $A = P + Prt$ for A, given that $P = \$850$, $r = 9\frac{1}{2}\%$, and $t = 10$ years.

11. Solve $A = P + Prt$ for r, given that $A = \$1372$, $P = \$700$, and $t = 12$ years. Express r as a percent.

12. Solve $A = P + Prt$ for r, given that $A = \$516$, $P = \$300$, and $t = 8$ years. Express r as a percent.

13. Solve $A = P + Prt$ for P, given that $A = \$326$, $r = 7\%$, and $t = 9$ years.

14. Solve $A = P + Prt$ for P, given that $A = \$720$, $r = 8\%$, and $t = 10$ years.

15. Use the formula $A = \frac{1}{2}h(b_1 + b_2)$ and complete the following chart.

A	98	104	49	162	$16\frac{1}{2}$	$38\frac{1}{2}$	Square feet
h	14	8	7	9	3	11	Feet
b_1	8	12	4	16	4	5	Feet
b_2	?	?	?	?	?	?	Feet

A = area, h = height, b_1 = one base,
b_2 = other base

16. Use the formula $P = 2l + 2w$ and complete the following chart. (You may want to change the form of the formula.)

P	28	18	12	34	68	Centimeters
w	6	3	2	7	14	Centimeters
l	?	?	?	?	?	Centimeters

P = perimeter, w = width, l = length

Solve each of the following for the indicated variable.

17. $V = Bh$ for h (Volume of a prism)

18. $A = lw$ for l (Area of a rectangle)

19. $V = \pi r^2 h$ for h (Volume of a circular cylinder)

20. $V = \dfrac{1}{3}Bh$ for B (Volume of a pyramid)

21. $C = 2\pi r$ for r (Circumference of a circle)

22. $A = 2\pi r^2 + 2\pi rh$ for h (Surface area of a circular cylinder)

23. $I = \dfrac{100M}{C}$ for C (Intelligence quotient)

24. $A = \dfrac{1}{2}h(b_1 + b_2)$ for h (Area of a trapezoid)

25. $F = \dfrac{9}{5}C + 32$ for C (Celsius to Fahrenheit)

26. $C = \dfrac{5}{9}(F - 32)$ for F (Fahrenheit to Celsius)

For Problems 27–36, solve each equation for x.

27. $y = mx + b$

28. $\dfrac{x}{a} + \dfrac{y}{b} = 1$

29. $y - y_1 = m(x - x_1)$

30. $a(x + b) = c$

31. $a(x + b) = b(x - c)$

32. $x(a - b) = m(x - c)$

33. $\dfrac{x - a}{b} = c$

34. $\dfrac{x}{a} - 1 = b$

35. $\dfrac{1}{3}x + a = \dfrac{1}{2}b$

36. $\dfrac{2}{3}x - \dfrac{1}{4}a = b$

For Problems 37–46, solve each equation for the indicated variable.

37. $2x - 5y = 7$ for x

38. $5x - 6y = 12$ for x

39. $-7x - y = 4$ for y

40. $3x - 2y = -1$ for y

41. $3(x - 2y) = 4$ for x

42. $7(2x + 5y) = 6$ for y

43. $\dfrac{y - a}{b} = \dfrac{x + b}{c}$ for x

44. $\dfrac{x - a}{b} = \dfrac{y - a}{c}$ for y

45. $(y + 1)(a - 3) = x - 2$ for y

46. $(y - 2)(a + 1) = x$ for y

Solve each of the following problems by setting up and solving an appropriate algebraic equation.

47. Suppose that the length of a certain rectangle is 2 meters less than four times its width. The perimeter of the rectangle is 56 meters. Find the length and width of the rectangle.

48. The perimeter of a triangle is 42 inches. The second side is 1 inch more than twice the first side and the third side is 1 inch less than three times the first side. Find the lengths of the three sides of the triangle.

49. How long will it take $500 to double itself at 9% simple interest?

50. How long will it take $700 to triple itself at 10% simple interest?

51. How long will it take P dollars to double itself at 9% simple interest?

52. How long will it take P dollars to triple itself at 10% simple interest?

53. Two airplanes leave Chicago at the same time and fly in opposite directions. If one travels at 450 miles per hour and the other at 550 miles per hour, how long will it take for them to be 4000 miles apart?

54. Look at Figure 2.5. Tyrone leaves city A on a moped traveling toward city B at 18 miles per hour. At the same time, Tina leaves city B on a bicycle traveling toward city A at 14 miles per hour. The distance between the two cities is 112 miles. How long will it take before Tyrone and Tina meet?

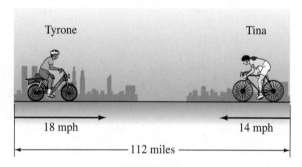

Tyrone   Tina

18 mph   14 mph

── 112 miles ──

FIGURE 2.5

55. Juan starts walking at 4 miles per hour. An hour and a half later Cathy starts jogging along the same route at 6 miles per hour. How long will it take Cathy to catch up with Juan?

56. A car leaves a town at 60 kilometers per hour. How long will it take a second car traveling at 75 kilometers per hour to catch the first car if it leaves 1 hour later?

57. Bret starts on a 70-mile bicycle ride at 20 miles per hour. After a time he becomes a little tired and slows down to 12 miles per hour for the rest of the trip. The entire trip of 70 miles took $4\frac{1}{2}$ hours. How far had Bret ridden when he reduced his speed to 12 miles per hour?

58. How many cups of grapefruit juice must be added to 40 cups of punch that contains 5% grapefruit juice to obtain a punch that is 10% grapefruit juice?

59. How many milliliters of pure acid must be added to 150 milliliters of a 30% solution of acid to obtain a 40% solution?

60. How many gallons of a 12%-salt solution must be mixed with 6 gallons of a 20%-salt solution to obtain a 15%-salt solution?

61. Suppose that you have a supply of a 30% solution of alcohol and a 70% solution of alcohol. How many quarts of each should be mixed to produce 20 quarts that is 40% alcohol?

62. A 16-quart radiator contains a 50% solution of antifreeze. How much needs to be drained out and replaced with pure antifreeze to obtain a 60%-antifreeze solution?

THOUGHTS INTO WORDS

63. Some people subtract 32 and then divide by 2 to estimate the change from a Fahrenheit reading to a Celsius reading. Why does this give an estimate and how good is the estimate?

64. One of your classmates analyzes Problem 56 as follows: The first car has traveled 60 kilometers before the second car starts. Since the second car travels 15 kilometers per hour faster, it will take $\frac{60}{15} = 4$ hours for the second car to overtake the first car. How would you react to this analysis of the problem?

65. Summarize the new ideas relative to problem solving that you have acquired thus far in this course.

Further Investigations

For Problems 66–73, use your calculator to help solve each formula for the indicated variable.

66. Solve $i = Prt$ for i, given that $P = \$875$, $r = 12\frac{1}{2}\%$, and $t = 4$ years.

67. Solve $i = Prt$ for i, given that $P = \$1125$, $r = 13\frac{1}{4}\%$, and $t = 4$ years.

68. Solve $i = Prt$ for t, given that $i = \$453.25$, $P = \$925$, and $r = 14\%$.

69. Solve $i = Prt$ for t, given that $i = \$243.75$, $P = \$1250$, and $r = 13\%$.

70. Solve $i = Prt$ for r, given that $i = \$356.50$, $P = \$1550$, and $t = 2$ years. Express r as a percent.

71. Solve $i = Prt$ for r, given that $i = \$159.50$, $P = \$2200$, and $t = .5$ of a year. Express r as a percent.

72. Solve $A = P + Prt$ for P, given that $A = \$1423.50$, $r = 9\frac{1}{2}\%$, and $t = 1$ year.

73. Solve $A = P + Prt$ for P, given that $A = \$2173.75$, $r = 8\frac{3}{4}\%$, and $t = 2$ years.

2.5 Inequalities

We listed the basic inequality symbols in Section 1.2. With these symbols we can make various **statements of inequality** as follows.

$a < b$ means a is less than b,

$a \leq b$ means a is less than or equal to b,

$a > b$ means a is greater than b,

$a \geq b$ means a is greater than or equal to b

Here are some examples of **numerical statements of inequality**.

$$7 + 8 > 10, \qquad -4 + (-6) \geq -10,$$
$$-4 > -6, \qquad 7 - 9 \leq -2,$$
$$7 - 1 < 20, \qquad 3 + 4 > 12,$$
$$8(-3) < 5(-3), \qquad 7 - 1 < 0$$

Notice that only $3 + 4 > 12$ and $7 - 1 < 0$ are *false*; the other six are *true* numerical statements.

Algebraic inequalities contain one or more variables. The following are examples of algebraic inequalities.

$$x + 4 > 8, \qquad 3x + 2y \leq 4,$$
$$3x - 1 < 15, \qquad x^2 + y^2 + z^2 \geq 7,$$
$$y^2 + 2y - 4 \geq 0$$

An algebraic inequality such as $x + 4 > 8$ is neither true nor false as it stands and we call it an **open sentence**. For each numerical value we substitute for x, the algebraic inequality $x + 4 > 8$ becomes a numerical statement of inequality that is true or false. For example, if $x = -3$ then $x + 4 > 8$ becomes $-3 + 4 > 8$, which is false. If $x = 5$, then $x + 4 > 8$ becomes $5 + 4 > 8$, which is true. **Solving an inequality** refers to the process of finding the numbers that make an algebraic inequality a true numerical statement. We call such numbers the *solutions* of the inequality; the solutions *satisfy* the inequality.

   The general process for solving inequalities closely parallels the process for solving equations. We continue to replace the given inequality with *equivalent, but simpler*, inequalities. For example,

$$3x + 4 > 10 \tag{1}$$
$$3x > 6 \tag{2}$$
$$x > 2 \tag{3}$$

are all equivalent inequalities; that is, they all have the same solutions. By inspection we see that the solutions for (3) are *all numbers greater than 2*. Thus, (1) has the same solutions.

   The exact procedure for simplifying inequalities so that we can determine the solutions is primarily based on two properties. The first of these is the addition property of inequality.

Addition Property of Inequality

For all real numbers a, b, and c,

   $a > b$ if and only if $a + c > b + c$.

The addition property of inequality states that *we can add any number to both sides of an inequality to produce an equivalent inequality*. We have stated the property in terms of $>$, but analogous properties exist for $<$, $\geq$, and $\leq$.

   Before we state the multiplication property of inequality let's look at some numerical examples.

$2 < 5$	Multiply by 4	$\underline{4}(2) < \underline{4}(5)$
$-3 > -7$	Multiply by 2	$\underline{2}(-3) > \underline{2}(-7)$
$-4 < 6$	Multiply by 10	$\underline{10}(-4) < \underline{10}(6)$
$4 < 8$	Multiply by -3	$\underline{-3}(4) > \underline{-3}(8)$
$3 > -2$	Multiply by -4	$\underline{-4}(3) < \underline{-4}(-2)$
$-4 < -1$	Multiply by -2	$\underline{-2}(-4) > \underline{-2}(-1)$

Notice in the first three examples that when we multiply both sides of an inequality by a *positive number* we get an inequality of the *same sense*. That means that if the original inequality is *less than*, then the new inequality is *less than*; and if the original inequality is *greater than*, then the new inequality is *greater than*. The last three examples illustrate that when we multiply both sides of an inequality by a *negative number* we get an inequality of the *opposite sense*.

We can state the multiplication property of inequality as follows.

Multiplication Property of Inequality

(a) For all real numbers a, b, and c, with $c > 0$,

$$a > b \quad \text{if and only if} \quad ac > bc.$$

(b) For all real numbers a, b, and c, with $c < 0$,

$$a > b \quad \text{if and only if} \quad ac < bc.$$

Similar properties hold if we reverse each inequality or if we replace $>$ with $\geq$ and $<$ with $\leq$. For example, if $a \leq b$ and $c < 0$, then $ac \geq bc$.

Now let's use the addition and multiplication properties of inequality to help solve some inequalities.

Example 1

Solve $3x - 4 > 8$.

Solution

$$3x - 4 > 8$$

$$3x - 4 + 4 > 8 + 4 \qquad \text{Add 4 to both sides.}$$

$$3x > 12$$

$$\frac{1}{3}(3x) > \frac{1}{3}(12) \qquad \text{Multiply both sides by } \frac{1}{3}.$$

$$x > 4$$

The solution set is $\{x \mid x > 4\}$. (Remember that we read the set $\{x \mid x > 4\}$ as "the set of all x such that x is greater than 4.") ▲

In Example 1, once we obtained the simple inequality $x > 4$, the solution set $\{x \mid x > 4\}$ became obvious. We can also express solution sets for inequalities on a number line graph. Figure 2.6 shows the graph of the solution set for Example 1. The left-hand parenthesis at 4 indicates that 4 is *not* a solution and the thickened portion to the right of 4 indicates that all numbers greater than 4 are solutions.

FIGURE 2.6

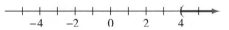

It is also convenient to express solution sets of inequalities using **interval notation**. For example, the notation $(4, \infty)$ also refers to the set of real numbers greater than 4. As in Figure 2.6, the left-hand parenthesis indicates that 4 is not to be included. The infinity symbol, ∞, along with the right-hand parenthesis, indicates that there is no right-hand endpoint. Following is a partial list of interval notations along with the sets of graphs they represent (Figure 2.7). We will add to this list in the next section.

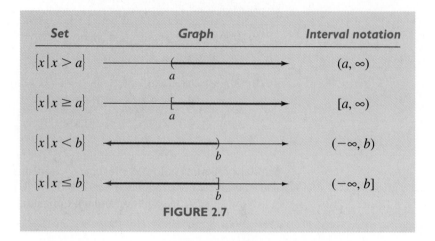

Set	Graph	Interval notation
$\{x \mid x > a\}$		(a, ∞)
$\{x \mid x \geq a\}$		$[a, \infty)$
$\{x \mid x < b\}$		$(-\infty, b)$
$\{x \mid x \leq b\}$		$(-\infty, b]$

FIGURE 2.7

Notice the use of square brackets to include endpoints. From now on, we will express the solution sets of inequalities using interval notation.

Example 2

Solve $-2x + 1 > 5$ and graph the solutions.

Solution

$$-2x + 1 > 5$$
$$-2x + 1 + (-1) > 5 + (-1) \qquad \text{Add } -1 \text{ to both sides.}$$
$$-2x > 4$$
$$-\frac{1}{2}(-2x) < -\frac{1}{2}(4) \qquad \text{Multiply both sides by } -\frac{1}{2}.$$
$$x < -2 \qquad \qquad \begin{array}{l} \text{Notice that the sense} \\ \text{of the inequality has} \\ \text{been reversed.} \end{array}$$

The solution set is $(-\infty, -2)$, which can be illustrated on a number line as in Figure 2.8.

FIGURE 2.8

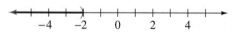

Many of the same techniques used to solve equations, such as removing parentheses and combining similar terms, may be used to solve inequalities. However, we must be extremely careful when using the multiplication property of inequality. Study each of the following examples very carefully. The format we used highlights the major steps of a solution.

Example 3

Solution

Solve $-3x + 5x - 2 \geq 8x - 7 - 9x$.

$$-3x + 5x - 2 \geq 8x - 7 - 9x$$

$$2x - 2 \geq -x - 7 \qquad \text{Combine similar terms on both sides.}$$

$$3x - 2 \geq -7 \qquad \text{Add } x \text{ to both sides.}$$

$$3x \geq -5 \qquad \text{Add 2 to both sides.}$$

$$\frac{1}{3}(3x) \geq \frac{1}{3}(-5) \qquad \text{Multiply both sides by } \frac{1}{3}.$$

$$x \geq -\frac{5}{3}$$

The solution set is $\left[-\dfrac{5}{3}, \infty\right)$.

Example 4

Solution

Solve $-5(x - 1) \leq 10$ and graph the solutions.

$$-5(x - 1) \leq 10$$

$$-5x + 5 \leq 10 \qquad \text{Apply the distributive property on left side.}$$

$$-5x \leq 5 \qquad \text{Add } -5 \text{ to both sides.}$$

$$-\frac{1}{5}(-5x) \geq -\frac{1}{5}(5) \qquad \begin{array}{l}\text{Multiply both sides by } -\dfrac{1}{5}\text{, which reverses} \\ \text{the inequality.}\end{array}$$

$$x \geq -1$$

The solution set is $[-1, \infty)$ and it can be graphed as in Figure 2.9.

FIGURE 2.9

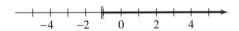

Example 5

Solution

Solve $4(x - 3) > 9(x + 1)$.

$$4(x - 3) > 9(x + 1)$$

$$4x - 12 > 9x + 9 \qquad \text{Apply the distributive property.}$$

$$-5x - 12 > 9 \qquad \text{Add } -9x \text{ to both sides.}$$

$$-5x > 21 \qquad \text{Add 12 to both sides.}$$

$$-\frac{1}{5}(-5x) < -\frac{1}{5}(21) \qquad \begin{array}{l}\text{Multiply both sides by } -\dfrac{1}{5}\text{, which} \\ \text{reverses the inequality.}\end{array}$$

$$x < -\frac{21}{5}$$

The solution set is $\left(-\infty, -\dfrac{21}{5}\right)$.

The next example will solve the inequality without indicating the justification for each step. Be sure that you can supply the reasons for the steps.

Example 6

Solve $3(2x + 1) - 2(2x + 5) < 5(3x - 2)$.

Solution

$$3(2x + 1) - 2(2x + 5) < 5(3x - 2)$$
$$6x + 3 - 4x - 10 < 15x - 10$$
$$2x - 7 < 15x - 10$$
$$-13x - 7 < -10$$
$$-13x < -3$$
$$-\frac{1}{13}(-13x) > -\frac{1}{13}(-3)$$
$$x > \frac{3}{13}$$

The solution set is $\left(\frac{3}{13}, \infty\right)$.

Checking solutions for an inequality presents a problem. Obviously, we cannot check all of the infinitely many solutions for a particular inequality. However, by checking at least one solution, especially when the multiplication property has been used, we might catch the common mistake of forgetting to change the sense of an inequality. In Example 6 we are claiming that *all numbers greater than* $\frac{3}{13}$ will satisfy the original inequality. Let's check one such number, say 1.

$$3(2x + 1) - 2(2x + 5) < 5(3x - 2)$$
$$3(2(1) + 1) - 2(2(1) + 5) \overset{?}{<} 5(3(1) - 2)$$
$$3(3) - 2(7) \overset{?}{<} 5(1)$$
$$9 - 14 \overset{?}{<} 5$$
$$-5 < 5$$

Thus, 1 satisfies the original inequality. Had we forgotten to switch the sense of the inequality when both sides were multiplied by $-\frac{1}{13}$, our answer would have been $x < \frac{3}{13}$, and we would have detected such an error by the check.

Problem Set 2.5

For Problems 1–8, express the given inequality in interval notation and sketch a graph of the interval.

1. $x > 1$
2. $x > -2$
3. $x \geq -1$
4. $x \geq 3$
5. $x < -2$
6. $x < 1$
7. $x \leq 2$
8. $x \leq 0$

For Problems 9–16, express each interval as an inequality using the variable of x. For example, we can express the interval $[5, \infty)$ as $x \geq 5$.

9. $(-\infty, 4)$ 10. $(-\infty, -2)$
11. $(-\infty, -7]$ 12. $(-\infty, 9]$
13. $(8, \infty)$ 14. $(-5, \infty)$
15. $[-7, \infty)$ 16. $[10, \infty)$

For Problems 17–40, solve each of the inequalities and graph the solution set on a number line.

17. $x - 3 > -2$
18. $x + 2 < 1$
19. $-2x \geq 8$
20. $-3x \leq -9$
21. $5x \leq -10$
22. $4x \geq -4$
23. $2x + 1 < 5$
24. $2x + 2 > 4$
25. $3x - 2 > -5$
26. $5x - 3 < -3$
27. $-7x - 3 \leq 4$
28. $-3x - 1 \geq 8$
29. $2 + 6x > -10$
30. $1 + 6x > -17$
31. $5 - 3x < 11$
32. $4 - 2x < 12$
33. $15 < 1 - 7x$
34. $12 < 2 - 5x$
35. $-10 \leq 2 + 4x$
36. $-9 \leq 1 + 2x$
37. $3(x + 2) > 6$
38. $2(x - 1) < -4$
39. $5x + 2 \geq 4x + 6$
40. $6x - 4 \leq 5x - 4$

For Problems 41–70, solve each inequality and express the solution sets using interval notation.

41. $2x - 1 > 6$
42. $3x - 2 < 12$

43. $-5x - 2 < -14$
44. $5 - 4x > -2$
45. $-3(2x + 1) \geq 12$
46. $-2(3x + 2) \leq 18$
47. $4(3x - 2) \geq -3$
48. $3(4x - 3) \leq -11$
49. $6x - 2 > 4x - 14$
50. $9x + 5 < 6x - 10$
51. $2x - 7 < 6x + 13$
52. $2x - 3 > 7x + 22$
53. $4(x - 3) \leq -2(x + 1)$
54. $3(x - 1) \geq -(x + 4)$
55. $5(x - 4) - 6(x + 2) < 4$
56. $3(x + 2) - 4(x - 1) < 6$
57. $-3(3x + 2) - 2(4x + 1) \geq 0$
58. $-4(2x - 1) - 3(x + 2) \geq 0$
59. $-(x - 3) + 2(x - 1) < 3(x + 4)$
60. $3(x - 1) - (x - 2) > -2(x + 4)$
61. $7(x + 1) - 8(x - 2) < 0$
62. $5(x - 6) - 6(x + 2) < 0$
63. $-5(x - 1) + 3 > 3x - 4 - 4x$
64. $3(x + 2) + 4 < -2x + 14 + x$
65. $3(x - 2) - 5(2x - 1) \geq 0$
66. $4(2x - 1) - 3(3x + 4) \geq 0$
67. $-5(3x + 4) < -2(7x - 1)$
68. $-3(2x + 1) > -2(x + 4)$
69. $-3(x + 2) > 2(x - 6)$
70. $-2(x - 4) < 5(x - 1)$

THOUGHTS INTO WORDS

71. Do the *less than* and *greater than* relations possess a symmetric property similar to the symmetric property of equality? Defend your answer.

72. Give a step-by-step description of how you would solve the inequality $-3 > 5 - 2x$.

73. How would you convince someone why it is necessary to reverse the inequality symbol when multiplying both sides of an inequality by a negative number?

(b) $3x - 4 < 3x + 7$
(c) $4(x + 1) < 2(2x + 5)$
(d) $-2(x - 1) > 2(x + 7)$
(e) $3(x - 2) < -3(x + 1)$
(f) $2(x + 1) + 3(x + 2) < 5(x - 3)$

Further Investigations

74. Solve each of the following inequalities.
(a) $5x - 2 > 5x + 3$

2.6 More on Inequalities

To solve equations involving fractions we found that **clearing the equation of all fractions** is frequently an effective technique. To accomplish this, multiply both sides of the equation by the least common denominator of all the denominators in the equation. This same basic approach also works very well with inequalities that involve fractions, as the next examples demonstrate.

Example 1

Solve $\frac{2}{3}x - \frac{1}{2}x > \frac{3}{4}$.

Solution

$$\frac{2}{3}x - \frac{1}{2}x > \frac{3}{4}$$

$$12\left(\frac{2}{3}x - \frac{1}{2}x\right) > 12\left(\frac{3}{4}\right)$$ Multiply both sides by 12, which is the LCD of 3, 2, and 4.

$$12\left(\frac{2}{3}x\right) - 12\left(\frac{1}{2}x\right) > 12\left(\frac{3}{4}\right)$$ Apply the distributive property.

$$8x - 6x > 9$$
$$2x > 9$$
$$x > \frac{9}{2}$$

The solution set is $\left(\frac{9}{2}, \infty\right)$.

Example 2

Solve $\frac{x + 2}{4} + \frac{x - 3}{8} < 1$.

Solution

$$\frac{x + 2}{4} + \frac{x - 3}{8} < 1$$

$$8\left(\frac{x + 2}{4} + \frac{x - 3}{8}\right) < 8(1)$$ Multiply both sides by 8, which is the LCD of 4 and 8.

$$8\left(\frac{x+2}{4}\right) + 8\left(\frac{x-3}{8}\right) < 8(1)$$

$$2(x+2) + (x-3) < 8$$

$$2x + 4 + x - 3 < 8$$

$$3x + 1 < 8$$

$$3x < 7$$

$$x < \frac{7}{3}$$

The solution set is $\left(-\infty, \frac{7}{3}\right)$.   ▲

Example 3

Solve $\dfrac{x}{2} - \dfrac{x-1}{5} \geq \dfrac{x+2}{10} - 4$.

Solution

$$\frac{x}{2} - \frac{x-1}{5} \geq \frac{x+2}{10} - 4$$

$$10\left(\frac{x}{2} - \frac{x-1}{5}\right) \geq 10\left(\frac{x+2}{10} - 4\right)$$

$$10\left(\frac{x}{2}\right) - 10\left(\frac{x-1}{5}\right) \geq 10\left(\frac{x+2}{10}\right) - 10(4)$$

$$5x - 2(x-1) \geq x + 2 - 40$$

$$5x - 2x + 2 \geq x - 38$$

$$3x + 2 \geq x - 38$$

$$2x + 2 \geq -38$$

$$2x \geq -40$$

$$x \geq -20$$

The solution set is $[-20, \infty)$.   ▲

The idea of **clearing all decimals** also works with inequalities in much the same way as it does with equations. We can multiply both sides of an inequality by an appropriate power of ten and then proceed to solve in the usual way. The next two examples illustrate this procedure.

Example 4

Solve $x \geq 1.6 + .2x$.

Solution

$$x \geq 1.6 + .2x$$

$$10(x) \geq 10(1.6 + .2x) \qquad \text{Multiply both sides by 10.}$$

$$10x \geq 16 + 2x$$

$$8x \geq 16$$

$$x \geq 2$$

The solution set is $[2, \infty)$.   ▲

Example 5

Solution

Solve $.08x + .09(x + 100) \geq 43$.

$$.08x + .09(x + 100) \geq 43$$
$$100(.08x + .09(x + 100)) \geq 100(43) \qquad \text{Multiply both sides by 100.}$$
$$8x + 9(x + 100) \geq 4300$$
$$8x + 9x + 900 \geq 4300$$
$$17x + 900 \geq 4300$$
$$17x \geq 3400$$
$$x \geq 200$$

The solution set is $[200, \infty)$.   ▲

Compound Statements

We use the words <u>and</u> and <u>or</u> in mathematics to form **compound statements**. The following are examples of some compound numerical statements that use <u>and</u>. We call such statements **conjunctions**. We agree to call a conjunction true only if all of its component parts are true. Statements 1 and 2 below are true, but 3, 4, and 5 are false.

    1. $3 + 4 = 7$ and $-4 < -3$.   True

    2. $-3 < -2$ and $-6 > -10$.   True

    3. $6 > 5$ and $-4 > -8$.   False

    4. $4 < 2$ and $0 < 10$.   False

    5. $-3 + 2 = 1$ and $5 + 4 = 8$.   False

We call compound statements that use <u>or</u> **disjunctions**. The following are some examples of disjunctions that involve numerical statements.

    6. $.14 > .13$ or $.235 < .237$.   True

    7. $\frac{3}{4} > \frac{1}{2}$ or $-4 + (-3) = 10$.   True

    8. $-\frac{2}{3} > \frac{1}{3}$ or $(.4)(.3) = .12$.   True

    9. $\frac{2}{5} < -\frac{2}{5}$ or $7 + (-9) = 16$.   False

A disjunction is true if at least one of its component parts is true. In other words, disjunctions are false only if all of the component parts are false. In the statements above, 6, 7, and 8 are true, but 9 is false.

Now let's consider finding solutions for some compound statements that involve algebraic inequalities. Keep in mind that our previous agreements for labeling conjunctions and disjunctions true or false form the basis for our reasoning.

Example 6

Graph the solution set for the conjunction $x > -1$ <u>and</u> $x < 3$.

Solution

The key word is <u>and</u>, so we need to satisfy both inequalities. Thus, all numbers between -1 and 3 are solutions and we can indicate this on a number line as in Figure 2.10.

FIGURE 2.10

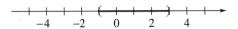

Using interval notation we can represent the interval enclosed in parentheses in Figure 2.10 by $(-1, 3)$. Using set builder notation we can express the same interval as $\{x \mid -1 < x < 3\}$, where the statement $-1 < x < 3$ is read "negative one is less than x and x is less than three." In other words, x is between -1 and 3.   ▲

Example 6 represents another concept that pertains to sets. The set of all elements common to two sets is called the **intersection** of the two sets. Thus, in Example 6 we found the intersection of the two sets $\{x \mid x > -1\}$ and $\{x \mid x < 3\}$ to be the set $\{x \mid -1 < x < 3\}$. In general, we define the intersection of two sets as follows.

DEFINITION 2.1

The **intersection** of two sets A and B (written $A \cap B$) is the set of all elements that are in both A and in B. Using set builder notation we can write

$$A \cap B = \{x \mid x \in A \quad \underline{\text{and}} \quad x \in B\}.$$

Example 7

Solve the conjunction $3x + 1 > -5$ <u>and</u> $2x + 5 > 7$ and graph its solution set on a number line.

Solution

First, let's simplify both inequalities.

$$3x + 1 > -5 \quad \text{and} \quad 2x + 5 > 7$$
$$3x > -6 \quad \text{and} \quad 2x > 2$$
$$x > -2 \quad \text{and} \quad x > 1$$

Since it is a conjunction we must satisfy both inequalities. Thus, all numbers greater than 1 are solutions and the solution set is $(1, \infty)$. We indicate the graph of the solution set in Figure 2.11.

FIGURE 2.11

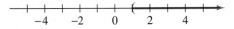

We can solve a conjunction such as $3x + 1 > -3$ and $3x + 1 < 7$, in which the same algebraic expression (in this case $3x + 1$) is contained in both inequalities, by using the **compact form** $-3 < 3x + 1 < 7$ as follows.

$$-3 < 3x + 1 < 7$$
$$-4 < 3x < 6 \qquad \text{Add } -1 \text{ to the left side, middle, and right side.}$$
$$-\frac{4}{3} < x < 2 \qquad \text{Multiply through by } \frac{1}{3}.$$

The solution set is $\left(-\frac{4}{3}, 2\right)$.

The word <u>and</u> ties the concept of a conjunction to the set concept of intersection. In a like manner, the word <u>or</u> links the idea of a disjunction to the set concept of **union**. We define the union of two sets as follows.

DEFINITION 2.2

The **union** of two sets A and B (written $A \cup B$) is the set of all elements that are in A or in B, or in both. Using set builder notation we can write

$$A \cup B = \{x \,|\, x \in A \quad \underline{or} \quad x \in B\}.$$

Example 8

Graph the solution set for the disjunction $x < -1$ <u>or</u> $x > 2$, and express it using interval notation.

Solution

The key word is <u>or</u>; so all numbers that satisfy either inequality (or both) are solutions. Thus, all numbers less than -1, along with all numbers greater than 2, are the solutions. The graph of the solution set is shown in Figure 2.12.

FIGURE 2.12

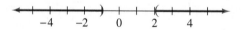

Using interval notation and the set concept of union, we can express the solution set as $(-\infty, -1) \cup (2, \infty)$.   ▲

Example 8 illustrates that in terms of set vocabulary, the solution set of a disjunction is the union of the solution sets of the component parts of the disjunction. Note that there is *no compact form* for writing $x < -1$ or $x > 2$ *or for any disjunction.*

Example 9

Solve the disjunction $2x - 5 < -11$ <u>or</u> $5x + 1 \geq 6$ and graph its solution set on a number line.

Solution

First, let's simplify both inequalities.

$$\begin{array}{ccc} 2x - 5 < -11 & \text{or} & 5x + 1 \geq 6 \\ 2x < -6 & \text{or} & 5x \geq 5 \\ x < -3 & \text{or} & x \geq 1 \end{array}$$

Because this is a disjunction, all numbers less than -3, along with all numbers greater than or equal to 1, will satisfy it. Thus, the solution set is $(-\infty, -3) \cup [1, \infty)$ and its graph is shown in Figure 2.13.

FIGURE 2.13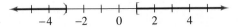

In summary, to solve a compound sentence involving an inequality, proceed as follows.

1. Solve separately each inequality in the compound sentence.

2. If it is a *conjunction*, the solution set is the *intersection* of the solution sets of each inequality.

3. If it is a *disjunction*, the solution set is the *union* of the solution sets of each inequality.

The following agreements on the use of interval notation should be added to the list on page 82 (Figure 2.14).

Set	Graph	Interval notation
$\{x \mid a < x < b\}$		(a, b)
$\{x \mid a \le x < b\}$		$[a, b)$
$\{x \mid a < x \le b\}$		$(a, b]$
$\{x \mid a \le x \le b\}$		$[a, b]$

FIGURE 2.14

Problem Solving

We will conclude this section with some word problems that contain inequality statements.

Problem I

Sari had scores of 94, 84, 86, and 88 on her first four exams of the semester. What score must she obtain on the 5th exam to have an average of 90 or better for the five exams?

Solution

Let s represent the score Sari needs on the 5th exam. Since the average is computed by adding all scores and dividing by the number of scores, we have the following inequality to solve.

$$\frac{94 + 84 + 86 + 88 + s}{5} \ge 90$$

Solving this inequality we obtain

$$\frac{352 + s}{5} \ge 90$$

$$5\left(\frac{352 + s}{5}\right) \ge 5(90) \qquad \text{Multiply both sides by 5.}$$

$$352 + s \ge 450$$

$$s \ge 98.$$

Sari must receive a score of 98 or better.   ▲

Problem 2

An investor has $1000 to invest. Suppose she invests $500 at 8% interest. At what rate must she invest the other $500 so that the two investments together yield more than $100 of yearly interest?

Solution

Let r represent the unknown rate of interest. We can use the following guideline to set up an inequality.

Interest from 8% investment	+	Interest from r percent investment	>	$100
↓		↓		↓
(8%)($500)	+	r($500)	>	$100

Solving this inequality yields

$$40 + 500r > 100$$

$$500r > 60$$

$$r > \frac{60}{500}$$

$$r > .12. \qquad \text{Change to a decimal.}$$

She must invest the other $500 at a rate greater than 12%.   ▲

Problem 3

If the temperature for a 24-hour period ranged between 41°F and 59°F, inclusive, (that is, $41 \le F \le 59$) what was the range in Celsius degrees?

Solution

Use the formula $F = \frac{9}{5}C + 32$, to solve the following compound inequality.

$$41 \le \frac{9}{5}C + 32 \le 59$$

Solving this yields

$$9 \le \frac{9}{5}C \le 27 \qquad \text{Add } -32.$$

$$\frac{5}{9}(9) \le \frac{5}{9}\left(\frac{9}{5}C\right) \le \frac{5}{9}(27) \qquad \text{Multiply by } \frac{5}{9}.$$

$$5 \le C \le 15.$$

The range is between 5°C and 15°C, inclusive.

Problem Set 2.6

For Problems 1–18, solve each of the inequalities and express the solution sets in interval notation.

1. $\frac{2}{5}x + \frac{1}{3}x > \frac{44}{15}$

2. $\frac{1}{4}x - \frac{4}{3}x < -13$

3. $x - \frac{5}{6} < \frac{x}{2} + 3$

4. $x + \frac{2}{7} > \frac{x}{2} - 5$

5. $\frac{x-2}{3} + \frac{x+1}{4} \ge \frac{5}{2}$

6. $\frac{x-1}{3} + \frac{x+2}{5} \le \frac{3}{5}$

7. $\frac{3-x}{6} + \frac{x+2}{7} \le 1$

8. $\frac{4-x}{5} + \frac{x+1}{6} \ge 2$

9. $\frac{x+3}{8} - \frac{x+5}{5} \ge \frac{3}{10}$

10. $\frac{x-4}{6} - \frac{x-2}{9} \le \frac{5}{18}$

11. $\frac{4x-3}{6} - \frac{2x-1}{12} < -2$

12. $\frac{3x+2}{9} - \frac{2x+1}{3} > -1$

13. $.06x + .08(250 - x) \ge 19$

14. $.08x + .09(2x) \ge 130$

15. $.09x + .1(x + 200) > 77$

16. $.07x + .08(x + 100) > 38$

17. $x \ge 3.4 + .15x$

18. $x \ge 2.1 + .3x$

For Problems 19–34, graph the solution set for each compound inequality and express the solution sets in interval notation.

19. $x > -1$ and $x < 2$

20. $x > 1$ and $x < 4$

21. $x \le 2$ and $x > -1$

22. $x \le 4$ and $x \ge -2$

23. $x > 2$ or $x < -1$

24. $x > 1$ or $x < -4$

25. $x \le 1$ or $x > 3$

26. $x < -2$ or $x \ge 1$

27. $x > 0$ and $x > -1$

28. $x > -2$ and $x > 2$

29. $x < 0$ and $x > 4$

30. $x > 1$ or $x < 2$

31. $x > -2$ or $x < 3$

32. $x > 3$ and $x < -1$

33. $x > -1$ or $x > 2$

34. $x < -2$ or $x < 1$

For Problems 35–44, solve each compound inequality and graph the solution sets. Express solution sets in interval notation.

35. $x - 2 > -1$ and $x - 2 < 1$

36. $x + 3 > -2$ and $x + 3 < 2$

37. $x + 2 < -3$ or $x + 2 > 3$

38. $x - 4 < -2$ or $x - 4 > 2$

39. $2x - 1 \geq 5$ and $x > 0$

40. $3x + 2 > 17$ and $x \geq 0$

41. $5x - 2 < 0$ and $3x - 1 > 0$

42. $x + 1 > 0$ and $3x - 4 < 0$

43. $3x + 2 < -1$ or $3x + 2 > 1$

44. $5x - 2 < -2$ or $5x - 2 > 2$

For Problems 45–56, solve each compound inequality using the compact form. Express the solution sets in interval notation.

45. $-3 < 2x + 1 < 5$

46. $-7 < 3x - 1 < 8$

47. $-17 \leq 3x - 2 \leq 10$

48. $-25 \leq 4x + 3 \leq 19$

49. $1 < 4x + 3 < 9$

50. $0 < 2x + 5 < 12$

51. $-6 < 4x - 5 < 6$

52. $-2 < 3x + 4 < 2$

53. $-4 \leq \dfrac{x - 1}{3} \leq 4$

54. $-1 \leq \dfrac{x + 2}{4} \leq 1$

55. $-3 < 2 - x < 3$

56. $-4 < 3 - x < 4$

For Problems 57–67, solve each problem by setting up and solving an appropriate inequality.

57. Suppose that Lance has $500 to invest. If he invests $300 at 9% interest, at what rate must he invest the remaining $200 so that the two investments yield more than $47 in yearly interest?

58. Mona invests $100 at 8% yearly interest. How much does she have to invest at 9% so that the total yearly interest from the two investments exceeds $26?

59. The average height of the two forwards and the center of a basketball team is 6 feet and 8 inches. What must the average height of the two guards be so that the team average is at least 6 feet and 4 inches?

60. Thanh has scores of 52, 84, 65, and 74 on his first four math exams. What score must he make on the 5th exam to have an average of 70 or better for the five exams?

61. Marsha bowled 142 and 170 in her first two games. What must she bowl in the third game to have an average of at least 160 for the three games?

62. Candace had scores of 95, 82, 93, and 84 on her first four exams of the semester. What score must she obtain on the fifth exam to have an average of 90 or better for the five exams?

63. Suppose that Derwin shot rounds of 82, 84, 78, and 79 on the first four days of a golf tournament. What must he shoot on the fifth day of the tournament to average 80 or less for the five days?

64. The temperatures for a 24-hour period range between $-4°F$ and $23°F$, inclusive. What was the range in Celsius degrees? $\left(\text{Use } F = \dfrac{9}{5}C + 32.\right)$

65. Oven temperatures for baking various foods usually range between $325°F$ and $425°F$, inclusive. Express this range in Celsius degrees. (Round answers to the nearest degree.)

66. A person's intelligence quotient (I) is found by dividing mental age (M), as indicated by standard tests, by the chronological age (C), and then multiplying this ratio by 100. The formula $I = \dfrac{100M}{C}$ can be used. If the I range of a group of 11-year-olds is given by $80 \leq I \leq 140$, find the mental age range of this group.

67. Repeat Problem 66 for an I range of 70 to 125, inclusive, for a group of 9-year-olds.

68. Explain the difference between a conjunction and a disjunction. Give an example of each (outside the field of mathematics).

69. How do you know by inspection that the solution set of the inequality $x + 3 > x + 2$ is the entire set of real numbers?

70. Find the solution set for each of the following compound statements and in each case explain your reasoning.

(a) $x < 3$ and $5 > 2$
(b) $x < 3$ or $5 > 2$
(c) $x < 3$ and $6 < 4$
(d) $x < 3$ or $6 < 4$

2.7 Equations and Inequalities Involving Absolute Value

In Section 1.2 we defined the absolute value of a real number by

$$|a| = \begin{cases} a, & \text{if } a \geq 0 \\ -a, & \text{if } a < 0. \end{cases}$$

We also interpreted the absolute value of any real number to be the distance between the number and zero on a number line. For example, $|6| = 6$ translates to 6 units between 6 and 0. Likewise, $|-8| = 8$ translates to 8 units between -8 and 0.

The interpretation of absolute value as distance on a number line provides a straightforward approach to solving a variety of equations and inequalities involving absolute value. First, let's consider some equations.

Example 1

Solve $|x| = 2$.

Solution

Think in terms of *distance between the number and zero* and you will see that x must be 2 or -2. That is, the equation $|x| = 2$ is equivalent to

$$x = -2 \qquad \text{or} \qquad x = 2.$$

The solution set is $\{-2, 2\}$.   ▲

Example 2

Solve $|x + 2| = 5$.

Solution

The number, $x + 2$, must be -5 or 5. Thus, $|x + 2| = 5$ is equivalent to

$$x + 2 = -5 \qquad \text{or} \qquad x + 2 = 5.$$

Solving each equation of the disjunction yields

$$x + 2 = -5 \qquad \text{or} \qquad x + 2 = 5$$
$$x = -7 \qquad \text{or} \qquad x = 3.$$

The solution set is $\{-7, 3\}$.

Check

$$|x + 2| = 5 \qquad |x + 2| = 5$$
$$|-7 + 2| \overset{?}{=} 5 \qquad |3 + 2| \overset{?}{=} 5$$
$$|-5| \overset{?}{=} 5 \qquad |5| \overset{?}{=} 5$$
$$5 = 5 \qquad 5 = 5$$

The following general property should seem reasonable from the distance interpretation of absolute value.

PROPERTY 2.1 $|ax + b| = k$ is equivalent to $ax + b = -k$ or $ax + b = k$, where k is a positive number.

Example 3 demonstrates our format for solving equations of the form $|ax + b| = k$.

Example 3

Solve $|5x + 3| = 7$.

Solution

$$|5x + 3| = 7$$
$$5x + 3 = -7 \qquad \text{or} \qquad 5x + 3 = 7$$
$$5x = -10 \qquad \text{or} \qquad 5x = 4$$
$$x = -2 \qquad \text{or} \qquad x = \frac{4}{5}$$

The solution set is $\left\{-2, \frac{4}{5}\right\}$. Check these solutions!

The *distance interpretation* for absolute value also provides a good basis for solving some inequalities that involve absolute value. Consider the following examples.

Example 4

Solve $|x| < 2$ and graph the solution set.

Solution

The number, x, must be *less than two units away from zero*. Thus, $|x| < 2$ is equivalent to

$$x > -2 \qquad \text{and} \qquad x < 2.$$

The solution set is $(-2, 2)$ and its graph is shown in Figure 2.15.

FIGURE 2.15

Example 5

Solve and graph the solutions for $|x + 3| < 1$.

Solution

Let's continue to think in terms of *distance* on a number line. The number, $x + 3$, must be *less than one unit away from zero*. Thus, $|x + 3| < 1$ is equivalent to

$$x + 3 > -1 \quad \text{and} \quad x + 3 < 1.$$

Solving this conjunction yields

$$x + 3 > -1 \quad \text{and} \quad x + 3 < 1$$
$$x > -4 \quad \text{and} \quad x < -2.$$

The solution set is $(-4, -2)$ and its graph is shown in Figure 2.16.

FIGURE 2.16

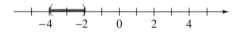

Take another look at Examples 4 and 5. The following general property should seem reasonable.

PROPERTY 2.2

$|ax + b| < k$ is equivalent to $ax + b > -k$ and $ax + b < k$, where k is a positive number.

Remember that we can write a conjunction such as $ax + b > -k$ and $ax + b < k$ in the compact form $-k < ax + b < k$. The compact form provides a very convenient format for solving inequalities such as $|3x - 1| < 8$, as Example 6 illustrates.

Example 6

Solve and graph the solutions for $|3x - 1| < 8$.

Solution

$$|3x - 1| < 8$$
$$-8 < 3x - 1 < 8$$
$$-7 < 3x < 9 \qquad \text{Add 1 to left side, middle, and right side.}$$
$$\frac{1}{3}(-7) < \frac{1}{3}(3x) < \frac{1}{3}(9) \qquad \text{Multiply through by } \frac{1}{3}.$$
$$-\frac{7}{3} < x < 3$$

The solution set is $\left(-\frac{7}{3}, 3\right)$ and its graph is shown in Figure 2.17.

FIGURE 2.17

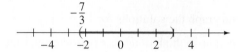

The distance interpretation also clarifies a property that pertains to *greater than* situations involving absolute value. Consider the following examples.

Example 7

Solve and graph the solutions for $|x| > 1$.

Solution

The number, x, must be *more than one unit away from zero*. Thus, $|x| > 1$ is equivalent to

$$x < -1 \quad \text{or} \quad x > 1.$$

The solution set is $(-\infty, -1) \cup (1, \infty)$ and its graph is shown in Figure 2.18.

FIGURE 2.18

Example 8

Solve and graph the solutions for $|x - 1| > 3$.

Solution

The number, $x - 1$, must be *more than three units away from zero*. Thus, $|x - 1| > 3$ is equivalent to

$$x - 1 < -3 \quad \text{or} \quad x - 1 > 3.$$

Solving this disjunction yields

$$x - 1 < -3 \quad \text{or} \quad x - 1 > 3$$
$$x < -2 \quad \text{or} \quad x > 4.$$

The solution set is $(-\infty, -2) \cup (4, \infty)$ and its graph is shown in Figure 2.19.

FIGURE 2.19

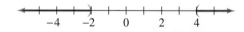

Examples 7 and 8 illustrate the following general property.

PROPERTY 2.3

$|ax + b| > k$ is equivalent to $ax + b < -k$ or $ax + b > k$, where k is a positive number.

Therefore, solving inequalities of the form $|ax + b| > k$ can take on the format in Example 9.

Example 9

Solve and graph the solutions for $|3x - 1| > 2$.

Solution

$$|3x - 1| > 2$$

$$3x - 1 < -2 \quad \text{or} \quad 3x - 1 > 2$$

$$3x < -1 \quad \text{or} \quad 3x > 3$$

$$x < -\frac{1}{3} \quad \text{or} \quad x > 1$$

The solution set is $\left(-\infty, -\frac{1}{3}\right) \cup (1, \infty)$ and its graph is shown in Figure 2.20.

FIGURE 2.20

Properties 2.1, 2.2, and 2.3 provide the basis for solving a variety of equations and inequalities that involve absolute value. However, if at any time you become doubtful as to what property applies, don't forget the distance interpretation. Furthermore, note that in each of the properties, k is a positive number. If k is a nonpositive number, we can determine the solution sets by inspection, as indicated by the following examples.

The solution set of $|x + 3| = 0$ is $\{-3\}$ because the number $x + 3$ has to be 0.

$|2x - 5| = -3$ has *no solutions* because the absolute value (distance) cannot be negative. (The solution set is $\varnothing$, the null set.)

$|x - 7| < -4$ has *no solutions* because we cannot obtain an absolute value less than -4. (The solution set is $\varnothing$.)

$|2x - 1| > -1$ is *satisfied by all real numbers* because the absolute value of $(2x - 1)$, regardless of what number is substituted for x, will always be greater than -1. (The solution set is the set of all real numbers, which we can express in interval notation as $(-\infty, \infty)$).

Problem Set 2.7

Solve and graph the solutions for each of the following.

1. $|x| < 5$

2. $|x| < 1$

3. $|x| \leq 2$

4. $|x| \leq 4$

5. $|x| > 2$

6. $|x| > 3$

7. $|x - 1| < 2$

8. $|x - 2| < 4$

9. $|x + 2| \leq 4$

10. $|x + 1| \leq 1$

11. $|x + 2| > 1$

12. $|x + 1| > 3$

13. $|x - 3| \geq 2$

14. $|x - 2| \geq 1$

Solve each of the following.

15. $|x - 1| = 8$

16. $|x + 2| = 9$

17. $|x - 2| > 6$

18. $|x - 3| > 9$

19. $|x + 3| < 5$

20. $|x + 1| < 8$

21. $|2x - 4| = 6$

22. $|3x - 4| = 14$

23. $|2x - 1| \leq 9$

24. $|3x + 1| \leq 13$

25. $|4x + 2| \geq 12$

26. $|5x - 2| \geq 10$

27. $|3x + 4| = 11$

28. $|5x - 7| = 14$

29. $|4 - 2x| = 6$

30. $|3 - 4x| = 8$

31. $|2 - x| > 4$

32. $|4 - x| > 3$

33. $|1 - 2x| < 2$

34. $|2 - 3x| < 5$

35. $|5x + 9| \leq 16$

36. $|7x - 6| \geq 22$

37. $\left| x - \dfrac{3}{4} \right| = \dfrac{2}{3}$

38. $\left| x + \dfrac{1}{2} \right| = \dfrac{3}{5}$

39. $|-2x + 7| \leq 13$

40. $|-3x - 4| \leq 15$

41. $\left| \dfrac{x - 3}{4} \right| < 2$

42. $\left| \dfrac{x + 2}{3} \right| < 1$

43. $\left| \dfrac{2x + 1}{2} \right| > 1$

44. $\left| \dfrac{3x - 1}{4} \right| > 3$

45. $|2x - 3| + 2 = 5$

46. $|3x - 1| - 1 = 9$

47. $|x + 7| - 3 \geq 4$

48. $|x - 2| + 4 \geq 10$

49. $|2x - 1| + 1 \leq 6$

50. $|4x + 3| - 2 \leq 5$

Solve each of the following *by inspection*. Don't forget the distance interpretation for absolute value.

51. $|2x + 1| = -4$

52. $|5x - 1| = -2$

53. $|3x - 1| > -2$

54. $|4x + 3| < -4$

55. $|5x - 2| = 0$

56. $|3x - 1| = 0$

57. $|4x - 6| < -1$

58. $|x + 9| > -6$

59. $|x + 4| < 0$

60. $|x + 6| > 0$

THOUGHTS INTO WORDS

61. Explain how you would solve the inequality $|2x + 5| > -3$.

62. Why is 2 the only solution for $|x - 2| \leq 0$?

63. Explain how you would solve the equation $|2x - 3| = 0$.

Further Investigations

Solve each of the following equations.

64. $|3x + 1| = |2x + 3|$ [*Hint*: $3x + 1 = 2x + 3$ or $3x + 1 = -(2x + 3)$]

65. $|-2x - 3| = |x + 1|$

66. $|2x - 1| = |x - 3|$

67. $|x - 2| = |x + 6|$

68. $|x + 1| = |x - 4|$

69. $|x + 1| = |x - 1|$

70. Use the definition of absolute value to help prove Property 2.1.

71. Use the definition of absolute value to help prove Property 2.2.

72. Use the definition of absolute value to help prove Property 2.3.

SUMMARY

(2.1) **Solving an algebraic equation** refers to the process of finding the number (or numbers) that makes the algebraic equation a true numerical statement. We call such numbers the **solutions** or **roots** of the equation that **satisfy** the equation. We call the set of all solutions of an equation the **solution set**. The general procedure for solving an equation is to continue replacing the given equation with **equivalent, but simpler,** equations until we arrive at one that can be solved by inspection. Two properties of equality play an important role in the process of solving equations.

Addition Property of Equality $a = b$ if and only if $a + c = b + c$.

Multiplication Property of Equality   For $c \neq 0$, $a = b$ if and only if $ac = bc$.

(2.2) To solve an equation involving fractions, first **clear the equation of all fractions**. It is usually easiest to begin by multiplying both sides of the equation by the least common multiple of all of the denominators in the equation (by the *least common denominator* or LCD).

Keep the following suggestions in mind as you solve word problems.

   1. Read the problem carefully.

   2. Sketch any figure, diagram, or chart that might be helpful.

   3. Choose a meaningful variable.

   4. Look for a guideline.

   5. Form an equation or inequality.

   6. Solve the equation or inequality.

   7. Check your answers.

(2.3) To solve equations that contain decimals, you can **clear the equation of all decimals** by multiplying both sides by an appropriate power of ten.

(2.4) We use equations to put rules in symbolic form; we call these rules **formulas**. We can solve a formula such as $P = 2l + 2w$ for $l\left(l = \dfrac{P - 2w}{2}\right)$ or for $w\left(w = \dfrac{P - 2l}{2}\right)$ by applying the addition and multiplication properties of equality.

We often use formulas as **guidelines** for solving word problems.

(2.5) **Solving an algebraic inequality** refers to the process of finding the numbers that make the algebraic inequality a true numerical statement. We call such numbers the **solutions** and we call the set of all solutions the **solution set**.

The general procedure for solving an inequality is to continue replacing the given inequality with **equivalent, but simpler**, inequalities until we arrive at one that we can solve by inspection. The following properties form the basis for solving algebraic inequalities.

1. $a > b$ if and only if $a + c > b + c$. (Addition property)
2. (a) For $c > 0$, $a > b$ if and only if $ac > bc$.
   (b) For $c < 0$, $a > b$ if and only if $ac < bc$. (Multiplication properties)

(2.6) To solve compound sentences that involve inequalities, we proceed as follows.

1. Solve separately each inequality in the compound sentence.
2. If it is a **conjunction**, the solution set is the **intersection** of the solution sets of each inequality.
3. If it is a **disjunction**, the solution set is the **union** of the solution sets of each inequality.

We define the intersection and union of two sets as follows.

Intersection $\quad A \cap B = \{x \mid x \in A \underline{\quad\text{and}\quad} x \in B\}$.

Union $\quad A \cup B = \{x \mid x \in A \underline{\quad\text{or}\quad} x \in B\}$.

The following are some examples of solution sets we examined in Sections 2.5 and 2.6 (Figure 2.21).

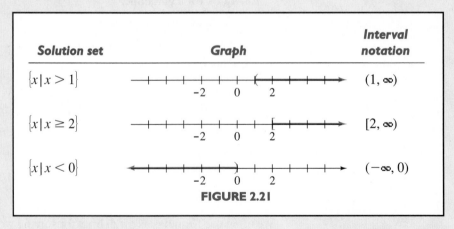

Solution set	Graph	Interval notation
$\{x \mid x > 1\}$		$(1, \infty)$
$\{x \mid x \geq 2\}$		$[2, \infty)$
$\{x \mid x < 0\}$		$(-\infty, 0)$

FIGURE 2.21

(continued on next page)

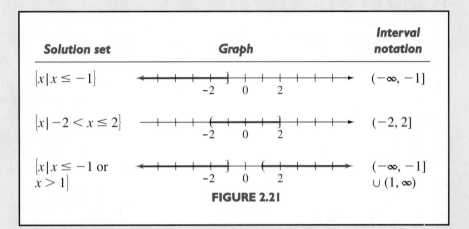

FIGURE 2.21

(2.7) We can interpret the **absolute value** of a number on the number line as the distance between that number and zero. The following properties form the basis for solving equations and inequalities involving absolute value.

1. $|ax + b| = k$ is equivalent to $ax + b = -k$ or $ax + b = k$
2. $|ax + b| < k$ is equivalent to $-k < ax + b < k$ $k > 0$
3. $|ax + b| > k$ is equivalent to $ax + b < -k$ or $ax + b > k$

Chapter 2 Review Problem Set

For Problems 1–15, solve each of the equations.

1. $5(x - 6) = 3(x + 2)$

2. $2(2x + 1) - (x - 4) = 4(x + 5)$

3. $-(2n - 1) + 3(n + 2) = 7$

4. $2(3n - 4) + 3(2n - 3) = -2(n + 5)$

5. $\dfrac{3t - 2}{4} = \dfrac{2t + 1}{3}$

6. $\dfrac{x + 6}{5} + \dfrac{x - 1}{4} = 2$

7. $1 - \dfrac{2x - 1}{6} = \dfrac{3x}{8}$

8. $\dfrac{2x + 1}{3} + \dfrac{3x - 1}{5} = \dfrac{1}{10}$

9. $\dfrac{3n - 1}{2} - \dfrac{2n + 3}{7} = 1$

10. $|3x - 1| = 11$

11. $.06x + .08(x + 100) = 15$

12. $.4(t - 6) = .3(2t + 5)$

13. $.1(n + 300) = .09n + 32$

14. $.2(x - .5) - .3(x + 1) = .4$

15. $|2n + 3| = 4$

For Problems 16–20, solve each equation for x.

16. $ax - b = b + 2$

17. $ax = bx + c$

18. $m(x + a) = p(x + b)$

19. $5x - 7y = 11$

20. $\dfrac{x - a}{b} = \dfrac{y + 1}{c}$

For Problems 21–24, solve each of the formulas for the indicated variable.

21. $A = \pi r^2 + \pi rs$ for s

22. $A = \dfrac{1}{2}h(b_1 + b_2)$ for b_2

23. $R = \dfrac{R_1 R_2}{R_1 + R_2}$ for R_1

24. $\dfrac{1}{R} = \dfrac{1}{R_1} + \dfrac{1}{R_2}$ for R

For Problems 25–36, solve each of the inequalities.

25. $5x - 2 \geq 4x - 7$

26. $3 - 2x < -5$

27. $2(3x - 1) - 3(x - 3) > 0$

28. $3(x + 4) \leq 5(x - 1)$

29. $\dfrac{5}{6}n - \dfrac{1}{3}n < \dfrac{1}{6}$

30. $\dfrac{n - 4}{5} + \dfrac{n - 3}{6} > \dfrac{7}{15}$

31. $s \geq 4.5 + .25s$

32. $.07x + .09(500 - x) \geq 43$

33. $|2x - 1| < 11$

34. $|3x + 1| > 10$

35. $-3(2t - 1) - (t + 2) > -6(t - 3)$

36. $\dfrac{2}{3}(x - 1) + \dfrac{1}{4}(2x + 1) < \dfrac{5}{6}(x - 2)$

For Problems 37–44, graph the solutions of each compound inequality.

37. $x > -1$ and $x < 1$

38. $x > 2$ or $x \leq -3$

39. $x > 2$ and $x > 3$

40. $x < 2$ or $x > -1$

41. $2x + 1 > 3$ or $2x + 1 < -3$

42. $2 \leq x + 4 \leq 5$

43. $-1 < 4x - 3 \leq 9$

44. $x + 1 > 3$ and $x - 3 < -5$

Solve each of the following problems by setting up and solving an appropriate equation or inequality.

45. The width of a rectangle is 2 meters more than one-third of the length. The perimeter of the rectangle is 44 meters. Find the length and width of the rectangle.

46. A total of $500 was invested, part of it at 7% interest and the remainder at 8%. If the total yearly interest from both investments amounted to $38, how much was invested at each rate?

47. Susan's average score for her first three psychology exams is 84. What must she get on the 4th exam so that her average for the four exams is 85 or better?

48. Find three consecutive integers such that the sum of one-half of the smallest and one-third of the largest is one less than the other integer.

49. Pat is paid time-and-a-half for each hour he works over 36 hours in a week. Last week he worked 42 hours for a total of $472.50. What is his normal hourly rate?

50. Marcela has a collection of nickels, dimes, and quarters worth $24.75. The number of dimes is 10 more than twice the number of nickels and the number of quarters is 25 more than the number of dimes. How many coins of each kind does she have?

51. If the complement of an angle is one-tenth of the supplement of the angle, find the measure of the angle.

52. A retailer has some sweaters that cost her $38 each. She wants to sell them at a profit of 20% of her cost. What price should she charge for the sweaters?

53. Nora scored 16, 22, 18, and 14 points for each of the first four basketball games. How many points does she need to score in the 5th game so that her average for the first five games is at least 20 points per game?

54. Gladys leaves a town driving at a rate of 40 miles per hour. Two hours later, Reena leaves from the same place traveling the same route and catches Gladys in 5 hours and 20 minutes. How fast was Reena traveling?

55. In $1\frac{1}{4}$ hours more time, Rita, riding her bicycle at 12 miles per hour, rode 2 miles farther than Sonya,

who was riding her bicycle at 16 miles per hour. How long did each girl ride?

56. How many cups of orange juice must be added to 50 cups of a punch that is 10% orange juice to obtain a punch that is 20% orange juice?

CHAPTER 2 TEST

For Problems 1–10, solve each equation.

1. $5x - 2 = 2x - 11$

2. $6(n - 2) - 4(n + 3) = -14$

3. $-3(x + 4) = 3(x - 5)$

4. $3(2x - 1) - 2(x + 5) = -(x - 3)$

5. $\dfrac{3t - 2}{4} = \dfrac{5t + 1}{5}$

6. $\dfrac{5x + 2}{3} - \dfrac{2x + 4}{6} = -\dfrac{4}{3}$

7. $|4x - 3| = 9$

8. $\dfrac{1 - 3x}{4} + \dfrac{2x + 3}{3} = 1$

9. $2 - \dfrac{3x - 1}{5} = -4$

10. $.05x + .06(1500 - x) = 83.5$

11. Solve $\dfrac{2}{3}x - \dfrac{3}{4}y = 2$ for y.

12. Solve $S = 2\pi r(r + h)$ for h.

For Problems 13–20, solve each inequality and express the solution sets using interval notation.

13. $7x - 4 > 5x - 8$

14. $-3x - 4 \le x + 12$

15. $2(x - 1) - 3(3x + 1) \ge -6(x - 5)$

16. $\dfrac{3}{5}x - \dfrac{1}{2}x < 1$

17. $\dfrac{x - 2}{6} - \dfrac{x + 3}{9} > -\dfrac{1}{2}$

18. $.05x + .07(800 - x) \ge 52$

19. $|6x - 4| < 10$

20. $|4x + 5| \ge 6$

For Problems 21–25, solve each problem by setting up and solving an appropriate equation or inequality.

21. Dela bought a dress at a 20% discount sale for $57.60. Find the original price of the dress.

22. The length of a rectangle is one centimeter more than three times its width. If the perimeter of the rectangle is 50 centimeters, find the length of the rectangle.

23. How many cups of grapefruit juice must be added to 30 cups of a punch that contains 8% grapefruit juice to obtain a punch that is 10% grapefruit juice?

(continued on next page)

CHAPTER 2 TEST (*continued*)

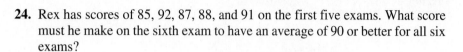

24. Rex has scores of 85, 92, 87, 88, and 91 on the first five exams. What score must he make on the sixth exam to have an average of 90 or better for all six exams?

25. If the complement of an angle is $\frac{2}{11}$ of the supplement of the angle, find the measure of the angle.

Polynomials

A strip of uniform width cut off of both sides and both ends of an 8-inch by 11-inch sheet of paper reduces the size of the paper to an area of 40 square inches. Find the width of the strip. With the equation $(11 - 2x)(8 - 2x) = 40$ you can determine that the strip should be 1.5 inches wide.

The main thrust of this text is to develop algebraic skills, to use these skills to solve equations and inequalities, and to use equations and inequalities to solve word problems. The work in this chapter will center around a class of algebraic expressions called polynomials.

3.1 Polynomials: Sums and Differences

Recall that algebraic expressions such as $5x$, $-6y^2$, $7xy$, $14a^2b$, and $-17ab^2c^3$ are called terms. A **term** is an indicated product and may contain any number of factors. The variables in a term are called **literal factors** and the numerical factor is called the **numerical coefficient**. Thus, in $7xy$ the x and y are literal factors, 7 is the numerical coefficient, and the term is *in two variables* (x and y).

Terms that contain variables with only nonnegative integers as exponents are called **monomials**. The previously listed terms, $5x$, $-6y^2$, $7xy$, $14a^2b$, and $-17ab^2c^3$, are all monomials. (We shall work with some algebraic expressions later such as $7x^{-1}y^{-1}$ and $6a^{-2}b^{-3}$, which are not monomials.)

The **degree** of a monomial is the sum of the exponents of the literal factors.

$7xy$ is of degree 2, $5x$ is of degree 1,

$14a^2b$ is of degree 3, $-6y^2$ is of degree 2,

$-17ab^2c^3$ is of degree 6

If the monomial contains only one variable, then the exponent of the variable is the degree of the monomial. The two examples on the right illustrate this point. We say that any nonzero constant term is of degree zero.

A **polynomial** is a monomial or a finite sum (or difference) of monomials. Thus,

$$4x^2, \qquad 3x^2 - 2x - 4, \qquad 7x^4 - 6x^3 + 4x^2 + x - 1,$$

$$3x^2y - 2xy^2, \qquad \frac{1}{5}a^2 - \frac{2}{3}b^2, \qquad \text{and} \qquad 14$$

are examples of polynomials. In addition to calling a polynomial with one term a **monomial**, we also classify polynomials with two terms as **binomials** and those with three terms as **trinomials**.

The **degree of a polynomial** is the degree of the term with the highest degree in the polynomial. The following examples help illustrate some of this terminology.

The polynomial $4x^3y^4$ is a monomial in two variables of degree 7.

The polynomial $4x^2y - 2xy$ is a binomial in two variables of degree 3.

The polynomial $9x^2 - 7x + 1$ is a trinomial in one variable of degree 2.

Combining Similar Terms

Remember that *similar* or *like terms* are terms that have the same literal factors. In the preceding chapters we have frequently simplified algebraic expressions by combining similar terms, as the next examples illustrate.

$$2x + 3y + 7x + 8y = \boxed{2x + 7x + 3y + 8y}$$
$$= \boxed{(2 + 7)x + (3 + 8)y}$$
$$= 9x + 11y.$$

Steps in dashed
boxes are usually
done mentally.

$$4a - 7 - 9a + 10 = \boxed{4a + (-7) + (-9a) + 10}$$
$$= \boxed{4a + (-9a) + (-7) + 10}$$
$$= \boxed{(4 + (-9))a + (-7) + 10}$$
$$= -5a + 3.$$

Both addition and subtraction of polynomials rely on basically the same ideas. The commutative, associative, and distributive properties provide the basis for rearranging, regrouping, and combining similar terms. Let's consider some examples.

Example 1

Add $4x^2 + 5x + 1$ and $7x^2 - 9x + 4$.

Solution

We usually use the horizontal format for such work. Thus,

$$(4x^2 + 5x + 1) + (7x^2 - 9x + 4) = (4x^2 + 7x^2) + (5x - 9x) + (1 + 4)$$
$$= 11x^2 - 4x + 5. \quad \blacktriangle$$

Example 2

Add $5x - 3$, $3x + 2$, and $8x + 6$.

Solution

$$(5x - 3) + (3x + 2) + (8x + 6) = (5x + 3x + 8x) + (-3 + 2 + 6)$$
$$= 16x + 5 \quad \blacktriangle$$

Example 3

Find the indicated sum: $(-4x^2y + xy^2) + (7x^2y - 9xy^2) + (5x^2y - 4xy^2)$.

Solution

$$(-4x^2y + xy^2) + (7x^2y - 9xy^2) + (5x^2y - 4xy^2)$$
$$= (-4x^2y + 7x^2y + 5x^2y) + (xy^2 - 9xy^2 - 4xy^2)$$
$$= 8x^2y - 12xy^2 \quad \blacktriangle$$

The idea of subtraction as *adding the opposite* ($a - b = a + (-b)$) extends to polynomials in general. We can form the opposite of a polynomial by taking the opposite of each term. For example, the opposite of $3x^2 - 7x + 1$ is $-3x^2 + 7x - 1$. Symbolically we express this as

$$-(3x^2 - 7x + 1) = -3x^2 + 7x - 1.$$

Now consider the following subtraction problems.

Example 4

Subtract $3x^2 + 7x - 1$ from $7x^2 - 2x - 4$.

Solution

Use the horizontal format to obtain

$$(7x^2 - 2x - 4) - (3x^2 + 7x - 1) = (7x^2 - 2x - 4) + (-3x^2 - 7x + 1)$$
$$= (7x^2 - 3x^2) + (-2x - 7x) + (-4 + 1)$$
$$= 4x^2 - 9x - 3.$$ ▲

Example 5

Subtract $-3y^2 + y - 2$ from $4y^2 + 7$.

Solution

$$(4y^2 + 7) - (-3y^2 + y - 2) = (4y^2 + 7) + (3y^2 - y + 2)$$
$$= (4y^2 + 3y^2) + (-y) + (7 + 2)$$
$$= 7y^2 - y + 9.$$ ▲

The next example demonstrates the use of the vertical format for this work.

Example 6

Subtract $4x^2 - 7xy + 5y^2$ from $3x^2 - 2xy + y^2$.

Solution

$$\begin{array}{l} 3x^2 - 2xy + y^2 \\ \underline{4x^2 - 7xy + 5y^2} \end{array}$$ Notice which polynomial goes on the bottom and how the similar terms are aligned.

Now we can *mentally form the opposite of the bottom polynomial* and add.

$$\begin{array}{l} 3x^2 - 2xy + y^2 \\ \underline{4x^2 - 7xy + 5y^2} \\ -x^2 + 5xy - 4y^2 \end{array}$$ The opposite of $4x^2 - 7xy + 5y^2$ is $-4x^2 + 7xy - 5y^2$. ▲

The distributive property and the properties $a = 1(a)$ and $-a = -1(a)$ can also be used when adding and subtracting polynomials. The next examples illustrate this approach.

Example 7

Perform the indicated operations: $(5x - 2) + (2x - 1) - (3x + 4)$.

Solution

$$(5x - 2) + (2x - 1) - (3x + 4) = 1(5x - 2) + 1(2x - 1) - 1(3x + 4)$$
$$= 1(5x) - 1(2) + 1(2x) - 1(1) - 1(3x) - 1(4)$$
$$= 5x - 2 + 2x - 1 - 3x - 4$$
$$= 5x + 2x - 3x - 2 - 1 - 4$$
$$= 4x - 7$$ ▲

Certainly you can do some of the steps mentally and simplify our format in the next two examples.

Example 8

Perform the indicated operations: $(5a^2 - 2b) - (2a^2 + 4) + (-7b - 3)$.

Solution

$$(5a^2 - 2b) - (2a^2 + 4) + (-7b - 3) = 5a^2 - 2b - 2a^2 - 4 - 7b - 3$$
$$= 3a^2 - 9b - 7 \qquad \blacktriangle$$

Example 9

Simplify $(4t^2 - 7t - 1) - (t^2 + 2t - 6)$.

Solution

$$(4t^2 - 7t - 1) - (t^2 + 2t - 6) = 4t^2 - 7t - 1 - t^2 - 2t + 6$$
$$= 3t^2 - 9t + 5 \qquad \blacktriangle$$

Remember that a polynomial in parentheses preceded by a negative sign can be written without the parentheses by replacing each term with its opposite. Thus, in Example 9, $-(t^2 + 2t - 6) = -t^2 - 2t + 6$. Finally, let's consider a simplification problem that contains grouping symbols within grouping symbols.

Example 10

Simplify $7x + [3x - (2x + 7)]$.

Solution

$$7x + [3x - (2x + 7)] = 7x + [3x - 2x - 7] \qquad \text{Remove the innermost parentheses first.}$$

$$= 7x + [x - 7]$$
$$= 7x + x - 7$$
$$= 8x - 7 \qquad \blacktriangle$$

Sometimes we encounter polynomials in a geometric setting. For example, we can find a polynomial that represents the total surface area of the rectangular solid in Figure 3.1 as follows.

FIGURE 3.1

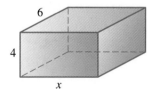

$4x$	+	$4x$	+	$6x$	+	$6x$	+	24	+	24
↑		↑		↑		↑		↑		↑
Area of front		Area of back		Area of top		Area of bottom		Area of left side		Area of right side

Simplifying $4x + 4x + 6x + 6x + 24 + 24$ we obtain the polynomial $20x + 48$, which represents the total surface area of the rectangular solid. Furthermore, by evaluating the polynomial $20x + 48$ for different positive values of x, we can determine the total surface area of any rectangular solid for which two dimensions are 4 and 6. The following chart contains some specific rectangular solids.

x	4 by 6 by x rectangular solid	Total surface area (20x + 48)
2	4 by 6 by 2	20(2) + 48 = 88
4	4 by 6 by 4	20(4) + 48 = 128
5	4 by 6 by 5	20(5) + 48 = 148
7	4 by 6 by 7	20(7) + 48 = 188
12	4 by 6 by 12	20(12) + 48 = 288

▼ Problem Set 3.1

For Problems 1–10, determine the degree of the given polynomials.

1. $7xy + 6y$

2. $-5x^2y^2 + 6xy^2 + x$

3. $-x^2y + 2xy^2 - xy$

4. $5x^3y^2 - 6x^3y^3$ **5.** $5x^2 - 7x - 2$

6. $7x^3 - 2x + 4$ **7.** $8x^6 + 9$

8. $5y^6 + y^4 - 2y^2 - 8$

9. -12 **10.** $7x - 2y$

For Problems 11–20, add the given polynomials.

11. $3x - 7$ and $7x + 4$

12. $9x + 6$ and $5x - 3$

13. $-5t - 4$ and $-6t + 9$

14. $-7t + 14$ and $-3t - 6$

15. $3x^2 - 5x - 1$ and $-4x^2 + 7x - 1$

16. $6x^2 + 8x + 4$ and $-7x^2 - 7x - 10$

17. $12a^2b^2 - 9ab$ and $5a^2b^2 + 4ab$

18. $15a^2b^2 - ab$ and $-20a^2b^2 - 6ab$

19. $2x - 4, -7x + 2,$ and $-4x + 9$

20. $-x^2 - x - 4, 2x^2 - 7x + 9,$ and $-3x^2 + 6x - 10$

For Problems 21–30, subtract the polynomials using the horizontal format.

21. $5x - 2$ from $3x + 4$

22. $7x + 5$ from $2x - 1$

23. $-4a - 5$ from $6a + 2$

24. $5a + 7$ from $-a - 4$

25. $3x^2 - x + 2$ from $7x^2 + 9x + 8$

26. $5x^2 + 4x - 7$ from $3x^2 + 2x - 9$

27. $2a^2 - 6a - 4$ from $-4a^2 + 6a + 10$

28. $-3a^2 - 6a + 3$ from $3a^2 + 6a - 11$

29. $2x^3 + x^2 - 7x - 2$ from $5x^3 + 2x^2 + 6x - 13$

30. $6x^3 + x^2 + 4$ from $9x^3 - x - 2$

For Problems 31–40, subtract the polynomials using the vertical format.

31. $5x - 2$ from $12x + 6$

32. $3x - 7$ from $2x + 1$

33. $-4x + 7$ from $-7x - 9$

34. $-6x - 2$ from $5x + 6$

35. $2x^2 + x + 6$ from $4x^2 - x - 2$

36. $4x^2 - 3x - 7$ from $-x^2 - 6x + 9$

37. $x^3 + x^2 - x - 1$ from $-2x^3 + 6x^2 - 3x + 8$

38. $2x^3 - x + 6$ from $x^3 + 4x^2 + 1$

39. $-5x^2 + 6x - 12$ from $2x - 1$

40. $2x^2 - 7x - 10$ from $-x^3 - 12$

For Problems 41–46, perform the operations as described.

41. Subtract $2x^2 - 7x - 1$ from the sum of $x^2 + 9x - 4$ and $-5x^2 - 7x + 10$.

42. Subtract $4x^2 + 6x + 9$ from the sum of $-3x^2 - 9x + 6$ and $-2x^2 + 6x - 4$.

43. Subtract $-x^2 - 7x - 1$ from the sum of $4x^2 + 3$ and $-7x^2 + 2x$.

44. Subtract $-4x^2 + 6x - 3$ from the sum of $-3x + 4$ and $9x^2 - 6$.

45. Subtract the sum of $5n^2 - 3n - 2$ and $-7n^2 + n + 2$ from $-12n^2 - n + 9$.

46. Subtract the sum of $-6n^2 + 2n - 4$ and $4n^2 - 2n + 4$ from $-n^2 - n + 1$.

For Problems 47–56, perform the indicated operations.

47. $(5x + 2) + (7x - 1) + (-4x - 3)$

48. $(-3x + 1) + (6x - 2) + (9x - 4)$

49. $(12x - 9) - (-3x + 4) - (7x + 1)$

50. $(6x + 4) - (4x - 2) - (-x - 1)$

51. $(2x^2 - 7x - 1) + (-4x^2 - x + 6) + (-7x^2 - 4x - 1)$

52. $(5x^2 + x + 4) + (-x^2 + 2x + 4) + (-14x^2 - x + 6)$

53. $(7x^2 - x - 4) - (9x^2 - 10x + 8) + (12x^2 + 4x - 6)$

54. $(-6x^2 + 2x + 5) - (4x^2 + 4x - 1) + (7x^2 + 4)$

55. $(n^2 - 7n - 9) - (-3n + 4) - (2n^2 - 9)$

56. $(6n^2 - 4) - (5n^2 + 9) - (6n + 4)$

For Problems 57–70, simplify by removing the inner parentheses first and working outward.

57. $3x - [5x - (x + 6)]$

58. $7x - [2x - (-x - 4)]$

59. $2x^2 - [-3x^2 - (x^2 - 4)]$

60. $4x^2 - [-x^2 - (5x^2 - 6)]$

61. $-2n^2 - [n^2 - (-4n^2 + n + 6)]$

62. $-7n^2 - [3n^2 - (-n^2 - n + 4)]$

63. $[4t^2 - (2t + 1) + 3] - [3t^2 + (2t - 1) - 5]$

64. $-(3n^2 - 2n + 4) - [2n^2 - (n^2 + n + 3)]$

65. $[2n^2 - (2n^2 - n + 5)] + [3n^2 + (n^2 - 2n - 7)]$

66. $3x^2 - [4x^2 - 2x - (x^2 - 2x + 6)]$

67. $[7xy - (2x - 3xy + y)] - [3x - (x - 10xy - y)]$

68. $[9xy - (4x + xy - y)] - [4y - (2x - xy + 6y)]$

69. $[4x^3 - (2x^2 - x - 1)] - [5x^3 - (x^2 + 2x - 1)]$

70. $[x^3 - (x^2 - x + 1)] - [-x^3 + (7x^2 - x + 10)]$

71. Find a polynomial that represents the perimeter of each of the following figures (Figures 3.2, 3.3, and 3.4).

(a)

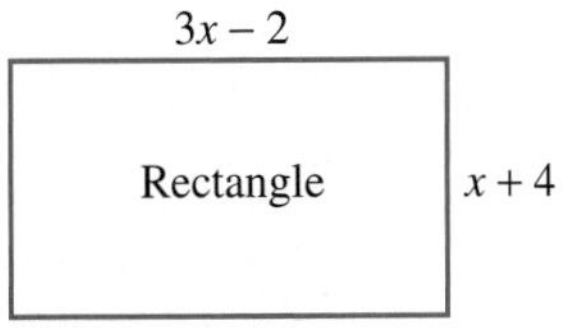

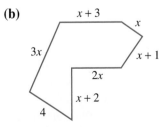

FIGURE 3.2

(b)

FIGURE 3.3

(c)

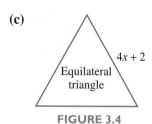

4x + 2

Equilateral
triangle

FIGURE 3.4

72. Find a polynomial that represents the total surface area of the rectangular solid in Figure 3.5.

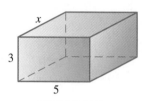

x

3

5

FIGURE 3.5

Now use that polynomial to determine the total surface area of each of the following rectangular solids.
(a) 3 by 5 by 4
(b) 3 by 5 by 7
(c) 3 by 5 by 11
(d) 3 by 5 by 13

73. Find a polynomial that represents the total surface area of the right circular cylinder in Figure 3.6. Now use that polynomial to determine the total sur-

face area of each of the following right circular cylinders that has a base with a radius of 4. Use 3.14 for π and express the answers to the nearest tenth.

(a) $h = 5$
(b) $h = 7$
(c) $h = 14$
(d) $h = 18$

4

h

FIGURE 3.6

THOUGHTS INTO WORDS

74. Explain how to subtract the polynomial $-3x^2 + 2x - 4$ from $4x^2 + 6$.

75. Is the sum of two binomials always another binomial? Defend your answer.

76. Explain how to simplify the expression $7x - [3x - (2x - 4) + 2] -x$.

3.2   Products and Quotients of Monomials

Suppose that we want to find the product of two monomials such as $3x^2y$ and $4x^3y^2$. To proceed, use the properties of real numbers and keep in mind that exponents indicate repeated multiplication.

$$(3x^2y)(4x^3y^2) = (3 \cdot x \cdot x \cdot y)(4 \cdot x \cdot x \cdot x \cdot y \cdot y)$$
$$= 3 \cdot 4 \cdot x \cdot x \cdot x \cdot x \cdot x \cdot y \cdot y \cdot y$$
$$= 12x^5y^3$$

You can use such an approach to find the product of any two monomials. However, there are some basic properties of exponents that make the process of multiplying monomials a much easier task. Let's consider each of these properties and illustrate its use when multiplying monomials. The following examples motivate the first property.

$$x^2 \cdot x^3 = (x \cdot x)(x \cdot x \cdot x) = x^5,$$
$$a^4 \cdot a^2 = (a \cdot a \cdot a \cdot a)(a \cdot a) = a^6,$$
$$b^3 \cdot b^4 = (b \cdot b \cdot b)(b \cdot b \cdot b \cdot b) = b^7,$$

In general,

$$b^n \cdot b^m = \underbrace{(b \cdot b \cdot b \cdots b)}_{\substack{n \text{ factors} \\ \text{of } b}}\underbrace{(b \cdot b \cdot b \cdots b)}_{\substack{m \text{ factors} \\ \text{of } b}}$$

$$= \underbrace{b \cdot b \cdot b \cdots b}_{(n + m) \text{ factors of } b}$$

$$= b^{n+m}.$$

We can state the first property as follows.

PROPERTY 3.1

If b is any real number and n and m are positive integers, then

$$b^n \cdot b^m = b^{n+m}.$$

Property 3.1 states *to find the product of two positive integral powers of the same base, add the exponents and use this sum as the exponent of the common base.*

$$x^7 \cdot x^8 = x^{7+8} = x^{15}, \qquad\qquad y^6 \cdot y^4 = y^{6+4} = y^{10},$$
$$2^3 \cdot 2^8 = 2^{3+8} = 2^{11}, \qquad\qquad (-3)^4 \cdot (-3)^5 = (-3)^{4+5} = (-3)^9,$$
$$\left(\frac{2}{3}\right)^7 \cdot \left(\frac{2}{3}\right)^5 = \left(\frac{2}{3}\right)^{7+5} = \left(\frac{2}{3}\right)^{12}$$

The following examples illustrate the use of Property 3.1 along with the commutative and associative properties of multiplication to form the basis for multiplying monomials. The steps enclosed in the dashed boxes could be performed mentally.

Example 1

$$(3x^2y)(4x^3y^2) = \boxed{3 \cdot 4 \cdot x^2 \cdot x^3 \cdot y \cdot y^2}$$
$$= \boxed{12x^{2+3}y^{1+2}}$$
$$= 12x^5y^3 \qquad \blacktriangle$$

Example 2

$$(-5a^3b^4)(7a^2b^5) = \boxed{-5 \cdot 7 \cdot a^3 \cdot a^2 \cdot b^4 \cdot b^5}$$
$$= \boxed{-35a^{3+2}b^{4+5}}$$
$$= -35a^5b^9 \qquad \blacktriangle$$

Example 3

$$\left(\frac{3}{4}xy\right)\left(\frac{1}{2}x^5y^6\right) = \frac{3}{4} \cdot \frac{1}{2} \cdot x \cdot x^5 \cdot y \cdot y^6$$

$$= \frac{3}{8}x^{1+5}y^{1+6}$$

$$= \frac{3}{8}x^6y^7$$

▲

Example 4

$$(-ab^2)(-5a^2b) = (-1)(-5)(a)(a^2)(b^2)(b)$$

$$= 5a^{1+2}b^{2+1}$$

$$= 5a^3b^3$$

▲

Example 5

$$(2x^2y^2)(3x^2y)(4y^3) = 2 \cdot 3 \cdot 4 \cdot x^2 \cdot x^2 \cdot y^2 \cdot y \cdot y^3$$

$$= 24x^{2+2}y^{2+1+3}$$

$$= 24x^4y^6$$

▲

The following examples demonstrate another useful property of exponents.

$$(x^2)^3 = x^2 \cdot x^2 \cdot x^2 = x^{2+2+2} = x^6,$$

$$(a^3)^2 = a^3 \cdot a^3 = a^{3+3} = a^6,$$

$$(b^4)^3 = b^4 \cdot b^4 \cdot b^4 = b^{4+4+4} = b^{12},$$

In general,

$$(b^n)^m = \underbrace{b^n \cdot b^n \cdot b^n \cdots b^n}_{m \text{ factors of } b^n}$$

$$= \overbrace{b^{n+n+n+\cdots+n}}^{m \text{ of these}}$$

$$= b^{mn}.$$

We can state the property as follows.

PROPERTY 3.2   If b is any real number and m and n are positive integers, then

$$(b^n)^m = b^{mn}.$$

Use Property 3.2 to find "the power of a power" as follows.

$$(x^4)^5 = x^{5(4)} = x^{20}, \qquad (y^6)^3 = y^{3(6)} = y^{18},$$
$$(2^3)^7 = 2^{7(3)} = 2^{21}$$

A third property of exponents pertains to raising a monomial to a power. Consider the following examples, which we use to introduce the property.

$$(3x)^2 = (3x)(3x) = 3 \cdot 3 \cdot x \cdot x = 3^2 \cdot x^2,$$
$$(4y^2)^3 = (4y^2)(4y^2)(4y^2) = 4 \cdot 4 \cdot 4 \cdot y^2 \cdot y^2 \cdot y^2 = (4)^3(y^2)^3,$$
$$(-2a^3b^4)^2 = (-2a^3b^4)(-2a^3b^4) = (-2)(-2)(a^3)(a^3)(b^4)(b^4)$$
$$= (-2)^2(a^3)^2(b^4)^2$$

In general,

$$(ab)^n = \underbrace{(ab)(ab)(ab) \cdots (ab)}_{n \text{ factors of } ab}$$

$$= \underbrace{(a \cdot a \cdot a \cdot a \cdots a)}_{\substack{n \text{ factors} \\ \text{of } a}}\underbrace{(b \cdot b \cdot b \cdots b)}_{\substack{n \text{ factors} \\ \text{of } b}}$$

$$= a^n b^n.$$

We can formally state Property 3.3 as follows.

PROPERTY 3.3

If a and b are real numbers and n is a positive integer, then
$$(ab)^n = a^n b^n.$$

Property 3.3 and Property 3.2 form the basis for raising a monomial to a power, as in the next examples.

Example 6

$$(x^2y^3)^4 = (x^2)^4(y^3)^4 \qquad \text{Use } (ab)^n = a^n b^n.$$
$$= x^8 y^{12} \qquad \text{Use } (b^n)^m = b^{mn}.$$

▲

Example 7

$$(3a^5)^3 = (3)^3(a^5)^3$$
$$= 27a^{15}$$

▲

Example 8

$$(-2xy^4)^5 = (-2)^5(x)^5(y^4)^5$$
$$= -32x^5 y^{20}$$

▲

Dividing Monomials

To develop an effective process for dividing by a monomial we need yet another property of exponents. This property is a direct consequence of the definition of an exponent. Study the following examples.

$$\frac{x^4}{x^3} = \frac{\cancel{x} \cdot \cancel{x} \cdot \cancel{x} \cdot x}{\cancel{x} \cdot \cancel{x} \cdot \cancel{x}} = x, \qquad \frac{x^3}{x^3} = \frac{\cancel{x} \cdot \cancel{x} \cdot \cancel{x}}{\cancel{x} \cdot \cancel{x} \cdot \cancel{x}} = 1,$$

$$\frac{a^5}{a^2} = \frac{\cancel{a} \cdot \cancel{a} \cdot a \cdot a \cdot a}{\cancel{a} \cdot \cancel{a}} = a^3, \qquad \frac{y^5}{y^5} = \frac{\cancel{y} \cdot \cancel{y} \cdot \cancel{y} \cdot \cancel{y} \cdot \cancel{y}}{\cancel{y} \cdot \cancel{y} \cdot \cancel{y} \cdot \cancel{y} \cdot \cancel{y}} = 1,$$

$$\frac{y^8}{y^4} = \frac{\cancel{y} \cdot \cancel{y} \cdot \cancel{y} \cdot \cancel{y} \cdot y \cdot y \cdot y \cdot y}{\cancel{y} \cdot \cancel{y} \cdot \cancel{y} \cdot \cancel{y}} = y^4$$

We can state the general property as follows.

PROPERTY 3.4

If b is any nonzero real number and m and n are positive integers, then

1. $\dfrac{b^n}{b^m} = b^{n-m}$, when $n > m$

2. $\dfrac{b^n}{b^m} = 1$, when $n = m$.

Apply Property 3.4 to the previous examples to yield

$$\frac{x^4}{x^3} = x^{4-3} = x^1 = x, \qquad \frac{x^3}{x^3} = 1,$$

$$\frac{a^5}{a^2} = a^{5-2} = a^3, \qquad \frac{y^5}{y^5} = 1,$$

$$\frac{y^8}{y^4} = y^{8-4} = y^4.$$

(We will discuss the situation when $n < m$ in a later chapter.)

Property 3.4, along with our knowledge of dividing integers, provides the basis for dividing monomials. The following examples demonstrate the process.

$$\frac{24x^5}{3x^2} = 8x^{5-2} = 8x^3, \qquad \frac{-36a^{13}}{-12a^5} = 3a^{13-5} = 3a^8,$$

$$\frac{-56x^9}{7x^4} = -8x^{9-4} = -8x^5, \qquad \frac{72b^5}{8b^5} = 9 \ \left(\frac{b^5}{b^5} = 1\right),$$

$$\frac{48y^7}{-12y} = -4y^{7-1} = -4y^6, \qquad \frac{12x^4y^7}{2x^2y^4} = 6x^{4-2}y^{7-4} = 6x^2y^3$$

Problem Set 3.2

Find each of the following products.

1. $(4x^3)(9x)$ **2.** $(6x^3)(7x^2)$

3. $(-2x^2)(6x^3)$

4. $(2xy)(-4x^2y)$

5. $(-a^2b)(-4ab^3)$

6. $(-8a^2b^2)(-3ab^3)$

7. $(x^2yz^2)(-3xyz^4)$

8. $(-2xy^2z^2)(-x^2y^3z)$

9. $(5xy)(-6y^3)$

10. $(-7xy)(4x^4)$

11. $(3a^2b)(9a^2b^4)$

12. $(-8a^2b^2)(-12ab^5)$

13. $(m^2n)(-mn^2)$

14. $(-x^3y^2)(xy^3)$

15. $\left(\frac{2}{5}xy^2\right)\left(\frac{3}{4}x^2y^4\right)$

16. $\left(\frac{1}{2}x^2y^6\right)\left(\frac{2}{3}xy\right)$

17. $\left(-\frac{3}{4}ab\right)\left(\frac{1}{5}a^2b^3\right)$

18. $\left(-\frac{2}{7}a^2\right)\left(\frac{3}{5}ab^3\right)$

19. $\left(-\frac{1}{2}xy\right)\left(\frac{1}{3}x^2y^3\right)$

20. $\left(\frac{3}{4}x^4y^5\right)(-x^2y)$

21. $(3x)(-2x^2)(-5x^3)$

22. $(-2x)(-6x^3)(x^2)$

23. $(-6x^2)(3x^3)(x^4)$

24. $(-7x^2)(3x)(4x^3)$

25. $(x^2y)(-3xy^2)(x^3y^3)$

26. $(xy^2)(-5xy)(x^2y^4)$

27. $(-3y^2)(-2y^2)(-4y^5)$

28. $(-y^3)(-6y)(-8y^4)$

29. $(4ab)(-2a^2b)(7a)$

30. $(3b)(-2ab^2)(7a)$

31. $(-ab)(-3ab)(-6ab)$

32. $(-3a^2b)(-ab^2)(-7a)$

33. $\left(\frac{2}{3}xy\right)(-3x^2y)(5x^4y^5)$

34. $\left(\frac{3}{4}x\right)(-4x^2y^2)(9y^3)$

35. $(12y)(-5x)\left(-\frac{5}{6}x^4y\right)$

36. $(-12x)(3y)\left(-\frac{3}{4}xy^6\right)$

Raise each of the following monomials to the indicated power.

37. $(3xy^2)^3$ **38.** $(4x^2y^3)^3$

39. $(-2x^2y)^5$

40. $(-3xy^4)^3$

41. $(-x^4y^5)^4$ **42.** $(-x^5y^2)^4$

43. $(ab^2c^3)^6$ **44.** $(a^2b^3c^5)^5$

45. $(2a^2b^3)^6$ **46.** $(2a^3b^2)^6$

47. $(9xy^4)^2$ **48.** $(8x^2y^5)^2$

49. $(-3ab^3)^4$ **50.** $(-2a^2b^4)^4$

51. $-(2ab)^4$ **52.** $-(3ab)^4$

53. $-(xy^2z^3)^6$

54. $-(xy^2z^3)^8$

55. $(-5a^2b^2c)^3$

56. $(-4abc^4)^3$

57. $(-xy^4z^2)^7$

58. $(-x^2y^4z^5)^5$

Find each of the following quotients.

59. $\dfrac{9x^4y^5}{3xy^2}$ **60.** $\dfrac{12x^2y^7}{6x^2y^3}$

61. $\dfrac{25x^5y^6}{-5x^2y^4}$ **62.** $\dfrac{56x^6y^4}{-7x^2y^3}$

63. $\dfrac{-54ab^2c^3}{-6abc}$ **64.** $\dfrac{-48a^3bc^5}{-6a^2c^4}$

65. $\dfrac{-18x^2y^2z^6}{xyz^2}$

66. $\dfrac{-32x^4y^5z^8}{x^2yz^3}$

67. $\dfrac{a^3b^4c^7}{-abc^5}$

68. $\dfrac{-a^4b^5c}{a^2b^4c}$

69. $\dfrac{-72x^2y^4}{-8x^2y^4}$

70. $\dfrac{-96x^4y^5}{12x^4y^4}$

71. $\dfrac{14ab^3}{-14\,ab}$

72. $\dfrac{-12abc^2}{12bc}$

73. $\dfrac{-36x^3y^5}{2y^5}$

74. $\dfrac{-48xyz^2}{2xz}$

Find each of the following products. Assume that the variables in the exponents represent positive integers. For example,

$$(x^{2n})(x^{3n}) = x^{2n+3n} = x^{5n}.$$

75. $(2x^n)(3x^{2n})$

76. $(3x^{2n})(x^{3n-1})$

77. $(a^{2n-1})(a^{3n+4})$

78. $(a^{5n-1})(a^{5n+1})$

79. $(x^{3n-2})(x^{n+2})$

80. $(x^{n-1})(x^{4n+3})$

81. $(a^{5n-2})(a^3)$

82. $(x^{3n-4})(x^4)$

83. $(2x^n)(-5x^n)$

84. $(4x^{2n-1})(-3x^{n+1})$

85. $(-3a^2)(-4a^{n+2})$

86. $(-5x^{n-1})(-6x^{2n+4})$

87. $(x^n)(2x^{2n})(3x^2)$

88. $(2x^n)(3x^{3n-1})(-4x^{2n+5})$

89. $(3x^{n-1})(x^{n+1})(4x^{2-n})$

90. $(-5x^{n+2})(x^{n-2})(4x^{3-2n})$

91. Find a polynomial that represents the total surface area of the rectangular solid in Figure 3.7. Also find a polynomial that represents the volume.

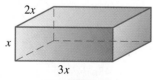

FIGURE 3.7

92. Find a polynomial that represents the total surface area of the rectangular solid in Figure 3.8. Also find a polynomial that represents the volume.

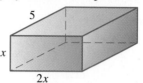

FIGURE 3.8

93. Find a polynomial that represents the area of the shaded region in Figure 3.9. The length of a radius of the larger circle is r units and the length of a radius of the smaller circle is 6 units.

FIGURE 3.9

THOUGHTS INTO WORDS

94. How would you convince someone that $x^6 \div x^2$ is x^4 and not x^3?

95. Your friend simplifies $2^3 \cdot 2^2$ as follows:

$$2^3 \cdot 2^2 = 4^{3+2} = 4^5 = 1024$$

What has she done incorrectly and how would you help her?

▼3.3 Multiplying Polynomials

We usually state the distributive property as $a(b + c) = ab + ac$; however, we can extend it as follows.

$$a(b + c + d) = ab + ac + ad$$
$$a(b + c + d + e) = ab + ac + ad + ae \qquad \text{etc.}$$

The commutative and associative properties, the properties of exponents, and the distributive property work together to find the product of a monomial and a polynomial. The following examples illustrate this idea.

Example 1

$$3x^2(2x^2 + 5x + 3) = 3x^2(2x^2) + 3x^2(5x) + 3x^2(3)$$
$$= 6x^4 + 15x^3 + 9x^2 \qquad \blacktriangle$$

Example 2

$$-2xy(3x^3 - 4x^2y - 5xy^2 + y^3) = -2xy(3x^3) - (-2xy)(4x^2y)$$
$$-(-2xy)(5xy^2) + (-2xy)(y^3)$$
$$= -6x^4y + 8x^3y^2 + 10x^2y^3 - 2xy^4 \qquad \blacktriangle$$

Now let's consider the product of two polynomials neither of which is a monomial. Consider the following examples.

Example 3

$$(x + 2)(y + 5) = x(y + 5) + 2(y + 5)$$
$$= x(y) + x(5) + 2(y) + 2(5)$$
$$= xy + 5x + 2y + 10 \qquad \blacktriangle$$

Notice that each term of the first polynomial is multiplied times each term of the second polynomial.

Example 4

$$(x - 3)(y + z + 3) = x(y + z + 3) - 3(y + z + 3)$$
$$= xy + xz + 3x - 3y - 3z - 9 \qquad \blacktriangle$$

Multiplying polynomials will often produce similar terms that can be combined to simplify the resulting polynomial.

Example 5

$$(x + 5)(x + 7) = x(x + 7) + 5(x + 7)$$
$$= x^2 + 7x + 5x + 35$$
$$= x^2 + 12x + 35 \qquad \blacktriangle$$

Example 6

$$\begin{aligned}(x-2)(x^2-3x+4) &= x(x^2-3x+4)-2(x^2-3x+4)\\ &= x^3-3x^2+4x-2x^2+6x-8\\ &= x^3-5x^2+10x-8\end{aligned}$$

▲

Example 7

$$\begin{aligned}(3x-2y)(x^2+xy-y^2) &= 3x(x^2+xy-y^2)-2y(x^2+xy-y^2)\\ &= 3x^3+3x^2y-3xy^2-2x^2y-2xy^2+2y^3\\ &= 3x^3+x^2y-5xy^2+2y^3\end{aligned}$$

▲

It helps to be able to find the product of two binomials without showing all of the intermediate steps. This is quite easy to do with a *three-step shortcut pattern* as demonstrated by Figures 3.10 and 3.11 in the following examples.

Example 8

FIGURE 3.10

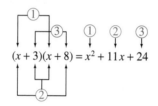

①
③
① ② ③
(x + 3)(x + 8) = x² + 11x + 24
②

STEP ①. Multiply $x \cdot x$.

STEP ②. Multiply $3 \cdot x$ and $8 \cdot x$ and combine.

STEP ③. Multiply $3 \cdot 8$.

▲

Example 9

FIGURE 3.11

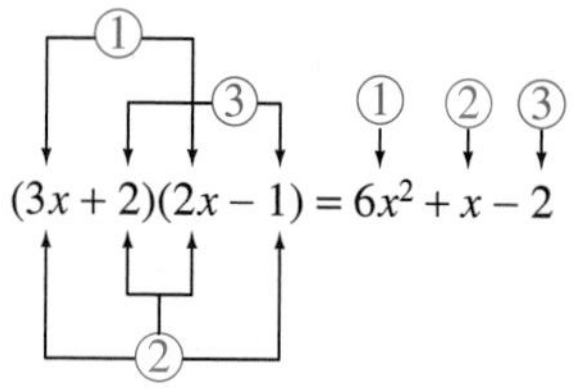

①
③
① ② ③
(3x + 2)(2x – 1) = 6x² + x – 2
②

▲

Now see if you can use the pattern to find the following products.

$(x+2)(x+6) = ?$

$(x-3)(x+5) = ?$

$(2x+5)(3x+7) = ?$

$(3x-1)(4x-3) = ?$

Your answers should be $x^2 + 8x + 12$, $x^2 + 2x - 15$, $6x^2 + 29x + 35$, and $12x^2 - 13x + 3$. Keep in mind that this shortcut pattern applies only to finding the product of two binomials.

We can use exponents to indicate repeated multiplication of polynomials. For example, $(x + 3)^2$ means $(x + 3)(x + 3)$; and $(x + 4)^3$ means $(x + 4)(x + 4) \cdot (x + 4)$. To square a binomial we can simply write it as the product of two equal binomials and apply the shortcut pattern. Thus,

$$(x + 3)^2 = (x + 3)(x + 3) = x^2 + 6x + 9,$$

$$(x - 6)^2 = (x - 6)(x - 6) = x^2 - 12x + 36, \quad \text{and}$$

$$(3x - 4)^2 = (3x - 4)(3x - 4) = 9x^2 - 24x + 16.$$

When squaring binomials, be careful not to forget the middle term. That is to say, $(x + 3)^2 \neq x^2 + 3^2$; instead, $(x + 3)^2 = x^2 + 6x + 9$.

When multiplying binomials, there are some special patterns that you should recognize. We can use these patterns to find products and later we will use some of them when factoring polynomials.

Pattern

$$(a + b)^2 = (a + b)(a + b) = a^2 + 2ab + b^2$$

Square of
first term   +
of binomial

Twice the
product of
the two terms
of binomial

+ Square of
second term
of binomial

Examples

$$(x + 4)^2 = x^2 + 8x + 16,$$

$$(2x + 3y)^2 = 4x^2 + 12xy + 9y^2,$$

$$(5a + 7b)^2 = 25a^2 + 70ab + 49b^2$$

▲

Pattern

$$(a - b)^2 = (a - b)(a - b) = a^2 - 2ab + b^2$$

Square of
first term   −
of binomial

Twice the
product of
the two terms
of binomial

+ Square of
second term
of binomial

Examples

$$(x - 8)^2 = x^2 - 16x + 64,$$

$$(3x - 4y)^2 = 9x^2 - 24xy + 16y^2,$$

$$(4a - 9b)^2 = 16a^2 - 72ab + 81b^2$$

▲

Pattern

$$(a + b)(a - b) = a^2 - b^2.$$

Square of | Square of
first term − second term
of binomials | of binomials

Examples

$$(x + 7)(x - 7) = x^2 - 49,$$
$$(2x + y)(2x - y) = 4x^2 - y^2,$$
$$(3a - 2b)(3a + 2b) = 9a^2 - 4b^2$$

Now suppose that we want to cube a binomial. One approach is as follows.

$$
\begin{aligned}
(x + 4)^3 &= (x + 4)(x + 4)(x + 4) \\
&= (x + 4)(x^2 + 8x + 16) \\
&= x(x^2 + 8x + 16) + 4(x^2 + 8x + 16) \\
&= x^3 + 8x^2 + 16x + 4x^2 + 32x + 64 \\
&= x^3 + 12x^2 + 48x + 64
\end{aligned}
$$

Another approach is to cube a general binomial and then use the resulting pattern.

Pattern

$$
\begin{aligned}
(a + b)^3 &= (a + b)(a + b)(a + b) \\
&= (a + b)(a^2 + 2ab + b^2) \\
&= a(a^2 + 2ab + b^2) + b(a^2 + 2ab + b^2) \\
&= a^3 + 2a^2b + ab^2 + a^2b + 2ab^2 + b^3 \\
&= a^3 + 3a^2b + 3ab^2 + b^3
\end{aligned}
$$

Let's use the pattern $(a + b)^3 = a^3 + 3a^2b + 3ab^2 + b^3$ to cube the binomial $x + 4$.

$$
\begin{aligned}
(x + 4)^3 &= x^3 + 3x^2(4) + 3x(4)^2 + 4^3 \\
&= x^3 + 12x^2 + 48x + 64
\end{aligned}
$$

Since $a - b = a + (-b)$ we can easily develop a pattern for cubing $a - b$.

Pattern

$$
\begin{aligned}
(a - b)^3 &= [a + (-b)]^3 \\
&= a^3 + 3a^2(-b) + 3a(-b)^2 + (-b)^3 \\
&= a^3 - 3a^2b + 3ab^2 - b^3
\end{aligned}
$$

Now let's use the pattern $(a - b)^3 = a^3 - 3a^2b + 3ab^2 - b^3$ to cube the binomial $3x - 2y$.

$$(3x - 2y)^3 = (3x)^3 - 3(3x)^2(2y) + 3(3x)(2y)^2 - (2y)^3$$
$$= 27x^3 - 54x^2y + 36xy^2 - 8y^3$$

Finally, we need to realize that if the patterns are forgotten or do not apply, then we can revert back to applying the distributive property.

$$(2x - 1)(x^2 - 4x + 6) = 2x(x^2 - 4x + 6) - 1(x^2 - 4x + 6)$$
$$= 2x^3 - 8x^2 + 12x - x^2 + 4x - 6$$
$$= 2x^3 - 9x^2 + 16x - 6 \qquad \blacktriangle$$

Back to the Geometry Connection

As you might expect, there are geometric interpretations for many of the algebraic concepts we present in this section. We will give you the opportunity to make some of these connections between algebra and geometry in the next problem set. Let's conclude this section with a problem that allows us to use some algebra and geometry.

Example 10

A rectangular piece of tin is 16 inches long and 12 inches wide as shown in Figure 3.12. From each corner a square piece x inches on a side is cut out. The flaps are then turned up to form an open box. Find polynomials that represent the volume and outside surface area of the box.

FIGURE 3.12

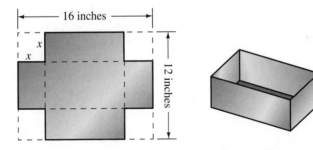

Solution

The length of the box will be $16 - 2x$, the width $12 - 2x$, and the height x. Using the volume formula $V = lwh$, the polynomial $(16 - 2x)(12 - 2x)(x)$, which simplifies to $4x^3 - 56x^2 + 192x$, represents the volume.

The outside surface area of the box is the area of the original piece of tin minus the four corners that were cut off. Therefore, the polynomial $16(12) - 4x^2$ or $192 - 4x^2$ represents the outside surface area of the box. $\qquad \blacktriangle$

REMARK

Recall in Section 3.1 that we found the total surface area of a rectangular solid by adding the areas of the sides, top, and bottom. Use this approach for the open box in Example 10 to check our answer of $192 - 4x^2$. Keep in mind, the box has no top. $\qquad \triangle$

Problem Set 3.3

Find the following indicated products. Remember the shortcut for multiplying binomials and the other special patterns we discussed in this section.

1. $2xy(5xy^2 + 3x^2y^3)$
2. $3x^2y(6y^2 - 5x^2y^4)$
3. $-3a^2b(4ab^2 - 5a^3)$
4. $-7ab^2(2b^3 - 3a^2)$
5. $8a^3b^4(3ab - 2ab^2 + 4a^2b^2)$

6. $9a^3b(2a - 3b + 7ab)$
7. $-x^2y(6xy^2 + 3x^2y^3 - x^3y)$

8. $-ab^2(5a + 3b - 6a^2b^3)$

9. $(a + 2b)(x + y)$
10. $(t - s)(x + y)$
11. $(a - 3b)(c + 4d)$
12. $(a - 4b)(c - d)$
13. $(x + 6)(x + 10)$
14. $(x + 2)(x + 10)$
15. $(y - 5)(y + 11)$
16. $(y - 3)(y + 9)$
17. $(n + 2)(n - 7)$
18. $(n + 3)(n - 12)$
19. $(x + 6)(x - 6)$
20. $(t + 8)(t - 8)$
21. $(x - 6)^2$
22. $(x - 2)^2$
23. $(x - 6)(x - 8)$
24. $(x - 3)(x - 13)$
25. $(x + 1)(x - 2)(x - 3)$
26. $(x - 1)(x + 4)(x - 6)$
27. $(x - 3)(x + 3)(x - 1)$
28. $(x - 5)(x + 5)(x - 8)$
29. $(t + 9)^2$
30. $(t + 13)^2$

31. $(y - 7)^2$
32. $(y - 4)^2$
33. $(4x + 5)(x + 7)$
34. $(6x + 5)(x + 3)$
35. $(3y - 1)(3y + 1)$
36. $(5y - 2)(5y + 2)$
37. $(7x - 2)(2x + 1)$
38. $(6x - 1)(3x + 2)$
39. $(1 + t)(5 - 2t)$
40. $(3 - t)(2 + 4t)$
41. $(3t + 7)^2$
42. $(4t + 6)^2$
43. $(2 - 5x)(2 + 5x)$
44. $(6 - 3x)(6 + 3x)$
45. $(7x - 4)^2$
46. $(5x - 7)^2$
47. $(6x + 7)(3x - 10)$
48. $(4x - 7)(7x + 4)$
49. $(2x - 5y)(x + 3y)$
50. $(x - 4y)(3x + 7y)$
51. $(5x - 2a)(5x + 2a)$
52. $(9x - 2y)(9x + 2y)$
53. $(t + 3)(t^2 - 3t - 5)$
54. $(t - 2)(t^2 + 7t + 2)$
55. $(x - 4)(x^2 + 5x - 4)$
56. $(x + 6)(2x^2 - x - 7)$
57. $(2x - 3)(x^2 + 6x + 10)$
58. $(3x + 4)(2x^2 - 2x - 6)$
59. $(4x - 1)(3x^2 - x + 6)$
60. $(5x - 2)(6x^2 + 2x - 1)$
61. $(x^2 + 2x + 1)(x^2 + 3x + 4)$

62. $(x^2 - x + 6)(x^2 - 5x - 8)$

63. $(2x^2 + 3x - 4)(x^2 - 2x - 1)$

64. $(3x^2 - 2x + 1)(2x^2 + x - 2)$

65. $(x + 2)^3$
66. $(x + 1)^3$
67. $(x - 4)^3$
68. $(x - 5)^3$
69. $(2x + 3)^3$
70. $(3x + 1)^3$
71. $(4x - 1)^3$
72. $(3x - 2)^3$
73. $(5x + 2)^3$
74. $(4x - 5)^3$

For Problems 75–84, find the indicated products. Assume all variables that appear as exponents represent positive integers.

75. $(x^n - 4)(x^n + 4)$
76. $(x^{3a} - 1)(x^{3a} + 1)$
77. $(x^a + 6)(x^a - 2)$
78. $(x^a + 4)(x^a - 9)$
79. $(2x^n + 5)(3x^n - 7)$
80. $(3x^n + 5)(4x^n - 9)$
81. $(x^{2a} - 7)(x^{2a} - 3)$
82. $(x^{2a} + 6)(x^{2a} - 4)$
83. $(2x^n + 5)^2$
84. $(3x^n - 7)^2$

85. Explain how Figure 3.13 can be used to demonstrate geometrically that $(x + 2)(x + 6) = x^2 + 8x + 12$.

FIGURE 3.13

86. Find a polynomial that represents the sum of the areas of the two rectangles shown in Figure 3.14.

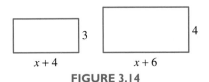

FIGURE 3.14

87. Find a polynomial that represents the area of the shaded region in Figure 3.15.

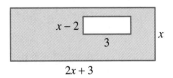

FIGURE 3.15

88. Explain how Figure 3.16 can be used to demonstrate geometrically that $(x + 7)(x - 3) = x^2 + 4x - 21$.

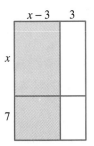

FIGURE 3.16

89. A square piece of cardboard is 16 inches on a side. From each corner a square piece x inches on a side is cut out. The flaps are then turned up to form an open box. Find polynomials that represent the volume and outside surface area of the box.

THOUGHTS INTO WORDS

90. How would you simplify $(2^3 + 2^2)^2$? Explain your reasoning.

91. Describe the process of multiplying two polynomials.

92. Determine the number of terms in the product of $(x + y)$ and $(a + b + c + d)$ without doing the multiplication. Explain how you arrived at your answer.

Further Investigations

93. We have used the following two multiplication patterns.

$$(a + b)^2 = a^2 + 2ab + b^2,$$

$$(a + b)^3 = a^3 + 3a^2b + 3ab^2 + b^3$$

By multiplying, we can extend these patterns as follows.

$$(a + b)^4 = a^4 + 4a^3b + 6a^2b^2 + 4ab^3 + b^4,$$

$$(a + b)^5 = a^5 + 5a^4b + 10a^3b^2 + 10a^2b^3 + 5a^4 + b^5$$

Based on these results, see if you can determine a pattern that will allow you to complete each of the following without using the long multiplication process.

(a) $(a + b)^6$

(b) $(a + b)^7$

(c) $(a + b)^8$

(d) $(a + b)^9$

94. Find each of the following indicated products. These patterns will be used again in Section 3.5.
(a) $(x - 1)(x^2 + x + 1)$
(b) $(x + 1)(x^2 - x + 1)$
(c) $(x + 3)(x^2 - 3x + 9)$
(d) $(x - 4)(x^2 + 4x + 16)$
(e) $(2x - 3)(4x^2 + 6x + 9)$
(f) $(3x + 5)(9x^2 - 15x + 25)$

3.4 Factoring: Use of the Distributive Property

Recall that 2 and 3 are said to be *factors* of 6 because the product of 2 and 3 is 6. Likewise, in an indicated product such as $7ab$, the 7, a, and b are called factors of the product. If a positive integer greater than 1 has no factors that are positive integers other than itself and 1, then it is called a **prime number**. Thus, the prime numbers less than 20 are 2, 3, 5, 7, 11, 13, 17, and 19. A positive integer greater than 1 that is not a prime number is called a **composite number**. The composite numbers less than 20 are 4, 6, 8, 9, 10, 12, 14, 15, 16, and 18. Every composite number is the product of prime numbers. Consider the following examples.

$$4 = 2 \cdot 2 \qquad 63 = 3 \cdot 3 \cdot 7$$
$$12 = 2 \cdot 2 \cdot 3 \qquad 121 = 11 \cdot 11$$
$$35 = 5 \cdot 7$$

The indicated product form that contains only prime factors is called the **prime factorization form** of a number. Thus, the prime factorization form of 63 is $3 \cdot 3 \cdot 7$. We also say that the number has been **completely factored** when it is in the prime factorization form.

In general, factoring is the reverse of multiplication. Previously, we have used the distributive property to find the product of a monomial and a polynomial, as in the next examples.

$$3(x + 2) = 3(x) + 3(2) = 3x + 6$$
$$5(2x - 1) = 5(2x) - 5(1) = 10x - 5$$
$$x(x^2 + 6x - 4) = x(x^2) + x(6x) - x(4) = x^3 + 6x^2 - 4x$$

We shall also use the distributive property (in the form, $ab + ac = a(b + c)$) to reverse the process, that is, to factor a given polynomial. Consider the following examples. (The steps in the dashed boxes can be done mentally.)

$$3x + 6 = \boxed{3(x) + 3(2)} = 3(x + 2),$$

$$10x - 5 = \boxed{5(2x) - 5(1)} = 5(2x - 1),$$

$$x^3 + 6x^2 - 4x = \boxed{x(x^2) + x(6x) - x(4)} = x(x^2 + 6x - 4)$$

Note that in each example a given polynomial has been factored into the product of a monomial and a polynomial. Obviously, polynomials could be factored in a variety of ways. Consider some factorizations of $3x^2 + 12x$.

$$3x^2 + 12x = 3x(x + 4) \quad \text{or} \quad 3x^2 + 12x = 3(x^2 + 4x) \quad \text{or}$$

$$3x^2 + 12x = x(3x + 12) \quad \text{or} \quad 3x^2 + 12x = \frac{1}{2}(6x^2 + 24x)$$

We are, however, primarily interested in the first of the previous factorization forms, which we refer to as the **completely factored form**. A polynomial with integral coefficients is in completely factored form if:

1. It is expressed as a product of polynomials with *integral coefficients*;
2. No polynomial, other than a monomial, within the factored form can be further factored into polynomials with integral coefficients.

Do you see why only the first of the above factored forms of $3x^2 + 12x$ is said to be in completely factored form? In each of the other three forms the polynomial inside the parentheses can be further factored. Furthermore, in the last form, $\frac{1}{2}(6x^2 + 24x)$, the condition of using only integral coefficients is violated.

The factoring process that we discuss in this section ($ab + ac = a(b + c)$) is often referred to as **factoring out the highest common monomial factor**. The key idea in this process is to recognize the monomial factor that is common to all terms. For example, observe that each term of the polynomial $2x^3 + 4x^2 + 6x$ has a factor of $2x$. Thus, we write

$$2x^3 + 4x^2 + 6x = 2x(\qquad)$$

and insert within the parentheses the appropriate polynomial factor. We determine the terms of this polynomial factor by dividing each term of the original polynomial by the factor of $2x$. The final completely factored form is

$$2x^3 + 4x^2 + 6x = 2x(x^2 + 2x + 3).$$

The following examples further demonstrate this process of factoring out the highest common monomial factor.

$$12x^3 + 16x^2 = 4x^2(3x + 4) \qquad\qquad 6x^2y^3 + 27xy^4 = 3xy^3(2x + 9y)$$

$$8ab - 18b = 2b(4a - 9) \qquad\qquad 8y^3 + 4y^2 = 4y^2(2y + 1)$$

$$30x^3 + 42x^4 - 24x^5 = 6x^3 \cdot (5 + 7x - 4x^2)$$

Note that in each example the common monomial factor itself is not in a completely factored form. For example, $4x^2(3x + 4)$ is not written as $2 \cdot 2x \cdot x \cdot (3x + 4)$.

Sometimes there may be a common binomial factor rather than a common monomial factor. For example, each of the two terms of the expression $x(y + 2) + z(y + 2)$ has a binomial factor of $(y + 2)$. Thus, we can factor $(y + 2)$ from each term and our result is as follows.

$$x(y + 2) + z(y + 2) = (y + 2)(x + z)$$

Consider a few more examples that involve a common binomial factor.

$$a^2(b + 1) + 2(b + 1) = (b + 1)(a^2 + 2)$$

$$x(2y - 1) - y(2y - 1) = (2y - 1)(x - y)$$

$$x(x + 2) + 3(x + 2) = (x + 2)(x + 3)$$

It may be that the original polynomial exhibits no apparent common monomial or binomial factor, which is the case with $ab + 3a + bc + 3c$. However, by factoring a from the first two terms and c from the last two terms we get

$$ab + 3a + bc + 3c = a(b + 3) + c(b + 3).$$

Now a common binomial factor of $(b + 3)$ is obvious and we can proceed as before.

$$a(b + 3) + c(b + 3) = (b + 3)(a + c)$$

We refer to this factoring process as **factoring by grouping**. Let's consider a few more examples of this type.

$$ab^2 - 4b^2 + 3a - 12 = b^2(a - 4) + 3(a - 4) \qquad \text{Factor } b^2 \text{ from first two terms and 3 from last two terms}$$

$$= (a - 4)(b^2 + 3), \qquad \text{Factor common binomial from both terms}$$

$$x^2 - x + 5x - 5 = x(x - 1) + 5(x - 1) \qquad \text{Factor } x \text{ from first two terms and 5 from last two terms}$$

$$= (x - 1)(x + 5), \qquad \text{Factor common binomial from both terms}$$

$$x^2 + 2x - 3x - 6 = x(x + 2) - 3(x + 2)$$

Factor x from first two terms and -3 from last two terms

$$= (x + 2)(x - 3)$$

Factor common binomial factor from both terms

It may be necessary to rearrange some terms first before applying the distributive property. Terms that contain common factors need to be grouped together and this may be done in more than one way. The next example illustrates this idea.

$$4a^2 - bc^2 - a^2b + 4c^2 = 4a^2 - a^2b + 4c^2 - bc^2$$
$$= a^2(4 - b) + c^2(4 - b)$$
$$= (4 - b)(a^2 + c^2), \qquad \text{or}$$
$$4a^2 - bc^2 - a^2b + 4c^2 = 4a^2 + 4c^2 - bc^2 - a^2b$$
$$= 4(a^2 + c^2) - b(c^2 + a^2)$$
$$= 4(a^2 + c^2) - b(a^2 + c^2)$$
$$= (a^2 + c^2)(4 - b)$$

Equations and Problem Solving

One reason that factoring is an important algebraic skill is that it extends our techniques for solving equations. Each time that we examine a factoring technique we will then use it to help solve certain types of equations.

We need another property of equality before we consider some equations where the highest-common-factor technique is useful. Suppose that the product of two numbers is zero. Can we conclude that at least one of the number must be zero? Let's state a property that formalizes this idea.

PROPERTY 3.5

Let a and b be real numbers,

$$ab = 0 \quad \text{if and only if } a = 0 \text{ or } b = 0.$$

Property 3.5, along with the highest-common-factor pattern, provides us with another technique for solving equations.

Example 1

Solve $x^2 + 6x = 0$.

Solution

$$x^2 + 6x = 0$$
$$x(x + 6) = 0 \qquad \text{Factor the left side.}$$
$$x = 0 \quad \text{or} \quad x + 6 = 0 \qquad \text{Use Property 3.5.}$$
$$x = 0 \quad \text{or} \quad x = -6$$

Thus, both 0 and -6 will satisfy the original equation and the solution set is $\{-6, 0\}$.   ▲

Example 2   Solve $a^2 = 11a$.

Solution
$$a^2 = 11a$$
$$a^2 - 11a = 0 \qquad \text{Add } -11a \text{ to both sides.}$$
$$a(a - 11) = 0 \qquad \text{Factor the left side.}$$
$$a = 0 \quad \text{or} \quad a - 11 = 0 \qquad \text{Use Property 3.5.}$$
$$a = 0 \quad \text{or} \quad a = 11$$

The solution set is $\{0, 11\}$.   ▲

REMARK   Notice in Example 2 we *did not* divide both sides of the equation by a. This would cause us to lose the solution of 0.   △

Example 3   Solve $3n^2 - 5n = 0$.

Solution
$$3n^2 - 5n = 0$$
$$n(3n - 5) = 0$$
$$n = 0 \quad \text{or} \quad 3n - 5 = 0$$
$$n = 0 \quad \text{or} \quad 3n = 5$$
$$n = 0 \quad \text{or} \quad n = \frac{5}{3}$$

The solution set is $\left\{0, \dfrac{5}{3}\right\}$.   ▲

Example 4   Solve $3ax^2 + bx = 0$ for x.

Solution
$$3ax^2 + bx = 0$$
$$x(3ax + b) = 0$$
$$x = 0 \quad \text{or} \quad 3ax + b = 0$$
$$x = 0 \quad \text{or} \quad 3ax = -b$$
$$x = 0 \quad \text{or} \quad x = -\frac{b}{3a}$$

The solution set is $\left\{0, -\dfrac{b}{3a}\right\}$.   ▲

Many of the problems that we are going to solve in the next few sections have a geometric setting. So let's briefly review a few formulas from geometry that we can put to use in a problem-solving situation (Figures 3.17–3.27).

$A = l\,w \qquad P = 2l + 2w$

    A area
    P perimeter
    l length
    w width

Rectangle

FIGURE 3.17

Square

$A = s^2 \quad P = 4s$

A area
P perimeter
s length of a side

FIGURE 3.18

Triangle

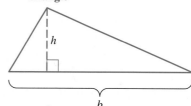

$A = \dfrac{1}{2}\,bh$

  A area
  b base
  h altitude (height)

FIGURE 3.19

Parallelogram

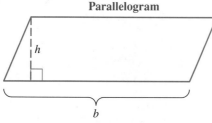

$A = bh$

A area
b base
h altitude (height)

FIGURE 3.20

Trapezoid

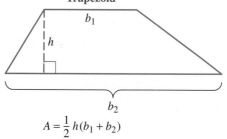

$A = \dfrac{1}{2}\,h(b_1 + b_2)$

    A area
b_1, b_2 bases
    h altitude

FIGURE 3.21

Circle

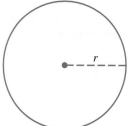

$A = \pi r^2 \qquad C = 2\pi r$

A area
C circumference
r radius

FIGURE 3.22

Sphere

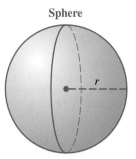

$$V = \frac{4}{3}\pi r^3 \quad S = 4\pi r^2$$

S   surface area
V   volume
r   radius

FIGURE 3.23

Right Circular Cylinder

$$V = \pi r^2 h \quad S = 2\pi r^2 + 2\pi rh$$

V   volume
S   total surface area
r   radius
h   altitude (height)

FIGURE 3.24

Right Circular Cone

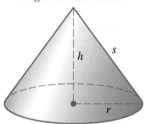

$$V = \frac{1}{3}\pi r^2 h \quad S = \pi r^2 + \pi rs$$

V   volume
S   total surface area
r   radius
h   altitude (height)
s   slant height

FIGURE 3.25

Prism

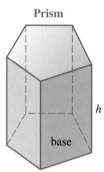

$$V = Bh$$

V   volume
B   area of base
h   altitude (height)

FIGURE 3.26

Pyramid

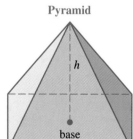

$$V = \frac{1}{3}Bh$$

V   volume
B   area of base
h   altitude (height)

FIGURE 3.27

Problem 1

The area of a square is three times its perimeter. Find the length of a side of the square.

Solution

Let s represent the length of a side of the square (Figure 3.28). The area is represented by s^2 and the perimeter by $4s$. Thus,

$$s^2 = 3(4s) \qquad \text{The area is to be three times the perimeter.}$$

$$s^2 = 12s$$

$$s^2 - 12s = 0$$

$$s(s - 12) = 0$$

$$s = 0 \quad \text{or} \quad s = 12.$$

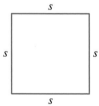

FIGURE 3.28

Since 0 is not a reasonable solution, it must be a 12-by-12 square. (Be sure to check this answer in the *original statement of the problem*!) ▲

Problem 2

Suppose that the volume of a right circular cylinder is numerically equal to the total surface area of the cylinder. If the height of the cylinder is equal to the length of a radius of the base, find the height.

Solution

Since $r = h$, the formula for volume $V = \pi r^2 h$ becomes $V = \pi r^3$ and the formula for the total surface area $S = 2\pi r^2 + 2\pi rh$ becomes $S = 2\pi r^2 + 2\pi r^2$ or $S = 4\pi r^2$. Therefore, we can set up and solve the following equation.

$$\pi r^3 = 4\pi r^2$$
$$\pi r^3 - 4\pi r^2 = 0$$
$$\pi r^2(r - 4) = 0$$
$$\pi r^2 = 0 \quad \text{or} \quad r - 4 = 0$$
$$r = 0 \quad \text{or} \quad r = 4$$

Since 0 is not a reasonable answer, the height must be 4 units. ▲

Problem Set 3.4

For problems 1–10, classify each number as prime or composite.

1. 63

2. 81

3. 59

4. 83

5. 51

6. 69

7. 91

8. 119

9. 71

10. 101

For Problems 11–20, factor each of the composite numbers into the product of prime numbers. For example, $30 = 2 \cdot 3 \cdot 5$.

11. 28

12. 39

13. 44

14. 49

15. 56

16. 64

17. 72

18. 84

19. 87

20. 91

For Problems 21–46, factor completely each of the following.

21. $6x + 3y$

22. $12x + 8y$

23. $6x^2 + 14x$

24. $15x^2 + 6x$

25. $28y^2 - 4y$

26. $42y^2 - 6y$

27. $20xy - 15x$

28. $27xy - 36y$

29. $7x^3 + 10x^2$

30. $12x^3 - 10x^2$

31. $18a^2b + 27ab^2$

32. $24a^3b^2 + 36a^2b$

33. $12x^3y^4 - 39x^4y^3$

34. $15x^4y^2 - 45x^5y^4$

35. $8x^4 + 12x^3 - 24x^2$

36. $6x^5 - 18x^3 + 24x$

37. $5x + 7x^2 + 9x^4$

38. $9x^2 - 17x^4 + 21x^5$

39. $15x^2y^3 + 20xy^2 + 35x^3y^4$

40. $8x^5y^3 - 6x^4y^5 + 12x^2y^3$

41. $x(y + 2) + 3(y + 2)$

42. $x(y - 1) + 5(y - 1)$

43. $3x(2a + b) - 2y(2a + b)$

44. $5x(a - b) + y(a - b)$

45. $x(x + 2) + 5(x + 2)$

46. $x(x - 1) - 3(x - 1)$

For Problems 47–64, factor each of the following by grouping.

47. $ax + 4x + ay + 4y$

48. $ax - 2x + ay - 2y$

49. $ax - 2bx + ay - 2by$

50. $2ax - bx + 2ay - by$

51. $3ax - 3bx - ay + by$

52. $5ax - 5bx - 2ay + 2by$

53. $2ax + 2x + ay + y$

54. $3bx + 3x + by + y$

55. $ax^2 - x^2 + 2a - 2$

56. $ax^2 - 2x^2 + 3a - 6$

57. $2ac + 3bd + 2bc + 3ad$

58. $2bx + cy + cx + 2by$

59. $ax - by + bx - ay$

60. $2a^2 - 3bc - 2ab + 3ac$

61. $x^2 + 9x + 6x + 54$

62. $x^2 - 2x + 5x - 10$

63. $2x^2 + 8x + x + 4$

64. $3x^2 + 18x - 2x - 12$

For Problems 65–80, solve each of the equations.

65. $x^2 + 7x = 0$

66. $x^2 + 9x = 0$

67. $x^2 - x = 0$

68. $x^2 - 14x = 0$

69. $a^2 = 5a$　　　**70.** $b^2 = -7b$

71. $-2y = 4y^2$

72. $-6x = 2x^2$

73. $3x^2 + 7x = 0$

74. $-4x^2 + 9x = 0$

75. $4x^2 = 5x$　　　**76.** $3x = 11x^2$

77. $x - 4x^2 = 0$

78. $x - 6x^2 = 0$

79. $12a = -a^2$

80. $-5a = -a^2$

For Problems 81–86, solve each equation for the indicated variable.

81. $5bx^2 - 3ax = 0$　for x

82. $ax^2 + bx = 0$　for x

83. $2by^2 = -3ay$　for y

84. $3ay^2 = by$　for y

85. $y^2 - ay + 2by - 2ab = 0$　for y

86. $x^2 + ax + bx + ab = 0$　for x

For Problems 87–96, set up an equation and solve each of the following problems.

87. The square of a number equals seven times the number. Find the number.

88. Suppose that the area of a square is six times its perimeter. Find the length of a side of the square.

89. The area of a circular region is numerically equal to three times the circumference of the circle. Find the length of a radius of the circle.

90. Find the length of a radius of a circle such that the circumference of the circle is numerically equal to the area of the circle.

91. Suppose that the area of a circle is numerically equal to the perimeter of a square and that the length of a radius of the circle is equal to the length of a side of the square. Find the length of a side of the square. Express your answer in terms of π.

92. Find the length of a radius of a sphere such that the surface area of the sphere is numerically equal to the volume of the sphere.

93. Suppose that the area of a square lot is twice the area of an adjoining rectangular plot of ground. If the rectangular plot is 50 feet wide and its length is the same as the length of a side of the square lot, find the dimensions of both the square and the rectangle.

94. The area of a square is one-fourth as large as the area of a triangle. One side of the triangle is 16 inches long and the altitude to that side is the same length as a side of the square. Find the length of a side of the square.

95. Suppose that the volume of a sphere is numerically equal to twice the surface area of the sphere. Find the length of a radius of the sphere.

96. Suppose that a radius of a sphere is equal in length to a radius of a circle. If the volume of the sphere is numerically equal to four times the area of the circle, find the length of a radius for both the sphere and the circle.

THOUGHTS INTO WORDS

97. Is $2 \cdot 3 \cdot 5 \cdot 7 \cdot 11 + 7$ a prime or composite number? Defend your answer.

98. Suppose that your friend factors $36x^2y + 48xy^2$ as follows.

$$36x^2y + 48xy^2 = (4xy)(9x + 12y)$$
$$= (4xy)(3)(3x + 4y)$$
$$= 12xy(3x + 4y)$$

Is this a correct approach? Would you have any suggestion to offer your friend?

99. Your classmate solves the equation $3ax + bx = 0$ for x as follows.

$$3ax + bx = 0$$
$$3ax = -bx$$
$$x = \frac{-bx}{3a}$$

How should he know that the solution is incorrect? How would you help him obtain the correct solution?

Further Investigations

100. The total surface area of a right circular cylinder is given by the formula $A = 2\pi r^2 + 2\pi rh$, where r represents the radius of a base and h represents the height of the cylinder. For computational purposes it may be more convenient to change the form of the right side of the formula by factoring it.

$$A = 2\pi r^2 + 2\pi rh$$
$$= 2\pi r(r + h)$$

Use $A = 2\pi r(r + h)$ to find the total surface area of each of the following cylinders. Also, use $\frac{22}{7}$ as an approximation for π.

(a) $r = 7$ centimeters and $h = 12$ centimeters;

(b) $r = 14$ meters and $h = 20$ meters;

(c) $r = 3$ feet and $h = 4$ feet;
(d) $r = 5$ yards and $h = 9$ yards.

For Problems 101–106, factor each expression and assume all variables that appear as exponents represent positive integers.

101. $2x^{2a} - 3x^a$

102. $6x^{2a} + 8x^a$

103. $y^{3m} + 5y^{2m}$

104. $3y^{5m} - y^{4m} - y^{3m}$

105. $2x^{6a} - 3x^{5a} + 7x^{4a}$

106. $6x^{3a} - 10x^{2a}$

3.5   Factoring: Difference of Two Squares and Sum or Difference of Two Cubes

In Section 3.3 we examined some special multiplication patterns. One of these patterns was the following.

$$(a + b)(a - b) = a^2 - b^2$$

This same pattern, viewed as a factoring pattern, we referred to as:

Difference of Two Squares

$$a^2 - b^2 = (a + b)(a - b).$$

Applying the pattern is a fairly simple process as these next examples demonstrate. Again the steps in dashed boxes are usually performed mentally.

$$x^2 - 16 = \boxed{(x)^2 - (4)^2} = (x + 4)(x - 4),$$

$$4x^2 - 25 = \boxed{(2x)^2 - (5)^2} = (2x + 5)(2x - 5),$$

$$16x^2 - 9y^2 = \boxed{(4x)^2 - (3y)^2} = (4x + 3y)(4x - 3y),$$

$$1 - a^2 = \boxed{(1)^2 - (a)^2} = (1 + a)(1 - a)$$

Since multiplication is commutative, the order of writing the factors is not important. For example, $(x + 4)(x - 4)$ can also be written as $(x - 4)(x + 4)$.

You must be careful not to assume an analogous factoring pattern for the *sum* of two squares; *it does not exist*. For example, $x^2 + 4 \neq (x + 2)(x + 2)$ because $(x + 2)(x + 2) = x^2 + 4x + 4$. We say that a polynomial such as $x^2 + 4$ is a **prime polynomial** or that it is *not factorable using integers*.

Sometimes the difference-of-two-squares pattern can be applied more than once as the next examples illustrate.

$$x^4 - y^4 = (x^2 + y^2)(x^2 - y^2) = (x^2 + y^2)(x + y)(x - y)$$

$$16x^4 - 81y^4 = (4x^2 + 9y^2)(4x^2 - 9y^2) = (4x^2 + 9y^2)(2x + 3y)(2x - 3y)$$

It may also be that the squares are other than simple monomial squares as in these next three examples.

$$(x + 3)^2 - y^2 = ((x + 3) + y)((x + 3) - y) = (x + 3 + y)(x + 3 - y)$$

$$4x^2 - (2y + 1)^2 = (2x + (2y + 1))(2x - (2y + 1))$$
$$= (2x + 2y + 1)(2x - 2y - 1)$$

$$(x - 1)^2 - (x + 4)^2 = ((x - 1) + (x + 4))((x - 1) - (x + 4))$$
$$= (x - 1 + x + 4)(x - 1 - x - 4)$$
$$= (2x + 3)(-5)$$

It is possible to apply both the technique of *factoring out a common monomial factor* and the pattern of the *difference of two squares* to the same problem. *In general, it is best to first look for a common monomial factor.* Consider the following examples.

$$2x^2 - 50 = 2(x^2 - 25)$$
$$= 2(x + 5)(x - 5)$$
$$48y^3 - 27y = 3y(16y^2 - 9)$$
$$= 3y(4y + 3)(4y - 3)$$
$$9x^2 - 36 = 9(x^2 - 4)$$
$$= 9(x + 2)(x - 2)$$

WORD OF CAUTION   The polynomial $9x^2 - 36$ can be factored as follows.

$$9x^2 - 36 = (3x + 6)(3x - 6)$$
$$= 3(x + 2)(3)(x - 2)$$
$$= 9(x + 2)(x - 2)$$

However, when taking this approach there seems to be a tendency to stop at the step $(3x + 6)(3x - 6)$. Therefore, remember the suggestion to *first look for a common monomial factor.*   △

The following examples should help you summarize all of our factoring techniques thus far.

$$7x^2 + 28 = 7(x^2 + 4)$$
$$4x^2y - 14xy^2 = 2xy(2x - 7y)$$
$$x^2 - 4 = (x + 2)(x - 2)$$
$$18 - 2x^2 = 2(9 - x^2) = 2(3 + x)(3 - x)$$
$y^2 + 9$ is not factorable using integers
$5x + 13y$ is not factorable using integers
$$x^4 - 16 = (x^2 + 4)(x^2 - 4) = (x^2 + 4)(x + 2)(x - 2)$$

Sum and Difference of Two Cubes

As we pointed out before, no sum-of-squares pattern exists analogous to the difference-of-squares factoring pattern. That is to say, a polynomial such as $x^2 + 9$ is not factorable using integers. However, there do exist patterns for both the *sum and difference of two cubes*. These patterns are as follows.

Sum and Difference of Two Cubes

$$a^3 + b^3 = (a + b)(a^2 - ab + b^2)$$
$$a^3 - b^3 = (a - b)((a^2 + ab + b^2)$$

Note how we apply these patterns in the next four examples.

$$x^3 + 27 = (x)^3 + (3)^3 = (x + 3)(x^2 - 3x + 9),$$
$$8a^3 + 125b^3 = (2a)^3 + (5b)^3 = (2a + 5b)(4a^2 - 10ab + 25b^2),$$
$$x^3 - 1 = (x)^3 - (1)^3 = (x - 1)(x^2 + x + 1),$$
$$27y^3 - 64x^3 = (3y)^3 - (4x)^3 = (3y - 4x)(9y^2 + 12xy + 16x^2)$$

Equations and Problem Solving

Remember that each time we pick up a new factoring technique we also develop more power for solving equations. Let's consider how we can use the difference-of-two-squares factoring pattern to help solve certain types of equations.

Example I

Solve $x^2 = 16$.

Solution

$$x^2 = 16$$
$$x^2 - 16 = 0$$
$$(x + 4)(x - 4) = 0$$
$$x + 4 = 0 \quad \text{or} \quad x - 4 = 0$$
$$x = -4 \quad \text{or} \quad x = 4$$

The solution set is $\{-4, 4\}$. (Be sure to check these solutions in the original equation!)   ▲

Example 2

Solve $9x^2 = 64$.

Solution

$$9x^2 = 64$$
$$9x^2 - 64 = 0$$
$$(3x + 8)(3x - 8) = 0$$
$$3x + 8 = 0 \quad \text{or} \quad 3x - 8 = 0$$
$$3x = -8 \quad \text{or} \quad 3x = 8$$
$$x = -\frac{8}{3} \quad \text{or} \quad x = \frac{8}{3}$$

The solution set is $\left\{-\frac{8}{3}, \frac{8}{3}\right\}$.   ▲

Example 3

Solution

Solve $7x^2 - 7 = 0$.

$$7x^2 - 7 = 0$$
$$7(x^2 - 1) = 0$$
$$x^2 - 1 = 0 \quad \text{Multiply both sides by } \tfrac{1}{7}.$$
$$(x + 1)(x - 1) = 0$$
$$x + 1 = 0 \quad \text{or} \quad x - 1 = 0$$
$$x = -1 \quad \text{or} \quad x = 1$$

The solution set is $\{-1, 1\}$. ▲

    In the previous examples we have been using the property $ab = 0$ if and only if $a = 0$ or $b = 0$. This property can be extended to any number of factors whose product is zero. Thus, for three factors the property could be stated $abc = 0$ if and only if $a = 0$ or $b = 0$ or $c = 0$. The next two examples illustrate this idea.

Example 4

Solution

Solve $x^4 - 16 = 0$.

$$x^4 - 16 = 0$$
$$(x^2 + 4)(x^2 - 4) = 0$$
$$(x^2 + 4)(x + 2)(x - 2) = 0$$
$$x^2 + 4 = 0 \quad \text{or} \quad x + 2 = 0 \quad \text{or} \quad x - 2 = 0$$
$$x^2 = -4 \quad \text{or} \quad x = -2 \quad \text{or} \quad x = 2$$

The solution set is $\{-2, 2\}$. (Since no real numbers when squared will produce -4, the equation $x^2 = -4$ yields no additional real solutions.) ▲

Example 5

Solution

Solve $x^3 - 49x = 0$.

$$x^3 - 49x = 0$$
$$x(x^2 - 49) = 0$$
$$x(x + 7)(x - 7) = 0$$
$$x = 0 \quad \text{or} \quad x + 7 = 0 \quad \text{or} \quad x - 7 = 0$$
$$x = 0 \quad \text{or} \quad x = -7 \quad \text{or} \quad x = 7$$

The solution set is $\{-7, 0, 7\}$. ▲

    The more that we know about solving equations the better off we are for solving word problems.

Problem 1

The combined area of two squares is 40 square centimeters. Each side of one square is three times as long as a side of the other square. Find the dimensions of each of the squares.

Solution

Let s represent the length of a side of the smaller square. Then $3s$ represents the length of a side of the larger square (Figure 3.29).

$$s^2 + (3s)^2 = 40$$
$$s^2 + 9s^2 = 40$$
$$10s^2 = 40$$
$$s^2 = 4$$
$$s^2 - 4 = 0$$
$$(s + 2)(s - 2) = 0$$
$$s + 2 = 0 \quad \text{or} \quad s - 2 = 0$$
$$s = -2 \quad \text{or} \quad s = 2$$

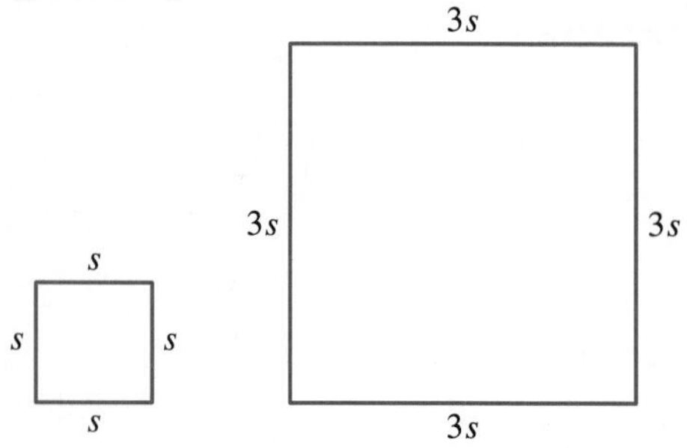

Since s represents the length of a side of a square, the solution -2 has to be disregarded. Thus, the length of a side of the small square is 2 centimeters and the large square has sides of $3(2) = 6$ centimeters.

FIGURE 3.29

Problem Set 3.5

For Problems 1–20, use the difference-of-squares pattern to factor each of the following.

1. $x^2 - 1$

2. $x^2 - 9$

3. $16x^2 - 25$

4. $4x^2 - 49$

5. $9x^2 - 25y^2$

6. $x^2 - 64y^2$

7. $25x^2y^2 - 36$

8. $x^2y^2 - a^2b^2$

9. $4x^2 - y^4$

10. $x^6 - 9y^2$

11. $1 - 144n^2$

12. $25 - 49n^2$

13. $(x + 2)^2 - y^2$

14. $(3x + 5)^2 - y^2$

15. $4x^2 - (y + 1)^2$

16. $x^2 - (y - 5)^2$

17. $9a^2 - (2b + 3)^2$

18. $16s^2 - (3t + 1)^2$

19. $(x + 2)^2 - (x + 7)^2$

20. $(x - 1)^2 - (x - 8)^2$

For Problems 21–44, factor each of the following completely. Indicate any that are not factorable using integers. Don't forget to first look for a common monomial factor.

21. $9x^2 - 36$

22. $8x^2 - 72$

23. $5x^2 + 5$

24. $7x^2 + 28$

25. $8y^2 - 32$

26. $5y^2 - 80$

27. $a^3b - 9ab$

28. $x^3y^2 - xy^2$

29. $16x^2 + 25$

30. $x^4 - 16$

31. $n^4 - 81$

32. $4x^2 + 9$

33. $3x^3 + 27x$

34. $20x^3 + 45x$

35. $4x^3y - 64xy^3$

36. $12x^3 - 27xy^2$

37. $6x - 6x^3$

38. $1 - 16x^4$

39. $1 - x^4y^4$

40. $20x - 5x^3$

41. $4x^2 - 64y^2$

42. $9x^2 - 81y^2$

43. $3x^4 - 48$

44. $2x^5 - 162x$

For Problems 45–56, use the sum or difference-of-cubes patterns to factor each of the following.

45. $a^3 - 64$

46. $a^3 - 27$

47. $x^3 + 1$

48. $x^3 + 8$

49. $27x^3 + 64y^3$

50. $8x^3 + 27y^3$

51. $1 - 27a^3$

52. $1 - 8x^3$

53. $x^3y^3 - 1$

54. $125x^3 + 27y^3$

55. $x^6 - y^6$

56. $x^6 + y^6$

For Problems 57–70, find all real number solutions for each equation.

57. $x^2 - 25 = 0$

58. $x^2 - 1 = 0$

59. $9x^2 - 49 = 0$

60. $4y^2 = 25$

61. $8x^2 - 32 = 0$

62. $3x^2 - 108 = 0$

63. $3x^3 = 3x$

64. $4x^3 = 64x$

65. $20 - 5x^2 = 0$

66. $54 - 6x^2 = 0$

67. $x^4 - 81 = 0$

68. $x^5 - x = 0$

69. $6x^3 + 24x = 0$

70. $4x^3 + 12x = 0$

For Problems 71–80, set up an equation and solve each of the following problems.

71. The cube of a number equals nine times the same number. Find the number.

72. The cube of a number equals the square of the same number. Find the number.

73. The combined area of two circles is 80π square centimeters. The length of a radius of one circle is twice the length of a radius of the other circle. Find the length of the radius of each circle.

74. The combined area of two squares is 26 square meters. The sides of the larger square are five times as long as the sides of the smaller square. Find the dimensions of each of the squares.

75. A rectangle is twice as long as it is wide and its area is 50 square meters. Find the length and the width of the rectangle.

76. Suppose that the length of a rectangle is one and one-third times as long as its width. The area of the rectangle is 48 square centimeters. Find the length and width of the rectangle.

77. The total surface area of a right circular cylinder is 54π square inches. If the altitude of the cylinder is twice the length of a radius, find the altitude of the cylinder.

78. The total surface area of a right circular cone is 108π square feet. If the slant height of the cone is twice the length of a radius of the base, find the length of a radius.

79. The sum of the areas of a circle and a square is $(16\pi + 64)$ square yards. If a side of the square is twice the length of a radius of the circle, find the length of a side of the square.

80. The length of an altitude of a triangle is one-third the length of the side to which it is drawn. If the area of the triangle is 6 square centimeters, find the length of that altitude.

81. Explain how you would solve the equation $4x^3 = 64x$.

82. What is wrong with the following factoring process?

$$25x^2 - 100 = (5x + 10)(5x - 10)$$

How would you correct the error?

83. Consider the following solution:

$$6x^2 - 24 = 0$$
$$6(x^2 - 4) = 0$$
$$6(x + 2)(x - 2) = 0$$

$6 = 0$ or $x + 2 = 0$ or $x - 2 = 0$

$6 = 0$ or $x = -2$ or $x = 2$

The solution set is $\{-2, 2\}$.

Is this a correct solution? Would you have any suggestion to offer the person who used this approach?

3.6   Factoring Trinomials

One of the most common types of factoring used in algebra is the expression of a trinomial as the product of two binomials. To develop a factoring technique we first look at some multiplication ideas. Let's consider the product $(x + a)(x + b)$ and use the distributive property to show how each term of the resulting trinomial is formed.

$$(x + a)(x + b) = x(x + b) + a(x + b)$$
$$= x(x) + x(b) + a(x) + a(b)$$
$$= x^2 + (a + b)x + ab$$

Notice that the coefficient of the middle term is the *sum* of a and b and the last term is the *product* of a and b. These two relationships can be used to factor trinomials. Let's consider some examples.

Example 1

Factor $x^2 + 8x + 12$.

Solution

We need to complete the following with two integers whose sum is 8 and whose product is 12.

$$x^2 + 8x + 12 = (x + \underline{\hspace{0.5cm}})(x + \underline{\hspace{0.5cm}})$$

The possible pairs of factors of 12 are 1(12), 2(6), and 3(4). Since $6 + 2 = 8$, we can complete the factoring as follows.

$$x^2 + 8x + 12 = (x + 6)(x + 2)$$

To *check* our answer we find the product of $(x + 6)$ and $(x + 2)$.   ▲

Example 2

Factor $x^2 - 10x + 24$.

Solution

We need two integers whose product is 24 and whose sum is -10. Let's use a small table to organize our thinking.

Product	Sum
$(-1)(-24) = 24$	$-1 + (-24) = -25$
$(-2)(-12) = 24$	$-2 + (-12) = -14$
$(-3)(-8) = 24$	$-3 + (-8) = -11$
$(-4)(-6) = 24$	$-4 + (-6) = -10$

The bottom line contains the numbers that we need. Thus,

$$x^2 - 10x + 24 = (x - 4)(x - 6).$$ ▲

Example 3

Factor $x^2 + 7x - 30$.

Solution

We need two integers whose product is -30 and whose sum is 7.

Product	Sum
$(-1)(30) = -30$	$-1 + 30 = 29$
$1(-30) = -30$	$1 + (-30) = -29$
$2(-15) = -30$	$2 + (-15) = -13$
$-2(15) = -30$	$-2 + 15 = 13$
$-3(10) = -30$	$-3 + 10 = 7$

No need to search any further

The numbers that we need are -3 and 10 and we can complete the factoring.

$$x^2 + 7x - 30 = (x + 10)(x - 3)$$ ▲

Example 4

Factor $x^2 + 7x + 16$.

Solution

We need two integers whose product is 16 and whose sum is 7.

Product	Sum
$1(16) = 16$	$1 + 16 = 17$
$2(8) = 16$	$2 + 8 = 10$
$4(4) = 16$	$4 + 4 = 8$

Since we have exhausted all possible pairs of factors of 16 and no two factors have a sum of 7, we conclude that $x^2 + 7x + 16$ *is not factorable using integers*.

▲

The tables in Examples 2, 3, and 4 were used to illustrate one way of organizing your thoughts for such problems. Normally you would probably factor such problems mentally without taking the time to formulate a table. Notice, however, that in Example 4 the table helped us to be absolutely sure that we tried all the possibilities. Whether or not you use the table, keep in mind that the key ideas are the product and sum relationships.

Example 5

Factor $n^2 - n - 72$.

Solution

Notice that the coefficient of the middle term is -1. So, we are looking for two integers whose product is -72, and since their sum is -1, the absolute value of the negative number must be one larger than the positive number. The numbers are -9 and 8, and we can complete the factoring.

$$n^2 - n - 72 = (n - 9)(n + 8)$$

▲

Example 6

Factor $t^2 + 2t - 168$.

Solution

We need two integers whose product is -168 and whose sum is 2. Since the absolute value of the constant term is rather large, it might help to look at it in prime factored form.

$$168 = 2 \cdot 2 \cdot 2 \cdot 3 \cdot 7$$

Now we can mentally form two numbers by using all of these factors in different combinations. Using two 2s and a 3 in one number, and the other 2 and the 7 in the second number produces $2 \cdot 2 \cdot 3 = 12$ and $2 \cdot 7 = 14$. Since the coefficient of the middle term of the trinomial is 2, we know that we must use 14 and -12. Thus, we obtain

$$t^2 + 2t - 168 = (t + 14)(t - 12).$$

▲

Trinomials of the Form $ax^2 + bx + c$

We have been factoring trinomials of the form $x^2 + bx + c$, that is, trinomials where the coefficient of the squared term is one. Now let's consider factoring trinomials where the coefficient of the squared term is not one. First, let's illustrate an informal trial and error technique that works quite well for certain types of trinomials. This technique is based on our knowledge of multiplication of binomials.

Example 7

Factor $2x^2 + 11x + 5$.

Solution

By looking at the first term, $2x^2$, and the positive signs of the other two terms, we know that the binomials are of the form

$$(x + \underline{\hspace{1cm}})(2x + \underline{\hspace{1cm}}).$$

Since the factors of the last term, 5, are 1 and 5, we have only the following two possibilities to try.

$$(x + 1)(2x + 5) \qquad \text{or} \qquad (x + 5)(2x + 1)$$

By checking the middle term formed in each of these products, we find that the second possibility yields the correct middle term of $11x$. Therefore,

$$2x^2 + 11x + 5 = (x + 5)(2x + 1). \qquad \blacktriangle$$

Example 8

Factor $10x^2 - 17x + 3$.

Solution

First, observe that $10x^2$ can be written as $x \cdot 10x$ or $2x \cdot 5x$. Secondly, since the middle term of the trinomial is negative and the last term is positive, we know that the binomials are of the form

$$(x - \underline{\hspace{1cm}})(10x - \underline{\hspace{1cm}}) \qquad \text{or} \qquad (2x - \underline{\hspace{1cm}})(5x - \underline{\hspace{1cm}}).$$

Since the factors of the last term, 3, are 1 and 3, the following possibilities exist.

$$(x - 1)(10x - 3) \qquad (2x - 1)(5x - 3)$$
$$(x - 3)(10x - 1) \qquad (2x - 3)(5x - 1)$$

By checking the middle term formed in each of these products, we find that the product $(2x - 3)(5x - 1)$ yields the desired middle term of $-17x$. Therefore,

$$10x^2 - 17x + 3 = (2x - 3)(5x - 1). \qquad \blacktriangle$$

Example 9

Factor $4x^2 + 6x + 9$.

Solution

The first term, $4x^2$, and the positive signs of the middle and last terms indicate that the binomials are of the form

$$(x + \underline{\hspace{1cm}})(4x + \underline{\hspace{1cm}}) \qquad \text{or} \qquad (2x + \underline{\hspace{1cm}})(2x + \underline{\hspace{1cm}}).$$

Since the factors of 9 are 1 and 9 or 3 and 3, we have the following five possibilities to try.

$$(x + 1)(4x + 9) \qquad (2x + 1)(2x + 9)$$
$$(x + 9)(4x + 1) \qquad (2x + 3)(2x + 3)$$
$$(x + 3)(4x + 3)$$

When we try all of these possibilities we find that none of them yields a middle term of $6x$. Therefore, $4x^2 + 6x + 9$ is *not factorable using integers*. $\qquad \blacktriangle$

By now it is obvious that factoring trinomials of the form $ax^2 + bx + c$ can be tedious. The key idea is to organize your work so that you consider all possibilities. We suggested one possible format in the previous three examples. As you practice such problems, you may come across a format of your own. Whatever works best for you is the right approach.

There is another more systematic technique that you may wish to use with some trinomials. It is an extension of the technique we used at the beginning of this section. To see the basis of this technique, let's look at the following product.

$$(px + r)(qx + s) = px(qx) + px(s) + r(qx) + r(s)$$
$$= (pq)x^2 + (ps + rq)x + rs$$

Notice that the product of the coefficient of the x^2 term and the constant term is $pqrs$. Likewise, the product of the two coefficients of x, ps and rq, is also $pqrs$. Therefore, when factoring the trinomial $(pq)x^2 + (ps + rq)x + rs$, the two coefficients of x must have a sum of $(ps) + (rq)$ and a product of $pqrs$. Let's see how this works in some examples.

Example 10

Factor $4x^2 - 4x - 15$.

Solution

$$4x^2 - 4x - 15 \qquad \text{Sum of } -4$$

Product of $4(-15) = -60$

We need two integers whose sum is -4 and whose product is -60. The integers -10 and 6 satisfy these conditions. Thus, we can express the middle term, $-4x$, as $-10x + 6x$ and we can factor as follows.

$$4x^2 - 4x - 15 = 4x^2 - 10x + 6x - 15$$
$$= 2x(2x - 5) + 3(2x - 5)$$
$$= (2x - 5)(2x + 3) \qquad \blacktriangle$$

Example 11

Factor $20x^2 + 39x + 18$.

Solution

$$20x^2 + 39x + 18 \qquad \text{Sum of } 39$$

Product of $20(18) = 360$

We need two integers whose sum is 39 and whose product is 360. To help find these integers, let's prime factor 360.

$$360 = 2 \cdot 2 \cdot 2 \cdot 3 \cdot 3 \cdot 5$$

Now by grouping these factors in various ways we find that $2 \cdot 2 \cdot 2 \cdot 3 = 24$ and $3 \cdot 5 = 15$ and $24 + 15 = 39$. So the numbers are 15 and 24 and the middle term, $39x$, can be written as $15x + 24x$. Thus, we can factor as follows.

$$20x^2 + 39x + 18 = 20x^2 + 15x + 24x + 18$$
$$= 5x(4x + 3) + 6(4x + 3)$$
$$= (4x + 3)(5x + 6) \qquad \blacktriangle$$

Perhaps at this time it would be helpful for you to go back and take another look at Examples 7 through 11. Both of the techniques we used have their strengths and weaknesses. The more that you work with them, the more comfortable you will feel using them.

Summary of Factoring Techniques

Before we summarize our work with factoring techniques let's look at two more special factoring patterns. In Section 3.3 we used the following two patterns to square binomials.

$$(a + b)^2 = a^2 + 2ab + b^2 \qquad \text{and} \qquad (a - b)^2 = a^2 - 2ab + b^2$$

These patterns can also be used for factoring purposes.

$$a^2 + 2ab + b^2 = (a + b)^2 \qquad \text{and} \qquad a^2 - 2ab + b^2 = (a - b)^2$$

The trinomials on the left sides are called **perfect square trinomials**; they are the result of squaring a binomial. We can always factor perfect square trinomials using the usual techniques for factoring trinomials. However, they are easily recognized by the nature of their terms. For example, $4x^2 + 12x + 9$ is a perfect square trinomial because

1. The first term is a perfect square; $(2x)^2$
2. The last term is a perfect square; $(3)^2$
3. The middle term is twice the product of the quantities $2(2x)(3)$
   being squared in the first and last terms.

Likewise, $9x^2 - 30x + 25$ is a perfect square trinomial because

1. The first term is a perfect square; $(3x)^2$
2. The last term is a perfect square; $(5)^2$
3. The middle term is the negative of twice the product of the $-2(3x)(5)$
   quantities being squared in the first and last terms.

Once we know that we have a perfect square trinomial, then the factors follow immediately from the two basic patterns. Thus,

$$4x^2 + 12x + 9 = (2x + 3)^2, \qquad 9x^2 - 30x + 25 = (3x - 5)^2.$$

Here are some additional examples of perfect square trinomials and their factored forms.

$$x^2 + 14x + 49 = (x)^2 + 2(x)(7) + (7) = (x + 7)^2,$$
$$n^2 - 16n + 64 = (n)^2 - 2(n)(8) + (8)^2 = (n - 8)^2,$$
$$36a^2 + 60ab + 25b^2 = (6a)^2 + 2(6a)(5b) + (5b)^2 = (6a + 5b)^2,$$
$$16x^2 - 8xy + y^2 = (4x)^2 - 2(4x)(y) + (y)^2 = (4x - y)^2$$

Perhaps you will want to do this
step mentally after you feel
comfortable with the process.

As we have indicated, factoring is an important algebraic skill. We learned some basic factoring techniques one at a time, but you must be able to apply them when the situation presents itself. So let's review the techniques and consider a variety of examples that demonstrate their use.

In this chapter we have discussed:

1. Factoring by using the distributive property to factor out a common monomial (or binomial) factor;

2. Factoring by applying the difference-of-two-squares pattern;

3. Factoring by applying the sum or difference-of-two-cubes patterns;

4. Factoring of trinomials into the product of two binomials (the *perfect square trinomial* pattern is a special case of this technique).

As a general guideline, *always look for a common monomial factor first and then proceed with the other techniques.* Study the following examples carefully and be sure that you agree with the indicated factors.

$$2x^2 + 20x + 48 = 2(x^2 + 10x + 24)$$
$$= 2(x + 4)(x + 6),$$
$$3x^3y^3 + 27xy = 3xy(x^2y^2 + 9),$$
$$2x^3 - 16 = 2(x^3 - 8)$$
$$= 2(x - 2)(x^2 + 2x + 4),$$
$$30n^2 - 31n + 5 = (5n - 1)(6n - 5)$$

$$16a^2 - 64 = 16(a^2 - 4)$$
$$= 16(a + 2)(a - 2),$$
$x^2 + 3x - 21$ is not factorable using integers,
$$t^4 + 3t^2 + 2 = (t^2 + 2)(t^2 + 1)$$

Problem Set 3.6

Factor completely each of the following. Indicate any that are not factorable using integers.

1. $x^2 + 9x + 20$
2. $x^2 + 11x + 24$
3. $x^2 - 11x + 28$
4. $x^2 - 8x + 12$
5. $a^2 + 5a - 36$
6. $a^2 + 6a - 40$
7. $y^2 + 20y + 84$
8. $y^2 + 21y + 98$

9. $x^2 - 5x - 14$

10. $x^2 - 3x - 54$

11. $x^2 + 9x + 12$

12. $35 - 2x - x^2$

13. $6 + 5x - x^2$

14. $x^2 + 8x - 24$

15. $x^2 + 15xy + 36y^2$

16. $x^2 - 14xy + 40y^2$

17. $a^2 - ab - 56b^2$

18. $a^2 + 2ab - 63b^2$

19. $15x^2 + 23x + 6$

20. $9x^2 + 30x + 16$

21. $12x^2 - x - 6$

22. $20x^2 - 11x - 3$

23. $8a^2 - 6a - 27$

24. $12a^2 + 4a - 5$

25. $12n^2 - 43n - 20$

26. $16n^2 + 43n - 15$

27. $3x^2 + 10x + 4$

28. $20n^2 - 47n + 21$

29. $20n^2 - 64n - 21$

30. $4x^2 - x + 6$

31. $16x^2 + 62x - 45$

32. $12x^2 + 4x - 33$

33. $6 - 29x - 42x^2$

34. $4 + 11x - 20x^2$

35. $20y^2 + 31y - 9$

36. $8y^2 + 22y - 21$

37. $24n^2 - 2n - 5$

38. $12n^2 - n - 35$

39. $5n^2 + 33n + 18$

40. $7n^2 + 31n + 12$

41. $x^2 + 25x + 150$

42. $x^2 + 21x + 108$

43. $n^2 - 36n + 320$

44. $n^2 - 26n + 168$

45. $t^2 + 3t - 180$

46. $t^2 - 2t - 143$

47. $t^4 - 5t^2 + 6$

48. $t^4 + 10t^2 + 24$

49. $10x^4 + 3x^2 - 4$

50. $3x^4 + 7x^2 - 6$

51. $x^4 - 9x^2 + 8$

52. $x^4 - x^2 - 12$

53. $18n^4 + 25n^2 - 3$

54. $4n^4 + 3n^2 - 27$

55. $x^4 - 17x^2 + 16$

56. $x^4 - 13x^2 + 36$

The following problems should help you pull together all of the factoring techniques of this chapter. Factor completely each of the following; indicate any that are not factorable using integers.

57. $2t^2 - 8$

58. $14w^2 - 29w - 15$

59. $12x^2 + 7xy - 10y^2$

60. $8x^2 + 2xy - y^2$

61. $18n^3 + 39n^2 - 15n$

62. $n^2 + 18n + 77$

63. $n^2 - 17n + 60$

64. $(x + 5)^2 - y^2$

65. $36a^2 - 12a + 1$

66. $2n^2 - n - 5$

67. $6x^2 + 54$

68. $x^5 - x$

69. $3x^2 + x - 5$

70. $10x^2 + 39x - 27$

71. $x^2 - (y - 7)^2$

72. $2n^3 + 6n^2 + 10n$

73. $1 - 16x^4$

74. $9a^2 - 42a + 49$

75. $12n^2 + 59n + 72$

76. $x^3 - 9x$

77. $n^3 - 49n$

78. $4x^2 + 16$

79. $x^2 - 7x - 8$

80. $x^2 + 3x - 54$

81. $3x^4 - 81x$

82. $x^3 + 125$

83. $x^4 + 6x^2 + 9$

84. $18x^2 - 12x + 2$

85. $x^4 - 5x^2 - 36$

86. $6x^4 - 5x^2 - 21$

87. $18w^2 + 9w - 35$

88. $10x^3 + 15x^2 + 20x$

89. $25n^2 + 64$

90. $4x^2 - 37x + 40$

91. $2n^3 + 14n^2 - 20n$

92. $25t^2 - 100$

93. $2xy + 6x + y + 3$

94. $3xy + 15x - 2y - 10$

THOUGHTS INTO WORDS

95. How can you determine that $x^2 + 5x + 12$ is not factorable using integers?

96. Explain your thought process when factoring $30x^2 + 13x - 56$.

97. Consider the following approach to factoring $12x^2 + 54x + 60$:

$$12x^2 + 54x + 60 = (3x + 6)(4x + 10)$$
$$= 3(x + 2)(2)(2x + 5)$$
$$= 6(x + 2)(2x + 5)$$

Is this a correct factoring process? Do you have any suggestion for the person using this approach?

Further Investigations

For Problems 98–103, factor each trinomial and assume that all variables that appear as exponents represent positive integers.

98. $x^{2a} + 2x^a - 24$

99. $x^{2a} + 10x^a + 21$

100. $6x^{2a} - 7x^a + 2$

101. $4x^{2a} + 20x^a + 25$

102. $12x^{2n} + 7x^n - 12$

103. $20x^{2n} + 21x^n - 5$

Consider the following approach to factoring $(x - 2)^2 + 3(x - 2) - 10$.

> Replace $x - 2$ with y.
>
> Factor.
>
> Replace y with $x - 2$.

$$(x - 2)^2 + 3(x - 2) - 10 = y^2 + 3y - 10$$
$$= (y + 5)(y - 2)$$
$$= (x - 2 + 5)(x - 2 - 2)$$
$$= (x + 3)(x - 4)$$

Use this approach to factor Problems 104–109.

104. $(x - 3)^2 + 10(x - 3) + 24$

105. $(x + 1)^2 - 8(x + 1) + 15$

106. $(2x + 1)^2 + 3(2x + 1) - 28$

107. $(3x - 2)^2 - 5(3x - 2) - 36$

108. $6(x - 4)^2 + 7(x - 4) - 3$

109. $15(x + 2)^2 - 13(x + 2) + 2$

3.7   Equations and Problem Solving

The techniques for factoring trinomials we presented in the previous section provide us with more power to solve equations. That is to say, the property $ab = 0$ if and only if $a = 0$ or $b = 0$ continues to play an important role as we solve equations that involve factorable trinomials. Let's consider some examples.

Example 1

Solve $x^2 - 11x - 12 = 0$.

Solution

$$x^2 - 11x - 12 = 0$$
$$(x - 12)(x + 1) = 0$$
$$x - 12 = 0 \quad \text{or} \quad x + 1 = 0$$
$$x = 12 \quad \text{or} \quad x = -1$$

The solution set is $\{-1, 12\}$. (Check these solutions in the original equation.)

▲

Example 2

Solve $20x^2 + 7x - 3 = 0$.

Solution

$$20x^2 + 7x - 3 = 0$$
$$(4x - 1)(5x + 3) = 0$$
$$4x - 1 = 0 \quad \text{or} \quad 5x + 3 = 0$$
$$4x = 1 \quad \text{or} \quad 5x = -3$$
$$x = \frac{1}{4} \quad \text{or} \quad x = -\frac{3}{5}$$

The solution set is $\left\{-\dfrac{3}{5}, \dfrac{1}{4}\right\}$.

▲

Example 3

Solve $-2n^2 - 10n + 12 = 0$.

Solution

$$-2n^2 - 10n + 12 = 0$$
$$-2(n^2 + 5n - 6) = 0$$
$$n^2 + 5n - 6 = 0 \qquad \text{Multiply both sides by } -\frac{1}{2}.$$
$$(n + 6)(n - 1) = 0$$
$$n + 6 = 0 \quad \text{or} \quad n - 1 = 0$$
$$n = -6 \quad \text{or} \quad n = 1$$

The solution set is $\{-6, 1\}$.

▲

Example 4

Solve $16x^2 - 56x + 49 = 0$

Solution

$$16x^2 - 56x + 49 = 0$$
$$(4x - 7)^2 = 0$$
$$(4x - 7)(4x - 7) = 0$$
$$4x - 7 = 0 \quad \text{or} \quad 4x - 7 = 0$$
$$4x = 7 \quad \text{or} \quad 4x = 7$$
$$x = \frac{7}{4} \quad \text{or} \quad x = \frac{7}{4}$$

The only solution is $\dfrac{7}{4}$; thus, the solution set is $\left\{\dfrac{7}{4}\right\}$.

▲

Sometimes the original form of an equation needs to be changed before a factoring technique becomes apparent. The next two examples demonstrate situations of this type.

Example 5

Solution

Solve $9a(a + 1) = 4$.

$$9a(a + 1) = 4$$
$$9a^2 + 9a = 4$$
$$9a^2 + 9a - 4 = 0$$
$$(3a + 4)(3a - 1) = 0$$
$$3a + 4 = 0 \quad \text{or} \quad 3a - 1 = 0$$
$$3a = -4 \quad \text{or} \quad 3a = 1$$
$$a = -\frac{4}{3} \quad \text{or} \quad a = \frac{1}{3}$$

The solution set is $\left\{-\frac{4}{3}, \frac{1}{3}\right\}$. (Perhaps you should check these solutions in the original equation.) ▲

Example 6

Solution

Solve $(x - 1)(x + 9) = 11$.

$$(x - 1)(x + 9) = 11$$
$$x^2 + 8x - 9 = 11$$
$$x^2 + 8x - 20 = 0$$
$$(x + 10)(x - 2) = 0$$
$$x + 10 = 0 \quad \text{or} \quad x - 2 = 0$$
$$x = -10 \quad \text{or} \quad x = 2$$

The solution set is $\{-10, 2\}$. ▲

Problem Solving

As you might expect, the increase in our power to solve equations broadens our base for solving problems. Now we are ready to tackle some problems using equations of the types presented in this section.

Problem 1

Solution

A room contains 78 chairs. The number of chairs per row is one more than twice the number of rows. Find the number of rows and the number of chairs per row.

Let r represent the number of rows. Then $2r + 1$ represents the number of chairs per row.

$$=r(2r + 1) = 78$$ The number of rows times the number of chairs per row yields the total number of chairs.

$$2r^2 + r = 78$$
$$2r^2 + r - 78 = 0$$
$$(2r + 13)(r - 6) = 0$$
$$2r + 13 = 0 \quad \text{or} \quad r - 6 = 0$$
$$2r = -13 \quad \text{or} \quad r = 6$$
$$r = -\frac{13}{2} \quad \text{or} \quad r = 6$$

The solution $-\dfrac{13}{2}$ must be disregarded, so there are 6 rows and $2r + 1$ or $2(6) + 1 = 13$ chairs per row.   ▲

Problem 2

A strip of uniform width cut off of both sides and both ends of an 8-inch by 11-inch sheet of paper reduces the size of the paper to an area of 40 square inches. Find the width of the strip.

Solution

Let x represent the width of the strip, as indicated in the following diagram (Figure 3.30).

FIGURE 3.30

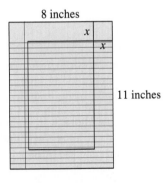

8 inches

x

x

11 inches

The length of the paper after the strips of width x are cut from both ends and both sides will be $11 - 2x$ and the width of the newly formed rectangle will be $8 - 2x$. Since the area ($A = lw$) is to be 40 square inches, we can set up and solve the following equation.

$$(11 - 2x)(8 - 2x) = 40$$
$$88 - 38x + 4x^2 = 40$$
$$4x^2 - 38x + 48 = 0$$
$$2x^2 - 19x + 24 = 0$$
$$(2x - 3)(x - 8) = 0$$

$$2x - 3 = 0 \quad \text{or} \quad x - 8 = 0$$
$$2x = 3 \quad \text{or} \quad x = 8$$
$$x = \frac{3}{2} \quad \text{or} \quad x = 8$$

The solution of 8 must be discarded since the width of the original sheet is only 8 inches. Therefore, the strip to be cut off of all four sides must be $1\frac{1}{2}$ inches wide. (Check this answer!) ▲

The Pythagorean theorem, an important theorem pertaining to right triangles, can sometimes serve as a guideline for solving problems that deal with right triangles (see Figure 3.31). The Pythagorean theorem states that "in any right triangle, the square of the longest side (called the hypotenuse) is equal to the sum of the squares of the other two sides (called legs)." Let's use this relationship to help solve a problem.

FIGURE 3.31

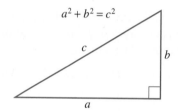

$a^2 + b^2 = c^2$

Problem 3

One leg of a right triangle is 2 centimeters more than twice the length of the other leg. The hypotenuse is 1 centimeter longer than the longer of the two legs. Find the lengths of the three sides of the right triangle.

Solution

Let l represent the length of the shortest leg. Then $2l + 2$ represents the length of the other leg and $2l + 3$ represents the length of the hypotenuse. Use the Pythagorean theorem as a guideline to set up and solve the following equation.

$$l^2 + (2l + 2)^2 = (2l + 3)^2$$
$$l^2 + 4l^2 + 8l + 4 = 4l^2 + 12l + 9$$
$$l^2 - 4l - 5 = 0$$
$$(l - 5)(l + 1) = 0$$
$$l - 5 = 0 \quad \text{or} \quad l + 1 = 0$$
$$l = 5 \quad \text{or} \quad l = -1$$

The negative solution must be discarded, so the length of one leg is 5 centimeters; the other leg is $2(5) + 2 = 12$ centimeters long and the hypotenuse is $2(5) + 3 = 13$ centimeters long. ▲

Problem Set 3.7

Solve each of the following equations. You will need to use factoring techniques that we discussed throughout this chapter.

1. $x^2 + 4x + 3 = 0$
2. $x^2 + 7x + 10 = 0$
3. $x^2 + 18x + 72 = 0$
4. $n^2 + 20n + 91 = 0$
5. $n^2 - 13n + 36 = 0$
6. $n^2 - 10n + 16 = 0$
7. $x^2 + 4x - 12 = 0$
8. $x^2 + 7x - 30 = 0$
9. $w^2 - 4w = 5$
10. $s^2 - 4s = 21$
11. $n^2 + 25n + 156 = 0$
12. $n(n - 24) = -128$
13. $3t^2 + 14t - 5 = 0$
14. $4t^2 - 19t - 30 = 0$
15. $6x^2 + 25x + 14 = 0$
16. $25x^2 + 30x + 8 = 0$
17. $3t(t - 4) = 0$
18. $1 - x^4 = 0$
19. $-6n^2 + 13n - 2 = 0$
20. $(x + 1)^2 - 4 = 0$
21. $2n^3 = 72n$
22. $a(a - 1) = 2$
23. $(x - 5)(x + 3) = 9$
24. $3w^3 - 24w^2 + 36w = 0$
25. $16 - x^4 = 0$
26. $16t^2 - 72t + 81 = 0$
27. $n^2 + 7n - 44 = 0$
28. $2x^3 = 50x$
29. $3x^2 = 75$
30. $x^2 + x - 2 = 0$
31. $15x^2 + 34x + 15 = 0$
32. $20x^2 + 41x + 20 = 0$

33. $8n^2 - 47n - 6 = 0$
34. $7x^2 + 62x - 9 = 0$
35. $28n^2 - 47n + 15 = 0$
36. $24n^2 - 38n + 15 = 0$
37. $35n^2 - 18n - 8 = 0$
38. $8n^2 - 6n - 5 = 0$
39. $-3x^2 - 19x + 14 = 0$
40. $5x^2 = 43x - 24$
41. $n(n + 2) = 360$
42. $n(n + 1) = 182$
43. $9x^4 - 37x^2 + 4 = 0$
44. $4x^4 - 13x^2 + 9 = 0$
45. $3x^4 - 46x^2 - 32 = 0$
46. $x^4 - 9x^2 = 0$
47. $2x^4 + x^2 - 3 = 0$
48. $x^4 + 5x^2 - 36 = 0$
49. $12x^3 + 46x^2 + 40x = 0$
50. $5x(3x - 2) = 0$
51. $(3x - 1)^2 - 16 = 0$
52. $(x + 8)(x - 6) = -24$
53. $4a(a + 1) = 3$
54. $-18n^2 - 15n + 7 = 0$

Set up an equation and solve each of the following problems.

55. Find two consecutive integers whose product is 72.

56. Find two consecutive even whole numbers whose product is 224.

57. Find two integers whose product is 105 such that one of the integers is one more than twice the other integer.

58. Find two integers whose product is 104 such that one of the integers is three less than twice the other integer.

59. The perimeter of a rectangle is 32 inches and the area is 60 square inches. Find the length and width of the rectangle.

60. Suppose that the length of a certain rectangle is two centimeters more than three times its width. If the area of the rectangle is 56 square centimeters, find its length and width.

61. The sum of the squares of two consecutive integers is 85. Find the integers.

62. The sum of the areas of two circles is 65π square feet. The length of a radius of the larger circle is one foot less than twice the length of a radius of the smaller circle. Find the length of a radius of each circle.

63. The combined area of a square and a rectangle is 64 square centimeters. The width of the rectangle is 2 centimeters more than the length of a side of the square and the length of the rectangle is 2 centimeters more than its width. Find the dimensions of the square and the rectangle.

64. The Ortegas have an apple orchard that contains 90 trees. The number of trees in each row is 3 more than twice the number of rows. Find the number of rows and the number of trees per row.

65. The lengths of the three sides of a right triangle are represented by consecutive whole numbers. Find the lengths of the three sides.

66. The area of the floor of a rectangular room is 175 square feet. The length of the room is $1\frac{1}{2}$ feet longer than the width. Find the length of the room.

Area = 175 square feet

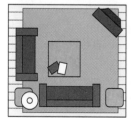

FIGURE 3.32

67. Suppose that the length of one leg of a right triangle is 3 inches more than the length of the other leg. If the length of the hypotenuse is 15 inches, find the lengths of the two legs.

68. The lengths of the three sides of a right triangle are represented by consecutive even whole numbers. Find the lengths of the three sides.

69. The area of a triangular sheet of paper is 28 square inches. One side of the triangle is two inches more than three times the length of the altitude to that side. Find the length of that side and the altitude to the side.

70. A strip of uniform width is shaded along both sides and both ends of a rectangular poster that is 12 inches by 16 inches. How wide is the shaded strip if one-half of the poster is shaded?

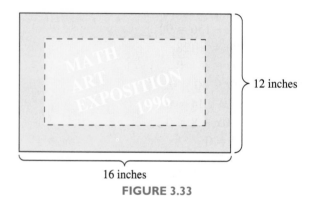

12 inches

16 inches

FIGURE 3.33

71. Discuss the role that factoring plays in solving equations.

72. Explain how you would solve the equation $(x + 6)$ $\cdot (x - 4) = 0$ and also how you would solve $(x + 6)$ $\cdot (x - 4) = -16$.

73. Consider the following two solutions for the equation $(x + 3)(x - 4) = (x + 3)(2x - 1)$.

Solution A

$$(x + 3)(x - 4) = (x + 3)(2x - 1)$$
$$(x + 3)(x - 4) - (x + 3)(2x - 1) = 0$$
$$(x + 3)[x - 4 - (2x - 1)] = 0$$
$$(x + 3)(x - 4 - 2x + 1) = 0$$
$$(x + 3)(-x - 3) = 0$$

$$x + 3 = 0 \quad \text{or} \quad -x - 3 = 0$$
$$x = -3 \quad \text{or} \quad -x = 3$$
$$x = -3 \quad \text{or} \quad x = -3$$

The solution set is $\{-3\}$.

Solution B

$$(x + 3)(x - 4) = (x + 3)(2x - 1)$$
$$x^2 - x - 12 = 2x^2 + 5x - 3$$
$$0 = x^2 + 6x + 9$$
$$0 = (x + 3)^2$$
$$x + 3 = 0$$
$$x = -3$$

The solution set is $\{-3\}$.

Are both approaches correct? Which approach would you use and why?

SUMMARY

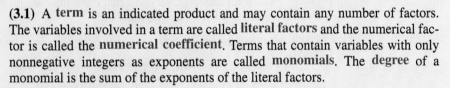

(3.1) A **term** is an indicated product and may contain any number of factors. The variables involved in a term are called **literal factors** and the numerical factor is called the **numerical coefficient**. Terms that contain variables with only nonnegative integers as exponents are called **monomials**. The **degree** of a monomial is the sum of the exponents of the literal factors.

A **polynomial** is a monomial or a finite sum (or difference) of monomials. We classify polynomials as follows.

Polynomial with one term $\longrightarrow$ Monomial

Polynomial with two terms $\longrightarrow$ Binomial

Polynomial with three terms $\longrightarrow$ Trinomial

Similar or **like terms** have the same literal factors. The commutative, associative, and distributive properties provide the basis for rearranging, regrouping, and combining similar terms.

(3.2) The following properties provide the basis for multiplying and dividing monomials.

1. $b^n \cdot b^m = b^{n+m}$

2. $(b^n)^m = b^{mn}$

3. $(ab)^n = a^n b^n$

4. (a) $\dfrac{b^n}{b^m} = b^{n-m}$ if $n > m$

   (b) $\dfrac{b^n}{b^m} = 1$ if $n = m$

(3.3) The commutative and associative properties, the properties of exponents, and the distributive property work together to form a basis for multiplying polynomials. The following can be used as multiplication patterns.

$$(a + b)^2 = a^2 + 2ab + b^2$$
$$(a - b)^2 = a^2 - 2ab + b^2$$
$$(a + b)(a - b) = a^2 - b^2$$
$$(a + b)^3 = a^3 + 3a^2b + 3ab^2 + b^3$$
$$(a - b)^3 = a^3 - 3a^2b + 3ab^2 - b^3$$

(3.4) If a positive integer greater than 1 has no factors that are positive integers other than itself and 1, then it is called a **prime number**. A positive integer greater than 1 that is not a prime number is called a **composite number**.

The indicated product form that contains only prime factors is called the **prime factorization form** of a number.

An expression such as $ax + bx + ay + by$ can be factored as follows.

$$ax + bx + ay + by = x(a + b) + y(a + b)$$
$$= (a + b)(x + y)$$

This is called **factoring by grouping**.

The distributive property in the form, $ab + ac = a(b + c)$, forms the basis for **factoring out the highest common monomial factor**.

Expressing polynomials in factored form and then applying the property, $ab = 0$ if and only if $a = 0$ or $b = 0$, provides us with another technique for solving equations.

(3.5) The factoring pattern

$$a^2 - b^2 = (a + b)(a - b)$$

is called the **difference of two squares.**

The difference-of-two-squares factoring pattern, along with the property, $ab = 0$ if and only if $a = 0$ or $b = 0$, provides us with another technique for solving equations. The factoring patterns

$$a^3 + b^3 = (a + b)(a^2 - ab + b^2), \quad \text{and}$$
$$a^3 - b^3 = (a - b)(a^2 + ab + b^2)$$

are called the **sum and difference of two cubes.**

(3.6) Expressing a trinomial (for which the coefficient of the squared term is one) as a product of two binomials is based on the following relationship.

$$(x + a)(x + b) = x^2 + (a + b)x + ab$$

The coefficient of the middle term is the *sum* of a and b, and the last term is the *product* of a and b.

If the coefficient of the squared term of a trinomial does not equal one, then the following relationship holds.

$$(px + r)(qx + s) = (pq)x^2 + (ps + rq)x + rs$$

The two coefficients of x, ps and rq, must have a sum of $(ps) + (rq)$ and a product of $pqrs$. Thus, to factor something like $6x^2 + 7x - 3$, we need to find two integers whose product is $6(-3) = -18$ and whose sum is 7. The integers are 9 and -2 and we can factor as follows.

$$6x^2 + 7x - 3 = 6x^2 + 9x - 2x - 3$$
$$= 3x(2x + 3) - 1(2x + 3)$$
$$= (2x + 3)(3x - 1)$$

A **perfect square trinomial** is the result of squaring a binomial. There are two basic perfect square trinomial factoring patterns as follows.

$$a^2 + 2ab + b^2 = (a + b)^2,$$
$$a^2 - 2ab + b^2 = (a - b)^2$$

(3.7) The factoring techniques we discussed in this chapter along with the property, $ab = 0$ if and only if $a = 0$ or $b = 0$, provide the basis for expanding our equation solving processes.

The ability to solve more types of equations increases our capabilities for problem solving.

Chapter 3 Review Problem Set

Perform the indicated operations and simplify each of the following.

1. $(3x - 2) + (4x - 6) + (-2x + 5)$
2. $(8x^2 + 9x - 3) - (5x^2 - 3x - 1)$

3. $(6x^2 - 2x - 1) + (4x^2 + 2x + 5) - (-2x^2 + x - 1)$
4. $(-5x^2y^3)(4x^3y^4)$
5. $(-2a^2)(3ab^2)(a^2b^3)$
6. $5a^2(3a^2 - 2a - 1)$
7. $(4x - 3y)(6x + 5y)$
8. $(x + 4)(3x^2 - 5x - 1)$
9. $(4x^2y^3)^4$
10. $(3x - 2y)^2$
11. $(-2x^2y^3z)^3$
12. $\dfrac{-39x^3y^4}{3xy^3}$
13. $[3x - (2x - 3y + 1)] - [2y - (x - 1)]$

14. $(x^2 - 2x - 5)(x^2 + 3x - 7)$

15. $(7 - 3x)(3 + 5x)$
16. $-(3ab)(2a^2b^3)^2$
17. $\left(\dfrac{1}{2}ab\right)(8a^3b^2)(-2a^3)$
18. $(7x - 9)(x + 4)$
19. $(3x + 2)(2x^2 - 5x + 1)$
20. $(3x^{n+1})(2x^{3n-1})$
21. $(2x + 5y)^2$
22. $(x - 2)^3$
23. $(2x + 5)^3$

Factor completely each of the following. Indicate any that are not factorable using integers.

24. $x^2 + 3x - 28$
25. $2t^2 - 18$

26. $4n^2 + 9$
27. $12n^2 - 7n + 1$
28. $x^6 - x^2$
29. $x^3 - 6x^2 - 72x$
30. $6a^3b + 4a^2b^2 - 2a^2bc$
31. $x^2 - (y - 1)^2$
32. $8x^2 + 12$
33. $12x^2 + x - 35$
34. $16n^2 - 40n + 25$
35. $4n^2 - 8n$
36. $3w^3 + 18w^2 - 24w$
37. $20x^2 + 3xy - 2y^2$
38. $16a^2 - 64a$
39. $3x^3 - 15x^2 - 18x$
40. $n^2 - 8n - 128$
41. $t^4 - 22t^2 - 75$
42. $35x^2 - 11x - 6$
43. $15 - 14x + 3x^2$
44. $64n^3 - 27$
45. $16x^3 + 250$

Solve each of the following equations.

46. $4x^2 - 36 = 0$
47. $x^2 + 5x - 6 = 0$
48. $49n^2 - 28n + 4 = 0$
49. $(3x - 1)(5x + 2) = 0$
50. $(3x - 4)^2 - 25 = 0$
51. $6a^3 = 54a$
52. $x^5 = x$
53. $-n^2 + 2n + 63 = 0$
54. $7n(7n + 2) = 8$
55. $30w^2 - w - 20 = 0$
56. $5x^4 - 19x^2 - 4 = 0$
57. $9n^2 - 30n + 25 = 0$

58. $n(2n + 4) = 96$

59. $7x^2 + 33x - 10 = 0$

60. $(x + 1)(x + 2) = 42$

61. $x^2 + 12x - x - 12 = 0$

62. $2x^4 + 9x^2 + 4 = 0$

63. $30 - 19x + 5x^2 = 0$

64. $3t^3 - 27t^2 + 24t = 0$

65. $-4n^2 - 39n + 10 = 0$

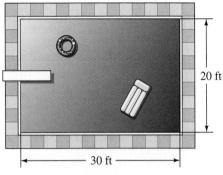

20 ft

30 ft

FIGURE 3.34

Set up an equation and solve each of the following problems.

66. Find three consecutive integers such that the product of the smallest and largest is 1 less than 9 times the middle integer.

67. Find two integers whose sum is 2 and whose product is -48.

68. Find two consecutive odd whole numbers whose product is 195.

69. Two cars leave an intersection at the same time, one traveling north and the other traveling east. Some time later, they are 20 miles apart and the car going east had traveled 4 miles further than the other car. How far had each car traveled?

70. The perimeter of a rectangle is 32 meters and its area is 48 square meters. Find the length and width of the rectangle.

71. A room contains 144 chairs. The number of chairs per row is 2 less than twice the number of rows. Find the number of rows and the number of chairs per row.

72. The area of a triangle is 39 square feet. The length of one side is 1 foot more than twice the altitude to that side. Find the length of that side and the altitude to the side.

73. A rectangular-shaped pool 20 feet by 30 feet has a sidewalk of uniform width around the pool (see Figure 3.34). The area of the sidewalk is 336 square feet. Find the width of the sidewalk.

74. The sum of the areas of two squares is 89 square centimeters. The length of a side of the larger square is 3 centimeters more than the length of a side of the smaller square. Find the dimensions of each square.

75. The total surface area of a right circular cylinder is 32π square inches. If the altitude of the cylinder is three times the length of a radius, find the altitude of the cylinder.

CHAPTER 3 TEST

For Problems 1–8, perform the indicated operations and simplify each expression.

1. $(-3x - 1) + (9x - 2) - (4x + 8)$

2. $(-6xy^2)(8x^3y^2)$ 3. $(-3x^2y^4)^3$

4. $(5x - 7)(4x + 9)$

5. $(3n - 2)(2n - 3)$

6. $(x - 4y)^3$

7. $(x + 6)(2x^2 - x - 5)$

8. $\dfrac{-70x^4y^3}{5xy^2}$

For Problems 9–14, factor each expression completely.

9. $6x^2 + 19x - 20$ 10. $12x^2 - 3$

11. $64 + t^3$

12. $30x + 4x^2 - 16x^3$

13. $x^2 - xy + 4x - 4y$

14. $24n^2 + 55n - 24$

For Problems 15–22, solve each equation.

15. $x^2 + 8x - 48 = 0$ 16. $4n^2 = n$

17. $4x^2 - 12x + 9 = 0$ 18. $(n - 2)(n + 7) = -18$

19. $3x^3 + 21x^2 - 54x = 0$

20. $12 + 13x - 35x^2 = 0$ 21. $n(3n - 5) = 2$

22. $9x^2 - 36 = 0$

For Problems 23–25, set up an equation and solve each problem.

23. The perimeter of a rectangle is 30 inches and its area is 54 square inches. Find the length of the longest side of the rectangle.

24. A room contains 105 chairs arranged in rows. The number of rows is one more than twice the number of chairs per row. Find the number of rows.

25. The combined area of a square and a rectangle is 57 square feet. The width of the rectangle is 3 feet more than the length of a side of the square and the length of the rectangle is 5 feet more than the length of a side of the square. Find the length of the rectangle.

Cumulative Review Problem Set

For Problems 1–10, evaluate each algebraic expression for the given values of the variables. Don't forget that in some cases it may be helpful to simplify the algebraic expression before evaluating it.

1. $x^2 - 2xy + y^2$ for $x = -2$ and $y = -4$
2. $-n^2 + 2n - 4$ for $n = -3$
3. $2x^2 - 5x + 6$ for $x = 3$
4. $3(2x - 1) - 2(x + 4) - 4(2x - 7)$ for $x = -1$
5. $-(2n - 1) + 5(2n - 3) - 6(3n + 4)$ for $n = 4$
6. $2(a - 4) - (a - 1) + (3a - 6)$ for $a = -5$
7. $(3x^2 - 4x - 7) - (4x^2 - 7x + 8)$ for $x = -4$
8. $-2(3x - 5y) - 4(x + 2y) + 3(-2x - 3y)$ for $x = 2$ and $y = -3$
9. $5(-x^2 - x + 3) - (2x^2 - x + 6) - 2(x^2 + 4x - 6)$ for $x = 2$
10. $3(x^2 - 4xy + 2y^2) - 2(x^2 - 6xy - y^2)$ for $x = -5$ and $y = -2$

For Problems 11–18, perform the indicated operations and express your answers in simplest form.

11. $4(3x - 2) - 2(4x - 1) - (2x + 5)$
12. $(-6ab^2)(2ab)(-3b^3)$
13. $(5x - 7)(6x + 1)$
14. $(-2x - 3)(x + 4)$
15. $(-4a^2b^3)^3$
16. $(x + 2)(5x - 6)(x - 2)$
17. $(x - 3)(x^2 - x - 4)$
18. $(x^2 + x + 4)(2x^2 - 3x - 7)$

For Problems 19–28, factor each of the algebraic expressions completely.

19. $7x^2 - 7$
20. $4a^2 - 4ab + b^2$
21. $3x^2 - 17x - 56$
22. $1 - x^3$
23. $xy - 5x + 2y - 10$
24. $3x^2 - 24x + 48$
25. $4n^4 - n^2 - 3$
26. $32x^4 + 108x$
27. $4x^2 + 36$
28. $20x^2 + x - 30$

For Problems 29–32, solve each equation for the indicated variable.

29. $5x - 2y = 6$ for x
30. $3x + 4y = 12$ for y
31. $V = 2\pi rh + 2\pi r^2$ for h
32. $\dfrac{1}{R} = \dfrac{1}{R_1} + \dfrac{1}{R_2}$ for R_1

For Problems 33–50, solve each of the equations.

33. $(x - 2)(x + 5) = 8$
34. $(5n - 2)(3n + 7) = 0$
35. $-2(n - 1) + 3(2n + 1) = -11$
36. $x^2 + 7x - 18 = 0$
37. $8x^2 - 8 = 0$
38. $\dfrac{3}{4}(x - 2) - \dfrac{2}{5}(2x - 3) = \dfrac{1}{5}$
39. $.1(x - .1) - .4(x + 2) = -5.31$
40. $\dfrac{2x - 1}{2} - \dfrac{5x + 2}{3} = 3$
41. $|3n - 2| = 7$
42. $|2x - 1| = |x + 4|$
43. $.08(x + 200) = .07x + 20$
44. $2x^2 - 12x - 80 = 0$
45. $x^3 = 16x$
46. $x(x + 2) - 3(x + 2) = 0$
47. $-12n^2 - 29n + 8 = 0$
48. $3y(y + 1) = 90$

49. $2x^3 + 6x^2 - 20x = 0$

50. $(3n - 1)(2n + 3) = (n + 4)(6n - 5)$

For Problems 51–58, solve each of the inequalities.

51. $-5(3n + 4) < -2(7n - 1)$

52. $7(x + 1) - 8(x - 2) < 0$

53. $|2x - 1| > 7$

54. $|3x + 7| < 14$

55. $.09x + .1(x + 200) > 77$

56. $\dfrac{2x - 1}{4} - \dfrac{x - 2}{6} \leq \dfrac{3}{8}$

57. $-(x - 1) + 2(3x - 1) \geq 2(x + 4) - (x - 1)$

58. $\dfrac{1}{4}(x - 2) + \dfrac{3}{7}(2x - 1) < \dfrac{3}{14}$

For Problems 59–72, solve each problem by setting up and solving an appropriate equation or inequality.

59. Find three consecutive odd integers such that three times the first minus the second is one more than the third.

60. Inez has a collection of 48 coins consisting of nickels, dimes, and quarters. The number of dimes is one less than twice the number of nickels, and the number of quarters is ten larger than the number of dimes. How many coins of each denomination are there in the collection?

61. The sum of the present ages of Joey and his mother is 46 years. In four years, Joey will be 3 years less than one-half as old as his mother at that time. Find the present ages of Joey and his mother.

62. The difference of the measures of two supplementary angles is 56°. Find the measure of each angle.

63. Norm invested a certain amount of money at 8% interest and $200 more than that amount at 9%. His total yearly interest was $86. How much did he invest at each rate?

64. Sanchez has a collection of pennies, nickels, and dimes worth $9.35. He has 5 more nickels than pennies and twice as many dimes as pennies. How many coins of each kind does he have?

65. Sandy starts with her bicycle at 8 miles per hour. Fifty minutes later, Billie starts riding along the same route at 12 miles per hour. How long will it take Billie to overtake Sandy?

66. How many milliliters of pure acid must be added to 150 milliliters of a 30% solution of acid to obtain a 40% solution?

67. Doreen bowled 152 and 174 in her first two games. What must she bowl in the third game to have an average of at least 160 for the three games?

68. Brad had scores of 88, 92, 93, and 89 on his first four algebra tests. What score must he obtain on the fifth test to have an average better than 90 for the five tests?

69. Suppose that the area of a square is one-half the area of a triangle. One side of the triangle is 16 inches long and the altitude to that side is the same length as a side of the square. Find the length of a side of the square.

70. A rectangle is twice as long as it is wide and its area is 98 square meters. Find the length and width of the rectangle.

71. A room contains 96 chairs. The number of chairs per row is 4 more than the number of rows. Find the number of rows and the number of chairs per row.

72. One leg of a right triangle is 3 feet longer than the other leg. The hypotenuse is 3 feet longer than the longer leg. Find the lengths of the three sides of the right triangle.

Rational Expressions

It takes Pat 12 hours to complete a task. After he had been working for 3 hours, he was joined by his brother, Liam, and together they finished the job in 5 hours. How long would it take Liam to do the job by himself? **We can use the** fractional equation $\frac{5}{12} + \frac{5}{h} = \frac{3}{4}$ to determine that Liam could do the entire job by himself in 15 hours.

Rational expressions are to algebra what rational numbers are to arithmetic. Most of the work we will do with rational expressions in this chapter parallels the work you have previously done with arithmetic fractions. The same basic properties we use to explain reducing, adding, subtracting, multiplying, and dividing arithmetic fractions will serve as a basis for our work with rational expressions. The techniques of factoring we studied in Chapter 3 will also play an important role in our discussions. At the end of this chapter we will work with some fractional equations that contain rational expressions.

Simplifying Rational Expressions

Any number that can be written in the form $\frac{a}{b}$, where a and b are integers and b is not zero, is called a **rational number**. The following are examples of rational numbers.

$$\frac{1}{2}, \quad \frac{3}{4}, \quad \frac{15}{7}, \quad \frac{-5}{6}, \quad \frac{7}{-8}, \quad \frac{-12}{-17}$$

Numbers such as 6, -4, 0, $4\frac{1}{2}$, $.7$, and $.21$ are also rational because we can express them as the indicated quotient of two integers. For example,

$$6 = \frac{6}{1} = \frac{12}{2} = \frac{18}{3} \quad \text{and so on,} \qquad\qquad 4\frac{1}{2} = \frac{9}{2},$$

$$-4 = \frac{4}{-1} = \frac{-4}{1} = \frac{8}{-2} \quad \text{and so on,} \qquad .7 = \frac{7}{10},$$

$$0 = \frac{0}{1} = \frac{0}{2} = \frac{0}{3} \quad \text{and so on,} \qquad\qquad .21 = \frac{21}{100}.$$

Our work with division of integers helps with the next examples.

$$\frac{8}{-2} = \frac{-8}{2} = -\frac{8}{2} = -4, \qquad \frac{12}{3} = \frac{-12}{-3} = 4$$

Observe the following general properties.

PROPERTY 4.1

1. $\dfrac{-a}{b} = \dfrac{a}{-b} = -\dfrac{a}{b}$

2. $\dfrac{-a}{-b} = \dfrac{a}{b}$

Therefore, a rational number such as $\dfrac{-2}{5}$ can also be written as $\dfrac{2}{-5}$ or $-\dfrac{2}{5}$.

We use the following property, often referred to as the **fundamental principle of fractions**, to reduce fractions to lowest terms or express fractions in simplest or reduced form.

PROPERTY 4.2

If b and k are nonzero integers and a is any integer, then

$$\frac{a \cdot k}{b \cdot k} = \frac{a}{b}.$$

Let's apply Properties 4.1 and 4.2 to the following examples.

Example 1

Reduce $\dfrac{18}{24}$ to lowest terms.

Solution

$$\frac{18}{24} = \frac{3 \cdot 6}{4 \cdot 6} = \frac{3}{4}$$

▲

Example 2

Change $\dfrac{40}{48}$ to simplest form.

Solution

$$\frac{\overset{5}{\cancel{40}}}{\underset{6}{\cancel{48}}} = \frac{5}{6}$$ A common factor of 8 was divided out of both numerator and denominator.

▲

Example 3

Express $\dfrac{-36}{63}$ in reduced form.

Solution

$$\frac{-36}{63} = -\frac{36}{63} = -\frac{4 \cdot 9}{7 \cdot 9} = -\frac{4}{7}$$

▲

Example 4

Reduce $\dfrac{72}{-90}$ to simplest form.

Solution

$$\frac{72}{-90} = -\frac{72}{90} = -\frac{\cancel{2} \cdot 2 \cdot 2 \cdot \cancel{3} \cdot \cancel{3}}{\cancel{2} \cdot \cancel{3} \cdot \cancel{3} \cdot 5} = -\frac{4}{5}$$

▲

In Example 4, notice the use of the prime factored forms to help identify the common factors of the numerator and denominator.

Rational Expressions

A **rational expression** is the indicated quotient of two polynomials. The following are examples of rational expressions.

$$\frac{3x^2}{5}, \quad \frac{x - 2}{x + 3}, \quad \frac{x^2 + 5x - 1}{x^2 - 9}, \quad \frac{xy^2 + x^2y}{xy}, \quad \frac{a^3 - 3a^2 - 5a - 1}{a^4 + a^3 + 6}$$

Because we must avoid division by zero, no values that create a denominator of zero can be assigned to variables. Thus, the rational expression $\dfrac{x - 2}{x + 3}$ is meaningful for all values of x except for $x = -3$. Rather than making restrictions for each individual expression, we will merely assume that all denominators represent nonzero real numbers.

Property 4.2 $\left(\dfrac{a \cdot k}{b \cdot k} = \dfrac{a}{b} \right)$ serves as the basis for simplifying rational expressions, as the next examples illustrate.

Example 5

Simplify $\dfrac{15xy}{25y}$.

Solution

$$\dfrac{15xy}{25y} = \dfrac{3 \cdot \cancel{5} \cdot x \cdot \cancel{y}}{\cancel{5} \cdot 5 \cdot \cancel{y}} = \dfrac{3x}{5}$$

▲

Example 6

Simplify $\dfrac{-9}{18x^2y}$.

Solution

$$\dfrac{-9}{18x^2y} = -\dfrac{\overset{1}{\cancel{9}}}{\underset{2}{\cancel{18}}x^2y} = -\dfrac{1}{2x^2y}$$ A common factor of 9 was divided out of numerator and denominator.

▲

Example 7

Simplify $\dfrac{-28a^2b^2}{-63a^2b^3}$.

Solution

$$\dfrac{-28a^2b^2}{-63a^2b^3} = \dfrac{4 \cdot \cancel{7} \cdot \cancel{a^2} \cdot \cancel{b^2}}{9 \cdot \cancel{7} \cdot \cancel{a^2} \cdot \underset{b}{\cancel{b^3}}} = \dfrac{4}{9b}$$

▲

The factoring techniques from Chapter 3 can be used to factor numerators and/or denominators so that we can apply the property $\dfrac{a \cdot k}{b \cdot k} = \dfrac{a}{b}$. Examples 8–12 should clarify this process.

Example 8

Simplify $\dfrac{x^2 + 4x}{x^2 - 16}$.

Solution

$$\dfrac{x^2 + 4x}{x^2 - 16} = \dfrac{x(\cancel{x + 4})}{(x - 4)(\cancel{x + 4})} = \dfrac{x}{x - 4}$$

▲

Example 9

Simplify $\dfrac{4a^2 + 12a + 9}{2a + 3}$.

Solution

$$\dfrac{4a^2 + 12a + 9}{2a + 3} = \dfrac{(\cancel{2a + 3})(2a + 3)}{1(\cancel{2a + 3})} = \dfrac{2a + 3}{1} = 2a + 3$$

▲

Example 10

Simplify $\dfrac{5n^2 + 6n - 8}{10n^2 - 3n - 4}$.

Solution

$$\dfrac{5n^2 + 6n - 8}{10n^2 - 3n - 4} = \dfrac{(\cancel{5n - 4})(n + 2)}{(\cancel{5n - 4})(2n + 1)} = \dfrac{n + 2}{2n + 1}$$

▲

Example 11

Simplify $\dfrac{6x^3y - 6xy}{x^3 + 5x^2 + 4x}$.

Solution

$$\dfrac{6x^3y - 6xy}{x^3 + 5x^2 + 4x} = \dfrac{6xy(x^2 - 1)}{x(x^2 + 5x + 4)} = \dfrac{6\cancel{x}y(\cancel{x + 1})(x - 1)}{\cancel{x}(\cancel{x + 1})(x + 4)} = \dfrac{6y(x - 1)}{x + 4}$$

▲

Note that in Example 11 we left the numerator of the final fraction in factored form. This is often done if expressions other than monomials are involved. Either $\dfrac{6y(x-1)}{x+4}$ or $\dfrac{6xy-6y}{x+4}$ is an acceptable answer.

Remember that the quotient of any nonzero real number and its opposite is -1. For example, $\dfrac{6}{-6} = -1$ and $\dfrac{-8}{8} = -1$. Likewise, the indicated quotient of any polynomial and its opposite is equal to -1. For example,

$$\frac{a}{-a} = -1 \quad \text{because } a \text{ and } -a \text{ are opposites,}$$

$$\frac{a-b}{b-a} = -1 \quad \text{because } a-b \text{ and } b-a \text{ are opposites,}$$

$$\frac{x^2-4}{4-x^2} = -1 \quad \text{because } x^2-4 \text{ and } 4-x^2 \text{ are opposites.}$$

Example 12 puts this idea to use when simplifying rational expressions.

Example 12

Simplify $\dfrac{6a^2 - 7a + 2}{10a - 15a^2}$.

Solution

$$\frac{6a^2 - 7a + 2}{10a - 15a^2} = \frac{(2a-1)(3a-2)}{5a(2-3a)} \qquad \frac{3a-2}{2-3a} = -1$$

$$= (-1)\left(\frac{2a-1}{5a}\right)$$

$$= -\frac{2a-1}{5a} \qquad \text{or} \qquad \frac{1-2a}{5a}$$

Problem Set 4.1

Express each of the following rational numbers in reduced form.

1. $\dfrac{27}{36}$ **2.** $\dfrac{14}{21}$ **3.** $\dfrac{45}{54}$

4. $\dfrac{-14}{42}$ **5.** $\dfrac{24}{-60}$ **6.** $\dfrac{45}{-75}$

7. $\dfrac{-16}{-56}$ **8.** $\dfrac{-30}{-42}$

Simplify each of the following.

9. $\dfrac{12xy}{42y}$ **10.** $\dfrac{21xy}{35x}$

11. $\dfrac{18a^2}{45ab}$ **12.** $\dfrac{48ab}{84b^2}$

13. $\dfrac{-14y^3}{56xy^2}$ **14.** $\dfrac{-14x^2y^3}{63xy^2}$

15. $\dfrac{54c^2d}{-78cd^2}$ **16.** $\dfrac{60x^3z}{-64xyz^2}$

17. $\dfrac{-40x^3y}{-24xy^4}$ **18.** $\dfrac{-30x^2y^2z^2}{-35xz^3}$

19. $\dfrac{x^2-4}{x^2+2x}$ **20.** $\dfrac{xy+y^2}{x^2-y^2}$

21. $\dfrac{18x+12}{12x-6}$ **22.** $\dfrac{20x+50}{15x-30}$

23. $\dfrac{a^2 + 7a + 10}{a^2 - 7a - 18}$

24. $\dfrac{a^2 + 4a - 32}{3a^2 + 26a + 16}$

25. $\dfrac{2n^2 + n - 21}{10n^2 + 33n - 7}$

26. $\dfrac{4n^2 - 15n - 4}{7n^2 - 30n + 8}$

27. $\dfrac{5x^2 + 7}{10x}$

28. $\dfrac{12x^2 + 11x - 15}{20x^2 - 23x + 6}$

29. $\dfrac{6x^2 + x - 15}{8x^2 - 10x - 3}$

30. $\dfrac{4x^2 + 8x}{x^3 + 8}$

31. $\dfrac{3x^2 - 12x}{x^3 - 64}$

32. $\dfrac{x^2 - 14x + 49}{6x^2 - 37x - 35}$

33. $\dfrac{3x^2 + 17x - 6}{9x^2 - 6x + 1}$

34. $\dfrac{9y^2 - 1}{3y^2 + 11y - 4}$

35. $\dfrac{2x^3 + 3x^2 - 14x}{x^2y + 7xy - 18y}$

36. $\dfrac{3x^3 + 12x}{9x^2 + 18x}$

37. $\dfrac{5y^2 + 22y + 8}{25y^2 - 4}$

38. $\dfrac{16x^3y + 24x^2y^2 - 16xy^3}{24x^2y + 12xy^2 - 12y^3}$

39. $\dfrac{15x^3 - 15x^2}{5x^3 + 5x}$

40. $\dfrac{5n^2 + 18n - 8}{3n^2 + 13n + 4}$

41. $\dfrac{4x^2y + 8xy^2 - 12y^3}{18x^3y - 12x^2y^2 - 6xy^3}$

42. $\dfrac{3 + x - 2x^2}{2 + x - x^2}$

43. $\dfrac{3n^2 + 14n - 24}{7n^2 + 44n + 12}$

44. $\dfrac{x^4 - 2x^2 - 15}{2x^4 + 9x^2 + 9}$

45. $\dfrac{8 + 18x - 5x^2}{10 + 31x + 15x^2}$

46. $\dfrac{6x^4 - 11x^2 + 4}{2x^4 + 17x^2 - 9}$

47. $\dfrac{27x^4 - x}{6x^3 + 10x^2 - 4x}$

48. $\dfrac{64x^4 + 27x}{12x^3 - 27x^2 - 27x}$

49. $\dfrac{-40x^3 + 24x^2 + 16x}{20x^3 + 28x^2 + 8x}$

50. $\dfrac{-6x^3 - 21x^2 + 12x}{-18x^3 - 42x^2 + 120x}$

Simplify each of the following. You will need to use factoring by grouping.

51. $\dfrac{xy + ay + bx + ab}{xy + ay + cx + ac}$

52. $\dfrac{xy + 2y + 3x + 6}{xy + 2y + 4x + 8}$

53. $\dfrac{ax - 3x + 2ay - 6y}{2ax - 6x + ay - 3y}$

54. $\dfrac{x^2 - 2x + ax - 2a}{x^2 - 2x + 3ax - 6a}$

55. $\dfrac{5x^2 + 5x + 3x + 3}{5x^2 + 3x - 30x - 18}$

56. $\dfrac{x^2 + 3x + 4x + 12}{2x^2 + 6x - x - 3}$

57. $\dfrac{2st - 30 - 12s + 5t}{3st - 6 - 18s + t}$

58. $\dfrac{nr - 6 - 3n + 2r}{nr + 10 + 2r + 5n}$

Simplify each of the following. You may want to refer to Example 12 in this section.

59. $\dfrac{5x - 7}{7 - 5x}$

60. $\dfrac{4a - 9}{9 - 4a}$

61. $\dfrac{n^2 - 49}{7 - n}$

62. $\dfrac{9 - y}{y^2 - 81}$

63. $\dfrac{2y - 2xy}{x^2 y - y}$

64. $\dfrac{3x - x^2}{x^2 - 9}$

65. $\dfrac{2x^3 - 8x}{4x - x^3}$

66. $\dfrac{x^2 - (y-1)^2}{(y-1)^2 - x^2}$

67. $\dfrac{n^2 - 5n - 24}{40 + 3n - n^2}$

68. $\dfrac{x^2 + 2x - 24}{20 - x - x^2}$

THOUGHTS INTO WORDS

69. Compare the concept of a rational number in arithmetic to the concept of a rational expression in algebra.

70. What role does factoring play when simplifying rational expressions?

71. Why is the rational expression $\dfrac{x+3}{x^2-4}$ undefined for $x = 2$ and $x = -2$ but defined for $x = -3$?

4.2 Multiplying and Dividing Rational Expressions

We define multiplication of rational numbers in common fraction form as follows.

DEFINITION 4.1

If a, b, c, and d are integers with b and d not equal to zero, then

$$\frac{a}{b} \cdot \frac{c}{d} = \frac{a \cdot c}{b \cdot d} = \frac{ac}{bd}.$$

To multiply rational numbers in common fraction form we merely **multiply numerators and multiply denominators** as the next examples demonstrate. (The steps in the dashed boxes are usually done mentally.)

$$\frac{2}{3} \cdot \frac{4}{5} = \frac{2 \cdot 4}{3 \cdot 5} = \frac{8}{15},$$

$$\frac{-3}{4} \cdot \frac{5}{7} = \frac{-3 \cdot 5}{4 \cdot 7} = \frac{-15}{28} = -\frac{15}{28},$$

$$-\frac{5}{6} \cdot \frac{13}{3} = \frac{-5}{6} \cdot \frac{13}{3} = \frac{-5 \cdot 13}{6 \cdot 3} = \frac{-65}{18} = -\frac{65}{18}$$

We also agree, when multiplying rational numbers, to express the final product in reduced form. The following examples show some different formats to *multiply and simplify* rational numbers.

$$\frac{3}{4} \cdot \frac{4}{7} = \frac{3 \cdot \cancel{4}}{\cancel{4} \cdot 7} = \frac{3}{7},$$

$$\frac{\overset{1}{\cancel{8}}}{\underset{1}{\cancel{9}}} \cdot \frac{\overset{3}{\cancel{27}}}{\underset{4}{\cancel{32}}} = \frac{3}{4},$$

A common factor of 9 was divided out of 9 and 27, and a common factor of 8 was divided out of 8 and 32.

$$\left(-\frac{28}{25}\right)\left(-\frac{65}{78}\right) = \frac{\cancel{2} \cdot 2 \cdot 7 \cdot \cancel{5} \cdot \cancel{13}}{\cancel{5} \cdot 5 \cdot \cancel{2} \cdot 3 \cdot \cancel{13}} = \frac{14}{15}.$$

We should recognize that a *negative times a negative is positive*. Also, notice the use of prime factors to help us recognize common factors.

Multiplication of rational expressions follows the same basic pattern as multiplication of rational numbers in common fraction form. That is to say, *we multiply numerators and multiply denominators and express the final product in simplified or reduced form.* Let's consider some examples.

$$\frac{3x}{4y} \cdot \frac{8y^2}{9x} = \frac{\cancel{3} \cdot \overset{2}{\cancel{8}} \cdot \cancel{x} \cdot \overset{y}{\cancel{y^2}}}{\underset{3}{\cancel{4} \cdot \cancel{9} \cdot \cancel{x} \cdot \cancel{y}}} = \frac{2y}{3},$$

Notice that we use the commutative property of multiplication to rearrange the factors in a form that allows us to identify common factors of the numerator and denominator.

$$\frac{-4a}{6a^2b^2} \cdot \frac{9ab}{12a^2} = -\frac{\cancel{4} \cdot \overset{\cancel{3}}{\cancel{9}} \cdot \cancel{a^2} \cdot \cancel{b}}{\underset{2}{\cancel{6}} \cdot \underset{\cancel{3}}{\cancel{12}} \cdot \underset{a^2}{\cancel{a^4}} \cdot \underset{b}{\cancel{b^2}}} = -\frac{1}{2a^2b},$$

$$\frac{12x^2y}{-18xy} \cdot \frac{-24xy^2}{56y^3} = \frac{\overset{2}{\cancel{12}} \cdot \overset{\cancel{3}}{\cancel{24}} \cdot \overset{x^2}{\cancel{x^3}} \cdot \cancel{y^3}}{\underset{\cancel{3}}{\cancel{18}} \cdot \underset{7}{\cancel{56}} \cdot \cancel{x} \cdot \underset{y}{\cancel{y^4}}} = \frac{2x^2}{7y}$$

You should recognize that the first fraction is equivalent to $-\dfrac{12x^2y}{18xy}$ and the second to $-\dfrac{24xy^2}{56y^3}$; thus, the product is positive.

If the rational expressions contain polynomials (other than monomials) that are factorable, then our work may take on the following format.

Example 1

Multiply and simplify $\dfrac{y}{x^2 - 4} \cdot \dfrac{x + 2}{y^2}$.

Solution

$$\frac{y}{x^2 - 4} \cdot \frac{x + 2}{y^2} = \frac{\cancel{y}(\cancel{x + 2})}{\underset{y}{\cancel{y^2}}(\cancel{x + 2})(x - 2)} = \frac{1}{y(x - 2)}$$

▲

In Example 1, notice that we combined the steps of multiplying numerators and denominators and factoring the polynomials. Also, notice that we left the final answer in factored form. Either $\dfrac{1}{y(x-2)}$ or $\dfrac{1}{xy-2y}$ would be an acceptable answer.

Example 2

Multiply and simplify $\dfrac{x^2-x}{x+5}\cdot\dfrac{x^2+5x+4}{x^4-x^2}$.

Solution

$$\frac{x^2-x}{x+5}\cdot\frac{x^2+5x+4}{x^4-x^2}=\frac{\cancel{x}(\cancel{x-1})(\cancel{x+1})(x+4)}{(x+5)(\cancel{x^2})(\cancel{x-1})(\cancel{x+1})}=\frac{x+4}{x(x+5)}$$
$$\underset{x}{}$$

▲

Example 3

Multiply and simplify $\dfrac{6n^2+7n-5}{n^2+2n-24}\cdot\dfrac{4n^2+21n-18}{12n^2+11n-15}$.

Solution

$$\frac{6n^2+7n-5}{n^2+2n-24}\cdot\frac{4n^2+21n-18}{12n^2+11n-15}$$
$$=\frac{(\cancel{3n+5})(2n-1)(\cancel{4n-3})(\cancel{n+6})}{(\cancel{n+6})(n-4)(\cancel{3n+5})(\cancel{4n-3})}=\frac{2n-1}{n-4}$$

▲

Dividing Rational Expressions

We define division of rational numbers in common fraction form as follows.

DEFINITION 4.2

If a, b, c, and d are integers with b, c, and d not equal to zero, then

$$\frac{a}{b}\div\frac{c}{d}=\frac{a}{b}\cdot\frac{d}{c}=\frac{ad}{bc}.$$

Definition 4.2 states that to divide two rational numbers in fractional form we **invert the divisor and multiply**. We call the numbers $\dfrac{c}{d}$ and $\dfrac{d}{c}$ **reciprocals** or **multiplicative inverses** of each other because their product is one. Thus, we can describe division as **to divide by a fraction, multiply by its reciprocal**. The following examples demonstrate the use of Definition 4.2.

$$\frac{7}{8}\div\frac{5}{6}=\frac{7}{\underset{4}{\cancel{8}}}\cdot\frac{\overset{3}{\cancel{6}}}{5}=\frac{21}{20},\qquad\frac{-5}{9}\div\frac{15}{18}=-\frac{\cancel{5}}{\cancel{9}}\cdot\frac{\overset{2}{\cancel{18}}}{\cancel{15}}=-\frac{2}{3},$$
$$\qquad\qquad\qquad\qquad\qquad\qquad\qquad\qquad\qquad\underset{3}{}$$

$$\frac{14}{-19}\div\frac{21}{-38}=\left(-\frac{14}{19}\right)\div\left(-\frac{21}{38}\right)=\left(-\frac{\overset{2}{\cancel{14}}}{\cancel{19}}\right)\left(-\frac{\overset{2}{\cancel{38}}}{\cancel{21}}\right)=\frac{4}{3}$$
$$\qquad\qquad\qquad\qquad\qquad\qquad\qquad\qquad\qquad\qquad\underset{3}{}$$

We define division of algebraic rational expressions in the same way that we define division of rational numbers. That is, the quotient of two rational expressions is the product of the first expression times the reciprocal of the second. Consider the following examples.

Example 4

Divide and simplify $\dfrac{16x^2y}{24xy^3} \div \dfrac{9xy}{8x^2y^2}$.

Solution

$$\frac{16x^2y}{24xy^3} \div \frac{9xy}{8x^2y^2} = \frac{16x^2y}{24xy^3} \cdot \frac{8x^2y^2}{9xy} = \frac{16 \cdot \overset{2}{8} \cdot \overset{x^2}{x^4} \cdot \overset{}{y^3}}{\underset{3}{24} \cdot 9 \cdot x^2 \cdot \underset{y}{y^4}} = \frac{16x^2}{27y}$$

Example 5

Divide and simplify $\dfrac{3a^2 + 12}{3a^2 - 15a} \div \dfrac{a^4 - 16}{a^2 - 3a - 10}$.

Solution

$$\frac{3a^2 + 12}{3a^2 - 15a} \div \frac{a^4 - 16}{a^2 - 3a - 10} = \frac{3a^2 + 12}{3a^2 - 15a} \cdot \frac{a^2 - 3a - 10}{a^4 - 16}$$

$$= \frac{3(a^2 + 4)(a - 5)(a + 2)}{3a(a - 5)(a^2 + 4)(a + 2)(a - 2)}$$

$$= \frac{1}{a(a - 2)}$$

Example 6

Divide and simplify $\dfrac{28t^3 - 51t^2 - 27t}{49t^2 + 42t + 9} \div (4t - 9)$.

Solution

$$\frac{28t^3 - 51t^2 - 27t}{49t^2 + 42t + 9} \div \frac{4t - 9}{1} = \frac{28t^3 - 51t^2 - 27t}{49t^2 + 42t + 9} \cdot \frac{1}{4t - 9}$$

$$= \frac{t(7t + 3)(4t - 9)}{(7t + 3)(7t + 3)(4t - 9)}$$

$$= \frac{t}{7t + 3}$$

In a problem such as Example 6, it may be helpful to write the divisor with a denominator of 1. Thus, we write $4t - 9$ as $\dfrac{4t - 9}{1}$; then its reciprocal is obviously $\dfrac{1}{4t - 9}$.

Let's consider one final example that involves both multiplication and division.

Example 7

Perform the indicated operations and simplify.

$$\frac{x^2 + 5x}{3x^2 - 4x - 20} \cdot \frac{x^2y + y}{2x^2 + 11x + 5} \div \frac{xy^2}{6x^2 - 17x - 10}$$

Solution

$$\frac{x^2 + 5x}{3x^2 - 4x - 20} \cdot \frac{x^2y + y}{2x^2 + 11x + 5} \div \frac{xy^2}{6x^2 - 17x - 10}$$

$$= \frac{x^2 + 5x}{3x^2 - 4x - 20} \cdot \frac{x^2y + y}{2x^2 + 11x + 5} \cdot \frac{6x^2 - 17x - 10}{xy^2}$$

$$= \frac{\cancel{x}(x+5)(\cancel{y})(x^2 + 1)(2x+1)(3x-10)}{(3x-10)(x + 2)(2x+1)(x+5)(\cancel{x})(\cancel{y^2})} = \frac{x^2 + 1}{y(x + 2)}$$

Problem Set 4.2

Perform the following indicated operations involving rational numbers. Express final answers in reduced form.

1. $\dfrac{7}{12} \cdot \dfrac{6}{35}$

2. $\dfrac{5}{8} \cdot \dfrac{12}{20}$

3. $\dfrac{-4}{9} \cdot \dfrac{18}{30}$

4. $\dfrac{-6}{9} \cdot \dfrac{36}{48}$

5. $\dfrac{3}{-8} \cdot \dfrac{-6}{12}$

6. $\dfrac{-12}{16} \cdot \dfrac{18}{-32}$

7. $\left(-\dfrac{5}{7}\right) \div \dfrac{6}{7}$

8. $\left(-\dfrac{5}{9}\right) \div \dfrac{10}{3}$

9. $\dfrac{-9}{5} \div \dfrac{27}{10}$

10. $\dfrac{4}{7} \div \dfrac{16}{-21}$

11. $\dfrac{4}{9} \cdot \dfrac{6}{11} \div \dfrac{4}{15}$

12. $\dfrac{2}{3} \cdot \dfrac{6}{7} \div \dfrac{8}{3}$

Perform the following indicated operations involving rational expressions. Express final answers in simplest form.

13. $\dfrac{6xy}{9y^4} \cdot \dfrac{30x^3y}{-48x}$

14. $\dfrac{-14xy^4}{18y^2} \cdot \dfrac{24x^2y^3}{35y^2}$

15. $\dfrac{5a^2b^2}{11ab} \cdot \dfrac{22a^3}{15ab^2}$

16. $\dfrac{10a^2}{5b^2} \cdot \dfrac{15b^3}{2a^4}$

17. $\dfrac{5xy}{8y^2} \cdot \dfrac{18x^2y}{15}$

18. $\dfrac{4x^2}{5y^2} \cdot \dfrac{15xy}{24x^2y^2}$

19. $\dfrac{5x^4}{12x^2y^3} \div \dfrac{9}{5xy}$

20. $\dfrac{7x^2y}{9xy^3} \div \dfrac{3x^4}{2x^2y^2}$

21. $\dfrac{9a^2c}{12bc^2} \div \dfrac{21ab}{14c^3}$

22. $\dfrac{3ab^3}{4c} \div \dfrac{21ac}{12bc^3}$

23. $\dfrac{9x^2y^3}{14x} \cdot \dfrac{21y}{15xy^2} \cdot \dfrac{10x}{12y^3}$

24. $\dfrac{5xy}{7a} \cdot \dfrac{14a^2}{15x} \cdot \dfrac{3a}{8y}$

25. $\dfrac{3x + 6}{5y} \cdot \dfrac{x^2 + 4}{x^2 + 10x + 16}$

26. $\dfrac{5xy}{x + 6} \cdot \dfrac{x^2 - 36}{x^2 - 6x}$

27. $\dfrac{5a^2 + 20a}{a^3 - 2a^2} \cdot \dfrac{a^2 - a - 12}{a^2 - 16}$

28. $\dfrac{2a^2 + 6}{a^2 - a} \cdot \dfrac{a^3 - a^2}{8a - 4}$

29. $\dfrac{3n^2 + 15n - 18}{3n^2 + 10n - 48} \cdot \dfrac{12n^2 - 17n - 40}{8n^2 + 2n - 10}$

30. $\dfrac{10n^2 + 21n - 10}{5n^2 + 33n - 14} \cdot \dfrac{2n^2 + 6n - 56}{2n^2 - 3n - 20}$

31. $\dfrac{9y^2}{x^2 + 12x + 36} \div \dfrac{12y}{x^2 + 6x}$

32. $\dfrac{7xy}{x^2 - 4x + 4} \div \dfrac{14y}{x^2 - 4}$

33. $\dfrac{x^2 - 4xy + 4y^2}{7xy^2} \div \dfrac{4x^2 - 3xy - 10y^2}{20x^2y + 25xy^2}$

34. $\dfrac{x^2 + 5xy - 6y^2}{xy^2 - y^3} \cdot \dfrac{2x^2 + 15xy + 18y^2}{xy + 4y^2}$

35. $\dfrac{5 - 14n - 3n^2}{1 - 2n - 3n^2} \cdot \dfrac{9 + 7n - 2n^2}{27 - 15n + 2n^2}$

36. $\dfrac{6 - n - 2n^2}{12 - 11n + 2n^2} \cdot \dfrac{24 - 26n + 5n^2}{2 + 3n + n^2}$

37. $\dfrac{3x^4 + 2x^2 - 1}{3x^4 + 14x^2 - 5} \cdot \dfrac{x^4 - 2x^2 - 35}{x^4 - 17x^2 + 70}$

38. $\dfrac{2x^4 + x^2 - 3}{2x^4 + 5x^2 + 2} \cdot \dfrac{3x^4 + 10x^2 + 8}{3x^4 + x^2 - 4}$

39. $\dfrac{6x^2 - 35x + 25}{4x^2 - 11x - 45} \div \dfrac{18x^2 + 9x - 20}{24x^2 + 74x + 45}$

40. $\dfrac{21t^2 + 22t - 8}{5t^2 - 43t - 18} \div \dfrac{12t^2 + 7t - 12}{20t^2 - 7t - 6}$

41. $\dfrac{10t^3 + 25t}{20t + 10} \cdot \dfrac{2t^2 - t - 1}{t^5 - t}$

42. $\dfrac{t^4 - 81}{t^2 - 6t + 9} \cdot \dfrac{6t^2 - 11t - 21}{5t^2 + 8t - 21}$

43. $\dfrac{4t^2 + t - 5}{t^3 - t^2} \cdot \dfrac{t^4 + 6t^3}{16t^2 + 40t + 25}$

44. $\dfrac{9n^2 - 12n + 4}{n^2 - 4n - 32} \cdot \dfrac{n^2 + 4n}{3n^3 - 2n^2}$

45. $\dfrac{nr + 3n + 2r + 6}{nr + 3n - 3r - 9} \cdot \dfrac{n^2 - 9}{n^3 - 4n}$

46. $\dfrac{xy + xc + ay + ac}{xy - 2xc + ay - 2ac} \cdot \dfrac{2x^3 - 8x}{12x^3 + 20x^2 - 8x}$

47. $\dfrac{x^2 - x}{4y} \cdot \dfrac{10xy^2}{2x - 2} \div \dfrac{3x^2 + 3x}{15x^2y^2}$

48. $\dfrac{4xy^2}{7x} \cdot \dfrac{14x^3y}{12y} \div \dfrac{7y}{9x^3}$

49. $\dfrac{a^2 - 4ab + 4b^2}{6a^2 - 4ab} \cdot \dfrac{3a^2 + 5ab - 2b^2}{6a^2 + ab - b^2} \div \dfrac{a^2 - 4b^2}{8a + 4b}$

50. $\dfrac{2x^2 + 3x}{2x^3 - 10x^2} \cdot \dfrac{x^2 - 8x + 15}{3x^3 - 27x} \div \dfrac{14x + 21}{x^2 - 6x - 27}$

THOUGHTS INTO WORDS

51. Explain in your own words how to divide two rational expressions.

52. Suppose that your friend missed class the day this section was discussed. How could you use her background in arithmetic to explain how to multiply and divide rational expressions?

53. Give a step-by-step description of how to do the following multiplication problem.

$$\dfrac{x^2 + 5x + 6}{x^2 - 2x - 8} \cdot \dfrac{x^2 - 16}{16 - x^2}$$

4.3 Adding and Subtracting Rational Expressions

We can define addition and subtraction of rational numbers as follows.

DEFINITION 4.3

If a, b, and c are integers and b is not zero, then

$$\frac{a}{b} + \frac{c}{b} = \frac{a+c}{b} \qquad \text{Addition}$$

$$\frac{a}{b} - \frac{c}{b} = \frac{a-c}{b} \qquad \text{Subtraction}$$

We can add or subtract rational numbers with a common denominator by adding or subtracting the numerators and placing the result over the common denominator. The following examples illustrate Definition 4.3.

$$\frac{2}{9} + \frac{3}{9} = \frac{2+3}{9} = \frac{5}{9},$$

$$\frac{7}{8} - \frac{3}{8} = \frac{7-3}{8} = \frac{4}{8} = \frac{1}{2}, \qquad \text{Don't forget to reduce!}$$

$$\frac{4}{6} + \frac{-5}{6} = \frac{4+(-5)}{6} = \frac{-1}{6} = -\frac{1}{6},$$

$$\frac{7}{10} + \frac{4}{-10} = \frac{7}{10} + \frac{-4}{10} = \frac{7+(-4)}{10} = \frac{3}{10}$$

We use this same *common denominator* approach when adding or subtracting rational expressions, as in these next examples.

$$\frac{3}{x} + \frac{9}{x} = \frac{3+9}{x} = \frac{12}{x},$$

$$\frac{8}{x-2} - \frac{3}{x-2} = \frac{8-3}{x-2} = \frac{5}{x-2},$$

$$\frac{9}{4y} + \frac{5}{4y} = \frac{9+5}{4y} = \frac{14}{4y} = \frac{7}{2y}, \qquad \text{Don't forget to simplify the final answer!}$$

$$\frac{n^2}{n-1} - \frac{1}{n-1} = \frac{n^2-1}{n-1} = \frac{(n+1)(\cancel{n-1})}{\cancel{n-1}} = n+1,$$

$$\frac{6a^2}{2a+1} + \frac{13a+5}{2a+1} = \frac{6a^2+13a+5}{2a+1} = \frac{(\cancel{2a+1})(3a+5)}{\cancel{2a+1}} = 3a+5$$

In each of the previous examples that involve rational expressions, we should technically restrict the variables to exclude division by zero. For example, $\frac{3}{x} + \frac{9}{x} = \frac{12}{x}$ is true for all real number values for x, *except* $x = 0$. Likewise, $\frac{8}{x-2} - \frac{3}{x-2} = \frac{5}{x-2}$ as long as x does not equal 2. Rather than taking the time and space to write down restrictions for each problem we will merely assume that such restrictions exist.

If rational numbers that do not have a common denominator are to be added or subtracted, then we apply the fundamental principle of fractions $\left(\frac{a}{b} = \frac{ak}{bk}\right)$ to obtain equivalent fractions with a common denominator. Equivalent fractions are fractions such as $\frac{1}{2}$ and $\frac{2}{4}$ that name the same number. Consider the following example.

$$\frac{1}{2} + \frac{1}{3} = \frac{3}{6} + \frac{2}{6} = \frac{3+2}{6} = \frac{5}{6}$$

$\left(\begin{array}{c}\frac{1}{2}\text{ and }\frac{3}{6}\\ \text{are equivalent}\\ \text{fractions.}\end{array}\right)\left(\begin{array}{c}\frac{1}{3}\text{ and }\frac{2}{6}\\ \text{are equivalent}\\ \text{fractions.}\end{array}\right)$

Notice that we chose 6 as our common denominator and 6 is the **least common multiple** of the original denominators 2 and 3. (The least common multiple of a set of whole numbers is the smallest nonzero whole number divisible by each of the numbers.) In general, we use the least common multiple of the denominators of the fractions to be added or subtracted as a **least common denominator** (LCD).

A least common denominator may be found by inspection or by using the prime factored forms of the numbers. Let's consider some examples and use each of these techniques.

Example 1

Subtract $\frac{5}{6} - \frac{3}{8}$.

Solution

By inspection we can see that the LCD is 24. Thus, both fractions can be changed to equivalent fractions, each with a denominator of 24.

$$\frac{5}{6} - \frac{3}{8} = \left(\frac{5}{6}\right)\left(\frac{4}{4}\right) - \left(\frac{3}{8}\right)\left(\frac{3}{3}\right) = \frac{20}{24} - \frac{9}{24} = \frac{11}{24}$$

$$\uparrow \qquad \uparrow$$
$$\text{Form of 1 Form of 1}$$

In Example 1, notice that the fundamental principle of fractions, $\frac{a}{b} = \frac{a \cdot k}{b \cdot k}$, can be written as $\frac{a}{b} = \left(\frac{a}{b}\right)\left(\frac{k}{k}\right)$. This latter form emphasizes the fact that one is the multiplication identity element.

Example 2

Perform the indicated operations $\frac{3}{5} + \frac{1}{6} - \frac{13}{15}$.

Solution

Again by inspection we can determine that the LCD is 30. Thus, we can proceed as follows.

$$\frac{3}{5} + \frac{1}{6} - \frac{13}{15} = \left(\frac{3}{5}\right)\left(\frac{6}{6}\right) + \left(\frac{1}{6}\right)\left(\frac{5}{5}\right) - \left(\frac{13}{15}\right)\left(\frac{2}{2}\right)$$

$$= \frac{18}{30} + \frac{5}{30} - \frac{26}{30} = \frac{18 + 5 - 26}{30}$$

$$= \frac{-3}{30} = -\frac{1}{10} \qquad \text{Don't forget to reduce!} \qquad \blacktriangle$$

Example 3

Add $\frac{7}{18} + \frac{11}{24}$.

Solution

Let's use the prime factored forms of the denominators to help find the LCD.

$$18 = 2 \cdot 3 \cdot 3, \qquad 24 = 2 \cdot 2 \cdot 2 \cdot 3$$

The LCD must contain three factors of 2 since 24 contains three 2s. The LCD must also contain two factors of 3 since 18 has two 3s. Thus, the LCD $= 2 \cdot 2 \cdot 2 \cdot 3 \cdot 3 = 72$. Now we can proceed as usual.

$$\frac{7}{18} + \frac{11}{24} = \left(\frac{7}{18}\right)\left(\frac{4}{4}\right) + \left(\frac{11}{24}\right)\left(\frac{3}{3}\right) = \frac{28}{72} + \frac{33}{72} = \frac{61}{72} \qquad \blacktriangle$$

To add and subtract rational expressions with different denominators, follow the same basic routine as you follow when you add or subtract rational numbers with different denominators. Study the following examples carefully and notice the similarity to our previous work with rational numbers.

Example 4

Add $\frac{x + 2}{4} + \frac{3x + 1}{3}$.

Solution

By inspection we see that the LCD is 12.

$$\frac{x + 2}{4} + \frac{3x + 1}{3} = \left(\frac{x + 2}{4}\right)\left(\frac{3}{3}\right) + \left(\frac{3x + 1}{3}\right)\left(\frac{4}{4}\right)$$

$$= \frac{3(x + 2)}{12} + \frac{4(3x + 1)}{12}$$

$$= \frac{3x + 6 + 12x + 4}{12}$$

$$= \frac{15x + 10}{12} \qquad \blacktriangle$$

Notice the final result in Example 4. The numerator, $15x + 10$, could be factored as $5(3x + 2)$. However, since this produces no common factors with the denominator, the fraction cannot be simplified. Thus, the final answer can be left as $\dfrac{15x + 10}{12}$, or it would also be acceptable to express it as $\dfrac{5(3x + 2)}{12}$.

Example 5

Subtract $\dfrac{a - 2}{2} - \dfrac{a - 6}{6}$.

Solution

By inspection we see that the LCD is 6.

$$\frac{a - 2}{2} - \frac{a - 6}{6} = \left(\frac{a - 2}{2}\right)\left(\frac{3}{3}\right) - \frac{a - 6}{6}$$

$$= \frac{3(a - 2) - (a - 6)}{6}$$ Be careful with this sign as you move to the next step!

$$= \frac{3a - 6 - a + 6}{6}$$

$$= \frac{2a}{6} = \frac{a}{3}$$ Don't forget to simplify.   ▲

Example 6

Perform the indicated operations $\dfrac{x + 3}{10} + \dfrac{2x + 1}{15} - \dfrac{x - 2}{18}$.

Solution

If you cannot determine the LCD by inspection, then use the prime factored forms of the denominators.

$$10 = 2 \cdot 5, \qquad 15 = 3 \cdot 5, \qquad 18 = 2 \cdot 3 \cdot 3$$

The LCD must contain one factor of 2, two factors of 3, and one factor of 5. Thus, the LCD is $2 \cdot 3 \cdot 3 \cdot 5 = 90$.

$$\frac{x + 3}{10} + \frac{2x + 1}{15} - \frac{x - 2}{18} = \left(\frac{x + 3}{10}\right)\left(\frac{9}{9}\right) + \left(\frac{2x + 1}{15}\right)\left(\frac{6}{6}\right) - \left(\frac{x - 2}{18}\right)\left(\frac{5}{5}\right)$$

$$= \frac{9(x + 3) + 6(2x + 1) - 5(x - 2)}{90}$$

$$= \frac{9x + 27 + 12x + 6 - 5x + 10}{90}$$

$$= \frac{16x + 43}{90}$$ ▲

A denominator that contains variables does not create any serious difficulties; our approach remains basically the same.

Example 7

Add $\dfrac{3}{2x} + \dfrac{5}{3y}$.

Solution

Using an LCD of $6xy$ we can proceed as follows.

$$\frac{3}{2x} + \frac{5}{3y} = \left(\frac{3}{2x}\right)\left(\frac{3y}{3y}\right) + \left(\frac{5}{3y}\right)\left(\frac{2x}{2x}\right)$$

$$= \frac{9y}{6xy} + \frac{10x}{6xy}$$

$$= \frac{9y + 10x}{6xy}$$

Example 8

Subtract $\dfrac{7}{12ab} - \dfrac{11}{15a^2}$.

Solution

We can prime factor the numerical coefficients of the denominators to help find the LCD.

$$\left.\begin{array}{l} 12ab = 2 \cdot 2 \cdot 3 \cdot a \cdot b \\ 15a^2 = 3 \cdot 5 \cdot a^2 \end{array}\right\} \longrightarrow \text{LCD} = 2 \cdot 2 \cdot 3 \cdot 5 \cdot a^2 \cdot b = 60a^2b$$

$$\frac{7}{12ab} - \frac{11}{15a^2} = \left(\frac{7}{12ab}\right)\left(\frac{5a}{5a}\right) - \left(\frac{11}{15a^2}\right)\left(\frac{4b}{4b}\right)$$

$$= \frac{35a}{60a^2b} - \frac{44b}{60a^2b}$$

$$= \frac{35a - 44b}{60a^2b}$$

Example 9

Add $\dfrac{x}{x - 3} + \dfrac{4}{x}$.

Solution

By inspection the LCD is $x(x - 3)$.

$$\frac{x}{x - 3} + \frac{4}{x} = \left(\frac{x}{x - 3}\right)\left(\frac{x}{x}\right) + \left(\frac{4}{x}\right)\left(\frac{x - 3}{x - 3}\right)$$

$$= \frac{x^2}{x(x - 3)} + \frac{4(x - 3)}{x(x - 3)}$$

$$= \frac{x^2 + 4x - 12}{x(x - 3)} \quad \text{or} \quad \frac{(x + 6)(x - 2)}{x(x - 3)}$$

Example 10

Subtract $\dfrac{2x}{x + 1} - 3$.

Solution

$$\frac{2x}{x + 1} - 3 = \frac{2x}{x + 1} - 3\left(\frac{x + 1}{x + 1}\right)$$

$$= \frac{2x}{x + 1} - \frac{3(x + 1)}{x + 1}$$

$$= \frac{2x - 3x - 3}{x + 1}$$

$$= \frac{-x - 3}{x + 1}$$

Problem Set 4.3

Perform the following indicated operations involving rational numbers. Be sure to express answers in reduced form.

1. $\dfrac{1}{4} + \dfrac{5}{6}$

2. $\dfrac{3}{5} + \dfrac{1}{6}$

3. $\dfrac{7}{8} - \dfrac{3}{5}$

4. $\dfrac{7}{9} - \dfrac{1}{6}$

5. $\dfrac{6}{5} + \dfrac{1}{-4}$

6. $\dfrac{7}{8} + \dfrac{5}{-12}$

7. $\dfrac{8}{15} + \dfrac{3}{25}$

8. $\dfrac{5}{9} - \dfrac{11}{12}$

9. $\dfrac{1}{5} + \dfrac{5}{6} - \dfrac{7}{15}$

10. $\dfrac{2}{3} - \dfrac{7}{8} + \dfrac{1}{4}$

11. $\dfrac{1}{3} - \dfrac{1}{4} - \dfrac{3}{14}$

12. $\dfrac{5}{6} - \dfrac{7}{9} - \dfrac{3}{10}$

Add or subtract the following rational expressions as indicated. Be sure to express answers in simplest form.

13. $\dfrac{2x}{x-1} + \dfrac{4}{x-1}$

14. $\dfrac{3x}{2x+1} - \dfrac{5}{2x+1}$

15. $\dfrac{4a}{a+2} + \dfrac{8}{a+2}$

16. $\dfrac{6a}{a-3} - \dfrac{18}{a-3}$

17. $\dfrac{3(y-2)}{7y} + \dfrac{4(y-1)}{7y}$

18. $\dfrac{2x-1}{4x^2} + \dfrac{3(x-2)}{4x^2}$

19. $\dfrac{x-1}{2} + \dfrac{x+3}{3}$

20. $\dfrac{x-2}{4} + \dfrac{x+6}{5}$

21. $\dfrac{2a-1}{4} + \dfrac{3a+2}{6}$

22. $\dfrac{a-4}{6} + \dfrac{4a-1}{8}$

23. $\dfrac{n+2}{6} - \dfrac{n-4}{9}$

24. $\dfrac{2n+1}{9} - \dfrac{n+3}{12}$

25. $\dfrac{3x-1}{3} - \dfrac{5x+2}{5}$

26. $\dfrac{4x-3}{6} - \dfrac{8x-2}{12}$

27. $\dfrac{x-2}{5} - \dfrac{x+3}{6} + \dfrac{x+1}{15}$

28. $\dfrac{x+1}{4} + \dfrac{x-3}{6} - \dfrac{x-2}{8}$

29. $\dfrac{3}{8x} + \dfrac{7}{10x}$

30. $\dfrac{5}{6x} - \dfrac{3}{10x}$

31. $\dfrac{5}{7x} - \dfrac{11}{4y}$

32. $\dfrac{5}{12x} - \dfrac{9}{8y}$

33. $\dfrac{4}{3x} + \dfrac{5}{4y} - 1$

34. $\dfrac{7}{3x} - \dfrac{8}{7y} - 2$

35. $\dfrac{7}{10x^2} + \dfrac{11}{15x}$

36. $\dfrac{7}{12a^2} - \dfrac{5}{16a}$

37. $\dfrac{10}{7n} - \dfrac{12}{4n^2}$

38. $\dfrac{6}{8n^2} - \dfrac{3}{5n}$

39. $\dfrac{3}{n^2} - \dfrac{2}{5n} + \dfrac{4}{3}$

40. $\dfrac{1}{n^2} + \dfrac{3}{4n} - \dfrac{5}{6}$

41. $\dfrac{3}{x} - \dfrac{5}{3x^2} - \dfrac{7}{6x}$

42. $\dfrac{7}{3x^2} - \dfrac{9}{4x} - \dfrac{5}{2x}$

43. $\dfrac{6}{5t^2} - \dfrac{4}{7t^3} + \dfrac{9}{5t^3}$

44. $\dfrac{5}{7t} + \dfrac{3}{4t^2} + \dfrac{1}{14t}$

45. $\dfrac{5b}{24a^2} - \dfrac{11a}{32b}$

46. $\dfrac{9}{14x^2y} - \dfrac{4x}{7y^2}$

47. $\dfrac{7}{9xy^3} - \dfrac{4}{3x} + \dfrac{5}{2y^2}$

48. $\dfrac{7}{16a^2b} + \dfrac{3a}{20b^2}$

49. $\dfrac{2x}{x-1} + \dfrac{3}{x}$

50. $\dfrac{3x}{x-4} - \dfrac{2}{x}$

51. $\dfrac{a-2}{a} - \dfrac{3}{a+4}$

52. $\dfrac{a+1}{a} - \dfrac{2}{a+1}$

53. $\dfrac{-3}{4n+5} - \dfrac{8}{3n+5}$

54. $\dfrac{-2}{n-6} - \dfrac{6}{2n+3}$

55. $\dfrac{-1}{x+4} + \dfrac{4}{7x-1}$

56. $\dfrac{-3}{4x+3} + \dfrac{5}{2x-5}$

57. $\dfrac{7}{3x-5} - \dfrac{5}{2x+7}$

58. $\dfrac{5}{x-1} - \dfrac{3}{2x-3}$

59. $\dfrac{5}{3x-2} + \dfrac{6}{4x+5}$

60. $\dfrac{3}{2x+1} + \dfrac{2}{3x+4}$

61. $\dfrac{3x}{2x+5} + 1$

62. $2 + \dfrac{4x}{3x-1}$

63. $\dfrac{4x}{x-5} - 3$

64. $\dfrac{7x}{x+4} - 2$

65. $-1 - \dfrac{3}{2x+1}$

66. $-2 - \dfrac{5}{4x-3}$

67. Recall that the indicated quotient of a polynomial and its opposite is -1. For example, $\dfrac{x-2}{2-x}$ simplifies to -1. Keep this idea in mind as you add or subtract the following rational expressions.

(a) $\dfrac{1}{x-1} - \dfrac{x}{x-1}$

(b) $\dfrac{3}{2x-3} - \dfrac{2x}{2x-3}$

(c) $\dfrac{4}{x-4} - \dfrac{x}{x-4} + 1$

(d) $-1 + \dfrac{2}{x-2} - \dfrac{x}{x-2}$

68. Consider the addition problem $\dfrac{8}{x-2} + \dfrac{5}{2-x}$. Notice that the denominators are opposites of each other. If the property $\dfrac{a}{-b} = -\dfrac{a}{b}$ is applied to the second fraction, we have $\dfrac{5}{2-x} = -\dfrac{5}{x-2}$. Thus, we proceed as follows.

$$\dfrac{8}{x-2} + \dfrac{5}{2-x} = \dfrac{8}{x-2} - \dfrac{5}{x-2} = \dfrac{8-5}{x-2} = \dfrac{3}{x-2}$$

Use this approach to do the following problems.

(a) $\dfrac{7}{x-1} + \dfrac{2}{1-x}$

(b) $\dfrac{5}{2x-1} + \dfrac{8}{1-2x}$

(c) $\dfrac{4}{a-3} - \dfrac{1}{3-a}$

(d) $\dfrac{10}{a-9}-\dfrac{5}{9-a}$

(e) $\dfrac{x^2}{x-1}-\dfrac{2x-3}{1-x}$

(f) $\dfrac{x^2}{x-4}-\dfrac{3x-28}{4-x}$

THOUGHTS INTO WORDS

69. What is the difference between the concept of least common multiple and the concept of least common denominator?

70. A classmate tells you that she finds the least common multiple of two counting numbers by list-

ing the multiples of each number and then choosing the smallest number that appears in both lists. Is this a correct procedure? What is the weakness of this procedure?

71. For which real numbers does $\dfrac{x}{x-3}+\dfrac{4}{x}$ equal $\dfrac{(x+6)(x-2)}{x(x-3)}$? Explain your answer.

72. Suppose that your friend does an addition problem as follows:

$$\dfrac{5}{8}+\dfrac{7}{12}=\dfrac{5(12)+8(7)}{8(12)}=\dfrac{60+56}{96}=\dfrac{116}{96}=\dfrac{29}{24}$$

Is this answer correct? What advice would you offer your friend?

4.4   More on Rational Expressions and Complex Fractions

In this section we expand our work with adding and subtracting rational expressions and we discuss the process of simplifying complex fractions. Before we begin, however, this seems like an appropriate time to offer a bit of advice regarding your study of algebra. Success in algebra depends upon having a good understanding of the concepts as well as the ability to perform the various computations. As for the computational work, you should adopt a carefully organized format that shows as many steps as *you need* in order to minimize the chances of making careless errors. Don't be eager to find shortcuts for certain computations before you have a thorough understanding of the steps involved in the process. This advice is especially appropriate at the beginning of this section.

Study the following examples very carefully. Notice the same basic procedure for each problem: (1) find the LCD, (2) change each fraction to an equivalent fraction that has the LCD as its denominator, (3) add or subtract numerators and place this result over the LCD, and (4) look for possibilities to simplify the resulting fraction.

Example 1

Add $\dfrac{8}{x^2-4x}+\dfrac{2}{x}$.

Solution

$$\left.\begin{array}{r} x^2-4x=x(x-4) \\ x=x \end{array}\right\} \longrightarrow \text{LCD is } x(x-4)$$

$$\dfrac{8}{x^2-4x}+\dfrac{2}{x}=\dfrac{8}{x(x-4)}+\dfrac{2}{x}$$

$$= \frac{8}{x(x-4)} + \left(\frac{2}{x}\right)\left(\frac{x-4}{x-4}\right)$$

$$= \frac{8}{x(x-4)} + \frac{2(x-4)}{x(x-4)}$$

$$= \frac{8+2x-8}{x(x-4)}$$

$$= \frac{2x}{x(x-4)} = \frac{2}{x-4}$$

Example 2

Subtract $\dfrac{a}{a^2-4} - \dfrac{3}{a+2}$.

Solution

$$\left.\begin{array}{l} a^2 - 4 = (a+2)(a-2) \\ a+2 = a+2 \end{array}\right\} \longrightarrow \text{LCD is } (a+2)(a-2)$$

$$\frac{a}{a^2-4} - \frac{3}{a+2} = \frac{a}{(a+2)(a-2)} - \frac{3}{a+2}$$

$$= \frac{a}{(a+2)(a-2)} - \left(\frac{3}{a+2}\right)\left(\frac{a-2}{a-2}\right)$$

$$= \frac{a}{(a+2)(a-2)} - \frac{3(a-2)}{(a+2)(a-2)}$$

$$= \frac{a-3a+6}{(a+2)(a-2)}$$

$$= \frac{-2a+6}{(a+2)(a-2)} \quad \text{or} \quad \frac{-2(a-3)}{(a+2)(a-2)}$$

Example 3

Add $\dfrac{3n}{n^2+6n+5} + \dfrac{4}{n^2-7n-8}$.

Solution

$$\left.\begin{array}{l} n^2 + 6n + 5 = (n+5)(n+1) \\ n^2 - 7n - 8 = (n-8)(n+1) \end{array}\right\} \longrightarrow \text{LCD is } (n+1)(n+5)(n-8)$$

$$\frac{3n}{n^2+6n+5} + \frac{4}{n^2-7n-8} = \frac{3n}{(n+5)(n+1)} + \frac{4}{(n-8)(n+1)}$$

$$= \left(\frac{3n}{(n+5)(n+1)}\right)\left(\frac{n-8}{n-8}\right) + \left(\frac{4}{(n-8)(n+1)}\right)\left(\frac{n+5}{n+5}\right)$$

$$= \frac{3n(n-8)}{(n+5)(n+1)(n-8)} + \frac{4(n+5)}{(n+5)(n+1)(n-8)}$$

$$= \frac{3n^2 - 24n + 4n + 20}{(n+5)(n+1)(n-8)}$$

$$= \frac{3n^2 - 20n + 20}{(n+5)(n+1)(n-8)}$$

Example 4

Perform the indicated operations.

$$\frac{2x^2}{x^4 - 1} + \frac{x}{x^2 - 1} - \frac{1}{x - 1}.$$

Solution

$$\left.\begin{array}{l} x^4 - 1 = (x^2 + 1)(x + 1)(x - 1) \\ x^2 - 1 = (x + 1)(x - 1) \\ x - 1 = x - 1 \end{array}\right\} \longrightarrow \text{LCD is } (x^2 + 1)(x + 1)(x - 1)$$

$$\frac{2x^2}{x^4 - 1} + \frac{x}{x^2 - 1} - \frac{1}{x - 1} = \frac{2x^2}{(x^2 + 1)(x + 1)(x - 1)} + \frac{x}{(x + 1)(x - 1)} - \frac{1}{x - 1}$$

$$= \frac{2x^2}{(x^2 + 1)(x + 1)(x - 1)} + \left(\frac{x}{(x + 1)(x - 1)}\right)\left(\frac{x^2 + 1}{x^2 + 1}\right) - \left(\frac{1}{x - 1}\right)\left(\frac{(x^2 + 1)(x + 1)}{(x^2 + 1)(x + 1)}\right)$$

$$= \frac{2x^2}{(x^2 + 1)(x + 1)(x - 1)} + \frac{x(x^2 + 1)}{(x^2 + 1)(x + 1)(x - 1)} - \frac{(x^2 + 1)(x + 1)}{(x^2 + 1)(x + 1)(x - 1)}$$

$$= \frac{2x^2 + x^3 + x - x^3 - x^2 - x - 1}{(x^2 + 1)(x + 1)(x - 1)}$$

$$= \frac{x^2 - 1}{(x^2 + 1)(x + 1)(x - 1)}$$

$$= \frac{(x + 1)(x - 1)}{(x^2 + 1)(x + 1)(x - 1)}$$

$$= \frac{1}{x^2 + 1}$$

▲

Complex Fractions

Complex fractions are fractional forms that contain rational numbers or rational expressions in the numerators and/or denominators. The following are examples of complex fractions.

$$\frac{\dfrac{4}{x}}{\dfrac{2}{xy}}, \qquad \frac{\dfrac{1}{2} + \dfrac{3}{4}}{\dfrac{5}{6} - \dfrac{3}{8}}, \qquad \frac{\dfrac{3}{x} + \dfrac{2}{y}}{\dfrac{5}{x} - \dfrac{6}{y^2}}, \qquad \frac{\dfrac{1}{x} + \dfrac{1}{y}}{2}, \qquad \frac{-3}{\dfrac{2}{x} - \dfrac{3}{y}}$$

It is often necessary to **simplify** a complex fraction. We will take each of the five examples above and examine some techniques for simplifying complex fractions.

Example 5

Simplify $\dfrac{\dfrac{4}{x}}{\dfrac{2}{xy}}$.

Solution

This type of problem is a simple division problem.

$$\frac{\dfrac{4}{x}}{\dfrac{2}{xy}} = \frac{4}{x} \div \frac{2}{xy}$$

$$= \frac{\overset{2}{\cancel{4}}}{\cancel{x}} \cdot \frac{\cancel{x}y}{\cancel{2}} = 2y$$

▲

Example 6

Simplify $\dfrac{\dfrac{1}{2} + \dfrac{3}{4}}{\dfrac{5}{6} - \dfrac{3}{8}}$.

Solution A

Let's look at two possible "attacks" for such a problem.

$$\frac{\dfrac{1}{2} + \dfrac{3}{4}}{\dfrac{5}{6} - \dfrac{3}{8}} = \frac{\dfrac{2}{4} + \dfrac{3}{4}}{\dfrac{20}{24} - \dfrac{9}{24}}$$

$$= \frac{\dfrac{5}{4}}{\dfrac{11}{24}} = \frac{5}{\cancel{4}} \cdot \frac{\overset{6}{\cancel{24}}}{11}$$

$$= \frac{30}{11}$$

Solution B

The LCD of all four denominators (2, 4, 6, and 8) is 24. Multiply the entire complex fraction by a form of 1, namely, $\dfrac{24}{24}$.

$$\frac{\dfrac{1}{2} + \dfrac{3}{4}}{\dfrac{5}{6} - \dfrac{3}{8}} = \left(\frac{24}{24}\right) \frac{\dfrac{1}{2} + \dfrac{3}{4}}{\dfrac{5}{6} - \dfrac{3}{8}}$$

$$= \frac{24\left(\dfrac{1}{2} + \dfrac{3}{4}\right)}{24\left(\dfrac{5}{6} - \dfrac{3}{8}\right)}$$

$$= \frac{24\left(\frac{1}{2}\right) + 24\left(\frac{3}{4}\right)}{24\left(\frac{5}{6}\right) - 24\left(\frac{3}{8}\right)}$$

$$= \frac{12 + 18}{20 - 9} = \frac{30}{11}$$

▲

Example 7

Simplify $\dfrac{\dfrac{3}{x} + \dfrac{2}{y}}{\dfrac{5}{x} - \dfrac{6}{y^2}}$.

Solution A

$$\frac{\dfrac{3}{x} + \dfrac{2}{y}}{\dfrac{5}{x} - \dfrac{6}{y^2}} = \frac{\left(\dfrac{3}{x}\right)\left(\dfrac{y}{y}\right) + \left(\dfrac{2}{y}\right)\left(\dfrac{x}{x}\right)}{\left(\dfrac{5}{x}\right)\left(\dfrac{y^2}{y^2}\right) - \left(\dfrac{6}{y^2}\right)\left(\dfrac{x}{x}\right)}$$

$$= \frac{\dfrac{3y}{xy} + \dfrac{2x}{xy}}{\dfrac{5y^2}{xy^2} - \dfrac{6x}{xy^2}}$$

$$= \frac{\dfrac{3y + 2x}{xy}}{\dfrac{5y^2 - 6x}{xy^2}}$$

$$= \frac{3y + 2x}{xy} \div \frac{5y^2 - 6x}{xy^2}$$

$$= \frac{3y + 2x}{\cancel{xy}} \cdot \frac{\overset{y}{\cancel{xy^2}}}{5y^2 - 6x}$$

$$= \frac{y(3y + 2x)}{5y^2 - 6x}$$

Solution B

The LCD of all four denominators (x, y, x, and y^2) is xy^2. Multiply the entire complex fraction by a form of 1, namely, $\dfrac{xy^2}{xy^2}$.

$$\frac{\dfrac{3}{x} + \dfrac{2}{y}}{\dfrac{5}{x} - \dfrac{6}{y^2}} = \left(\frac{xy^2}{xy^2}\right) \frac{\dfrac{3}{x} + \dfrac{2}{y}}{\dfrac{5}{x} - \dfrac{6}{y^2}}$$

$$= \frac{xy^2\left(\dfrac{3}{x} + \dfrac{2}{y}\right)}{xy^2\left(\dfrac{5}{x} - \dfrac{6}{y^2}\right)}$$

$$= \frac{xy^2\left(\dfrac{3}{x}\right) + xy^2\left(\dfrac{2}{y}\right)}{xy^2\left(\dfrac{5}{x}\right) - xy^2\left(\dfrac{6}{y^2}\right)}$$

$$= \frac{3y^2 + 2xy}{5y^2 - 6x} \quad \text{or} \quad \frac{y(3y + 2x)}{5y^2 - 6x} \quad \blacktriangle$$

Certainly either approach (Solution A or Solution B) will work with problems such as Examples 6 and 7. Examine Solution B in both examples carefully. This approach works effectively with complex fractions where the LCD of all the denominators is easy to find. (Don't be misled by the length of Solution B for Example 6; we were especially careful to show every step.)

Example 8

Simplify $\dfrac{\dfrac{1}{x} + \dfrac{1}{y}}{2}$.

Solution

The number 2 can be written as $\dfrac{2}{1}$; thus, the LCD of all three denominators (x, y, and 1) is xy. So let's multiply the entire complex fraction by a form of 1, namely, $\dfrac{xy}{xy}$.

$$\left(\frac{\dfrac{1}{x} + \dfrac{1}{y}}{\dfrac{2}{1}}\right)\left(\frac{xy}{xy}\right) = \frac{xy\left(\dfrac{1}{x}\right) + xy\left(\dfrac{1}{y}\right)}{2xy} = \frac{y + x}{2xy} \quad \blacktriangle$$

Example 9

Simplify $\dfrac{-3}{\dfrac{2}{x} - \dfrac{3}{y}}$.

Solution

$$\left(\frac{\dfrac{-3}{1}}{\dfrac{2}{x} - \dfrac{3}{y}}\right)\left(\frac{xy}{xy}\right) = \frac{-3(xy)}{xy\left(\dfrac{2}{x}\right) - xy\left(\dfrac{3}{y}\right)} = \frac{-3xy}{2y - 3x} \quad \blacktriangle$$

Let's conclude this section with an example that has a complex fraction as part of an algebraic expression.

Example 10

Simplify $1 - \dfrac{n}{1 - \dfrac{1}{n}}$.

Solution

First simplify the complex fraction $\dfrac{n}{1 - \dfrac{1}{n}}$ by multiplying by $\dfrac{n}{n}$.

$$\left(\dfrac{n}{1 - \dfrac{1}{n}}\right)\left(\dfrac{n}{n}\right) = \dfrac{n^2}{n - 1}$$

Now we can perform the subtraction.

$$1 - \dfrac{n^2}{n - 1} = \left(\dfrac{n - 1}{n - 1}\right)\left(\dfrac{1}{1}\right) - \dfrac{n^2}{n - 1}$$

$$= \dfrac{n - 1}{n - 1} - \dfrac{n^2}{n - 1}$$

$$= \dfrac{n - 1 - n^2}{n - 1} \quad \text{or} \quad \dfrac{-n^2 + n - 1}{n - 1}$$

Problem Set 4.4

Perform the indicated operations and express your answers in simplest form.

1. $\dfrac{2x}{x^2 + 4x} + \dfrac{5}{x}$

2. $\dfrac{3x}{x^2 - 6x} + \dfrac{4}{x}$

3. $\dfrac{4}{x^2 + 7x} - \dfrac{1}{x}$

4. $\dfrac{-10}{x^2 - 9x} - \dfrac{2}{x}$

5. $\dfrac{x}{x^2 - 1} + \dfrac{5}{x + 1}$

6. $\dfrac{2x}{x^2 - 16} + \dfrac{7}{x - 4}$

7. $\dfrac{6a + 4}{a^2 - 1} - \dfrac{5}{a - 1}$

8. $\dfrac{4a - 4}{a^2 - 4} - \dfrac{3}{a + 2}$

9. $\dfrac{2n}{n^2 - 25} - \dfrac{3}{4n + 20}$

10. $\dfrac{3n}{n^2 - 36} - \dfrac{2}{5n + 30}$

11. $\dfrac{5}{x} - \dfrac{5x - 30}{x^2 + 6x} + \dfrac{x}{x + 6}$

12. $\dfrac{3}{x + 1} + \dfrac{x + 5}{x^2 - 1} - \dfrac{3}{x - 1}$

13. $\dfrac{3}{x^2 + 9x + 14} + \dfrac{5}{2x^2 + 15x + 7}$

14. $\dfrac{6}{x^2 + 11x + 24} + \dfrac{4}{3x^2 + 13x + 12}$

15. $\dfrac{1}{a^2 - 3a - 10} - \dfrac{4}{a^2 + 4a - 45}$

16. $\dfrac{6}{a^2 - 3a - 54} - \dfrac{10}{a^2 + 5a - 6}$

17. $\dfrac{3a}{20a^2 - 11a - 3} + \dfrac{1}{12a^2 + 7a - 12}$

18. $\dfrac{2a}{6a^2 + 11a - 10} + \dfrac{a}{2a^2 - 3a - 20}$

19. $\dfrac{5}{x^2 + 3} - \dfrac{2}{x^2 + 4x - 21}$

20. $\dfrac{7}{x^2 + 1} - \dfrac{3}{x^2 + 7x - 60}$

21. $\dfrac{2}{y^2 + 6y - 16} - \dfrac{4}{y + 8} - \dfrac{3}{y - 2}$

22. $\dfrac{7}{y - 6} - \dfrac{10}{y + 12} + \dfrac{4}{y^2 + 6y - 72}$

23. $x - \dfrac{x^2}{x - 2} + \dfrac{3}{x^2 - 4}$

24. $x + \dfrac{5}{x^2 - 25} - \dfrac{x^2}{x + 5}$

25. $\dfrac{x + 3}{x + 10} + \dfrac{4x - 3}{x^2 + 8x - 20} + \dfrac{x - 1}{x - 2}$

26. $\dfrac{2x - 1}{x + 3} + \dfrac{x + 4}{x - 6} + \dfrac{3x - 1}{x^2 - 3x - 18}$

27. $\dfrac{n}{n - 6} + \dfrac{n + 3}{n + 8} + \dfrac{12n + 26}{n^2 + 2n - 48}$

28. $\dfrac{n - 1}{n + 4} + \dfrac{n}{n + 6} + \dfrac{2n + 18}{n^2 + 10n + 24}$

29. $\dfrac{4x - 3}{2x^2 + x - 1} - \dfrac{2x + 7}{3x^2 + x - 2} - \dfrac{3}{3x - 2}$

30. $\dfrac{2x + 5}{x^2 + 3x - 18} - \dfrac{3x - 1}{x^2 + 4x - 12} + \dfrac{5}{x - 2}$

31. $\dfrac{n}{n^2 + 1} + \dfrac{n^2 + 3n}{n^4 - 1} - \dfrac{1}{n - 1}$

32. $\dfrac{2n^2}{n^4 - 16} - \dfrac{n}{n^2 - 4} + \dfrac{1}{n + 2}$

33. $\dfrac{15x^2 - 10}{5x^2 - 7x + 2} - \dfrac{3x + 4}{x - 1} - \dfrac{2}{5x - 2}$

34. $\dfrac{32x + 9}{12x^2 + x - 6} - \dfrac{3}{4x + 3} - \dfrac{x + 5}{3x - 2}$

35. $\dfrac{t + 3}{3t - 1} + \dfrac{8t^2 + 8t + 2}{3t^2 - 7t + 2} - \dfrac{2t + 3}{t - 2}$

36. $\dfrac{t - 3}{2t + 1} + \dfrac{2t^2 + 19t - 46}{2t^2 - 9t - 5} - \dfrac{t + 4}{t - 5}$

Simplify each of the following complex fractions.

37. $\dfrac{\dfrac{1}{2} - \dfrac{1}{4}}{\dfrac{5}{8} + \dfrac{3}{4}}$

38. $\dfrac{\dfrac{3}{8} + \dfrac{3}{4}}{\dfrac{5}{8} - \dfrac{7}{12}}$

39. $\dfrac{\dfrac{3}{28} - \dfrac{5}{14}}{\dfrac{5}{7} + \dfrac{1}{4}}$

40. $\dfrac{\dfrac{5}{9} + \dfrac{7}{36}}{\dfrac{3}{18} - \dfrac{5}{12}}$

41. $\dfrac{\dfrac{5}{6y}}{\dfrac{10}{3xy}}$

42. $\dfrac{\dfrac{9}{8xy^2}}{\dfrac{5}{4x^2}}$

43. $\dfrac{\dfrac{3}{x} - \dfrac{2}{y}}{\dfrac{4}{y} - \dfrac{7}{xy}}$

44. $\dfrac{\dfrac{9}{x} + \dfrac{7}{x^2}}{\dfrac{5}{y} + \dfrac{3}{y^2}}$

45. $\dfrac{\dfrac{6}{a} - \dfrac{5}{b^2}}{\dfrac{12}{a^2} + \dfrac{2}{b}}$

46. $\dfrac{\dfrac{4}{ab} - \dfrac{3}{b^2}}{\dfrac{1}{a} + \dfrac{3}{b}}$

47. $\dfrac{\dfrac{2}{x} - 3}{\dfrac{3}{y} + 4}$

48. $\dfrac{1 + \dfrac{3}{x}}{1 - \dfrac{6}{x}}$

49. $\dfrac{3 + \dfrac{2}{n+4}}{5 - \dfrac{1}{n+4}}$

50. $\dfrac{4 + \dfrac{6}{n-1}}{7 - \dfrac{4}{n-1}}$

51. $\dfrac{5 - \dfrac{2}{n-3}}{4 - \dfrac{1}{n-3}}$

52. $\dfrac{\dfrac{3}{n-5} - 2}{1 - \dfrac{4}{n-5}}$

53. $\dfrac{\dfrac{-1}{y-2} + \dfrac{5}{x}}{\dfrac{3}{x} - \dfrac{4}{xy-2x}}$

54. $\dfrac{\dfrac{-2}{x} - \dfrac{4}{x+2}}{\dfrac{3}{x^2+2x} + \dfrac{3}{x}}$

55. $\dfrac{\dfrac{2}{x-3} - \dfrac{3}{x+3}}{\dfrac{5}{x^2-9} - \dfrac{2}{x-3}}$

56. $\dfrac{\dfrac{2}{x-y} + \dfrac{3}{x+y}}{\dfrac{5}{x+y} - \dfrac{1}{x^2-y^2}}$

57. $\dfrac{3a}{2 - \dfrac{1}{a}} - 1$

58. $\dfrac{a}{\dfrac{1}{a} + 4} + 1$

59. $2 - \dfrac{x}{3 - \dfrac{2}{x}}$

60. $1 + \dfrac{x}{1 + \dfrac{1}{x}}$

THOUGHTS INTO WORDS

61. Which of the two techniques presented in the text would you use to simplify $\dfrac{\dfrac{1}{4} + \dfrac{1}{3}}{\dfrac{3}{4} - \dfrac{1}{6}}$? Which technique would you use to simplify $\dfrac{\dfrac{3}{8} - \dfrac{5}{7}}{\dfrac{7}{9} + \dfrac{6}{25}}$? Explain your choice for each problem.

62. Give a step-by-step description of how to do the following addition problem.

$$\dfrac{3x+4}{8} + \dfrac{5x-2}{12}$$

4.5 Dividing Polynomials

In Chapter 3 we saw how the property $\frac{b^n}{b^m} = b^{n-m}$ along with our knowledge of dividing integers was used to divide monomials. For example,

$$\frac{12x^3}{3x} = 4x^2, \qquad \frac{-36x^4y^5}{4xy^2} = -9x^3y^3$$

In Section 4.3, we used $\frac{a}{b} + \frac{c}{b} = \frac{a+c}{b}$ and $\frac{a}{b} - \frac{c}{b} = \frac{a-c}{b}$ as the basis for adding and subtracting rational expressions. These same equalities, viewed as $\frac{a+b}{c} = \frac{a}{c} + \frac{b}{c}$ and $\frac{a-c}{b} = \frac{a}{b} - \frac{c}{b}$, along with our knowledge of dividing monomials, provide the basis for dividing polynomials by monomials. Consider the following examples.

$$\frac{18x^3 + 24x^2}{6x} = \frac{18x^3}{6x} + \frac{24x^2}{6x} = 3x^2 + 4x,$$

$$\frac{35x^2y^3 - 55x^3y^4}{5xy^2} = \frac{35x^2y^3}{5xy^2} - \frac{55x^3y^4}{5xy^2} = 7xy - 11x^2y^2$$

To divide a polynomial by a monomial we divide each term of the polynomial by the monomial. As with many skills, once you feel comfortable with the process you may then want to perform some of the steps mentally. Your work could take on the following format.

$$\frac{40x^4y^5 + 72x^5y^7}{8x^2y} = 5x^2y^4 + 9x^3y^6, \qquad \frac{36a^3b^4 - 45a^4b^6}{-9a^2b^3} = -4ab + 5a^2b^3$$

In Section 4.1 we saw that a fraction like $\frac{3x^2 + 11x - 4}{x + 4}$ can be simplified as follows.

$$\frac{3x^2 + 11x - 4}{x + 4} = \frac{(3x - 1)(x + 4)}{x + 4} = 3x - 1$$

We can obtain the same result by using a dividing process similar to long division in arithmetic. The process is as follows.

STEP 1 Use the conventional long division format and arrange both the dividend and the divisor in descending powers of the variable.

$$x + 4)\overline{3x^2 + 11x - 4}$$

STEP 2 Find the first term of the quotient by dividing the first term of the dividend by the first term of the divisor.

$$\frac{3x}{x + 4)\overline{3x^2 + 11x - 4}}$$

STEP 3   Multiply the entire divisor by the term of the quotient found in Step 2 and position the product to be subtracted from the dividend.

$$\begin{array}{r} 3x \\ x + 4\overline{)3x^2 + 11x - 4} \\ 3x^2 + 12x \end{array}$$

STEP 4   Subtract

$$\begin{array}{r} 3x \\ x + 4\overline{)3x^2 + 11x - 4} \\ 3x^2 + 12x \end{array}$$

Remember to add the opposite! $\longrightarrow$

$(3x^2 + 11x - 4) - (3x^2 + 12x) = -x - 4$ $\longrightarrow$ $-x - 4$

STEP 5   Repeat the process beginning with Step 2; use the polynomial that resulted from the subtraction in Step 4 as a new dividend.

$$\begin{array}{r} 3x \quad - 1 \\ x + 4\overline{)3x^2 + 11x - 4} \\ 3x^2 + 12x \\ \hline -x - 4 \\ -x - 4 \end{array}$$

In the next example let's **think** in terms of the previous step-by-step procedure but arrange our work in a more compact form.

Example 1   Divide $5x^2 + 6x - 8$ by $x + 2$.

Solution

$$\begin{array}{r} 5x \quad - \quad 4 \\ x + 2\overline{)5x^2 + \quad 6x - 8} \\ 5x^2 + 10x \\ \hline - \quad 4x - 8 \\ - \quad 4x - 8 \\ \hline 0 \end{array}$$

Think Steps

1. $\dfrac{5x^2}{x} = 5x$.

2. $5x(x + 2) = 5x^2 + 10x$.

3. $(5x^2 + 6x - 8) - (5x^2 + 10x) = -4x - 8$.

4. $\dfrac{-4x}{x} = -4$.

5. $-4(x + 2) = -4x - 8$.   ▲

Recall that to check a division problem we can multiply the divisor times the quotient and add the remainder. In other words,

Dividend = (Divisor)(Quotient) + (Remainder).

Sometimes the remainder is expressed as a fractional part of the divisor. The relationship then becomes

$$\frac{\text{Dividend}}{\text{Divisor}} = \text{Quotient} + \frac{\text{Remainder}}{\text{Divisor}}.$$

Example 2   Divide $2x^2 - 3x + 1$ by $x - 5$.

Solution

$$\begin{array}{r} 2x + 7 \\ x - 5\overline{\smash{\big)}\ 2x^2 - 3x + 1} \\ \underline{2x^2 - 10x} \\ 7x + 1 \\ \underline{7x - 35} \\ 36 \end{array}$$ ⟵——————— Remainder

Thus,

$$\frac{2x^2 - 3x + 1}{x - 5} = 2x + 7 + \frac{36}{x - 5}, \qquad x \neq 5.$$

✔ *Check*

$$(x - 5)(2x + 7) + 36 \overset{?}{=} 2x^2 - 3x + 1$$
$$2x^2 - 3x - 35 + 36 \overset{?}{=} 2x^2 - 3x + 1$$
$$2x^2 - 3x + 1 = 2x^2 - 3x + 1 \qquad\qquad ▲$$

Each of the next two examples illustrates another point regarding the division process. Study them carefully and then you should be ready to work the exercises in the next problem set.

Example 3

Divide $t^3 - 8$ by $t - 2$.

Solution

$$\begin{array}{r} t^2 + 2t + 4 \\ t - 2\overline{\smash{\big)}\ t^3 + 0t^2 + 0t - 8} \\ \underline{t^3 - 2t^2} \\ 2t^2 + 0t - 8 \\ \underline{2t^2 - 4t} \\ 4t - 8 \\ \underline{4t - 8} \\ 0 \end{array}$$ ⟵————— Notice the insertion of a
"t squared" and a "t-term"
with zero coefficients

Check this result! ▲

Example 4

Divide $y^3 + 3y^2 - 2y - 1$ by $y^2 + 2y$.

Solution

$$\begin{array}{r} y + 1 \\ y^2 + 2y\overline{\smash{\big)}\ y^3 + 3y^2 - 2y - 1} \\ \underline{y^3 + 2y^2} \\ y^2 - 2y - 1 \\ \underline{y^2 + 2y} \\ -4y - 1 \end{array}$$ ⟵————— Remainder of $-4y - 1$

(The division process is complete when the degree of the remainder is less than the degree of the divisor.) Thus,

$$\frac{y^3 + 3y^2 - 2y - 1}{y^2 + 2y} = y + 1 + \frac{-4y - 1}{y^2 + 2y}.$$

▲

REMARK   If the divisor is of the form $x - k$, where the coefficient of the x-term is one, then the format of the division process described in this section can be simplified by a procedure called **synthetic division**. This procedure is outlined later in the text.

△

▼ Problem Set 4.5

Perform the following divisions of polynomials by monomials.

1. $\dfrac{9x^4 + 18x^3}{3x}$

2. $\dfrac{12x^3 - 24x^2}{6x^2}$

3. $\dfrac{-24x^6 + 36x^8}{4x^2}$

4. $\dfrac{-35x^5 - 42x^3}{-7x^2}$

5. $\dfrac{15a^3 - 25a^2 - 40a}{5a}$

6. $\dfrac{-16a^4 + 32a^3 - 56a^2}{-8a}$

7. $\dfrac{13x^3 - 17x^2 + 28x}{-x}$

8. $\dfrac{14xy - 16x^2y^2 - 20x^3y^4}{-xy}$

9. $\dfrac{-18x^2y^2 + 24x^3y^2 - 48x^2y^3}{6xy}$

10. $\dfrac{-27a^3b^4 - 36a^2b^3 + 72a^2b^5}{9a^2b^2}$

Perform the following divisions.

11. $\dfrac{x^2 - 7x - 78}{x + 6}$

12. $\dfrac{x^2 + 11x - 60}{x - 4}$

13. $(x^2 + 12x - 160) \div (x - 8)$

14. $(x^2 - 18x - 175) \div (x + 7)$

15. $\dfrac{2x^2 - x - 4}{x - 1}$

16. $\dfrac{3x^2 - 2x - 7}{x + 2}$

17. $\dfrac{15x^2 + 22x - 5}{3x + 5}$

18. $\dfrac{12x^2 - 32x - 35}{2x - 7}$

19. $\dfrac{3x^3 + 7x^2 - 13x - 21}{x + 3}$

20. $\dfrac{4x^3 - 21x^2 + 3x + 10}{x - 5}$

21. $(2x^3 + 9x^2 - 17x + 6) \div (2x - 1)$

22. $(3x^3 - 5x^2 - 23x - 7) \div (3x + 1)$

23. $(4x^3 - x^2 - 2x + 6) \div (x - 2)$

24. $(6x^3 - 2x^2 + 4x - 3) \div (x + 1)$

25. $(x^4 - 10x^3 + 19x^2 + 33x - 18) \div (x - 6)$

26. $(x^4 + 2x^3 - 16x^2 + x + 6) \div (x - 3)$

27. $\dfrac{x^3 - 125}{x - 5}$

28. $\dfrac{x^3 + 64}{x + 4}$

29. $(x^3 + 64) \div (x + 1)$

30. $(x^3 - 8) \div (x - 4)$

31. $(2x^3 - x - 6) \div (x + 2)$

32. $(5x^3 + 2x - 3) \div (x - 2)$

33. $\dfrac{4a^2 - 8ab + 4b^2}{a - b}$

34. $\dfrac{3x^2 - 2xy - 8y^2}{x - 2y}$

35. $\dfrac{4x^3 - 5x^2 + 2x - 6}{x^2 - 3x}$

36. $\dfrac{3x^3 + 2x^2 - 5x - 1}{x^2 + 2x}$

37. $\dfrac{8y^3 - y^2 - y + 5}{y^2 + y}$

38. $\dfrac{5y^3 - 6y^2 - 7y - 2}{y^2 - y}$

39. $(2x^3 + x^2 - 3x + 1) \div (x^2 + x - 1)$

40. $(3x^3 - 4x^2 + 8x + 8) \div (x^2 - 2x + 4)$

41. $(4x^3 - 13x^2 + 8x - 15) \div (4x^2 - x + 5)$

42. $(5x^3 + 8x^2 - 5x - 2) \div (5x^2 - 2x - 1)$

43. $(5a^3 + 7a^2 - 2a - 9) \div (a^2 + 3a - 4)$

44. $(4a^3 - 2a^2 + 7a - 1) \div (a^2 - 2a + 3)$

45. $(2n^4 + 3n^3 - 2n^2 + 3n - 4) \div (n^2 + 1)$

46. $(3n^4 + n^3 - 7n^2 - 2n + 2) \div (n^2 - 2)$

47. $(x^5 - 1) \div (x - 1)$

48. $(x^5 + 1) \div (x + 1)$

49. $(x^4 - 1) \div (x + 1)$

50. $(x^4 - 1) \div (x - 1)$

51. $(3x^4 + x^3 - 2x^2 - x + 6) \div (x^2 - 1)$

52. $(4x^3 - 2x^2 + 7x - 5) \div (x^2 + 2)$

THOUGHTS INTO WORDS

53. Describe the process of long division of polynomials.

54. Give a step-by-step description of how you would do the following division problem.

$$(4 - 3x - 7x^3) \div (x + 6)$$

55. How do you know by inspection that $3x^2 + 5x + 1$ cannot be the correct answer for the division problem $(3x^3 - 7x^2 - 22x + 8) \div (x - 4)$?

4.6 Fractional Equations

The fractional equations used in this text are of two basic types. One type has only constants as denominators and the other type contains variables in the denominators.

In Chapter 2 we considered fractional equations that involve only constants in the denominators. Let's briefly review our approach to solving such equations as we will be using that same basic technique to solve any type of fractional equation.

Example 1

Solve $\dfrac{x-2}{3} + \dfrac{x+1}{4} = \dfrac{1}{6}$.

Solution

$$\frac{x-2}{3} + \frac{x+1}{4} = \frac{1}{6}$$

$$12\left(\frac{x-2}{3} + \frac{x+1}{4}\right) = 12\left(\frac{1}{6}\right) \qquad \text{Multiply both sides by 12, which is the LCD of all of the denominators.}$$

$$4(x-2) + 3(x+1) = 2$$

$$4x - 8 + 3x + 3 = 2$$

$$7x - 5 = 2$$

$$7x = 7$$

$$x = 1$$

The solution set is $\{1\}$. Check it!   ▲

    If an equation contains a variable (or variables) in one or more denominators, then we proceed in essentially the same way as in Example 1 above, **except we must avoid any value of the variable that makes a denominator zero.** Consider the following examples.

Example 2

Solve $\dfrac{5}{n} + \dfrac{1}{2} = \dfrac{9}{n}$.

Solution

First, we need to realize that n **cannot equal zero.** (Let's indicate this restriction so that it is not forgotten!) Then we can proceed as follows.

$$\frac{5}{n} + \frac{1}{2} = \frac{9}{n}, \qquad n \neq 0$$

$$2n\left(\frac{5}{n} + \frac{1}{2}\right) = 2n\left(\frac{9}{n}\right) \qquad \text{Multiply both sides by the LCD, which is } 2n.$$

$$10 + n = 18$$

$$n = 8$$

The solution set is $\{8\}$. Check it!   ▲

Example 3

Solve $\dfrac{35-x}{x} = 7 + \dfrac{3}{x}$.

Solution

$$\frac{35-x}{x} = 7 + \frac{3}{x}, \qquad x \neq 0$$

$$x\left(\frac{35-x}{x}\right) = x\left(7 + \frac{3}{x}\right) \qquad \text{Multiply both sides by } x.$$

$$35 - x = 7x + 3$$

$$32 = 8x$$
$$4 = x$$

The solution set is $\{4\}$. ▲

Example 4

Solve $\dfrac{3}{a-2} = \dfrac{4}{a+1}$.

Solution

$$\frac{3}{a-2} = \frac{4}{a+1}, \qquad a \neq 2 \text{ and } a \neq -1$$

$$(a-2)(a+1)\left(\frac{3}{a-2}\right) = (a-2)(a+1)\left(\frac{4}{a+1}\right) \qquad \text{Multiply both sides by } (a-2)(a+1).$$

$$3(a+1) = 4(a-2)$$
$$3a + 3 = 4a - 8$$
$$11 = a$$

The solution set is $\{11\}$. ▲

Keep in mind that listing the restrictions at the beginning of a problem does not replace *checking* the potential solutions. In Example 4, 11 needs to be checked in the original equation.

Example 5

Solve $\dfrac{a}{a-2} + \dfrac{2}{3} = \dfrac{2}{a-2}$.

Solution

$$\frac{a}{a-2} + \frac{2}{3} = \frac{2}{a-2}, \qquad a \neq 2$$

$$3(a-2)\left(\frac{a}{a-2} + \frac{2}{3}\right) = 3(a-2)\left(\frac{2}{a-2}\right) \qquad \text{Multiply both sides by } 3(a-2).$$

$$3a + 2(a-2) = 6$$
$$3a + 2a - 4 = 6$$
$$5a = 10$$
$$a = 2$$

Because our initial restriction was $a \neq 2$, we conclude that this equation *has no solution*. Thus, the solution set is $\varnothing$. ▲

REMARK Example 5 demonstrates the importance of recognizing the restrictions that must be made to exclude division by zero. △

Ratio and Proportion

A **ratio** is the comparison of two numbers by division. We often use the fractional form to express ratios. For example, we can write the ratio of a to b as $\dfrac{a}{b}$. A statement of equality between two ratios is called a **proportion**. Thus, if $\dfrac{a}{b}$

and $\dfrac{c}{d}$ are two equal ratios, we can form the proportion $\dfrac{a}{b} = \dfrac{c}{d}(b \neq 0$ and $d \neq 0)$. We deduce an important property of proportions as follows.

$$\frac{a}{b} = \frac{c}{d}, \qquad b \neq 0 \text{ and } d \neq 0$$

$$bd\left(\frac{a}{b}\right) = bd\left(\frac{c}{d}\right) \qquad \text{Multiply both sides by } bd.$$

$$ad = bc$$

Cross-Multiplication Property of Proportions

If $\dfrac{a}{b} = \dfrac{c}{d}(b \neq 0$ and $d \neq 0)$, then $ad = bc$.

We can treat some fractional equations as proportions and solve them by using the cross-multiplication idea, as in the next examples.

Example 6

Solve $\dfrac{5}{x+6} = \dfrac{7}{x-5}$.

Solution

$$\frac{5}{x+6} = \frac{7}{x-5},$$

$$5(x-5) = 7(x+6) \qquad \text{Apply the cross-multiplication property.}$$

$$5x - 25 = 7x + 42$$

$$-67 = 2x$$

$$-\frac{67}{2} = x$$

The solution set is $\left\{-\dfrac{67}{2}\right\}$. ▲

Example 7

Solve $\dfrac{x}{7} = \dfrac{4}{x+3}$.

Solution

$$\frac{x}{7} = \frac{4}{x+3}, \qquad x \neq -3$$

$$x(x+3) = 7(4) \qquad \text{Cross-multiplication property}$$

$$x^2 + 3x = 28$$

$$x^2 + 3x - 28 = 0$$

$$(x+7)(x-4) = 0$$

$$x + 7 = 0 \quad \text{or} \quad x - 4 = 0$$
$$x = -7 \quad \text{or} \quad x = 4$$

The solution set is $\{-7, 4\}$. Check these solutions in the original equation. ▲

Problem Solving

The ability to solve fractional equations broadens our base for solving word problems. We are now ready to tackle some word problems that translate into fractional equations.

Problem 1

The sum of a number and its reciprocal is $\dfrac{10}{3}$. Find the number.

Solution

Let n represent the number. Then $\dfrac{1}{n}$ represents its reciprocal.

$$n + \frac{1}{n} = \frac{10}{3}, \quad n \neq 0$$
$$3n\left(n + \frac{1}{n}\right) = 3n\left(\frac{10}{3}\right)$$
$$3n^2 + 3 = 10n$$
$$3n^2 - 10n + 3 = 0$$
$$(3n - 1)(n - 3) = 0$$
$$3n - 1 = 0 \quad \text{or} \quad n - 3 = 0$$
$$3n = 1 \quad \text{or} \quad n = 3$$
$$n = \frac{1}{3} \quad \text{or} \quad n = 3$$

If the number is $\dfrac{1}{3}$, then its reciprocal is $\dfrac{1}{\frac{1}{3}} = 3$. If the number is 3, then its reciprocal is $\dfrac{1}{3}$. ▲

Now let's consider a problem where we can use the relationship

$$\frac{\text{Dividend}}{\text{Divisor}} = \text{Quotient} + \frac{\text{Remainder}}{\text{Divisor}}$$

as a guideline.

Problem 2

The sum of two numbers is 52. If the larger is divided by the smaller, the quotient is 9 and the remainder is 2. Find the numbers.

Solution

Let n represent the smaller number. Then $52 - n$ represents the larger number. Let's use the relationship we discussed previously as a guideline and proceed as follows.

$$\frac{\text{Dividend}}{\text{Divisor}} = \text{Quotient} + \frac{\text{Remainder}}{\text{Divisor}}$$

$$\frac{52 - n}{n} = 9 + \frac{2}{n}, \qquad n \neq 0$$

$$n\left(\frac{52 - n}{n}\right) = n\left(9 + \frac{2}{n}\right)$$

$$52 - n = 9n + 2$$

$$50 = 10n$$

$$5 = n$$

If $n = 5$, then $52 - n$ equals 47. The numbers are 5 and 47.   ▲

We can conveniently set up some problems and solve them using the concepts of ratio and proportion. Let's conclude this section with two such examples.

Problem 3

On a certain map $1\frac{1}{2}$ inches represents 25 miles. If two cities are $5\frac{1}{4}$ inches apart on the map, find the number of miles between the cities (see Figure 4.1).

FIGURE 4.1

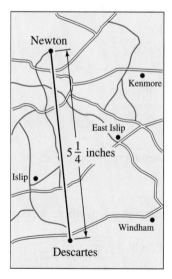

Solution

Let m represent the number of miles between the two cities. Set up the following proportion and solve the problem.

$$\frac{1\frac{1}{2}}{25} = \frac{5\frac{1}{4}}{m}, \qquad m \neq 0$$

$$\frac{\frac{3}{2}}{25} = \frac{\frac{21}{4}}{m}$$

$$\frac{3}{2}m = 25\left(\frac{21}{4}\right) \qquad \text{Cross-multiplication property}$$

$$\frac{2}{3}\left(\frac{3}{2}m\right) = \frac{\cancel{2}}{\cancel{3}}(25)\left(\frac{\overset{7}{\cancel{21}}}{\underset{2}{\cancel{4}}}\right) \qquad \text{Multiply both sides by } \frac{2}{3}.$$

$$m = \frac{175}{2} = 87\frac{1}{2}$$

The distance between the two cities is $87\frac{1}{2}$ miles. ▲

Problem 4

A sum of $750 is to be divided between two people in the ratio of 2 to 3. How much does each person receive?

Solution

Let d represent the amount of money that one person receives. Then $750 - d$ represents the amount for the other person.

$$\frac{d}{750 - d} = \frac{2}{3}, \qquad d \neq 750$$

$$3d = 2(750 - d)$$

$$3d = 1500 - 2d$$

$$5d = 1500$$

$$d = 300$$

If $d = 300$, then $750 - d$ equals 450. Therefore, one person receives $300 and the other person receives $450. ▲

Problem Set 4.6

Solve each of the following equations.

1. $\dfrac{x+1}{4} + \dfrac{x-2}{6} = \dfrac{3}{4}$

2. $\dfrac{x+2}{5} + \dfrac{x-1}{6} = \dfrac{3}{5}$

3. $\dfrac{x+3}{2} - \dfrac{x-4}{7} = 1$

4. $\dfrac{x+4}{3} - \dfrac{x-5}{9} = 1$

5. $\dfrac{5}{n} + \dfrac{1}{3} = \dfrac{7}{n}$

6. $\dfrac{3}{n} + \dfrac{1}{6} = \dfrac{11}{3n}$

7. $\dfrac{7}{2x} + \dfrac{3}{5} = \dfrac{2}{3x}$

8. $\dfrac{9}{4x} + \dfrac{1}{3} = \dfrac{5}{2x}$

9. $\dfrac{3}{4x} + \dfrac{5}{6} = \dfrac{4}{3x}$

10. $\dfrac{5}{7x} - \dfrac{5}{6} = \dfrac{1}{6x}$

11. $\dfrac{47 - n}{n} = 8 + \dfrac{2}{n}$

12. $\dfrac{45 - n}{n} = 6 + \dfrac{3}{n}$

13. $\dfrac{n}{65 - n} = 8 + \dfrac{2}{65 - n}$

14. $\dfrac{n}{70 - n} = 7 + \dfrac{6}{70 - n}$

15. $n + \dfrac{1}{n} = \dfrac{17}{4}$

16. $n + \dfrac{1}{n} = \dfrac{37}{6}$

17. $n - \dfrac{2}{n} = \dfrac{23}{5}$

18. $n - \dfrac{3}{n} = \dfrac{26}{3}$

19. $\dfrac{5}{7x - 3} = \dfrac{3}{4x - 5}$

20. $\dfrac{3}{2x - 1} = \dfrac{5}{3x + 2}$

21. $\dfrac{-2}{x - 5} = \dfrac{1}{x + 9}$

22. $\dfrac{5}{2a - 1} = \dfrac{-6}{3a + 2}$

23. $\dfrac{x}{x + 1} - 2 = \dfrac{3}{x - 3}$

24. $\dfrac{x}{x - 2} + 1 = \dfrac{8}{x - 1}$

25. $\dfrac{a}{a + 5} - 2 = \dfrac{3a}{a + 5}$

26. $\dfrac{a}{a - 3} - \dfrac{3}{2} = \dfrac{3}{a - 3}$

27. $\dfrac{5}{x + 6} = \dfrac{6}{x - 3}$

28. $\dfrac{3}{x - 1} = \dfrac{4}{x + 2}$

29. $\dfrac{3x - 7}{10} = \dfrac{2}{x}$

30. $\dfrac{x}{-4} = \dfrac{3}{12x - 25}$

31. $\dfrac{x}{x - 6} - 3 = \dfrac{6}{x - 6}$

32. $\dfrac{x}{x + 1} + 3 = \dfrac{4}{x + 1}$

33. $\dfrac{3s}{s + 2} + 1 = \dfrac{35}{2(3s + 1)}$

34. $\dfrac{s}{2s - 1} - 3 = \dfrac{-32}{3(s + 5)}$

35. $2 - \dfrac{3x}{x - 4} = \dfrac{14}{x + 7}$

36. $-1 + \dfrac{2x}{x + 3} = \dfrac{-4}{x + 4}$

37. $\dfrac{n + 6}{27} = \dfrac{1}{n}$

38. $\dfrac{n}{5} = \dfrac{10}{n - 5}$

39. $\dfrac{3n}{n - 1} - \dfrac{1}{3} = \dfrac{-40}{3n - 18}$

40. $\dfrac{n}{n + 1} + \dfrac{1}{2} = \dfrac{-2}{n + 2}$

41. $\dfrac{-3}{4x + 5} = \dfrac{2}{5x - 7}$

42. $\dfrac{7}{x + 4} = \dfrac{3}{x - 8}$

43. $\dfrac{2x}{x - 2} + \dfrac{15}{x^2 - 7x + 10} = \dfrac{3}{x - 5}$

44. $\dfrac{x}{x - 4} - \dfrac{2}{x + 3} = \dfrac{20}{x^2 - x - 12}$

Set up an algebraic equation and solve each of the following problems.

45. A sum of \$1750 is to be divided between two people in the ratio of 3 to 4. How much does each person receive?   \$750 and \$1000

46. A blueprint has a scale of 1 inch represents 5 feet. Find the dimensions of a rectangular room that measures $3\dfrac{1}{2}$ inches by $5\dfrac{3}{4}$ inches on the blueprint.

47. One angle of a triangle has a measure of 60° and the measures of the other two angles are in a ratio of 2 to 3. Find the measures of the other two angles.

48. The ratio of the complement of an angle to its supplement is 1 to 4. Find the measure of the angle.

49. The sum of a number and its reciprocal is $\frac{53}{14}$. Find the number.

50. The sum of two numbers is 80. If the larger is divided by the smaller, the quotient is 7 and the remainder is 8. Find the numbers.

51. If a home valued at $50,000 is assessed $900 in real estate taxes, at the same rate how much are the taxes on a home valued at $60,000?

52. The ratio of male students to female students at a certain university is 5 to 7. If there is a total of 16,200 students, find the number of male students and the number of female students.

53. Suppose that, together, Laura and Tammy sold $120.75 worth of candy for the annual school fair. If the ratio of Tammy's sales to Laura's sales was 4 to 3, how much did each sell?

54. The total value of a house and a lot is $68,000. If the ratio of the value of the house to the value of the lot is 7 to 1, find the value of the house.

55. The sum of two numbers is 90. If the larger is divided by the smaller, the quotient is 10 and the remainder is 2. Find the numbers.

56. What number must be added to the numerator and denominator of $\frac{2}{5}$ to produce a rational number that is equivalent to $\frac{7}{8}$?

57. A 20-foot board is to be cut into two pieces whose lengths are in the ratio of 7 to 3. Find the lengths of the two pieces.

58. An inheritance of $300,000 is to be divided between a son and the local heart fund in the ratio of 3 to 1. How much money will the son receive?

59. Suppose that in a certain precinct, 1150 people voted in the last presidential election. If the ratio of female voters to male voters was 3 to 2, how many females and how many males voted?

60. The perimeter of a rectangle is 114 centimeters. If the ratio of its width to its length is 7 to 12, find the dimensions of the rectangle.

61. How could you do Problem 57 without using algebra?

62. Now do Problem 59 using the same approach that you used in Problem 61. What difficulties do you encounter?

63. How can you tell by inspection that the equation $\frac{x}{x+2} = \frac{-2}{x+2}$ has no solution?

64. How would you help someone solve the equation $\frac{3}{x} - \frac{4}{x} = \frac{-1}{x}$?

4.7 More Fractional Equations and Applications

Let's begin this section by considering a few more fractional equations. We will continue to solve them using the same basic techniques as in the previous section. That is, we will multiply both sides of the equation by the least common denominator of all of the denominators in the equation, with the necessary restrictions to avoid division by zero. Some of the denominators in these problems will require factoring before we can determine a least common denominator.

Example 1

Solve $\dfrac{x}{2x-8} + \dfrac{16}{x^2-16} = \dfrac{1}{2}$.

Solution

$$\frac{x}{2x-8} + \frac{16}{x^2-16} = \frac{1}{2}$$

$$\frac{x}{2(x-4)} + \frac{16}{(x+4)(x-4)} = \frac{1}{2}, \qquad x \neq 4 \text{ and } x \neq -4$$

$$2(x-4)(x+4)\left(\frac{x}{2(x-4)} + \frac{16}{(x+4)(x-4)}\right) = 2(x+4)(x-4)\left(\frac{1}{2}\right)$$

$$x(x+4) + 2(16) = (x+4)(x-4)$$

$$x^2 + 4x + 32 = x^2 - 16$$

$$4x = -48$$

$$x = -12$$

The solution set is $\{-12\}$. Perhaps you should check it!   ▲

   In Example 1, notice that the restrictions were not indicated until the denominators were expressed in factored form. It is usually easier to determine the necessary restrictions at this step.

Example 2

Solve $\dfrac{3}{n-5} - \dfrac{2}{2n+1} = \dfrac{n+3}{2n^2-9n-5}$.

Solution

$$\frac{3}{n-5} - \frac{2}{2n+1} = \frac{n+3}{2n^2-9n-5}$$

$$\frac{3}{n-5} - \frac{2}{2n+1} = \frac{n+3}{(2n+1)(n-5)}, \qquad n \neq -\frac{1}{2} \text{ and } n \neq 5$$

$$(2n+1)(n-5)\left(\frac{3}{n-5} - \frac{2}{2n+1}\right) = (2n+1)(n-5)\left(\frac{n+3}{(2n+1)(n-5)}\right)$$

$$3(2n+1) - 2(n-5) = n+3$$

$$6n + 3 - 2n + 10 = n + 3$$

$$4n + 13 = n + 3$$

$$3n = -10$$

$$n = -\frac{10}{3}$$

The solution set is $\left\{-\dfrac{10}{3}\right\}$.   ▲

Example 3

Solve $2 + \dfrac{4}{x-2} = \dfrac{8}{x^2-2x}$.

Solution

$$2 + \frac{4}{x - 2} = \frac{8}{x^2 - 2x}$$

$$2 + \frac{4}{x - 2} = \frac{8}{x(x - 2)}, \qquad x \ne 0 \text{ and } x \ne 2$$

$$x(x - 2)\left(2 + \frac{4}{x - 2}\right) = x(x - 2)\left(\frac{8}{x(x - 2)}\right)$$

$$2x(x - 2) + 4x = 8$$

$$2x^2 - 4x + 4x = 8$$

$$2x^2 = 8$$

$$x^2 = 4$$

$$x^2 - 4 = 0$$

$$(x + 2)(x - 2) = 0$$

$$x + 2 = 0 \qquad \text{or} \qquad x - 2 = 0$$

$$x = -2 \qquad \text{or} \qquad x = 2$$

Since our initial restriction indicated that $x \ne 2$, then the *only solution* is -2. Thus, the solution set is $\{-2\}$. ▲

In Section 2.4, we discussed using the properties of equality to change the form of various formulas. For example, we considered the simple interest formula $A = P + Prt$ and changed its form by solving for P as follows.

$$A = P + Prt$$

$$A = P(1 + rt)$$

$$\frac{A}{1 + rt} = P \qquad \text{Multiply both sides by } \frac{1}{1 + rt}.$$

If the formula is in the form of a fractional equation, then the techniques of these last two sections are applicable. Consider the following example.

Example 4

If the original cost of some business property is C dollars and it is depreciated linearly over N years, its value, V, at the end of T years is given by

$$V = C\left(1 - \frac{T}{N}\right).$$

Solve this formula for N in terms of V, C, and T.

Solution

$$V = C\left(1 - \frac{T}{N}\right)$$

$$V = C - \frac{CT}{N}$$

$$N(V) = N\left(C - \frac{CT}{N}\right) \qquad \text{Multiply both sides by } N.$$

$$NV = NC - CT$$
$$NV - NC = -CT$$
$$N(V - C) = -CT$$
$$N = \frac{-CT}{V - C}$$
$$N = -\frac{CT}{V - C}.$$

▲

Problem Solving

In Section 2.4 we solved some uniform motion problems. The formula $d = rt$ was used in the analysis of the problems and we used guidelines that involve distance relationships. Now let's consider some uniform motion problems where guidelines that involve either times or rates are appropriate. These problems will generate fractional equations to solve.

Problem I

An airplane travels 2050 miles in the same time that a car travels 260 miles. If the rate of the plane is 358 miles per hour greater than the rate of the car, find the rate of each.

Solution

Let r represent the rate of the car. Then $r + 358$ represents the rate of the plane. The fact that the times are equal can be a guideline.

Time of plane	Equals	Time of car
↓		↓
$\dfrac{\text{Distance of plane}}{\text{Rate of plane}}$	$=$	$\dfrac{\text{Distance of car}}{\text{Rate of car}}$

$$\frac{2050}{r + 358} = \frac{260}{r}$$
$$2050r = 260(r + 358)$$
$$2050r = 260r + 93{,}080$$
$$1790r = 93{,}080$$
$$r = 52$$

If $r = 52$, then $r + 358$ equals 410. Thus, the rate of the car is 52 miles per hour and the rate of the plane is 410 miles per hour.

▲

Problem 2

It takes a freight train 2 hours longer to travel 300 miles than it takes an express train to travel 280 miles. The rate of the express train is 20 miles per hour greater than the rate of the freight train. Find the times and rates of both trains.

Solution

Let t represent the time of the express train. Then $t + 2$ represents the time of the freight train. Let's record the information of this problem in a table as follows.

	Distance	Time	$r = \dfrac{d}{t}$
Express train	280	t	$\dfrac{280}{t}$
Freight train	300	$t + 2$	$\dfrac{300}{t + 2}$

The fact that the rate of the express train is 20 miles per hour greater than the rate of the freight train can be a guideline.

Rate of express   Equals   Rate of freight train plus 20

$$\frac{280}{t} = \frac{300}{t + 2} + 20$$

$$t(t + 2)\left(\frac{280}{t}\right) = t(t + 2)\left(\frac{300}{t + 2} + 20\right)$$

$$280(t + 2) = 300t + 20t(t + 2)$$

$$280t + 560 = 300t + 20t^2 + 40t$$

$$280t + 560 = 340t + 20t^2$$

$$0 = 20t^2 + 60t - 560$$

$$0 = t^2 + 3t - 28$$

$$0 = (t + 7)(t - 4)$$

$$t + 7 = 0 \quad \text{or} \quad t - 4 = 0$$

$$t = -7 \quad \text{or} \quad t = 4$$

The negative solution must be discarded, so the time of the express train (t) is 4 hours and the time of the freight train ($t + 2$) is 6 hours. The rate of the express train $\left(\dfrac{280}{t}\right)$ is $\dfrac{280}{4} = 70$ miles per hour and the rate of the freight train $\left(\dfrac{300}{t + 2}\right)$ is $\dfrac{300}{6} = 50$ miles per hour. ▲

REMARK   Note that to solve Problem 1 we went directly to a guideline without the use of a table, but for Problem 2 we used a table. Again, remember that this is a personal preference; we are merely exposing you to a variety of techniques. △

     Uniform motion problems are a special case of a larger group of problems we refer to as **rate-time problems**. For example, if a certain machine can produce 150 items in 10 minutes, then we say that the machine is producing at a rate of $\dfrac{150}{10} = 15$ items per minute. Likewise, if a person can do a certain job in 3 hours, then, assuming a constant rate of work we say that the person is working at a rate of $\frac{1}{3}$ of the job per hour. In general, if Q is the quantity of something done in t units of time, then the rate, r, is given by $r = \frac{Q}{t}$. We state the rate in terms of *so much quantity per unit of time*. (In uniform motion problems the "quantity" is distance.) Let's consider some examples of rate-time problems.

Problem 3

If Jim can mow a lawn in 50 minutes and his son, Todd, can mow the same lawn in 40 minutes, how long will it take them to mow the lawn if they work together?

Solution

Jim's rate is $\frac{1}{50}$ of the lawn per minute and Todd's rate is $\frac{1}{40}$ of the lawn per minute. If we let m represent the number of minutes that they work together, then $\frac{1}{m}$ represents their rate when working together. Therefore, since the sum of the individual rates must equal the rate when working together, we can set up and solve the following equation.

$$\overset{\text{Jim's rate}}{\frac{1}{50}} + \overset{\text{Todd's rate}}{\frac{1}{40}} = \overset{\text{Combined rate}}{\frac{1}{m}}$$

$$200m\left(\frac{1}{50} + \frac{1}{40}\right) = 200m\left(\frac{1}{m}\right)$$

$$4m + 5m = 200$$

$$9m = 200$$

$$m = \frac{200}{9} = 22\frac{2}{9}$$

It should take them $22\frac{2}{9}$ minutes.   ▲

Problem 4

Working together, Linda and Kathy can type a term paper in $3\frac{3}{5}$ hours. Linda can type the paper by herself in 6 hours. How long would it take Kathy to type the paper by herself?

Solution

Their rate working together is $\dfrac{1}{3\frac{3}{5}} = \dfrac{1}{\frac{18}{5}} = \dfrac{5}{18}$ of the job per hour and Linda's rate is $\frac{1}{6}$ of the job per hour. If we let h represent the number of hours that it would take Kathy by herself, then her rate is $\frac{1}{h}$ of the job per hour. Thus, we have

Linda's rate   Kathy's rate   Combined rate

$$\frac{1}{6} \quad + \quad \frac{1}{h} \quad = \quad \frac{5}{18}.$$

Solving this equation yields

$$18h\left(\frac{1}{6}+\frac{1}{h}\right)=18h\left(\frac{5}{18}\right)$$
$$3h+18=5h$$
$$18=2h$$
$$9=h.$$

It would take Kathy 9 hours to type the paper by herself. ▲

One final example of this section illustrates another approach that some people find meaningful for rate-time problems. For this approach, think in terms of fractional parts of the job. For example, if a person can do a certain job in 5 hours, then at the end of 2 hours he or she has done $\frac{2}{5}$ of the job. (Again, assume a constant rate of work.) At the end of 4 hours, he or she has finished $\frac{4}{5}$ of the job and, in general, at the end of h hours, he or she has done $\frac{h}{5}$ of the job. Let's see how this works in a problem.

Problem 5

It takes Pat 12 hours to complete a task. After he had been working for 3 hours, he was joined by his brother, Mike, and together they finished the task in 5 hours. How long would it take Mike to do the job by himself?

Solution

Let h represent the number of hours that it would take Mike by himself. Since Pat has been working for 3 hours, he has done $\frac{3}{12}=\frac{1}{4}$ of the job before Mike joins him. Thus, there is $\frac{3}{4}$ of the original job to be done while working together. We can set up and solve the following equation.

Fractional part of the remaining $\frac{3}{4}$ of the job that Pat does   Fractional part of the remaining $\frac{3}{4}$ of the job that Mike does

$$\frac{5}{12} \quad + \quad \frac{5}{h} = \frac{3}{4}$$
$$12h\left(\frac{5}{12}+\frac{5}{h}\right)=12h\left(\frac{3}{4}\right)$$
$$5h+60=9h$$

$$60 = 4h$$
$$15 = h$$

It would take Mike 15 hours to do the entire job by himself.

▲

▼ Problem Set 4.7

Solve each of the following equations.

1. $\dfrac{x}{4x-4} + \dfrac{5}{x^2-1} = \dfrac{1}{4}$

2. $\dfrac{x}{3x-6} + \dfrac{4}{x^2-4} = \dfrac{1}{3}$

3. $3 + \dfrac{6}{t-3} = \dfrac{6}{t^2-3t}$

4. $2 + \dfrac{4}{t-1} = \dfrac{4}{t^2-t}$

5. $\dfrac{3}{n-5} + \dfrac{4}{n+7} = \dfrac{2n+11}{n^2+2n-35}$

6. $\dfrac{2}{n+3} + \dfrac{3}{n-4} = \dfrac{2n-1}{n^2-n-12}$

7. $\dfrac{5x}{2x+6} - \dfrac{4}{x^2-9} = \dfrac{5}{2}$

8. $\dfrac{3x}{5x+5} - \dfrac{2}{x^2-1} = \dfrac{3}{5}$

9. $1 + \dfrac{1}{n-1} = \dfrac{1}{n^2-n}$

10. $3 + \dfrac{9}{n-3} = \dfrac{27}{n^2-3n}$

11. $\dfrac{2}{n-2} - \dfrac{n}{n+5} = \dfrac{10n+15}{n^2+3n-10}$

12. $\dfrac{n}{n+3} + \dfrac{1}{n-4} = \dfrac{11-n}{n^2-n-12}$

13. $\dfrac{2}{2x-3} - \dfrac{2}{10x^2-13x-3} = \dfrac{x}{5x+1}$

14. $\dfrac{1}{3x+4} + \dfrac{6}{6x^2+5x-4} = \dfrac{x}{2x-1}$

15. $\dfrac{2x}{x+3} - \dfrac{3}{x-6} = \dfrac{29}{x^2-3x-18}$

16. $\dfrac{x}{x-4} - \dfrac{2}{x+8} = \dfrac{63}{x^2+4x-32}$

17. $\dfrac{a}{a-5} + \dfrac{2}{a-6} = \dfrac{2}{a^2-11a+30}$

18. $\dfrac{a}{a+2} + \dfrac{3}{a+4} = \dfrac{14}{a^2+6a+8}$

19. $\dfrac{-1}{2x-5} + \dfrac{2x-4}{4x^2-25} = \dfrac{5}{6x+15}$

20. $\dfrac{-2}{3x+2} + \dfrac{x-1}{9x^2-4} = \dfrac{3}{12x-8}$

21. $\dfrac{7y+2}{12y^2+11y-15} - \dfrac{1}{3y+5} = \dfrac{2}{4y-3}$

22. $\dfrac{5y-4}{6y^2+y-12} - \dfrac{2}{2y+3} = \dfrac{5}{3y-4}$

23. $\dfrac{2n}{6n^2+7n-3} - \dfrac{n-3}{3n^2+11n-4} =$
$\dfrac{5}{2n^2+11n+12}$

24. $\dfrac{x+1}{2x^2+7x-4} - \dfrac{x}{2x^2-7x+3} =$
$\dfrac{1}{x^2+x-12}$

25. $\dfrac{1}{2x^2-x-1} + \dfrac{3}{2x^2+x} = \dfrac{2}{x^2-1}$

26. $\dfrac{2}{n^2+4n} + \dfrac{3}{n^2-3n-28} = \dfrac{5}{n^2-6n-7}$

27. $\dfrac{x+1}{x^3-9x} - \dfrac{1}{2x^2+x-21} = \dfrac{1}{2x^2+13x+21}$

28. $\dfrac{x}{2x^2+5x} - \dfrac{x}{2x^2+7x+5} = \dfrac{2}{x^2+x}$

29. $\dfrac{4t}{4t^2 - t - 3} + \dfrac{2 - 3t}{3t^2 - t - 2} = \dfrac{1}{12t^2 + 17t + 6}$

30. $\dfrac{2t}{2t^2 + 9t + 10} + \dfrac{1 - 3t}{3t^2 + 4t - 4} =$
$\dfrac{4}{6t^2 + 11t - 10}$

For Problems 31–44, solve each equation for the indicated variable.

31. $y = \dfrac{5}{6}x + \dfrac{2}{9}$ for x

32. $y = \dfrac{3}{4}x - \dfrac{2}{3}$ for x

33. $\dfrac{-2}{x - 4} = \dfrac{5}{y - 1}$ for y

34. $\dfrac{7}{y - 3} = \dfrac{3}{x + 1}$ for y

35. $I = \dfrac{100M}{C}$ for M

36. $V = C\left(1 - \dfrac{T}{N}\right)$ for T

37. $\dfrac{R}{S} = \dfrac{T}{S + T}$ for R

38. $\dfrac{1}{R} = \dfrac{1}{S} + \dfrac{1}{T}$ for R

39. $\dfrac{y - 1}{x - 3} = \dfrac{b - 1}{a - 3}$ for y

40. $y = -\dfrac{a}{b}x + \dfrac{c}{d}$ for x

41. $\dfrac{x}{a} + \dfrac{y}{b} = 1$ for y

42. $\dfrac{y - b}{x} = m$ for y

43. $\dfrac{y - 1}{x + 6} = \dfrac{-2}{3}$ for y

44. $\dfrac{y + 5}{x - 2} = \dfrac{3}{7}$ for y

Set up an equation and solve each of the following problems.

45. Kent drives his Mazda 270 miles in the same time that Dave drives his Datsun 250 miles. If Kent averages 4 miles per hour faster than Dave, find their rates.

46. Suppose that Wendy rides her bicycle 30 miles in the same time that it takes Kim to ride her bicycle 20 miles. If Wendy rides 5 miles per hour faster than Kim, find the rate of each.

47. An inlet pipe can fill a tank in 10 minutes. A drain can empty the tank in 12 minutes. If the tank is empty and both the pipe and drain are open, how long will it take before the tank overflows?

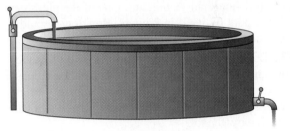

FIGURE 4.2

48. Barry can do a certain job in 3 hours, while it takes Sanchez 5 hours to do the same job. How long would it take them to do the job working together?

49. Connie can type 600 words in 5 minutes less than it takes Katie to type 600 words. If Connie types at a rate of 20 words per minute faster than Katie types, find the typing rate of each woman.

50. Walt can mow a lawn in 1 hour, while his son, Malik, can mow the same lawn in 50 minutes. One day Malik started mowing the lawn by himself and worked for 30 minutes. Then Walt joined him and they finished the lawn. How long did it take them to finish mowing the lawn after Walt started to help?

51. Plane A can travel 1400 miles in one hour less time than it takes plane B to travel 2000 miles. The rate of plane B is 50 miles per hour faster than the rate of plane A. Find the times and rates of both planes.

52. To travel 60 miles, it takes Sue, riding a moped, 2 hours less time than it takes Doreen to travel 50 miles riding a bicycle. Sue travels 10 miles per hour faster than Doreen. Find the times and rates of both girls.

53. It takes Amy twice as long to deliver papers as it does Nancy. How long would it take each if they can deliver the papers together in 40 minutes?

54. If two inlet pipes are both open, they can fill a pool in 1 hour and 12 minutes. One of the pipes can fill the pool by itself in 2 hours. How long would it take the other pipe to fill the pool by itself?

55. Dan agreed to mow a vacant lot for $12. It took him an hour longer than what he had anticipated, so he earned $1 per hour less than he originally calculated. How long had he anticipated that it would take him to mow the lot?

56. Last week Al bought some golf balls for $20. The next day they were on sale for $.50 per ball less and he bought $22.50 worth of balls. If he purchased 5 more balls on the second day than he did on the first day, how many did he buy each day and at what price per ball?

57. Debbie rode her bicycle out into the country for a distance of 24 miles. On the way back, she took a much shorter route of 12 miles and made the return trip in one-half hour less time. If her rate out into the country was 4 miles per hour faster than her rate on the return trip, find both rates.

58. Calvin jogs for 10 miles and then walks another 10 miles. He jogs $2\frac{1}{2}$ miles per hour faster than he walks and the entire distance of 20 miles takes 6 hours. Find the rate that he walks and the rate that he jogs.

THOUGHTS INTO WORDS

59. Why is it important to consider more than one way to do a problem?

60. Write a paragraph or two summarizing the new ideas about problem solving you have acquired thus far in this course.

SUMMARY

(4.1) Any number that can be written in the form $\frac{a}{b}$, where a and b are integers and $b \neq 0$, is called a **rational number**.

A **rational expression** is defined as the indicated quotient of two polynomials. The following properties pertain to rational numbers and rational expressions.

$$\textbf{1.} \ \frac{-a}{b} = \frac{a}{-b} = -\frac{a}{b}$$

2. $\dfrac{-a}{-b} = \dfrac{a}{b}$

3. $\dfrac{a \cdot k}{b \cdot k} = \dfrac{a}{b}$ Fundamental principle of fractions

(4.2) Multiplication and division of rational expressions are based upon the following.

1. $\dfrac{a}{b} \cdot \dfrac{c}{d} = \dfrac{ac}{bd}$ Multiplication

2. $\dfrac{a}{b} \div \dfrac{c}{d} = \dfrac{a}{b} \cdot \dfrac{d}{c} = \dfrac{ad}{bc}$ Division

(4.3) Addition and subtraction of rational expressions are based upon the following.

1. $\dfrac{a}{b} + \dfrac{c}{b} = \dfrac{a+c}{b}$ Addition

2. $\dfrac{a}{b} - \dfrac{c}{b} = \dfrac{a-c}{b}$ Subtraction

(4.4) The following basic procedure is used to add or subtract rational expressions.

1. Find the LCD of all denominators.
2. Change each fraction to an equivalent fraction that has the LCD as its denominator.
3. Add or subtract numerators and place this result over the LCD.
4. Look for possibilities to simplify the resulting fraction.

Fractional forms that contain rational numbers or rational expressions in the numerators and/or denominators are called **complex fractions**. The fundamental principle of fractions serves as a basis for simplifying complex fractions.

(4.5) To divide a polynomial by a monomial we divide each term of the polynomial by the monomial.

The procedure for dividing a polynomial by a polynomial, rather than a monomial, resembles the long division process in arithmetic. (See the examples in Section 4.5.)

(4.6) To solve a fractional equation, it is often easiest to begin by multiplying both sides of the equation by the LCD of all of the denominators in the equation. If an equation contains a variable in one or more denominators, then we must be careful to avoid any value of the variable that makes the denominator zero.

A **ratio** is the comparison of two numbers by division. A statement of equality between two ratios is a **proportion**.

We can treat some fractional equations as proportions and we can solve them by applying the following property. This property is often called the **cross-multiplication** property.

$$\text{If } \frac{a}{b} = \frac{c}{d}, \text{ then } ad = bc.$$

(4.7) Those techniques we use to solve fractional equations can also be used to change the form of formulas containing rational expressions so that we can use those formulas to solve problems.

Chapter 4 Review Problem Set

For Problems 1–6, simplify each of the rational expressions.

1. $\dfrac{26x^2y^3}{39x^4y^2}$

2. $\dfrac{a^2 - 9}{a^2 + 3a}$

3. $\dfrac{n^2 - 3n - 10}{n^2 + n - 2}$

4. $\dfrac{x^4 - 1}{x^3 - x}$

5. $\dfrac{8x^3 - 2x^2 - 3x}{12x^2 - 9x}$

6. $\dfrac{x^4 - 7x^2 - 30}{2x^4 + 7x^2 + 3}$

For Problems 7–10, simplify each complex fraction.

7. $\dfrac{\dfrac{5}{8} - \dfrac{1}{2}}{\dfrac{1}{6} + \dfrac{3}{4}}$

8. $\dfrac{\dfrac{3}{2x} + \dfrac{5}{3y}}{\dfrac{4}{x} - \dfrac{3}{4y}}$

9. $\dfrac{\dfrac{3}{x - 2} - \dfrac{4}{x^2 - 4}}{\dfrac{2}{x + 2} + \dfrac{1}{x - 2}}$

10. $1 - \dfrac{1}{2 - \dfrac{1}{x}}$

For Problems 11–22, perform the indicated operations and express answers in simplest form.

11. $\dfrac{6xy^2}{7y^3} \div \dfrac{15x^2y}{5x^2}$

12. $\dfrac{9ab}{3a + 6} \cdot \dfrac{a^2 - 4a - 12}{a^2 - 6a}$

13. $\dfrac{n^2 + 10n + 25}{n^2 - n} \cdot \dfrac{5n^3 - 3n^2}{5n^2 + 22n - 15}$

14. $\dfrac{x^2 - 2xy - 3y^2}{x^2 + 9y^2} \div \dfrac{2x^2 + xy - y^2}{2x^2 - xy}$

15. $\dfrac{2x + 1}{5} + \dfrac{3x - 2}{4}$

16. $\dfrac{3}{2n} + \dfrac{5}{3n} - \dfrac{1}{9}$

17. $\dfrac{3x}{x+7} - \dfrac{2}{x}$

18. $\dfrac{10}{x^2 - 5x} + \dfrac{2}{x}$

19. $\dfrac{3}{n^2 - 5n - 36} + \dfrac{2}{n^2 + 3n - 4}$

20. $\dfrac{3}{2y+3} + \dfrac{5y-2}{2y^2 - 9y - 18} - \dfrac{1}{y-6}$

21. $(18x^2 + 9x - 2) \div (3x + 2)$

22. $(3x^3 + 5x^2 - 6x - 2) \div (x + 4)$

For Problems 23–32, solve each equation.

23. $\dfrac{4x+5}{3} + \dfrac{2x-1}{5} = 2$

24. $\dfrac{3}{4x} + \dfrac{4}{5} = \dfrac{9}{10x}$

25. $\dfrac{a}{a-2} - \dfrac{3}{2} = \dfrac{2}{a-2}$

26. $\dfrac{4}{5y-3} = \dfrac{2}{3y+7}$

27. $n + \dfrac{1}{n} = \dfrac{53}{14}$

28. $\dfrac{1}{2x-7} + \dfrac{x-5}{4x^2 - 49} = \dfrac{4}{6x - 21}$

29. $\dfrac{x}{2x+1} - 1 = \dfrac{-4}{7(x-2)}$

30. $\dfrac{2x}{-5} = \dfrac{3}{4x - 13}$

31. $\dfrac{2n}{2n^2 + 11n - 21} - \dfrac{n}{n^2 + 5n - 14} = \dfrac{3}{n^2 + 5n - 14}$

32. $\dfrac{2}{t^2 - t - 6} + \dfrac{t+1}{t^2 + t - 12} = \dfrac{t}{t^2 + 6t + 8}$

33. Solve $\dfrac{y-6}{x+1} = \dfrac{3}{4}$ for y.

34. Solve $\dfrac{x}{a} - \dfrac{y}{b} = 1$ for y.

For Problems 35–40, set up an equation and solve the problem.

35. A sum of $1400 is to be divided between two people in the ratio of $\dfrac{3}{5}$. How much does each person receive?

36. Working together, Dan and Julio can mow a lawn in 12 minutes. Julio can mow the lawn by himself in 10 minutes less time than it takes Dan by himself. How long does it take each of them to mow the lawn alone?

37. Suppose that car A can travel 250 miles in 3 hours less time than it takes car B to travel 440 miles. The rate of car B is 5 miles per hour faster than car A. Find the rates of both cars.

38. Mark can overhaul an engine in 20 hours and Phil can do the same job by himself in 30 hours. If they both work together for a time and then Mark finishes the job by himself in 5 hours, how long did they work together?

39. Kelly contracted to paint a house for $640. It took him 20 hours longer than he had anticipated, so he earned $1.60 per hour less than he had calculated. How long had he anticipated that it would take him to paint the house?

40. Nasser rode his bicycle 66 miles in $4\dfrac{1}{2}$ hours. For the first 40 miles he averaged a certain rate and then for the last 26 miles he reduced his rate by 3 miles per hour. Find his rate for the last 26 miles.

CHAPTER 4 TEST

For Problems 1–4, simplify each rational expression.

1. $\dfrac{39x^2y^3}{72x^3y}$

2. $\dfrac{3x^2 + 17x - 6}{x^3 - 36x}$

3. $\dfrac{6n^2 - 5n - 6}{3n^2 + 14n + 8}$

4. $\dfrac{2x - 2x^2}{x^2 - 1}$

For Problems 5–13, perform the indicated operations and express answers in simplest form.

5. $\dfrac{5x^2y}{8x} \cdot \dfrac{12y^2}{20xy}$

6. $\dfrac{5a + 5b}{20a + 10b} \cdot \dfrac{a^2 - ab}{2a^2 + 2ab}$

7. $\dfrac{3x^2 + 10x - 8}{5x^2 + 19x - 4} \div \dfrac{3x^2 - 23x + 14}{x^2 - 3x - 28}$

8. $\dfrac{3x - 1}{4} + \dfrac{2x + 5}{6}$

9. $\dfrac{5x - 6}{3} - \dfrac{x - 12}{6}$

10. $\dfrac{3}{5n} + \dfrac{2}{3} - \dfrac{7}{3n}$

11. $\dfrac{3x}{x - 6} + \dfrac{2}{x}$

12. $\dfrac{9}{x^2 - x} - \dfrac{2}{x}$

13. $\dfrac{3}{2n^2 + n - 10} + \dfrac{5}{n^2 + 5n - 14}$

14. Divide $3x^3 + 10x^2 - 9x - 4$ by $x + 4$.

15. Simplify the complex fraction $\dfrac{\dfrac{3}{2x} - \dfrac{1}{6}}{\dfrac{2}{3x} + \dfrac{3}{4}}$.

16. Solve $\dfrac{x + 2}{y - 4} = \dfrac{3}{4}$ for y.

For Problems 17–22, solve each equation.

17. $\dfrac{x - 1}{2} - \dfrac{x + 2}{5} = -\dfrac{3}{5}$

18. $\dfrac{5}{4x} + \dfrac{3}{2} = \dfrac{7}{5x}$

19. $\dfrac{-3}{4n - 1} = \dfrac{-2}{3n + 11}$

20. $n - \dfrac{5}{n} = 4$

(continued on next page)

CHAPTER 4 TEST *(continued)*

21. $\dfrac{6}{x-4} - \dfrac{4}{x+3} = \dfrac{8}{x-4}$

22. $\dfrac{1}{3x-1} + \dfrac{x-2}{9x^2-1} = \dfrac{7}{6x-2}$

For Problems 23–25, set up an equation and solve the problem.

23. The denominator of a rational number is 9 less than three times the numerator. The number in simplest form is $\dfrac{3}{8}$. Find the number.

24. It takes Jodi three times as long to deliver papers as it does Jannie. Together they can deliver the papers in 15 minutes. How long would it take Jodi by herself?

25. René can ride her bike 60 miles in one hour less time than it takes Sue to ride 60 miles. René's rate is 3 miles per hour faster than Sue's rate. Find René's rate.

Exponents and Radicals

How long will it take a pendulum that is 1.5 feet long to swing from one side to the other side and back? The formula $T = 2\pi\sqrt{\frac{L}{32}}$ can be used to determine that it will take approximately 1.4 seconds.

It is not uncommon in mathematics to find two separately developed concepts that are closely related to each other. In this chapter we will first develop the concepts of exponent and root individually and then show how they merge to become even more functional as a unified idea.

5.1 Using Integers as Exponents

Thus far in the text we have used only positive integers as exponents. In Chapter 1 the expression b^n, where b is any real number and n is a positive integer, was defined by

$$b^n = b \cdot b \cdot b \cdots b \qquad n \text{ factors of } b.$$

Then in Chapter 3 some of the parts of the following property served as a basis for manipulation with polynomials.

PROPERTY 5.1

If m and n are positive integers and a and b are real numbers (except $b \neq 0$ whenever it appears in a denominator), then

1. $b^n \cdot b^m = b^{n+m}$

2. $(b^n)^m = b^{mn}$

3. $(ab)^n = a^n b^n$

4. $\left(\dfrac{a}{b}\right)^n = \dfrac{a^n}{b^n}$

5. $\dfrac{b^n}{b^m} = b^{n-m}$ when $n > m$

$\dfrac{b^n}{b^m} = 1$ when $n = m$

$\dfrac{b^n}{b^m} = \dfrac{1}{b^{m-n}}$ when $n < m$

We are now ready to extend the concept of an exponent to include the use of zero and the negative integers as exponents.

First, let's consider the use of zero as an exponent. We want to use zero in such a way that the previously listed properties continue to hold. If $b^n \cdot b^m = b^{n+m}$ is to hold, then $x^4 \cdot x^0 = x^{4+0} = x^4$. In other words, x^0 *acts like* 1 because $x^4 \cdot x^0 = x^4$. This line of reasoning suggests the following definition.

DEFINITION 5.1

If b is a nonzero real number, then

$$b^0 = 1.$$

According to Definition 5.1 the following statements are all true.

$5^0 = 1,$ $\qquad\qquad (-413)^0 = 1,$

$\left(\dfrac{3}{11}\right)^0 = 1,$ $\qquad\qquad n^0 = 1, \quad n \neq 0,$

$(x^3 y^4)^0 = 1, \quad x \neq 0, y \neq 0$

We can use a similar line of reasoning to motivate a definition for the use of negative integers as exponents. Consider the example $x^4 \cdot x^{-4}$. If $b^n \cdot b^m = b^{n+m}$ is to hold, then $x^4 \cdot x^{-4} = x^{4+(-4)} = x^0 = 1$. Thus, x^{-4} must be the reciprocal of x^4, since their product is 1. That is to say,

$$x^{-4} = \frac{1}{x^4}.$$

This suggests the following general definition.

DEFINITION 5.2

If n is a positive integer and b is a nonzero real number, then

$$b^{-n} = \frac{1}{b^n}.$$

According to Definition 5.2 the following statements are true.

$$x^{-5} = \frac{1}{x^5}, \qquad\qquad 2^{-4} = \frac{1}{2^4} = \frac{1}{16},$$

$$10^{-2} = \frac{1}{10^2} = \frac{1}{100} \text{ or } .01, \qquad \frac{2}{x^{-3}} = \frac{2}{\frac{1}{x^3}} = (2)\left(\frac{x^3}{1}\right) = 2x^3,$$

$$\left(\frac{3}{4}\right)^{-2} = \frac{1}{\left(\frac{3}{4}\right)^2} = \frac{1}{\frac{9}{16}} = \frac{16}{9}$$

It can be verified (although it is beyond the scope of this text) that all of the parts of Property 5.1 hold for *all integers*. In fact, the following equality can replace the three separate statements for part (5).

$$\frac{b^n}{b^m} = b^{n-m} \quad \text{for all integers } n \text{ and } m$$

Let's restate Property 5.1 as it holds for all integers and include, at the right, a *name tag* for easy reference.

PROPERTY 5.2

If m and n are integers and a and b are real numbers, except $b \neq 0$ whenever it appears in a denominator, then

1. $b^n \cdot b^m = b^{n+m}$ Product of two powers
2. $(b^n)^m = b^{mn}$ Power of a power
3. $(ab)^n = a^n b^n$ Power of a product

4. $\left(\dfrac{a}{b}\right)^n = \dfrac{a^n}{b^n}$ Power of a quotient

5. $\dfrac{b^n}{b^m} = b^{n-m}$ Quotient of two powers

Having the use of all integers as exponents allows us to work with a large variety of numerical and algebraic expressions. Let's consider some examples that illustrate the use of the various parts of Property 5.2.

Example 1

Simplify each of the following numerical expressions.

   (a) $10^{-3} \cdot 10^2$ (b) $(2^{-3})^{-2}$ (c) $(2^{-1} \cdot 3^2)^{-1}$

   (d) $\left(\dfrac{2^{-3}}{3^{-2}}\right)^{-1}$ (e) $\dfrac{10^{-2}}{10^{-4}}$

Solution

   (a) $10^{-3} \cdot 10^2 = 10^{-3+2}$ Product of two powers

$$= 10^{-1}$$

$$= \frac{1}{10^1} = \frac{1}{10}$$

   (b) $(2^{-3})^{-2} = 2^{(-2)(-3)}$ Power of a power

$$= 2^6 = 64$$

   (c) $(2^{-1} \cdot 3^2)^{-1} = (2^{-1})^{-1}(3^2)^{-1}$ Power of a product

$$= 2^1 \cdot 3^{-2}$$

$$= \frac{2^1}{3^2} = \frac{2}{9}$$

   (d) $\left(\dfrac{2^{-3}}{3^{-2}}\right)^{-1} = \dfrac{(2^{-3})^{-1}}{(3^{-2})^{-1}}$ Power of a quotient

$$= \frac{2^3}{3^2} = \frac{8}{9}$$

   (e) $\dfrac{10^{-2}}{10^{-4}} = 10^{-2-(-4)}$ Quotient of two powers

$$= 10^2 = 100$$

▲

Example 2

Simplify each of the following; express final results without using zero or negative integers as exponents.

   (a) $x^2 \cdot x^{-5}$ (b) $(x^{-2})^4$ (c) $(x^2 y^{-3})^{-4}$

   (d) $\left(\dfrac{a^3}{b^{-5}}\right)^{-2}$ (e) $\dfrac{x^{-4}}{x^{-2}}$

Solution

(a) $x^2 \cdot x^{-5} = x^{2+(-5)}$ Product of two powers

$$= x^{-3}$$

$$= \frac{1}{x^3}$$

(b) $(x^{-2})^4 = x^{4(-2)}$ Power of a power

$$= x^{-8}$$

$$= \frac{1}{x^8}$$

(c) $(x^2 y^{-3})^{-4} = (x^2)^{-4}(y^{-3})^{-4}$ Power of a product

$$= x^{-4(2)} y^{-4(-3)}$$

$$= x^{-8} y^{12}$$

$$= \frac{y^{12}}{x^8}$$

(d) $\left(\dfrac{a^3}{b^{-5}}\right)^{-2} = \dfrac{(a^3)^{-2}}{(b^{-5})^{-2}}$ Power of a quotient

$$= \frac{a^{-6}}{b^{10}}$$

$$= \frac{1}{a^6 b^{10}}$$

(e) $\dfrac{x^{-4}}{x^{-2}} = x^{-4-(-2)}$ Quotient of two powers

$$= x^{-2}$$

$$= \frac{1}{x^2}$$

▲

Example 3

Find the indicated products and quotients; express your results using positive integral exponents only.

(a) $(3x^2 y^{-4})(4x^{-3} y)$ (b) $\dfrac{12a^3 b^2}{-3a^{-1} b^5}$ (c) $\left(\dfrac{15x^{-1} y^2}{5xy^{-4}}\right)^{-1}$

Solution

(a) $(3x^2 y^{-4})(4x^{-3} y) = 12x^{2+(-3)} y^{-4+1}$

$$= 12x^{-1} y^{-3}$$

$$= \frac{12}{xy^3}$$

(b) $\dfrac{12a^3 b^2}{-3a^{-1} b^5} = -4a^{3-(-1)} b^{2-5}$

$$= -4a^4 b^{-3}$$

$$= -\frac{4a^4}{b^3}$$

$$(c) \ \left(\frac{15x^{-1}y^2}{5xy^{-4}}\right)^{-1} = (3x^{-1-1}y^{2-(-4)})^{-1}$$

Notice that we are first simplifying inside the parentheses.

$$= (3x^{-2}y^6)^{-1}$$
$$= 3^{-1}x^2y^{-6}$$
$$= \frac{x^2}{3y^6}.$$

The final examples of this section show the simplification of numerical and algebraic expressions that involve sums and differences. In such cases, we use Definition 5.2 to change from negative to positive exponents so that we can proceed in the usual way.

Example 4

Simplify $2^{-3} + 3^{-1}$.

Solution

$$2^{-3} + 3^{-1} = \frac{1}{2^3} + \frac{1}{3^1}$$
$$= \frac{1}{8} + \frac{1}{3}$$
$$= \frac{3}{24} + \frac{8}{24}$$
$$= \frac{11}{24}$$

Example 5

Simplify $(4^{-1} - 3^{-2})^{-1}$.

Solution

$$(4^{-1} - 3^{-2})^{-1} = \left(\frac{1}{4^1} - \frac{1}{3^2}\right)^{-1}$$

Apply $b^{-n} = \frac{1}{b^n}$ to 4^{-1} and to 3^{-2}.

$$= \left(\frac{1}{4} - \frac{1}{9}\right)^{-1}$$
$$= \left(\frac{9}{36} - \frac{4}{36}\right)^{-1}$$
$$= \left(\frac{5}{36}\right)^{-1}$$
$$= \frac{1}{\left(\frac{5}{36}\right)^1}$$

Apply $b^{-n} = \frac{1}{b^n}$.

$$= \frac{1}{\frac{5}{36}} = \frac{36}{5}$$

Example 6

Express $a^{-1} + b^{-2}$ as a single fraction involving positive exponents only.

Solution

$$a^{-1} + b^{-2} = \frac{1}{a^1} + \frac{1}{b^2}$$

$$= \left(\frac{1}{a}\right)\left(\frac{b^2}{b^2}\right) + \left(\frac{1}{b^2}\right)\left(\frac{a}{a}\right) \qquad \text{Use } ab^2 \text{ as the LCD.}$$

$$= \frac{b^2}{ab^2} + \frac{a}{ab^2}$$

$$= \frac{b^2 + a}{ab^2}$$

▲

Problem Set 5.1

Simplify each of the following numerical expressions.

1. 3^{-3}

2. 2^{-4}

3. -10^{-2}

4. 10^{-3}

5. $\dfrac{1}{3^{-4}}$

6. $\dfrac{1}{2^{-6}}$

7. $-\left(\dfrac{1}{3}\right)^{-3}$

8. $\left(\dfrac{1}{2}\right)^{-3}$

9. $\left(-\dfrac{1}{2}\right)^{-3}$

10. $\left(\dfrac{2}{7}\right)^{-2}$

11. $\left(-\dfrac{3}{4}\right)^{0}$

12. $\dfrac{1}{\left(\dfrac{4}{5}\right)^{-2}}$

13. $\dfrac{1}{\left(\dfrac{3}{7}\right)^{-2}}$

14. $-\left(\dfrac{5}{6}\right)^{0}$

15. $2^7 \cdot 2^{-3}$

16. $3^{-4} \cdot 3^6$

17. $10^{-5} \cdot 10^2$

18. $10^4 \cdot 10^{-6}$

19. $10^{-1} \cdot 10^{-2}$

20. $10^{-2} \cdot 10^{-2}$

21. $(3^{-1})^{-3}$

22. $(2^{-2})^{-4}$

23. $(5^3)^{-1}$

24. $(3^{-1})^3$

25. $(2^3 \cdot 3^{-2})^{-1}$

26. $(2^{-2} \cdot 3^{-1})^{-3}$

27. $(4^2 \cdot 5^{-1})^2$

28. $(2^{-3} \cdot 4^{-1})^{-1}$

29. $\left(\dfrac{2^{-1}}{5^{-2}}\right)^{-1}$

30. $\left(\dfrac{2^{-4}}{3^{-2}}\right)^{-2}$

31. $\left(\dfrac{2^{-1}}{3^{-2}}\right)^{2}$

32. $\left(\dfrac{3^2}{5^{-1}}\right)^{-1}$

33. $\dfrac{3^3}{3^{-1}}$

34. $\dfrac{2^{-2}}{2^3}$

35. $\dfrac{10^{-2}}{10^2}$

36. $\dfrac{10^{-2}}{10^{-5}}$

37. $2^{-2} + 3^{-2}$

38. $2^{-4} + 5^{-1}$

39. $\left(\dfrac{1}{3}\right)^{-1} - \left(\dfrac{2}{5}\right)^{-1}$

40. $\left(\dfrac{3}{2}\right)^{-1} - \left(\dfrac{1}{4}\right)^{-1}$

41. $(2^{-3} + 3^{-2})^{-1}$

42. $(5^{-1} - 2^{-3})^{-1}$

Simplify each of the following; express final results without using zero or negative integers as exponents.

43. $x^2 \cdot x^{-8}$

44. $x^{-3} \cdot x^{-4}$

45. $a^3 \cdot a^{-5} \cdot a^{-1}$

46. $b^{-2} \cdot b^3 \cdot b^{-6}$

47. $(a^{-4})^2$

48. $(b^4)^{-3}$

49. $(x^2 y^{-6})^{-1}$

50. $(x^5 y^{-1})^{-3}$

51. $(ab^3 c^{-2})^{-4}$

52. $(a^3 b^{-3} c^{-2})^{-5}$

53. $(2x^3 y^{-4})^{-3}$

54. $(4x^5 y^{-2})^{-2}$

55. $\left(\dfrac{x^{-1}}{y^{-4}}\right)^{-3}$

56. $\left(\dfrac{y^3}{x^{-4}}\right)^{-2}$

57. $\left(\dfrac{3a^{-2}}{2b^{-1}}\right)^{-2}$ **58.** $\left(\dfrac{2xy^2}{5a^{-1}b^{-2}}\right)^{-1}$

59. $\dfrac{x^{-6}}{x^{-4}}$ **60.** $\dfrac{a^{-2}}{a^2}$

61. $\dfrac{a^3b^{-2}}{a^{-2}b^{-4}}$ **62.** $\dfrac{x^{-3}y^{-4}}{x^2y^{-1}}$

Find the indicated products and quotients; express results using positive integral exponents only.

63. $(2xy^{-1})(3x^{-2}y^4)$

64. $(-4x^{-1}y^2)(6x^3y^{-4})$

65. $(-7a^2b^{-5})(-a^{-2}b^7)$

66. $(-9a^{-3}b^{-6})(-12a^{-1}b^4)$

67. $\dfrac{28x^{-2}y^{-3}}{4x^{-3}y^{-1}}$ **68.** $\dfrac{63x^2y^{-4}}{7xy^{-4}}$

69. $\dfrac{-72a^2b^{-4}}{6a^3b^{-7}}$ **70.** $\dfrac{108a^{-5}b^{-4}}{9a^{-2}b}$

71. $\left(\dfrac{35x^{-1}y^{-2}}{7x^4y^3}\right)^{-1}$

72. $\left(\dfrac{-48ab^2}{-6a^3b^5}\right)^{-2}$

73. $\left(\dfrac{-36a^{-1}b^{-6}}{4a^{-1}b^4}\right)^{-2}$

74. $\left(\dfrac{8xy^3}{-4x^4y}\right)^{-3}$

Express each of the following as a single fraction involving positive exponents only.

75. $x^{-2} + x^{-3}$

76. $x^{-1} + x^{-5}$

77. $x^{-3} - y^{-1}$

78. $2x^{-1} - 3y^{-2}$

79. $3a^{-2} + 4b^{-1}$

80. $a^{-1} + a^{-1}b^{-3}$

81. $x^{-1}y^{-2} - xy^{-1}$

82. $x^2y^{-2} - x^{-1}y^{-3}$

83. $2x^{-1} - 3x^{-2}$

84. $5x^{-2}y + 6x^{-1}y^{-2}$

THOUGHTS INTO WORDS

85. Is the following simplification process correct?

$$(3^{-2})^{-1} = \left(\dfrac{1}{3^2}\right)^{-1} = \left(\dfrac{1}{9}\right)^{-1} = \dfrac{1}{\left(\dfrac{1}{9}\right)^1} = 9$$

Could you suggest a better way to do the problem?

86. Explain how to simplify $(2^{-1} \cdot 3^{-2})^{-1}$ and also how to simplify $(2^{-1} + 3^{-2})^{-1}$.

Further Investigations

87. Use a calculator to check your answers for Problems 1–42.

88. Use a calculator to simplify each of the following numerical expressions. Express your answers to the nearest hundredth.

(a) $(2^{-3} + 3^{-3})^{-2}$

(b) $(4^{-3} - 2^{-1})^{-2}$

(c) $(5^{-3} - 3^{-5})^{-1}$

(d) $(6^{-2} + 7^{-4})^{-2}$

(e) $(7^{-3} - 2^{-4})^{-2}$

(f) $(3^{-4} + 2^{-3})^{-3}$

5.2 Roots and Radicals

To **square a number** means to raise it to the second power, that is, to use the number as a factor twice.

$$4^2 = 4 \cdot 4 = 16 \qquad \text{Read "four squared equals sixteen"}$$

$$10^2 = 10 \cdot 10 = 100$$

$$\left(\frac{1}{2}\right)^2 = \frac{1}{2} \cdot \frac{1}{2} = \frac{1}{4}$$

$$(-3)^2 = (-3)(-3) = 9$$

A **square root of a number** is one of its two equal factors. Thus, 4 is a square root of 16 because $4 \cdot 4 = 16$. Likewise, -4 is also a square root of 16 because $(-4)(-4) = 16$. In general, a is a square root of b if $a^2 = b$. The following generalizations are a direct consequence of the previous statement.

1. Every positive real number has two square roots; one is positive and the other is negative. They are opposites of each other.

2. Negative real numbers have no real number square roots because any nonzero real number is positive when squared.

3. The square root of 0 is 0.

The symbol $\sqrt{}$, called a **radical sign**, is used to designate the nonnegative square root. The number under the radical sign is called the **radicand**. The entire expression, such as $\sqrt{16}$, is called a **radical**.

$\sqrt{16} = 4$ $\sqrt{16}$ indicates the nonnegative or **principal square root** of 16.

$-\sqrt{16} = -4$ $-\sqrt{16}$ indicates the negative square root of 16.

$\sqrt{0} = 0$ Zero has only one square root. Technically, we could write $-\sqrt{0} = -0 = 0$.

$\sqrt{-4}$ is not a real number.

$-\sqrt{-4}$ is not a real number.

In general, the following definition is useful.

DEFINITION 5.3

If $a \geq 0$ and $b \geq 0$, then $\sqrt{b} = a$ if and only if $a^2 = b$; a is called the **principal square root** of b.

To **cube a number** means to raise it to the third power, that is, to use the number as a factor three times.

$$2^3 = 2 \cdot 2 \cdot 2 = 8 \qquad \text{Read "two cubed equals eight"}$$

$$4^4 = 4 \cdot 4 \cdot 4 = 64$$

$$\left(\frac{2}{3}\right)^3 = \frac{2}{3} \cdot \frac{2}{3} \cdot \frac{2}{3} = \frac{8}{27}$$

$$(-2)^3 = (-2)(-2)(-2) = -8$$

A **cube root of a number** is one of its three equal factors. Thus, 2 is a cube root of 8 because $2 \cdot 2 \cdot 2 = 8$. (In fact, 2 is the only real number that is a cube root of 8.) Furthermore, -2 is a cube root of -8 because $(-2)(-2)(-2) = -8$. (In fact, -2 is the only real number that is a cube root of -8.)

In general, a is a cube root of b if $a^3 = b$. The following generalizations are a direct consequence of the previous statement.

1. Every positive real number has one positive real number cube root.

2. Every negative real number has one negative real number cube root.

3. The cube root of 0 is 0.

REMARK Technically, every nonzero real number has three cube roots, but only one of them is a real number. The other two roots are classified as complex numbers. We are restricting our work at this time to the set of real numbers. △

The symbol $\sqrt[3]{}$ designates the cube root of a number. Thus, we can write

$$\sqrt[3]{8} = 2, \qquad \sqrt[3]{\frac{1}{27}} = \frac{1}{3},$$

$$\sqrt[3]{-8} = -2, \qquad \sqrt[3]{-\frac{1}{27}} = -\frac{1}{3}$$

In general, the following definition is useful.

DEFINITION 5.4 $\sqrt[3]{b} = a$ if and only if $a^3 = b$.

In Definition 5.4, if $b \geq 0$ then $a \geq 0$ whereas if $b < 0$ then $a < 0$. The number a is called **the principal cube root of b** or simply **the cube root of b**.

The concept of root can be extended to fourth roots, fifth roots, sixth roots, and, in general, nth roots. We can make the following generalizations.

If n is an even positive integer, then the following statements are true.

1. Every positive real number has exactly two real nth roots—one positive and one negative. For example, the real fourth roots of 16 are 2 and -2.

2. Negative real numbers do not have real nth roots. For example, there are no real fourth roots of -16.

If n is an odd positive integer greater than one, then the following statements are true.

1. Every real number has exactly one real nth root.
2. The real nth root of a positive number is positive. For example, the fifth root of 32 is 2.
3. The real nth root of a negative number is negative. For example, the fifth root of -32 is -2.

In general, the following definition is useful.

DEFINITION 5.5 $\sqrt[n]{b} = a$ if and only if $a^n = b$.

In Definition 5.5, if n is an even positive integer, then a and b are both nonnegative. If n is an odd positive integer greater than one, then a and b are both nonnegative or both negative. The symbol $\sqrt[n]{}$ designates the **principal nth root**. Consider the following examples.

$$\sqrt[4]{81} = 3 \quad \text{because } 3^4 = 81$$
$$\sqrt[5]{32} = 2 \quad \text{because } 2^5 = 32$$
$$\sqrt[5]{-32} = -2 \quad \text{because } (-2)^5 = -32$$

To complete our terminology, the n in the radical $\sqrt[n]{b}$ is called the **index** of the radical. If $n = 2$, we commonly write $\sqrt{b}$ instead of $\sqrt[2]{b}$. In the future as we use symbols, such as $\sqrt[n]{b}$, $\sqrt[m]{y}$, and $\sqrt[r]{x}$, we will assume the previous agreements relative to the existence of real roots (without listing the various restrictions) unless a special restriction is necessary.

The following property is a direct consequence of Definition 5.5.

PROPERTY 5.3   1. $(\sqrt[n]{b})^n = b$ n is any positive integer greater than one.

2. $\sqrt[n]{b^n} = b$ n is any positive integer greater than one if $b \geq 0$; n is an odd positive integer greater than one if $b < 0$.

The following examples demonstrate the use of Property 5.3.

$$\sqrt{16^2} = (\sqrt{16})^2 = 4^2 = 16$$
$$\sqrt[3]{64^3} = (\sqrt[3]{64})^3 = 4^3 = 64$$
$$\sqrt[3]{(-8)^3} = (\sqrt[3]{-8})^3 = (-2)^3 = -8$$
but $\sqrt{(-16)^2} \neq (\sqrt{-16})^2$ because $\sqrt{-16}$ is not a real number.

Let's use some examples to lead into the next very useful property of radicals.

$$\sqrt{4 \cdot 9} = \sqrt{36} = 6 \qquad \text{and} \qquad \sqrt{4} \cdot \sqrt{9} = 2 \cdot 3 = 6,$$

$$\sqrt{16 \cdot 25} = \sqrt{400} = 20 \qquad \text{and} \qquad \sqrt{16} \cdot \sqrt{25} = 4 \cdot 5 = 20,$$

$$\sqrt[3]{8 \cdot 27} = \sqrt[3]{216} = 6 \qquad \sqrt[3]{8} \cdot \sqrt[3]{27} = 2 \cdot 3 = 6,$$

$$\sqrt[3]{(-8)(27)} = \sqrt[3]{-216} = -6 \qquad \text{and} \qquad \sqrt[3]{-8} \cdot \sqrt[3]{27} = (-2)(3) = -6$$

In general, we can state the following property.

PROPERTY 5.4

$$\sqrt[n]{bc} = \sqrt[n]{b}\sqrt[n]{c} \qquad \sqrt[n]{b} \text{ and } \sqrt[n]{c} \text{ are real numbers}$$

Property 5.4 states that **the nth root of a product is equal to the product of the nth roots**.

Simplest Radical Form

The definition of nth root, along with Property 5.4, provides the basis for changing radicals to simplest radical form. The concept of **simplest radical form** takes on additional meaning as we encounter more complicated expressions, but for now it simply means that the radicand is not to contain any perfect powers of the index. Let's consider some examples to clarify this idea.

Example 1

Express each of the following in simplest radical form.

    (a) $\sqrt{8}$ *(b)* $\sqrt{45}$ *(c)* $\sqrt[3]{24}$ *(d)* $\sqrt[3]{54}$

Solution

(a) $\sqrt{8} = \sqrt{4 \cdot 2} = \sqrt{4}\sqrt{2} = 2\sqrt{2}$

$\uparrow$

4 is a
perfect
square.

(b) $\sqrt{45} = \sqrt{9 \cdot 5} = \sqrt{9}\sqrt{5} = 3\sqrt{5}$

$\uparrow$

9 is a
perfect
square.

(c) $\sqrt[3]{24} = \sqrt[3]{8 \cdot 3} = \sqrt[3]{8}\sqrt[3]{3} = 2\sqrt[3]{3}$

$\uparrow$

8 is a
perfect
cube.

(d) $\sqrt[3]{54} = \sqrt[3]{27 \cdot 2} = \sqrt[3]{27}\,\sqrt[3]{2} = 3\sqrt[3]{2}$

27 is a
perfect
cube.

▲

The first step in each example is to express the radicand of the given radical as the product of two factors, one of which must be a perfect nth power other than 1. Also, observe the radicands of the final radicals. In each case, the radicand *cannot* be the product of two factors; one must be a perfect nth power other than 1. We say that the final radicals $2\sqrt{2}$, $3\sqrt{5}$, $2\sqrt[3]{3}$, and $3\sqrt[3]{2}$ are in **simplest radical form**.

You may vary the steps somewhat in changing to simplest radical form, but the final result should be the same. Consider some different approaches to change $\sqrt{72}$ to simplest form.

$$\sqrt{72} = \sqrt{9}\sqrt{8} = 3\sqrt{8} = 3\sqrt{4}\sqrt{2} = 3 \cdot 2\sqrt{2} = 6\sqrt{2}, \quad \text{or}$$
$$\sqrt{72} = \sqrt{4}\sqrt{18} = 2\sqrt{18} = 2\sqrt{9}\sqrt{2} = 2 \cdot 3\sqrt{2} = 6\sqrt{2}, \quad \text{or}$$
$$\sqrt{72} = \sqrt{36}\sqrt{2} = 6\sqrt{2}$$

Another variation of the technique for changing radicals to simplest form is to prime factor the radicand and then to look for perfect nth powers in exponential form. The following example illustrates the use of this technique.

Example 2

Express each of the following in simplest radical form.

(a) $\sqrt{50}$ *(b)* $3\sqrt{80}$ *(c)* $\sqrt[3]{108}$

Solution

(a) $\sqrt{50} = \sqrt{2 \cdot 5 \cdot 5} = \sqrt{5^2}\sqrt{2} = 5\sqrt{2}$

(b) $3\sqrt{80} = 3\sqrt{2 \cdot 2 \cdot 2 \cdot 2 \cdot 5} = 3\sqrt{2^4}\sqrt{5} = 3 \cdot 2^2\sqrt{5} = 12\sqrt{5}$

(c) $\sqrt[3]{108} = \sqrt[3]{2 \cdot 2 \cdot 3 \cdot 3 \cdot 3} = \sqrt[3]{3^3}\sqrt[3]{4} = 3\sqrt[3]{4}$

▲

Another property of nth roots is demonstrated by the following examples.

$$\sqrt{\frac{36}{9}} = \sqrt{4} = 2 \quad \text{and} \quad \frac{\sqrt{36}}{\sqrt{9}} = \frac{6}{3} = 2,$$

$$\sqrt[3]{\frac{64}{8}} = \sqrt[3]{8} = 2 \quad \text{and} \quad \frac{\sqrt[3]{64}}{\sqrt[3]{8}} = \frac{4}{2} = 2,$$

$$\sqrt[3]{\frac{-8}{64}} = \sqrt[3]{-\frac{1}{8}} = -\frac{1}{2} \quad \text{and} \quad \frac{\sqrt[3]{-8}}{\sqrt[3]{64}} = \frac{-2}{4} = -\frac{1}{2}$$

In general, we can state the following property.

PROPERTY 5.5 $\sqrt[n]{\dfrac{b}{c}} = \dfrac{\sqrt[n]{b}}{\sqrt[n]{c}}$ $\sqrt[n]{b}$ and $\sqrt[n]{c}$ are real numbers and $c \neq 0$.

Property 5.5 states that **the nth root of a quotient is equal to the quotient of the nth roots.**

To evaluate radicals such as $\sqrt{\dfrac{4}{25}}$ and $\sqrt[3]{\dfrac{27}{8}}$, for which the numerator and denominator of the fractional radicand are perfect nth powers, you may use Property 5.5 or merely rely on the definition of nth root.

$$\sqrt{\dfrac{4}{25}} = \dfrac{\sqrt{4}}{\sqrt{25}} = \dfrac{2}{5} \qquad \text{or} \qquad \sqrt{\dfrac{4}{25}} = \dfrac{2}{5} \quad \text{because } \dfrac{2}{5} \cdot \dfrac{2}{5} = \dfrac{4}{25},$$

Property 5.5 ↑ ↓   Definition of nth root ↑ ↓

$$\sqrt[3]{\dfrac{27}{8}} = \dfrac{\sqrt[3]{27}}{\sqrt[3]{8}} = \dfrac{3}{2} \qquad \text{or} \qquad \sqrt[3]{\dfrac{27}{8}} = \dfrac{3}{2} \quad \text{because } \dfrac{3}{2} \cdot \dfrac{3}{2} \cdot \dfrac{3}{2} = \dfrac{27}{8}$$

Radicals such as $\sqrt{\dfrac{28}{9}}$ and $\sqrt[3]{\dfrac{24}{27}}$ in which only the denominators of the radicand are perfect nth powers can be simplified as follows.

$$\sqrt{\dfrac{28}{9}} = \dfrac{\sqrt{28}}{\sqrt{9}} = \dfrac{\sqrt{28}}{3} = \dfrac{\sqrt{4}\sqrt{7}}{3} = \dfrac{2\sqrt{7}}{3}$$

$$\sqrt[3]{\dfrac{24}{27}} = \dfrac{\sqrt[3]{24}}{\sqrt[3]{27}} = \dfrac{\sqrt[3]{24}}{3} = \dfrac{\sqrt[3]{8}\sqrt[3]{3}}{3} = \dfrac{2\sqrt[3]{3}}{3}$$

Before we consider more examples, let's summarize some ideas that pertain to the simplifying of radicals. A radical is said to be in **simplest radical form** if the following conditions are satisfied.

1. No fraction appears with a radical sign. $\sqrt{\dfrac{3}{4}}$ violates this condition.

2. No radical appears in the denominator. $\dfrac{\sqrt{2}}{\sqrt{3}}$ violates this condition.

3. No radicand, when expressed in prime factored form, contains a factor raised to a power equal to or greater than the index. $\sqrt{2^3 \cdot 5}$ violates this condition.

Now let's consider an example in which neither the numerator nor the denominator of the radicand is a perfect nth power.

Example 3

Simplify $\sqrt{\dfrac{2}{3}}$.

Solution

$$\sqrt{\dfrac{2}{3}} = \dfrac{\sqrt{2}}{\sqrt{3}} = \dfrac{\sqrt{2}}{\sqrt{3}} \cdot \dfrac{\sqrt{3}}{\sqrt{3}} = \dfrac{\sqrt{6}}{3}$$

↑
Form of 1

▲

We refer to the process we used to simplify the radical in Example 3 as **rationalizing the denominator**. Notice that the denominator becomes a rational number. The process of rationalizing the denominator can often be accomplished in more than one way, as we will see in the next example.

Example 4

Simplify $\dfrac{\sqrt{5}}{\sqrt{8}}$.

Solution A

$$\dfrac{\sqrt{5}}{\sqrt{8}} = \dfrac{\sqrt{5}}{\sqrt{8}} \cdot \dfrac{\sqrt{8}}{\sqrt{8}} = \dfrac{\sqrt{40}}{8} = \dfrac{\sqrt{4}\sqrt{10}}{8} = \dfrac{2\sqrt{10}}{8} = \dfrac{\sqrt{10}}{4}$$

Solution B

$$\dfrac{\sqrt{5}}{\sqrt{8}} = \dfrac{\sqrt{5}}{\sqrt{8}} \cdot \dfrac{\sqrt{2}}{\sqrt{2}} = \dfrac{\sqrt{10}}{\sqrt{16}} = \dfrac{\sqrt{10}}{4}$$

Solution C

$$\dfrac{\sqrt{5}}{\sqrt{8}} = \dfrac{\sqrt{5}}{\sqrt{4}\sqrt{2}} = \dfrac{\sqrt{5}}{2\sqrt{2}} = \dfrac{\sqrt{5}}{2\sqrt{2}} \cdot \dfrac{\sqrt{2}}{\sqrt{2}} = \dfrac{\sqrt{10}}{4}$$

▲

The three approaches to Example 4 again illustrate the need to think first and then push the pencil. You may find one approach easier than another. To conclude this section, study the following examples and check the final radicals according to the three conditions previously listed for **simplest radical form**.

Example 5

Simplify each of the following.

(a) $\dfrac{3\sqrt{2}}{5\sqrt{3}}$ (b) $\dfrac{3\sqrt{7}}{2\sqrt{18}}$ (c) $\sqrt[3]{\dfrac{5}{9}}$ (d) $\dfrac{\sqrt[3]{5}}{\sqrt[3]{16}}$

Solution

(a) $\dfrac{3\sqrt{2}}{5\sqrt{3}} = \dfrac{3\sqrt{2}}{5\sqrt{3}} \cdot \dfrac{\sqrt{3}}{\sqrt{3}} = \dfrac{3\sqrt{6}}{5\sqrt{9}} = \dfrac{3\sqrt{6}}{15} = \dfrac{\sqrt{6}}{5}$

↑
Form of 1

(b) $\dfrac{3\sqrt{7}}{2\sqrt{18}} = \dfrac{3\sqrt{7}}{2\sqrt{18}} \cdot \dfrac{\sqrt{2}}{\sqrt{2}} = \dfrac{3\sqrt{14}}{2\sqrt{36}} = \dfrac{3\sqrt{14}}{12} = \dfrac{\sqrt{14}}{4}$

↑
Form of 1

(c) $\sqrt[3]{\dfrac{5}{9}} = \dfrac{\sqrt[3]{5}}{\sqrt[3]{9}} = \dfrac{\sqrt[3]{5}}{\sqrt[3]{9}} \cdot \dfrac{\sqrt[3]{3}}{\sqrt[3]{3}} = \dfrac{\sqrt[3]{15}}{\sqrt[3]{27}} = \dfrac{\sqrt[3]{15}}{3}$

$\uparrow$

Form of I

(d) $\dfrac{\sqrt[3]{5}}{\sqrt[3]{16}} = \dfrac{\sqrt[3]{5}}{\sqrt[3]{16}} \cdot \dfrac{\sqrt[3]{4}}{\sqrt[3]{4}} = \dfrac{\sqrt[3]{20}}{\sqrt[3]{64}} = \dfrac{\sqrt[3]{20}}{4}$

$\uparrow$

Form of I   ▲

Applications of Radicals

Many real world applications involve radical expressions. For example, police often use the formula $S = \sqrt{30Df}$ to estimate the speed of a car based on the length of skid marks. In this formula, S represents the speed of the car in miles per hour, D the length of skid marks measured in feet, and f represents a coefficient of friction. For a particular situation, the coefficient of friction is a constant depending on the type and condition of the road surface.

Example 6

Using .35 as a coefficient of friction, determine how fast a car was traveling if it skidded 325 feet.

Solution

Substitute .35 for f and 325 for D in the formula.

$$S = \sqrt{30Df} = \sqrt{30(325)(.35)} = 58, \quad \text{to the nearest whole number}$$

The car was traveling at approximately 58 miles per hour.   ▲

The **period** of a pendulum is the time it takes to swing from one side to the other side and back. The formula

$$T = 2\pi\sqrt{\dfrac{L}{32}},$$

where T represents the time in seconds and L the length in feet, can be used to determine the period of a pendulum (see Figure 5.1).

FIGURE 5.1

Example 7

Find the period, to the nearest tenth of a second, of a pendulum of length 3.5 feet.

Solution

Let's use 3.14 as an approximation for π and substitute 3.5 for L in the formula.

$$T = 2\pi\sqrt{\frac{L}{32}}, = 2(3.14)\sqrt{\frac{3.5}{32}} = 2.1, \quad \text{to the nearest tenth}$$

The period is approximately 2.1 seconds. ▲

Radical expressions are also used in some geometric applications. For example, if a, b, and c represent the lengths of the three sides of a triangle, the formula $K = \sqrt{s(s - a)(s - b)(s - c)}$, known as Heron's formula, can be used to determine the area (K) of the triangle. The letter s represents the semiperimeter of the triangle or $s = \dfrac{a + b + c}{2}$.

Example 8

Find the area of a triangular piece of sheet metal that has sides of lengths 17 inches, 19 inches, and 26 inches.

Solution

First, let's find the value of s.

$$s = \frac{17 + 19 + 26}{2} = 31$$

Now we can use Heron's formula.

$$\begin{aligned} K = \sqrt{s(s - a)(s - b)(s - c)} &= \sqrt{31(31 - 17)(31 - 19)(31 - 26)} \\ &= \sqrt{31(14)(12)(5)} \\ &= \sqrt{20{,}640} \\ &= 161.4, \quad \text{to the nearest tenth} \end{aligned}$$

Thus, the area of the piece of sheet metal is approximately 161.4 square inches. ▲

REMARK Notice in Examples 6–8 that we did not simplify the radicals. When using a calculator to approximate the square roots, there is no need to simplify first. △

▼ Problem Set 5.2

Evaluate each of the following: for example, $\sqrt{25} = 5$.

1. $\sqrt{64}$

2. $\sqrt{49}$

3. $-\sqrt{100}$

4. $-\sqrt{81}$

5. $\sqrt[3]{27}$

6. $\sqrt[3]{216}$

7. $\sqrt[3]{-64}$

8. $\sqrt[3]{-125}$

9. $\sqrt[4]{81}$

10. $-\sqrt[4]{16}$

11. $\sqrt{\dfrac{16}{25}}$

12. $\sqrt{\dfrac{25}{64}}$

13. $-\sqrt{\dfrac{36}{49}}$

14. $\sqrt{\dfrac{16}{64}}$

15. $\sqrt{\dfrac{9}{36}}$

16. $\sqrt{\dfrac{144}{36}}$

17. $\sqrt[3]{\dfrac{27}{64}}$

18. $\sqrt[3]{-\dfrac{8}{27}}$

19. $\sqrt[3]{8^3}$

20. $\sqrt[4]{16^4}$

Change each of the following radicals to simplest radical form.

21. $\sqrt{27}$

22. $\sqrt{48}$

23. $\sqrt{32}$

24. $\sqrt{98}$

25. $\sqrt{80}$

26. $\sqrt{125}$

27. $\sqrt{160}$

28. $\sqrt{112}$

29. $4\sqrt{18}$

30. $5\sqrt{32}$

31. $-6\sqrt{20}$

32. $-4\sqrt{54}$

33. $\dfrac{2}{5}\sqrt{75}$

34. $\dfrac{1}{3}\sqrt{90}$

35. $\dfrac{3}{2}\sqrt{24}$

36. $\dfrac{3}{4}\sqrt{45}$

37. $-\dfrac{5}{6}\sqrt{28}$

38. $-\dfrac{2}{3}\sqrt{96}$

39. $\sqrt{\dfrac{19}{4}}$

40. $\sqrt{\dfrac{22}{9}}$

41. $\sqrt{\dfrac{27}{16}}$

42. $\sqrt{\dfrac{8}{25}}$

43. $\sqrt{\dfrac{75}{81}}$

44. $\sqrt{\dfrac{24}{49}}$

45. $\sqrt{\dfrac{2}{7}}$

46. $\sqrt{\dfrac{3}{8}}$

47. $\sqrt{\dfrac{2}{3}}$

48. $\sqrt{\dfrac{7}{12}}$

49. $\dfrac{\sqrt{5}}{\sqrt{12}}$

50. $\dfrac{\sqrt{3}}{\sqrt{7}}$

51. $\dfrac{\sqrt{11}}{\sqrt{24}}$

52. $\dfrac{\sqrt{5}}{\sqrt{48}}$

53. $\dfrac{\sqrt{18}}{\sqrt{27}}$

54. $\dfrac{\sqrt{10}}{\sqrt{20}}$

55. $\dfrac{\sqrt{35}}{\sqrt{7}}$

56. $\dfrac{\sqrt{42}}{\sqrt{6}}$

57. $\dfrac{2\sqrt{3}}{\sqrt{7}}$

58. $\dfrac{3\sqrt{2}}{\sqrt{6}}$

59. $-\dfrac{4\sqrt{12}}{\sqrt{5}}$

60. $\dfrac{-6\sqrt{5}}{\sqrt{18}}$

61. $\dfrac{3\sqrt{2}}{4\sqrt{3}}$

62. $\dfrac{6\sqrt{5}}{5\sqrt{12}}$

63. $\dfrac{-8\sqrt{18}}{10\sqrt{50}}$

64. $\dfrac{4\sqrt{45}}{-6\sqrt{20}}$

65. $\sqrt[3]{16}$

66. $\sqrt[3]{40}$

67. $2\sqrt[3]{81}$

68. $-3\sqrt[3]{54}$

69. $\dfrac{2}{\sqrt[3]{9}}$

70. $\dfrac{3}{\sqrt[3]{3}}$

71. $\dfrac{\sqrt[3]{27}}{\sqrt[3]{4}}$

72. $\dfrac{\sqrt[3]{8}}{\sqrt[3]{16}}$

73. $\dfrac{\sqrt[3]{6}}{\sqrt[3]{4}}$

74. $\dfrac{\sqrt[3]{4}}{\sqrt[3]{2}}$

75. Use a coefficient of friction of .4 in the formula from Example 6 and find the speeds of cars leaving skid marks of lengths 150 feet, 200 feet, and 350 feet. Express your answers to the nearest mile per hour.

76. Use the formula from Example 7 and find the periods of pendulums of lengths 2 feet, 3 feet, and 4.5 feet. Express your answers to the nearest tenth of a second.

77. Find the area, to the nearest square centimeter, of a triangle that measures 14 centimeters by 16 centimeters by 18 centimeters.

78. Find the area, to the nearest square yard, of a triangular plot of ground that measures 45 yards by 60 yards by 75 yards.

79. Find the area of an equilateral triangle, each of whose sides is 18 inches long. Express the area to the nearest square inch.

80. Find the area, to the nearest square inch, of the quadrilateral in Figure 5.2.

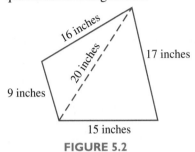

9 inches

FIGURE 5.2

estimate, and then use your calculator to see how well you estimated.

(a) $3\sqrt{10} - 4\sqrt{24} + 6\sqrt{65}$

(b) $9\sqrt{27} + 5\sqrt{37} - 3\sqrt{80}$

(c) $12\sqrt{5} + 13\sqrt{18} + 9\sqrt{47}$

(d) $3\sqrt{98} - 4\sqrt{83} - 7\sqrt{120}$

(e) $4\sqrt{170} + 2\sqrt{198} + 5\sqrt{227}$

(f) $-3\sqrt{256} - 6\sqrt{287} + 11\sqrt{321}$

THOUGHTS INTO WORDS

81. Why is $\sqrt{-9}$ not a real number?

82. Why do we say that 25 has two square roots (5 and −5), but we write $\sqrt{25} = 5$?

83. How is the multiplication property of one used when simplifying radicals?

84. How could you find a whole number approximation for $\sqrt{2750}$ if you did not have a calculator or table available?

Further Investigations

85. Use your calculator to find a rational approximation, to the nearest thousandth, for (a) through (i).

(a) $\sqrt{2}$ **(b)** $\sqrt{75}$

(c) $\sqrt{156}$ **(d)** $\sqrt{691}$

(e) $\sqrt{3249}$ **(f)** $\sqrt{45,123}$

(g) $\sqrt{.14}$ **(h)** $\sqrt{.023}$

(i) $\sqrt{.8649}$

86. Sometimes a fairly good estimate can be made of a radical expression by using whole number approximations. For example, $5\sqrt{35} + 7\sqrt{50}$ is approximately $5(6) + 7(7) = 79$. Using a calculator we find that $5\sqrt{35} + 7\sqrt{50} = 79.1$ to the nearest tenth. In this case our whole number estimate is very good. For (a) through (f), first make a whole number

5.3 Combining Radicals and Simplifying Radicals That Contain Variables

Recall our use of the distributive property as the basis for combining similar terms. For example,

$$3x + 2x = (3 + 2)x = 5x$$

$$8y - 5y = (8 - 5)y = 3y$$

$$\frac{2}{3}a^2 + \frac{3}{4}a^2 = \left(\frac{2}{3} + \frac{3}{4}\right)a^2 = \left(\frac{8}{12} + \frac{9}{12}\right)a^2 = \frac{17}{12}a^2.$$

In a like manner, expressions that contain radicals can often be simplified by using the distributive property as follows.

$$3\sqrt{2} + 5\sqrt{2} = (3 + 5)\sqrt{2} = 8\sqrt{2}$$

$$7\sqrt[3]{5} - 3\sqrt[3]{5} = (7 - 3)\sqrt[3]{5} = 4\sqrt[3]{5}$$

$$4\sqrt{7} + 5\sqrt{7} + 6\sqrt{11} - 2\sqrt{11} = (4 + 5)\sqrt{7} + (6 - 2)\sqrt{11} = 9\sqrt{7} + 4\sqrt{11}$$

Notice that *to add or subtract radicals they must have the same index and the same radicand*. Thus, we cannot simplify an expression such as $5\sqrt{2} + 7\sqrt{11}$.

Simplifying by combining radicals sometimes requires that you first express the given radicals in simplest form and then apply the distributive property. The following examples illustrate this idea.

Example 1

Simplify $3\sqrt{8} + 2\sqrt{18} - 4\sqrt{2}$.

Solution

$$3\sqrt{8} + 2\sqrt{18} - 4\sqrt{2} = 3\sqrt{4}\sqrt{2} + 2\sqrt{9}\sqrt{2} - 4\sqrt{2}$$

$$= 6\sqrt{2} + 6\sqrt{2} - 4\sqrt{2}$$

$$= (6 + 6 - 4)\sqrt{2} = 8\sqrt{2} \qquad \blacktriangle$$

Example 2

Simplify $\frac{1}{4}\sqrt{45} + \frac{1}{3}\sqrt{20}$.

Solution

$$\frac{1}{4}\sqrt{45} + \frac{1}{3}\sqrt{20} = \frac{1}{4}\sqrt{9}\sqrt{5} + \frac{1}{3}\sqrt{4}\sqrt{5}$$

$$= \frac{1}{4} \cdot 3 \cdot \sqrt{5} + \frac{1}{3} \cdot 2 \cdot \sqrt{5}$$

$$= \frac{3}{4}\sqrt{5} + \frac{2}{3}\sqrt{5} = \left(\frac{3}{4} + \frac{2}{3}\right)\sqrt{5}$$

$$= \left(\frac{9}{12} + \frac{8}{12}\right)\sqrt{5} = \frac{17}{12}\sqrt{5} \qquad \blacktriangle$$

Example 3

Simplify $5\sqrt[3]{2} - 2\sqrt[3]{16} - 6\sqrt[3]{54}$.

Solution

$$5\sqrt[3]{2} - 2\sqrt[3]{16} - 6\sqrt[3]{54} = 5\sqrt[3]{2} - 2\sqrt[3]{8}\sqrt[3]{2} - 6\sqrt[3]{27}\sqrt[3]{2}$$
$$= 5\sqrt[3]{2} - 2 \cdot 2 \cdot \sqrt[3]{2} - 6 \cdot 3 \cdot \sqrt[3]{2}$$
$$= 5\sqrt[3]{2} - 4\sqrt[3]{2} - 18\sqrt[3]{2}$$
$$= (5 - 4 - 18)\sqrt[3]{2}$$
$$= -17\sqrt[3]{2} \qquad \blacktriangle$$

Radicals That Contain Variables

Before we discuss the process of simplifying *radicals that contain variables*, there is one technicality that we should call to your attention. Let's look at some examples to clarify the point. Consider the radical $\sqrt{x^2}$ as follows.

Let $x = 3$,　then $\sqrt{x^2} = \sqrt{3^2} = \sqrt{9} = 3$.

Let $x = -3$,　then $\sqrt{x^2} = \sqrt{(-3)^2} = \sqrt{9} = 3$.

Thus, if $x \geq 0$, then $\sqrt{x^2} = x$, *but if* $x < 0$, then $\sqrt{x^2} = -x$. Using the concept of absolute value we can state that *for all real numbers,* $\sqrt{x^2} = |x|$.

Now consider the radical $\sqrt{x^3}$. Since x^3 is negative when x is negative, we need to restrict x to the nonnegative reals when working with $\sqrt{x^3}$. Thus, we can write: If $x \geq 0$, then $\sqrt{x^3} = \sqrt{x^2}\sqrt{x} = x\sqrt{x}$, and no absolute value sign is necessary. Finally, let's consider the radical $\sqrt[3]{x^3}$.

Let $x = 2$,　then $\sqrt[3]{x^3} = \sqrt[3]{2^3} = \sqrt[3]{8} = 2$.

Let $x = -2$,　then $\sqrt[3]{x^3} = \sqrt[3]{(-2)^3} = \sqrt[3]{-8} = -2$

Thus, it is correct to write: $\sqrt[3]{x^3} = x$ for all real numbers, and again no absolute value sign is necessary.

The previous discussion indicates that technically every radical expression involving variables in the radicand needs to be analyzed individually as to the necessary restrictions imposed on the variables. However, to avoid considering such restrictions on a problem-to-problem basis we shall merely *assume that all variables represent positive real numbers*.

Let's consider the process of simplifying radicals that contain variables in the radicand. Study the following examples and notice that it is the same basic approach we used in Section 5.2.

Example 4

Simplify each of the following.

(a) $\sqrt{8x^3}$ 　　　*(b)* $\sqrt{45x^3y^7}$ 　　*(c)* $\sqrt{180a^4b^3}$ 　　*(d)* $\sqrt[3]{40x^4y^8}$

Solution

(a) $\sqrt{8x^3} = \sqrt{4x^2}\sqrt{2x} = 2x\sqrt{2x}$

                $4x^2$ is a
                perfect square.

(b) $\sqrt{45x^3y^7} = \sqrt{9x^2y^6}\sqrt{5xy} = 3xy^3\sqrt{5xy}$

                $9x^2y^6$ is a
                perfect square.

(c) If the numerical coefficient of the radicand is quite large, you may want to look at it in prime factored form.

$$\sqrt{180a^4b^3} = \sqrt{2\cdot 2\cdot 3\cdot 3\cdot 5\cdot a^4\cdot b^3}$$
$$= \sqrt{36\cdot 5\cdot a^4\cdot b^3}$$
$$= \sqrt{36a^4b^2}\sqrt{5b}$$
$$= 6a^2b\sqrt{5b}$$

(d) $\sqrt[3]{40x^4y^8} = \sqrt[3]{8x^3y^6}\sqrt[3]{5xy^2} = 2xy^2\sqrt[3]{5xy^2}$

                $8x^3y^6$ is a
                perfect cube.   ▲

Before we consider more examples, let's restate (so as to include radicands containing variables) the conditions necessary for a radical to be in *simplest radical form.*

1. A radicand contains no polynomial factor raised to a power equal to or greater than the index of the radical. $\sqrt{x^3}$ violates this condition.

2. No fraction appears within a radical sign. $\sqrt{\dfrac{2x}{3y}}$ violates this condition.

3. No radical appears in the denominator. $\dfrac{3}{\sqrt[3]{4x}}$ violates this condition.

Example 5

Express each of the following in simplest radical form.

(a) $\sqrt{\dfrac{2x}{3y}}$ (b) $\dfrac{\sqrt{5}}{\sqrt{12a^3}}$ (c) $\dfrac{\sqrt{8x^2}}{\sqrt{27y^5}}$

(d) $\dfrac{3}{\sqrt[3]{4x}}$ (e) $\dfrac{\sqrt[3]{16x^2}}{\sqrt[3]{9y^5}}$

Solution

(a) $\sqrt{\dfrac{2x}{3y}} = \dfrac{\sqrt{2x}}{\sqrt{3y}} = \dfrac{\sqrt{2x}}{\sqrt{3y}} \cdot \dfrac{\sqrt{3y}}{\sqrt{3y}} = \dfrac{\sqrt{6xy}}{3y}$

Form of 1

(b) $\dfrac{\sqrt{5}}{\sqrt{12a^3}} = \dfrac{\sqrt{5}}{\sqrt{12a^3}} \cdot \dfrac{\sqrt{3a}}{\sqrt{3a}} = \dfrac{\sqrt{15a}}{\sqrt{36a^4}} = \dfrac{\sqrt{15a}}{6a^2}$

Form of 1

(c) $\dfrac{\sqrt{8x^2}}{\sqrt{27y^5}} = \dfrac{\sqrt{4x^2}\sqrt{2}}{\sqrt{9y^4}\sqrt{3y}} = \dfrac{2x\sqrt{2}}{3y^2\sqrt{3y}} = \dfrac{2x\sqrt{2}}{3y^2\sqrt{3y}} \cdot \dfrac{\sqrt{3y}}{\sqrt{3y}}$

$= \dfrac{2x\sqrt{6y}}{(3y^2)(3y)} = \dfrac{2x\sqrt{6y}}{9y^3}$

(d) $\dfrac{3}{\sqrt[3]{4x}} = \dfrac{3}{\sqrt[3]{4x}} \cdot \dfrac{\sqrt[3]{2x^2}}{\sqrt[3]{2x^2}} = \dfrac{3\sqrt[3]{2x^2}}{\sqrt[3]{8x^3}} = \dfrac{3\sqrt[3]{2x^2}}{2x}$

(e) $\dfrac{\sqrt[3]{16x^2}}{\sqrt[3]{9y^5}} = \dfrac{\sqrt[3]{16x^2}}{\sqrt[3]{9y^5}} \cdot \dfrac{\sqrt[3]{3y}}{\sqrt[3]{3y}} = \dfrac{\sqrt[3]{48x^2y}}{\sqrt[3]{27y^6}} = \dfrac{\sqrt[3]{8}\sqrt[3]{6x^2y}}{3y^2} = \dfrac{2\sqrt[3]{6x^2y}}{3y^2}$

▲

Notice that in *(c)* we did some simplifying first before rationalizing the denominator, whereas in *(b)* we proceeded immediately to rationalize the denominator. This is an individual choice and you should probably do it both ways a few times to help determine your preference.

▼ Problem Set 5.3

Use the distributive property to help simplify each of the following. For example,

$3\sqrt{8} - \sqrt{32} = 3\sqrt{4}\sqrt{2} - \sqrt{16}\sqrt{2}$

$= 3(2)\sqrt{2} - 4\sqrt{2}$

$= 6\sqrt{2} - 4\sqrt{2}$

$= (6 - 4)\sqrt{2} = 2\sqrt{2}.$

1. $5\sqrt{18} - 2\sqrt{2}$

2. $7\sqrt{12} + 4\sqrt{3}$

3. $7\sqrt{12} + 10\sqrt{48}$

4. $6\sqrt{8} - 5\sqrt{18}$

5. $-2\sqrt{50} - 5\sqrt{32}$

6. $-2\sqrt{20} - 7\sqrt{45}$

7. $3\sqrt{20} - \sqrt{5} - 2\sqrt{45}$

8. $6\sqrt{12} + \sqrt{3} - 2\sqrt{48}$

9. $-9\sqrt{24} + 3\sqrt{54} - 12\sqrt{6}$

10. $13\sqrt{28} - 2\sqrt{63} - 7\sqrt{7}$

11. $\dfrac{3}{4}\sqrt{7} - \dfrac{2}{3}\sqrt{28}$

12. $\frac{3}{5}\sqrt{5} - \frac{1}{4}\sqrt{80}$

13. $\frac{3}{5}\sqrt{40} + \frac{5}{6}\sqrt{90}$

14. $\frac{3}{8}\sqrt{96} - \frac{2}{3}\sqrt{54}$

15. $\frac{3\sqrt{18}}{5} - \frac{5\sqrt{72}}{6} + \frac{3\sqrt{98}}{4}$

16. $\frac{-2\sqrt{20}}{3} + \frac{3\sqrt{45}}{4} - \frac{5\sqrt{80}}{6}$

17. $5\sqrt[3]{3} + 2\sqrt[3]{24} - 6\sqrt[3]{81}$

18. $-3\sqrt[3]{2} - 2\sqrt[3]{16} + \sqrt[3]{54}$

19. $-\sqrt[3]{16} + 7\sqrt[3]{54} - 9\sqrt[3]{2}$

20. $4\sqrt[3]{24} - 6\sqrt[3]{3} + 13\sqrt[3]{81}$

Express each of the following in simplest radical form. All variables represent positive real numbers.

21. $\sqrt{32x}$

22. $\sqrt{50y}$

23. $\sqrt{75x^2}$

24. $\sqrt{108y^2}$

25. $\sqrt{20x^2y}$

26. $\sqrt{80xy^2}$

27. $\sqrt{64x^3y^7}$

28. $\sqrt{36x^5y^6}$

29. $\sqrt{54a^4b^3}$

30. $\sqrt{96a^7b^8}$

31. $\sqrt{63x^6y^8}$

32. $\sqrt{28x^4y^{12}}$

33. $2\sqrt{40a^3}$

34. $4\sqrt{90a^5}$

35. $\frac{2}{3}\sqrt{96xy^3}$

36. $\frac{4}{5}\sqrt{125x^4y}$

37. $\sqrt{\frac{2x}{5y}}$

38. $\sqrt{\frac{3x}{2y}}$

39. $\sqrt{\frac{5}{12x^4}}$

40. $\sqrt{\frac{7}{8x^2}}$

41. $\frac{5}{\sqrt{18y}}$

42. $\frac{3}{\sqrt{12x}}$

43. $\frac{\sqrt{7x}}{\sqrt{8y^5}}$

44. $\frac{\sqrt{5y}}{\sqrt{18x^3}}$

45. $\frac{\sqrt{18y^3}}{\sqrt{16x}}$

46. $\frac{\sqrt{2x^3}}{\sqrt{9y}}$

47. $\frac{\sqrt{24a^2b^3}}{\sqrt{7ab^6}}$

48. $\frac{\sqrt{12a^2b}}{\sqrt{5a^3b^3}}$

49. $\sqrt[3]{24y}$

50. $\sqrt[3]{16x^2}$

51. $\sqrt[3]{16x^4}$

52. $\sqrt[3]{54x^3}$

53. $\sqrt[3]{56x^6y^8}$

54. $\sqrt[3]{81x^5y^6}$

55. $\sqrt[3]{\frac{7}{9x^2}}$

56. $\sqrt[3]{\frac{5}{2x}}$

57. $\frac{\sqrt[3]{3y}}{\sqrt[3]{16x^4}}$

58. $\frac{\sqrt[3]{2y}}{\sqrt[3]{3x}}$

59. $\frac{\sqrt[3]{12xy}}{\sqrt[3]{3x^2y^5}}$

60. $\frac{5}{\sqrt[3]{9xy^2}}$

61. $\sqrt{8x + 12y}$ [*Hint*: $\sqrt{8x + 12y}$ $= \sqrt{4(2x + 3y)}$]

62. $\sqrt{4x + 4y}$

63. $\sqrt{16x + 48y}$

64. $\sqrt{27x + 18y}$

Use the distributive property to help simplify each of the following. All variables represent positive real numbers.

65. $-3\sqrt{4x} + 5\sqrt{9x} + 6\sqrt{16x}$

66. $-2\sqrt{25x} - 4\sqrt{36x} + 7\sqrt{64x}$

67. $2\sqrt{18x} - 3\sqrt{8x} - 6\sqrt{50x}$

68. $4\sqrt{20x} + 5\sqrt{45x} - 10\sqrt{80x}$

69. $5\sqrt{27n} - \sqrt{12n} - 6\sqrt{3n}$

70. $4\sqrt{8n} + 3\sqrt{18n} - 2\sqrt{72n}$

71. $7\sqrt{4ab} - \sqrt{16ab} - 10\sqrt{25ab}$

72. $4\sqrt{ab} - 9\sqrt{36ab} + 6\sqrt{49ab}$

73. $-3\sqrt{2x^3} + 4\sqrt{8x^3} - 3\sqrt{32x^3}$

74. $2\sqrt{40x^5} - 3\sqrt{90x^5} + 5\sqrt{160x^5}$

THOUGHTS INTO WORDS

75. Is the expression $3\sqrt{2} + \sqrt{50}$ in simplest radical form? Defend your answer.

76. Your friend simplified $\dfrac{\sqrt{6}}{\sqrt{8}}$ as follows:

$$\frac{\sqrt{6}}{\sqrt{8}} \cdot \frac{\sqrt{8}}{\sqrt{8}} = \frac{\sqrt{48}}{8} = \frac{\sqrt{16}\sqrt{3}}{8} = \frac{4\sqrt{3}}{8} = \frac{\sqrt{3}}{2}$$

Is this a correct procedure? Can you show her a better way to do this problem?

77. Does $\sqrt{x + y}$ equal $\sqrt{x} + \sqrt{y}$? Defend your answer.

Further Investigations

78. Use your calculator and evaluate each expression in Problems 1–16. Then evaluate the simplified expression that you obtained when doing these problems. Your two results for each problem should be the same.

5.4 Products and Quotients Involving Radicals

As we have seen, Property 5.4 ($\sqrt[n]{bc} = \sqrt[n]{b}\sqrt[n]{c}$) is used to express one radical as the product of two radicals and also to express the product of two radicals as one radical. In fact, we have used the property for both purposes within the framework of simplifying radicals. For example,

$$\frac{\sqrt{3}}{\sqrt{32}} = \frac{\sqrt{3}}{\sqrt{16}\sqrt{2}} = \frac{\sqrt{3}}{4\sqrt{2}} = \frac{\sqrt{3}}{4\sqrt{2}} \cdot \frac{\sqrt{2}}{\sqrt{2}} = \frac{\sqrt{6}}{8}.$$

$$\uparrow \qquad\qquad \uparrow \qquad\qquad\qquad\qquad \uparrow \qquad\quad \uparrow$$

$$\sqrt[n]{bc} = \sqrt[n]{b}\sqrt[n]{c} \qquad\qquad\qquad \sqrt[n]{b}\sqrt[n]{c} = \sqrt[n]{bc}$$

The following examples demonstrate the use of Property 5.4 to multiply radicals and to express the product in simplest form.

Example 1

Multiply and simplify where possible.

(a) $(2\sqrt{3})(3\sqrt{5})$ (b) $(3\sqrt{8})(5\sqrt{2})$

(c) $(7\sqrt{6})(3\sqrt{8})$ (d) $(2\sqrt[3]{6})(5\sqrt[3]{4})$

Solution

(a) $(2\sqrt{3})(3\sqrt{5}) = 2 \cdot 3 \cdot \sqrt{3} \cdot \sqrt{5} = 6\sqrt{15}.$

(b) $(3\sqrt{8})(5\sqrt{2}) = 3 \cdot 5 \cdot \sqrt{8} \cdot \sqrt{2} = 15\sqrt{16} = 15 \cdot 4 = 60.$

(c) $(7\sqrt{6})(3\sqrt{8}) = 7 \cdot 3 \cdot \sqrt{6} \cdot \sqrt{8} = 21\sqrt{48} = 21\sqrt{16}\sqrt{3}$
$= 21 \cdot 4 \cdot \sqrt{3} = 84\sqrt{3}$

(d) $(2\sqrt[3]{6})(5\sqrt[3]{4}) = 2 \cdot 5 \cdot \sqrt[3]{6} \cdot \sqrt[3]{4} = 10\sqrt[3]{24}$
$= 10\sqrt[3]{8}\sqrt[3]{3}$
$= 10 \cdot 2 \cdot \sqrt[3]{3}$
$= 20\sqrt[3]{3}$ ▲

Recall the use of the distributive property when finding the product of a monomial and a polynomial. For example, $3x^2(2x + 7) = 3x^2(2x) + 3x^2(7)$ $= 6x^3 + 21x^2$. In a similar manner, the distributive property, along with Property 5.4, provide the basis for finding certain special products that involve radicals. The following examples illustrate this idea.

Example 2

Multiply and simplify where possible.

(a) $\sqrt{3}(\sqrt{6} + \sqrt{12})$ (b) $2\sqrt{2}(4\sqrt{3} - 5\sqrt{6})$

(c) $\sqrt{6x}(\sqrt{8x} + \sqrt{12xy}$ (d) $\sqrt[3]{2}(5\sqrt[3]{4} - 3\sqrt[3]{16})$

Solution

(a) $\sqrt{3}(\sqrt{6} + \sqrt{12}) = \sqrt{3}\sqrt{6} + \sqrt{3}\sqrt{12}$
$= \sqrt{18} + \sqrt{36}$
$= \sqrt{9}\sqrt{2} + 6$
$= 3\sqrt{2} + 6$

(b) $2\sqrt{2}(4\sqrt{3} - 5\sqrt{6}) = (2\sqrt{2})(4\sqrt{3}) - (2\sqrt{2})(5\sqrt{6})$
$= 8\sqrt{6} - 10\sqrt{12}$
$= 8\sqrt{6} - 10\sqrt{4}\sqrt{3}$
$= 8\sqrt{6} - 20\sqrt{3}$

(c) $\sqrt{6x}(\sqrt{8x} + \sqrt{12xy}) = (\sqrt{6x})(\sqrt{8x}) + (\sqrt{6x})(\sqrt{12xy})$
$= \sqrt{48x^2} + \sqrt{72x^2y}$
$= \sqrt{16x^2}\sqrt{3} + \sqrt{36x^2}\sqrt{2y}$
$= 4x\sqrt{3} + 6x\sqrt{2y}$

(d) $\sqrt[3]{2}(5\sqrt[3]{4} - 3\sqrt[3]{16}) = (\sqrt[3]{2})(5\sqrt[3]{4}) - (\sqrt[3]{2})(3\sqrt[3]{16})$
$= 5\sqrt[3]{8} - 3\sqrt[3]{32}$
$= 5 \cdot 2 - 3\sqrt[3]{8}\sqrt[3]{4}$
$= 10 - 6\sqrt[3]{4}$ ▲

The distributive property also plays a central role in determining the product of two binomials. For example, $(x + 2)(x + 3) = x(x + 3) + 2(x + 3) = x^2 + 3x + 2x + 6 = x^2 + 5x + 6$. Finding the product of two binomial expressions that involve radicals can be handled in a similar fashion, as in the next examples.

Example 3

Find the following products and simplify.

(a) $(\sqrt{3} + \sqrt{5})(\sqrt{2} + \sqrt{6})$ (b) $(2\sqrt{2} - \sqrt{7})(3\sqrt{2} + 5\sqrt{7})$

(c) $(\sqrt{8} + \sqrt{6})(\sqrt{8} - \sqrt{6})$ (d) $(\sqrt{x} + \sqrt{y})(\sqrt{x} - \sqrt{y})$

Solution

(a) $(\sqrt{3} + \sqrt{5})(\sqrt{2} + \sqrt{6}) = \sqrt{3}(\sqrt{2} + \sqrt{6}) + \sqrt{5}(\sqrt{2} + \sqrt{6})$

$\qquad\qquad\qquad\qquad\quad = \sqrt{3}\sqrt{2} + \sqrt{3}\sqrt{6} + \sqrt{5}\sqrt{2} + \sqrt{5}\sqrt{6}$

$\qquad\qquad\qquad\qquad\quad = \sqrt{6} + \sqrt{18} + \sqrt{10} + \sqrt{30}$

$\qquad\qquad\qquad\qquad\quad = \sqrt{6} + 3\sqrt{2} + \sqrt{10} + \sqrt{30}$

(b) $(2\sqrt{2} - \sqrt{7})(3\sqrt{2} + 5\sqrt{7}) = 2\sqrt{2}(3\sqrt{2} + 5\sqrt{7}) - \sqrt{7}(3\sqrt{2}+5\sqrt{7})$

$\qquad\qquad\qquad\qquad\qquad = (2\sqrt{2})(3\sqrt{2}) + (2\sqrt{2})(5\sqrt{7})$

$\qquad\qquad\qquad\qquad\qquad\quad - (\sqrt{7})(3\sqrt{2}) - (\sqrt{7})(5\sqrt{7})$

$\qquad\qquad\qquad\qquad\qquad = 12 + 10\sqrt{14} - 3\sqrt{14} - 35$

$\qquad\qquad\qquad\qquad\qquad = -23 + 7\sqrt{14}$

(c) $(\sqrt{8} + \sqrt{6})(\sqrt{8} - \sqrt{6}) = \sqrt{8}(\sqrt{8} - \sqrt{6}) + \sqrt{6}(\sqrt{8} - \sqrt{6})$

$\qquad\qquad\qquad\qquad\qquad = \sqrt{8}\sqrt{8} - \sqrt{8}\sqrt{6} + \sqrt{6}\sqrt{8} - \sqrt{6}\sqrt{6}$

$\qquad\qquad\qquad\qquad\qquad = 8 - \sqrt{48} + \sqrt{48} - 6$

$\qquad\qquad\qquad\qquad\qquad = 2$

(d) $(\sqrt{x} + \sqrt{y})(\sqrt{x} - \sqrt{y}) = \sqrt{x}(\sqrt{x} - \sqrt{y}) + \sqrt{y}(\sqrt{x} - \sqrt{y})$

$\qquad\qquad\qquad\qquad\qquad = \sqrt{x}\sqrt{x} - \sqrt{x}\sqrt{y} + \sqrt{y}\sqrt{x} - \sqrt{y}\sqrt{y}$

$\qquad\qquad\qquad\qquad\qquad = x - \sqrt{xy} + \sqrt{xy} - y$

$\qquad\qquad\qquad\qquad\qquad = x - y$ ▲

Notice parts (c) and (d) of Example 3; they fit the special product pattern $(a + b)(a - b) = a^2 - b^2$. Furthermore, in each case the final product is in rational form. This suggests a way of rationalizing the denominator in an expression that contains a binomial denominator with radicals. Consider the following example.

Example 4

Simplify $\dfrac{4}{\sqrt{5} + \sqrt{2}}$ by rationalizing the denominator.

Solution

$$\frac{4}{\sqrt{5} + \sqrt{2}} = \frac{4}{\sqrt{5} + \sqrt{2}} \cdot \left(\frac{\sqrt{5} - \sqrt{2}}{\sqrt{5} - \sqrt{2}}\right) \quad \text{A form of I}$$

$$= \frac{4(\sqrt{5} - \sqrt{2})}{(\sqrt{5} + \sqrt{2})(\sqrt{5} - \sqrt{2})} = \frac{4(\sqrt{5} - \sqrt{2})}{5 - 2}$$

$$= \frac{4(\sqrt{5} - \sqrt{2})}{3} \quad \text{or} \quad \frac{4\sqrt{5} - 4\sqrt{2}}{3}$$

Either answer
is acceptable

The next examples further illustrate the process of rationalizing and simplifying expressions that contain binomial denominators.

Example 5

For each of the following, rationalize the denominator and simplify.

(a) $\dfrac{\sqrt{3}}{\sqrt{6} - 9}$

(b) $\dfrac{7}{3\sqrt{5} + 2\sqrt{3}}$

(c) $\dfrac{\sqrt{x} + 2}{\sqrt{x} - 3}$

(d) $\dfrac{2\sqrt{x} - 3\sqrt{y}}{\sqrt{x} + \sqrt{y}}$

Solution

(a) $\dfrac{\sqrt{3}}{\sqrt{6} - 9} = \dfrac{\sqrt{3}}{\sqrt{6} - 9} \cdot \dfrac{\sqrt{6} + 9}{\sqrt{6} + 9}$

$$= \frac{\sqrt{3}(\sqrt{6} + 9)}{(\sqrt{6} - 9)(\sqrt{6} + 9)}$$

$$= \frac{\sqrt{18} + 9\sqrt{3}}{6 - 81}$$

$$= \frac{3\sqrt{2} + 9\sqrt{3}}{-75}$$

$$= \frac{3(\sqrt{2} + 3\sqrt{3})}{(-3)(25)}$$

$$= -\frac{\sqrt{2} + 3\sqrt{3}}{25} \quad \text{or} \quad \frac{-\sqrt{2} - 3\sqrt{3}}{25}$$

(b) $\dfrac{7}{3\sqrt{5} + 2\sqrt{3}} = \dfrac{7}{3\sqrt{5} + 2\sqrt{3}} \cdot \dfrac{3\sqrt{5} - 2\sqrt{3}}{3\sqrt{5} - 2\sqrt{3}}$

$$= \frac{7(3\sqrt{5} - 2\sqrt{3})}{(3\sqrt{5} + 2\sqrt{3})(3\sqrt{5} - 2\sqrt{3})}$$

$$= \frac{7(3\sqrt{5} - 2\sqrt{3})}{45 - 12}$$

$$= \frac{7(3\sqrt{5} - 2\sqrt{3})}{33} \quad \text{or} \quad \frac{21\sqrt{5} - 14\sqrt{3}}{33}$$

(c) $\dfrac{\sqrt{x} + 2}{\sqrt{x} - 3} = \dfrac{\sqrt{x} + 2}{\sqrt{x} - 3} \cdot \dfrac{\sqrt{x} + 3}{\sqrt{x} + 3} = \dfrac{(\sqrt{x} + 2)(\sqrt{x} + 3)}{(\sqrt{x} - 3)(\sqrt{x} + 3)}$

$$= \frac{x + 3\sqrt{x} + 2\sqrt{x} + 6}{x - 9}$$

$$= \frac{x + 5\sqrt{x} + 6}{x - 9}$$

(d) $\dfrac{2\sqrt{x} - 3\sqrt{y}}{\sqrt{x} + \sqrt{y}} = \dfrac{2\sqrt{x} - 3\sqrt{y}}{\sqrt{x} + \sqrt{y}} \cdot \dfrac{\sqrt{x} - \sqrt{y}}{\sqrt{x} - \sqrt{y}}$

$$= \frac{(2\sqrt{x} - 3\sqrt{y})(\sqrt{x} - \sqrt{y})}{(\sqrt{x} + \sqrt{y})(\sqrt{x} - \sqrt{y})}$$

$$= \frac{2x - 2\sqrt{xy} - 3\sqrt{xy} + 3y}{x - y}$$

$$= \frac{2x - 5\sqrt{xy} + 3y}{x - y}$$

Problem Set 5.4

Multiply and simplify where possible.

1. $\sqrt{6}\sqrt{12}$ **2.** $\sqrt{8}\sqrt{6}$

3. $(3\sqrt{3})(2\sqrt{6})$

4. $(5\sqrt{2})(3\sqrt{12})$

5. $(4\sqrt{2})(-6\sqrt{5})$

6. $(-7\sqrt{3})(2\sqrt{5})$

7. $(-3\sqrt{3})(-4\sqrt{8})$

8. $(-5\sqrt{8})(-6\sqrt{7})$

9. $(5\sqrt{6})(4\sqrt{6})$ **10.** $(3\sqrt{7})(2\sqrt{7})$

11. $(2\sqrt[3]{4})(6\sqrt[3]{2})$ **12.** $(4\sqrt[3]{3})(5\sqrt[3]{9})$

13. $(4\sqrt[3]{6})(7\sqrt[3]{4})$

14. $(9\sqrt[3]{6})(2\sqrt[3]{9})$

Find the following products and express answers in simplest radical form. All variables represent non-negative real numbers.

15. $\sqrt{2}(\sqrt{3} + \sqrt{5})$

16. $\sqrt{3}(\sqrt{7} + \sqrt{10})$

17. $3\sqrt{5}(2\sqrt{2} - \sqrt{7})$

18. $5\sqrt{6}(2\sqrt{5} - 3\sqrt{11})$

19. $2\sqrt{6}(3\sqrt{8} - 5\sqrt{12})$

20. $4\sqrt{2}(3\sqrt{12} + 7\sqrt{6})$

21. $-4\sqrt{5}(2\sqrt{5} + 4\sqrt{12})$

22. $-5\sqrt{3}(3\sqrt{12} - 9\sqrt{8})$

23. $3\sqrt{x}(5\sqrt{2} + \sqrt{y})$

24. $\sqrt{2x}(3\sqrt{y} - 7\sqrt{5})$

25. $\sqrt{xy}(5\sqrt{xy} - 6\sqrt{x})$

26. $4\sqrt{x}(2\sqrt{xy} + 2\sqrt{x})$

27. $\sqrt{5y}(\sqrt{8x} + \sqrt{12y^2})$

28. $\sqrt{2x}(\sqrt{12xy} - \sqrt{8y})$

29. $5\sqrt{3}(2\sqrt{8} - 3\sqrt{18})$

30. $2\sqrt{2}(3\sqrt{12} - \sqrt{27})$

31. $(\sqrt{3} + 4)(\sqrt{3} - 7)$

32. $(\sqrt{2} + 6)(\sqrt{2} - 2)$

33. $(\sqrt{5} - 6)(\sqrt{5} - 3)$

34. $(\sqrt{7} - 2)(\sqrt{7} - 8)$

35. $(3\sqrt{5} - 2\sqrt{3})(2\sqrt{7} + \sqrt{2})$

36. $(\sqrt{2} + \sqrt{3})(\sqrt{5} - \sqrt{7})$

37. $(2\sqrt{6} + 3\sqrt{5})(\sqrt{8} - 3\sqrt{12})$

38. $(5\sqrt{2} - 4\sqrt{6})(2\sqrt{8} + \sqrt{6})$

39. $(2\sqrt{6} + 5\sqrt{5})(3\sqrt{6} - \sqrt{5})$

40. $(7\sqrt{3} - \sqrt{7})(2\sqrt{3} + 4\sqrt{7})$

41. $(3\sqrt{2} - 5\sqrt{3})(6\sqrt{2} - 7\sqrt{3})$

42. $(\sqrt{8} - 3\sqrt{10})(2\sqrt{8} - 6\sqrt{10})$

43. $(\sqrt{6} + 4)(\sqrt{6} - 4)$

44. $(\sqrt{7} - 2)(\sqrt{7} + 2)$

45. $(\sqrt{2} + \sqrt{10})(\sqrt{2} - \sqrt{10})$

46. $(2\sqrt{3} + \sqrt{11})(2\sqrt{3} - \sqrt{11})$

47. $(\sqrt{2x} + \sqrt{3y})(\sqrt{2x} - \sqrt{3y})$

48. $(2\sqrt{x} - 5\sqrt{y})(2\sqrt{x} + 5\sqrt{y})$

49. $2\sqrt[3]{3}(5\sqrt[3]{4} + \sqrt[3]{6})$

50. $2\sqrt[3]{2}(3\sqrt[3]{6} - 4\sqrt[3]{5})$

51. $3\sqrt[3]{4}(2\sqrt[3]{2} - 6\sqrt[3]{4})$

52. $3\sqrt[3]{3}(4\sqrt[3]{9} + 5\sqrt[3]{7})$

For each of the following, rationalize the denominator and simplify. All variables represent positive real numbers.

53. $\dfrac{2}{\sqrt{7} + 1}$

54. $\dfrac{6}{\sqrt{5} + 2}$

55. $\dfrac{3}{\sqrt{2} - 5}$

56. $\dfrac{-4}{\sqrt{6} - 3}$

57. $\dfrac{1}{\sqrt{2} + \sqrt{7}}$

58. $\dfrac{3}{\sqrt{3} + \sqrt{10}}$

59. $\dfrac{\sqrt{2}}{\sqrt{10} - \sqrt{3}}$

60. $\dfrac{\sqrt{3}}{\sqrt{7} - \sqrt{2}}$

61. $\dfrac{\sqrt{3}}{2\sqrt{5} + 4}$

62. $\dfrac{\sqrt{7}}{3\sqrt{2} - 5}$

63. $\dfrac{6}{3\sqrt{7} - 2\sqrt{6}}$

64. $\dfrac{5}{2\sqrt{5} + 3\sqrt{7}}$

65. $\dfrac{\sqrt{6}}{3\sqrt{2} + 2\sqrt{3}}$

66. $\dfrac{3\sqrt{6}}{5\sqrt{3} - 4\sqrt{2}}$

67. $\dfrac{2}{\sqrt{x} + 4}$

68. $\dfrac{3}{\sqrt{x} + 7}$

69. $\dfrac{\sqrt{x}}{\sqrt{x} - 5}$

70. $\dfrac{\sqrt{x}}{\sqrt{x}-1}$

71. $\dfrac{\sqrt{x}-2}{\sqrt{x}+6}$

72. $\dfrac{\sqrt{x}+1}{\sqrt{x}-10}$

73. $\dfrac{\sqrt{x}}{\sqrt{x}+2\sqrt{y}}$

74. $\dfrac{\sqrt{y}}{2\sqrt{x}-\sqrt{y}}$

75. $\dfrac{3\sqrt{y}}{2\sqrt{x}-3\sqrt{y}}$

76. $\dfrac{2\sqrt{x}}{3\sqrt{x}+5\sqrt{y}}$

THOUGHTS INTO WORDS

77. How would you help someone rationalize the denominator and simplify $\dfrac{4}{\sqrt{8}+\sqrt{12}}$?

78. Discuss how the distributive property has been used thus far in this chapter.

79. How would you simplify the expression $\dfrac{\sqrt{8}+\sqrt{12}}{\sqrt{2}}$?

Further Investigations

80. Use your calculator to evaluate each expression in Problems 53–66. Then evaluate the results you obtained when you did the problems.

5.5 Equations Involving Radicals

We often refer to equations that contain radicals with variables in a radicand as **radical equations**. In this section we discuss techniques for solving such equations that contain one or more radicals. To solve radical equations we need the following property of equality.

PROPERTY 5.6

Let a and b be real numbers and n be a positive integer.

If $a = b$, then $a^n = b^n$.

Property 5.6 states that we can *raise both sides of an equation to a positive integral power*. However, raising both sides of an equation to a positive integral power sometimes produces results that do not satisfy the original equation. Let's consider two examples to illustrate this point.

Example 1

Solution

Solve $\sqrt{2x-5} = 7$.

$$\sqrt{2x-5} = 7$$
$$(\sqrt{2x-5})^2 = 7^2 \quad \text{Square both sides.}$$
$$2x - 5 = 49$$
$$2x = 54$$
$$x = 27$$

Exponents and Radicals

 Check

$$\sqrt{2x - 5} = 7$$
$$\sqrt{2(27) - 5} \overset{?}{=} 7$$
$$\sqrt{49} \overset{?}{=} 7$$
$$7 = 7$$

The solution set for $\sqrt{2x - 5} = 7$ is $\{27\}$.

Example 2

Solution

Solve $\sqrt{3a + 4} = -4$.

$$\sqrt{3a + 4} = -4$$
$$(\sqrt{3a + 4})^2 = (-4)^2 \qquad \text{Square both sides.}$$
$$3a + 4 = 16$$
$$3a = 12$$
$$a = 4$$

 Check

$$\sqrt{3a + 4} = -4$$
$$\sqrt{3(4) + 4} \overset{?}{=} -4$$
$$\sqrt{16} \overset{?}{=} -4$$
$$4 \neq -4$$

Since 4 does not check, the original equation *has no real number solution.* Thus, the solution set is $\varnothing$.

In general, raising both sides of an equation to a positive integral power produces an equation that has all of the solutions of the original equation; it may also have some extra solutions that will not satisfy the original equation. Such extra solutions are called **extraneous solutions**. Therefore, when using Property 5.6 you *must* check each potential solution in the original equation.

Let's consider some examples to illustrate different situations that arise when solving radical equations.

Example 3

Solution

Solve $\sqrt{2t - 4} = t - 2$.

$$\sqrt{2t - 4} = t - 2$$
$$(\sqrt{2t - 4})^2 = (t - 2)^2 \qquad \text{Square both sides.}$$
$$2t - 4 = t^2 - 4t + 4$$
$$0 = t^2 - 6t + 8$$
$$0 = (t - 2)(t - 4) \qquad \text{Factor the right side.}$$
$$t - 2 = 0 \quad \text{or} \quad t - 4 = 0 \qquad \text{Apply: } ab = 0 \text{ if and}$$
$$t = 2 \quad \text{or} \quad t = 4 \qquad \text{only if } a = 0 \text{ or } b = 0.$$

✔ *Check*

$$\sqrt{2t - 4} = t - 2 \qquad\qquad \sqrt{2t - 4} = t - 2$$
$$\sqrt{2(2) - 4} \stackrel{?}{=} 2 - 2 \quad \text{or} \quad \sqrt{2(4) - 4} \stackrel{?}{=} 4 - 2$$
$$\sqrt{0} \stackrel{?}{=} 0 \qquad\qquad\qquad \sqrt{4} \stackrel{?}{=} 2$$
$$0 = 0 \qquad\qquad\qquad\qquad 2 = 2$$

The solution set is $\{2, 4\}$.

▲

Example 4

Solve $\sqrt{y} + 6 = y$.

Solution

$$\sqrt{y} + 6 = y$$
$$\sqrt{y} = y - 6$$
$$(\sqrt{y})^2 = (y - 6)^2 \qquad\qquad \text{Square both sides.}$$
$$y = y^2 - 12y + 36$$
$$0 = y^2 - 13y + 36$$
$$0 = (y - 4)(y - 9) \qquad\qquad \text{Factor the right side.}$$
$$y - 4 = 0 \quad \text{or} \quad y - 9 = 0 \qquad \text{Apply: } ab = 0 \text{ if and}$$
$$y = 4 \quad \text{or} \quad y = 9 \qquad\qquad \text{only if } a = 0 \text{ or } b = 0.$$

✔ *Check*

$$\sqrt{y} + 6 = y \qquad\qquad \sqrt{y} + 6 = y$$
$$\sqrt{4} + 6 \stackrel{?}{=} 4 \quad \text{or} \quad \sqrt{9} + 6 \stackrel{?}{=} 9$$
$$2 + 6 \stackrel{?}{=} 4 \qquad\qquad 3 + 6 \stackrel{?}{=} 9$$
$$8 \neq 4 \qquad\qquad\qquad 9 = 9$$

The only solution is 9; the solution set is $\{9\}$.

▲

In Example 4, note that we changed the form of the original equation $\sqrt{y} + 6 = y$ to $\sqrt{y} = y - 6$ before we squared both sides. Squaring both sides of $\sqrt{y} + 6 = y$ produces $y + 12\sqrt{y} + 36 = y^2$, which is a much more complex equation that still contains a radical. So, again it pays to think ahead a few steps before carrying out the details. Now let's consider an example involving a cube root.

Example 5

Solve $\sqrt[3]{n^2 - 1} = 2$.

Solution

$$\sqrt[3]{n^2 - 1} = 2$$
$$(\sqrt[3]{n^2 - 1})^3 = 2^3 \qquad\qquad \text{Cube both sides.}$$
$$n^2 - 1 = 8$$
$$n^2 - 9 = 0$$
$$(n + 3)(n - 3) = 0$$
$$n + 3 = 0 \quad \text{or} \quad n - 3 = 0$$
$$n = -3 \quad \text{or} \quad n = 3$$

✔ **Check**

$$\sqrt[3]{n^2 - 1} = 2 \qquad\qquad \sqrt[3]{n^2 - 1} = 2$$

$$\sqrt[3]{(-3)^2 - 1} \stackrel{?}{=} 2 \qquad \text{or} \qquad \sqrt[3]{3^2 - 1} \stackrel{?}{=} 2$$

$$\sqrt[3]{8} \stackrel{?}{=} 2 \qquad\qquad \sqrt[3]{8} \stackrel{?}{=} 2$$

$$2 = 2 \qquad\qquad\qquad 2 = 2$$

The solution set is $\{-3, 3\}$.   ▲

It may be necessary to square both sides of an equation, simplify the resulting equation, and then square both sides again. The next example illustrates this type of problem.

Example 6

Solution

Solve $\sqrt{x + 2} = 7 - \sqrt{x + 9}$.

$$\sqrt{x + 2} = 7 - \sqrt{x + 9}$$

$$(\sqrt{x + 2})^2 = (7 - \sqrt{x + 9})^2 \qquad \text{Square both sides.}$$

$$x + 2 = 49 - 14\sqrt{x + 9} + x + 9$$

$$x + 2 = x + 58 - 14\sqrt{x + 9}$$

$$-56 = -14\sqrt{x + 9}$$

$$4 = \sqrt{x + 9}$$

$$(4)^2 = (\sqrt{x + 9})^2 \qquad \text{Square both sides.}$$

$$16 = x + 9$$

$$7 = x$$

✔ **Check**

$$\sqrt{x + 2} = 7 - \sqrt{x + 9}$$

$$\sqrt{7 + 2} \stackrel{?}{=} 7 - \sqrt{7 + 9}$$

$$\sqrt{9} \stackrel{?}{=} 7 - \sqrt{16}$$

$$3 \stackrel{?}{=} 7 - 4$$

$$3 = 3$$

The solution set is $\{7\}$.   ▲

Another Look at Applications

In Section 5.1 we used the formula $S = \sqrt{30Df}$ to approximate how fast a car was traveling based on the length of skid marks. (Remember that S represents the speed of the car in miles per hour, D the length of the skid marks measured in feet, and f represents a coefficient of friction.) This same formula can be used to estimate the length of skid marks that are produced by cars traveling at different rates on various types of road surfaces. To use the formula for this purpose, let's change the form of the equation by solving for D.

$$\sqrt{30Df} = S$$
$$30Df = S^2 \qquad \text{The result of squaring both sides of the original equation}$$
$$D = \frac{S^2}{30f} \qquad \text{D, S, and f are positive numbers, so this final equation and the original one are equivalent.}$$

Example 7

Suppose that for a particular road surface the coefficient of friction is .35. How far will a car skid when applying the brakes at 60 miles per hour?

Solution

We can substitute .35 for f and 60 for S in the formula $D = \dfrac{S^2}{30f}$.

$$D = \frac{60^2}{30(.35)} = 343 \quad \text{to the nearest whole number}$$

The car will skid approximately 343 feet.   ▲

REMARK   Pause for a moment and think about the result in Example 7. The coefficient of friction of .35 refers to a wet concrete road surface. Note that a car traveling at 60 miles per hour will skid more than the length of a football field.   △

Problem Set 5.5

Solve each of the following equations. Don't forget to check each of your potential solutions.

1. $\sqrt{5x} = 10$

2. $\sqrt{3x} = 9$

3. $\sqrt{2x} + 4 = 0$

4. $\sqrt{4x} + 5 = 0$

5. $2\sqrt{n} = 5$

6. $5\sqrt{n} = 3$

7. $3\sqrt{n} - 2 = 0$

8. $2\sqrt{n} - 7 = 0$

9. $\sqrt{3y + 1} = 4$

10. $\sqrt{2y - 3} = 5$

11. $\sqrt{4y - 3} - 6 = 0$

12. $\sqrt{3y + 5} - 2 = 0$

13. $\sqrt{2x - 5} = -1$

14. $\sqrt{4x - 3} = -4$

15. $\sqrt{5x + 2} = \sqrt{6x + 1}$

16. $\sqrt{4x + 2} = \sqrt{3x + 4}$

17. $\sqrt{3x + 1} = \sqrt{7x - 5}$

18. $\sqrt{6x + 5} = \sqrt{2x + 10}$

19. $\sqrt{3x - 2} - \sqrt{x + 4} = 0$

20. $\sqrt{7x - 6} - \sqrt{5x + 2} = 0$

21. $5\sqrt{t - 1} = 6$

22. $4\sqrt{t + 3} = 6$

23. $\sqrt{x^2 + 7} = 4$

24. $\sqrt{x^2 + 3} - 2 = 0$

25. $\sqrt{x^2 + 13x + 37} = 1$

26. $\sqrt{x^2 + 5x - 20} = 2$

27. $\sqrt{x^2 - x + 1} = x + 1$

28. $\sqrt{n^2 - 2n - 4} = n$

29. $\sqrt{x^2 + 3x + 7} = x + 2$

30. $\sqrt{x^2 + 2x + 1} = x + 3$

31. $\sqrt{-4x + 17} = x - 3$

32. $\sqrt{2x - 1} = x - 2$

33. $\sqrt{n + 4} = n + 4$

34. $\sqrt{n + 6} = n + 6$

35. $\sqrt{3y} = y - 6$

36. $2\sqrt{n} = n - 3$

37. $4\sqrt{x} + 5 = x$

38. $\sqrt{-x} - 6 = x$

39. $\sqrt[3]{x - 2} = 3$

40. $\sqrt[3]{x + 1} = 4$

41. $\sqrt[3]{2x + 3} = -3$

42. $\sqrt[3]{3x - 1} = -4$

43. $\sqrt[3]{2x + 5} = \sqrt[3]{4 - x}$

44. $\sqrt[3]{3x - 1} = \sqrt[3]{2 - 5x}$

45. $\sqrt{x + 19} - \sqrt{x + 28} = -1$

46. $\sqrt{x + 4} = \sqrt{x - 1} + 1$

47. $\sqrt{3x + 1} + \sqrt{2x + 4} = 3$

48. $\sqrt{2x - 1} - \sqrt{x + 3} = 1$

49. $\sqrt{n - 4} + \sqrt{n + 4} = 2\sqrt{n - 1}$

50. $\sqrt{n - 3} + \sqrt{n + 5} = 2\sqrt{n}$

51. $\sqrt{t + 3} - \sqrt{t - 2} = \sqrt{7 - t}$

52. $\sqrt{t + 7} - 2\sqrt{t - 8} = \sqrt{t - 5}$

53. Use the formula given in Example 7 with a coefficient of friction of .95. How far will a car skid at 40 miles per hour? at 55 miles per hour? at 65 miles per hour? Express the answers to the nearest foot.

54. Solve the formula $T = 2\pi\sqrt{\dfrac{L}{32}}$ for L. (Remember that in this formula, which was used in Section 5.2, T represents the period of a pendulum expressed in seconds, and L represents the length of the pendulum in feet.)

55. In Problem 54, you should have obtained the equation $L = \dfrac{8T^2}{\pi^2}$. What is the length of a pendulum that has a period of 2 seconds? of 2.5 seconds? of 3 seconds? Express your answers to the nearest tenth of a foot.

THOUGHTS INTO WORDS

56. Explain the concept of extraneous solutions.

57. Explain why possible solutions for radical equations *must* be checked.

58. Your friend attempts to solve the equation $3 + 2\sqrt{x} = x$ as follows.

$$(3 + 2\sqrt{x})^2 = x^2$$
$$9 + 12\sqrt{x} + 4x = x^2$$

At this step he stops and doesn't know how to proceed. What help would you give him?

5.6 Merging Exponents and Roots

Recall that the basic properties of positive integral exponents led to a definition for the use of negative integers as exponents. In this section, the properties of integral exponents are used to form definitions for the use of rational numbers as exponents. These definitions will tie together the concepts of *exponent* and *root*.

Let's consider the following comparisons.

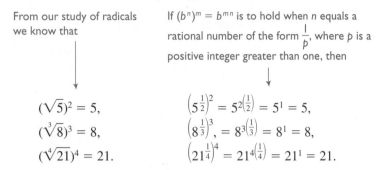

From our study of radicals we know that

$$(\sqrt{5})^2 = 5,$$
$$(\sqrt[3]{8})^3 = 8,$$
$$(\sqrt[4]{21})^4 = 21.$$

If $(b^n)^m = b^{mn}$ is to hold when n equals a rational number of the form $\frac{1}{p}$, where p is a positive integer greater than one, then

$$\left(5^{\frac{1}{2}}\right)^2 = 5^{2\left(\frac{1}{2}\right)} = 5^1 = 5,$$
$$\left(8^{\frac{1}{3}}\right)^3, = 8^{3\left(\frac{1}{3}\right)} = 8^1 = 8,$$
$$\left(21^{\frac{1}{4}}\right)^4 = 21^{4\left(\frac{1}{4}\right)} = 21^1 = 21.$$

It would seem reasonable to make the following definition.

DEFINITION 5.6

If b is a real number, n is a positive integer greater than one, and $\sqrt[n]{b}$ exists, then

$$b^{\frac{1}{n}} = \sqrt[n]{b}.$$

Definition 5.6 states that $b^{\frac{1}{n}}$ means the nth root of b. We shall assume that b and n are chosen so that $\sqrt[n]{b}$ exists. For example, $(-25)^{\frac{1}{2}}$ is not meaningful at this time because $\sqrt{-25}$ is not a real number. Consider the following examples that demonstrate the use of Definition 5.6.

$$25^{\frac{1}{2}} = \sqrt{25} = 5, \qquad\qquad 16^{\frac{1}{4}} = \sqrt[4]{16} = 2,$$
$$8^{\frac{1}{3}} = \sqrt[3]{8} = 2, \qquad\qquad \left(\frac{36}{49}\right)^{\frac{1}{2}} = \sqrt{\frac{36}{49}} = \frac{6}{7},$$
$$(-27)^{\frac{1}{3}} = \sqrt[3]{-27} = -3$$

The following definition provides the basis for the use of *all* rational numbers as exponents.

DEFINITION 5.7

If $\frac{m}{n}$ is a rational number, where n is a positive integer greater than one, and b is a real number such that $\sqrt[n]{b}$ exists, then

$$b^{\frac{m}{n}} = \sqrt[n]{b^m} = (\sqrt[n]{b})^m.$$

In Definition 5.7, notice that the denominator of the exponent is the index of the radical; and the numerator of the exponent is either the exponent of the radicand or the exponent of the root.

Whether we use the form $\sqrt[n]{b^m}$ or $(\sqrt[n]{b})^m$ for computational purposes depends somewhat on the magnitude of the problem. Let's use both forms on two problems to illustrate this point.

$$8^{\frac{2}{3}} = \sqrt[3]{8^2} \qquad \text{or} \qquad 8^{\frac{2}{3}} = (\sqrt[3]{8})^2$$
$$= \sqrt[3]{64} \qquad\qquad\qquad = 2^2$$
$$= 4 \qquad\qquad\qquad\quad = 4$$

$$27^{\frac{2}{3}} = \sqrt[3]{27^2} \qquad \text{or} \qquad 27^{\frac{2}{3}} = (\sqrt[3]{27})^2$$
$$= \sqrt[3]{729} \qquad\qquad\qquad = 3^2$$
$$= 9 \qquad\qquad\qquad\quad = 9$$

To compute $8^{\frac{2}{3}}$, either form seems to work about well as the other one. However, to compute $27^{\frac{2}{3}}$, it should be obvious that $(\sqrt[3]{27})^2$ is much easier to handle than $\sqrt[3]{27^2}$.

Example 1

Simplify each of the following numerical expressions.

(a) $25^{\frac{3}{2}}$ (b) $16^{\frac{3}{4}}$ (c) $(32)^{-\frac{2}{5}}$

(d) $(-64)^{\frac{2}{3}}$ (e) $-8^{\frac{1}{3}}$

Solution

(a) $25^{\frac{3}{2}} = (\sqrt{25})^3 = 5^3 = 125$

(b) $16^{\frac{3}{4}} = (\sqrt[4]{16})^3 = 2^3 = 8$

(c) $(32)^{-\frac{2}{5}} = \dfrac{1}{(32)^{\frac{2}{5}}} = \dfrac{1}{(\sqrt[5]{32})^2} = \dfrac{1}{2^2} = \dfrac{1}{4}$

(d) $(-64)^{\frac{2}{3}} = (\sqrt[3]{-64})^2 = (-4)^2 = 16$

(e) $-8^{\frac{1}{3}} = -\sqrt[3]{8} = -2$ ▲

The basic laws of exponents we stated in Property 5.2 are true for all rational exponents. Therefore, from now on we will use Property 5.2 for rational as well as integral exponents.

Some problems can be handled better in exponential form and others in radical form. Thus, we must be able to switch forms with a certain amount of ease. Let's consider some examples where we switch from one form to the other.

Example 2

Write each of the following expressions in radical form.

(a) $x^{\frac{3}{4}}$ (b) $3y^{\frac{2}{5}}$

(c) $x^{\frac{1}{4}}y^{\frac{3}{4}}$ (d) $(x + y)^{\frac{2}{3}}$

Solution

(a) $x^{\frac{3}{4}} = \sqrt[4]{x^3}$

(b) $3y^{\frac{2}{5}} = 3\sqrt[5]{y^2}$

(c) $x^{\frac{1}{4}}y^{\frac{3}{4}} = (xy^3)^{\frac{1}{4}} = \sqrt[4]{xy^3}$

(d) $(x + y)^{\frac{2}{3}} = \sqrt[3]{(x + y)^2}$ ▲

Example 3

Write each of the following using positive rational exponents.

(a) $\sqrt{xy}$ (b) $\sqrt[4]{a^3b}$ (c) $4\sqrt[3]{x^2}$ (d) $\sqrt[5]{(x + y)^4}$

Solution

(a) $\sqrt{xy} = (xy)^{\frac{1}{2}} = x^{\frac{1}{2}}y^{\frac{1}{2}}$

(b) $\sqrt[4]{a^3b} = (a^3b)^{\frac{1}{4}} = a^{\frac{3}{4}}b^{\frac{1}{4}}$

(c) $4\sqrt[3]{x^2} = 4x^{\frac{2}{3}}$

(d) $\sqrt[5]{(x + y)^4} = (x + y)^{\frac{4}{5}}$ ▲

The basic properties of exponents provide the basis for simplifying algebraic expressions that contain rational exponents, as these next examples illustrate.

Example 4

Simplify each of the following. Express final results using positive exponents only.

(a) $\left(3x^{\frac{1}{2}}\right)\left(4x^{\frac{2}{3}}\right)$ (b) $\left(5a^{\frac{1}{3}}b^{\frac{1}{2}}\right)^2$ (c) $\dfrac{12y^{\frac{1}{3}}}{6y^{\frac{1}{2}}}$ (d) $\left(\dfrac{3x^{\frac{2}{5}}}{2y^{\frac{2}{3}}}\right)^4$

Solution

(a) $\left(3x^{\frac{1}{2}}\right)\left(4x^{\frac{2}{3}}\right) = 3 \cdot 4 \cdot x^{\frac{1}{2}} \cdot x^{\frac{2}{3}}$

$= 12x^{\frac{1}{2}+\frac{2}{3}}$ $b^n \cdot b^m = b^{n+m}$

$= 12x^{\frac{3}{6}+\frac{4}{6}}$

$= 12x^{\frac{7}{6}}$

(b) $\left(5a^{\frac{1}{3}}b^{\frac{1}{2}}\right)^2 = 5^2 \cdot \left(a^{\frac{1}{3}}\right)^2 \cdot \left(b^{\frac{1}{2}}\right)^2$ $(ab)^n = a^nb^n$

$= 25a^{\frac{2}{3}}b$ $(b^n)^m = b^{mn}$

(c) $\dfrac{12y^{\frac{1}{3}}}{6y^{\frac{1}{2}}} = 2y^{\frac{1}{3}-\frac{1}{2}}$ $\dfrac{b^n}{b^m} = b^{n-m}$

$= 2y^{\frac{2}{6}-\frac{3}{6}}$

$= 2y^{-\frac{1}{6}}$

$= \dfrac{2}{y^{\frac{1}{6}}}$

(d) $\left(\dfrac{3x^{\frac{2}{5}}}{2y^{\frac{2}{3}}}\right)^4 = \dfrac{\left(3x^{\frac{2}{5}}\right)^4}{\left(2y^{\frac{2}{3}}\right)^4}$ $\left(\dfrac{a}{b}\right) - \dfrac{a^n}{b^n}$

$\qquad = \dfrac{3^4 \cdot \left(x^{\frac{2}{5}}\right)^4}{2^4 \cdot \left(y^{\frac{2}{3}}\right)^4}$ $(ab)^n = a^n b^n$

$\qquad = \dfrac{81x^{\frac{8}{5}}}{16y^{\frac{8}{3}}}$ $(b^n)^m = b^{mn}$

The link between exponents and roots also provides a basis for multiplying and dividing some radicals even if they have a different index. The general procedure is to change from radical form to exponential form, apply the properties of exponents, and then change back to radical form. The three parts of Example 5 will illustrate this process.

Example 5

Perform the indicated operations and express the answer in simplest radical form.

(a) $\sqrt{2}\sqrt[3]{2}$ (b) $\dfrac{\sqrt{5}}{\sqrt[3]{5}}$ (c) $\dfrac{\sqrt{4}}{\sqrt[3]{2}}$

Solution

(a) $\sqrt{2}\sqrt[3]{2} = 2^{\frac{1}{2}} \cdot 2^{\frac{1}{3}}$

$\qquad = 2^{\frac{1}{2}+\frac{1}{3}}$

$\qquad = 2^{\frac{3}{6}+\frac{2}{6}}$

$\qquad = 2^{\frac{5}{6}}$

$\qquad = \sqrt[6]{2^5} = \sqrt[6]{32}$

(b) $\dfrac{\sqrt{5}}{\sqrt[3]{5}} = \dfrac{5^{\frac{1}{2}}}{5^{\frac{1}{3}}}$

$\qquad = 5^{\frac{1}{2}-\frac{1}{3}}$

$\qquad = 5^{\frac{3}{6}-\frac{2}{6}}$

$\qquad = 5^{\frac{1}{6}} = \sqrt[6]{5}$

(c) $\dfrac{\sqrt{4}}{\sqrt[3]{2}} = \dfrac{4^{\frac{1}{2}}}{2^{\frac{1}{3}}}$

$\qquad = \dfrac{(2^2)^{\frac{1}{2}}}{2^{\frac{1}{3}}}$

$\qquad = \dfrac{2^1}{2^{\frac{1}{3}}}$

$\qquad = 2^{1-\frac{1}{3}}$

$\qquad = 2^{\frac{2}{3}} = \sqrt[3]{2^2} = \sqrt[3]{4}$

Problem Set 5.6

Simplify each of the following numerical expressions.

1. $81^{\frac{1}{2}}$ **2.** $64^{\frac{1}{2}}$ **3.** $27^{\frac{1}{3}}$

4. $(-32)^{\frac{1}{5}}$ **5.** $(-8)^{\frac{1}{3}}$

6. $\left(-\frac{27}{8}\right)^{\frac{1}{3}}$ **7.** $-25^{\frac{1}{2}}$

8. $-64^{\frac{1}{3}}$ **9.** $36^{-\frac{1}{2}}$

10. $81^{-\frac{1}{2}}$ **11.** $\left(\frac{1}{27}\right)^{-\frac{1}{3}}$

12. $\left(-\frac{8}{27}\right)^{-\frac{1}{3}}$ **13.** $4^{\frac{3}{2}}$

14. $64^{\frac{2}{3}}$ **15.** $27^{\frac{4}{3}}$

16. $4^{\frac{7}{2}}$ **17.** $(-1)^{\frac{7}{3}}$

18. $(-8)^{\frac{4}{3}}$ **19.** $-4^{\frac{5}{2}}$

20. $-16^{\frac{3}{2}}$ **21.** $\left(\frac{27}{8}\right)^{\frac{4}{3}}$

22. $\left(\frac{8}{125}\right)^{\frac{2}{3}}$ **23.** $\left(\frac{1}{8}\right)^{-\frac{2}{3}}$

24. $\left(-\frac{1}{27}\right)^{-\frac{2}{3}}$ **25.** $64^{-\frac{7}{6}}$

26. $32^{-\frac{4}{5}}$ **27.** $-25^{\frac{3}{2}}$

28. $-16^{\frac{3}{4}}$ **29.** $125^{\frac{4}{3}}$

30. $81^{\frac{5}{4}}$

Write each of the following in radical form. For example, $3x^{\frac{2}{3}} = 3\sqrt[3]{x^2}$.

31. $x^{\frac{4}{3}}$ **32.** $x^{\frac{2}{5}}$

33. $3x^{\frac{1}{2}}$ **34.** $5x^{\frac{1}{4}}$

35. $(2y)^{\frac{1}{3}}$ **36.** $(3xy)^{\frac{1}{2}}$

37. $(2x - 3y)^{\frac{1}{2}}$

38. $(5x + y)^{\frac{1}{3}}$

39. $(2a - 3b)^{\frac{2}{3}}$

40. $(5a + 7b)^{\frac{3}{5}}$

41. $x^{\frac{2}{3}}y^{\frac{1}{3}}$ **42.** $x^{\frac{3}{7}}y^{\frac{5}{7}}$

43. $-3x^{\frac{1}{5}}y^{\frac{2}{5}}$ **44.** $-4x^{\frac{3}{4}}y^{\frac{1}{4}}$

Write each of the following using positive rational exponents. For example,
$$\sqrt{ab} = (ab)^{\frac{1}{2}} = a^{\frac{1}{2}}b^{\frac{1}{2}}.$$

45. $\sqrt{5y}$ **46.** $\sqrt{2xy}$

47. $3\sqrt{y}$ **48.** $5\sqrt{ab}$

49. $\sqrt[3]{xy^2}$ **50.** $\sqrt[5]{x^2y^4}$

51. $\sqrt[4]{a^2b^3}$ **52.** $\sqrt[6]{ab^5}$

53. $\sqrt[5]{(2x - y)^3}$

54. $\sqrt[7]{(3x - y)^4}$

55. $5x\sqrt{y}$

56. $4y\sqrt[3]{x}$

57. $-\sqrt[3]{x + y}$

58. $-\sqrt[5]{(x - y)^2}$

Simplify each of the following. Express final results using positive exponents only. For example,
$$\left(2x^{\frac{1}{2}}\right)\left(3x^{\frac{1}{3}}\right) = 6x^{\frac{5}{6}}.$$

59. $\left(2x^{\frac{2}{5}}\right)\left(6x^{\frac{1}{4}}\right)$ **60.** $\left(3x^{\frac{1}{4}}\right)\left(5x^{\frac{1}{3}}\right)$

61. $\left(y^{\frac{2}{3}}\right)\left(y^{-\frac{1}{4}}\right)$ **62.** $\left(y^{\frac{3}{4}}\right)\left(y^{-\frac{1}{2}}\right)$

63. $\left(x^{\frac{2}{5}}\right)\left(4x^{-\frac{1}{2}}\right)$ **64.** $\left(2x^{\frac{1}{3}}\right)\left(x^{-\frac{1}{2}}\right)$

65. $\left(4x^{\frac{1}{2}}y\right)^2$ **66.** $\left(3x^{\frac{1}{4}}y^{\frac{1}{5}}\right)^3$

67. $(8x^6y^3)^{\frac{1}{3}}$ **68.** $(9x^2y^4)^{\frac{1}{2}}$

69. $\dfrac{24x^{\frac{3}{5}}}{6x^{\frac{1}{3}}}$ **70.** $\dfrac{18x^{\frac{1}{2}}}{9x^{\frac{1}{3}}}$

71. $\dfrac{48b^{\frac{1}{3}}}{12b^{\frac{3}{4}}}$ **72.** $\dfrac{56a^{\frac{1}{6}}}{8a^{\frac{1}{4}}}$

73. $\left(\dfrac{6x^{\frac{2}{5}}}{7y^{\frac{2}{3}}}\right)^2$

74. $\left(\dfrac{2x^{\frac{1}{3}}}{3y^{\frac{1}{4}}}\right)^4$

75. $\left(\dfrac{x^2}{y^3}\right)^{-\frac{1}{2}}$

76. $\left(\dfrac{a^3}{b^{-2}}\right)^{-\frac{1}{3}}$

77. $\left(\dfrac{18x^{\frac{1}{3}}}{9x^{\frac{1}{4}}}\right)^2$

78. $\left(\dfrac{72x^{\frac{3}{4}}}{6x^{\frac{1}{2}}}\right)^2$

79. $\left(\dfrac{60a^{\frac{1}{5}}}{15a^{\frac{3}{4}}}\right)^2$

80. $\left(\dfrac{64a^{\frac{1}{3}}}{16a^{\frac{5}{9}}}\right)^3$

Perform the indicated operations and express answers in simplest radical form. (See Example 5.)

81. $\sqrt[3]{3}\sqrt{3}$

82. $\sqrt{2}\sqrt[4]{2}$

83. $\sqrt[4]{6}\sqrt{6}$

84. $\sqrt[3]{5}\sqrt{5}$

85. $\dfrac{\sqrt[3]{3}}{\sqrt[4]{3}}$

86. $\dfrac{\sqrt{2}}{\sqrt[3]{2}}$

87. $\dfrac{\sqrt[3]{8}}{\sqrt[4]{4}}$

88. $\dfrac{\sqrt{9}}{\sqrt[3]{3}}$

89. $\dfrac{\sqrt[4]{27}}{\sqrt{3}}$

90. $\dfrac{\sqrt[3]{16}}{\sqrt[6]{4}}$

THOUGHTS INTO WORDS

91. Your friend keeps getting an error message when evaluating $-4^{\frac{5}{2}}$ on his calculator. What error is he probably making?

92. Explain how you would evaluate $27^{\frac{2}{3}}$ without a calculator.

Further Investigations

93. Use your calculator to evaluate each of the following.

(a) $\sqrt[3]{1728}$

(b) $\sqrt[3]{5832}$

(c) $\sqrt[4]{2401}$

(d) $\sqrt[4]{65,536}$

(e) $\sqrt{161,051}$

(f) $\sqrt[5]{6,436,343}$

94. Definition 5.7 states that
$$b^{\frac{m}{n}} = \sqrt[n]{b^m} = (\sqrt[n]{b})^m.$$
Use your calculator to verify each of the following.

(a) $\sqrt[3]{27^2} = (\sqrt[3]{27})^2$ **(b)** $\sqrt[3]{8^5} = (\sqrt[3]{8})^5$

(c) $\sqrt[4]{16^3} = (\sqrt[4]{16})^3$ **(d)** $\sqrt[3]{16^2} = (\sqrt[3]{16})^2$

(e) $\sqrt[5]{9^4} = (\sqrt[5]{9})^4$ **(f)** $\sqrt[3]{12^4} = (\sqrt[3]{12})^4$

95. Use your calculator to evaluate each of the following.

(a) $16^{\frac{5}{2}}$ **(b)** $25^{\frac{7}{2}}$

(c) $16^{\frac{9}{4}}$ **(d)** $27^{\frac{5}{3}}$

(e) $343^{\frac{2}{3}}$ **(f)** $512^{\frac{4}{3}}$

96. Use your calculator to estimate each of the following to the nearest one-thousandth.

(a) $7^{\frac{4}{3}}$ **(b)** $10^{\frac{4}{5}}$

(c) $12^{\frac{3}{5}}$ **(d)** $19^{\frac{2}{5}}$

(e) $7^{\frac{3}{4}}$ **(f)** $10^{\frac{5}{4}}$

97. (a) Since $\dfrac{4}{5} = .8$, we can evaluate $10^{\frac{4}{5}}$ by evaluating $10^{.8}$, which involves a shorter sequence of "calculator steps." Evaluate parts (b), (c), (d), (e), and (f) of Problem 96 and take advantage of decimal exponents.

(b) What problem is created when we try to evaluate $7^{\frac{4}{3}}$ by changing the exponent to decimal form?

Scientific Notation

Many applications of mathematics involve the use of very large and very small numbers. For example,

1. A light year—the distance that a ray of light travels in one year—is approximately 5,900,000,000,000 miles;

2. In 1982 the national debt was approximately 950,000,000,000 dollars;

3. In the metric system, a millimicron equals .000000001 of a meter;

4. The weight of an oxygen molecule is approximately .0000000000000000000000053 of a gram.

Working with numbers of this type in standard form is quite cumbersome. It is much more convenient to represent very small and very large numbers in **scientific notation**, sometimes called **scientific form**. We express a number in scientific notation when we write it as a product of a number between 1 and 10 and an integral power of 10. Symbolically, a number written in scientific notation has the form

$$(N)(10)^k$$

where N is a number between 1 and 10, written in decimal form, and k is an integer. Consider the following examples that show a comparison between ordinary notation and scientific notation.

Ordinary notation	Scientific notation
2.14	$(2.14)(10)^0$
31.78	$(3.178)(10)^1$
412.9	$(4.129)(10)^2$
8,000,000	$(8)(10)^6$
.14	$(1.4)(10)^{-1}$
.0379	$(3.79)(10)^{-2}$
.00000049	$(4.9)(10)^{-7}$

To switch from ordinary notation to scientific notation, you can use the following procedure.

Write the given number as the product of a number between 1 and 10 and a power of 10. Determine the exponent of 10 by counting the number of places that the decimal point was moved when going from the original number to the number between 1 and 10. This exponent is (a) negative if the original number is less than 1, (b) positive if the original number is greater than 10, and (c) 0 if the original number itself is between 1 and 10.

Thus, we can write

$$.00467 = (4.67)(10)^{-3},$$
$$87,000 = (8.7)(10)^4,$$
$$3.1416 = (3.1416)(10)^0.$$

To switch from scientific notation to ordinary notation you can use the following procedure.

Move the decimal point the number of places indicated by the exponent of 10. The decimal point is moved to the right if the exponent is positive and to the left if it is negative.

Thus, we can write

$$(4.78)(10)^4 = 47,800,$$
$$(8.4)(10)^{-3} = .0084.$$

Scientific notation can frequently be used to simplify numerical calculations. We merely change the numbers to scientific notation and use the appropriate properties of exponents. Consider the following examples.

Example 1

Perform the indicated operations.

(a) $(.00024)(20,000)$

(b) $\dfrac{7,800,000}{.0039}$

(c) $\dfrac{(.00069)(.0034)}{(.0000017)(.023)}$

(d) $\sqrt{.000004}$

Solution

$$
\begin{aligned}
(a) \quad (.00024)(20,000) &= (2.4)(10)^{-4}(2)(10)^4 \\
&= (2.4)(2)(10)^{-4}(10)^4 \\
&= (4.8)(10)^0 \\
&= (4.8)(1) \\
&= 4.8
\end{aligned}
$$

(b) $\dfrac{7,800,000}{.0039} = \dfrac{(7.8)(10)^6}{(3.9)(10)^{-3}}$

$= (2)(10)^9$

$= 2,000,000,000$

(c) $\dfrac{(.00069)(.0034)}{(.0000017)(.023)} = \dfrac{(6.9)(10)^{-4}(3.4)(10)^{-3}}{(1.7)(10)^{-6}(2.3)(10)^{-2}}$

$= \dfrac{(\cancel{6.9})^3(\cancel{3.4})^2(10)^{-7}}{(\cancel{1.7})(\cancel{2.3})(10)^{-8}}$

$= (6)(10)^1$

$= 60$

(d) $\sqrt{.000004} = \sqrt{(4)(10)^{-6}}$

$= ((4)(10)^{-6})^{\frac{1}{2}}$

$= 4^{\frac{1}{2}}((10)^{-6})^{\frac{1}{2}}$

$= (2)(10)^{-3}$

$= .002$ ▲

Many calculators are equipped to display numbers in scientific notation. The display panel shows the number between 1 and 10 and the appropriate exponent of 10. For example, evaluating $(3,800,000)^2$ yields

| 1.444 | 13 |

Thus, $(3,800,000)^2 = (1.444)(10)^{13} = 14,440,000,000,000.$

Similarly, the answer for $(.000168)^2$ is displayed as

| 2.8224 | -08 |

Thus, $(.000168)^2 = (2.8224)(10)^{-8} = .000000028224.$

Calculators vary as to the number of digits displayed in the number between 1 and 10 when using scientific notation. For example, we used two different calculators to estimate $(6729)^6$ and obtained the following results.

| 9.2833 | 22 |

| 9.283316768 | 22 |.

Obviously, you need to know the capabilities of your calculator when working with problems in scientific notation. Many calculators also allow the entry of a number in scientific notation. Such calculators are equipped with an enter-the-exponent key (often labeled as $\boxed{\text{EE}}$ or $\boxed{\text{E EX}}$). Thus, a number such as $(3.14)(10)^8$ might be entered as follows.

Enter	Press	Display
3.14	$\boxed{\text{EE}}$	3.14 00
8		3.14 08

Furthermore, it may be that your calculator will perform the switch from ordinary notation to scientific notation with a routine such as the following.

Enter	Press	Display
4721	$\boxed{\text{EE}}$	4721 00
	$\boxed{=}$	4.721 03

Be sure that you know the routine that will accomplish this switch on your calculator.

It should be evident from this brief discussion that even when using a calculator, you need to have a thorough understanding of scientific notation.

Problem Set 5.7

Write each of the following in scientific notation. For example, $27,800 = (2.78)(10)^4$.

1. 89   **2.** 117

3. 4290   **4.** 812,000

5. 6,120,000

6. 72,400,000

7. 40,000,000

8. 500,000,000

9. 376.4

10. 9126.21

11. .347   **12.** .2165

13. .0214   **14.** .0037

15. .00005

16. .00000082

17. .00000000194

18. .00000000003

Write each of the following in ordinary notation. For example, $(3.18)(10)^2 = 318$.

19. $(2.3)(10)^1$

20. $(1.62)(10)^2$

21. $(4.19)(10)^3$

22. $(7.631)(10)^4$

23. $(5)(10)^8$

24. $(7)(10)^9$
25. $(3.14)(10)^{10}$
26. $(2.04)(10)^{12}$
27. $(4.3)(10)^{-1}$
28. $(5.2)(10)^{-2}$
29. $(9.14)(10)^{-4}$
30. $(8.76)(10)^{-5}$
31. $(5.123)(10)^{-8}$
32. $(6)(10)^{-9}$

Use scientific notation and the properties of exponents to help perform the following operations.

33. $(.0037)(.00002)$
34. $(.00003)(.00025)$
35. $(.00007)(11,000)$
36. $(.000004)(120,000)$
37. $\dfrac{360,000,000}{.0012}$
38. $\dfrac{66,000,000,000}{.022}$
39. $\dfrac{.000064}{16,000}$
40. $\dfrac{.00072}{.0000024}$
41. $\dfrac{(60,000)(.006)}{(.0009)(400)}$
42. $\dfrac{(.00063)(960,000)}{(3,200)(.0000021)}$
43. $\dfrac{(.0045)(60000)}{(1800)(.00015)}$
44. $\dfrac{(.00016)(300)(.028)}{.064}$
45. $\sqrt{9,000,000}$
46. $\sqrt{.00000009}$
47. $\sqrt[3]{8000}$
48. $\sqrt[3]{.001}$
49. $(90,000)^{\frac{3}{2}}$
50. $(8000)^{\frac{2}{3}}$

51. Explain the importance of scientific notation.
52. Why do we still need scientific notation when using calculators and computers?

Further Investigations

53. Sometimes it is more convenient to express a number as a product of a power of 10 and a number that is not between 1 and 10. For example, suppose that we want to calculate $\sqrt{640,000}$. We can proceed as follows.

$$\begin{aligned}\sqrt{640,000} &= \sqrt{(64)(10)^4}\\ &= ((64)(10)^4)^{\frac{1}{2}}\\ &= (64)^{\frac{1}{2}}(10^4)^{\frac{1}{2}}\\ &= (8)(10)^2\\ &= 8(100) = 800\end{aligned}$$

Compute each of the following.

(a) $\sqrt{49,000,000}$
(b) $\sqrt{.0025}$ **(c)** $\sqrt{14,400}$
(d) $\sqrt{.000121}$ **(e)** $\sqrt[3]{27,000}$
(f) $\sqrt[3]{.000064}$

54. Use your calculator to evaluate each of the following. Express final answers in ordinary notation.
(a) $(27,000)^2$
(b) $(450,000)^2$
(c) $(14,800)^2$
(d) $(1700)^3$
(e) $(900)^4$
(f) $(60)^5$
(g) $(.0213)^2$
(h) $(.000213)^2$
(i) $(.000198)^2$
(j) $(.000009)^3$

55. Use your calculator to estimate each of the following. Express final answers in scientific notation with the number between 1 and 10 rounded to the nearest one-thousandth.

(a) $(4576)^4$

(b) $(719)^{10}$

(c) $(28)^{12}$

(d) $(8619)^6$

(e) $(314)^5$

(f) $(145,723)^2$

56. Use your calculator to estimate each of the following. Express final answers in ordinary notation rounded to the nearest one-thousandth.

(a) $(1.09)^5$

(b) $(1.08)^{10}$

(c) $(1.14)^7$

(d) $(1.12)^{20}$

(e) $(.785)^4$

(f) $(.492)^5$

SUMMARY

(5.1) The following properties form the basis for manipulating with exponents.

1. $b^n \cdot b^m = b^{n+m}$ Product of two powers

2. $(b^n)^m = b^{mn}$ Power of a power

3. $(ab)^n = a^n b^n$ Power of a product

4. $\left(\dfrac{a}{b}\right)^n = \dfrac{a^n}{b^n}$ Power of a quotient

5. $\dfrac{b^n}{b^m} = b^{n-m}$ Quotient of two powers

(5.2) and (5.3) The **principal nth root of b** is designated by $\sqrt[n]{b}$, where n is the **index** and b is the **radicand**.

A radical expression is in **simplest radical form** if:

1. A radicand contains no polynomial factor raised to a power equal to or greater than the index of the radical.

2. No fraction appears within a radical sign.

3. No radical appears in the denominator.

The following properties are used to express radicals in simplest form.

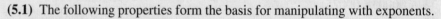

$$\sqrt[n]{bc} = \sqrt[n]{b}\,\sqrt[n]{c}, \qquad \sqrt[n]{\dfrac{b}{c}} = \dfrac{\sqrt[n]{b}}{\sqrt[n]{c}}$$

Simplifying by combining radicals sometimes requires that you first express the given radicals in simplest form and then apply the distributive property.

(5.4) The distributive property and the property $\sqrt[n]{b}\,\sqrt[n]{c} = \sqrt[n]{bc}$ are used to find products of expressions that involve radicals.

The special product pattern $(a + b)(a - b) = a^2 - b^2$ suggests a procedure for **rationalizing the denominator** of an expression that contains a binomial denominator with radicals.

(5.5) Equations that contain radicals with variables in a radicand are called **radical equations**. The property, if $a = b$, then $a^n = b^n$, forms the basis for solving radical equations. Raising both sides of an equation to a positive integral power may produce **extraneous solutions**, that is, solutions that do not satisfy the original equation. Therefore, you **must check** each potential solution.

(5.6) If b is a real number, n is a positive integer greater than one, and $\sqrt[n]{b}$ exists, then

$$b^{\frac{1}{n}} = \sqrt[n]{b}.$$

Thus, $b^{\frac{1}{n}}$ means **the nth root of b**.

If $\dfrac{m}{n}$ is a rational number, when n is a positive integer greater than one, and b is a real number such that $\sqrt[n]{b}$ exists, then

$$b^{\frac{m}{n}} = \sqrt[n]{b^m} = (\sqrt[n]{b})^m.$$

Both $\sqrt[n]{b^m}$ and $(\sqrt[n]{n})^m$ can be used for computational purposes.
We need to be able to switch back and forth between **exponential** and **radical form**. The link between exponents and roots provides a basis for multiplying and dividing some radicals even if they have different indices.

(5.7) The **scientific form** of a number is expressed as

$$(N)(10)^k$$

where N is a number between 1 and 10, written in decimal form, and k is an integer. Scientific notation is often convenient to use with very small and very large numbers. For example, .000046 can be expressed as $(4.6)(10^{-5})$ and 92,000,000 can be written as $(9.2)(10)^7$.

Scientific notation can often be used to simplify numerical calculations. For example,

$$(.000016)(30000) = (1.6)(10)^{-5}(3)(10)^4$$
$$= (4.8)(10)^{-1} = .48.$$

Chapter 5 Review Problem Set

Evaluate each of the following numerical expressions.

1. 4^{-3}

2. $\left(\dfrac{2}{3}\right)^{-2}$

3. $(3^2 \cdot 3^{-3})^{-1}$

4. $\sqrt[3]{-8}$

5. $\sqrt[4]{\dfrac{16}{81}}$

6. $4^{\frac{5}{2}}$

7. $(-1)^{\frac{2}{3}}$

8. $\left(\dfrac{8}{27}\right)^{\frac{2}{3}}$

9. $-16^{\frac{3}{2}}$

10. $\dfrac{2^3}{2^{-2}}$

11. $(4^{-2} \cdot 4^2)^{-1}$

12. $\left(\dfrac{3^{-1}}{3^2}\right)^{-1}$

Express each of the following radicals in simplest radical form.

13. $\sqrt{54}$

14. $\sqrt{48x^3y}$

15. $\dfrac{4\sqrt{3}}{\sqrt{6}}$

16. $\sqrt{\dfrac{5}{12x^3}}$

17. $\sqrt[3]{56}$

18. $\dfrac{\sqrt[3]{2}}{\sqrt[3]{9}}$

19. $\sqrt{\dfrac{9}{5}}$

20. $\sqrt{\dfrac{3x^3}{7}}$

21. $\sqrt[3]{108x^4y^8}$

22. $\dfrac{3}{4}\sqrt{150}$

23. $\dfrac{2}{3}\sqrt{45xy^3}$

24. $\dfrac{\sqrt{8x^2}}{\sqrt{2x}}$

Multiply and simplify.

25. $(3\sqrt{8})(4\sqrt{5})$

26. $(5\sqrt[3]{2})(6\sqrt[3]{4})$

27. $3\sqrt{2}(4\sqrt{6} - 2\sqrt{7})$

28. $(\sqrt{x} + 3)(\sqrt{x} - 5)$

29. $(2\sqrt{5} - \sqrt{3})(2\sqrt{5} + \sqrt{3})$

30. $(3\sqrt{2} + \sqrt{6})(5\sqrt{2} - 3\sqrt{6})$

31. $(2\sqrt{a} + \sqrt{b})(3\sqrt{a} - 4\sqrt{b})$

32. $(4\sqrt{8} - \sqrt{2})(\sqrt{8} + 3\sqrt{2})$

Rationalize the denominator and simplify.

33. $\dfrac{4}{\sqrt{7} - 1}$

34. $\dfrac{\sqrt{3}}{\sqrt{8} + \sqrt{5}}$

35. $\dfrac{3}{2\sqrt{3} + 3\sqrt{5}}$

36. $\dfrac{3\sqrt{2}}{2\sqrt{6} - \sqrt{10}}$

Simplify each of the following and express the final results using positive exponents.

37. $(x^{-3}y^4)^{-2}$

38. $\left(\dfrac{2a^{-1}}{3b^4}\right)^{-3}$

39. $\left(4x^{\frac{1}{2}}\right)\left(5x^{\frac{1}{5}}\right)$

40. $\dfrac{42a^{\frac{3}{4}}}{6a^{\frac{1}{3}}}$

41. $\left(\dfrac{x^3}{y^4}\right)^{-\frac{1}{3}}$

42. $\left(\dfrac{6x^{-2}}{2x^4}\right)^{-2}$

Use the distributive property to help simplify each of the following.

43. $3\sqrt{45} - 2\sqrt{20} - \sqrt{80}$

44. $4\sqrt[3]{24} + 3\sqrt[3]{3} - 2\sqrt[3]{81}$

45. $3\sqrt{24} - \dfrac{2\sqrt{54}}{5} + \dfrac{\sqrt{96}}{4}$

46. $-2\sqrt{12x} + 3\sqrt{27x} - 5\sqrt{48x}$

Express each of the following as a single fraction involving positive exponents only.

47. $x^{-2} + y^{-1}$

48. $a^{-2} - 2a^{-1}b^{-1}$

Solve each of the following equations.

49. $\sqrt{7x - 3} = 4$

50. $\sqrt{2y + 1} = \sqrt{5y - 11}$

51. $\sqrt{2x} = x - 4$

52. $\sqrt{n^2 - 4n - 4} = n$

53. $\sqrt[3]{2x - 1} = 3$

54. $\sqrt{t^2 + 9t - 1} = 3$

55. $\sqrt{x^2 + 3x - 6} = x$

56. $\sqrt{x + 1} - \sqrt{2x} = -1$

Use scientific notation and the properties of exponents to help perform the following calculations.

57. $(.00002)(.0003)$

58. $(120,000)(300,000)$

59. $(.000015)(400,000)$

60. $\dfrac{.000045}{.0003}$

61. $\dfrac{(.00042)(.0004)}{.006}$

62. $\sqrt{.000004}$

63. $\sqrt[3]{.000000008}$

64. $(4,000,000)^{\frac{3}{2}}$

CHAPTER 5 TEST

For Problems 1–4, simplify each of the numerical expressions.

1. $(4)^{-\frac{5}{2}}$

2. $-16^{\frac{5}{4}}$

3. $\left(\dfrac{2}{3}\right)^{-4}$

4. $\left(\dfrac{2^{-1}}{2^{-2}}\right)^{-2}$

For Problems 5–9, express each radical expression in simplest radical form.

5. $\sqrt{63}$

6. $\sqrt[3]{108}$

7. $\sqrt{52x^4y^3}$

8. $\dfrac{5\sqrt{18}}{3\sqrt{12}}$

9. $\sqrt{\dfrac{7}{24x^3}}$

10. Multiply and simplify: $(4\sqrt{6})(3\sqrt{12})$.

11. Multiply and simplify: $(3\sqrt{2} + \sqrt{3})(\sqrt{2} - 2\sqrt{3})$.

12. Simplify by combining similar radicals: $2\sqrt{50} - 4\sqrt{18} - 9\sqrt{32}$.

13. Rationalize the denominator and simplify: $\dfrac{3\sqrt{2}}{4\sqrt{3} - \sqrt{8}}$.

14. Simplify and express answer using positive exponents: $\left(\dfrac{2x^{-1}}{3y}\right)^{-2}$.

15. Simplify and express answer using positive exponents: $\dfrac{-84a^{\frac{1}{2}}}{7a^{\frac{4}{5}}}$

16. Express $x^{-1} + y^{-3}$ as a single fraction involving positive exponents.

17. Multiply and express answer using positive exponents: $(3x^{-\frac{1}{2}})(-4x^{\frac{3}{4}})$.

18. Multiply and simplify: $(3\sqrt{5} - 2\sqrt{3})(3\sqrt{5} + 2\sqrt{3})$.

For Problems 19 and 20, use scientific notation and the properties of exponents to help with the calculations.

19. $\dfrac{(.00004)(300)}{.00002}$

20. $\sqrt{.000009}$

(continued on next page)

CHAPTER 5 TEST (*continued*)

For Problems 21–25, solve each equation.

21. $\sqrt{3x + 1} = 3$

22. $\sqrt[3]{3x + 2} = 2$

23. $\sqrt{x} = x - 2$

24. $\sqrt{5x - 2} = \sqrt{3x + 8}$

25. $\sqrt{x^2 - 10x + 28} = 2$

Cumulative Review Problem Set

For Problems 1–5, evaluate each algebraic expression for the given values of the variables.

1. $\dfrac{4a^2 b^3}{12 a^3 b}$ for $a = 5$ and $b = -8$

2. $\dfrac{\frac{1}{x} + \frac{1}{y}}{\frac{1}{x} - \frac{1}{y}}$ for $x = 4$ and $y = 7$

3. $\dfrac{3}{n} + \dfrac{5}{2n} - \dfrac{4}{3n}$ for $n = 25$

4. $\dfrac{4}{x - 1} - \dfrac{2}{x + 2}$ for $x = \dfrac{1}{2}$

5. $2\sqrt{2x + y} - 5\sqrt{3x - y}$ for $x = 5$ and $y = 6$

For Problems 6–17, perform the indicated operations and express answers in simplified form.

6. $(3a^2 b)(-2ab)(4ab^3)$

7. $(x + 3)(2x^2 - x - 4)$

8. $\dfrac{6xy^2}{14y} \cdot \dfrac{7x^2 y}{8x}$

9. $\dfrac{a^2 + 6a - 40}{a^2 - 4a} \div \dfrac{2a^2 + 19a - 10}{a^3 + a^2}$

10. $\dfrac{3x + 4}{6} - \dfrac{5x - 1}{9}$

11. $\dfrac{4}{x^2 + 3x} + \dfrac{5}{x}$

12. $\dfrac{3n^2 + n}{n^2 + 10n + 16} \cdot \dfrac{2n^2 - 8}{3n^3 - 5n^2 - 2n}$

13. $\dfrac{3}{5x^2 + 3x - 2} - \dfrac{2}{5x^2 - 22x + 8}$

14. $\dfrac{y^3 - 7y^2 + 16y - 12}{y - 2}$

15. $(4x^3 - 17x^2 + 7x + 10) \div (4x - 5)$

16. $(3\sqrt{2} + 2\sqrt{5})(5\sqrt{2} - \sqrt{5})$

17. $(\sqrt{x} - 3\sqrt{y})(2\sqrt{x} + 4\sqrt{y})$

For Problems 18–25, evaluate each of the numerical expressions.

18. $-\sqrt{\dfrac{9}{64}}$

19. $\sqrt[3]{-\dfrac{8}{27}}$

20. $\sqrt[3]{.008}$

21. $32^{-\frac{1}{5}}$

22. $3^0 + 3^{-1} + 3^{-2}$

23. $-9^{\frac{3}{2}}$

24. $\left(\dfrac{3}{4}\right)^{-2}$

25. $\dfrac{1}{\left(\frac{2}{3}\right)^{-3}}$

For Problems 26–31, factor each of the algebraic expressions completely.

26. $3x^4 + 81x$

27. $6x^2 + 19x - 20$

28. $12 + 13x - 14x^2$

29. $9x^4 + 68x^2 - 32$

30. $2ax - ay - 2bx + by$

31. $27x^3 - 8y^3$

For Problems 32–49, solve each of the equations.

32. $3(x - 2) - 2(3x + 5) = 4(x - 1)$

33. $.06n + .08(n + 50) = 25$

34. $4\sqrt{x} + 5 = x$

35. $\sqrt[3]{n^2 - 1} = -1$

36. $6x^2 - 24 = 0$

37. $a^2 + 14a + 49 = 0$

38. $3n^2 + 14n - 24 = 0$

39. $\dfrac{2}{5x - 2} = \dfrac{4}{6x + 1}$

40. $\sqrt{2x - 1} - \sqrt{x + 2} = 0$

41. $5x - 4 = \sqrt{5x - 4}$

42. $|3x - 1| = 11$

43. $(3x - 2)(4x - 1) = 0$

44. $(2x + 1)(x - 2) = 7$

45. $\dfrac{5}{6x} - \dfrac{2}{3} = \dfrac{7}{10x}$

46. $\dfrac{3}{y + 4} + \dfrac{2y - 1}{y^2 - 16} = \dfrac{-2}{y - 4}$

47. $6x^4 - 23x^2 - 4 = 0$

48. $3n^3 + 3n = 0$

49. $n^2 - 13n - 114 = 0$

For Problems 50–55, solve each of the inequalities.

50. $6 - 2x \geq 10$

51. $4(2x - 1) < 3(x + 5)$

52. $\dfrac{n + 1}{4} + \dfrac{n - 2}{12} > \dfrac{1}{6}$

53. $|2x - 1| < 5$

54. $|3x + 2| > 11$

55. $\dfrac{1}{2}(3x - 1) - \dfrac{2}{3}(x + 4) \leq \dfrac{3}{4}(x - 1)$

For Problems 56–61, solve each problem by setting up and solving an appropriate equation.

56. How many liters of a 60% acid solution must be added to 14 liters of a 10% acid solution to produce a 25% acid solution?

57. A sum of $2250 is to be divided between two people in the ratio of 2 to 3. How much does each person receive?

58. The length of a picture without its border is 7 inches less than twice its width. If the border is 1 inch wide and its area is 62 square inches, what are the dimensions of the picture alone?

59. Lolita and Doug working together can paint a shed in 3 hours and 20 minutes. If Doug can paint the shed by himself in 10 hours, how long would it take Lolita to paint the shed by herself?

60. Angie bought some golf balls for $14. If each ball had cost $.25 less, she could have purchased one more ball for the same amount of money. How many golf balls did Angie buy?

61. A jogger who can run an 8-minute mile starts a half mile ahead of a jogger who can run a 6-minute mile. How long will it take the faster jogger to catch the slower jogger?

Quadratic Equations and Inequalities

A page for a magazine contains 70 square inches of type. The height of the page is twice the width. If the margin around the type is 2 inches uniformly, what are the dimensions of a page? **We can use the quadratic equation $(x - 4)(2x - 4) = 70$ to determine that the page measures 9 inches by 18 inches.**

Solving equations is one of the central themes of this text. Let's pause for a moment and reflect back on the different types of equations that we have solved in the last five chapters.

Type of equation	Examples
First-degree equations in one variable	$3x + 2x = x - 4$; $5(x + 4) = 12$; $\dfrac{x + 2}{3} + \dfrac{x - 1}{4} = 2$.
Second-degree equations in one variable *that are factorable*	$x^2 + 5x = 0$; $x^2 + 5x + 6 = 0$; $x^2 - 9 = 0$; $x^2 - 10x + 25 = 0$.
Fractional equations	$\dfrac{2}{x} + \dfrac{3}{x} = 4$; $\dfrac{5}{a - 1} = \dfrac{6}{a - 2}$; $\dfrac{2}{x^2 - 9} + \dfrac{3}{x + 3} = \dfrac{4}{x - 3}$.
Radical equations	$\sqrt{x} = 2$; $\sqrt{3x - 2} = 5$; $\sqrt{5y + 1} = \sqrt{3y + 4}$.

As this chart shows, we have solved second-degree equations in one variable, but only those for which the polynomial is factorable. In this chapter we will expand our work to include more general types of second-degree equations as well as inequalities in one variable.

6.1   Complex Numbers

Since the square of any real number is nonnegative, a simple equation such as $x^2 = -4$ has no solutions in the set of real numbers. To handle this situation, we can expand the set of real numbers into a larger set called the **complex numbers**. In this section we will instruct you on how to manipulate complex numbers.

To provide a solution for the equation $x^2 + 1 = 0$, we use the number i, such that

$$i^2 = -1.$$

The number i is not a real number and is often called the **imaginary unit**, but the number i^2 is the real number -1. The imaginary unit i is used to define a complex number as follows.

DEFINITION 6.1

> A **complex number** is any number that can be expressed in the form
>
> $a + bi$,
>
> where a and b are real numbers.

The form $a + bi$ is called the **standard form** of a complex number. The real number a is called the **real part** of the complex number and b is called the **imaginary part**. (Note that b is a real number even though it is called the imaginary part.) The following list exemplifies this terminology.

1. The number $7 + 5i$ is a complex number that has a real part of 7 and an imaginary part of 5.

2. The number $\frac{2}{3} + i\sqrt{2}$ is a complex number that has a real part of $\frac{2}{3}$ and an imaginary part of $\sqrt{2}$. (It is easy to mistake $\sqrt{2i}$ for $\sqrt{2}\,i$. Thus, it is common to write $i\sqrt{2}$ instead of $\sqrt{2}\,i$ to avoid any difficulties with the radical sign.)

3. The number $-4 - 3i$ can be written in the standard form $-4 + (-3i)$ and therefore is a complex number that has a real part of -4 and an imaginary part of -3. (The form $-4 - 3i$ is often used but we know that it means $-4 + (-3i)$.)

4. The number $-9i$ can be written as $0 + (-9i)$; thus, it is a complex number that has a real part of 0 and an imaginary part of -9. (Complex numbers, such as $-9i$, for which $a = 0$ and $b \neq 0$ are called **pure imaginary numbers**.)

5. The real number 4 can be written as $4 + 0i$ and is thus a complex number that has a real part of 4 and an imaginary part of 0.

Look at number 5. We see that the set of real numbers is a subset of the set of complex numbers. The following diagram indicates the organizational format of the complex numbers.

Complex numbers ($a + bi$,   where a and b are real numbers)

Real numbers   **Imaginary numbers**
($a + bi$,   where $b = 0$)   ($a + bi$,   where $b \neq 0$)

Pure imaginary numbers
($a + bi$,   where $a = 0$ and $b \neq 0$)

Two complex numbers $a + bi$ and $c + di$ are said to be **equal** if and only if $a = c$ and $b = d$.

Adding and Subtracting Complex Numbers

To **add complex numbers**, we simply add their real parts and add their imaginary parts. Thus,

$$(a + bi) + (c + di) = (a + c) + (b + d)i.$$

The following examples show addition of two complex numbers.

1. $(4 + 3i) + (5 + 9i) = (4 + 5) + (3 + 9)i = 9 + 12i.$

2. $(-6 + 4i) + (8 - 7i) = (-6 + 8) + (4 - 7)i$
$$= 2 - 3i.$$

3. $\left(\dfrac{1}{2} + \dfrac{3}{4}i\right) + \left(\dfrac{2}{3} + \dfrac{1}{5}i\right) = \left(\dfrac{1}{2} + \dfrac{2}{3}\right) + \left(\dfrac{3}{4} + \dfrac{1}{5}\right)i$

$$= \left(\dfrac{3}{6} + \dfrac{4}{6}\right) + \left(\dfrac{15}{20} + \dfrac{4}{20}\right)i$$

$$= \dfrac{7}{6} + \dfrac{19}{20}i.$$

The set of complex numbers is closed with respect to addition; that is, the sum of two complex numbers is a complex number. Furthermore, the commutative and associative properties of addition hold for all complex numbers. The addition identity element is $0 + 0i$ (or simply the real number 0). The additive inverse of $a + bi$ is $-a - bi$, because

$$(a + bi) + (-a - bi) = 0.$$

To **subtract complex numbers**, $c + di$ from $a + bi$, add the additive inverse of $c + di$. Thus,

$$(a + bi) - (c + di) = (a + bi) + (-c - di)$$
$$= (a - c) + (b - d)i.$$

In other words, we subtract the real parts and subtract the imaginary parts, as in the next examples.

1. $(9 + 8i) - (5 + 3i) = (9 - 5) + (8 - 3)i$
$$= 4 + 5i.$$

2. $(3 - 2i) - (4 - 10i) = (3 - 4) + (-2 - (-10))i$
$$= -1 + 8i.$$

Products and Quotients of Complex Numbers

Since $i^2 = -1$, i is a square root of negative one; so we let $i = \sqrt{-1}$. It should also be evident that $-i$ is a square root of negative one since

$$(-i)^2 = (-i)(-i) = i^2 = -1.$$

Thus, in the set of complex numbers, -1 has two square roots, i and $-i$. We express these symbolically as

$$\sqrt{-1} = i \qquad \text{and} \qquad -\sqrt{-1} = -i.$$

Let us extend our definition so that in the set of complex numbers every negative real number has two square roots. We simply define $\sqrt{-b}$, where b is a positive real number, to be the number whose square is $-b$. Thus,

$$(\sqrt{-b})^2 = -b, \quad \text{for } b > 0.$$

Furthermore, since $(i\sqrt{b})(i\sqrt{b}) = i^2(b) = -1(b) = -b$ we see that

$$\sqrt{-b} = i\sqrt{b}.$$

In other words, a square root of any negative real number can be represented as the product of a real number and the imaginary unit i. Consider the following examples.

$$\sqrt{-4} = i\sqrt{4} = 2i,$$
$$\sqrt{-17} = i\sqrt{17},$$
$$\sqrt{-24} = i\sqrt{24} = i\sqrt{4}\sqrt{6} = 2i\sqrt{6} \qquad \text{Notice that we simplified the radical } \sqrt{24} \text{ to } 2\sqrt{6}.$$

We should also observe that $-\sqrt{-b}$, where $b > 0$, is a square root of $-b$ since

$$(-\sqrt{-b})^2 = (-i\sqrt{b})^2 = i^2(b) = -1(b) = -b.$$

Thus, in the set of complex numbers, $-b$ (where $b > 0$) has two square roots, $i\sqrt{b}$ and $-i\sqrt{b}$. We express these symbolically as

$$\sqrt{-b} = i\sqrt{b} \qquad \text{and} \qquad -\sqrt{-b} = -i\sqrt{b}.$$

We must be very careful with the use of the symbol $\sqrt{-b}$, where $b > 0$. Some relationships that are true in the set of real numbers that involve the square root symbol do not hold if the square root symbol does not represent a real number. For example, $\sqrt{a}\sqrt{b} = \sqrt{ab}$ *does not hold* if a and b are both negative numbers.

Correct $\quad \sqrt{-4}\sqrt{-9} = (2i)(3i) = 6i^2 = 6(-1) = -6.$

Incorrect $\quad \sqrt{-4}\,\sqrt{-9} = \sqrt{(-4)(-9)} = \sqrt{36} = 6.$

To avoid difficulty with this idea, you should rewrite all expressions of the form $\sqrt{-b}$, where $b > 0$, in the form $i\sqrt{b}$ before doing *any computations*. The following examples further demonstrate this point.

1. $\sqrt{-5}\sqrt{-7} = (i\sqrt{5})(i\sqrt{7}) = i^2\sqrt{35} = (-1)\sqrt{35} = -\sqrt{35}.$

2. $\sqrt{-2}\sqrt{-8} = (i\sqrt{2})(i\sqrt{8}) = i^2\sqrt{16} = (-1)(4) = -4.$

3. $\sqrt{-6}\sqrt{-8} = (i\sqrt{6})(i\sqrt{8}) = i^2\sqrt{48} = (-1)\sqrt{16}\sqrt{3} = -4\sqrt{3}.$

4. $\dfrac{\sqrt{-75}}{\sqrt{-3}} = \dfrac{i\sqrt{75}}{i\sqrt{3}} = \dfrac{\sqrt{75}}{\sqrt{3}} = \sqrt{\dfrac{75}{3}} = \sqrt{25} = 5.$

5. $\dfrac{\sqrt{-48}}{\sqrt{12}} = \dfrac{i\sqrt{48}}{\sqrt{12}} = i\sqrt{\dfrac{48}{12}} = i\sqrt{4} = 2i.$

Since complex numbers have a *binomial form*, we find the *product* of two complex numbers in the same way that we find the product of two binomials. Then by replacing i^2 with -1, we are able to simplify and express the final result in standard form. Consider the following examples.

6. $(2 + 3i)(4 + 5i) = 2(4 + 5i) + 3i(4 + 5i)$
$$= 8 + 10i + 12i + 15i^2$$
$$= 8 + 22i + 15i^2$$
$$= 8 + 22i + 15(-1) = -7 + 22i.$$

7. $(-3 + 6i)(2 - 4i) = -3(2 - 4i) + 6i(2 - 4i)$
$$= -6 + 12i + 12i - 24i^2$$
$$= -6 + 24i - 24(-1)$$
$$= -6 + 24i + 24 = 18 + 24i.$$

8. $(1 - 7i)^2 = (1 - 7i)(1 - 7i)$
$$= 1(1 - 7i) - 7i(1 - 7i)$$
$$= 1 - 7i - 7i + 49i^2$$
$$= 1 - 14i + 49(-1)$$
$$= 1 - 14i - 49$$
$$= -48 - 14i.$$

9. $(2 + 3i)(2 - 3i) = 2(2 - 3i) + 3i(2 - 3i)$
$$= 4 - 6i + 6i - 9i^2$$
$$= 4 - 9(-1)$$
$$= 4 + 9$$
$$= 13.$$

Example 9 illustrates an important situation: The complex numbers $2 + 3i$ and $2 - 3i$ are conjugates of each other. In general, two complex numbers $a + bi$ and $a - bi$ are called **conjugates** of each other. *The product of a complex number and its conjugate is always a real number*, which can be shown as follows.

$$(a + bi)(a - bi) = a(a - bi) + bi(a - bi)$$
$$= a^2 - abi + abi - b^2i^2$$
$$= a^2 - b^2(-1)$$
$$= a^2 + b^2$$

We use conjugates to *simplify expressions* such as $\dfrac{3i}{5 + 2i}$ that *indicate the quotient* of two complex numbers. To eliminate i in the denominator and change the indicated quotient to the standard form of a complex number, we can multiply both the numerator and the denominator by the conjugate of the denominator as follows.

$$\frac{3i}{5 + 2i} = \frac{3i(5 - 2i)}{(5 + 2i)(5 - 2i)}$$
$$= \frac{15i - 6i^2}{25 - 4i^2}$$
$$= \frac{15i - 6(-1)}{25 - 4(-1)}$$
$$= \frac{15i + 6}{29}$$
$$= \frac{6}{29} + \frac{15}{29}i$$

The following examples further clarify the process of *dividing* complex numbers.

10. $\dfrac{2 - 3i}{4 - 7i} = \dfrac{(2 - 3i)(4 + 7i)}{(4 - 7i)(4 + 7i)}$ $4 + 7i$ is the conjugate of $4 - 7i$.
$$= \frac{8 + 14i - 12i - 21i^2}{16 - 49i^2}$$
$$= \frac{8 + 2i - 21(-1)}{16 - 49(-1)}$$
$$= \frac{8 + 2i + 21}{16 + 49}$$

$$= \frac{29 + 2i}{65}$$

$$= \frac{29}{65} + \frac{2}{65}i.$$

11. $\dfrac{4 - 5i}{2i} = \dfrac{(4 - 5i)(-2i)}{(2i)(-2i)}$ $-2i$ is the conjugate of $2i$.

$$= \frac{-8i + 10i^2}{-4i^2}$$

$$= \frac{-8i + 10(-1)}{-4(-1)}$$

$$= \frac{-8i - 10}{4}$$

$$= -\frac{5}{2} - 2i$$

In Example 11 where the denominator is a pure imaginary number, we can change to standard form by choosing a multiplier other than the conjugate. Consider the following alternate approach for Example 11.

$$\frac{4 - 5i}{2i} = \frac{(4 - 5i)(i)}{(2i)(i)}$$

$$= \frac{4i - 5i^2}{2i^2}$$

$$= \frac{4i - 5(-1)}{2(-1)}$$

$$= \frac{4i + 5}{-2}$$

$$= -\frac{5}{2} - 2i$$

Problem Set 6.1

Label each of the following statements true or false.

1. Every complex number is a real number.

2. Every real number is a complex number.

3. The real part of the complex number $6i$ is 0.

4. Every complex number is a pure imaginary number.

5. The sum of two complex numbers is always a complex number.

6. The imaginary part of the complex number 7 is 0.

7. The sum of two complex numbers is sometimes a real number.

8. The sum of two pure imaginary numbers is always a pure imaginary number.

Add or subtract as indicated.

9. $(6 + 3i) + (4 + 5i)$

10. $(5 + 2i) + (7 + 10i)$

11. $(-8 + 4i) + (2 + 6i)$

12. $(5 - 8i) + (-7 + 2i)$

13. $(3 + 2i) - (5 + 7i)$

14. $(1 + 3i) - (4 + 9i)$

15. $(-7 + 3i) - (5 - 2i)$

16. $(-8 + 4i) - (9 - 4i)$

17. $(-3 - 10i) + (2 - 13i)$

18. $(-4 - 12i) + (-3 + 16i)$

19. $(4 - 8i) - (8 - 3i)$

20. $(12 - 9i) - (14 - 6i)$

21. $(-1 - i) - (-2 - 4i)$

22. $(-2 - 3i) - (-4 - 14i)$

23. $\left(\dfrac{3}{2} + \dfrac{1}{3}i\right) + \left(\dfrac{1}{6} - \dfrac{3}{4}i\right)$

24. $\left(\dfrac{2}{3} - \dfrac{1}{5}i\right) + \left(\dfrac{3}{5} - \dfrac{3}{4}i\right)$

25. $\left(-\dfrac{5}{9} + \dfrac{3}{5}i\right) - \left(\dfrac{4}{3} - \dfrac{1}{6}i\right)$

26. $\left(\dfrac{3}{8} - \dfrac{5}{2}i\right) - \left(\dfrac{5}{6} + \dfrac{1}{7}i\right)$

Write each of the following in terms of i and simplify. For example, $\sqrt{-20} = i\sqrt{20} = i\sqrt{4}\sqrt{5} = 2i\sqrt{5}$.

27. $\sqrt{-81}$

28. $\sqrt{-49}$

29. $\sqrt{-14}$

30. $\sqrt{-33}$

31. $\sqrt{-\dfrac{16}{25}}$

32. $\sqrt{-\dfrac{64}{36}}$

33. $\sqrt{-18}$

34. $\sqrt{-84}$

35. $\sqrt{-75}$

36. $\sqrt{-63}$

37. $3\sqrt{-28}$

38. $5\sqrt{-72}$

39. $-2\sqrt{-80}$

40. $-6\sqrt{-27}$

41. $12\sqrt{-90}$

42. $9\sqrt{-40}$

Write each of the following in terms of i, perform the indicated operations, and simplify. For example,

$$\sqrt{-3}\sqrt{-8} = (i\sqrt{3})(i\sqrt{8})$$
$$= i^2\sqrt{24}$$
$$= (-1)\sqrt{4}\sqrt{6}$$
$$= -2\sqrt{6}.$$

43. $\sqrt{-4}\sqrt{-16}$

44. $\sqrt{-81}\sqrt{-25}$

45. $\sqrt{-3}\sqrt{-5}$

46. $\sqrt{-7}\sqrt{-10}$

47. $\sqrt{-9}\sqrt{-6}$

48. $\sqrt{-8}\sqrt{-16}$

49. $\sqrt{-15}\sqrt{-5}$

50. $\sqrt{-2}\sqrt{-20}$

51. $\sqrt{-2}\sqrt{-27}$

52. $\sqrt{-3}\sqrt{-15}$

53. $\sqrt{6}\sqrt{-8}$

54. $\sqrt{-75}\sqrt{3}$

55. $\dfrac{\sqrt{-25}}{\sqrt{-4}}$

56. $\dfrac{\sqrt{-81}}{\sqrt{-9}}$

57. $\dfrac{\sqrt{-56}}{\sqrt{-7}}$

58. $\dfrac{\sqrt{-72}}{\sqrt{-6}}$

59. $\dfrac{\sqrt{-24}}{\sqrt{6}}$

60. $\dfrac{\sqrt{-96}}{\sqrt{2}}$

Find each of the following products and express answers in the standard form of a complex number.

61. $(5i)(4i)$

62. $(-6i)(9i)$

63. $(7i)(-6i)$

64. $(-5i)(-12i)$

65. $(3i)(2 - 5i)$

66. $(7i)(-9 + 3i)$

67. $(-6i)(-2 - 7i)$

68. $(-9i)(-4 - 5i)$

69. $(3 + 2i)(5 + 4i)$

70. $(4 + 3i)(6 + i)$

71. $(6 - 2i)(7 - i)$

72. $(8 - 4i)(7 - 2i)$

73. $(-3 - 2i)(5 + 6i)$

74. $(-5 - 3i)(2 - 4i)$

75. $(9 + 6i)(-1 - i)$

76. $(10 + 2i)(-2 - i)$

77. $(4 + 5i)^2$

78. $(5 - 3i)^2$

79. $(-2 - 4i)^2$

80. $(-3 - 6i)^2$

81. $(6 + 7i)(6 - 7i)$

82. $(5 - 7i)(5 + 7i)$

83. $(-1 + 2i)(-1 - 2i)$

84. $(-2 - 4i)(-2 + 4i)$

Find each of the following quotients and express answers in the standard form of a complex number.

85. $\dfrac{3i}{2 + 4i}$

86. $\dfrac{4i}{5 + 2i}$

87. $\dfrac{-2i}{3 - 5i}$

88. $\dfrac{-5i}{2 - 4i}$

89. $\dfrac{-2 + 6i}{3i}$

90. $\dfrac{-4 - 7i}{6i}$

91. $\dfrac{2}{7i}$

92. $\dfrac{3}{10i}$

93. $\dfrac{2 + 6i}{1 + 7i}$

94. $\dfrac{5 + i}{2 + 9i}$

95. $\dfrac{3 + 6i}{4 - 5i}$

96. $\dfrac{7 - 3i}{4 - 3i}$

97. $\dfrac{-2 + 7i}{-1 + i}$

98. $\dfrac{-3 + 8i}{-2 + i}$

99. $\dfrac{-1 - 3i}{-2 - 10i}$

100. $\dfrac{-3 - 4i}{-4 - 11i}$

THOUGHTS INTO WORDS

101. Why is the set of real numbers a subset of the set of complex numbers?

102. Can the sum of two nonreal complex numbers be a real number? Defend your answer.

103. Can the product of two nonreal complex numbers be a real number? Defend your answer.

6.2 Quadratic Equations

A second-degree equation in one variable contains the variable with an exponent of two, but no higher power. Such equations are also called **quadratic equations**. The following are examples of quadratic equations.

$$x^2 = 36, \qquad y^2 + 4y = 0, \qquad x^2 + 5x - 2 = 0,$$

$$3n^2 + 2n - 1 = 0, \qquad 5x^2 + x + 2 = 3x^2 - 2x - 1$$

A quadratic equation in the variable x can also be defined as any equation that can be written in the form

$$ax^2 + bx + c = 0,$$

where a, b, and c are real numbers and $a \neq 0$. The form $ax^2 + bx + c = 0$ is called the **standard form** of a quadratic equation.

In previous chapters you solved quadratic equations (the term *quadratic* was not used at that time) by factoring and applying the property, $ab = 0$ if and only if $a = 0$ or $b = 0$. Let's review a few such examples.

Example 1

Solve $3n^2 + 14n - 5 = 0$.

Solution

$$3n^2 + 14n - 5 = 0$$
$$(3n - 1)(n + 5) = 0 \qquad \text{Factor the left side.}$$
$$3n - 1 = 0 \qquad \text{or} \qquad n + 5 = 0 \qquad \text{Apply: } ab = 0 \text{ if and}$$
$$\text{only if } a = 0 \text{ or } b = 0.$$
$$3n = 1 \qquad \text{or} \qquad n = -5$$
$$n = \frac{1}{3} \qquad \text{or} \qquad n = -5$$

The solution set is $\left\{-5, \dfrac{1}{3}\right\}$. ▲

Example 2

Solve $x^2 + 3kx - 10k^2 = 0$ for x.

Solution

$$x^2 + 3kx - 10k^2 = 0$$
$$(x + 5k)(x - 2k) = 0 \qquad \text{Factor the left side.}$$
$$x + 5k = 0 \qquad \text{or} \qquad x - 2k = 0 \qquad \text{Apply: } ab = 0 \text{ if and}$$
$$\text{only if } a = 0 \text{ or } b = 0.$$
$$x = -5k \qquad \text{or} \qquad x = 2k$$

The solution set is $\{-5k, 2k\}$. ▲

Example 3

Solve $2\sqrt{x} = x - 8$.

Solution

$$2\sqrt{x} = x - 8$$
$$(2\sqrt{x})^2 = (x - 8)^2 \qquad \text{Square both sides.}$$
$$4x = x^2 - 16x + 64$$
$$0 = x^2 - 20x + 64$$
$$0 = (x - 16)(x - 4) \qquad \text{Factor the right side.}$$
$$x - 16 = 0 \qquad \text{or} \qquad x - 4 = 0 \qquad \text{Apply: } ab = 0 \text{ if and}$$
$$\text{only if } a = 0 \text{ or } b = 0.$$
$$x = 16 \qquad \text{or} \qquad x = 4$$

✔ *Check*

$$2\sqrt{x} = x - 8 \qquad\qquad 2\sqrt{x} = x - 8$$
$$2\sqrt{16} \overset{?}{=} 16 - 8 \quad \text{or} \quad 2\sqrt{4} \overset{?}{=} 4 - 8$$
$$2(4) \overset{?}{=} 8 \qquad\qquad 2(2) \overset{?}{=} -4$$
$$8 = 8 \qquad\qquad 4 \neq -4$$

The solution set is $\{16\}$. ▲

We should make two comments about Example 3. First, remember that applying the property, if $a = b$, then $a^n = b^n$, might produce extraneous solutions. Therefore, we *must* check all potential solutions. Secondly, the equation $2\sqrt{x} = x - 8$ is said to be of **quadratic form** because it can be written as $2x^{\frac{1}{2}} = \left(x^{\frac{1}{2}}\right)^2 - 8$. More will be said about the phrase, *quadratic form*, later.

Let's consider quadratic equations of the form $x^2 = a$, where x is the variable and a is any real number. We can solve $x^2 = a$ as follows.

$$x^2 = a$$
$$x^2 - a = 0$$
$$x^2 - (\sqrt{a})^2 = 0 \qquad\qquad a = (\sqrt{a})^2$$
$$(x - \sqrt{a})(x + \sqrt{a}) = 0 \qquad\qquad \text{Factor the left side.}$$
$$x - \sqrt{a} = 0 \quad \text{or} \quad x + \sqrt{a} = 0 \qquad \text{Apply: } ab = 0 \text{ if and}$$
$$\text{only if } a = 0 \text{ or } b = 0.$$
$$x = \sqrt{a} \quad \text{or} \quad x = -\sqrt{a}.$$

The solutions are $\sqrt{a}$ and $-\sqrt{a}$.

We can state the previous result as a general property and use it to solve certain types of quadratic equations.

PROPERTY 6.1

For any real number a,

$$x^2 = a \quad \text{if and only if } x = \sqrt{a} \text{ or } x = -\sqrt{a}.$$

(The statement $x = \sqrt{a}$ or $x = -\sqrt{a}$ can be written as $x = \pm\sqrt{a}$.)

Property 6.1, along with our knowledge of square roots, makes it very easy to solve quadratic equations of the form $x^2 = a$.

Example 4

Solution

Solve $x^2 = 45$.

$$x^2 = 45$$
$$x = \pm\sqrt{45}$$
$$x = \pm 3\sqrt{5} \qquad \sqrt{45} = \sqrt{9}\sqrt{5} = 3\sqrt{5}$$

The solution set is $\{\pm 3\sqrt{5}\}$. ▲

Example 5

Solution

Solve $x^2 = -9$.

$$x^2 = -9$$
$$x = \pm\sqrt{-9}$$
$$x = \pm 3i. \qquad \sqrt{-9} = i\sqrt{9} = 3i$$

Thus, the solution set is $\{\pm 3i\}$.   ▲

Example 6

Solution

Solve $7n^2 = 12$.

$$7n^2 = 12$$
$$n^2 = \frac{12}{7}$$
$$n = \pm\sqrt{\frac{12}{7}}$$
$$n = \pm\frac{2\sqrt{21}}{7}. \qquad \sqrt{\frac{12}{7}} = \frac{\sqrt{12}}{\sqrt{7}} \cdot \frac{\sqrt{7}}{\sqrt{7}} = \frac{\sqrt{84}}{7} = \frac{\sqrt{4}\sqrt{21}}{7} = \frac{2\sqrt{21}}{7}$$

The solution set is $\left\{\pm\dfrac{2\sqrt{21}}{7}\right\}$.   ▲

Example 7

Solution

Solve $(3n + 1)^2 = 25$.

$$(3n + 1)^2 = 25$$
$$(3n + 1) = \pm\sqrt{25}$$
$$3n + 1 = \pm 5$$
$$3n + 1 = 5 \qquad \text{or} \qquad 3n + 1 = -5$$
$$3n = 4 \qquad \text{or} \qquad 3n = -6$$
$$n = \frac{4}{3} \qquad \text{or} \qquad n = -2$$

The solution set is $\left\{-2, \dfrac{4}{3}\right\}$.   ▲

Example 8

Solution

Solve $(x - 3)^2 = -10$.

$$(x - 3)^2 = -10$$
$$x - 3 = \pm\sqrt{-10}$$
$$x - 3 = \pm i\sqrt{10}$$
$$x = 3 \pm i\sqrt{10}$$

Thus, the solution set is $\left\{3 \pm i\sqrt{10}\right\}$.   ▲

Sometimes it may be necessary to change the form before we can apply Property 6.1. Let's consider one example to illustrate this idea.

Example 9

Solve $3(2x - 3)^2 + 8 = 44$.

Solution

$$3(2x - 3)^2 + 8 = 44$$
$$3(2x - 3)^2 = 36$$
$$(2x - 3)^2 = 12$$
$$2x - 3 = \pm\sqrt{12}$$
$$2x - 3 = \pm 2\sqrt{3}$$
$$2x = 3 \pm 2\sqrt{3}$$
$$x = \frac{3 \pm 2\sqrt{3}}{2}$$

The solution set is $\left\{\dfrac{3 \pm 2\sqrt{3}}{2}\right\}$.

▲

Back to the Pythagorean Theorem

Our work with radicals, Property 6.1, and the Pythagorean theorem form a basis for solving a variety of problems that pertain to right triangles.

Example 10

A 50-foot rope hangs from the top of a flagpole. When pulled taut to its full length, the rope reaches a point on the ground 18 feet from the base of the pole. Find the height of the pole to the nearest tenth of a foot.

Solution

Let's make a sketch (Figure 6.1) and record the given information.

FIGURE 6.1

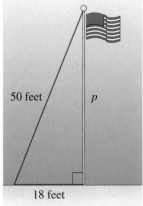

50 feet

p

18 feet

p represents the height of the flagpole.

Use the Pythagorean theorem to solve for p as follows.

$$p^2 + 18^2 = 50^2$$
$$p^2 + 324 = 2500$$
$$p^2 = 2176$$
$$p = \sqrt{2176} = 46.6, \quad \text{to the nearest tenth}$$

The height of the flagpole is approximately 46.6 feet.

▲

There are two special kinds of right triangles that we use extensively in later mathematics courses. The first is an **isosceles right triangle**, which is a right triangle that has both legs of the same length. Let's consider a problem that involves an isosceles right triangle.

Example 11

Find the length of each leg of an isosceles right triangle that has a hypotenuse of length 5 meters.

Solution

Let's sketch an isosceles right triangle and let x represent the length of each leg (Figure 6.2). Then we can apply the Pythagorean theorem as follows.

FIGURE 6.2

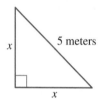

$$x^2 + x^2 = 5^2$$
$$2x^2 = 25$$
$$x^2 = \frac{25}{2}$$
$$x = \pm\sqrt{\frac{25}{2}} = \pm\frac{5}{\sqrt{2}} = \pm\frac{5\sqrt{2}}{2}$$

Each leg is $\dfrac{5\sqrt{2}}{2}$ meters long.

REMARK   In Example 10 we made no attempt to express $\sqrt{2176}$ in simplest radical form because the answer was to be given as a rational approximation to the nearest tenth. However, in Example 11 we left the final answer in radical form and therefore expressed it in simplest radical form.

The second special kind of right triangle that we use frequently is one that contains acute angles of 30° and 60°. In such a right triangle, which we refer to as a **30°–60° right triangle**, the side opposite the 30° angle is equal in length to one-half of the length of the hypotenuse. This relationship, along with the Pythagorean theorem, provides us with another problem solving technique.

Example 12

Suppose that a 20-foot ladder is leaning against a building and makes an angle of 60° with the ground. How far up the building does the top of the ladder reach? Express your answer to the nearest tenth of a foot.

Solution

Figure 6.3 depicts this situation. The side opposite the 30° angle equals one-half of the hypotenuse, so it is of length $\frac{1}{2}(20) = 10$ feet. Now we can apply the Pythagorean theorem.

FIGURE 6.3

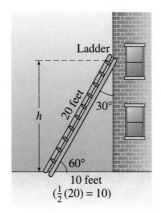

$$h^2 + 10^2 = 20^2$$
$$h^2 + 100 = 400$$
$$h^2 = 300$$
$$h = \sqrt{300} = 17.3, \quad \text{to the nearest tenth}$$

The top of the ladder touches the building at approximately 17.3 feet from the ground.

Problem Set 6.2

Solve each of the following quadratic equations by factoring and applying the property, $ab = 0$ if and only if $a = 0$ or $b = 0$. If necessary, return to Chapter 3 and review the factoring techniques that we studied.

1. $x^2 - 9x = 0$ **2.** $x^2 + 5x = 0$

3. $x^2 = -3x$ **4.** $x^2 = 15x$

5. $3y^2 + 12y = 0$

6. $6y^2 - 24y = 0$

7. $5n^2 - 9n = 0$

8. $4n^2 + 13n = 0$

9. $x^2 + x - 30 = 0$

10. $x^2 - 8x - 48 = 0$

11. $x^2 - 19x + 84 = 0$

12. $x^2 - 21x + 104 = 0$

13. $2x^2 + 19x + 24 = 0$

14. $4x^2 + 29x + 30 = 0$

15. $15x^2 + 29x - 14 = 0$

16. $24x^2 + x - 10 = 0$

17. $25x^2 - 30x + 9 = 0$

18. $16x^2 - 8x + 1 = 0$

19. $6x^2 - 5x - 21 = 0$

20. $12x^2 - 4x - 5 = 0$

Solve each of the following radical equations. Don't forget, you *must check* potential solutions.

21. $3\sqrt{x} = x + 2$

22. $3\sqrt{2x} = x + 4$

23. $\sqrt{2x} = x - 4$ **24.** $\sqrt{x} = x - 2$

25. $\sqrt{3x} + 6 = x$

26. $\sqrt{5x} + 10 = x$

Solve each of the following equations for x by factoring and applying the property, $ab = 0$ if and only if $a = 0$ or $b = 0$.

27. $x^2 - 5kx = 0$

28. $x^2 + 7kx = 0$

29. $x^2 = 16k^2x$

30. $x^2 = 25k^2x$

31. $x^2 - 12kx + 35k^2 = 0$

32. $x^2 - 3kx - 18k^2 = 0$

33. $2x^2 + 5kx - 3k^2 = 0$

34. $3x^2 - 20kx - 7k^2 = 0$

Use Property 6.1 to help solve each of the following quadratic equations.

35. $x^2 = 1$

36. $x^2 = 81$

37. $x^2 = -36$

38. $x^2 = -49$

39. $x^2 = 14$

40. $x^2 = 22$

41. $n^2 - 28 = 0$

42. $n^2 - 54 = 0$

43. $3t^2 = 54$

44. $4t^2 = 108$

45. $2t^2 = 7$

46. $3t^2 = 8$

47. $15y^2 = 20$

48. $14y^2 = 80$

49. $10x^2 + 48 = 0$

50. $12x^2 + 50 = 0$

51. $24x^2 = 36$

52. $12x^2 = 49$

53. $(x - 2)^2 = 9$

54. $(x + 1)^2 = 16$

55. $(x + 3)^2 = 25$

56. $(x - 2)^2 = 49$

57. $(x + 6)^2 = -4$

58. $(3x + 1)^2 = 9$

59. $(2x - 3)^2 = 1$

60. $(2x + 5)^2 = -4$

61. $(n - 4)^2 = 5$

62. $(n - 7)^2 = 6$

63. $(t + 5)^2 = 12$

64. $(t - 1)^2 = 18$

65. $(3y - 2)^2 = -27$

66. $(4y + 5)^2 = 80$

67. $3(x + 7)^2 + 4 = 79$

68. $2(x + 6)^2 - 9 = 63$

69. $2(5x - 2)^2 + 5 = 25$

70. $3(4x - 1)^2 + 1 = -17$

For Problems 71–76, a and b represent the lengths of the legs of a right triangle, and c represents the length of the hypotenuse. Express answers in simplest radical form.

71. Find c if $a = 4$ centimeters and $b = 6$ centimeters.

72. Find c if $a = 3$ meters and $b = 7$ meters.

73. Find a if $c = 12$ inches and $b = 8$ inches.

74. Find a if $c = 8$ feet and $b = 6$ feet.

75. Find b if $c = 17$ yards and $a = 15$ yards.

76. Find b if $c = 14$ meters and $a = 12$ meters.

For Problems 77–80, use the isosceles right triangle in Figure 6.4. Express your answers in simplest radical form.

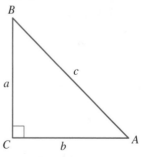

FIGURE 6.4

77. If $b = 6$ inches, find c.

78. If $a = 7$ centimeters, find c.

79. If $c = 8$ meters, find a and b.

80. If $c = 9$ feet, find a and b.

For Problems 81–86, use the triangle in Figure 6.5. Express your answers in simplest radical form.

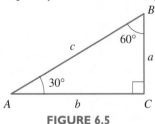

FIGURE 6.5

81. If $a = 3$ inches, find b and c.

82. If $a = 6$ feet, find b and c.

83. If $c = 14$ centimeters, find a and b.

84. If $c = 9$ centimeters, find a and b.

85. If $b = 10$ feet, find a and c.

86. If $b = 8$ meters, find a and c.

87. A 24-foot ladder resting against a house reaches a windowsill 16 feet above the ground. How far is the foot of the ladder from the foundation of the house? Express your answer to the nearest tenth of a foot.

88. A 62-foot guy-wire makes an angle of 60° with the ground and is attached to a telephone pole (see Figure 6.6). Find the distance from the base of the pole to the point on the pole where the wire is attached. Express your answer to the nearest tenth of a foot.

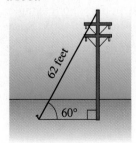

FIGURE 6.6

89. A rectangular plot measures 16 meters by 34 meters. Find the distance, to the nearest meter, from one corner of the plot to the corner diagonally opposite.

90. Consecutive bases of a square-shaped baseball diamond are 90 feet apart (see Figure 6.7). Find the distance from first base diagonally across the diamond to third base, to the nearest tenth of a foot.

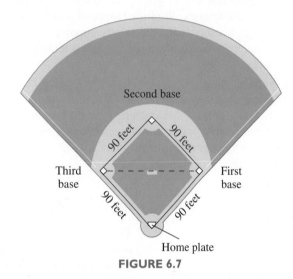

FIGURE 6.7

91. A diagonal of a square parking lot is 75 meters. Find, to the nearest meter, the length of a side of the lot.

THOUGHTS INTO WORDS

92. Explain why the equation $(x + 2)^2 + 5 = 1$ has no real number solutions.

93. Suppose that your friend solved the equation $(x + 3)^2 = 25$ as follows:

$$(x + 3)^2 = 25$$
$$x^2 + 6x + 9 = 25$$
$$x^2 + 6x - 16 = 0$$
$$(x + 8)(x - 2) = 0$$
$$x + 8 = 0 \quad \text{or} \quad x - 2 = 0$$
$$x = -8 \quad \text{or} \quad x = 2$$

Is this a correct approach to the problem? Would you offer any suggestion about an easier approach to the problem?

Further Investigations

94. Suppose that we are given a cube with edges of length 12 centimeters. Find the length of a diagonal from a lower corner to the diagonally opposite upper corner. Express your answer to the nearest tenth of a centimeter.

95. Suppose that we are given a rectangular box with a length of 8 centimeters, a width of 6 centimeters, and a height of 4 centimeters. Find the length of a diagonal from a lower corner to the upper corner diagonally opposite. Express your answer to the nearest tenth of a centimeter.

96. The converse of the Pythagorean theorem is also true. It states that *if the measures a, b, and c of the sides of a triangle are such that* $a^2 + b^2 = c^2$, *then the triangle is a right triangle with a and b the measures of the legs and c the measure of the hypotenuse.* Use the converse of the Pythagorean theorem to determine which of the triangles with sides of the following measures are right triangles.

(a) 9, 40, 41   (b) 20, 48, 52
(c) 19, 21, 26   (d) 32, 37, 49
(e) 65, 156, 169   (f) 21, 72, 75

97. Find the length of the hypotenuse (h) of an isosceles right triangle if each leg is s units long. Then use this relationship and redo Problems 77–80.

98. Suppose that the side opposite the 30° angle in a 30°–60° right triangle is s units long. Express the length of the hypotenuse and the length of the other leg in terms of s. Then use these relationships and redo Problems 81–86.

6.3 Completing the Square

Thus far we have solved quadratic equations by factoring and applying the property $ab = 0$ *if and only if* $a = 0$ *or* $b = 0$, or by applying the property, $x^2 = a$ *if and only if* $x = \pm\sqrt{a}$. In this section we examine another method called **completing the square**, which will give us the power to solve *any* quadratic equation.

A factoring technique we studied in Chapter 3 relied on recognizing **perfect square trinomials**. In each of the following, the perfect square trinomial on the right side is the result of squaring the binomial on the left side.

$$(x + 4)^2 = x^2 + 8x + 16, \qquad (x - 6)^2 = x^2 - 12x + 36,$$
$$(x + 7)^2 = x^2 + 14x + 49, \qquad (x - 9)^2 = x^2 - 18x + 81$$

Notice that in each of the square trinomials *the constant term is equal to the square of one-half of the coefficient of the x-term.* This relationship allows us to form a perfect square trinomial by adding a proper constant term. For example, suppose that we want to form a perfect square trinomial from $x^2 + 10x$. Since $\frac{1}{2}(10) = 5$ and $5^2 = 25$, the perfect square trinomial $x^2 + 10x + 25$ can be formed. Let's use the previous ideas to help solve some quadratic equations.

Example 1

Solve $x^2 + 10x - 2 = 0$.

Solution

$$x^2 + 10x - 2 = 0$$
$$x^2 + 10x = 2$$

$$x^2 + 10x + 25 = 2 + 25$$

We add 25 to the left side to form a perfect square trinomial; 25 must also be added to the right side.

$$(x + 5)^2 = 27$$

(Now we can proceed as in the last section.)

$$x + 5 = \pm\sqrt{27}$$

$$x + 5 = \pm 3\sqrt{3}$$

$$x = -5 \pm 3\sqrt{3}$$

The solution set is $\{-5 \pm 3\sqrt{3}\}$. ▲

Notice from Example 1 that the method of completing the square to solve a quadratic equation is merely what the name implies. A perfect square trinomial is formed, then the equation can be changed to the necessary form for applying the property, $x^2 = a$ *if and only if* $x = \pm\sqrt{a}$. Let's consider another example.

Example 2

Solve $x^2 + 4x + 7 = 0$.

Solution

$$x^2 + 4x + 7 = 0$$

$$x^2 + 4x = -7$$

$$x^2 + 4x + 4 = -7 + 4$$

$$(x + 2)^2 = -3$$

$$x + 2 = \pm\sqrt{-3}$$

$$x + 2 = \pm i\sqrt{3}$$

$$x = -2 \pm i\sqrt{3}$$

The solution set is $\{-2 \pm i\sqrt{3}\}$. ▲

REMARK   Even though we do not show a check for each problem, it is probably a good idea for you to check one now and then. Example 2 is a good one to check since it involves some manipulation with complex numbers. △

Example 3

Solve $x^2 - 3x + 1 = 0$.

Solution

$$x^2 - 3x + 1 = 0$$

$$x^2 - 3x = -1$$

$$x^2 - 3x + \frac{9}{4} = -1 + \frac{9}{4} \qquad \frac{1}{2}(3) = \frac{3}{2} \text{ and } \left(\frac{3}{2}\right)^2 = \frac{9}{4}$$

$$\left(x - \frac{3}{2}\right)^2 = \frac{5}{4}$$

$$x - \frac{3}{2} = \pm\sqrt{\frac{5}{4}}$$

$$x - \frac{3}{2} = \pm\frac{\sqrt{5}}{2}$$

$$x = \frac{3}{2} \pm \frac{\sqrt{5}}{2}$$

$$x = \frac{3 \pm \sqrt{5}}{2}$$

The solution set is $\left\{\dfrac{3 \pm \sqrt{5}}{2}\right\}$.

In Example 3 notice that since the coefficient of the x-term is odd, we are forced into the realm of fractions. The use of common fractions rather than decimals allows us to apply our previous work with radicals.

The relationship for a perfect square trinomial that states that *the constant term is equal to the square of one-half of the coefficient of the x-term* holds only if the coefficient of x^2 is 1. Thus, an adjustment needs to be made when solving quadratic equations that have a coefficient of x^2 other than 1. The next example shows how to make this adjustment.

Example 4

Solution

Solve $2x^2 + 12x - 5 = 0$.

$$2x^2 + 12x - 5 = 0$$

$$2x^2 + 12x = 5$$

$$x^2 + 6x = \frac{5}{2} \qquad \text{Multiply both sides by } \frac{1}{2}.$$

$$x^2 + 6x + 9 = \frac{5}{2} + 9$$

$$x^2 + 6x + 9 = \frac{23}{2}$$

$$(x + 3)^2 = \frac{23}{2}$$

$$x + 3 = \pm\sqrt{\frac{23}{2}}$$

$$x + 3 = \pm\frac{\sqrt{46}}{2} \qquad \sqrt{\frac{23}{2}} = \frac{\sqrt{23}}{\sqrt{2}} \cdot \frac{\sqrt{2}}{\sqrt{2}} = \frac{\sqrt{46}}{2}$$

$$x = -3 \pm \frac{\sqrt{46}}{2}$$

$$x = \frac{-6 \pm \sqrt{46}}{2}$$

The solution set is $\left\{\dfrac{-6 \pm \sqrt{46}}{2}\right\}$.

As we mentioned earlier, we can use the method of completing the square to solve *any* quadratic equation. To illustrate, let's use it to solve an equation that could also be solved by factoring.

Example 5

Solution

Solve $x^2 - 2x - 8 = 0$ by completing the square.

$$x^2 - 2x - 8 = 0$$
$$x^2 - 2x = 8$$
$$x^2 - 2x + 1 = 8 + 1$$
$$(x - 1)^2 = 9$$
$$x - 1 = \pm 3$$
$$x - 1 = 3 \quad\text{or}\quad x - 1 = -3$$
$$x = 4 \quad\text{or}\quad x = -2$$

The solution set is $\{-2, 4\}$.   ▲

We make no claim that using the method of completing the square with an equation such as the one in Example 5 is easier than the factoring technique. However, you should recognize that the method of completing the square will work with any quadratic equation.

Problem Set 6.3

Solve each of the following quadratic equations by using (a) the factoring method and (b) the method of completing the square.

1. $x^2 - 4x - 60 = 0$

2. $x^2 + 6x - 16 = 0$

3. $x^2 - 14x = -40$

4. $x^2 - 18x = -72$

5. $x^2 - 5x - 50 = 0$

6. $x^2 + 3x - 18 = 0$

7. $x(x + 7) = 8$

8. $x(x - 1) = 30$

9. $2n^2 - n - 15 = 0$

10. $3n^2 + n - 14 = 0$

11. $3n^2 + 7n - 6 = 0$

12. $2n^2 + 7n - 4 = 0$

13. $n(n + 6) = 160$

14. $n(n - 6) = 216$

Use the method of completing the square to solve each of the following quadratic equations.

15. $x^2 + 4x - 2 = 0$

16. $x^2 + 2x - 1 = 0$

17. $x^2 + 6x - 3 = 0$

18. $x^2 + 8x - 4 = 0$

19. $y^2 - 10y = 1$

20. $y^2 - 6y = -10$

21. $n^2 - 8n + 17 = 0$

22. $n^2 - 4n + 2 = 0$

23. $n(n + 12) = -9$

24. $n(n + 14) = -4$

25. $n^2 + 2n + 6 = 0$

26. $n^2 + n - 1 = 0$

27. $x^2 + 3x - 2 = 0$

28. $x^2 + 5x - 3 = 0$

29. $x^2 + 5x + 1 = 0$

30. $x^2 + 7x + 2 = 0$

31. $y^2 - 7y + 3 = 0$

32. $y^2 - 9y + 30 = 0$

33. $2x^2 + 4x - 3 = 0$

34. $2t^2 - 4t + 1 = 0$

35. $3n^2 - 6n + 5 = 0$

36. $3x^2 + 12x - 2 = 0$

37. $3x^2 + 5x - 1 = 0$

38. $2x^2 + 7x - 3 = 0$

Solve each of the following quadratic equations and use whatever method seems most appropriate.

39. $x^2 + 8x - 48 = 0$

40. $x^2 + 5x - 14 = 0$

41. $2n^2 - 8n = -3$

42. $3x^2 + 6x = 1$

43. $(3x - 1)(2x + 9) = 0$

44. $(5x + 2)(x - 4) = 0$

45. $(x + 2)(x - 7) = 10$

46. $(x - 3)(x + 5) = -7$

47. $(x - 3)^2 = 12$

48. $x^2 = 16x$

49. $3n^2 - 6n + 4 = 0$

50. $2n^2 - 2n - 1 = 0$

51. $n(n + 8) = 240$

52. $t(t - 26) = -160$

53. $3x^2 + 29x = -66$

54. $6x^2 - 13x = 28$

55. $6n^2 + 23n + 21 = 0$

56. $6n^2 + n - 2 = 0$

57. $x^2 + 12x = 4$

58. $x^2 + 6x = -11$

59. $12n^2 - 7n + 1 = 0$

60. $5(x + 2)^2 + 1 = 16$

61. Use the method of completing the square to solve $ax^2 + bx + c = 0$ for x, where a, b, and c are real numbers and $a \neq 0$.

62. Explain the process of *completing the square* to solve a quadratic equation.

63. Give a step-by-step description of how to solve $3x^2 + 9x - 4 = 0$ by completing the square.

Further Investigations

Solve Problems 64−67 for the indicated variable. Assume that all letters represent positive numbers.

64. $\dfrac{x^2}{a^2} - \dfrac{y^2}{b^2} = 1$ for y

65. $\dfrac{x^2}{a^2} + \dfrac{y^2}{b^2} = 1$ for x

66. $s = \dfrac{1}{2}gt^2$ for t

67. $A = \pi r^2$ for r

Solve each of the following equations for x.

68. $x^2 + 8ax + 15a^2 = 0$

69. $x^2 - 5ax + 6a^2 = 0$

70. $10x^2 - 31ax - 14a^2 = 0$

71. $6x^2 + ax - 2a^2 = 0$

72. $4x^2 + 4bx + b^2 = 0$

73. $9x^2 - 12bx + 4b^2 = 0$

6.4 Quadratic Formula

As we saw in the last section, the method of completing the square can be used to solve *any* quadratic equation. Furthermore, the equation $ax^2 + bx + c = 0$, where a, b, and c are real numbers with $a \neq 0$, can be used to represent *any* quadratic equation. These two ideas merge to produce the **quadratic formula**, a formula that can be used to solve *any* quadratic equation. The merger is accomplished by using the method of completing the square to solve the equation $ax^2 + bx + c = 0$ as follows.

$$ax^2 + bx + c = 0$$

$$ax^2 + bx = -c$$

$$x^2 + \frac{b}{a}x = -\frac{c}{a} \qquad \text{Multiply both sides by } \frac{1}{a}.$$

$$x^2 + \frac{b}{a}x + \frac{b^2}{4a^2} = -\frac{c}{a} + \frac{b^2}{4a^2} \qquad \textit{Complete the square by adding } \frac{b^2}{4a^2} \text{ to both sides.}$$

$$\left(x + \frac{b}{2a}\right)^2 = \frac{b^2 - 4ac}{4a^2} \qquad \text{The right side is combined into a single fraction.}$$

$$x + \frac{b}{2a} = \pm\sqrt{\frac{b^2 - 4ac}{4a^2}}$$

$$x + \frac{b}{2a} = \pm\frac{\sqrt{b^2 - 4ac}}{\sqrt{4a^2}}$$

$$x + \frac{b}{2a} = \pm\frac{\sqrt{b^2 - 4ac}}{2a} \qquad \sqrt{4a^2} = |2a| \text{ but } 2a \text{ can be used because of the use of } \pm.$$

$$x + \frac{b}{2a} = \frac{\sqrt{b^2 - 4ac}}{2a} \qquad \text{or} \qquad x + \frac{b}{2a} = -\frac{\sqrt{b^2 - 4ac}}{2a}$$

$$x = -\frac{b}{2a} + \frac{\sqrt{b^2 - 4ac}}{2a} \qquad \text{or} \qquad x = -\frac{b}{2a} - \frac{\sqrt{b^2 - 4ac}}{2a}$$

$$x = \frac{-b + \sqrt{b^2 - 4ac}}{2a} \qquad \text{or} \qquad x = \frac{-b - \sqrt{b^2 - 4ac}}{2a}$$

The quadratic formula is usually stated as follows.

Quadratic Formula

$$x = \frac{-b \pm \sqrt{b^2 - 4ac}}{2a}, \qquad a \neq 0$$

We can use this formula to solve any quadratic equation by expressing the equation in the standard form, $ax^2 + bx + c = 0$, and substituting the values for a, b, and c into the formula. Let's consider some examples.

Example 1

Solve $x^2 + 5x + 2 = 0$.

Solution

The given equation is in standard form, so $a = 1$, $b = 5$, and $c = 2$. Substituting these values into the formula and simplifying, we obtain

$$x = \frac{-b \pm \sqrt{b^2 - 4ac}}{2a}$$

$$x = \frac{-5 \pm \sqrt{5^2 - 4(1)(2)}}{2(1)}$$

$$x = \frac{-5 \pm \sqrt{25 - 8}}{2}$$

$$x = \frac{-5 \pm \sqrt{17}}{2}.$$

The solution set is $\left\{ \dfrac{-5 \pm \sqrt{17}}{2} \right\}$.

▲

Example 2

Solve $x^2 - 2x - 4 = 0$.

Solution

We need to think of $x^2 - 2x - 4 = 0$ as $x^2 + (-2x) + (-4) = 0$ to determine the values $a = 1$, $b = -2$, $c = -4$. Substitute these values into the quadratic formula and simplify, to obtain

$$x = \frac{-b \pm \sqrt{b^2 - 4ac}}{2a}$$

$$x = \frac{-(-2) \pm \sqrt{(-2)^2 - 4(1)(-4)}}{2(1)}$$

$$x = \frac{2 \pm \sqrt{4 + 16}}{2}$$

$$x = \frac{2 \pm \sqrt{20}}{2}$$

$$x = \frac{2 \pm 2\sqrt{5}}{2}$$

$$= \frac{\cancel{2}(1 \pm \sqrt{5})}{\cancel{2}}.$$

The solution set is $\left\{ 1 \pm \sqrt{5} \right\}$.

▲

Example 3

Solve $x^2 - 2x + 19 = 0$.

Solution

$$x^2 - 2x + 19 = 0$$

$$x = \frac{-(-2) \pm \sqrt{(-2)^2 - 4(1)(19)}}{2(1)}$$

$$x = \frac{2 \pm \sqrt{4 - 76}}{2}$$

$$x = \frac{2 \pm \sqrt{-72}}{2}$$

$$x = \frac{2 \pm 6i\sqrt{2}}{2} \qquad \sqrt{-72} = i\sqrt{72} = i\sqrt{36}\sqrt{2} = 6i\sqrt{2}$$

$$x = 1 \pm 3i\sqrt{2}.$$

The solution set is $\left\{1 \pm 3i\sqrt{2}\right\}$.

▲

Example 4

Solve $2x^2 + 4x - 3 = 0$.

Solution

Substitute $a = 2$, $b = 4$, and $c = -3$ into the quadratic formula and simplify to obtain

$$x = \frac{-b \pm \sqrt{b^2 - 4ac}}{2a}$$

$$x = \frac{-4 \pm \sqrt{4^2 - 4(2)(-3)}}{2(2)}$$

$$x = \frac{-4 \pm \sqrt{16 + 24}}{4}$$

$$x = \frac{-4 \pm \sqrt{40}}{4}$$

$$x = \frac{-4 \pm 2\sqrt{10}}{4}$$

$$x = \frac{-2 \pm \sqrt{10}}{2}.$$

The solution set is $\left\{\dfrac{-2 \pm \sqrt{10}}{2}\right\}$.

▲

Example 5

Solve $n(3n - 10) = 25$.

Solution

First, we need to change the equation to the standard form, $an^2 + bn + c = 0$.

$$n(3n - 10) = 25$$

$$3n^2 - 10n = 25$$

$$3n^2 - 10n - 25 = 0$$

Now substituting $a = 3$, $b = -10$, and $c = -25$ into the quadratic formula we obtain

$$n = \frac{-b \pm \sqrt{b^2 - 4ac}}{2a}$$

$$n = \frac{-(-10) \pm \sqrt{(-10)^2 - 4(3)(-25)}}{2(3)}$$

$$n = \frac{10 \pm \sqrt{100 + 300}}{2(3)}$$

$$n = \frac{10 \pm \sqrt{400}}{6}$$

$$n = \frac{10 \pm 20}{6}$$

$$n = \frac{10 + 20}{6} \quad \text{or} \quad n = \frac{10 - 20}{6}$$

$$n = 5 \quad \text{or} \quad n = -\frac{5}{3}.$$

The solution set is $\left\{-\frac{5}{3}, 5\right\}$.

In Example 5, notice that we used the variable n. The quadratic formula is usually stated in terms of x, but it certainly can be applied to quadratic equations in other variables. Also note in Example 5 that the polynomial $3n^2 - 10n - 25$ can be factored as $(3n + 5)(n - 5)$. Therefore, we could also solve the equation $3n^2 - 10n - 25 = 0$ by using the factoring approach. We will give you some guidance in the next section as to which approach to use for a particular equation.

Nature of Roots

The quadratic formula makes it easy to determine the nature of the roots of a quadratic equation without completely solving the equation. The number

$$b^2 - 4ac,$$

which appears under the radical sign in the quadratic formula, is called the **discriminant** of the quadratic equation. It is the indicator as to the kind of roots of the equation. For example, suppose that we start to solve the equation $x^2 - 4x + 7 = 0$ as follows.

$$x = \frac{-b \pm \sqrt{b^2 - 4ac}}{2a}$$

$$x = \frac{-(-4) \pm \sqrt{(-4)^2 - 4(1)(7)}}{2(1)}$$

$$x = \frac{4 \pm \sqrt{16 - 28}}{2}$$

$$x = \frac{4 \pm \sqrt{-12}}{2}$$

At this stage you should be able to look ahead and realize that you will obtain two complex solutions for the equation. (By the way, observe that these solutions are complex conjugates.) In other words, the discriminant, -12, indicates the type of roots you will obtain.

We make the following general statements relative to the roots of a quadratic equation of the form $ax^2 + bx + c = 0$.

1. If $b^2 - 4ac < 0$, then the equation has two nonreal complex solutions.
2. If $b^2 - 4ac = 0$, then the equation has one real solution.
3. If $b^2 - 4ac > 0$, then the equation has two real solutions.

The following examples illustrate each of these situations. (You may want to solve the equations completely to verify the conclusions.)

Equation	Discriminant	Nature of roots
$x^2 - 3x + 7 = 0$	$b^2 - 4ac = (-3)^2 - 4(1)(7)$ $= 9 - 28$ $= -19$	Two nonreal complex solutions
$9x^2 - 12x + 4 = 0$	$b^2 - 4ac = (-12)^2 - 4(9)(4)$ $= 144 - 144$ $= 0$	One real solution
$2x^2 + 5x - 3 = 0$	$b^2 - 4ac = (5)^2 - 4(2)(-3)$ $= 25 + 24$ $= 49$	Two real solutions

There is another very useful relationship that involves the roots of a quadratic equation and the numbers a, b, and c of the general form $ax^2 + bx + c = 0$. Suppose that we let x_1 and x_2 be the two roots generated by the quadratic formula. Thus, we have

$$x_1 = \frac{-b + \sqrt{b^2 - 4ac}}{2a} \quad \text{and} \quad x_2 = \frac{-b - \sqrt{b^2 - 4ac}}{2a}.$$

REMARK A point of clarification should be made at this time. Previously, we made the statement that if $b^2 - 4ac = 0$, then the equation has one real solution. Technically, such an equation has two solutions but they are equal. For example, each factor of $(x - 2)(x - 2) = 0$ produces a solution but both solutions are the number 2. We sometimes refer to this as one real solution with a *multiplicity of two*. Using the idea of multiplicity of roots, we can say that every quadratic equation has two roots. △

Now let's consider the sum and product of the two roots.

Sum $x_1 + x_2 = \dfrac{-b + \sqrt{b^2 - 4ac}}{2a} + \dfrac{-b - \sqrt{b^2 - 4ac}}{2a} = \dfrac{-2b}{2a} = \boxed{-\dfrac{b}{a}}.$

Product $(x_1)(x_2) = \left(\dfrac{-b + \sqrt{b^2 - 4ac}}{2a}\right)\left(\dfrac{-b - \sqrt{b^2 - 4ac}}{2a}\right) = \dfrac{b^2 - (b^2 - 4ac)}{4a^2}$

$$= \dfrac{b^2 - b^2 + 4ac}{4a^2}$$

$$= \dfrac{4ac}{4a^2} = \boxed{\dfrac{c}{a}}.$$

These relationships provide another way of checking potential solutions when solving quadratic equations. For example, back in Example 3 we solved the equation $x^2 - 2x + 19 = 0$ and obtained solutions of $1 + 3i\sqrt{2}$ and $1 - 3i\sqrt{2}$. Let's check these solutions by using the sum and product relationships.

Check for Example 3

Sum of roots $(1 + 3i\sqrt{2}) + (1 - 3i\sqrt{2}) = 2$ and $-\dfrac{b}{a} = -\dfrac{-2}{1} = 2$

Product of roots $(1 + 3i\sqrt{2})(1 - 3i\sqrt{2}) = 1 - 18i^2 = 1 + 18 = 19$

and $\dfrac{c}{a} = \dfrac{19}{1} = 19$

Likewise, a check for Example 4 is as follows.

Check for Example 4

Sum of roots $\left(\dfrac{-2 + \sqrt{10}}{2}\right) + \left(\dfrac{-2 - \sqrt{10}}{2}\right) = -\dfrac{4}{2} = -2$ and $-\dfrac{b}{a} = -\dfrac{4}{2} = -2$

Product of roots $\left(\dfrac{-2 + \sqrt{10}}{2}\right)\left(\dfrac{-2 - \sqrt{10}}{2}\right) = -\dfrac{6}{4} = -\dfrac{3}{2}$ and $\dfrac{c}{a} = \dfrac{-3}{2} = -\dfrac{3}{2}$

Notice that for both Examples 3 and 4, it was much easier to check by using the sum and product relationships than it would have been by substituting back into the original equation. Don't forget that the values for a, b, and c come from a quadratic equation of the form $ax^2 + bx + c = 0$. Therefore, in Example 5 we must be certain that no errors were made when changing the given equation $n(3n - 10) = 25$ to the form $3n^2 - 10n - 25 = 0$ if we are going to check the potential solutions by using the sum and product relationships.

◢ Problem Set 6.4

For each quadratic equation in Problems 1–10, first use the discriminant to determine whether the equation has two nonreal complex solutions, one real solution with a multiplicity of two, or two real solutions, and then solve the equation.

1. $x^2 + 4x - 21 = 0$

2. $x^2 - 3x - 54 = 0$

3. $9x^2 - 6x + 1 = 0$

4. $4x^2 + 20x + 25 = 0$

5. $x^2 - 7x + 13 = 0$

6. $2x^2 - x + 5 = 0$

7. $15x^2 + 17x - 4 = 0$

8. $8x^2 + 18x - 5 = 0$

9. $3x^2 + 4x = 2$

10. $2x^2 - 6x = -1$

For Problems 11–50, use the quadratic formula to solve each of the quadratic equations. Check your solutions by using the *sum and product relationships*.

11. $x^2 + 2x - 1 = 0$

12. $x^2 + 4x - 1 = 0$

13. $n^2 + 5n - 3 = 0$

14. $n^2 + 3n - 2 = 0$

15. $a^2 - 8a = 4$

16. $a^2 - 6a = 2$

17. $n^2 + 5n + 8 = 0$

18. $2n^2 - 3n + 5 = 0$

19. $x^2 - 18x + 80 = 0$

20. $x^2 + 19x + 70 = 0$

21. $-y^2 = -9y + 5$

22. $-y^2 + 7y = 4$

23. $2x^2 + x - 4 = 0$

24. $2x^2 + 5x - 2 = 0$

25. $4x^2 + 2x + 1 = 0$

26. $3x^2 - 2x + 5 = 0$

27. $3a^2 - 8a + 2 = 0$

28. $2a^2 - 6a + 1 = 0$

29. $-2n^2 + 3n + 5 = 0$

30. $-3n^2 - 11n + 4 = 0$

31. $3x^2 + 19x + 20 = 0$

32. $2x^2 - 17x + 30 = 0$

33. $36n^2 - 60n + 25 = 0$

34. $9n^2 + 42n + 49 = 0$

35. $4x^2 - 2x = 3$

36. $6x^2 - 4x = 3$

37. $5x^2 - 13x = 0$

38. $7x^2 + 12x = 0$

39. $3x^2 = 5$

40. $4x^2 = 3$

41. $6t^2 + t - 3 = 0$

42. $2t^2 + 6t - 3 = 0$

43. $n^2 + 32n + 252 = 0$

44. $n^2 - 4n - 192 = 0$

45. $12x^2 - 73x + 110 = 0$

46. $6x^2 + 11x - 255 = 0$

47. $-2x^2 + 4x - 3 = 0$

48. $-2x^2 + 6x - 5 = 0$

49. $-6x^2 + 2x + 1 = 0$

50. $-2x^2 + 4x + 1 = 0$

51. Your friend states that the equation $-2x^2 + 4x - 1 = 0$ must be changed to $2x^2 - 4x + 1 = 0$ (by multiplying both sides by -1) before the quadratic formula can be applied. Is she right about this and if not, how would you convince her?

52. Another of your friends claims that the quadratic formula can be used to solve the equation $x^2 - 9 = 0$. How would you react to this claim?

53. Why must the equation $3x^2 - 2x = 4$ be changed to $3x^2 - 2x - 4 = 0$ before applying the quadratic formula?

Further Investigations

The solution set for $x^2 - 4x - 37 = 0$ is $\{2 \pm \sqrt{41}\}$. With a calculator, we found a rational approximation, to the nearest one-thousandth, for each of these solutions.

$$2 - \sqrt{41} = -4.403 \quad \text{and} \quad 2 + \sqrt{41} = 8.403.$$

Thus, the solution set is $\{-4.403, 8.403\}$, with the answers rounded to the nearest one-thousandth.

Solve each of the following equations and express solutions to the nearest one-thousandth.

54. $x^2 - 6x - 10 = 0$
55. $x^2 - 16x - 24 = 0$
56. $x^2 + 6x - 44 = 0$
57. $x^2 + 10x - 46 = 0$
58. $x^2 + 8x + 2 = 0$
59. $x^2 + 9x + 3 = 0$
60. $4x^2 - 6x + 1 = 0$
61. $5x^2 - 9x + 1 = 0$
62. $2x^2 - 11x - 5 = 0$
63. $3x^2 - 12x - 10 = 0$

For Problems 64–66, use the discriminant to help solve each problem.

64. Determine k so that the solutions of $x^2 - 2x + k = 0$ are complex but nonreal.
65. Determine k so that $4x^2 - kx + 1 = 0$ has two equal real solutions.
66. Determine k so that $3x^2 - kx - 2 = 0$ has real solutions.

6.5 More Quadratic Equations and Applications

Which method should be used to solve a particular quadratic equation? There is no definite answer to that question; it depends upon the *type* of equation and your personal preference. In the following examples we will state reasons for choosing a specific technique. However, keep in mind that usually this is a decision *you* must make as the need arises. So become familiar with the strengths and weaknesses of each method.

Example 1

Solve $2x^2 - 3x - 1 = 0$.

Solution

Because of the leading coefficient of 2 and the constant term of -1, there are very few factoring possibilities to consider. Therefore, with such problems, first try the factoring approach. Unfortunately, this particular polynomial is not factorable using integers. Thus, let's use the quadratic formula to solve the equation.

$$x = \frac{-b \pm \sqrt{b^2 - 4ac}}{2a}$$

$$x = \frac{-(-3) \pm \sqrt{(-3)^2 - 4(2)(-1)}}{2(2)}$$

$$x = \frac{3 \pm \sqrt{9 + 8}}{4}$$

$$x = \frac{3 \pm \sqrt{17}}{4}$$

✔ **Check**   We can use the *sum and product-of-roots* relationships for our checking purposes.

Sum of roots $\dfrac{3 + \sqrt{17}}{4} + \dfrac{3 - \sqrt{17}}{4} = \dfrac{6}{4} = \dfrac{3}{2}$ and $-\dfrac{b}{a} = -\dfrac{-3}{2} = \dfrac{3}{2}.$

Product of roots $\left(\dfrac{3 + \sqrt{17}}{4}\right)\left(\dfrac{3 - \sqrt{17}}{4}\right) = \dfrac{9 - 17}{16} = -\dfrac{8}{16} = -\dfrac{1}{2}$ and

$$\frac{c}{a} = \frac{-1}{2} = -\frac{1}{2}.$$

The solution set is $\left\{\dfrac{3 \pm \sqrt{17}}{4}\right\}.$ ▲

Example 2

Solve $\dfrac{3}{n} + \dfrac{10}{n + 6} = 1.$

Solution

$$\frac{3}{n} + \frac{10}{n + 6} = 1, \qquad n \neq 0 \text{ and } n \neq -6.$$

$$n(n + 6)\left(\frac{3}{n} + \frac{10}{n + 6}\right) = 1(n)(n + 6) \qquad \begin{array}{l}\text{Multiply both sides by } n(n + 6) \\ \text{which is the LCD.}\end{array}$$

$$3(n + 6) + 10n = n(n + 6)$$

$$3n + 18 + 10n = n^2 + 6n$$

$$13n + 18 = n^2 + 6n$$

$$0 = n^2 - 7n - 18$$

This equation is an easy one to consider the possibilities for factoring, and it factors as follows.

$$0 = (n - 9)(n + 2)$$

$$n - 9 = 0 \qquad \text{or} \qquad n + 2 = 0$$

$$n = 9 \qquad \text{or} \qquad n = -2$$

✔ *Check* Substituting 9 and -2 back into the original equation, we obtain

$$\frac{3}{n} + \frac{10}{n+6} = 1 \qquad\qquad \frac{3}{n} + \frac{10}{n+6} = 1$$

$$\frac{3}{9} + \frac{10}{9+6} \stackrel{?}{=} 1 \qquad\qquad \frac{3}{-2} + \frac{10}{-2+6} \stackrel{?}{=} 1$$

$$\frac{1}{3} + \frac{10}{15} \stackrel{?}{=} 1 \quad \text{or} \quad -\frac{3}{2} + \frac{10}{4} \stackrel{?}{=} 1$$

$$\frac{1}{3} + \frac{2}{3} \stackrel{?}{=} 1 \qquad\qquad -\frac{3}{2} + \frac{5}{2} \stackrel{?}{=} 1$$

$$1 = 1 \qquad\qquad \frac{2}{2} = 1$$

The solution set is $\{-2, 9\}$. ▲

We should make two comments about Example 2. First, notice the indication of the initial restrictions $n \neq 0$ and $n \neq -6$. Remember that we need to do this when solving fractional equations. Secondly, the *sum and product-of-roots* relationships were not used for checking purposes in this problem. Those relationships would only check the validity of our work from the step $0 = n^2 - 7n - 18$ to the finish. In other words, an error made in changing the original equation to quadratic form would not be detected by checking the sum and product of potential roots. Thus, with such a problem the only *absolute check* is to substitute the potential solutions back into the *original equation*.

Example 3

Solve $x^2 + 22x + 112 = 0$.

Solution

The size of the constant term makes the factoring approach a little cumbersome for this problem. Furthermore, since the leading coefficient is 1 and the coefficient of the x-term is even, the method of completing the square will work effectively as follows.

$$x^2 + 22x + 112 = 0$$
$$x^2 + 22x = -112$$
$$x^2 + 22x + 121 = -112 + 121$$
$$(x + 11)^2 = 9$$
$$x + 11 = \pm\sqrt{9}$$
$$x + 11 = \pm 3$$
$$x + 11 = 3 \quad \text{or} \quad x + 11 = -3$$
$$x = -8 \quad \text{or} \quad x = -14$$

✔ *Check* **Sum of roots** $\quad -8 + (-14) = -22 \quad$ and $\quad -\dfrac{b}{a} = -22.$

Product of roots $(-8)(-14) = 112$ and $\dfrac{c}{a} = 112.$

The solution set is $\{-14, -8\}$. ▲

Example 4   Solve $x^4 - 4x^2 - 96 = 0$.

Solution   An equation such as $x^4 - 4x^2 - 96 = 0$ is not a quadratic equation, but we can solve it using the techniques that we use on quadratic equations. That is to say, we can factor the polynomial and apply the property, $ab = 0$ if *and only if* $a = 0$ or $b = 0$, as follows.

$$x^4 - 4x^2 - 96 = 0$$
$$(x^2 - 12)(x^2 + 8) = 0$$

$x^2 - 12 = 0$ or $x^2 + 8 = 0$

$x^2 = 12$ or $x^2 = -8$

$x = \pm\sqrt{12}$ or $x = \pm\sqrt{-8}$

$x = \pm 2\sqrt{3}$ or $x = \pm 2i\sqrt{2}$

The solution set is $\{\pm 2\sqrt{3}, \pm 2i\sqrt{2}\}$. (We will leave the check for this problem for you to do!) ▲

REMARK   Another approach to Example 4 would be to substitute y for x^2 and y^2 for x^4. The equation $x^4 - 4x^2 - 96 = 0$ becomes the quadratic equation $y^2 - 4y - 96 = 0$. Thus, we say that $x^4 - 4x^2 - 96 = 0$ is of *quadratic form*. Then we could solve the quadratic equation $y^2 - 4y - 96 = 0$ and use the equation $y = x^2$ to determine the solutions for x. △

Applications

Before we conclude this section with some word problems that can be solved using quadratic equations, let's restate the suggestions for solving word problems we made in an earlier chapter.

Suggestions for Solving Word Problems
_____

1. Read the problem carefully and make certain that you understand the meanings of all the words. Be especially alert for any technical terms used in the statement of the problem.

2. Read the problem a second time (perhaps even a third time) to get an overview of the situation being described and to determine the known facts, as well as what is to be found.

3. Sketch any figure, diagram, or chart that might be helpful in analyzing the problem.

4. Choose a meaningful variable to represent an unknown quantity in the problem (perhaps *l*, if the length of a rectangle is an unknown quantity) and represent any other unknowns in terms of that variable.

5. Look for a *guideline* that you can use to set up an equation. A guideline might be a formula such as $A = lw$ or a relationship such as *the fractional part of a job done by Bill plus the fractional part of the job done by Mary equals the total job.*

6. Form an equation that contains the variable that translates the conditions of the guideline from English to algebra.

7. Solve the equation and use the solutions to determine all facts requested in the problem.

8. **Check all answers back into the original statement of the problem.**

Keep these suggestions in mind, as we now consider some word problems.

Problem 1

A page for a magazine contains 70 square inches of type. The height of a page is twice the width. If the margin around the type is to be 2 inches uniformly, what are the dimensions of a page?

Solution

Let x represent the width of a page. Then $2x$ represents the height of a page. Now let's draw and label a model of a page (Figure 6.8).

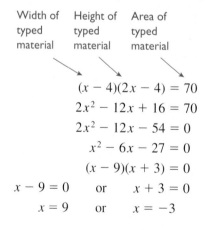

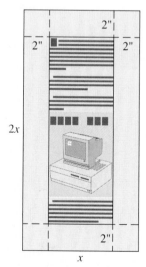

Width of typed material   Height of typed material   Area of typed material

$$(x - 4)(2x - 4) = 70$$
$$2x^2 - 12x + 16 = 70$$
$$2x^2 - 12x - 54 = 0$$
$$x^2 - 6x - 27 = 0$$
$$(x - 9)(x + 3) = 0$$
$$x - 9 = 0 \quad \text{or} \quad x + 3 = 0$$
$$x = 9 \quad \text{or} \quad x = -3$$

FIGURE 6.8

Disregard the negative solution; the page must be 9 inches wide and its height is $2(9) = 18$ inches.

Let's use our knowledge of quadratic equations to analyze some applications of the business world. For example, if P dollars is invested at r rate of interest compounded annually for t years, then the amount of money, A, accumulated at the end of t years is given by the formula

$$A = P(1 + r)^t.$$

This compound interest formula serves as a guideline for the next problem.

Problem 2

Suppose that $100 is invested at a certain rate of interest compounded annually for 2 years. If the accumulated value at the end of 2 years is $121, find the rate of interest.

Solution

Let r represent the rate of interest. Substitute the known values into the compound interest formula to yield

$$A = P(1 + r)^t$$
$$121 = 100(1 + r)^2.$$

Solving this equation, we obtain

$$\frac{121}{100} = (1 + r)^2$$

$$\pm\sqrt{\frac{121}{100}} = (1 + r)$$

$$\pm\frac{11}{10} = 1 + r$$

$$1 + r = \frac{11}{10} \quad \text{or} \quad 1 + r = -\frac{11}{10}$$

$$r = -1 + \frac{11}{10} \quad \text{or} \quad r = -1 - \frac{11}{10}$$

$$r = \frac{1}{10} \quad \text{or} \quad r = -\frac{21}{10}.$$

We must disregard the negative solution, so $r = \dfrac{1}{10}$ is the only solution. Change $\dfrac{1}{10}$ to a percent and the rate of interest is 10%.   ▲

Problem 3

A businesswoman bought a parcel of land on speculation for $120,000. She subdivided the land into lots and when she had sold all but 18 lots at a profit of $6000 per lot, she regained the entire cost of the land. How many lots were sold and at what price per lot?

Solution

Let x represent the number of lots sold. Then $x + 18$ represents the total number of lots. Therefore, $\dfrac{120,000}{x}$ represents the selling price per lot and $\dfrac{120,000}{x + 18}$ represents the cost per lot. The following equation represents the situation.

Selling price
per lot   Equals   Cost per lot   Plus $6000

$$\frac{120{,}000}{x} = \frac{120{,}000}{x + 18} + 6000$$

Solving this equation, we obtain

$$x(x + 18)\left(\frac{120{,}000}{x}\right) = \left(\frac{120{,}000}{x + 18} + 6000\right)(x)(x + 18)$$

$$120{,}000(x + 18) = 120{,}000x + 6000x(x + 18)$$

$$120{,}000x + 2{,}160{,}000 = 120{,}000x + 6000x^2 + 108{,}000x$$

$$0 = 6000x^2 + 108{,}000x - 2{,}160{,}000$$

$$0 = x^2 + 18x - 360.$$

The method of completing the square works very well with this equation.

$$x^2 + 18x = 360$$

$$x^2 + 18x + 81 = 441$$

$$(x + 9)^2 = 441$$

$$x + 9 = \pm\sqrt{441}$$

$$x + 9 = \pm 21$$

$$x + 9 = 21 \quad \text{or} \quad x + 9 = -21$$

$$x = 12 \quad \text{or} \quad x = -30$$

We discard the negative solution; thus, 12 lots were sold at $\dfrac{120{,}000}{x} = \dfrac{120{,}000}{12}$ = $10,000 per lot. ▲

Problem 4

Barry bought a number of shares of stock for $600. A week later the value of the stock increased $3 per share and he sold all but 10 shares and regained his original investment of $600. How many shares did he sell and at what price per share?

Solution

Let s represent the number of shares Barry sold. Then $s + 10$ represents the number of shares purchased. Therefore $\dfrac{600}{s}$ represents the selling price per share and $\dfrac{600}{s + 10}$ represents the cost per share.

Selling price
per share   Cost per share

$$\frac{600}{s} = \frac{600}{s + 10} + 3$$

Solving this equation yields

$$s(s + 10)\left(\frac{600}{s}\right) = \left(\frac{600}{s + 10} + 3\right)(s)(s + 10)$$

$$600(s + 10) = 600s + 3s(s + 10)$$

$$600s + 6000 = 600s + 3s^2 + 30s$$

$$0 = 3s^2 + 30s - 6000$$

$$0 = s^2 + 10s - 2000.$$

Use the quadratic formula to obtain

$$s = \frac{-10 \pm \sqrt{10^2 - 4(1)(-2000)}}{2(1)}$$

$$s = \frac{-10 \pm \sqrt{100 + 8000}}{2}$$

$$s = \frac{-10 \pm \sqrt{8100}}{2}$$

$$s = \frac{-10 \pm 90}{2}$$

$$s = \frac{-10 + 90}{2} \qquad \text{or} \qquad s = \frac{-10 - 90}{2}$$

$$s = 40 \qquad \text{or} \qquad s = -50.$$

We discard the negative solution and we know that 40 shares were sold at $\frac{600}{s} = \frac{600}{40} = \$15$ per share. ▲

This next problem set contains a large variety of word problems. Not only are there some business applications similar to those we discussed in this section, but there are also more problems of the types we discussed back in Chapters 3 and 4. Try to give them your best shot without referring back to examples in earlier chapters.

▼ Problem Set 6.5

Solve each of the following quadratic equations and use the method that seems most appropriate to you.

1. $x^2 - 4x - 6 = 0$

2. $x^2 - 8x - 4 = 0$

3. $3x^2 + 23x - 36 = 0$

4. $n^2 + 22n + 105 = 0$

5. $x^2 - 18x = 9$

6. $x^2 + 20x = 25$

7. $2x^2 - 3x + 4 = 0$

8. $3y^2 - 2y + 1 = 0$

9. $135 + 24n + n^2 = 0$

10. $28 - x - 2x^2 = 0$

11. $(x - 2)(x + 9) = -10$

12. $(x + 3)(2x + 1) = -3$

13. $2x^2 - 4x + 7 = 0$

14. $3x^2 - 2x + 8 = 0$

15. $x^2 - 18x + 15 = 0$

16. $x^2 - 16x + 14 = 0$

17. $20y^2 + 17y - 10 = 0$

18. $12x^2 + 23x - 9 = 0$

19. $4t^2 + 4t - 1 = 0$

20. $5t^2 + 5t - 1 = 0$

Solve each of the following equations.

21. $n + \dfrac{3}{n} = \dfrac{19}{4}$

22. $n - \dfrac{2}{n} = -\dfrac{7}{3}$

23. $\dfrac{3}{x} + \dfrac{7}{x - 1} = 1$

24. $\dfrac{2}{x} + \dfrac{5}{x + 2} = 1$

25. $\dfrac{12}{x - 3} + \dfrac{8}{x} = 14$

26. $\dfrac{16}{x + 5} - \dfrac{12}{x} = -2$

27. $\dfrac{3}{x - 1} - \dfrac{2}{x} = \dfrac{5}{2}$

28. $\dfrac{4}{x + 1} + \dfrac{2}{x} = \dfrac{5}{3}$

29. $\dfrac{6}{x} + \dfrac{40}{x + 5} = 7$

30. $\dfrac{12}{t} + \dfrac{18}{t + 8} = \dfrac{9}{2}$

31. $\dfrac{5}{n - 3} - \dfrac{3}{n + 3} = 1$

32. $\dfrac{3}{t + 2} + \dfrac{4}{t - 2} = 2$

33. $x^4 - 18x^2 + 72 = 0$

34. $x^4 - 21x^2 + 54 = 0$

35. $3x^4 - 35x^2 + 72 = 0$

36. $5x^4 - 32x^2 + 48 = 0$

37. $3x^4 + 17x^2 + 20 = 0$

38. $4x^4 + 11x^2 - 45 = 0$

39. $6x^4 - 29x^2 + 28 = 0$

40. $6x^4 - 31x^2 + 18 = 0$

Set up an equation and solve each of the following problems.

41. Find two consecutive whole numbers such that the sum of their squares is 145.

42. Find two consecutive odd whole numbers such that the sum of their squares is 74.

43. Two positive integers differ by 3, and their product is 108. Find the numbers.

44. Suppose that the sum of two numbers is 20 and the sum of their squares is 232. Find the numbers.

45. Find two numbers such that their sum is 10 and their product is 22.

46. Find two numbers such that their sum is 6 and their product is 7.

47. Suppose that the sum of two whole numbers is 9 and the sum of their reciprocals is $\dfrac{1}{2}$. Find the numbers.

48. The difference between two whole numbers is 8 and the difference between their reciprocals is $\dfrac{1}{6}$. Find the two numbers.

49. The sum of the lengths of the two legs of a right triangle is 21 inches. If the length of the hypotenuse is 15 inches, find the length of each leg.

50. The length of a rectangle floor is 1 meter less than twice its width. If a diagonal of the rectangle is 17 meters, find the length and width of the floor.

51. A rectangular plot of ground measuring 12 meters by 20 meters is surrounded by a sidewalk of a uniform width (see Figure 6.9). The area of the sidewalk is 68 square meters. Find the width of the walk.

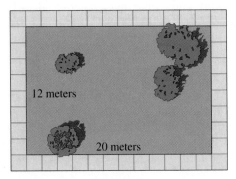

12 meters

20 meters

FIGURE 6.9

52. A 5-inch-by-7-inch picture is surrounded by a frame of uniform width. The area of the picture and frame together is 80 square inches. Find the width of the frame.

53. The perimeter of a rectangle is 44 inches and its area is 112 square inches. Find the length and width of the rectangle.

54. A rectangular piece of cardboard is 2 units longer than it is wide. From each of its corners a square piece 2 units on a side is cut out. The flaps are then turned up to form an open box that has a volume of 70 cubic units. Find the length and width of the original piece of cardboard.

55. Charlotte traveled 250 miles in one hour more time than it took Lorraine to travel 180 miles. Charlotte drove 5 miles per hour faster than Lorraine. How fast did each one travel?

56. Larry drove 156 miles in one hour more than it took Terrell to drive 108 miles. Terrell drove at an average rate of 2 miles per hour faster than Larry. How fast did each one travel?

57. On a 570-mile trip, Andy averaged 5 miles per hour faster for the last 240 miles than he did for the first 330 miles. The entire trip took 10 hours. How fast did he travel for the first 330 miles?

58. On a 135-mile bicycle excursion, Maria averaged 5 miles per hour faster for the first 60 miles than she did for the last 75 miles. The entire trip took 8 hours. Find her rate for the first 60 miles.

59. It takes Terry 2 hours longer to do a certain job than it takes Tom. They worked together for 3 hours; then Tom left and Terry finished the job in 1 hour. How long would it take each of them to do the job alone?

60. Suppose that Arlene can mow the entire lawn in 40 minutes less time with the power mower than she can with the push mower. One day the power mower broke down after she had been mowing for 30 minutes. She finished the lawn with the push mower in 20 minutes. How long does it take Arlene to mow the entire lawn with the power mower?

61. A man did a job for $360. It took him 6 hours longer than he expected and therefore he earned $2 per hour less than he anticipated. How long did he expect that it would take to do the job?

62. A group of students agreed to each chip in the same amount to pay for a party that would cost $100. Then they found 5 more students interested in the party and in sharing the expenses. This decreased the amount each had to pay by $1. How many students were involved in the party and how much did they each have to pay?

63. A group of customers agreed to each contribute the same amount to buy their favorite waitress a $100 birthday gift. At the last minute, 2 of the people decided not to chip in. This increased the amount that the remaining people had to pay by $2.50 per person. How many people actually contributed to the gift?

64. A retailer bought a number of special mugs for $48. Two of the mugs were broken in the store, but by selling each of the other mugs $3 above the original cost per mug she made a total profit of $22. How many mugs did she buy and at what price per mug did she sell them?

65. Tony bought a number of shares of stock for $720. A month later the value of the stock increased by $8 per share and he sold all but 20 shares and regained his original investment plus a profit of $80. How many shares did he sell and at what price per share?

66. The formula $D = \dfrac{n(n-3)}{2}$ yields the number of diagonals, D, in a polygon of n sides. Find the number of sides of a polygon that has 54 diagonals.

67. The formula $S = \dfrac{n(n+1)}{2}$ yields the sum, S, of the first n natural numbers $1, 2, 3, 4, \ldots$ How many consecutive natural numbers starting with 1 will give a sum of 1275?

68. At a point 16 yards from the base of a tower, the distance to the top of the tower is 4 yards more than the height of the tower (see Figure 6.10). Find the height of the tower.

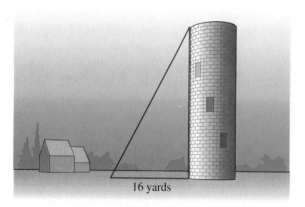

16 yards

FIGURE 6.10

69. Suppose that $500 is invested at a certain rate of interest compounded annually for 2 years. If the accumulated value at the end of 2 years is $594.05, find the rate of interest.

70. Suppose that $10,000 is invested at a certain rate of interest compounded annually for 2 years. If the accumulated value at the end of 2 years is $12,544, find the rate of interest.

71. How would you solve the equation $x^2 - 4x = 252$? Explain your choice of the method that you would use.

72. Explain how you would solve $(x - 2)(x - 7) = 0$ and also how you would solve $(x - 2)(x - 7) = 4$.

73. One of our problem solving suggestions is to *look for a guideline that can be used to help determine an equation.* What does this suggestion mean to you?

Further Investigations

Solve each of the following equations.

74. $x - 9\sqrt{x} + 18 = 0$ [*Hint*: Let $y = \sqrt{x}$.]

75. $x - 4\sqrt{x} + 3 = 0$

76. $x + \sqrt{x} - 2 = 0$

77. $x^{\frac{2}{3}} + x^{\frac{1}{3}} - 6 = 0$ [*Hint*: Let $y = x^{\frac{1}{3}}$.]

78. $6x^{\frac{2}{3}} - 5x^{\frac{1}{3}} - 6 = 0$

79. $x^{-2} + 4x^{-1} - 12 = 0$

80. $12x^{-2} - 17x^{-1} - 5 = 0$

6.6 **Quadratic Inequalities**

We refer to the equation $ax^2 + bx + c = 0$ as the standard form of a quadratic equation in one variable. Similarly, the following forms express **quadratic inequalities** in one variable.

$$ax^2 + bx + c > 0 \qquad\qquad ax^2 + bx + c < 0$$
$$ax^2 + bx + c \geq 0 \qquad\qquad ax^2 + bx + c \leq 0$$

We can use the number line very effectively to help solve quadratic inequalities where the quadratic polynomial is factorable. Let's consider some examples to illustrate the procedure.

Example 1

Solve and graph the solutions for $x^2 + 2x - 8 > 0$.

Solution

First, let's factor the polynomial.

$$x^2 + 2x - 8 > 0$$
$$(x + 4)(x - 2) > 0$$

FIGURE 6.11

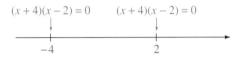

On a number line (Figure 6.11), we indicate that at $x = 2$ and $x = -4$ the product $(x + 4)(x - 2)$ equals zero. The numbers -4 and 2 divide the number line into three intervals: (1) the numbers less than -4, (2) the numbers between -4 and 2, and (3) the numbers greater than 2. We can choose a **test number** from each of these intervals and see how it affects the signs of the factors $x + 4$ and $x - 2$, and consequently, the sign of the product of these factors. For example, if $x < -4$ (try $x = -5$) then $x + 4$ is negative and $x - 2$ is negative; so their product is positive. If $-4 < x < 2$ (try $x = 0$), then $x + 4$ is positive and $x - 2$ is negative; so their product is negative. If $x > 2$ (try $x = 3$), then $x + 4$ is positive and $x - 2$ is positive; so their product is positive. This information can be conveniently arranged using a number line as follows (Figure 6.12). Note the open circles at -4 and 2 to indicate that they are not included in the solution set.

FIGURE 6.12

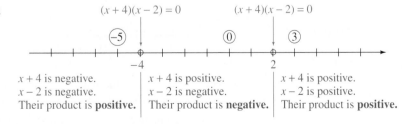

Therefore, the given inequality, $x^2 + 2x - 8 > 0$ is satisfied by numbers less than -4 along with numbers greater than 2. Using interval notation the solution set is $(-\infty, -4) \cup (2, \infty)$. These solutions can be shown on a number line (Figure 6.13).

FIGURE 6.13

We refer to numbers such as -4 and 2 in the preceding example (where the given polynomial or algebraic expression equals zero or is undefined) as **critical numbers**. Let's consider some additional examples that make use of critical numbers and test numbers.

Example 2

Solve and graph the solutions for $x^2 + 2x - 3 \leq 0$.

Solution

First, factor the polynomial.

$$x^2 + 2x - 3 \leq 0$$
$$(x + 3)(x - 1) \leq 0$$

Secondly, locate the values for which $(x + 3)(x - 1)$ equals zero. We put dots at -3 and 1 to remind ourselves that these two numbers are to be included in the solution set since the given statement includes equality. Now let's choose a test number from each of the three intervals and record the sign behavior of the factors $(x + 3)$ and $(x - 1)$ (Figure 6.14).

FIGURE 6.14

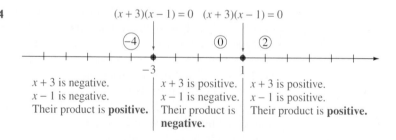

Therefore, the solution set is $[-3, 1]$ and it can be graphed as on Figure 6.15.

FIGURE 6.15

Examples 1 and 2 have indicated a systematic approach for solving quadratic inequalities where the polynomial is factorable. This same type of number line analysis can also be used to solve indicated quotients such as $\frac{x + 1}{x - 5} > 0$.

Example 3

Solve and graph the solutions for $\dfrac{x+1}{x-5} > 0$.

Solution

First, indicate that at $x = -1$ the given quotient equals zero, and at $x = 5$ the quotient is undefined. Second, choose test numbers from each of the three intervals and record the sign behavior of $(x + 1)$ and $(x - 5)$ as in Figure 6.16.

FIGURE 6.16

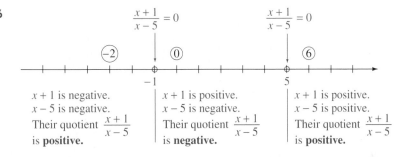

Therefore, the solution set is $(-\infty, -1) \cup (5, \infty)$ and its graph is shown in Figure 6.17.

FIGURE 6.17

Example 4

Solve $\dfrac{x+2}{x+4} \leq 0$.

Solution

The indicated quotient equals zero at $x = -2$ and is undefined at $x = -4$. (Note that -2 is to be included in the solution set, but -4 is not to be included.) Now let's choose some test numbers and record the sign behavior of $(x + 2)$ and $(x + 4)$ as in Figure 6.18.

FIGURE 6.18

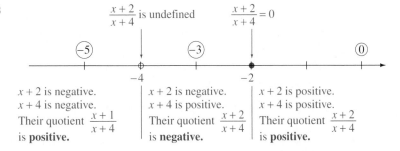

Therefore, the solution set is $(-4, -2]$.

The final example illustrates that sometimes we need to change the form of the given inequality before we use the number line analysis.

Example 5

Solve $\dfrac{x}{x+2} \geq 3$.

Solution

First, let's change the form of the given inequality as follows.

$$\frac{x}{x+2} \geq 3$$

$$\frac{x}{x+2} - 3 \geq 0 \qquad \text{Add } -3 \text{ to both sides.}$$

$$\frac{x - 3(x+2)}{x+2} \geq 0 \qquad \text{Express the left side over a common denominator.}$$

$$\frac{x - 3x - 6}{x+2} \geq 0$$

$$\frac{-2x - 6}{x+2} \geq 0.$$

Now we can proceed as we did with the previous examples. If $x = -3$, then $\dfrac{-2x-6}{x+2}$ equals zero; and if $x = -2$, then $\dfrac{-2x-6}{x+2}$ is undefined. Then choosing test numbers we can record the sign behavior of $(-2x - 6)$ and $(x + 2)$ as in Figure 6.19.

FIGURE 6.19

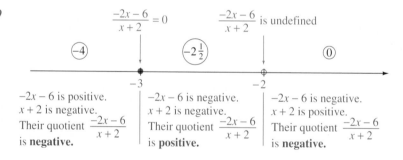

Therefore, the solution set is $[-3, -2)$. Perhaps you should check a few numbers from this solution set back into the original inequality! ▲

Problem Set 6.6

Solve each of the following inequalities and graph each solution set on a number line.

1. $(x + 2)(x - 1) > 0$

2. $(x - 2)(x + 3) > 0$

3. $(x + 1)(x + 4) < 0$

4. $(x - 3)(x - 1) < 0$

5. $(2x - 1)(3x + 7) \geq 0$

6. $(3x + 2)(2x - 3) \geq 0$

7. $(x + 2)(4x - 3) \leq 0$

8. $(x - 1)(2x - 7) \leq 0$

9. $(x + 1)(x - 1)(x - 3) > 0$

10. $(x + 2)(x + 1)(x - 2) > 0$

11. $x(x + 2)(x - 4) \leq 0$

12. $x(x + 3)(x - 3) \leq 0$

13. $\dfrac{x + 1}{x - 2} > 0$

14. $\dfrac{x - 1}{x + 2} > 0$

15. $\dfrac{x - 3}{x + 2} < 0$

16. $\dfrac{x + 2}{x - 4} < 0$

17. $\dfrac{2x - 1}{x} \geq 0$

18. $\dfrac{x}{3x + 7} \geq 0$

19. $\dfrac{-x + 2}{x - 1} \leq 0$

20. $\dfrac{3 - x}{x + 4} \leq 0$

Solve each of the following inequalities.

21. $x^2 + 2x - 35 < 0$

22. $x^2 + 3x - 54 < 0$

23. $x^2 - 11x + 28 > 0$

24. $x^2 + 11x + 18 > 0$

25. $3x^2 + 13x - 10 \leq 0$

26. $4x^2 - x - 14 \leq 0$

27. $8x^2 + 22x + 5 \geq 0$

28. $12x^2 - 20x + 3 \geq 0$

29. $x(5x - 36) > 32$

30. $x(7x + 40) < 12$

31. $x^2 - 14x + 49 \geq 0$

32. $(x + 9)^2 \geq 0$

33. $4x^2 + 20x + 25 \leq 0$

34. $9x^2 - 6x + 1 \leq 0$

35. $(x + 1)(x - 3)^2 > 0$

36. $(x - 4)^2(x - 1) \leq 0$

37. $\dfrac{2x}{x + 3} > 4$

38. $\dfrac{x}{x - 1} > 2$

39. $\dfrac{x - 1}{x - 5} \leq 2$

40. $\dfrac{x + 2}{x + 4} \leq 3$

41. $\dfrac{x + 2}{x - 3} > -2$

42. $\dfrac{x - 1}{x - 2} < -1$

43. $\dfrac{3x + 2}{x + 4} \leq 2$

44. $\dfrac{2x - 1}{x + 2} \geq -1$

45. $\dfrac{x + 1}{x - 2} < 1$

46. $\dfrac{x + 3}{x - 4} \geq 1$

THOUGHTS INTO WORDS

47. Explain how to solve the inequality $(x + 1) \cdot (x - 2)(x - 3) > 0$.

48. Explain how to solve the inequality $(x - 2)^2 > 0$ by inspection.

49. Your friend looks at the inequality $1 + \dfrac{1}{x} > 2$ and without any computation states that the solution set is all real numbers between 0 and 1. How can she do that?

Further Investigations

50. The product $(x - 2)(x + 3)$ is positive if both factors are negative *or* if both factors are positive. Therefore, we can solve $(x - 2)(x + 3) > 0$ as follows.

$$(x - 2 < 0 \text{ and } x + 3 < 0) \text{ or } (x - 2 > 0 \text{ and } x + 3 > 0)$$
$$(x < 2 \text{ and } x < -3) \text{ or } (x > 2 \text{ and } x > -3)$$
$$x < -3 \text{ or } x > 2$$

The solution set is $(-\infty, -3) \cup (2, \infty)$. Use this type of analysis to solve each of the following.

(a) $(x - 2)(x + 7) > 0$

(b) $(x - 3)(x + 9) \geq 0$

(c) $(x + 1)(x - 6) \leq 0$

(d) $(x + 4)(x - 8) < 0$

(e) $\dfrac{x + 4}{x - 7} > 0$

(f) $\dfrac{x - 5}{x + 8} \leq 0$

SUMMARY

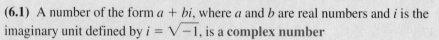

(6.1) A number of the form $a + bi$, where a and b are real numbers and i is the imaginary unit defined by $i = \sqrt{-1}$, is a **complex number**

Two complex numbers $a + bi$ and $c + di$ are said to be *equal* if and only if $a = c$ and $b = d$.

We describe addition and subtraction of complex numbers as follows.

$$(a + bi) + (c + di) = (a + c) + (b + d)i,$$
$$(a + bi) - (c + di) = (a - c) + (b - d)i$$

We can represent a square root of any negative real number as the product of a real number and the imaginary unit i. That is,

$$\sqrt{-b} = i\sqrt{b}, \quad \text{where } b \text{ is a positive real number.}$$

The product of two complex numbers conforms with the product of two binomials. The **conjugate** of $a + bi$ is $a - bi$. The product of a complex number and its conjugate is a real number. Therefore, conjugates are used to simplify expressions such as $\dfrac{4 + 3i}{5 - 2i}$, which indicate the quotient of two complex numbers.

(6.2) The **standard form for a quadratic equation** in one variable is

$$ax^2 + bx + c = 0,$$

where a, b, and c are real numbers and $a \neq 0$.

Some quadratic equations can be solved by *factoring* and applying the property, $ab = 0$ if and only if $a = 0$ or $b = 0$.

Don't forget that applying the property, *if $a = b$, then $a^n = b^n$*, might produce extraneous solutions. Therefore, we *must check* all potential solutions.

We can solve some quadratic equations by applying the property, $x^2 = a$ *if and only if $x = \pm\sqrt{a}$.*

(6.3) To solve a quadratic equation of the form $x^2 + bx = k$ by **completing the square**, we (1) add $\left(\dfrac{b}{2}\right)^2$ to both sides, (2) factor the left side, and (3) apply the property, $x^2 = a$ *if and only if* $x = \pm\sqrt{a}$.

(6.4) We can solve any quadratic equation of the form $ax^2 + bx + c = 0$ by the **quadratic formula**, which we usually state as

$$x = \frac{-b \pm \sqrt{b^2 - 4ac}}{2a}.$$

The **discriminant**, $b^2 - 4ac$, can be used to determine the nature of the roots of a quadratic equation as follows.

1. If $b^2 - 4ac < 0$, then the equation has two nonreal complex solutions.
2. If $b^2 - 4ac = 0$, then the equation has two equal real solutions.
3. If $b^2 - 4ac > 0$, then the equation has two unequal real solutions.

If x_1 and x_2 are roots of a quadratic equation, then the following relationships exist.

$$x_1 + x_2 = -\frac{b}{a} \qquad \text{and} \qquad (x_1)(x_2) = \frac{c}{a}$$

These **sum-and-product relationships** can be used to check potential solutions of quadratic equations.

(6.5) To review the strengths and weaknesses of the three basic methods for solving a quadratic equation (factoring, completing the square, the quadratic formula), go back over the examples in this section.

Keep the following suggestions in mind as you solve word problems.

1. Read the problem carefully.
2. Sketch any figure, diagram, or chart that might help organize and analyze the problem.
3. Choose a meaningful variable.
4. Look for a guideline that can be used to set up an equation.
5. Form an equation that translates the guideline from English to algebra.
6. Solve the equation and use the solutions to determine all facts requested in the problem.
7. Check all answers back into the original statement of the problem.

(6.6) The number line, along with **critical numbers** and **test numbers**, provides a good basis for solving **quadratic inequalities** where the polynomial is factorable. We can use this same basic approach to solve inequalities, such as $\frac{3x + 1}{x - 4} > 0$, which indicate quotients.

Chapter 6 Review Problem Set

For Problems 1–8, perform the indicated operations and express the answers in the standard form of a complex number.

1. $(-7 + 3i) + (9 - 5i)$

2. $(4 - 10i) - (7 - 9i)$

3. $5i(3 - 6i)$

4. $(5 - 7i)(6 + 8i)$

5. $(-2 - 3i)(4 - 8i)$

6. $(4 - 3i)(4 + 3i)$

7. $\dfrac{4 + 3i}{6 - 2i}$

8. $\dfrac{-1 - i}{-2 + 5i}$

For Problems 9–12, find the discriminant of each equation and determine whether the equation has (1) two nonreal complex solutions, (2) one real solution with a multiplicity of two, or (3) two real solutions. Do not solve the equations.

9. $4x^2 - 20x + 25 = 0$

10. $5x^2 - 7x + 31 = 0$

11. $7x^2 - 2x - 14 = 0$

12. $5x^2 - 2x = 4$

For Problems 13–31, solve each equation.

13. $x^2 - 17x = 0$

14. $(x - 2)^2 = 36$

15. $(2x - 1)^2 = -64$

16. $x^2 - 4x - 21 = 0$

17. $x^2 + 2x - 9 = 0$

18. $x^2 - 6x = -34$

19. $4\sqrt{x} = x - 5$

20. $3n^2 + 10n - 8 = 0$

21. $n^2 - 10n = 200$

22. $3a^2 + a - 5 = 0$

23. $x^2 - x + 3 = 0$

24. $2x^2 - 5x + 6 = 0$

25. $2a^2 + 4a - 5 = 0$

26. $t(t + 5) = 36$

27. $x^2 + 4x + 9 = 0$

28. $(x - 4)(x - 2) = 80$

29. $\dfrac{3}{x} + \dfrac{2}{x + 3} = 1$

30. $2x^4 - 23x^2 + 56 = 0$

31. $\dfrac{3}{n - 2} = \dfrac{n + 5}{4}$

For Problems 32–35, solve each inequality and indicate the solution set on a number line graph.

32. $x^2 + 3x - 10 > 0$

33. $2x^2 + x - 21 \le 0$

34. $\dfrac{x - 4}{x + 6} \ge 0$

35. $\dfrac{2x - 1}{x + 1} > 4$

For Problems 36–43, set up an equation and solve each problem.

36. Find two numbers whose sum is 6 and whose product is 2.

37. Sherry bought a number of shares of stock for $250. Six months later the value of the stock increased by $5 per share and she sold all but 5 shares and regained her original investment plus a profit of $50. How many shares did she sell and at what price per share?

38. Andre traveled 270 miles in one hour more time than it took Sandy to travel 260 miles. Sandy drove 7 miles per hour faster than Andre. How fast did each one travel?

39. The area of a square is numerically equal to twice its perimeter. Find the length of a side of the square.

40. Find two consecutive even whole numbers such that the sum of their squares is 164.

41. The perimeter of a rectangle is 38 inches and its area is 84 square inches. Find the length and width of the rectangle.

42. It takes Billy 2 hours longer to do a certain job than it takes Reena. They worked together for 2 hours; then Reena left and Billy finished the job in 1 hour. How long would it take each of them to do the job alone?

43. A company has a rectangular parking lot 40 meters wide and 60 meters long. They plan to increase the area of the lot by 1100 square meters by adding a strip of equal width to one side and one end. Find the width of the strip to be added.

CHAPTER 6 TEST

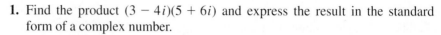

1. Find the product $(3 - 4i)(5 + 6i)$ and express the result in the standard form of a complex number.

2. Find the quotient $\dfrac{2 - 3i}{3 + 4i}$ and express the result in the standard form of a complex number.

For Problems 3–15, solve each equation.

3. $x^2 = 7x$ 4. $(x - 3)^2 = 16$

5. $x^2 + 3x - 18 = 0$ 6. $x^2 - 2x - 1 = 0$

7. $5x^2 - 2x + 1 = 0$

8. $x^2 + 30x = -224$

9. $(3x - 1)^2 + 36 = 0$

10. $(5x - 6)(4x + 7) = 0$ 11. $(2x + 1)(3x - 2) = 55$

12. $n(3n - 2) = 40$ 13. $x^4 + 12x^2 - 64 = 0$

14. $\dfrac{3}{x} + \dfrac{2}{x + 1} = 4$

15. $3x^2 - 2x - 3 = 0$

16. Does the equation $4x^2 + 20x + 25 = 0$ have (a) two nonreal complex solutions, (b) two equal real solutions, or (c) two unequal real solutions?

17. Does the equation $4x^2 - 3x = -5$ have (a) two nonreal complex solutions, (b) two equal real solutions, or (c) two unequal real solutions?

For Problems 18–20, solve each inequality and express the solution set using interval notation.

18. $x^2 - 3x - 54 \leq 0$ 19. $\dfrac{3x - 1}{x + 2} > 0$

20. $\dfrac{x - 2}{x + 6} \geq 3$

For Problems 21–25, set up an equation and solve each problem.

21. A 24-foot ladder leans against a building and makes an angle of 60° with the ground. How far up on the building does the top of the ladder reach? Express your answer to the nearest tenth of a foot.

(continued on next page)

CHAPTER 6 TEST (*continued*)

22. A rectangular plot of ground measures 16 meters by 24 meters. Find the distance, to the nearest meter, from one corner of the plot to the diagonally opposite corner.

23. Dana bought a number of shares of stock for a total of $3000. Three months later the stock had increased in value by $5 per share and she sold all but 50 shares and regained her original investment of $3000. How many shares did she sell?

24. The perimeter of a rectangle is 41 inches and its area is 91 square inches. Find the length of its shortest side.

25. The sum of two numbers is 6 and their product is 4. Find the larger of the two numbers.

Coordinate Geometry and Graphing Techniques

René Descartes, a French mathematician of the seventeenth century, transformed geometric problems into an algebraic setting so that he could use the tools of algebra to solve the problems. This connecting of algebraic and geometric ideas is the foundation of a branch of mathematics called analytic geometry, more commonly called coordinate geometry.

Basically, there are two kinds of problems in coordinate geometry: Given an algebraic equation, find its geometric graph; and given a set of conditions pertaining to a geometric graph, find its algebraic equation. We discuss problems of both types in this chapter.

7.1 Coordinate Geometry

Recall that the real number line exhibits a **one-to-one correspondence** between the set of real numbers and the points on a line. That is to say, to each real number there corresponds one and only one point on the line, and to each point on the line there corresponds one and only one real number. The number that corresponds to a particular point on the line is called the **coordinate** of that point.

Suppose that on the number line we want to determine the distance from -4 to 7. The "from-to" vocabulary implies a **directed distance**, which in this case is $7 - (-4) = 11$ units. In other words, it is 11 units in a positive direction from -4 to 7. Likewise, the distance from 5 to -8 is $-8 - 5 = -13$, that is, 13 units in a negative direction. In general, if x_1 and x_2 are the coordinates of two points on the number line, then the distance from x_1 to x_2 is given by $x_2 - x_1$, and the distance from x_2 to x_1 is given by $x_1 - x_2$.

Now suppose that we want to find the distance between -3 and 5. The "between" vocabulary implies distance without regard to direction. Thus we can find the distance between -3 and 5 by using either $|5 - (-3)| = 8$ or $|-3 - 5| = 8$. In general, if x_1 and x_2 are the coordinates of two points on the number line, we can find the distance between x_1 and x_2 by using either $|x_2 - x_1|$ or $|x_1 - x_2|$.

Rectangular Coordinate System

Consider two number lines, one vertical and one horizontal, perpendicular to each other at the point associated with zero on both lines (Figure 7.1). We refer to these number lines individually as the **horizontal** and **vertical axes** or together as the **coordinate axes**. These axes partition the plane into four regions called **quadrants**. The quadrants are numbered counterclockwise from I through IV as indicated in Figure 7.1. The point of intersection of the two axes is called the **origin**.

FIGURE 7.1

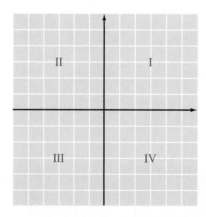

It is now possible to set up a one-to-one correspondence between **ordered pairs** of real numbers and the points in a plane. To each ordered pair of real numbers there corresponds a unique point in the plane, and to each point in the plane there corresponds a unique ordered pair of real numbers. We have illustrated a part of this correspondence in Figure 7.2. The ordered pair (3, 2) means that the point A is located 3 units to the right and two units up from the origin. (The ordered pair (0, 0) is associated with the origin.) The ordered pair $(-3, -5)$ means that the point D is located three units to the left and five units down from the origin.

FIGURE 7.2

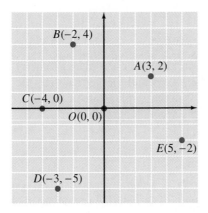

The notation $(-2, 4)$ was used earlier in this text to indicate an interval of the real number line. Now we are using the same notation to indicate an ordered pair of real numbers. This double meaning should not be confusing since the context of the material will definitely indicate the appropriate meaning. Throughout this chapter we will use the ordered-pair interpretation. △

In general, the real numbers a and b in the ordered pair (a, b) are associated with a point; they are referred to as the **coordinates of the point**. The first number, a, called the **abscissa**, is the directed distance of the point from the vertical axis measured parallel to the horizontal axis. The second number, b, called the **ordinate**, is the directed distance of the point from the horizontal axis measured parallel to the vertical axis (Figure 7.3(a)). Thus, in the first quadrant all points have a positive abscissa and a positive ordinate. In the second quadrant all points have a negative abscissa and a positive ordinate. We have indicated the sign situations for all four quadrants in Figure 7.3(b). This system of associating points in a plane with pairs of real numbers is called the **rectangular coordinate system** or the **Cartesian coordinate system**.

FIGURE 7.3

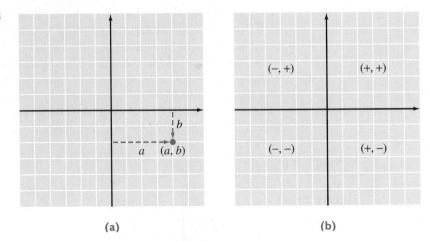

(a)   (b)

For most problems in coordinate geometry it is customary to label the horizontal axis the **x-axis** and the vertical axis the **y-axis**. Then, ordered pairs that represent points in the *xy*-plane are of the form (x, y); that is, x is the first coordinate, and y is the second coordinate.

Distance

As we work with the rectangular coordinate system, it is sometimes necessary to express the length of certain line segments. In other words, we need to be able to find the *distance between* two points. Let's first consider two specific examples and then develop the general distance formula.

Example I

Find the distance between the points $A(2, 2)$ and $B(5, 2)$, and between the points $C(-2, 5)$ and $D(-2, -4)$.

Solution

Let's plot the points and draw $\overline{AB}$ and $\overline{CD}$ as in Figure 7.4. Since $\overline{AB}$ is parallel to the x-axis, its length can be expressed as $|5 - 2|$ or $|2 - 5|$. (The absolute value symbol is used to ensure a nonnegative value.) Thus, the length of $\overline{AB}$ is 3 units. Likewise, since $\overline{CD}$ is parallel to the y-axis, the length of $\overline{CD}$ is $|5 - (-4)| = |-4 - 5| = 9$ units.

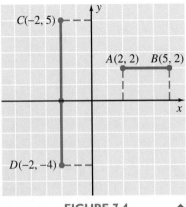

FIGURE 7.4   ▲

Example 2

Find the distance between the points $A(2, 3)$ and $B(5, 7)$.

Solution

Let's plot the points and form a right triangle as indicated in Figure 7.5. Notice that the coordinates of point D are $(5, 3)$. Because $\overline{AD}$ is parallel to the horizontal axis, we can easily determine its length to be 3 units. Likewise, $\overline{DB}$ is parallel to the vertical axis and its length is 4 units. Let d represent the length of $\overline{AB}$ and apply the Pythagorean theorem to obtain

$$d^2 = 3^2 + 4^2$$
$$d^2 = 9 + 16$$
$$d^2 = 25$$
$$d = \pm\sqrt{25} = \pm 5.$$

Since *distance between* is a nonnegative value, the length of $\overline{AB}$ is 5 units.

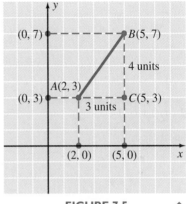

FIGURE 7.5

We can use the approach in Example 2 to develop a general distance formula for finding the distance between any two points in a coordinate plane. The development proceeds as follows.

1. Let $P_1(x_1, y_1)$ and $P_2(x_2, y_2)$ represent any two points in a coordinate plane.

2. Form a right triangle as indicated in Figure 7.6. The coordinates of the vertex of the right angle, point R, are (x_2, y_1).

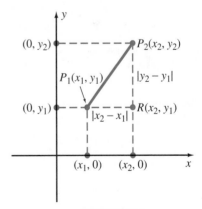

FIGURE 7.6

The length of $\overline{P_1R}$ is $|x_2 - x_1|$, and the length of $\overline{RP_2}$ is $|y_2 - y_1|$. (We use the absolute value symbol to ensure a nonnegative value.) Let d represent the length of $\overline{P_1P_2}$ and apply the Pythagorean theorem to yield

$$d^2 = |x_2 - x_1|^2 + |y_2 - y_1|^2.$$

Since $|a|^2 = a^2$, the distance formula can be stated as follows.

Distance Formula
$$d = \sqrt{(x_2 - x_1)^2 + (y_2 - y_1)^2}.$$

It makes no difference which point you call P_1 or P_2 when using the distance formula. Remember, if you forget the formula, don't panic; merely form a right triangle and apply the Pythagorean theorem as we did in Example 2.

Let's consider two examples that illustrate the use of the distance formula.

Example 3

Find the distance between $(-1, 4)$ and $(1, 2)$.

Solution

Let $(-1, 4)$ be P_1 and $(1, 2)$ be P_2. Using the distance formula, we obtain

$$d = \sqrt{[1 - (-1)]^2 + (2 - 4)^2}$$
$$= \sqrt{2^2 + (-2)^2}$$
$$= \sqrt{4 + 4}$$
$$= \sqrt{8} = 2\sqrt{2}. \quad \text{Express the answer in simplest radical form.}$$

The distance between the two points is $2\sqrt{2}$ units. ▲

In Example 3, we did not sketch a figure because of the simplicity of the problem. However, sometimes we can use a figure to organize the given information and aid in the analysis of the problem, as we see in the next example.

Example 4

Verify that the points $(-3, 6)$, $(3, 4)$, and $(1, -2)$ are vertices of an isosceles triangle. (An isosceles triangle has two sides of the same length.)

Solution

Let's plot the points and draw the triangle (Figure 7.7). Using the distance formula, we can find the lengths d_1, d_2, and d_3 as follows.

$$d_1 = \sqrt{[3 - 1]^2 + [4 - (-2)]^2}$$
$$= \sqrt{2^2 + 6^2} = \sqrt{40} = 2\sqrt{10};$$

$$d_2 = \sqrt{[-3 - 3]^2 + [6 - 4]^2}$$
$$= \sqrt{(-6)^2 + 2^2} = \sqrt{40} = 2\sqrt{10};$$

$$d_3 = \sqrt{[-3 - 1]^2 + [6 - (-2)]^2}$$
$$= \sqrt{(-4)^2 + 8^2} = \sqrt{80} = 4\sqrt{5}.$$

Since $d_1 = d_2$, we know that the triangle is an isosceles triangle.

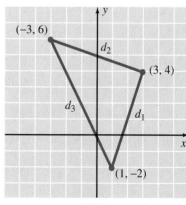

FIGURE 7.7 ▲

Slope of a Line

In coordinate geometry, we use the concept of **slope** to discuss the steepness of lines. The slope of a line is the ratio of the vertical change compared to the horizontal change as we move from one point on a line to another point. We use points P_1 and P_2 to illustrate slope in Figure 7.8. In Figure 7.9, consider the coordinates of the points P_1, P_2, and R for a precise definition of slope.

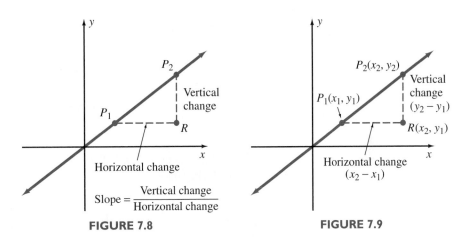

FIGURE 7.8   **FIGURE 7.9**

The horizontal change as we move from P_1 to P_2 is $x_2 - x_1$, and the vertical change is $y_2 - y_1$. Thus, the following definition for slope is given.

DEFINITION 7.1

If points P_1 and P_2 with coordinates (x_1, y_1) and (x_2, y_2), respectively, are any two different points on a line, then the slope of the line (denoted by m) is

$$m = \frac{y_2 - y_1}{x_2 - x_1}, \qquad x_2 \neq x_1.$$

Since $\frac{y_2 - y_1}{x_2 - x_1} = \frac{y_1 - y_2}{x_1 - x_2}$, how we designate P_1 and P_2 is not important. Let's use Definition 7.1 to find the slopes of some lines.

Example 5

Find the slope of the line determined by each of the following pairs of points, and graph the lines.

(*a*) $(-1, 1)$ and $(3, 2)$ (*b*) $(4, -2)$ and $(-1, 5)$
(*c*) $(2, -3)$ and $(-3, -3)$

Solutions

(*a*) Let $(-1, 1)$ be P_1, and let $(3, 2)$ be P_2 (Figure 7.10).

$$m = \frac{y_2 - y_1}{x_2 - x_1} = \frac{2 - 1}{3 - (-1)}$$

$$= \frac{1}{4}$$

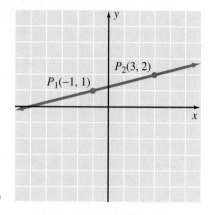

FIGURE 7.10

(*b*) Let $(4, -2)$ be P_1, and let $(-1, 5)$ be P_2 (Figure 7.11).

$$m = \frac{y_2 - y_1}{x_2 - x_1} = \frac{5 - (-2)}{-1 - 4}$$

$$= \frac{7}{-5} = -\frac{7}{5}$$

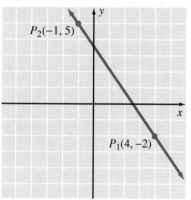

FIGURE 7.11

(*c*) Let $(2, -3)$ be P_1, and let $(-3, -3)$ be P_2 (Figure 7.12).

$$m = \frac{y_2 - y_1}{x_2 - x_1} = \frac{-3 - (-3)}{-3 - 2}$$

$$= \frac{0}{-5} = 0$$

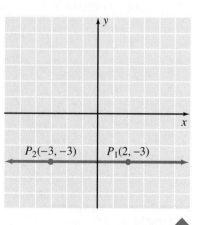

FIGURE 7.12

The three parts of Example 5 illustrate the three basic possibilities for slope; that is, the slope of a line can be positive, negative, or zero. A line that has a positive slope, as in Figure 7.10, rises as we move from left to right. A line that has a negative slope, as in Figure 7.11, falls as we move from left to right. A horizontal line, as in Figure 7.12, has a slope of zero. Finally, we need to realize that **the concept of slope is undefined for vertical lines.** This is due to the fact that for any vertical line the horizontal change as we move from one point on the line to another is zero. Thus, the ratio $\dfrac{y_2 - y_1}{x_2 - x_1}$ will have a denominator of zero and will therefore be undefined. So in Definition 7.1 the restriction $x_2 \neq x_1$ is made.

One final idea pertaining to the concept of slope needs to be emphasized. The slope of a line is a *ratio*, the ratio of vertical change compared to horizontal change. A slope of $\dfrac{2}{3}$ means that for every 2 units of vertical change there must be a corresponding 3 units of horizontal change. Thus, starting at some point on a line having a slope of $\dfrac{2}{3}$, we could locate other points on the line as follows.

$$\frac{2}{3} = \frac{4}{6}$$ By moving 4 units *up* and 6 units to the *right*

$$\frac{2}{3} = \frac{8}{12}$$ By moving 8 units *up* and 12 units to the *right*

$$\frac{2}{3} = \frac{-2}{-3}$$ By moving 2 units *down* and 3 units to the *left*

Likewise, if a line has a slope of $-\dfrac{3}{4}$, then by starting at some point on the line we could locate other points on the line as follows.

$$-\frac{3}{4} = \frac{-3}{4}$$ By moving 3 units *down* and 4 units to the *right*

$$-\frac{3}{4} = \frac{3}{-4}$$ By moving 3 units *up* and 4 units to the *left*

$$-\frac{3}{4} = \frac{-9}{12}$$ By moving 9 units *down* and 12 units to the *right*

$$-\frac{3}{4} = \frac{15}{-20}$$ By moving 15 units *up* and 20 units to the *left*

Problem Set 7.1

For Problems 1–12, find the following indicated distances on a number line.

1. From -3 to 12

2. From -7 to 19

3. From 7 to 2

4. From 12 to 6

5. From -4 to -13

6. From -2 to -10

7. From -14 to -3

8. From -9 to -3

9. Between -6 and 4
10. Between -8 and 9
11. Between -3 and -8
12. Between -2 and -10

For Problems 13–24, find the distance between each of the pairs of points. Express answers in simplest radical form.

13. $(3, 2)$ and $(9, 10)$
14. $(1, 3)$ and $(4, 7)$
15. $(-1, 3)$ and $(4, 2)$
16. $(5, -1)$ and $(2, 4)$
17. $(-2, -4)$ and $(2, 2)$
18. $(-1, -4)$ and $(0, -11)$
19. $(5, -3)$ and $(-7, -19)$
20. $(-2, 3)$ and $(6, -12)$
21. $(-2, -1)$ and $(-6, -9)$
22. $(-3, -5)$ and $(-9, -8)$
23. $(6, -4)$ and $(9, -7)$
24. $(-5, 2)$ and $(-1, 6)$
25. Verify that the points $(3, 4)$, $(5, -2)$, and $(-9, 0)$ are vertices of a right triangle. [*Hint:* If $a^2 + b^2 = c^2$, then it is a right triangle with the right angle opposite side c.]

26. Verify that the points $(0, 3)$, $(2, -3)$, and $(-4, -5)$ are vertices of an isosceles triangle. (An isosceles triangle has two sides of the same length.)

27. Verify that $(2, -1)$ is the midpoint of the line segment joining $(-2, -4)$ and $(6, 2)$.

28. Verify that the points $(2, -2)$ and $(6, -8)$ divide the line segment between $(-2, 4)$ and $(10, -14)$ into three segments of equal length.

29. Find the perimeter of the triangle whose vertices are $(4, 1)$, $(1, -3)$, and $(-5, 5)$.

30. Use the distance formula to verify that the points $(3, 3)$, $(1, -1)$, and $(-2, -7)$ lie on the same straight line.

31. Verify that $(-4, 9)$, $(8, 4)$, $(3, -8)$ and $(-9, -3)$ are vertices of a square.

32. Verify that the points $(4, -5)$, $(6, 7)$, and $(-8, -3)$ lie on a circle that has its center at $(-1, 2)$.

For Problems 33–46, find the slope of the line determined by each pair of points.

33. $(4, 2)$ and $(7, 4)$
34. $(7, 10)$ and $(1, 5)$
35. $(-1, 3)$ and $(3, -3)$
36. $(-4, 2)$ and $(5, -1)$
37. $(-1, -2)$ and $(-7, -10)$
38. $(-2, -3)$ and $(-6, -12)$
39. $(-2, 4)$ and $(7, 4)$
40. $(a, 0)$ and $(0, b)$
41. (a, b) and (c, d)
42. $(5, -3)$ and $(-7, -3)$
43. $(-2, 4)$ and $(4, 10)$
44. $(-2, 6)$ and $(-2, -4)$
45. $(-1, 5)$ and $(-1, -7)$
46. $(-2, 3)$ and $(-7, -2)$
47. Find x if the line through $(-2, 4)$ and $(x, 6)$ has a slope of $\dfrac{2}{9}$.

48. Find y if the line through $(1, y)$ and $(4, 2)$ has a slope of $\dfrac{5}{3}$.

49. Find y if the line through $(5, 2)$ and $(-3, y)$ has a slope of $-\dfrac{7}{8}$.

50. Find x if the line through $(x, 4)$ and $(2, -5)$ has a slope of $-\dfrac{9}{4}$.

For Problems 51–58, you are given one point on a line and the slope of the line. Find the coordinates of three other points on the line.

51. $(2, 4)$, $m = \dfrac{5}{6}$

52. $(3, 2), m = \dfrac{2}{5}$

53. $(-2, 1), m = 3$

54. $(-3, 4), m = 2$

55. $(-3, -2), m = -\dfrac{2}{3}$

56. $(-2, -4), m = -\dfrac{1}{4}$

57. $(4, -1), m = -\dfrac{3}{2}$

58. $(1, -3), m = -4$

59. The concept of slope is used for highway construction. The "grade" of a highway expressed as a percent means the number of feet that the highway changes in elevation for each 100 feet of horizontal change.
(a) A certain highway has a 2% grade. How many feet does it rise in a horizontal distance of 1 mile? (1 mile = 5280 feet)
(b) The grade of a highway up a hill is 30%. How much change in horizontal distance is there if the vertical height of the hill is 75 feet?

60. Slope is often expressed as the ratio of "rise to run" in construction of steps.
(a) If the ratio of rise to run is to be $\dfrac{3}{5}$ for some steps and the rise is 19 centimeters, find the measure of the run to the nearest centimeter.
(b) If the ratio of rise to run is to be $\dfrac{2}{3}$ for some steps and the run is 28 centimeters, find the rise to the nearest centimeter.

61. Suppose that a county ordinance requires a $2\dfrac{1}{4}\%$ "fall" for a sewage pipe from the house to the main pipe at the street. How much vertical drop must there be for a horizontal distance of 45 feet? Express the answer to the nearest tenth of a foot.

THOUGHTS INTO WORDS

62. Suppose that you forgot the distance formula. How would you find the distance between the points (3, 4) and (7, 10)?

63. How would you explain the concept of slope to someone who was absent from class the day it was discussed?

64. If one line has a slope of $\dfrac{2}{5}$ and another line has a slope of $\dfrac{3}{7}$, which line is steeper? Explain your answer.

65. Suppose that a line has a slope of $\dfrac{2}{3}$ and contains the point (4, 7). Are the points (7, 9) and (1, 3) also on the line? Explain your answer.

Further Investigations

66. We can use the tools of coordinate geometry to prove various geometric properties. For example, consider the following way of proving that the diagonals of a rectangle are equal in length. First, we draw a rectangle. Secondly, we choose the coordinate axes in a convenient position and label the vertices of the rectangle as indicated in Figure 7.13.

Finally, the distance formula can be used to find the lengths of the diagonals $\overline{AC}$ and $\overline{BD}$.

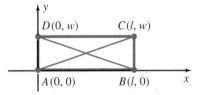

FIGURE 7.13

$$AC = \sqrt{(l - 0)^2 + (w - 0)^2} = \sqrt{l^2 + w^2}$$
$$BD = \sqrt{(0 - l)^2 + (w - 0)^2} = \sqrt{l^2 + w^2}$$

Therefore, $AC = BD$, and we have verified that the diagonals are equal in length.

Prove each of the following.
(a) The diagonals of an isosceles trapezoid are equal in length.
(b) The line segment that joins the midpoints of two sides of a triangle is equal in length to one-half of the third side.
(c) The midpoint of the hypotenuse of a right triangle is equally distant from all three vertices.

67. Sometimes it is necessary to find the coordinate of a point on a number line that is located somewhere between two given points. For example, suppose that we want to find the coordinate (x) of the point located two-thirds of the distance *from* 2 *to* 8.

Since the total distance from 2 to 8 is $8 - 2 = 6$ units, we can start at 2 and move $\frac{2}{3}(6) = 4$ units toward 8. Thus, $x = 2 + \frac{2}{3}(6) = 2 + 4 = 6$.

For each of the following, find the coordinate of the indicated point on a number line.

(a) Two-thirds of the distance from 1 to 10
(b) Three-fourths of the distance from -2 to 14
(c) One-third of the distance from -3 to 7
(d) Two-fifths of the distance from -5 to 6
(e) Three-fifths of the distance from -1 to -11
(f) Five-sixths of the distance from 3 to -7

68. Now suppose that we want to find the coordinates of point P, which is located two-thirds of the distance from $A(1, 2)$ to $B(7, 5)$ in a coordinate plane. In Figure 7.14 we plotted the given points A and B

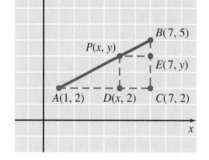

FIGURE 7.14

and completed a figure to help with the analysis of this problem. Point D is two-thirds of the distance from A to C because parallel lines cut off proportional segments on every transversal that intersects the lines. Thus, $\overline{AC}$ can be treated as a segment of a number line as shown in Figure 7.15. Therefore

FIGURE 7.15

$$x = 1 + \frac{2}{3}(7 - 1) = 1 + \frac{2}{3}(6) = 5.$$

Similarly, $\overline{CB}$ can be treated as a segment of a number line as shown in Figure 7.16. Therefore,

FIGURE 7.16

$$y = 2 + \frac{2}{3}(5 - 2) = 2 + \frac{2}{3}(3) = 4$$

The coordinates of point P are $(5, 4)$.

For each of the following, find the coordinates of the indicated point in the xy-plane.

(a) One-third of the distance from $(2, 3)$ to $(5, 9)$

(b) Two-thirds of the distance from $(1, 4)$ to $(7, 13)$

(c) Two-fifths of the distance from $(-2, 1)$ to $(8, 11)$

(d) Three-fifths of the distance from $(2, -3)$ to $(-3, 8)$

(e) Five-eighths of the distance from $(-1, -2)$ to $(4, -10)$

(f) Seven-eighths of the distance from $(-2, 3)$ to $(-1, -9)$

7.2 Graphing Techniques—Linear Equations and Inequalities

As you continue to study mathematics, you will find that the ability to quickly sketch the graph of an equation is an important skill. Thus, throughout precalculus and calculus courses, we discuss various curve sketching techniques. We will devote a good portion of this chapter to expanding your repertoire of graphing techniques.

First, let's briefly review some basic ideas by considering the solutions for the equation $y = x + 2$. A **solution** of an equation in two variables is an ordered pair of real numbers that satisfy the equation. When the variables x and y are used, the ordered pairs are of the form (x, y). We see that $(1, 3)$ is a solution for $y = x + 2$, because if x is replaced by 1 and y by 3, the true numerical statement $3 = 1 + 2$ is obtained. Likewise, $(-2, 0)$ is a solution because $0 = -2 + 2$ is a true statement. We can find an infinite number of pairs of real numbers that satisfy $y = x + 2$ when we arbitrarily choose values for x and then determine corresponding values for y. Let's use a table to record some of the solutions for $y = x + 2$.

Choose x	Determine y from $y = x + 2$	Solutions for $y = x + 2$
0	2	$(0, 2)$
1	3	$(1, 3)$
3	5	$(3, 5)$
5	7	$(5, 7)$
-2	0	$(-2, 0)$
-4	-2	$(-4, -2)$
-6	-4	$(-6, -4)$

Plot the points associated with the ordered pairs from the table above to produce Figure 7.17(a). The straight line that contains the points [Figure 7.17(b)] is called the **graph of the equation $y = x + 2$.**

FIGURE 7.17

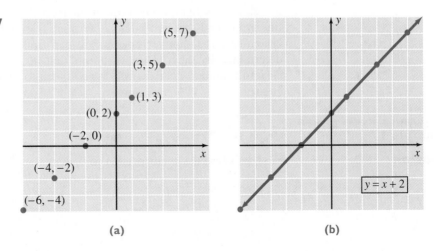

(a)   (b)

Graphing Linear Equations

It is very valuable to be able to recognize the kind of graph that a certain type of equation produces. For example, from previous mathematics courses you probably remember that any equation of the form $Ax + By = C$, where A, B, and C are constants (A and B not both zero) and x and y are variables, is a *linear equation* and *its graph is a straight line.* Two comments about this description of a linear equation should be made. First, the choice of x and y as variables is arbitrary; any two letters could be used to represent the variables. For example, an equation such as $3r + 2s = 9$ is also a linear equation in two variables. In order to avoid constantly changing the labeling of the coordinate axes when graphing equations, we will use the same two variables, x and y, in all equations. Second, the statement "any equation of the form $Ax + By = C$" technically means any equation of the form $Ax + By = C$ or *equivalent* to that form. For example, the equation $y = 2x - 1$ is equivalent to $-2x + y = -1$ and therefore it is linear and produces a straight line graph.

Before we graph some linear equations, let's define in general the **intercepts** of a graph.

> The x-coordinates of the points that a graph has in common with the x-axis are called the x-**intercepts** of the graph. (To compute the x-intercepts, let $y = 0$ and solve for x.)
>
> The y-coordinates of the points that a graph has in common with the y-axis are called the y-**intercepts** of the graph. (To compute the y-intercepts, let $x = 0$ and solve for y.)

Knowing that any equation of the form $Ax + By = C$ produces a straight line graph and that two points determine a straight line makes graphing linear equations a simple process. We can find two points and draw the line determined by them. The two points that involve the intercepts are usually easy to find, and it's generally a good idea to plot a third point to serve as a check.

Example I

Graph $3x - 2y = 6$.

Solution

First, let's find the intercepts. If $x = 0$, then

$$3(0) - 2y = 6$$
$$-2y = 6$$
$$y = -3.$$

Therefore, the point $(0, -3)$ is on the line. If $y = 0$, then

$$3x - 2(0) = 6$$
$$3x = 6$$
$$x = 2.$$

Thus, the point $(2, 0)$ is also on the line. Now let's find a check point. If $x = -2$, then

$$3(-2) - 2y = 6$$
$$-6 - 2y = 6$$
$$-2y = 12$$
$$y = -6.$$

So the point $(-2, -6)$ is also on the line.

In Figure 7.18 the three points are plotted and the graph of $3x - 2y = 6$ is drawn.

FIGURE 7.18

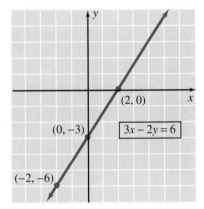

Notice in Example 1 that we did not solve the given equation for y in terms of x or for x in terms of y. Since we know it is a straight line, there is no need for an extensive table of values; thus, there is no need to change the form of the original equation. Furthermore, the point $(-2, -6)$ served as a check point. If it had not been on the line determined by the two intercepts, then we would have known that an error had been made in finding the intercepts.

Example 2

Graph $y = -2x$.

Solution

If $x = 0$, then $y = -2(0) = 0$; so the origin $(0, 0)$ is on the line. Since both intercepts are determined by the point $(0, 0)$, another point is necessary to form the line. Then a third point should be found as a check point. The graph of $y = -2x$ is shown in Figure 7.19.

x	y
0	0
1	−2
−1	2

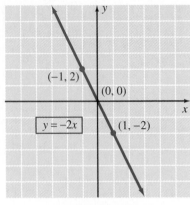

FIGURE 7.19

Example 2 illustrates the general concept that for the form $Ax + By = C$, if $C = 0$, then the line contains the origin. Stated another way, the graph of any equation of the form $y = kx$, where k is any real number, is a straight line that contains the origin.

Example 3

Graph $x = 2$.

Solution

Because we are considering linear equations *in two variables*, the equation $x = 2$ is equivalent to $x + 0(y) = 2$. Any value of y can be used, but the x-value must always be 2. Therefore, some of the solutions are $(2, 0)$, $(2, 1)$, $(2, 2)$, $(2, -1)$, and $(2, -2)$. The graph of $x = 2$ is the vertical line shown in Figure 7.20.

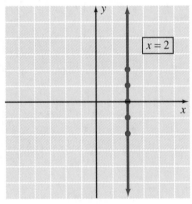

FIGURE 7.20

In general, the graph of any equation of the form $Ax + By = C$, where $A = 0$ or $B = 0$ (but not both), is a line parallel to one of the axes. More specifically, any equation of the form $x = a$, where a is any real number, is a line parallel to the y-axis (a vertical line) that has an x-intercept of a. Any equation of the form $y = b$, where b is a real number, is a line parallel to the x-axis (a horizontal line) that has a y-intercept of b.

Graphing Linear Inequalities

Linear inequalities in two variables are of the form $Ax + By > C$ or $Ax + By < C$, where A, B, and C are real numbers. (Combined linear equality and inequality statements are of the form $Ax + By \geq C$ or $Ax + By \leq C$.) Graphing linear inequalities is almost as easy as graphing linear equations. The following discussion leads to a simple, step-by-step process.

Let's consider the following equation and related inequalities.

$$x + y = 2, \qquad x + y > 2, \qquad x + y < 2$$

FIGURE 7.21

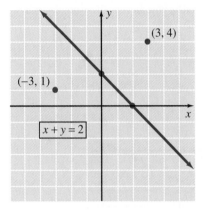

The straight line in Figure 7.21 is the graph of $x + y = 2$. The line divides the plane into two half planes, one above the line and one below the line. For each point in the half plane *above* the line, the ordered pair associated with the point satisfies the inequality $x + y > 2$. For example, the ordered pair (3, 4) produces the true statement $3 + 4 > 2$. Likewise, for each point in the half plane *below* the line, the ordered pair associated with the point satisfies the inequality $x + y < 2$. For example, (−3, 1) produces the true statement $-3 + 1 < 2$.

Now let's use the ideas from the previous discussion to graph some inequalities.

Example 4

Graph $x - 2y > 4$.

Solution

First, graph $x - 2y = 4$ as a dashed line since equality is not included in the expression $x - 2y > 4$ (Figure 7.22).

FIGURE 7.22

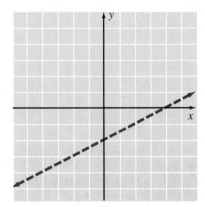

Second, since *all* of the points in a specific half plane satisfy either $x - 2y > 4$ or $x - 2y < 4$, let's try a **test point**. For example, try the origin. Then $x - 2y > 4$ becomes $0 - 2(0) > 4$, which is a false statement. Because the ordered pairs in the half plane containing the origin do not satisfy $x - 2y > 4$, the ordered pairs in the other half plane must satisfy it. Therefore, the graph of $x - 2y > 4$ is the half plane below the line, as indicated by the shaded portion in Figure 7.23.

FIGURE 7.23

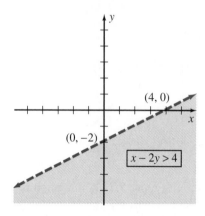

We suggest the following steps for graphing a linear inequality.

1. Graph the corresponding equality. Use a solid line if equality is included in the original statement and a dashed line if equality is not included.

2. Choose a test point not on the line, and substitute its coordinates into the inequality. (The origin is a convenient point if it is not on the line.)

3. The graph of the original inequality is

   (a) the half plane containing the test point if the inequality is satisfied by the point; or

   (b) the half plane not containing the test point if the inequality is not satisfied by the point.

Example 5

Solution

Graph $2x + 3y \geq -6$.

1. Graph $2x + 3y = -6$ as a solid line (Figure 7.24).

2. Choose the origin as a test point. Then $2x + 3y \geq -6$ becomes $2(0) + 3(0) \geq -6$, which is true.

3. Since the test point satisfies the given inequality, all points in the same half plane as the test point satisfy it. Thus, the graph of $2x + 3y \geq -6$ is the line and the half plane above the line (Figure 7.24).

FIGURE 7.24

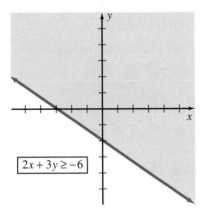

$$2x + 3y \geq -6$$

 Graphing Utilities

The term graphing utility is used in current literature to refer to either a graphics calculator (see Figure 7.25) or a computer with a graphics software package. (We will frequently use the phrase "use a graphics calculator to. . ." to mean a graphics calculator or a computer with the appropriate software.) These devices have a large range of capabilities that allow the user not only to obtain a quick sketch of a graph, but also to study various characteristics of it; for example, x-intercepts, y-intercepts, and turning points of a graph. We will introduce some of these features of graphing utilities as we need them in the text. Since there are so many different types of graphing utilities available, we will use mostly generic terminology and let you consult your user's manual for specific key punching instructions. We also suggest that you study the graphing utility examples in this text even if you do not have access to a graphics calculator or a computer. The examples were chosen to reinforce the concepts we are discussing.

FIGURE 7.25

Graphics Calculators T181 and T182
Texas Instruments Incorporated

Example 6

Use a graphing utility to obtain a graph of the line $2.1x + 5.3y = 7.9$.

Solution

It may be necessary to first solve the equation for y in terms of x.

$$2.1x + 5.3y = 7.9$$

$$5.3y = 7.9 - 2.1x$$

$$y = \frac{7.9 - 2.1x}{5.3}$$

Now, we can enter the expression $\dfrac{7.9 - 2.1x}{5.3}$ for Y_1 and obtain the graph as shown in Figure 7.26.

FIGURE 7.26

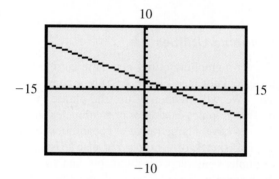

REMARK Note in the solution for Example 6 we said that it may be necessary to first solve the equation for y in terms of x. This is true for most graphics calculators but many computer software packages are designed to handle an equation such as $2.1x + 5.3y = 7.9$ without changing to y-form. △

▼ Problem Set 7.2

For Problems 1–20, graph each of the linear equations.

1. $x + y = 2$

2. $2x + y = 4$

3. $x - 2y = 6$

4. $x - y = -3$

5. $2x + 3y = -6$

6. $3x + 4y = 12$

7. $3x - y = 4$

8. $x - 4y = 3$

9. $2x + 4y = 7$

10. $4x - 5y = 10$

11. $y = -3x + 4$

12. $y = -2x - 1$

13. $y = 3x$

14. $y = -x$

15. $y = -2$

16. $x = 0$

17. $y = \frac{3}{4}x$

18. $y = -\frac{2}{3}x$

19. $-2x + 5y = -10$

20. $-3x + 4y = -12$

For Problems 21–36, graph each of the linear inequalities.

21. $x + y > 3$

22. $2x + y > 2$

23. $3x - y < 3$

24. $x - y < 4$

25. $2x + y \le 4$

26. $x + 3y \le 6$

27. $3x + 2y \ge 6$

28. $5x + 2y \ge 10$

29. $y > -x - 2$

30. $y > -2x + 4$

31. $y \le -x$

32. $y \le x$

33. $x > 2$

34. $y < 3$

35. $x - 2y < 0$

36. $3x - y > 0$

For Problems 37–48, find the coordinates of two points on the given line and then use those coordinates to find the slope of the line.

37. $4x + 5y = 20$ **38.** $5x - 4y = 40$

39. $3x - y = 15$ **40.** $x - 5y = 20$

41. $2x + 3y = 7$ **42.** $3x + 4y = 9$

43. $-3x + 4y = 12$ **44.** $-2x + 5y = -10$

45. $y = -3x - 1$ **46.** $y = -4x + 2$

47. $x = 2y - 1$ **48.** $x = -3y + 6$

THOUGHTS INTO WORDS

49. How do we know that the graph of $y = -3x$ is a straight line that contains the origin?

50. How do we know that the graphs of $2x - 3y = 6$ and $-2x + 3y = -6$ are the same line?

51. What is the graph of the conjunction $x = 2$ and $y = 4$? What is the graph of the disjunction $x = 2$ or $y = 4$? Explain your answers.

52. Why is the point $(-4, 1)$ not a good test point to use when graphing $5x - 2y > -22$?

Further Investigations

From our work with absolute value we know that $|x + y| = 4$ is equivalent to $x + y = 4$ or $x + y = -4$. Therefore, the graph of $|x + y| = 4$ consists of the two lines $x + y = 4$ and $x + y = -4$. Graph each of the following.

53. $|x + y| = 3$

54. $|x - y| = 2$

55. $|2x + y| = 4$

56. $|x + 3y| = 3$

57. $|x - y| < 4$

58. $|x + y| \le 2$

59. $|x + 2y| \ge 2$

60. $|x - 2y| > 4$

By the definition of absolute value, the equation $y = |x| + 2$ becomes $y = x + 2$ for $x \ge 0$ and $y = -x + 2$ for $x < 0$. Therefore, the graph of $y = |x| + 2$ is as shown in Figure 7.27.

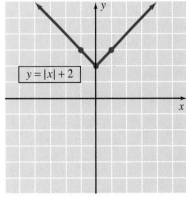

$y = |x| + 2$

FIGURE 7.27

Graph each of the following.

61. $y = |x| - 1$

62. $y = |x - 2|$

63. $|y| = x$

64. $|y| = |x|$

65. $y = 2|x|$

66. $|x| + |y| = 4$

Graphics Calculator Activities

This is the first of many appearances of a group of problems we call Graphics Calculator Activities. These problems were specifically designed for those of you who have access to a graphics calculator or a computer with an appropriate software package. Within the framework of these problems you will be given the opportunity to reinforce concepts we discussed in the text, lay groundwork for concepts we will introduce later in the text, predict shapes and locations of graphs based on your previous graphing experiences, solve problems that are unreasonable or perhaps impossible to solve without a graphing utility, and in general become familiar with the capabilities and limitations of your graphing utility.

This first set of activities will help you get started using your graphing utility. When you do these problems you will review some concepts of this section and lay groundwork for some concepts discussed in the next section.

67. **(a)** Graph $y = 3x + 4$, $y = 2x + 4$, $y = -4x + 4$, and $y = -2x + 4$ on the same set of axes.

(b) Graph $y = \frac{1}{2}x - 3$, $y = 5x - 3$, $y = .1x - 3$, and $y = -7x - 3$ on the same set of axes.

(c) What common characteristic would all lines of the form $y = ax + 2$, where a is any real number, possess?

68. **(a)** Graph $y = 2x - 3$, $y = 2x + 3$, $y = 2x - 6$, and $y = 2x + 5$ on the same set of axes.

(b) Graph $y = -3x + 1$, $y = -3x + 4$, $y = -3x - 2$, and $y = -3x - 5$ on the same set of axes.

(c) Graph $y = \frac{1}{2}x + 3$, $y = \frac{1}{2}x - 4$, $y = \frac{1}{2}x + 5$, and $y = \frac{1}{2}x - 2$ on the same set of axes.

(d) What relationship exists between all lines of the form $y = 3x + b$, where b is any real number?

69. **(a)** Graph $2x + 3y = 4$, $2x + 3y = -6$, $4x + 6y = 7$, and $8x + 12y = -1$ on the same set of axes.

(b) Graph $5x - 2y = 4$, $5x - 2y = -3$, $10x - 4y = 3$, and $15x - 6y = 30$ on the same set of axes.

(c) Graph $x + 4y = 8$, $2x + 8y = 3$, $x - 4y = 6$, and $3x + 12y = 10$ on the same set of axes.

(d) Graph $3x - 4y = 6$, $3x + 4y = 10$, $6x - 8y = 20$, and $6x - 8y = 24$ on the same set of axes.

(e) For each of the following pairs of lines, (a) predict whether or not they are parallel lines, and (b) graph each pair of lines to check your prediction.

(1) $5x - 2y = 10$ and $5x - 2y = -4$

(2) $x + y = 6$ and $x - y = 4$

(3) $2x + y = 8$ and $4x + 2y = 2$

(4) $y = .2x + 1$ and $y = .2x - 4$

(5) $3x - 2y = 4$ and $3x + 2y = 4$

(6) $4x - 3y = 8$ and $8x - 6y = 3$

(7) $2x - y = 10$ and $6x - 3y = 6$

(8) $x + 2y = 6$ and $3x - 6y = 6$

70. You may need to consult the user's manual to see how your graphing utility handles the graphing of inequalities. Then use your graphing utility to check your graphs for Problems 21–36.

7.3 Determining the Equation of a Line

As we stated earlier, there are basically two types of problems in coordinate geometry:

1. Given an algebraic equation, find its geometric graph.
2. Given a set of conditions pertaining to a geometric figure, find its algebraic equation.

Problems of type 1 were our primary concern in the previous section. At this time we want to introduce some problems of type 2 that specifically deal with straight lines. In other words, given certain facts about a line, we need to be able to determine its algebraic equation. Let's consider some examples.

Example 1

Find the equation of the line that has a slope of $\frac{2}{3}$ and contains the point $(1, 2)$.

Solution

First, let's draw the line and record the given information. Then we will choose a point (x, y) that represents any point on the line other than the given point $(1, 2)$. (See Figure 7.28.)

FIGURE 7.28

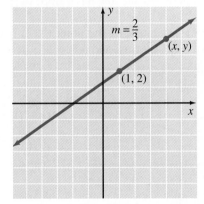

The slope determined by $(1, 2)$ and (x, y) is $\frac{2}{3}$. Thus,

$$\frac{y - 2}{x - 1} = \frac{2}{3}$$

$$2(x - 1) = 3(y - 2)$$

$$2x - 2 = 3y - 6$$

$$2x - 3y = -4.$$

Example 2

Find the equation of the line that contains $(3, 2)$ and $(-2, 5)$.

Solution

First, let's draw the line determined by the given points (Figure 7.29). Since two points are known, the slope can be found.

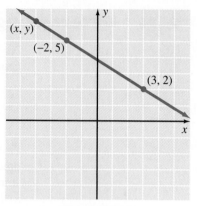

$$m = \frac{y_2 - y_1}{x_2 - x_1} = \frac{3}{-5} = -\frac{3}{5}$$

Now we can use the same approach as in Example 1. We form an equation using a variable point (x, y), one of the two given points, and the slope of $-\frac{3}{5}$.

$$\frac{y - 5}{x + 2} = \frac{3}{-5} \qquad -\frac{3}{5} = \frac{3}{-5}$$

$$3(x + 2) = -5(y - 5)$$

$$3x + 6 = -5y + 25$$

$$3x + 5y = 19$$

FIGURE 7.29

Example 3

Find the equation of the line that has a slope of $\frac{1}{4}$ and a y-intercept of 2.

Solution

A y-intercept of 2 means that the point $(0, 2)$ is on the line (Figure 7.30). We choose a variable point (x, y), and proceed as in the previous examples.

FIGURE 7.30

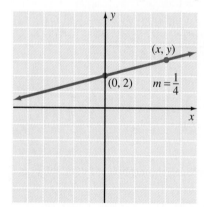

$$\frac{y - 2}{x - 0} = \frac{1}{4}$$

$$1(x - 0) = 4(y - 2)$$

$$x = 4y - 8$$

$$x - 4y = -8$$

Perhaps it would be helpful for you to pause a moment and look back over Examples 1, 2, and 3. Notice that we use the same basic approach in all three situations—that is, we choose a variable point (x, y) and use it to determine the equation that satisfies the conditions given in the problem. You should also recognize that the approach we took in the previous examples can be generalized to produce some special forms of equations of straight lines.

Point-Slope Form

Example 4

Find the equation of the line that has a slope of m and contains the point (x_1, y_1).

Solution

If we choose (x, y) to represent any other point on the line (Figure 7.31) then the slope of the line is given by

$$m = \frac{y - y_1}{x - x_1}, \qquad x \neq x_1$$

where we have

$$y - y_1 = m(x - x_1).$$

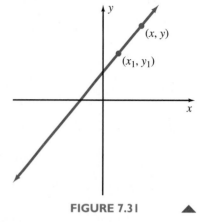

FIGURE 7.31 ▲

We refer to the equation

$$\boxed{y - y_1 = m(x - x_1)}$$

as the **point-slope form** of the equation of a straight line. Instead of using the approach we used in Example 1, we can use the point-slope form to write the equation of a line with a given slope and that contains a given point. For example, the equation of the line that has a slope of $\frac{3}{5}$ and contains the point $(2, 4)$ can be determined as follows. Substituting $(2, 4)$ for (x_1, y_1) and $\frac{3}{5}$ for m in

$$y - y_1 = m(x - x_1),$$

we have

$$y - 4 = \frac{3}{5}(x - 2)$$

$$5(y - 4) = 3(x - 2)$$

$$5y - 20 = 3x - 6$$

$$-14 = 3x - 5y.$$

Slope-Intercept Form

Now consider the equation of the line that has a slope of m and a y-intercept of b (Figure 7.32). A y-intercept of b means that the line contains the point $(0, b)$; therefore, we can use the point-slope form as follows.

FIGURE 7.32

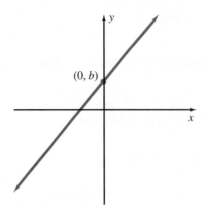

$$y - y_1 = m(x - x_1)$$
$$y - b = m(x - 0)$$
$$y - b = mx$$
$$y = mx + b$$

We refer to the equation

$$\boxed{y = mx + b}$$

as the **slope-intercept form** of the equation of a straight line. We use it for two primary purposes, as Examples 5 and 6 demonstrate.

Example 5

Find the equation of the line that has a slope of $\frac{1}{4}$ and a y-intercept of 2.

Solution

This is a restatement of Example 3, but this time we will use the slope-intercept form ($y = mx + b$) of a line to write its equation. Since $m = \frac{1}{4}$ and $b = 2$, we can substitute these values into $y = mx + b$.

$$y = mx + b$$

$$y = \frac{1}{4}x + 2$$

$4y = x + 8$ 　　　Multiply both sides by 4.

$x - 4y = -8$ 　　　Same result as in Example 3 　　　▲

Example 6

Find the slope of the line when the equation is $3x + 2y = 6$.

Solution

We can solve the equation for y in terms of x and then compare the resulting equation to the slope-intercept form to determine the slope. Thus,

$$3x + 2y = 6$$

$$2y = -3x + 6$$

$$y = -\frac{3}{2}x + 3$$

$$y = -\frac{3}{2}x + 3 \qquad y = mx + b$$

The slope of the line is $-\dfrac{3}{2}$. Furthermore, the y-intercept is 3. ▲

In general, if the equation of a nonvertical line is written in slope-intercept form ($y = mx + b$), the coefficient of x is the slope of the line and the constant term is the y-intercept. (Remember that the concept of slope is not defined for a vertical line.)

Parallel and Perpendicular Lines

Two important relationships between lines and their slopes can help solve certain kinds of problems. We can show that nonvertical parallel lines have the same slope, and that two nonvertical lines are perpendicular if the product of their slopes is -1. (Details for verifying these facts are left to another course.) In other words, if two lines have slopes m_1 and m_2, respectively, then

1. The two lines are parallel if and only if $m_1 = m_2$;
2. The two lines are perpendicular if and only if $(m_1)(m_2) = -1$.

The following examples illustrate the use of these properties.

Example 7

(a) Verify that the graphs of $2x + 3y = 7$ and $4x + 6y = 11$ are parallel lines.
(b) Verify that the graphs of $8x - 12y = 3$ and $3x + 2y = 2$ are perpendicular lines.

Solution

(a) Let's change each equation to slope-intercept form.

$$2x + 3y = 7 \longrightarrow 3y = -2x + 7$$

$$y = -\frac{2}{3}x + \frac{7}{3},$$

$$4x + 6y = 11 \longrightarrow 6y = -4x + 11$$

$$y = -\frac{4}{6}x + \frac{11}{6}$$

$$y = -\frac{2}{3}x + \frac{11}{6}$$

Both lines have a slope of $-\dfrac{2}{3}$ and different y-intercepts. Therefore, the two lines are parallel.

(b) Solving each equation for y in terms of x, we obtain

$$8x - 12y = 3 \longrightarrow -12y = -8x + 3$$

$$y = \frac{8}{12}x - \frac{3}{12}$$

$$y = \frac{2}{3}x - \frac{1}{4},$$

$$3x + 2y = 2 \longrightarrow 2y = -3x + 2$$

$$y = -\frac{3}{2}x + 1.$$

Because $\left(\frac{2}{3}\right)\left(-\frac{3}{2}\right) = -1$ (the product of the two slopes is -1), the lines are perpendicular. ▲

REMARK The statement "the product of two slopes is -1" is equivalent to saying that the two slopes are negative reciprocals of each other; that is, $m_1 = -\frac{1}{m_2}$. △

Example 8 Find the equation of the line that contains the point $(1, 4)$ and is parallel to the line determined by $x + 2y = 5$.

Solution First, let's draw a figure to help in our analysis of the problem (Figure 7.33). Since the line through $(1, 4)$ is to be parallel to the line determined by $x + 2y = 5$, it must have the same slope. So let's find the slope by changing $x + 2y = 5$ to the slope-intercept form.

$$x + 2y = 5$$

$$2y = -x + 5$$

$$y = -\frac{1}{2}x + \frac{5}{2}$$

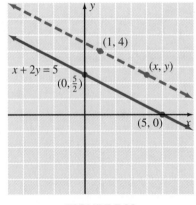

FIGURE 7.33

The slope of both lines is $-\frac{1}{2}$. Now we can choose a variable point (x, y) on the line through $(1, 4)$ and proceed as we did in earlier examples.

$$\frac{y - 4}{x - 1} = \frac{1}{-2}$$

$$1(x - 1) = -2(y - 4)$$

$$x - 1 = -2y + 8$$

$$x + 2y = 9$$

▲

Example 9

Find the equation of the line that contains the point $(-1, -2)$ and is perpendicular to the line determined by $2x - y = 6$.

Solution

Figure 7.34 will help in our analysis of the problem. Because the line through

FIGURE 7.34

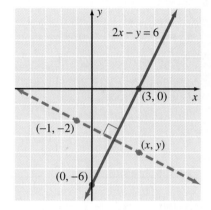

$(-1, -2)$ is to be perpendicular to the line determined by $2x - y = 6$, its slope must be the negative reciprocal of the slope of $2x - y = 6$. So let's find the slope of $2x - y = 6$ by changing it to the slope-intercept form.

$$2x - y = 6$$
$$-y = -2x + 6$$
$$y = 2x - 6$$

The slope is 2.

The slope of the desired line is $-\dfrac{1}{2}$ (the negative reciprocal of 2); we can use a variable point (x, y) and proceed as before.

$$\frac{y + 2}{x + 1} = \frac{1}{-2}$$
$$1(x + 1) = -2(y + 2)$$
$$x + 1 = -2y - 4$$
$$x + 2y = -5$$ ▲

Two forms of equations of straight lines are used extensively. We refer to them as the **standard form** and the **slope-intercept form** and we can describe them as follows.

Standard Form

$$Ax + By = C,$$

where B and C are integers and A is a nonnegative integer (A and B not both zero).

Slope-Intercept Form

$$y = mx + b,$$

where m is a real number that represents the slope and b is a real number that represents the y-intercept.

Problem Set 7.3

For Problems 1–8, write the equation of each line that has the indicated slope and contains the given point. Express final equations in standard form.

1. $m = \dfrac{1}{2}$, $(3, 4)$

2. $m = \dfrac{3}{4}$, $(2, 6)$

3. $m = -\dfrac{5}{6}$, $(-2, 7)$

4. $m = -\dfrac{2}{5}$, $(-1, 4)$

5. $m = 3$, $(-1, -4)$

6. $m = -2$, $(-2, -5)$

7. $m = -\dfrac{3}{2}$, $(2, -8)$

8. $m = \dfrac{4}{3}$, $(3, -4)$

For Problems 9–18, write the equation of each line that contains the given pair of points. Express final equations in standard form.

9. $(1, 3), (5, 7)$
10. $(2, 1), (9, 10)$
11. $(-2, 2), (3, -1)$
12. $(-1, -2), (6, -4)$

13. $(-7, -3), (-2, 4)$
14. $(3, 2), (-2, -7)$
15. $(2, -4), (7, -4)$
16. $(-1, 7), (-1, -4)$
17. $(-3, 0), (0, 6)$
18. $(0, -2), (7, 0)$

For Problems 19–26, write the equation of each line that has the given slope (m) and y-intercept (b). Express final equations in slope-intercept form.

19. $m = \dfrac{2}{7}, b = 3$

20. $m = \dfrac{3}{8}, b = 1$

21. $m = 3, b = -2$

22. $m = 2, b = -\dfrac{1}{2}$

23. $m = -\dfrac{1}{4}, b = -5$

24. $m = -\dfrac{3}{5}, b = 2$

25. $m = -\dfrac{1}{2}, b = \dfrac{2}{3}$

26. $m = -\dfrac{3}{2}, b = \dfrac{7}{5}$

For Problems 27–32, write the equation of each line that satisfies the given conditions. Express final equations in standard form.

27. x-intercept of -3 and y-intercept of 4

28. x-intercept of 2 and slope of $-\dfrac{7}{2}$

29. Contains the point $(3, -5)$ and is parallel to the y-axis

30. Contains the point $(-2, 4)$ and is parallel to the x-axis

31. Contains the point $(4, 7)$ and is perpendicular to the y-axis

32. Contains the origin and the point $\left(1, \dfrac{2}{3}\right)$

For Problems 33–44, determine whether each pair of lines is (a) parallel, (b) perpendicular, or (c) a pair of intersecting lines that are not perpendicular.

33. $y = \dfrac{5}{6}x - 2$

$y = \dfrac{5}{6}x + 8$

34. $y = 4x - 1$

$y = -\dfrac{1}{4}x + 2$

35. $5x - 7y = 8$
$7x + 5y = 9$

36. $x - 2y = 4$
$2x - 4y = 9$

37. $2x + 3y = 9$
$3x + 2y = 4$

38. $4x - 7y = 13$
$7x - 4y = 19$

39. $y = -2x - 5$
$6x + 3y = 14$

40. $x = -1$
$y = 4$

41. $y = 3x$
$y = -3x$

42. $y = x$
$x + y = 0$

43. $x = 2y + 4$
$y = -2x + 7$

44. $y = 2x + 1$
$y = -2x - 3$

For Problems 45–52, write the equation of each line that satisfies the stated conditions. Express final equations in standard form.

45. Contains $(2, 4)$ and is parallel to $5x - y = 12$

46. Contains $(-3, 1)$ and is parallel to $4x - 7y = 2$

47. Contains the origin and is parallel to $3x + 2y = 8$

48. Contains $(-2, -6)$ and is parallel to $5x + 3y = 1$

49. Contains $(2, -2)$ and is perpendicular to $2x + y = 6$

50. Contains $(3, 7)$ and is perpendicular to $x - 4y = 6$

51. Contains $(-4, -1)$ and is perpendicular to $y = -3x$

52. Contains the origin and is perpendicular to $y = \dfrac{1}{2}x + 4$

For Problems 53–56, use the slope-intercept form, $y = mx + b$, to help write the equation of each line that has the given slope and contains the given point. Express final equations in slope-intercept form.

53. $m = \dfrac{1}{5}$, $(2, 6)$

54. $m = \dfrac{3}{7}$, $(-2, 4)$

55. $m = -\dfrac{5}{4}$, $(-2, -3)$

56. $m = -\dfrac{2}{5}$, $(4, -7)$

57. The slope-intercept form of a line can also be used for graphing purposes. Suppose that we want to graph the equation $y = \frac{1}{4}x + 2$. Since the y-intercept is 2, the point $(0, 2)$ is on the line. Furthermore, since the slope is $\frac{1}{4}$, another point can be located by moving 1 unit *up* and 4 units to the *right*. Thus, the point $(4, 3)$ is also on the line. The two points $(0, 2)$ and $(4, 3)$ determine the line. Use the slope-intercept form to graph the following lines.

(a) $y = \frac{2}{3}x + 1$

(b) $y = \frac{3}{5}x - 2$

(c) $y = -\frac{1}{2}x - 3$

(d) $y = -\frac{1}{3}x + 1$

(e) $y = -2x$

(f) $y = 4x$

(g) $y = \frac{1}{2}x + \frac{2}{3}$

(h) $y = \frac{2}{3}x - \frac{3}{4}$

58. Use the concept of slope to verify that $(6, 6)$, $(2, -2)$, $(-8, -5)$, and $(-4, 3)$ are vertices of a parallelogram.

59. Use the concept of slope to verify that the quadrilateral whose vertices are $(0, 7)$, $(-2, -1)$, $(2, -2)$, and $(4, 6)$ is a rectangle.

60. Use the concept of slope to verify that the triangle determined by $(4, 3)$, $(5, 1)$, and $(3, 0)$ is a right triangle.

61. Use the concept of slope to verify that $(1, 1)$, $(3, 7)$, and $(-2, -8)$ lie on the same straight line.

62. We can use linear equations in two variables to describe some real world situations. If two pairs of values are known, then the equation can be determined by the approach we used in Example 2 of this section. For each of the following cases, assume that the relationship can be expressed as a linear equation in two variables, and use the given information to determine the equation. Express the equation in standard form.

(a) A company produces 10 fiberglass shower stalls for $2015 and 15 stalls for $3015. Let y be the cost and x the number of stalls.

(b) A company can produce 6 boxes of candy for $8 and 10 boxes of candy for $13. Let y represent the cost and x the number of boxes of candy.

(c) Two banks on opposite corners of a town square have signs that display the current temperature. One bank displays the temperature in Celsius degrees and the other in degrees Fahrenheit. One day a temperature of 10°C was displayed at the same time as a temperature of 50°F. On another day, a temperature of -5°C was displayed at the same time as a temperature of 23°F. Let y represent the temperature in Fahrenheit and x the temperature in Celsius.

THOUGHTS INTO WORDS

63. What does it mean to say that two points determine a line?

64. Do three points determine a line? Explain your answer.

65. How would you help a friend determine the equation of the line that is perpendicular to $x - 5y = 7$ and contains the point $(5, 4)$?

66. Explain how you would find the slope of the line $y = 4$.

Further Investigations

67. The equation of a line that contains the two points (x_1, y_1) and (x_2, y_2) is $\frac{y - y_1}{x - x_1} = \frac{y_2 - y_1}{x_2 - x_1}$. This form is often referred to as the **two-point form** of the equation of a straight line. Using the two-point form, write the equation of each of the lines that contains the indicated pair of points. Express final equations in standard form.

(a) $(1, 1)$ and $(5, 2)$

(b) $(2, 4)$ and $(-2, -1)$

(c) $(-3, 5)$ and $(3, 1)$

(d) $(-5, 1)$ and $(2, -7)$

68. Let $Ax + By = C$ and $A'x + B'y = C'$ represent two lines. Change each of these equations to slope-intercept form, and then verify each of the following properties.

(a) If $\dfrac{A}{A'} = \dfrac{B}{B'} \neq \dfrac{C}{C'}$, then the lines are parallel.

(b) If $AA' = -BB'$, then the lines are perpendicular.

69. The properties in Problem 68 provide us with another way to write the equation of a line parallel or perpendicular to a given line that contains a given point not on the line. For example, suppose that we want the equation of the line perpendicular to $3x + 4y = 6$ that contains the point $(1, 2)$. The form $4x - 3y = k$, where k is a constant, represents a family of lines perpendicular to $3x + 4y = 6$ because we satisfied the condition $AA' = -BB'$. Therefore, to find the specific line of the family containing $(1, 2)$, we substitute 1 for x and 2 for y to determine k.

$$4x - 3y = k$$
$$4(1) - 3(2) = k$$
$$-2 = k$$

Thus, the equation of the desired line is $4x - 3y = -2$. Use the properties from Problem 68 to write the equation of each of the following lines.

(a) Contains $(1, 8)$ and is parallel to $2x + 3y = 6$

(b) Contains $(-1, 4)$ and is parallel to $x - 2y = 4$

(c) Contains $(2, -7)$ and is perpendicular to $3x - 5y = 10$

(d) Contains $(-1, -4)$ and is perpendicular to $2x + 5y = 12$.

Graphics Calculator Activities

70. Graph $y = x^2$. In the next section we refer to this figure as the <u>basic parabola</u>.

71. **(a)** Graph $y = x^2$, $y = x^2 + 1$, $y = x^2 + 3$, and $y = x^2 + 6$ on the same set of axes.

(b) Graph $y = x^2$, $y = x^2 - 2$, $y = x^2 - 4$, and $y = x^2 - 5$ on the same set of axes.

(c) Make a statement about the graph of $y = x^2 + k$, where k is any nonzero real number.

72. **(a)** Graph $y = x^2$, $y = 2x^2$, $y = 3x^2$, and $y = 4x^2$ on the same set of axes.

(b) Graph $y = x^2$, $y = \dfrac{3}{4}x^2$, $y = \dfrac{1}{2}x^2$, and $y = \dfrac{1}{5}x^2$ on the same set of axes.

(c) Graph $y = x^2$, $y = -x^2$, $y = -3x^2$, and $y = -\dfrac{1}{4}x^2$ on the same set of axes.

(d) Make a statement about the graph of $y = ax^2$, where a is any nonzero real number except one.

73. **(a)** Graph $y = x^2$, $y = (x - 2)^2$, $y = (x - 3)^2$, and $y = (x - 5)^2$ on the same set of axes.

(b) Graph $y = x^2$, $y = (x + 1)^2$, $y = (x + 3)^2$, and $y = (x + 6)^2$ on the same set of axes.

(c) Make a statement about the graph of $y = (x - h)^2$, where h is any nonzero real number.

74. **(a)** Graph $y = x^2$, $y = (x - 2)^2 + 3$, $y = (x + 4)^2 - 2$, and $y = (x - 6)^2 - 4$ on the same set of axes.

(b) Graph $y = x^2$, $y = 2(x + 1)^2 + 4$, $y = 3(x - 1)^2 - 3$, and $y = \dfrac{1}{2}(x - 5)^2 + 2$ on the same set of axes.

(c) Graph $y = x^2$, $y = -(x - 4)^2 - 3$, $y = -2(x + 3)^2 - 1$, and $y = -\dfrac{1}{2}(x - 2)^2 + 6$ on the same set of axes.

(d) Make a statement about the graph of $y = a(x - h)^2 + k$, where a, h, and k are nonzero real numbers.

7.4   Graphing Parabolas

Let's return to the problem of sketching a graph of an algebraic equation in two variables. Suppose that we want the graph of the equation $y = x^2$. First we set up a table that indicates some of the solutions. Then we can plot the points associated with these solutions, as indicated in Figure 7.35(a). Finally, by connecting these points with a smooth curve, we produce Figure 7.35(b).

x	$y = x^2$	Solutions (x, y)
0	0	$(0, 0)$
1	1	$(1, 1)$
2	4	$(2, 4)$
3	9	$(3, 9)$
-1	1	$(-1, 1)$
-2	4	$(-2, 4)$
-3	9	$(-3, 9)$

FIGURE 7.35

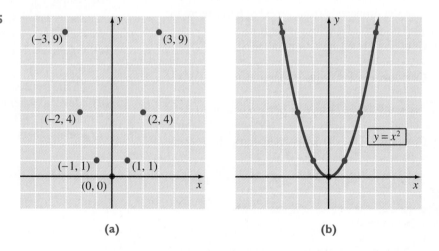

(a)   (b)

The curve shown in Figure 7.35(b) is called a **parabola**. The graph of any equation in the form $y = ax^2 + bx + c$, where a, b, and c are real numbers and $a \neq 0$, is a parabola. As we work with parabolas, we will use the vocabulary in Figure 7.36.

The parabola in Figure 7.35(b) is said to be **symmetric with respect to the y-axis**. That is to say, the y-axis is a line of symmetry. Each half of the curve is a

FIGURE 7.36

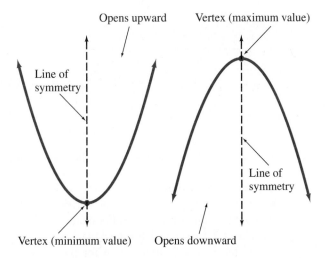

mirror image of the other half through the y-axis. Notice that in the table of values for the equation $y = x^2$, for each ordered pair (x, y) the ordered pair $(-x, y)$ is also a solution. A general test for y-axis symmetry can be stated as follows.

y-Axis Symmetry

The graph of an equation is symmetric with respect to the y-axis if replacing x with $-x$ results in an equivalent equation.

The equation $y = x^2$ exhibits y-axis symmetry because replacing x with $-x$ produces $y = (-x)^2 = x^2$. Likewise, the equations $y = x^2 + 6$, $y = x^4$, and $y = x^4 + 2x^2$ exhibit y-axis symmetry.

Graphing parabolas requires the ability to find the vertex, determine which way the parabola opens, and locate two points on opposite sides of the line of symmetry. Some of this information can be found by comparing the parabolas produced by various types of equations such as $y = x^2 + k$, $y = ax^2$, $y = (x - h)^2$, and $y = a(x - h)^2 + k$ to the **basic parabola** produced by the equation $y = x^2$. Let's first consider some equations of the form $y = x^2 + k$, where k is a constant.

Example 1

Graph $y = x^2 + 1$.

Solution

Let's set up a table of values to compare y-values for $y = x^2 + 1$ to corresponding y-values for $y = x^2$.

x	$y = x^2$	$y = x^2 + 1$
0	0	1
1	1	2
2	4	5
−1	1	2
−2	4	5

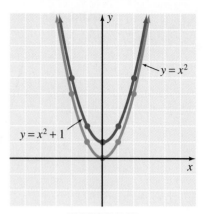

FIGURE 7.37

It should be evident that y-values for $y = x^2 + 1$ are *one larger* than corresponding y-values for $y = x^2$. For example, if $x = 2$, then $y = 4$ for the equation $y = x^2$ but if $x = 2$, then $y = 5$ for the equation $y = x^2 + 1$. Thus, the graph of $y = x^2 + 1$ is the same as the graph of $y = x^2$ but *moved up* 1 unit (Figure 7.37).

▲

Example 2

Graph $y = x^2 - 2$.

Solution

The y-values for $y = x^2 - 2$ are 2 *less than* the corresponding y-values for $y = x^2$, as indicated in the following table.

x	$y = x^2$	$y = x^2 - 2$
0	0	−2
1	1	−1
2	4	2
−1	1	−1
−2	4	2

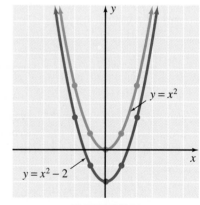

Thus, the graph of $y = x^2 - 2$ is the same as the graph of $y = x^2$, but *moved down* 2 units (Figure 7.38).

FIGURE 7.38 ▲

In general, the graph of a quadratic equation of the form $y = x^2 + k$ is the same as the graph of $y = x^2$ except moved up or down k units depending on whether k is positive or negative. We say that the graph of $y = x^2 + k$ is a **vertical translation** of the graph of $y = x^2$. Also note that equations of the form $y = x^2 + k$ exhibit y-axis symmetry.

Now let's consider some quadratic equations of the form $y = ax^2$, where a is a nonzero constant.

Example 3

Graph $y = 2x^2$.

Solution

Let's again use a table to make some comparisons of y-values.

x	$y = x^2$	$y = 2x^2$
0	0	0
1	1	2
2	4	8
-1	1	2
-2	4	8

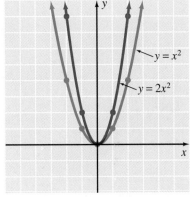

FIGURE 7.39

Obviously, the y-values for $y = 2x^2$ are *twice* the corresponding y-values for $y = x^2$. Thus, the parabola associated with $y = 2x^2$ has the same vertex (the origin) as the graph of $y = x^2$, but it is narrower (Figure 7.39). ▲

Example 4

Graph $y = \frac{1}{2}x^2$.

Solution

The following table indicates some comparisons of y-values.

x	$y = x^2$	$y = \frac{1}{2}x^2$
0	0	0
1	1	$\frac{1}{2}$
2	4	2
-1	1	$\frac{1}{2}$
-2	4	2

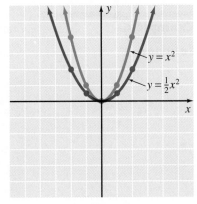

FIGURE 7.40

The y-values for $y = \frac{1}{2}x^2$ are *one-half* of the corresponding y-values for $y = x^2$. Therefore, the graph of $y = \frac{1}{2}x^2$ is wider than the basic parabola (Figure 7.40). ▲

Example 5

Solution

Graph $y = -x^2$.

x	$y = x^2$	$y = -x^2$
0	0	0
1	1	-1
2	4	-4
-1	1	-1
-2	4	-4

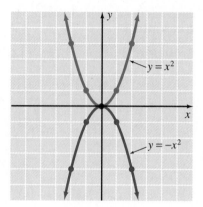

FIGURE 7.41

The y-values for $y = -x^2$ are the *opposites* of the corresponding y-values for $y = x^2$. Thus, the graph of $y = -x^2$ is a reflection across the x-axis of the basic parabola (Figure 7.41).

▲

In general, the graph of a quadratic equation of the form $y = ax^2$ has its vertex at the origin, exhibits y-axis symmetry, and opens upward if a is positive and downward if a is negative. The parabola is narrower than the basic parabola if $|a| > 1$ and wider if $|a| < 1$.

Let's continue our investigation of quadratic equations by considering those of the form $y = (x - h)^2$, where h is a nonzero constant.

Example 6

Solution

Graph $y = (x - 2)^2$.

A fairly extensive table of values illustrates a pattern.

x	$y = x^2$	$y = (x - 2)^2$
-2	4	16
-1	1	9
0	0	4
1	1	1
2	4	0
3	9	1
4	16	4
5	25	9

Notice that $y = (x - 2)^2$ and $y = x^2$ take on the same y-values, *but* for different values of x. More specifically, if $y = x^2$ achieves a certain y-value at x equals a constant, then $y = (x - 2)^2$ achieves that same y-value at x equals the *constant plus two*. In other words, the graph of $y = (x - 2)^2$ is the same as the graph of $y = x^2$ *but moved 2 units to the right* (Figure 7.42).

FIGURE 7.42

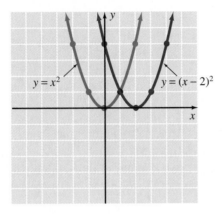

Example 7

Graph $y = (x + 3)^2$.

Solution

x	$y = x^2$	$y = (x + 3)^2$
-3	9	0
-2	4	1
-1	1	4
0	0	9
1	1	16
2	4	25
3	9	36

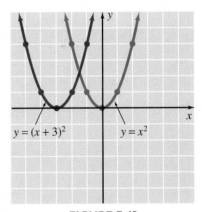

FIGURE 7.43

If $y = x^2$ achieves a certain y-value at x equals a constant, then $y = (x + 3)^2$ achieves the same y-value at x equals that *constant minus* 3. Therefore, the graph of $y = (x + 3)^2$ is the same as the graph of $y = x^2$ *but moved* 3 *units to the left* (Figure 7.43). ▲

In general, the graph of a quadratic equation of the form $y = (x - h)^2$ is the same as the graph of $y = x^2$ but moved to the right h units if h is positive or moved to the left $|h|$ units if h is negative. For example,

$$y = (x - 4)^2 \qquad \longleftarrow \text{Moved to the } \textit{right} \text{ 4 units}$$

$$y = (x + 2)^2 = [x - (-2)]^2. \longleftarrow \text{Moved to the } \textit{left} \text{ 2 units}$$

We say that the graph of $y = (x - h)^2$ is a **horizontal translation** of the graph of $y = x^2$. The following diagram summarizes our work thus far in graphing quadratic equations.

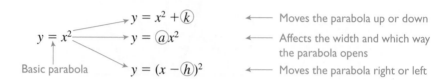

Equations of the form $y = x^2 + k$ and $y = ax^2$ are symmetric about the y-axis. The final two examples of this section put these ideas together to graph a quadratic equation of the form $y = a(x - h)^2 + k$.

Example 8

Graph $y = 2(x - 3)^2 + 1$.

Solution

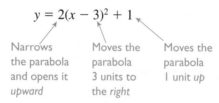

The parabola is shown in Figure 7.44. Two points in addition to the vertex are located to determine the parabola.

FIGURE 7.44

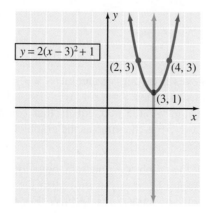

Example 9

Graph $y = -\dfrac{1}{2}(x + 1)^2 - 2$.

Solution

$$y = -\dfrac{1}{2}(x + 1)^2 - 2$$

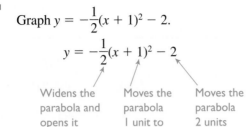

| Widens the parabola and opens it *downward* | Moves the parabola 1 unit to the *left* | Moves the parabola 2 units *down* |

The parabola is shown in Figure 7.45.

FIGURE 7.45

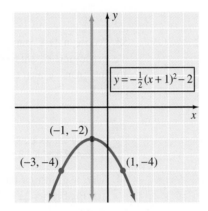

$$y = -\tfrac{1}{2}(x+1)^2 - 2$$

(−1, −2)

(−3, −4)   (1, −4)

Finally, using a graphing utility we can demonstrate some of the ideas of this section. Let's graph $y = x^2$, $y = -3(x - 7)^2 - 1$, $y = 2(x + 9)^2 + 5$, and $y = -.2(x + 8)^2 - 3.5$ on the same set of axes as shown in Figure 7.46. Certainly, Figure 7.46 is consistent with the ideas we presented in this section.

FIGURE 7.46

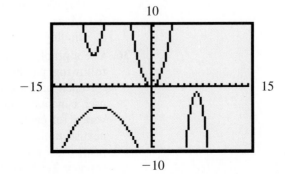

Problem Set 7.4

Graph each of the following parabolas.

1. $y = x^2 - 3$

2. $y = x^2 - 1$

3. $y = x^2 + 2$

4. $y = x^2 + 3$

5. $y = 3x^2$

6. $y = 4x^2$

7. $y = -2x^2$

8. $y = -3x^2$

9. $y = -\frac{1}{2}x^2$

10. $y = -\frac{1}{4}x^2$

11. $y = 2x^2 - 1$

12. $y = 2x^2 + 2$

13. $y = \frac{1}{4}x^2 + 1$

14. $y = \frac{1}{2}x^2 - 3$

15. $y = (x - 3)^2$

16. $y = (x - 1)^2$

17. $y = (x + 1)^2$

18. $y = (x + 4)^2$

19. $y = 2(x - 2)^2$

20. $y = 3(x + 2)^2$

21. $y = -(x + 4)^2$

22. $y = -\frac{1}{2}(x - 3)^2$

23. $y = (x - 1)^2 + 3$

24. $y = (x + 2)^2 - 4$

25. $y = -(x + 3)^2 + 4$

26. $y = -(x - 3)^2 - 1$

27. $y = 3(x - 2)^2 - 3$

28. $y = 2(x + 2)^2 + 1$

29. $y = \frac{1}{3}(x + 1)^2 - 2$

30. $y = \frac{1}{4}(x - 1)^2 + 2$

31. $y = -4(x - 3)^2 + 1$

32. $y = -3(x + 3)^2 - 1$

THOUGHTS INTO WORDS

33. Write a few paragraphs summarizing the ideas we presented in this section for someone who was absent from class the day this material was discussed.

Graphics Calculator Activities

34. Use a graphics calculator to check your graphs for Problems 22–32.

35. (a) Graph $y = x^2 - 12x + 41$ and $y = x^2 + 12x + 41$ on the same set of axes. What relationship seems to exist between the two graphs?

(b) Graph $y = x^2 - 8x + 22$ and $y = -x^2 + 8x - 22$ on the same set of axes. What relationship seems to exist between the two graphs?

(c) Graph $y = x^2 + 10x + 29$ and $y = -x^2 + 10x - 29$ on the same set of axes. What relationship seems to exist between the two graphs?

(d) Make a statement that summarizes your findings in parts (a) through (c).

36. Use a graphics calculator to graph each of the following parabolas and determine the vertex for each one. (You may need to consult your user's manual.) Draw a sketch and record the vertex for each one to use when you study the next section.

(a) $y = x^2 + 6x + 8$ **(b)** $y = x^2 - 3x - 1$

(c) $y = 2x^2 + 8x + 9$ **(d)** $y = -3x^2 + 6x - 5$

7.5 More Parabolas and Some Circles

We are now ready to graph quadratic equations of the form $y = ax^2 + bx + c$, where a, b, and c are real numbers and $a \neq 0$. The general approach is to change equations of the form $y = ax^2 + bx + c$ to the form $y = a(x - h)^2 + k$. Then we can proceed to graph them as we did in the previous section. The process of *completing the square* serves as the basis for making the change in the form of the equations. Let's consider some examples to indicate the details.

Example 1

Solution

Graph $y = x^2 + 6x + 8$.

$$y = x^2 + 6x + 8$$

$$y = (x^2 + 6x \qquad) + 8 \qquad \text{Add 9, which is the square of one-half of the coefficient of } x.$$

$$y = (x^2 + 6x + 9) + 8 - 9 \qquad \text{Subtract 9 to compensate for the } 9 \text{ that was added.}$$

$$\qquad\qquad\qquad\qquad\qquad\qquad x^2 + 6x + 9 = (x + 3)^2$$

$$y = (x + 3)^2 - 1$$

The graph of $y = (x + 3)^2 - 1$ is the basic parabola moved three units to the left and one unit down (Figure 7.47).

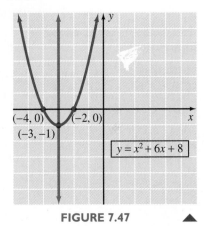

$(-4, 0)$ $(-2, 0)$

$(-3, -1)$

$y = x^2 + 6x + 8$

FIGURE 7.47 ▲

Example 2

Solution

Graph $y = x^2 - 3x - 1$.

$$y = x^2 - 3x - 1$$

$$y = (x^2 - 3x \qquad) - 1$$

$$y = \left(x^2 - 3x + \frac{9}{4}\right) - 1 - \frac{9}{4} \qquad \text{Add and subtract } \frac{9}{4}, \text{ which is the}$$

square of one-half of the coefficient of x.

$$y = \left(x - \frac{3}{2}\right)^2 - \frac{13}{4}$$

The graph of $y = \left(x - \dfrac{3}{2}\right)^2 - \dfrac{13}{4}$ is the basic parabola moved $1\dfrac{1}{2}$ units to the right and $3\dfrac{1}{4}$ units down (Figure 7.48).

FIGURE 7.48

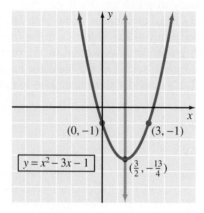

$(0, -1)$ $(3, -1)$

$y = x^2 - 3x - 1$

$\left(\dfrac{3}{2}, -\dfrac{13}{4}\right)$

If the coefficient of x^2 is not 1, then a slight adjustment has to be made before we apply the process of completing the square. The next two examples illustrate this situation.

Example 3

Graph $y = 2x^2 + 8x + 9$.

Solution

$y = 2x^2 + 8x + 9$

$y = 2(x^2 + 4x \qquad) + 9$ Factor a 2 from the first two terms on the right side.

$y = 2(x^2 + 4x + 4) + 9 - 8$ Add 4 inside the parentheses, which is the square of one-half of the coefficient of x.

Subtract 8 to compensate for the 4 added inside the parentheses times the factor of 2.

$y = 2(x + 2)^2 + 1$

See Figure 7.49 for the graph of

$y = 2(x + 2)^2 + 1.$

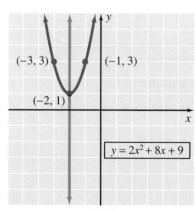

$(-3, 3)$ $(-1, 3)$

$(-2, 1)$

$y = 2x^2 + 8x + 9$

FIGURE 7.49

Example 4

Graph $y = -3x^2 + 6x - 5$.

Solution

$$y = -3x^2 + 6x - 5$$

$$y = -3(x^2 - 2x \qquad) - 5 \qquad \text{Factor } -3 \text{ from the first two terms on the right side.}$$

$$y = -3(x^2 - 2x + 1) - 5 + 3 \qquad \text{Add I inside the parentheses to complete the square.}$$

Add 3 to compensate for the I added inside the parentheses times the factor of -3.

$$y = -3(x - 1)^2 - 2$$

The graph of $y = -3(x - 1)^2 - 2$ is drawn in Figure 7.50.

FIGURE 7.50

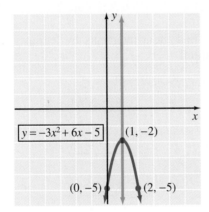

$y = -3x^2 + 6x - 5$ $(1, -2)$

$(0, -5)$ $(2, -5)$

Parabolas possess various useful properties. For example, if a parabola is rotated about its axis, a parabolic surface is formed and such surfaces are used for light and sound reflectors. A projectile fired into the air will follow the curvature of a parabola. The "trend line" of profit and cost functions sometimes follows a parabolic curve. (We will study some of these applications in more detail in the next chapter.) In most applications of the parabola, we are primarily interested in the x-intercepts and the vertex. Let's find the x-intercepts and the vertex in the next examples.

Example 5

Find the x-intercepts and the vertex for each of the following parabolas.

(**a**) $y = -x^2 + 11x - 18$
(**b**) $y = x^2 - 8x - 3$

Solutions

(**a**) To find the x-intercepts, let $y = 0$ and solve the resulting equation.

$$-x^2 + 11x - 18 = 0$$

$$x^2 - 11x + 18 = 0$$

$$(x - 2)(x - 9) = 0$$

$$x - 2 = 0 \quad \text{or} \quad x - 9 = 0$$

$$x = 2 \quad \text{or} \quad x = 9$$

To find the vertex, let's complete the square on x.

$$y = -x^2 + 11x - 18$$
$$= -(x^2 - 11x \quad\quad) - 18$$
$$= -\left(x^2 - 11x + \frac{121}{4}\right) - 18 + \frac{121}{4}$$
$$= -\left(x - \frac{11}{2}\right)^2 + \frac{49}{4}$$

Therefore, the x-intercepts are 2 and 9, and the vertex is at $\left(\dfrac{11}{2}, \dfrac{49}{4}\right)$.

(b) Let's find the x-intercepts by letting $y = 0$ and solving the resulting equation.

$$x^2 - 8x - 3 = 0$$

$$x = \frac{8 \pm \sqrt{64 + 12}}{2}$$

$$x = \frac{8 \pm \sqrt{76}}{2}$$

$$x = \frac{8 \pm 2\sqrt{19}}{2}$$

$$x = 4 \pm \sqrt{19}$$

To find the vertex, we need to complete the square on x.

$$y = x^2 - 8x - 3$$
$$= x^2 - 8x + 16 - 3 - 16$$
$$= (x - 4)^2 - 19$$

Therefore, the x-intercepts are $4 + \sqrt{19}$ and $4 - \sqrt{19}$, and the vertex is at $(4, -19)$. ▲

Circles

When we apply the distance formula that we developed in Section 7.1,

$$d = \sqrt{(x_2 - x_1)^2 + (y_2 - y_1)^2},$$

to the definition of a circle, we get what is known as the **standard equation of a circle**. We start with a precise definition of a circle.

DEFINITION 7.2

A **circle** is the set of all points in a plane equidistant from a given fixed point called the **center**. A line segment determined by the center and any point on the circle is called a **radius**.

Let's consider a circle that has a radius of length r and a center at (h, k) on a coordinate system (Figure 7.51). For any point P on the circle with coordinates (x, y), the length of a radius (denoted by r) can be expressed as

$$r = \sqrt{(x - h)^2 + (y - k)^2}.$$

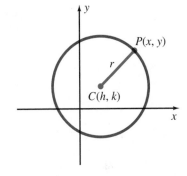

FIGURE 7.51

Thus, squaring both sides of the equation, we obtain the **standard form of the equation of a circle**,

$$(x - h)^2 + (y - k)^2 = r^2.$$

The standard form of the equation of a circle can be used to solve the two basic kinds of problems:

1. Given the coordinates of the center and the length of a radius of a circle, find its equation;
2. Given the equation of a circle, find its center and the length of a radius.

Let's look at some examples of each problem.

Example 6

Write the equation of a circle that has its center at $(3, -5)$ and a radius of length 6 units.

Solution

Let's substitute 3 for h, -5 for k, and 6 for r into the standard form.

$$(x - h)^2 + (y - k)^2 = r^2 \quad \text{becomes} \quad (x - 3)^2 + (y + 5)^2 = 6^2,$$

which can be simplified as follows.

$$(x - 3)^2 + (y + 5)^2 = 6^2$$
$$x^2 - 6x + 9 + y^2 + 10y + 25 = 36$$
$$x^2 + y^2 - 6x + 10y - 2 = 0$$

▲

Notice that in Example 6 we simplified the equation to the form $x^2 + y^2 + Dx + Ey + F = 0$, where D, E, and F are integers. This is another form commonly used in work with circles.

Example 7

Graph $x^2 + y^2 + 4x - 6y + 9 = 0$.

Solution

This equation is of the form $x^2 + y^2 + Dx + Ey + F = 0$, so its graph is a circle. We can change the given equation into the form $(x - h)^2 + (y - k)^2 = r^2$ by completing the square on x and on y as follows.

$$x^2 + y^2 + 4x - 6y + 9 = 0$$

$$(x^2 + 4x \qquad) + (y^2 - 6y \qquad) = -9$$

$$(x^2 + 4x + 4) + (y^2 - 6y + 9) = -9 + 4 + 9$$

Add 4 to complete the square on x.

Add 9 to complete the square on y.

Add 4 and 9 to compensate for the 4 and 9 added on the left side.

Then,

$$(x + 2)^2 + (y - 3)^2 = 4$$

$$(x - (-2))^2 + (y - 3)^2 = 2^2.$$

$$h \qquad k \qquad r$$

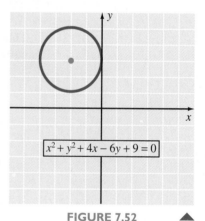

The center of the circle is at $(-2, 3)$, and the length of a radius is 2 (Figure 7.52).

$$x^2 + y^2 + 4x - 6y + 9 = 0$$

FIGURE 7.52   ▲

Examples 6 and 7 illustrate that both forms, $(x - h)^2 + (y - k)^2 = r^2$ and $x^2 + y^2 + Dx + Ey + F = 0$, play an important role in solving problems involving circles.

Finally, we need to recognize that the standard form of a circle that has its center at the origin is $x^2 + y^2 = r^2$. This is the result of letting $h = 0$ and $k = 0$ in the general standard form.

$$(x - h)^2 + (y - k)^2 = r^2$$

$$(x - 0)^2 + (y - 0)^2 = r^2$$

$$x^2 + y^2 = r^2$$

Thus, by inspection we can recognize that $x^2 + y^2 = 9$ is a circle with its center at the origin; the length of the radius is 3 units. Furthermore, the equation of a circle that has its center at the origin and a radius of length 6 units is $x^2 + y^2 = 36$.

When using a graphing utility to graph a circle we may need to solve the equation for y in terms of x. This will produce two equations that can be graphed on the same set of axes. Furthermore, as with any graph it may be necessary to change the boundaries on x or y (or both) to obtain a complete graph. Let's consider an example.

Example 8

Use a graphing utility to graph $x^2 - 40x + y^2 + 351 = 0$.

Solution

First, we need to solve for y in terms of x.

$$x^2 - 40x + y^2 + 351 = 0$$

$$y^2 = -x^2 + 40x - 351$$

$$y = \pm \sqrt{-x^2 + 40x - 351}$$

Now we can make the following assignments.

$$Y_1 = \sqrt{-x^2 + 40x - 351}$$

$$Y_2 = -Y_1$$

(Note that we assigned Y_2 in terms of Y_1. By doing this we avoid repetitive key strokes and hopefully we reduce the chance for errors. You may need to consult your user's manual for instructions on how to key stroke $-Y_1$.) Figure 7.53 shows the graph.

FIGURE 7.53

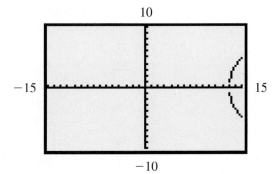

Since we know from the original equation that this graph should be a circle, we need to make some adjustments on the boundaries in order to get a complete graph. This can be done by completing the square on the original equation to change its form to $(x - 20)^2 + y^2 = 49$ or simply by a trial-and-error process. By changing the boundaries on x such that $-15 \leq x \leq 30$, we obtain Figure 7.54.

FIGURE 7.54

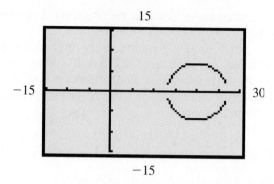

Problem Set 7.5

For Problems 1–30, graph each of the parabolas.

1. $y = x^2 - 2x + 4$

2. $y = x^2 - 4x + 3$

3. $y = x^2 + 4x$

4. $y = x^2 + 6x + 11$

5. $y = x^2 - 6x + 7$

6. $y = x^2 + 2x$

7. $y = 4x^2 - 1$

8. $y = 2x^2 - 4$

9. $y = x^2 + 7x + 14$

10. $y = x^2 - x - 1$

11. $y = x^2 - 5x + 3$

12. $y = x^2 + 3x + 1$

13. $y = 2x^2 - 4x - 1$

14. $y = 2x^2 - 12x + 19$

15. $y = 3x^2 + 12x + 14$

16. $y = 4x^2 + 32x + 62$

17. $y = x^2 + 6x + 9$

18. $y = x^2 - 10x + 25$

19. $y = 2x^2 - x + 2$

20. $y = 3x^2 + 2x + 1$

21. $y = 3x^2 - x - 1$

22. $y = 2x^2 + 3x + 1$

23. $y = -2x^2 + 8x - 5$

24. $y = -2x^2 - 4x - 3$

25. $y = -x^2 - 8x - 18$

26. $y = -x^2 + 10x - 21$

27. $y = -3x^2 - 6x$

28. $y = -3x^2 + 6x$

29. $y = 4x^2 + 24x + 32$

30. $y = 4x^2 + 40x + 97$

For each parabola in Problems 31–42, find the x-intercepts and the vertex.

31. $y = x^2 - 8x + 15$

32. $y = x^2 - 16x + 63$

33. $y = 2x^2 - 28x + 96$

34. $y = 3x^2 - 60x + 297$

35. $y = -x^2 + 10x - 24$

36. $y = -2x^2 + 36x - 160$

37. $y = x^2 - 14x + 44$

38. $y = x^2 - 18x + 68$

39. $y = -x^2 + 9x - 21$

40. $y = 2x^2 + 3x + 3$

41. $y = -4x^2 + 4x + 4$

42. $y = -2x^2 + 3x + 7$

For Problems 43–52, write the equation of each circle with the given center and length of a radius. Express the final equation in the form $x^2 + y^2 + Dx + Ey + F = 0$.

43. center at $(3, 5)$ and $r = 2$

44. center at $(4, 7)$ and $r = 1$

45. center at $(-2, 1)$ and $r = 3$

46. center at $(1, -1)$ and $r = 4$

47. center at $(0, 0)$ and $r = \sqrt{3}$
48. center at $(0, 0)$ and $r = 3\sqrt{2}$
49. center at $(-2, -5)$ and $r = 2\sqrt{3}$

50. center at $(-4, -2)$ and $r = \sqrt{5}$

51. center at $(0, -3)$ and $r = 6$
52. center at $(-4, 0)$ and $r = 4$

For Problems 53–64, find the center and the length of a radius of each circle.

53. $x^2 + y^2 - 2x - 8y + 8 = 0$
54. $x^2 + y^2 - 10x - 4y + 25 = 0$
55. $x^2 + y^2 - 6x + 4y - 23 = 0$
56. $x^2 + y^2 + 8x - 2y - 32 = 0$
57. $x^2 + y^2 = 24$
58. $x^2 + y^2 = 32$
59. $x^2 + y^2 + 4x + 12y + 15 = 0$
60. $x^2 + y^2 + 2x + 10y - 23 = 0$
61. $x^2 + y^2 + 4x - 3 = 0$
62. $x^2 + y^2 - 2y - 5 = 0$
63. $x^2 + y^2 + 6x - 8y = 0$
64. $x^2 + y^2 - 12x + 16y = 0$

65. Find the equation of the circle that passes through the origin and has its center at $(0, -5)$.

66. Find the equation of the circle that passes through the origin and has its center at $(6, 0)$.

67. Find the equation of the circle that passes through the origin and has its center at $(-4, -3)$.

68. Find the equation of the circle that passes through the origin and has its center at $(-8, 15)$.

69. What is the graph of $x^2 + y^2 = -4$? Explain your answer.

70. On which axis is the center of the circle $x^2 + y^2 - 8y + 7 = 0$? Defend your answer.

71. Give a step-by-step description of how you would help someone graph the parabola $y = 2x^2 - 12x + 9$.

72. The discriminant of the quadratic equation $x^2 + 4x + 5 = 0$ is -4. How do we know that the parabola $y = x^2 + 4x + 5$ has no x-intercepts?

Further Investigations

73. If an object is thrown upward with an initial velocity of 32 feet per second, then its height (h) at a time of t seconds is given by the following equation.

$$h = 32t - 16t^2$$

Label the horizontal axis t and the vertical axis h, and graph the equation

$$h = 32t - 16t^2.$$

74. The area of a rectangle that has a perimeter of 12 meters is given by the equation

$$A = l(6 - l),$$

where l represents the length in meters and A the area in square meters. Label the horizontal axis l and the vertical axis A, and graph the equation $A = l(6 - l)$.

75. The points (x, y) and (y, x) are mirror images of each other across the line $y = x$. Therefore, by interchanging x and y in the equation $y = ax^2 + bx + c$, we obtain the equation of its mirror image across the line $y = x$—namely, $x = ay^2 + by + c$. Thus to graph $x = y^2 + 2$, we can first graph $y = x^2 + 2$ and then reflect it across the line $y = x$ as indicated in Figure 7.55.

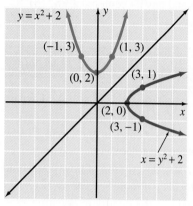

FIGURE 7.55

Graph each of the following parabolas.
(a) $x = y^2$
(b) $x = -y^2$
(c) $x = y^2 - 1$
(d) $x = -y^2 + 3$
(e) $x = -2y^2$
(f) $x = 3y^2$
(g) $x = y^2 + 4y + 7$
(h) $x = y^2 - 2y - 3$

76. Expanding $(x - h)^2 + (y - k)^2 = r^2$, we obtain $x^2 - 2hx + h^2 + y^2 - 2ky + k^2 - r^2 = 0$. Comparing this result to the form $x^2 + y^2 + Dx + Ey + F = 0$, we see that $D = -2h$, $E = -2k$, and $F = h^2 + k^2 - r^2$. Therefore, the center and length of a radius of a circle can be found by using $h = \dfrac{D}{-2}$, $k = \dfrac{E}{-2}$, and $r = \sqrt{h^2 + k^2 - F}$. Use these relationships to find the center and length of a radius of each of the following circles.
(a) $x^2 + y^2 - 2x - 8y + 8 = 0$
(b) $x^2 + y^2 + 4x - 14y + 49 = 0$
(c) $x^2 + y^2 + 12x + 8y - 12 = 0$
(d) $x^2 + y^2 - 16x + 20y + 115 = 0$
(e) $x^2 + y^2 - 12y - 45 = 0$
(f) $x^2 + y^2 + 14x = 0$

Graphics Calculator Activities

77. Use a graphics calculator to check your graphs for Problems 13–30.

78. Graph each of the following parabolas and circles. Be sure to set your boundaries so that you get a complete graph.
(a) $x^2 + 24x + y^2 + 135 = 0$
(b) $y = x^2 - 4x + 18$
(c) $x^2 + y^2 - 18y + 56 = 0$
(d) $x^2 + y^2 + 24x + 28y + 336 = 0$
(e) $y = -3x^2 - 24x - 58$
(f) $y = x^2 - 10x + 3$

79. For each of the following parabolas use a graphics calculator to graph the parabola, and use the "trace function" to help estimate the x-intercepts and the vertex. Then use the approach of Example 5 to find the x-intercepts and the vertex.
(a) $y = x^2 - 6x + 3$
(b) $y = x^2 - 18x + 66$

(c) $y = -x^2 + 8x - 3$
(d) $y = -x^2 + 24x - 129$

(e) $y = 14x^2 - 7x + 1$
(f) $y = -\dfrac{1}{2}x^2 + 5x - \dfrac{17}{2}$

7.6   Ellipses and Hyperbolas

In the previous section we found that the graph of the equation $x^2 + y^2 = 36$ is a circle of radius 6 units with its center at the origin. Generally, any equation of the form $Ax^2 + By^2 = C$, where $A = B$ and A, B, and C are nonzero constants having the same sign, is a circle with the center at the origin. For example, $3x^2 + 3y^2 = 12$ is equivalent to $x^2 + y^2 = 4$ (divide both sides of the given equation by 3), and thus it is a circle of radius 2 units with its center at the origin.

The general equation $Ax^2 + By^2 = C$ can be used to describe other geometric figures if the restrictions on A and B are changed. For example, if A, B, and C are of the same sign but $A \neq B$, then the graph of the equation $Ax^2 + By^2 = C$ is an ellipse. Let's consider two examples.

Example 1

Graph $4x^2 + 25y^2 = 100$.

Solution

Let's find the x- and y-intercepts. Let $x = 0$, then

$$4(0)^2 + 25y^2 = 100$$

$$25y^2 = 100$$

$$y^2 = 4$$

$$y = \pm 2.$$

Thus, the points $(0, 2)$ and $(0, -2)$ are on the graph. Let $y = 0$, then

$$4x^2 + 25(0)^2 = 100$$

$$4x^2 = 100$$

$$x^2 = 25$$

$$x = \pm 5.$$

Thus, the points $(5, 0)$ and $(-5, 0)$ are also on the graph. Plotting the four points we have and knowing that it is an ellipse, we can make a pretty good sketch of the figure (Figure 7.56).

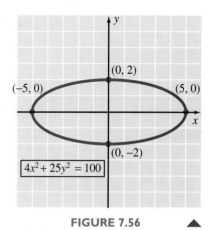

FIGURE 7.56   ▲

In Figure 7.56, the line segment with endpoints at $(-5, 0)$ and $(5, 0)$ is called the **major axis** of the ellipse. The shorter line segment with endpoints at $(0, -2)$ and $(0, 2)$ is called the **minor axis**. Establishing the endpoints of the major and minor axes provides a basis for sketching an ellipse.

Example 2

Graph $9x^2 + 4y^2 = 36$.

Solution

Again let's find the x- and y-intercepts. Let $x = 0$, then

$$9(0)^2 + 4y^2 = 36$$

$$4y^2 = 36$$

$$y^2 = 9$$

$$y = \pm 3.$$

Thus, the points $(0, 3)$ and $(0, -3)$ are on the graph. Let $y = 0$, then

$$9x^2 + 4(0)^2 = 36$$

$$9x^2 = 36$$

$$x^2 = 4$$

$$x = \pm 2.$$

Thus, the points $(2, 0)$ and $(-2, 0)$ are also on the graph. The ellipse is sketched in Figure 7.57.

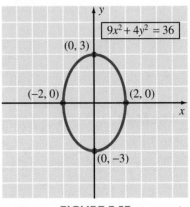

FIGURE 7.57 ▲

In Figure 7.57 the major axis has endpoints at $(0, -3)$ and $(0, 3)$, and the minor axis has endpoints at $(-2, 0)$ and $(2, 0)$. The ellipses in Figures 7.56 and 7.57 are symmetrical about the x-axis and about the y-axis. In other words, both the x-axis and the y-axis serve as lines of symmetry.

Hyperbolas

The graph of an equation of the form $Ax^2 + By^2 = C$, where A, B, and C are nonzero constants and A and B are of unlike signs, is a **hyperbola**. The next two examples illustrate the graphing of hyperbolas.

Example 3

Graph $x^2 - y^2 = 9$.

Solution

If we let $y = 0$, we obtain

$$x^2 - 0^2 = 9$$

$$x^2 = 9$$

$$x = \pm 3.$$

Thus, the points $(3, 0)$ and $(-3, 0)$ are on the graph. If we let $x = 0$, we obtain

$$0^2 - y^2 = 9$$

$$-y^2 = 9$$

$$y^2 = -9.$$

Since $y^2 = -9$ has no real number solutions, there are no points of the y-axis on this graph. That is to say, the graph does not intersect the y-axis. Now let's solve the given equation for y so as to have a more convenient form for finding other solutions.

$$x^2 - y^2 = 9$$
$$-y^2 = 9 - x^2$$
$$y^2 = x^2 - 9$$
$$y = \pm\sqrt{x^2 - 9}$$

Since the radicand, $x^2 - 9$, must be nonnegative, the values chosen for x must be greater than or equal to 3, or less than or equal to -3. With this in mind, we can form the following table of values.

x	y	
3	0	
-3	0	Intercepts
4	$\pm\sqrt{7}$	
-4	$\pm\sqrt{7}$	
5	± 4	
-5	± 4	Other points

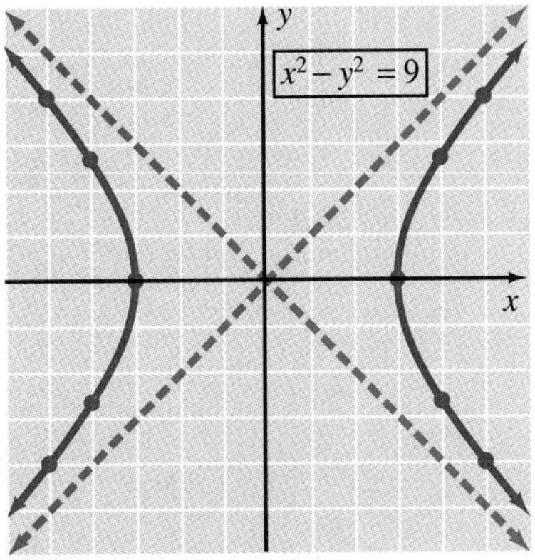

FIGURE 7.58

These points are plotted and the hyperbola is drawn in Figure 7.58. (This graph is also symmetrical about both axes.)

Notice the dashed lines in Figure 7.58; they are called **asymptotes**. Each branch of the hyperbola approaches one of these lines but does not intersect it. Therefore, the ability to sketch the asymptotes of a hyperbola is very helpful in graphing the hyperbola. Fortunately, the equations of the asymptotes are easy to determine. They can be found by replacing the constant term in the given equation of the hyperbola with 0 and solving for y. (The reason this works will become evident in a later course.) Thus, for the hyperbola in Example 3 we obtain

$$x^2 - y^2 = 0$$
$$y^2 = x^2$$
$$y = \pm x.$$

So, the two lines $y = x$ and $y = -x$ are the asymptotes indicated by the dashed lines in Figure 7.58.

Example 4

Solution

Graph $y^2 - 5x^2 = 4$.

If we let $x = 0$, we obtain

$$y^2 - 5(0)^2 = 4$$
$$y^2 = 4$$
$$y = \pm 2.$$

The points $(0, 2)$ and $(0, -2)$ are on the graph. If we let $y = 0$, we obtain

$$0^2 - 5x^2 = 4$$
$$-5x^2 = 4$$
$$x^2 = -\frac{4}{5}.$$

Since $x^2 = -\dfrac{4}{5}$ has no real number solutions, we know that this hyperbola does not intersect the x-axis. Solving the given equation for y yields

$$y^2 - 5x^2 = 4$$
$$y^2 = 5x^2 + 4$$
$$y = \pm\sqrt{5x^2 + 4}.$$

The following table shows some additional solutions for the equation.

x	y	
0	2	
0	-2	Intercepts
1	±3	
-1	±3	
2	$\pm\sqrt{24}$	
-2	$\pm\sqrt{24}$	Other points

The equations of the asymptotes are determined as follows.

$$y^2 - 5x^2 = 0$$
$$y^2 = 5x^2$$
$$y = \pm\sqrt{5}x$$

Sketching the asymptotes and plotting the points listed in the table of values helps to determine the hyperbola in Figure 7.59. (Notice that this hyperbola is also symmetrical about the x-axis and the y-axis.)

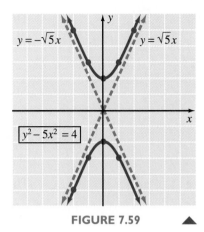

FIGURE 7.59

Other Ellipses and Hyperbolas

The graphing of ellipses and hyperbolas whose centers are not at the origin can be done in much the same way as we handled circles in Section 7.5. The final two examples of this section illustrate these ideas.

Example 5

Graph $4x^2 + 24x + 9y^2 - 36y + 36 = 0$.

Solution

Let's complete the square on x and y as follows.

$$4x^2 + 24x + 9y^2 - 36y + 36 = 0$$

$$4(x^2 + 6x + \underline{}) + 9(y^2 - 4y + \underline{}) = -36$$

$$4(x^2 + 6x + 9) + 9(y^2 - 4y + 4) = -36 + 36 + 36$$

$$4(x + 3)^2 + 9(y - 2)^2 = 36$$

$$4(x - (-3))^2 + 9(y - 2)^2 = 36$$

Since 4 and 9 are of the same sign, but not equal, the graph is an ellipse. The center of the ellipse is at $(-3, 2)$. The equation $4(x + 3)^2 + 9(y - 2)^2 = 36$ can be used to find the endpoints of the major and minor axes as follows. Let $y = 2$, then

$$4(x + 3)^2 + 9(2 - 2)^2 = 36$$

$$4(x + 3)^2 = 36$$

$$(x + 3)^2 = 9$$

$$x + 3 = \pm 3$$

$$x + 3 = 3 \quad \text{or} \quad x + 3 = -3$$

$$x = 0 \quad \text{or} \quad x = -6.$$

The endpoints of the major axis are at $(0, 2)$ and $(-6, 2)$. Let $x = -3$, then

$$4(-3 + 3)^2 + 9(y - 2)^2 = 36$$
$$9(y - 2)^2 = 36$$
$$(y - 2)^2 = 4$$
$$y - 2 = \pm 2$$
$$y - 2 = 2 \quad \text{or} \quad y - 2 = -2$$
$$y = 4 \quad \text{or} \quad y = 0.$$

The endpoints of the minor axis are at $(-3, 4)$ and $(-3, 0)$. The ellipse is sketched in Figure 7.60.

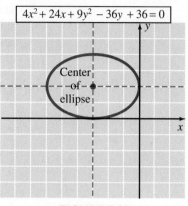

$$4x^2 + 24x + 9y^2 - 36y + 36 = 0$$

Center of ellipse

FIGURE 7.60   ▲

Example 6

Graph $4x^2 - 8x - y^2 - 4y - 16 = 0$.

Solution

Completing the square on x and y, we obtain

$$4x^2 - 8x - y^2 - 4y - 16 = 0$$
$$4(x^2 - 2x + \underline{\ \ }) - (y^2 + 4y + \underline{\ \ }) = 16$$
$$4(x^2 - 2x + 1) - (y^2 + 4y + 4) = 16 + 4 - 4$$
$$4(x - 1)^2 - (y + 2)^2 = 16$$
$$4(x - 1)^2 - 1(y - (-2))^2 = 16.$$

Since 4 and -1 are of opposite signs, the graph is a hyperbola. The center of the hyperbola is at $(1, -2)$.

Now using the equation $4(x - 1)^2 - (y + 2)^2 = 16$, we can proceed as follows. Let $y = -2$, then

$$4(x - 1)^2 - (-2 + 2)^2 = 16$$
$$4(x - 1)^2 = 16$$
$$(x - 1)^2 = 4$$
$$x - 1 = \pm 2$$
$$x - 1 = 2 \quad \text{or} \quad x - 1 = -2$$
$$x = 3 \quad \text{or} \quad x = -1.$$

Thus, the hyperbola intersects the horizontal line $y = -2$ at $(3, -2)$ and at $(-1, -2)$. Let $x = 1$, then

$$4(1 - 1)^2 - (y + 2)^2 = 16$$
$$-(y + 2)^2 = 16$$
$$(y + 2)^2 = -16.$$

Since $(y + 2)^2 = -16$ has no real number solutions, we know that the hyperbola does not intersect the vertical line $x = 1$. Replacing the constant term of $4(x - 1)^2 - (y + 2)^2 = 16$ with 0 and solving for y produces the equations of the asymptotes as follows.

$$4(x - 1)^2 - (y + 2)^2 = 0$$

The left side can be factored using the pattern of the difference of squares.

$$(2(x - 1) + (y + 2))(2(x - 1) - (y + 2)) = 0$$

$$(2x - 2 + y + 2)(2x - 2 - y - 2) = 0$$

$$(2x + y)(2x - y - 4) = 0$$

$$2x + y = 0 \qquad \text{or} \qquad 2x - y - 4 = 0$$

$$y = -2x \qquad \text{or} \qquad 2x - 4 = y$$

So, the equations of the asymptotes are $y = -2x$ and $y = 2x - 4$. Sketching the asymptotes and plotting the two points $(3, -2)$ and $(-1, -2)$, we can draw the hyperbola as in Figure 7.61.

FIGURE 7.61

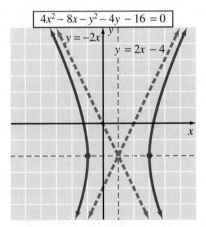

$$4x^2 - 8x - y^2 - 4y - 16 = 0$$

$y = -2x$

$y = 2x - 4$

As a way of summarizing, let's focus our attention on the continuity pattern we used in these last two sections. In Section 7.5 we used the definition of a circle to generate a standard form for the equation of a circle. Then, in Section 7.6 we discussed ellipses and hyperbolas not from a definition viewpoint, but by considering variations of the general equation of a circle with its center at the origin ($Ax^2 + By^2 = C$, where A, B, and C are of the same sign and $A = B$). In a subsequent mathematics course, parabolas, ellipses, and hyperbolas will also be developed from a definition viewpoint. That is to say, first each concept will be defined and then the definition will be used to generate a standard form of its equation.

One final comment should be made relative to the material in these last two sections. Parabolas, circles, ellipses, and hyperbolas can be formed by intersecting a plane with a conical surface as shown in Figure 7.62; these curves are often called **conic sections**

FIGURE 7.62

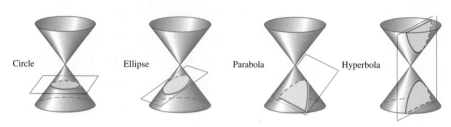

Circle   Ellipse   Parabola   Hyperbola

▼ Problem Set 7.6

Graph each of the following equations. If the equation represents a hyperbola, find its asymptotes and use them to help sketch the hyperbola.

1. $x^2 + 4y^2 = 36$
2. $x^2 + 4y^2 = 16$
3. $9x^2 + y^2 = 36$
4. $16x^2 + 9y^2 = 144$
5. $x^2 - y^2 = 1$
6. $x^2 - y^2 = 4$
7. $y^2 - 4x^2 = 9$
8. $4y^2 - x^2 = 16$
9. $4x^2 + 3y^2 = 12$
10. $5x^2 + 4y^2 = 20$
11. $5x^2 - 2y^2 = 20$
12. $9x^2 - 4y^2 = 9$
13. $25x^2 + 2y^2 = 50$
14. $12x^2 + y^2 = 36$
15. $y^2 - 16x^2 = 4$
16. $y^2 - 9x^2 = 16$
17. $-4x^2 + y^2 = -4$
18. $-9x^2 + y^2 = -36$
19. $25y^2 - 3x^2 = 75$
20. $16y^2 - 5x^2 = 80$

21. The graphs of equations of the form $xy = k$, where k is a nonzero constant, are also hyperbolas, sometimes referred to as rectangular hyperbolas. Graph each of the following.
(a) $xy = 3$
(b) $xy = 5$
(c) $xy = -2$
(d) $xy = -4$

22. What is the graph of $xy = 0$? Defend your answer.

23. We have graphed various equations of the form $Ax^2 + By^2 = C$, where C is a nonzero constant. Now graph each of the following.
(a) $x^2 + y^2 = 0$
(b) $2x^2 + 3y^2 = 0$
(c) $x^2 - y^2 = 0$
(d) $4y^2 - x^2 = 0$

24. A flashlight produces a *cone of light* that can be cut by the plane of a wall to illustrate the conic sections. Try shining a flashlight against a wall at different angles to produce a circle, an ellipse, a parabola, and one branch of a hyperbola. (You may find it difficult to distinguish between a parabola and a branch of a hyperbola.)

For Problems 25–32, graph each of the ellipses and hyperbolas.

25. $4x^2 + 8x + 16y^2 - 64y + 4 = 0$

26. $9x^2 - 36x + 4y^2 - 24y + 36 = 0$

27. $-4x^2 + 32x + 9y^2 - 18y - 91 = 0$

28. $x^2 - 4x - y^2 + 6y - 14 = 0$

29. $x^2 + 8x + 9y^2 + 36y + 16 = 0$

30. $4x^2 - 24x + y^2 + 4y + 24 = 0$

31. $-4x^2 + 24x + 16y^2 + 64y - 36 = 0$

32. $x^2 + 4x - 9y^2 + 54y - 113 = 0$

THOUGHTS INTO WORDS

33. Explain the concept of an asymptote.

34. Are the graphs of $x^2 - y^2 = 0$ and $y^2 - x^2 = 0$ identical? Are the graphs of $x^2 - y^2 = 4$ and $y^2 - x^2 = 4$ identical? Defend your answers.

35. Is the graph of $x^2 + 4y^2 = 16$ a circle or an ellipse? Defend your answer.

36. Explain how asymptotes can be used when graphing hyperbolas.

Graphics Calculator Activities

37. Use a graphics calculator to graph the ellipses in Examples 1 and 2 of this section.

38. To graph the hyperbola along with its asymptotes of Example 3 we can make the following assignments for the graphics calculator.

$$Y_1 = \sqrt{x^2 - 9}, \qquad Y_2 = -Y_1,$$
$$Y_3 = x, \qquad Y_4 = -Y_3$$

Do this and see if your graph agrees with Figure 7.58. Also graph the hyperbola and its asymptotes for Example 4.

39. Use a graphics calculator to check your graphs for Problems 13–20.

40. For each of the following equations, (1) predict the type and location of the graph, and (2) use your graphics calculator to check your prediction.

(a) $x^2 + y^2 = 100$
(b) $x^2 - y^2 = 100$
(c) $y^2 - x^2 = 100$
(d) $y = -x^2 + 9$
(e) $2x^2 + y^2 = 14$
(f) $x^2 + 2y^2 = 14$
(g) $x^2 + 2x + y^2 - 4 = 0$
(h) $x^2 + y^2 - 4y - 2 = 0$
(i) $y = x^2 + 16$
(j) $y^2 = x^2 + 16$
(k) $9x^2 - 4y^2 = 72$
(l) $4x^2 - 9y^2 = 72$
(m) $y^2 = -x^2 - 4x + 6$

More on Graphing

As you noticed in the previous section, it is very helpful to recognize that a certain type of equation produces a particular kind of graph. However, we also need to develop some general graphing techniques to use with equations when we do not recognize the graph.

In Section 7.4 we stated the following test for y-axis symmetry.

y-Axis Symmetry

The graph of an equation is symmetric with respect to the y-axis if replacing x with $-x$ results in an equivalent equation.

Figure 7.63 illustrates the reasoning that justifies this test. The points $A(x, y)$ and $B(-x, y)$ are y-axis reflections of each other. Thus, if an equation is satisfied by both (x, y) and $(-x, y)$, it exhibits y-axis symmetry. Likewise, as indicated in Figure 7.63 $A(x, y)$ and $C(x, -y)$ are x-axis reflections of each other. Therefore, the following test for x-axis symmetry can be stated.

FIGURE 7.63

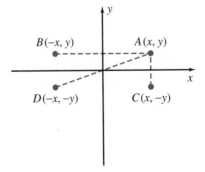

x-Axis Symmetry

The graph of an equation is symmetric with respect to the x-axis if replacing y with $-y$ results in an equivalent equation.

Again in Figure 7.63, we illustrated that $A(x, y)$ and $D(-x, -y)$ are origin reflections of each other. That is, the origin is the midpoint of the line segment $\overline{DA}$. Thus, the following test for origin symmetry can be stated.

The graph of an equation is symmetric with respect to the origin if replacing x with $-x$ and y with $-y$ results in an equivalent equation.

Following is a list of graphing suggestions. The order of the suggestions indicates the order in which we usually attack a "new" graphing problem—that is, one that we do not recognize from its equation.

1. Determine the type of symmetry that the equation exhibits.
2. Find the intercepts.
3. Solve the equation for y in terms of x or for x in terms of y if it is not already in such a form.
4. Determine the necessary restrictions so as to ensure real number solutions. (This will be explained in a moment.)
5. Set up a table of ordered pairs that satisfy the equation. The type of symmetry and the restrictions will affect your choice of values in the table.
6. Plot the points associated with the ordered pairs, and connect them with a smooth curve. Then, if appropriate, reflect this curve according to the symmetry shown by the graph.

The following examples illustrate the use of these suggestions.

Example 1

Graph $y = x^4$.

Solution

Replacing x with $-x$ produces $y = (-x)^4 = x^4$. Therefore, this graph has *y-axis symmetry*. If $x = 0$, then $y = 0$; so the graph contains the origin. Because of the y-axis symmetry, we can limit the table of values to positive values of x.

x	y
$\frac{1}{2}$	$\frac{1}{16}$
1	1
$\frac{3}{2}$	$\frac{81}{16}$
2	16

Plotting these points and connecting them with a smooth curve, we obtain the portion of the curve in the first quadrant, as shown in Figure 7.64(a). Then reflecting this portion of the curve across the y-axis produces the complete graph, as shown in Figure 7.64(b).

FIGURE 7.64

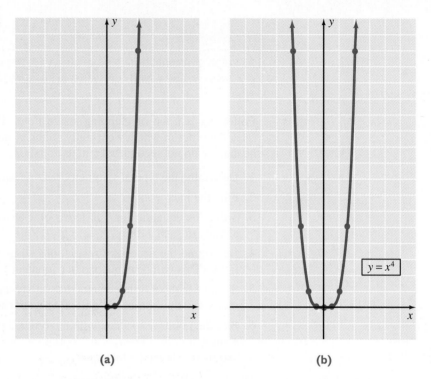

(a)   (b)

REMARK   The curve in Figure 7.64(b) is not a parabola even though its general shape resembles one. This curve is much flatter than a parabola at the bottom, and it rises much more rapidly.   △

Example 2   Graph $y = x^3$.

Solution   Replacing x with $-x$ and y with $-y$ produces $-y = (-x)^3 = -x^3$. This equation is equivalent to $y = x^3$, since multiplying both sides of $-y = -x^3$ by -1 produces $y = x^3$. Therefore, this graph has *origin symmetry*. If $x = 0$, then $y = 0$; so the graph contains the origin. Because of origin symmetry, we can limit the table of values to positive values of x.

x	y
$\dfrac{1}{2}$	$\dfrac{1}{8}$
1	1
$\dfrac{3}{2}$	$\dfrac{27}{8}$
2	8

Plotting these points and connecting them with a smooth curve, we obtain Figure 7.65(a). Then reflecting this portion of the curve through the origin produces the complete graph, as shown in Figure 7.65(b).

FIGURE 7.65

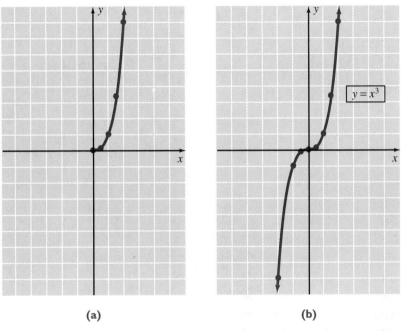

(a)   (b)   ▲

REMARK   From the symmetry tests we can see that if a curve has both x-axis and y-axis symmetry, it must have origin symmetry. However, it is possible for a curve to have origin symmetry and not be symmetrical to either axis. Figure 7.65(b) is an example of such a curve.   △

Example 3   Graph $y^2 = x^3 + 1$.

Solution   Replacing y with $-y$ produces $(-y)^2 = x^3 + 1$, which is equivalent to $y^2 = x^3 + 1$. Therefore, this graph is *symmetric with respect to the x-axis*. If $x = 0$, then $y^2 = 1$; so $y = \pm 1$ and the y-intercepts are 1 and -1. Thus, the points $(0, 1)$ and $(0, -1)$ are on the graph. If $y = 0$, then $x^3 + 1 = 0$; so $x^3 = -1$ and $x = -1$. (The other two solutions for $x^3 + 1 = 0$ are complex numbers.) Thus, the point $(-1, 0)$ is on the graph. Solving $y^2 = x^3 + 1$ for y produces $y = \pm \sqrt{x^3 + 1}$. Using $y = \sqrt{x^3 + 1}$, we can set up the following table of values.

x	y
$-\dfrac{1}{2}$	$\dfrac{\sqrt{14}}{4}$
$\dfrac{1}{2}$	$\dfrac{3\sqrt{2}}{4}$
1	$\sqrt{2}$
2	3
3	$2\sqrt{7}$

Plotting these points and connecting them with a smooth curve produces Figure 7.66(a). Then reflecting this portion of the curve across the x-axis produces the complete graph shown in Figure 7.66(b).

FIGURE 7.66

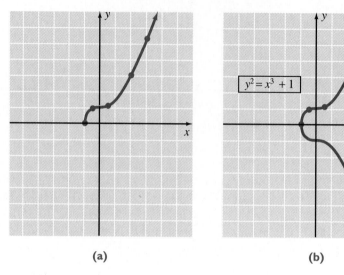

(a)   (b)

Example 4

Graph $x^2y^2 = 9$.

Solution

All three tests for symmetry are satisfied; therefore, this graph is *symmetric with respect to both axes and to the origin.* It should be evident from the equation that neither x nor y can equal zero. Therefore, no points of either axis are on the graph. Solving $x^2y^2 = 9$ for y produces $y = \pm\sqrt{\dfrac{9}{x^2}} = \pm\dfrac{3}{x}$. Using $y = \dfrac{3}{x}$, we ca set up the following table of values.

x	y
$\dfrac{1}{2}$	6
1	3
2	$\dfrac{3}{2}$
3	1
4	$\dfrac{3}{4}$

Plotting these points and connecting them with a smooth curve produces Figure 7.67(a). Then reflecting this portion of the curve through the x-axis, y-axis, and the origin produces the complete graph shown in Figure 7.67(b).

FIGURE 7.67

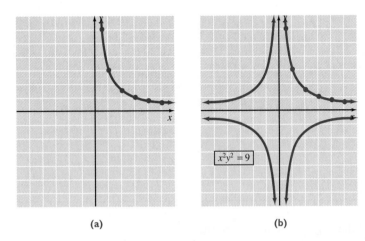

(a)   (b)

Another graphing consideration is that of *restricting a variable* to ensure real number solutions. The following example illustrates this point.

Example 5

Solution

Graph $y = \sqrt{x - 1}$.

The radicand, $x - 1$, must be nonnegative. Therefore,

$$x - 1 \geq 0$$
$$x \geq 1.$$

The restriction, $x \geq 1$, indicates that we need not look for a y-intercept. The x-intercept can be found as follows. If $y = 0$, then

$$0 = \sqrt{x - 1}$$
$$0 = x - 1$$
$$1 = x.$$

The point $(1, 0)$ is on the graph.

Now, keeping the restriction in mind, we can determine the following table of values.

x	y
2	1
3	$\sqrt{2}$
4	$\sqrt{3}$
5	2
10	3

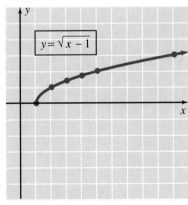

Plotting these points and connecting them with a smooth curve produces the curve in Figure 7.68.

FIGURE 7.68

Problem Set 7.7 has a section of graphing problems that includes a mixture of "old" and "new" graphs. That is, some of the graphs are straight lines, circles, parabolas, ellipses, or hyperbolas, but others are curves that you probably will not recognize from their equations. Each time you face a new curve, remember the suggestions we offered in this section. You will be exposed to more graphing ideas in Chapters 8 and 9.

Problem Set 7.7

For each of the following points, determine the points that are symmetric with respect to the (a) x-axis, (b) y-axis, and (c) origin (Problems 1–6).

1. $(3, 5)$

2. $(-1, 3)$

3. $(-2, -4)$

4. $(4, -3)$

5. $(0, -1)$

6. $(7, 0)$

For Problems 7–20, determine the type of symmetry (x-axis, y-axis, origin) possessed by each of the graphs. Do not sketch the graph.

7. $x = y^2 + 4$

8. $y = 2x^2 + 7$

9. $x^2 = y^3$

10. $x^3 = y^2$

11. $x^2 - 2x + 3y^2 = 4$

12. $2x^2 - y^2 + 4y - 6 = 0$

13. $y = x$

14. $y = -x$

15. $xy = 4$

16. $xy = -6$

17. $x^2 + 5y^2 = 9$

18. $4x^2 - 3y^2 = 12$

19. $y = x^4 + x^2$

20. $y = x^4 + 2$

For Problems 21–50, graph each of the equations. Remember that some of them are straight lines, circles, parabolas, ellipses, and hyperbolas, but some of them are "new" curves.

21. $y = -x^2 - 1$

22. $y = x^2 + 2$

23. $y = 3x - 6$

24. $y = -x + 4$

25. $y^3 = x^2$

26. $y^2 = x^3$

27. $x^2 - 4y^2 = 8$

28. $y^2 - x^2 = 4$

29. $y = \sqrt{x + 1}$

30. $y = -\sqrt{x}$

31. $xy = -1$

32. $xy = 1$

33. $y = x^3 + 2$

34. $y = -x^3$

35. $y = x^2 - 2x - 3$

36. $y = x^2 + 2x - 2$

37. $x^2 + 2y^2 = 16$

38. $3x^2 + y^2 = 9$

39. $y = -x^4$

40. $y = 4x^2 - x^4$

41. $x^2 + y^2 - 6x + 4y + 9 = 0$

42. $x^2 + y^2 + 4x - 6y + 4 = 0$

43. $y = x^3 - 4x$

44. $y^2 = x^3 + 8$

45. $xy^2 = 4$

46. $x^2y = 4$

47. $y = \dfrac{-2}{x^2 + 1}$

48. $y = \dfrac{4}{x^2 + 2}$

49. $x = y^2 + 2$

50. $x = -y^2 - 1$

51. How would you help someone who is having difficulty graphing $y = x^4 - 4x^2$?

52. Your friend claims that the graphs of $y = 3x - 6$ and $y = 6 - 3x$ are x-axis reflections of each other. Is she correct? Defend your answer.

Graphics Calculator Activities

53. Check your answers for Problems 7–20 by using a graphics calculator to graph each equation.

54. Based on previous work with radicals, predict the graph for each of the following. Then use a graphics calculator to check your prediction.
 (a) $y = \sqrt{x^2}$ **(b)** $y = \sqrt{x^3}$
 (c) $y = \sqrt{x^4}$ **(d)** $y = x\sqrt{x}$

55. Use your graphing experiences of this chapter to help predict the graph for each of the following. Then use a graphics calculator to check your prediction.
 (a) $x = y^3$ **(b)** $y = \sqrt{x + 4}$
 (c) $y = \dfrac{1}{x} + 4$ **(d)** $y = x^4 + x^2$
 (e) $y = x^4 - x^2$ **(f)** $y = x^2 - x^4$
 (g) $y = -\sqrt{x - 1}$ **(h)** $x^2y^2 = -9$
 (i) $y = 1 - x^3$ **(j)** $x = 4y^2$

SUMMARY

(7.1) The **Cartesian (or rectangular) coordinate system** is used to graph ordered pairs of real numbers. The first number, a, of the ordered pair (a, b) is called the **abscissa**, and the second number, b, is called the **ordinate**; together they are referred to as the **coordinates** of a point.

Two basic kinds of problems exist in coordinate geometry.

 1. Given an algebraic equation, find its geometric graph.
 2. Given a set of conditions pertaining to a geometric figure, find its algebraic equation.

The distance between any two points (x_1, y_1) and (x_2, y_2) is given by the **distance formula,**

$$d = \sqrt{(x_2 - x_1)^2 + (y_2 - y_1)^2}.$$

The **slope** (denoted by m) of a line determined by the points (x_1, y_1) and (x_2, y_2) is given by the slope formula,

$$m = \frac{y_2 - y_1}{x_2 - x_1}, \qquad x_2 \neq x_1.$$

(7.2) A **solution** of an equation in two variables is an ordered pair of real numbers that satisfies the equation.

Any equation of the form $Ax + By = C$, where A, B, and C are constants (A and B not both zero) and x and y are variables, is a **linear equation**, and its graph is a **straight line**.

Any equation of the form $Ax + By = C$, where $C = 0$, is a straight line that contains the **origin**.

Any equation of the form $x = a$, where a is a constant, is a line parallel to the y-axis that has an x-intercept of a.

Any equation of the form $y = b$, where b is a constant, is a line parallel to the x-axis that has a y-intercept of b.

Linear inequalities in two variables are of the form $Ax + By > C$ or $Ax + By < C$. The following steps are suggested for graphing a linear inequality.

1. First, graph the corresponding equality. Use a solid line if equality is included in the original statement and a dashed line if equality is not included.

2. Choose a test point not on the line, and substitute its coordinates into the inequality.

3. The graph of the original inequality is
   (a) the half plane containing the test point if the inequality is satisfied by that point, or
   (b) the half plane not containing the test point if the inequality is not satisfied by that point.

(7.3) Look back at Examples 1, 2, and 3 of this section to review the *general approach* to writing the equation of a line given certain facts about the line.

The equation

$$y - y_1 = m(x - x_1)$$

is called the **point-slope form** of the equation of a line.

The equation

$$y = mx + b$$

is called the **slope-intercept form** of the equation of a line. If the equation of a nonvertical line is written in this form, the coefficient of x is the slope of the line and the constant term is the y-intercept.

If two lines have slopes m_1 and m_2, respectively, then

    1. The two lines are parallel if and only if $m_1 = m_2$;

    2. The two lines are perpendicular if and only if $(m_1)(m_2) = -1$.

(7.4) and (7.5) The graph of any quadratic equation of the form $y = ax^2 + bx + c$, where a, b, and c are real numbers and $a \neq 0$, is a **parabola**

The following diagram summarizes the graphing of parabolas.

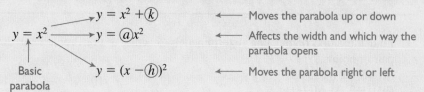

The **standard form of the equation of a circle** with its center at (h, k) and a radius of length r is

$$(x - h)^2 + (y - k)^2 = r^2.$$

The standard form of the equation of a circle with its center at the origin and a radius of length r is

$$x^2 + y^2 = r^2.$$

(7.6) The graph of an equation of the form

$$Ax^2 + By^2 = C,$$

where A, B, and C are nonzero constants of the same sign and $A \neq B$, is an **ellipse**.

The graph of an equation of the form

$$Ax^2 + By^2 = C,$$

where A, B, and C are nonzero constants and A and B are of unlike signs, is a **hyperbola**.

Circles, ellipses, parabolas, and hyperbolas are often referred to as **conic sections**.

(7.7) Probably the most effective graphing technique is to be able to recognize the kind of graph that is produced by a certain type of equation. However, for determining "new" curves, we offer the following suggestions.

1. Determine the type of symmetry that the equation exhibits.
2. Find the x- and y-intercepts.
3. Solve the equation for y in terms of x or for x in terms of y.
4. Determine the necessary restrictions to ensure real number solutions.
5. Set up a table of ordered pairs that satisfy the equation.
6. Plot the points associated with the ordered pairs.
7. Connect the points with a smooth curve. Then, if appropriate, reflect this portion of the curve according to the symmetry possessed by the graph.

Chapter 7 Review Problem Set

1. Find the slope of the line determined by each pair of points.
   (a) $(3, 4), (-2, -2)$
   (b) $(-2, 3), (4, -1)$

2. Find the slope of each of the following lines.
   (a) $4x + y = 7$
   (b) $2x - 7y = 3$

3. Find the lengths of the sides of a triangle whose vertices are at $(2, 3), (5, -1),$ and $(-4, -5)$.

For Problems 4–8, write the equation of each of the lines that satisfy the stated conditions. Express final equations in standard form.

4. Contains the points $(-1, 2)$ and $(3, -5)$

5. Has a slope of $-\dfrac{3}{7}$ and a y-intercept of 4

6. Contains the point $(-1, -6)$ and has a slope of $\dfrac{2}{3}$

7. Contains the point $(2, 5)$ and is parallel to the line $x - 2y = 4$

8. Contains the point $(-2, -6)$ and is perpendicular to the line $3x + 2y = 12$

For Problems 9–20, graph each of these equations.

9. $2x - y = 6$
10. $y = -2x^2 - 1$
11. $x^2 + y^2 = 1$
12. $4x^2 + y^2 = 16$
13. $y = -4x$
14. $2x^2 - y^2 = 8$
15. $y = x^2 + 4x - 1$
16. $y = 4x^2 - 8x + 2$
17. $xy^2 = -1$
18. $x^2 - 4x + 2y^2 - 4y - 2 = 0$
19. $y^2 + 4y - 4x^2 - 24x - 36 = 0$
20. $y = -2x^2 + 12x - 16$

For Problems 21 and 22, graph each of the inequalities.

21. $x + 2y \geq 4$
22. $2x - 3y < 6$
23. Find the center and length of a radius of the circle $x^2 + y^2 + 6x - 8y + 16 = 0$.
24. Find the coordinates of the vertex of the parabola $y = x^2 + 8x + 10$.
25. Find the equations of the asymptotes of the hyperbola $9x^2 - 4y^2 = 72$.
26. Find the length of the major axis of the ellipse $4x^2 + y^2 = 9$.

CHAPTER 7 TEST

1. Find the slope of the line determined by the points $(-2, 4)$ and $(3, -2)$.

2. Find the slope of the line determined by the equation $3x - 7y = 12$.

3. Find the length of the line segment whose endpoints are $(4, 2)$ and $(-3, -1)$. Express your answer in simplest radical form.

4. Find the equation of the line that has a slope of $-\dfrac{3}{2}$ and contains the point $(4, -5)$. Express the equation in standard form.

5. Find the equation of the line that contains the points $(-4, 2)$ and $(2, 1)$. Express the equation in slope-intercept form.

6. Find the equation of the line that is parallel to the line $5x + 2y = 7$ and contains the point $(-2, -4)$. Express the equation in standard form.

7. Find the equation of the line that is perpendicular to the line $x - 6y = 9$ and contains the point $(4, 7)$. Express the equation in standard form.

8. Find the center of the circle $x^2 + y^2 - 4x + 10y - 4 = 0$.

9. Find the length of a radius of the circle $x^2 + y^2 + 6x - 8y - 11 = 0$.

10. Find the vertex of the parabola $y = -x^2 - 10x - 27$.

11. Find the vertex of the parabola $y = 3x^2 - 18x + 23$.

12. Find the length of the major axis of the ellipse $3x^2 + y^2 = 16$.

13. Find the equations of the asymptotes of the hyperbola $16x^2 - 9y^2 = 12$.

For Problems 14–17, identify the graph of the given equation as a straight line, circle, ellipse, hyperbola, or parabola.

14. $x^2 - 3y^2 = -9$

15. $y = -5x^2 - 2$

16. $y = -5x - 2$

17. $2x^2 + 5x + 2y^2 - 9y - 4 = 0$

For Problems 18–23, graph each of the equations.

18. $y = -2x^2 - 4$

19. $y = -2x - 4$

20. $x^2 - 2y^2 = -2$

21. $2x^2 + y^2 = 18$

(continued on next page)

CHAPTER 7 TEST (*continued*)

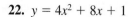

22. $y = 4x^2 + 8x + 1$

23. $x^2 + y^2 - 4y = 0$

For Problems 24 and 25, graph each inequality.

24. $2x - y < 4$

25. $3x + 2y \geq 6$

Cumulative Review Problem Set

For Problems 1–6, express each radical in simplest radical form.

1. $\sqrt{32xy^3}$

2. $\sqrt{48a^3b^4}$

3. $\dfrac{3\sqrt{2}}{2\sqrt{3}}$

4. $-\dfrac{2\sqrt{5}}{\sqrt{8}}$

5. $\sqrt[3]{48x^5}$

6. $\dfrac{3}{4}\sqrt{108}$

For Problems 7 and 8, rationalize the denominator and simplify.

7. $\dfrac{3}{2\sqrt{5} - 1}$

8. $\dfrac{2\sqrt{3}}{\sqrt{6} + 2\sqrt{2}}$

For Problems 9–18, perform the indicated operations and express answers in simplified form.

9. $(3\sqrt{6})(2\sqrt{8})$

10. $(\sqrt{x} + 2)(\sqrt{x} - 3)$

11. $(2\sqrt{3} + \sqrt{5})(\sqrt{3} - 2\sqrt{5})$

12. $(x - 1)(2x^2 - x + 7)$

13. $\dfrac{9y^2}{x^2 + 12x + 36} \div \dfrac{12y}{x^2 + 6x}$

14. $\dfrac{x^2 - x}{4y} \cdot \dfrac{10xy^2}{2x - 2} \div \dfrac{3x^2 + 3x}{15x^2y^2}$

15. $\dfrac{3}{5x} - \dfrac{2}{3x} + \dfrac{5x}{6}$

16. $\dfrac{5}{x - 9} + \dfrac{4}{x^2 - 3x - 54} - \dfrac{1}{x + 6}$

17. $(20x^2 - 39x + 18) \div (5x - 6)$

18. $\dfrac{4x^3 - 5x^2 + 2x - 6}{x^2 - 3x}$

For Problems 19–21, use scientific notation to help perform the indicated operations.

19. $\dfrac{(.00063)(960000)}{(3200)(.0000021)}$

20. $(8000)^{\frac{2}{3}}$

21. $\sqrt{.000009}$

For Problems 22–27, find each of the indicated products or quotients and express answers in standard form.

22. $(3 - 2i)(5 + i)$

23. $(-2 + 5i)(4 - 7i)$

24. $(2 - 6i)^2$

25. $\dfrac{2}{5i}$

26. $\dfrac{3 + i}{2 - 4i}$

27. $\dfrac{-1 - 4i}{-2 + 8i}$

For Problems 28–32, evaluate each of the numerical expressions.

28. $-8^{-\frac{1}{3}}$

29. $(-8)^{\frac{1}{3}}$

30. $\left(\dfrac{2}{5}\right)^{-2}$

31. $\sqrt[4]{16}$

32. $2^0 + 2^{-1} + 2^{-2} + 2^{-3}$

For Problems 33–37, factor each of the algebraic expressions completely.

33. $9x^2 + 12xy + 4y^2$

34. $27x^3 - 64y^3$

35. $4x^4 - 13x^2 + 9$

36. $3x^3 - 30x^2 - 72x$

37. $x^2 - 4xy + 4y^2 - 4$

For Problems 38–57, solve each of the equations. Some of these may have complex numbers as solutions.

38. $16n^2 - 40n + 25 = 0$

39. $x^3 = 8x$

40. $n^2 = -4n - 1$

41. $(y + 2)^2 = -24$

42. $|3x - 2| = |-2x - 4|$

43. $x^3 = 8$

44. $(2x - 7)(x + 4) = 0$

45. $(x - 5)(4x - 1) = 23$

46. $\dfrac{2n - 3}{3} + \dfrac{n + 1}{2} = 3$

47. $(x - 1)(x + 4) = (x + 1)(2x - 5)$

48. $.5(3x + .7) = 20.6$

49. $t^2 - 2t = -4$

50. $4x^2 + 23x - 6 = 0$

51. $x^2 - 4x = 192$

52. $\dfrac{3}{2x - 8} - \dfrac{x - 5}{x^2 - 2x - 8} = \dfrac{7}{x + 2}$

53. $\dfrac{3n}{n^2 + n - 6} + \dfrac{2}{n^2 + 4n + 3} = \dfrac{n}{n^2 - n - 2}$

54. $\sqrt{x + 6} = x$

55. $\sqrt{x + 4} = \sqrt{x - 1} + 1$

56. $x^4 + 5x^2 - 36 = 0$

57. $\dfrac{2}{x - 1} = \dfrac{x + 4}{3}$

For Problems 58–61, solve each equation for the indicated variable.

58. $3x - 5y = 10$ for y

59. $\dfrac{3}{4} = \dfrac{y - 1}{x - 2}$ for y

60. $f = \dfrac{1}{\dfrac{1}{a} + \dfrac{1}{b}}$ for b

61. $V = C\left(1 - \dfrac{T}{N}\right)$ for T

For Problems 62–70, solve each inequality and graph the solution set on a number line.

62. $|-2x - 1| > 3$

63. $|3x + 5| < 2$

64. $(x - 2)(x + 4) > 0$

65. $(x + 1)(2x - 3) < 0$

66. $\dfrac{x - 1}{4} - \dfrac{x + 2}{6} \leq \dfrac{1}{8}$

67. $\dfrac{x - 3}{x - 5} \geq 0$

68. $6x^2 + 13x - 5 > 0$

69. $\dfrac{2x}{x + 3} > 4$

70. $\dfrac{3x + 2}{x + 4} \leq 2$

For Problems 71–74, solve each problem by setting up and solving an appropriate equation.

71. Find two numbers whose sum is -2 and whose product is -35.

72. The sum of the lengths of the two legs of a right triangle is 9 centimeters. If the length of the hypotenuse is $3\sqrt{5}$ centimeters, find the length of each leg.

73. A 3- by 5-inch picture is surrounded by a frame of uniform width. The area of the picture and frame together is 24 square inches. Find the width of the frame.

74. It takes Kent 2 hours longer to do a certain job than it takes Maika. They worked together for 2 hours; then Maika left to go shopping and Kent finished the job in 1 hour. How long would it take each of them to do the job alone?

Functions

A golf pro-shop operator finds that she can sell 30 sets of golf clubs at $500 per set in a year. Furthermore, she predicts that for each $25 decrease in price, three extra sets of golf clubs could be sold. At what price should she sell the clubs to maximize gross income? **We can use the quadratic function $f(x) = (30 + 3x)(500 - 25x)$ to determine that the clubs should be sold at $375 per set.**

One of the fundamental concepts of mathematics is the function. Functions are used to unify mathematics and also to apply mathematics to many real world problems. Functions provide a means of studying quantities that vary with one another, that is, when a change in one quantity causes a corresponding change in another.

In this chapter we will (1) introduce the basic ideas that pertain to the function concept, (2) review and extend some concepts from Chapter 7, and (3) discuss some applications of functions.

8.1   Relations and Functions

Mathematically, a function is a special kind of **relation**, so we will begin our discussion with a simple definition of a relation.

DEFINITION 8.1   A **relation** is a set of ordered pairs.

Thus, a set of ordered pairs such as $\{(1, 2), (3, 7), (8, 14)\}$ is a relation. The set of all first components of the ordered pairs is the **domain** of the relation and the set of all second components is the **range** of the relation. The relation $\{(1, 2), (3, 7) (8, 14)\}$ has a domain of $\{1, 3, 8\}$ and a range of $\{2, 7, 14\}$.

The ordered pairs we refer to in Definition 8.1 may be generated by various means, such as a graph or a chart. However, one of the most common ways of generating ordered pairs is by the use of equations. Since the solution set of an equation in two variables is a set of ordered pairs, such an equation describes a relation. Each of the following equations describes a relation between the variables x and y. We have listed *some* of the infinitely many ordered pairs (x, y) of each relation.

1. $x^2 + y^2 = 4$ $\qquad (1, \sqrt{3}), (1, -\sqrt{3}), (0, 2), (0, -2)$
2. $y^2 = x^3$ $\qquad (0, 0), (1, 1), (1, -1), (4, 8), (4, -8)$
3. $y = x + 2$ $\qquad (0, 2), (1, 3), (2, 4), (-1, 1), (5, 7)$
4. $y = \dfrac{1}{x - 1}$ $\qquad (0, -1), (2, 1), \left(3, \dfrac{1}{2}\right), \left(-1, -\dfrac{1}{2}\right), \left(-2, -\dfrac{1}{3}\right)$
5. $y = x^2$ $\qquad (0, 0), (1, 1), (2, 4), (-1, 1), (-2, 4)$

Now we direct your attention to the ordered pairs associated with equations 3, 4, and 5 in the list above. Note that in each case no two ordered pairs have the same first component. Such a set of ordered pairs is called a **function**.

DEFINITION 8.2   A **function** is a relation where no two ordered pairs have the same first component.

Stated another way, Definition 8.2 means that a function is a relation where each member of the domain is assigned *one and only one* member of the range. Thus, it is easy to determine that each of the following sets of ordered pairs is a function.

$$f = \{(x, y) \,|\, y = x + 2\},$$

$$g = \left\{(x, y) \,\middle|\, y = \frac{1}{x - 1}\right\},$$

$$h = \{(x, y) \,|\, y = x^2\}$$

In each case there is one and only one value of y (an element of the range) associated with each value of x (an element of the domain).

Notice that we named the previous functions f, g, and h. It is common to name functions by means of a single letter and the letters f, g, and h are often used. We would suggest more meaningful choices when functions are used to portray real world situations. For example, if a problem involves a profit function, then naming the function p or even P would seem natural.

The symbol for a function can be used along with a variable that represents an element in the domain to represent the associated element in the range. For example, suppose that we have a function f specified in terms of the variable x. The symbol, $f(x)$, (read "f of x" or "the value of f at x") represents the element in the range associated with the element x from the domain. The function $f = \{(x, y) \,|\, y = x + 2\}$ can be written as $f = \{(x, f(x)) \,|\, f(x) = x + 2\}$ and this is usually shortened to read "f is the function determined by the equation $f(x) = x + 2$."

REMARK Be careful with the symbolism $f(x)$. As we stated above, it means the value of the function f at x. It does not mean f times x. △

This *function notation* is very convenient when computing and expressing various values of the function. For example, the value of the function $f(x) = 3x - 5$ at $x = 1$ is

$$f(1) = 3(1) - 5 = -2.$$

Likewise, the functional values for $x = 2$, $x = -1$, and $x = 5$ are

$$f(2) = 3(2) - 5 = 1,$$

$$f(-1) = 3(-1) - 5 = -8, \qquad \text{and}$$

$$f(5) = 3(5) - 5 = 10.$$

Thus, this function f contains the ordered pairs $(1, -2)$, $(2, 1)$, $(-1, -8)$, $(5, 10)$, and in general all ordered pairs of the form $(x, f(x))$, where $f(x) = 3x - 5$ and x is any real number.

It may be helpful for you to mentally picture the concept of a function in terms of a "function machine" as in Figure 8.1. Each time that a value of x is put into the machine, the equation $f(x) = x + 2$ is used to generate one and only one value for $f(x)$ to be ejected from the machine. For example, if 3 is put into this machine, then $f(3) = 3 + 2 = 5$, and 5 is ejected. Thus, the ordered pair $(3, 5)$ is one element of the function. The following examples will help us pull together some of the ideas about functions.

FIGURE 8.1

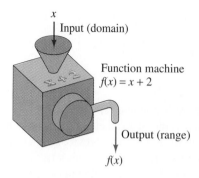

x

Input (domain)

Function machine
$f(x) = x + 2$

Output (range)

$f(x)$

Example 1

Determine whether the relation $\{(x, y) \mid y^2 = x\}$ is a function and specify its domain and range.

Solution

Because $y^2 = x$ is equivalent to $y = \pm\sqrt{x}$, to each value of x there are assigned *two* values for y. Therefore, this relation is not a function.
The expression $\sqrt{x}$ requires that x be nonnegative; therefore, the domain *(D)* is

$$D = \{x \mid x \geq 0\}.$$

To each nonnegative real number, the relation assigns two real numbers, $\sqrt{x}$ and $-\sqrt{x}$. Thus, the range *(R)* is

$$R = \{y \mid y \text{ is a real number}\}.$$

▲

Example 2

For the function $f(x) = x^2$,

    (a) specify its domain,   *(b)* determine its range, and
    (c) evaluate $f(-2), f(0)$, and $f(4)$.

Solution

    (a) Any real number can be squared; therefore, the domain *(D)* is

$$D = \{x \mid x \text{ is a real number}\}.$$

    (b) Squaring a real number always produces a nonnegative result. Thus, the range *(R)* is

$$R = \{f(x) \mid f(x) \geq 0\}.$$

    (c) $f(-2) = (-2)^2 = 4,$

$$f(0) = (0)^2 = 0,$$

$$f(4) = (4)^2 = 16$$

▲

For our purposes in this text, if the domain of a function is not specifically determined by a real world application, then we assume the domain to be the set of all **real numbers** that produce **real number** functional values. Consider the following examples.

Example 3

Specify the domain for each of the following.

(a) $f(x) = \dfrac{1}{x - 1}$ (b) $f(t) = \dfrac{1}{t^2 - 4}$ (c) $f(s) = \sqrt{s - 3}$

Solution

(a) We can replace x with any real number except 1, because 1 makes the denominator zero. Thus, the domain is given by

$$D = \{x \mid x \neq 1\}.$$

(b) We need to eliminate any value of t that will make the denominator zero. Thus, let's solve the equation $t^2 - 4 = 0$.

$$t^2 - 4 = 0$$

$$t^2 = 4$$

$$t = \pm 2$$

The domain is the set

$$D = \{t \mid t \neq -2 \text{ and } t \neq 2\}.$$

(c) The radicand, $s - 3$, must be nonnegative.

$$s - 3 \geq 0$$

$$s \geq 3$$

The domain is the set

$$D = \{s \mid s \geq 3\}.$$ ▲

REMARK   Certainly interval notation could be used to express the domains of functions in Example 3. However, we decided at this time to give you a little more experience with set-builder notation.   △

Example 4

If $f(x) = -2x + 7$ and $g(x) = x^2 - 5x + 6$, find $f(3), f(-4), g(2),$ and $g(-1)$.

Solution

$$f(x) = -2x + 7 \qquad\qquad g(x) = x^2 - 5x + 6$$

$$f(3) = -2(3) + 7 = 1, \qquad g(2) = 2^2 - 5(2) + 6 = 0,$$

$$f(-4) = -2(-4) + 7 = 15 \quad g(-1) = (-1)^2 - 5(-1) + 6 = 12 \quad ▲$$

In Example 4, notice that we worked with two different functions in the same problem. Thus, we used different names, f and g.

The quotient $\dfrac{f(a + h) - f(a)}{h}$ is often called a **difference quotient**; we use it extensively with functions when studying the limit concept in calculus. The next example illustrates finding the difference quotient for a specific function.

Example 5

If $f(x) = x^2 + 2x - 3$, find $\dfrac{f(a + h) - f(a)}{h}$.

Solution

$$f(a + h) = (a + h)^2 + 2(a + h) - 3$$
$$= a^2 + 2ah + h^2 + 2a + 2h - 3,$$

and

$$f(a) = a^2 + 2a - 3.$$

Therefore,

$$f(a + h) - f(a) = (a^2 + 2ah + h^2 + 2a + 2h - 3) - (a^2 + 2a - 3)$$
$$= a^2 + 2ah + h^2 + 2a + 2h - 3 - a^2 - 2a + 3$$
$$= 2ah + h^2 + 2h,$$

and

$$\frac{f(a + h) - f(a)}{h} = \frac{2ah + h^2 + 2h}{h}$$
$$= \frac{\cancel{h}(2a + h + 2)}{\cancel{h}}$$
$$= 2a + h + 2. \qquad \blacktriangle$$

Functions and functional notation provide the basis for describing many real world relationships. The next example illustrates this point.

Example 6

Suppose a factory determines that the overhead for producing a quantity of a certain item is $500 and the cost for each item is $25. Express the total expenses as a function of the number of items produced and compute the expenses for producing 12, 25, 50, 75, and 100 items.

Solution

Let n represent the number of items produced. Then $25n + 500$ represents the total expenses. Let's use E to represent the *expense function,* so that we have

$$E(n) = 25n + 500, \quad \text{where } n \text{ is a whole number,}$$

from which we obtain

$$E(12) = 25(12) + 500 = 800,$$
$$E(25) = 25(25) + 500 = 1125,$$
$$E(50) = 25(50) + 500 = 1750,$$
$$E(75) = 25(75) + 500 = 2375,$$
$$E(100) = 25(100) + 500 = 3000.$$

So the total expenses for producing 12, 25, 50, 75, and 100 items are $800, $1125, $1750, $2375, and $3000, respectively. $\qquad \blacktriangle$

Problem Set 8.1

For Problems 1–10, specify the domain and the range for each relation. Also state whether or not the relation is a function.

1. $\{(1, 5), (2, 8), (3, 11), (4, 14)\}$

2. $\{(0, 0), (2, 10), (4, 20), (6, 30), (8, 40)\}$

3. $\{(0, 5), (0, -5), (1, 2\sqrt{6}), (1, -2\sqrt{6})\}$

4. $\{(1, 1), (1, 2), (1, -1), (1, -2), (1, 3)\}$

5. $\{(1, 2), (2, 5), (3, 10), (4, 17), (5, 26)\}$

6. $\{(-1, 5), (0, 1), (1, -3), (2, -7)\}$

7. $\{(x, y) \mid 5x - 2y = 6\}$

8. $\{(x, y) \mid y = -3x\}$

9. $\{(x, y) \mid x^2 = y^3\}$

10. $\{(x, y) \mid x^2 - y^2 = 16\}$

For Problems 11–36, specify the domain for each of the functions.

11. $f(x) = 7x - 2$

12. $f(x) = x^2 + 1$

13. $f(x) = \dfrac{1}{x - 1}$

14. $f(x) = \dfrac{-3}{x + 4}$

15. $g(x) = \dfrac{3x}{4x - 3}$

16. $g(x) = \dfrac{5x}{2x + 7}$

17. $h(x) = \dfrac{2}{(x + 1)(x - 4)}$

18. $h(x) = \dfrac{-3}{(x - 6)(2x + 1)}$

19. $f(x) = \dfrac{14}{x^2 + 3x - 40}$

20. $f(x) = \dfrac{7}{x^2 - 8x - 20}$

21. $f(x) = \dfrac{-4}{x^2 + 6x}$

22. $f(x) = \dfrac{9}{x^2 - 12x}$

23. $f(t) = \dfrac{4}{t^2 + 9}$

24. $f(t) = \dfrac{8}{t^2 + 1}$

25. $f(t) = \dfrac{3t}{t^2 - 4}$

26. $f(t) = \dfrac{-2t}{t^2 - 25}$

27. $h(x) = \sqrt{x + 4}$

28. $h(x) = \sqrt{5x - 3}$

29. $f(s) = \sqrt{4s - 5}$

30. $f(s) = \sqrt{s - 2} + 5$

31. $f(x) = \sqrt{x^2 - 16}$

32. $f(x) = \sqrt{x^2 - 49}$

33. $f(x) = \sqrt{x^2 - 3x - 18}$

34. $f(x) = \sqrt{x^2 + 4x - 32}$

35. $f(x) = \sqrt{1 - x^2}$

36. $f(x) = \sqrt{9 - x^2}$

37. If $f(x) = 5x - 2$, find $f(0)$, $f(2)$, $f(-1)$, and $f(-4)$.

38. If $f(x) = -3x - 4$, find $f(-2), f(-1), f(3)$, and $f(5)$.

39. If $f(x) = \dfrac{1}{2}x - \dfrac{3}{4}$, find $f(-2)$, $f(0)$, $f\left(\dfrac{1}{2}\right)$, and

$f\left(\dfrac{2}{3}\right).$

40. If $g(x) = x^2 + 3x - 1$, find $g(1)$, $g(-1)$, $g(3)$, and $g(-4)$.

41. If $g(x) = 2x^2 - 5x - 7$, find $g(-1)$, $g(2)$, $g(-3)$, and $g(4)$.

42. If $h(x) = -x^2 - 3$, find $h(1)$, $h(-1)$, $h(-3)$, and $h(5)$.

43. If $h(x) = -2x^2 - x + 4$, find $h(-2)$, $h(-3)$, $h(4)$, and $h(5)$.

44. If $f(x) = \sqrt{x - 1}$, find $f(1)$, $f(5)$, $f(13)$, and $f(26)$.

45. If $f(x) = \sqrt{2x + 1}$, find $f(3)$, $f(4)$, $f(10)$, and $f(12)$.

46. If $f(x) = \dfrac{3}{x - 2}$, find $f(3)$, $f(0)$, $f(-1)$, and $f(-5)$.

47. If $f(x) = \dfrac{-4}{x + 3}$, find $f(1)$, $f(-1)$, $f(3)$, and $f(-6)$.

48. If $f(x) = 2x^2 - 7$ and $g(x) = x^2 + x - 1$, find $f(-2)$, $f(3)$, $g(-4)$, and $g(5)$.

49. If $f(x) = 5x^2 - 2x + 3$ and $g(x) = -x^2 + 4x - 5$, find $f(-2)$, $f(3)$, $g(-4)$, and $g(6)$.

50. If $f(x) = |3x - 2|$ and $g(x) = |x| + 2$, find $f(1)$, $f(-1)$, $g(2)$, and $g(-3)$.

51. If $f(x) = 3|x| - 1$ and $g(x) = -|x| + 1$, find $f(-2)$, $f(3)$, $g(-4)$, and $g(5)$.

For Problems 52–59, find $\dfrac{f(a + h) - f(a)}{h}$ for each of the given functions.

52. $f(x) = 5x - 4$ **53.** $f(x) = -3x + 6$

54. $f(x) = x^2 + 5$

55. $f(x) = -x^2 - 1$

56. $f(x) = x^2 - 3x + 7$

57. $f(x) = 2x^2 - x + 8$

58. $f(x) = -3x^2 + 4x - 1$ –

59. $f(x) = -4x^2 - 7x - 9$ –

60. Suppose that the cost function for producing a certain item is given by $C(n) = 3n + 5$, where n represents the number of items produced. Compute $C(150)$, $C(500)$, $C(750)$, and $C(1500)$.

61. The height of a projectile fired vertically into the air (neglecting air resistance) at an initial velocity of 64 feet per second is a function of the time (t) and is given by the equation

$$h(t) = 64t - 16t^2.$$

Compute $h(1)$, $h(2)$, $h(3)$, and $h(4)$.

62. The profit function for selling n items is given by $P(n) = -n^2 + 500n - 61{,}500$. Compute $P(200)$, $P(230)$, $P(250)$, and $P(260)$.

63. A car rental agency charges \$50 per day plus \$0.32 a mile. Therefore, the daily charge for renting a car is a function of the number of miles traveled *(m)* and can be expressed as $C(m) = 50 + .32m$. Compute $C(75)$, $C(150)$, $C(225)$, and $C(650)$.

64. The equation $A(r) = \pi r^2$ expresses the area of a circular region as a function of the length of a radius *(r)*. Use 3.14 as an approximation for π and compute $A(2)$, $A(3)$, $A(12)$, and $A(17)$.

65. The equation $I(r) = 500r$ expresses the amount of simple interest earned by an investment of \$500 for one year as a function of the rate of interest *(r)*. Compute $I(.11)$, $I(.12)$, $I(.135)$, and $I(.15)$.

THOUGHTS INTO WORDS

66. Are all functions also relations? Are all relations also functions? Defend your answers.

67. What does it mean to say that the domain of a function may be restricted if the function represents a real world situation? Give two or three examples of such situations.

68. Does $f(a + b) = f(a) + f(b)$ for all functions? Defend your answer.

69. Are there any functions for which $f(a + b) = f(a) + f(b)$? Defend your answer.

8.2 Functions: Graphs and Applications

In Section 7.1, we used phrases such as "the graph of the solution set of the equation $y = x - 1$," or simply "the graph of the equation $y = x - 1$" is a line that contains the points $(0, -1)$ and $(1, 0)$. Because the equation $y = x - 1$ (which can be written as $f(x) = x - 1$) can be used to specify a function, the line previously referred to is also called the **graph of the function specified by the equation** or simply the **graph of the function**. Generally speaking, the graph of any equation that determines a function is also called the graph of the function. Thus, the graphing techniques we discussed in Chapter 7 will continue to play an important role as we graph functions.

As we use the function concept in our study of mathematics, it is helpful to classify certain types of functions and become familiar with their equations, characteristics, and graphs. This section will cover two special types of functions: **linear** and **quadratic functions**. These functions are merely an outgrowth of our earlier study of linear and quadratic equations.

Linear Functions

Any function that can be written in the form

$$f(x) = ax + b,$$

where a and b are real numbers, is called a **linear function**. The following equations are examples of linear functions.

$$f(x) = -3x + 6, \qquad f(x) = 2x + 4, \qquad f(x) = -\frac{1}{2}x - \frac{3}{4}$$

Graphing linear functions is quite easy because the graph of every linear function is a straight line. Therefore, all we need to do is determine two points of the graph and draw the line determined by those two points. You may want to continue to use a third point as a check point.

Example 1

Graph the function $f(x) = -3x + 6$.

Solution

Because $f(0) = 6$, the point $(0, 6)$ is on the graph. Likewise, because $f(1) = 3$, the point $(1, 3)$ is on the graph. Plotting these two points and drawing the line determined by the two points produces Figure 8.2.

FIGURE 8.2

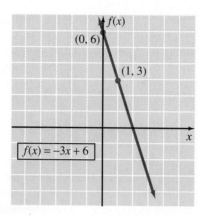

$f(x)$

$(0, 6)$

$(1, 3)$

x

$f(x) = -3x + 6$

REMARK   Note that in Figure 8.2 we labeled the vertical axis $f(x)$. It could also be labeled y, since $f(x) = -3x + 6$, and $y = -3x + 6$ mean the same thing. We will continue to use $f(x)$ in this chapter to help you adjust to the function notation.   △

Example 2   Graph the function $f(x) = x$.

Solution   The equation $f(x) = x$ can be written as $f(x) = 1x + 0$; thus, it is a linear function. Since $f(0) = 0$ and $f(2) = 2$, the points $(0, 0)$ and $(2, 2)$ determine the line in Figure 8.3. The function $f(x) = x$ is often called the **identity function**

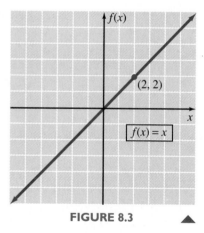

$f(x)$

$(2, 2)$

x

$f(x) = x$

FIGURE 8.3   ▲

As you use function notation to graph functions, it is often helpful to think of the ordinate of every point on the graph as the value of the function at a specific value of x. Geometrically, this functional value is the directed distance of the point from the x-axis as illustrated in Figure 8.4, with the function $f(x) = 2x - 4$. For example, consider the graph of the function $f(x) = 2$. The function $f(x) = 2$ means that every functional value is 2, or geometrically, that every point of the graph is 2 units above the x-axis. Thus, the graph is the horizontal line indicated in Figure 8.5. Any linear function of the form $f(x) = ax + b$, where $a = 0$, is called a **constant function** and its graph is a horizontal line.

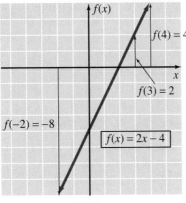

FIGURE 8.4

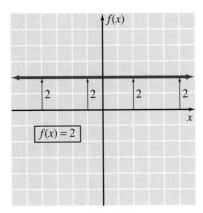

FIGURE 8.5

Quadratic Functions

Any function that can be written in the form

$$f(x) = ax^2 + bx + c,$$

where a, b, and c are real numbers, with $a \neq 0$, is called a **quadratic function**. The following are examples of quadratic functions.

$$f(x) = 3x^2,$$

$$f(x) = -2x^2 + 5x,$$

$$f(x) = 4x^2 - 7x + 1$$

The techniques we discussed in Chapter 7 relative to graphing quadratic equations of the form $y = ax^2 + bx + c$ provide the basis for graphing quadratic functions. Let's review some work from Chapter 7 with an example.

Example 3   Graph the function $f(x) = 2x^2 - 4x + 5$.

Solution
$$\begin{aligned}
f(x) &= 2x^2 - 4x + 5 \\
&= 2(x^2 - 2x + \quad\ ) + 5 \qquad \text{Recall the process of completing the square!} \\
&= 2(x^2 - 2x + 1) + 5 - 2 \\
&= 2(x - 1)^2 + 3
\end{aligned}$$

This is the general form of the equation for a parabola as presented in Section 7.4. From this form we can obtain the following information about the parabola.

$$f(x) = 2(x - 1)^2 + 3$$

Narrows the parabola and opens it *upward*

Moves the parabola I unit to the *right*

Moves the parabola 3 units *up*

Therefore, the vertex of the parabola is at (1, 3) and the line $x = 1$ is the axis of symmetry indicated in Figure 8.6.

FIGURE 8.6

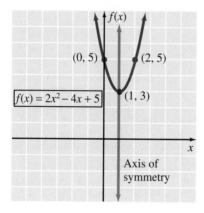

(0, 5)   (2, 5)

$f(x) = 2x^2 - 4x + 5$ (1, 3)

Axis of symmetry

In general, if we complete the square on

$$f(x) = ax^2 + bx + c,$$

we obtain

$$f(x) = a\left(x^2 + \frac{b}{a}x + \underline{\quad}\right) + c$$

$$= a\left(x^2 + \frac{b}{a}x + \frac{b^2}{4a^2}\right) + c - \frac{b^2}{4a}$$

$$= a\left(x + \frac{b}{2a}\right)^2 + \frac{4ac - b^2}{4a}.$$

Therefore, the parabola associated with $f(x) = ax^2 + bx + c$ has its vertex at

$$\left(-\frac{b}{2a}, \frac{4ac - b^2}{4a}\right)$$

and the equation of its axis of symmetry is $x = -\dfrac{b}{2a}$. These facts are illustrated in Figure 8.7.

FIGURE 8.7

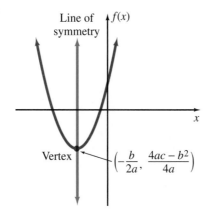

Line of symmetry

Vertex $\left(-\dfrac{b}{2a}, \dfrac{4ac - b^2}{4a}\right)$

By using the information from Figure 8.7 we now have another way of graphing quadratic functions of the form $f(x) = ax^2 + bx + c$, without completing the square, as indicated by the following steps.

1. Determine whether the parabola opens upward (if $a > 0$) or downward (if $a < 0$).

2. Find $-\dfrac{b}{2a}$, which is the x-coordinate of the vertex.

3. Find $f\left(-\dfrac{b}{2a}\right)$, which is the y-coordinate of the vertex. $\left(\text{You could also}\right.$ find the y-coordinate by evaluating $\left.\dfrac{4ac - b^2}{4a}\right)$.

4. Locate another point on the parabola and also locate its image across the line of symmetry, $x = -\dfrac{b}{2a}$.

The three points found in steps 2, 3, and 4 should determine the general shape of the parabola. Let's try two examples and use these steps.

Example 4

Graph $f(x) = 3x^2 - 6x + 5$.

Solution

STEP 1 Because $a = 3$, the parabola opens upward.

STEP 2 $-\dfrac{b}{2a} = -\dfrac{-6}{6} = 1$.

STEP 3 $f\left(-\dfrac{b}{2a}\right) = f(1) = 3 - 6 + 5 = 2$.
Thus, the vertex is at $(1, 2)$.

STEP 4 Letting $x = 2$, we obtain $f(2) = 12 - 12 + 5 = 5$. Thus $(2, 5)$ is on the graph and so is its reflection $(0, 5)$ across the line of symmetry $x = 1$.

The three points (1, 2), (2, 5), and (0, 5) are used to graph the parabola in Figure 8.8.

FIGURE 8.8

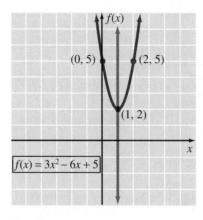

$f(x) = 3x^2 - 6x + 5$

Example 5

Graph $f(x) = -x^2 - 4x - 7$.

Solution

STEP 1 Since $a = -1$, the parabola opens downward.

STEP 2 $-\dfrac{b}{2a} = -\dfrac{-4}{-2} = -2$

STEP 3 $f\left(-\dfrac{b}{2a}\right) = f(-2) = -(-2)^2 - 4(-2) - 7 = -3$.
Thus, the vertex is at $(-2, -3)$.

STEP 4 Letting $x = 0$, we obtain $f(0) = -7$. Thus $(0, -7)$ is on the graph and so is its reflection $(-4, -7)$ across the line of symmetry $x = -2$.

The three points $(-2, -3)$, $(0, -7)$, and $(-4, -7)$ are used to draw the parabola in Figure 8.9.

FIGURE 8.9

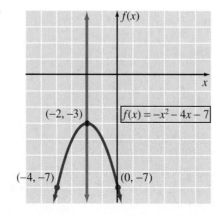

$f(x) = -x^2 - 4x - 7$

Our knowledge of parabolas can be helpful even when using a graphing utility to graph a quadratic function. Consider the following example.

Example 6

Use a graphing utility to obtain the graph of the quadratic function $f(x) = -x^2 + 37x - 311$.

Solution

First, we know from the equation that the parabola opens downward and its width is the same as that of the basic parabola $f(x) = x^2$. Furthermore, because of the size of the constant term we suspect that the boundaries of the viewing rectangle will need to be changed from their normal settings. We will use $\left(-\dfrac{b}{2a}, f\left(-\dfrac{b}{2a}\right)\right)$ to determine the vertex.

$$-\frac{b}{2a} = -\frac{37}{2(-1)} = 18.5 \qquad \text{and} \qquad f(18.5) = 31.25$$

Therefore, setting the boundaries of the viewing rectangle so that $-2 \le x \le 25$ and $-10 \le y \le 35$ we obtain the graph in Figure 8.10.

FIGURE 8.10

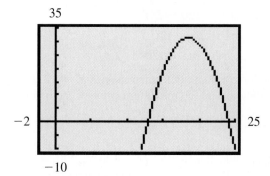

REMARK   The graph in Figure 8.10 is sufficient for most purposes since it shows the vertex and the x-intercepts of the parabola. Certainly other boundaries could be used that would also give this information.   △

Quadratic Functions and Problem Solving

As we have seen, the vertex of the graph of a quadratic function is either the lowest or the highest point on the graph. Thus, the vocabulary *minimum value* or *maximum value* of a function is often used in applications of the parabola. The x-value of the vertex indicates where the minimum or maximum occurs and $f(x)$ yields the minimum or maximum value of the function. Let's consider some examples that use these ideas.

Example 7

A farmer has 120 rods of fencing and wants to enclose a rectangular plot of land that requires fencing on only three sides, since it is bounded by a river on one side. Find the length and width of the plot that will maximize the area.

Solution

Let x represent the width; then $120 - 2x$ represents the length, as indicated in Figure 8.11.

FIGURE 8.11

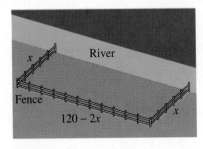

The function $A(x) = x(120 - 2x)$ represents the area of the plot in terms of the width x. Since

$$A(x) = x(120 - 2x)$$
$$= 120x - 2x^2$$
$$= -2x^2 + 120x,$$

we have a quadratic function with $a = -2$, $b = 120$, and $c = 0$. Therefore, the x-value where the maximum value of the function is obtained is

$$-\frac{b}{2a} = -\frac{120}{2(-2)} = 30.$$

If $x = 30$, then $120 - 2x = 120 - 2(30) = 60$. Thus, the farmer should make the plot 30 rods wide and 60 rods long to maximize the area at $(30)(60) = 1800$ square rods. ▲

Example 8

Find two numbers whose sum is 30, such that the sum of their squares is a minimum.

Solution

Let x represent one of the numbers; then $30 - x$ represents the other number. By expressing the sum of the squares as a function of x we obtain

$$f(x) = x^2 + (30 - x)^2,$$

which can be simplified to

$$f(x) = x^2 + 900 - 60x + x^2$$
$$= 2x^2 - 60x + 900.$$

This is a quadratic function with $a = 2$, $b = -60$, and $c = 900$. Therefore, the x-value where the minimum occurs is

$$-\frac{b}{2a} = -\frac{-60}{4} = 15.$$

If $x = 15$, then $30 - x = 30 - (15) = 15$. Thus, the two numbers should both be 15. ▲

Example 9

A golf pro-shop operator finds that she can sell 30 sets of golf clubs at $500 per set in a year. Furthermore, she predicts that for each $25 decrease in price, three extra sets of golf clubs could be sold. At what price should she sell the clubs to maximize gross income?

Solution

Sometimes when analyzing such a problem it helps to start setting up a table as follows.

	Number of sets	Price per set	Income
3 additional sets can be sold for a $25 decrease in price →	30	$500	$15,000
	33	$475	$15,675
	36	$450	$16,200

Let x represent the number of $25 decreases in price. Then the income can be expressed as a function of x as follows.

$$f(x) = (30 + 3x)(500 - 25x)$$

Number of sets   Price per set

Simplifying this, we obtain

$$f(x) = 15{,}000 - 750x + 1500x - 75x^2$$
$$= -75x^2 + 750x + 15{,}000.$$

Completing the square we obtain

$$f(x) = -75x^2 + 750x + 15{,}000$$
$$= -75(x^2 - 10x + \underline{\qquad}) + 15{,}000$$
$$= -75(x^2 - 10x + 25) + 15{,}000 + 1875$$
$$= -75(x - 5)^2 + 16{,}875.$$

From this form we know that the vertex of the parabola is at $(5, 16875)$. So 5 decreases of $25, that is, a $125 reduction in price, will give a maximum income of $16,875. The golf clubs should be sold at $375 per set. ▲

Other Functions

To graph a new function, that is, one we are unfamiliar with, we can use some of the graphing suggestions offered in Chapter 7. Let's restate those suggestions in terms of function vocabulary and notation. Pay special attention to steps 2 and 3 where we restated the concepts of intercepts and symmetry in terms of function notation.

1. Determine the domain of the function.

2. Determine any types of symmetry that the equation possesses. If $f(-x) = f(x)$, then the function exhibits y-axis symmetry. If $f(-x) = -f(x)$, then the function exhibits origin symmetry. (Note that the definition of a function rules out the possibility that the graph of a function has x-axis symmetry.)

3. Find the y-intercept (we are labeling the y-axis with $f(x)$) by evaluating $f(0)$. Find the x-intercept by finding the value(s) of x such that $f(x) = 0$.

4. Set up a table of ordered pairs that satisfy the equation. The type of symmetry and the domain will affect your choice of values of x in the table.

5. Plot the points associated with the ordered pairs and connect them with a smooth curve. Then, if appropriate, reflect this part of the curve according to any symmetries possessed by the graph.

Let's consider some examples in terms of these suggestions.

Example 10

Graph $f(x) = \sqrt{x}$.

Solution

Since the radicand must be nonnegative, the domain is the set of nonnegative real numbers. Thus, $D = \{x \mid x \geq 0\}$. Since $x \geq 0$, $f(-x)$ is not a real number. There is no symmetry for this graph. We see that $f(0) = 0$, so both intercepts are 0. That is to say, the origin $(0, 0)$ is a point of the graph.

Now let's set up a table of values and keep in mind that $x \geq 0$.

x	f(x)
0	0
1	1
4	2
9	3

Plotting these points and connecting them with a smooth curve produces Figure 8.12.

FIGURE 8.12

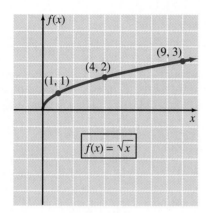

Sometimes a new function is defined in terms of old functions. In such cases, the definition plays an important role in the study of the new function. Consider the following example.

Example 11

Graph the function $f(x) = |x|$.

Solution

The concept of absolute value is defined for all real numbers as

$$|x| = x \quad \text{if } x \geq 0,$$

$$|x| = -x \quad \text{if } x < 0.$$

Therefore, the absolute value function can be expressed as

$$f(x) = |x| = \begin{cases} x & \text{if } x \geq 0 \\ -x & \text{if } x < 0 \end{cases}.$$

The graph of $f(x) = x$ for $x \geq 0$ is the ray in the first quadrant and the graph of $f(x) = -x$ for $x < 0$ is the half-line in the second quadrant as indicated in Figure 8.13.

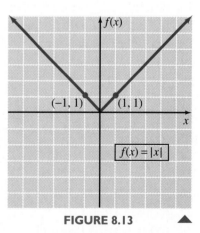

FIGURE 8.13

REMARK Note in Example 11 that the equation $f(x) = |x|$ does exhibit y-axis symmetry because $f(-x) = |-x| = |x|$. Even though we did not use the symmetry idea to sketch the curve, we should recognize that the symmetry does exist.

Problem Set 8.2

For Problems 1–30, graph each of the linear and quadratic functions.

1. $f(x) = 2x - 4$

2. $f(x) = 3x + 3$

3. $f(x) = -2x^2$

4. $f(x) = -4x^2$

5. $f(x) = -3x$

6. $f(x) = -4x$

7. $f(x) = -(x + 1)^2 - 2$

8. $f(x) = -(x - 2)^2 + 4$

9. $f(x) = -x + 3$

10. $f(x) = -2x - 4$

11. $f(x) = x^2 + 2x - 2$

12. $f(x) = x^2 - 4x - 1$

13. $f(x) = -x^2 + 6x - 8$

14. $f(x) = -x^2 - 8x - 15$

15. $f(x) = -3$

16. $f(x) = 1$

17. $f(x) = 2x^2 - 20x + 52$

18. $f(x) = 2x^2 + 12x + 14$

19. $f(x) = -3x^2 + 6x$

20. $f(x) = -4x^2 - 8x$

21. $f(x) = x^2 - x + 2$

22. $f(x) = x^2 + 3x + 2$

23. $f(x) = 2x^2 + 10x + 11$

24. $f(x) = 2x^2 - 10x + 15$

25. $f(x) = -2x^2 - 1$

26. $f(x) = -3x^2 + 2$

27. $f(x) = -3x^2 + 12x - 7$

28. $f(x) = -3x^2 - 18x - 23$

29. $f(x) = -2x^2 + 14x - 25$

30. $f(x) = -2x^2 - 10x - 14$

31. Suppose that the cost function for a particular item is given by the equation $C(x) = 2x^2 - 320x + 12,920$, where x represents the number of items. How many items should be produced to minimize the cost?

32. Suppose that the equation $p(x) = -2x^2 + 280x - 1000$, where x represents the number of items sold, describes the profit function for a certain business. How many items should be sold to maximize the profit?

33. Find two numbers whose sum is 30, such that the sum of the square of one number plus ten times the other number is a minimum.

34. The height of a projectile fired vertically into the air (neglecting air resistance) at an initial velocity of 96 feet per second is a function of the time and is given by the equation $f(x) = 96x - 16x^2$, where x represents the time. Find the highest point reached by the projectile.

35. Two hundred and forty meters of fencing is available to enclose a rectangular playground. What should be the dimensions of the playground to maximize the area?

36. Find two numbers whose sum is 50 and whose product is a maximum.

37. A Cable TV company has 1000 subscribers who each pay $15 per month. Based on a survey, they feel that for each decrease of $.25 on the monthly rate, they could obtain 20 additional subscribers. At what rate will maximum revenue be obtained and how many subscribers will it take at that rate?

38. A motel advertises that they will provide dinner, a dance, and drinks for $50 per couple for a New Year's Eve party. They must have a guarantee of 30 couples. Furthermore, they will agree that for each couple in excess of 30, they will reduce the price per couple for all attending by $.50. How many couples will it take to maximize the motel's revenue?

For Problems 39–48, graph each of the functions.

39. $f(x) = \sqrt{x + 3}$

40. $f(x) = (x - 2)^3$

41. $f(x) = \dfrac{x}{x-2}$

42. $f(x) = \dfrac{x-2}{x}$

43. $f(x) = x + |x|$

44. $f(x) = x - |x|$

45. $f(x) = |x| - x$

46. $f(x) = \dfrac{x}{|x|}$

47. $f(x) = \dfrac{3}{x^2}$

48. $f(x) = \dfrac{3}{x^2 + 1}$

THOUGHTS INTO WORDS

49. Give a step-by-step description of how you would use the ideas of this section to graph $f(x) = -4x^2 + 16x - 13$.

50. Suppose that your friend has no idea how to graph the function $f(x) = x^4 + x^2$. How would you help him?

Graphics Calculator Activities

51. Use a graphics calculator to check your graphs for Problems 39–48.

52. **(a)** Graph $f(x) = x^3$, $f(x) = x^3 + 2$, $f(x) = x^3 + 3$, and $f(x) = x^3 + 4$ on the same set of axes.
  (b) Graph $f(x) = x^3$, $f(x) = x^3 - 1$, $f(x) = x^3 - 3$, and $f(x) = x^3 - 4$ on the same set of axes.
  (c) Based on your results in parts (a) and (b), make a general statement about the graph of $f(x) = x^3 + k$, where k is any real number.

53. **(a)** Graph $f(x) = x^3$, $f(x) = (x-2)^3$, $f(x) = (x-4)^3$, and $f(x) = (x-6)^3$ on the same set of axes.
  (b) Graph $f(x) = x^3$, $f(x) = (x+1)^3$, $f(x) = (x+3)^3$ and $f(x) = (x+5)^3$ on the same set of axes.
  (c) Based on your results in parts (a) and (b), make a general statement about the graph of $f(x) = (x-h)^3$, where h is any real number.

54. **(a)** Graph $f(x) = x^3$, $f(x) = 2x^3$, $f(x) = 3x^3$, and $f(x) = 5x^3$ on the same set of axes.
  (b) Graph $f(x) = x^3$, $f(x) = .75x^3$, $f(x) = .5x^3$, and $f(x) = .25x^3$ on the same set of axes.
  (c) Graph $f(x) = x^3$, $f(x) = -x^3$, $f(x) = -4x^3$, and $f(x) = -.5x^3$ on the same set of axes.
  (d) Based on your results in parts (a), (b), and (c), make a general statement about the graph of $f(x) = ax^3$, where a is a nonzero real number.

8.3 Graphing Made Easy Via Transformations

In Section 7.4 we found that the graph of $f(x) = (x - 5)^2$ is the basic parabola, $f(x) = x^2$, translated five units to the right. Likewise, we know that the graph of $f(x) = -x^2 - 2$ is the basic parabola reflected across the x-axis and translated downward two units. Translations and reflections apply not only to parabolas but to curves in general. Therefore, if we know the shapes of a few basic curves, then numerous variations of these curves can be easily sketched using the concepts of translation and reflection.

From our previous graphing experiences, we can sketch the graphs of the basic functions $f(x) = x^3, f(x) = \sqrt{x}, f(x) = |x|$, and $f(x) = \dfrac{1}{x}$. Let's reproduce those figures at this time so that we have them at our fingertips for our work with transformations (Figures 8.14–8.17).

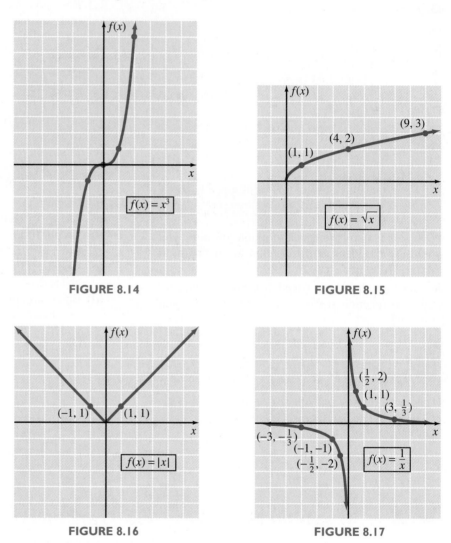

FIGURE 8.14

FIGURE 8.15

FIGURE 8.16

FIGURE 8.17

Translations of the Basic Curves

From our work in Chapter 7 we know that the graph of $f(x) = x^2 + 3$ is the graph of $f(x) = x^2$ moved up three units. Likewise, the graph of $f(x) = x^2 - 2$ is the graph of $f(x) = x^2$ moved down two units. Now we will describe in general the concept of **vertical translation**.

Vertical Translation

The graph of $y = f(x) + k$ is the graph of $y = f(x)$ shifted k units upward if $k > 0$ or shifted $|k|$ units downward if $k < 0$.

In Figure 8.18(a) the graph of $f(x) = \dfrac{1}{x} + 2$ is obtained by shifting the graph of $f(x) = \dfrac{1}{x}$ upward two units. Likewise, in Figure 8.18(b) the graph of $f(x) = \dfrac{1}{x} - 3$ is obtained by shifting the graph of $f(x) = \dfrac{1}{x}$ downward three units. (Remember that $f(x) = \dfrac{1}{x} - 3$ can be written as $f(x) = \dfrac{1}{x} + (-3)$.)

FIGURE 8.18

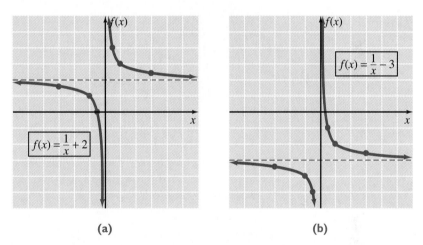

(a)   (b)

Horizontal translations of the basic parabola were also graphed in Chapter 7. For example, the graph of $f(x) = (x - 4)^2$ is the graph of $f(x) = x^2$ shifted four units to the right and the graph of $f(x) = (x + 5)^2$ is the graph of $f(x) = x^2$ shifted five units to the left. We can describe the general concept of a **horizontal translation** as follows.

Horizontal Translation

The graph of $y = f(x - h)$ is the graph of $y = f(x)$ shifted h units to the right if $h > 0$ or shifted $|h|$ units to the left if $h < 0$.

In Figure 8.19 the graph of $f(x) = (x - 3)^3$ is obtained by shifting the graph of $f(x) = x^3$ three units to the right. Likewise, the graph of $f(x) = (x + 2)^3$ is obtained by shifting the graph of $f(x) = x^3$ two units to the left.

FIGURE 8.19

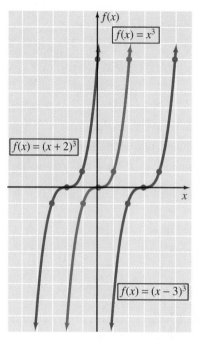

Reflections of the Basic Curves

From our work in Chapter 7 we know that the graph of $f(x) = -x^2$ is the graph of $f(x) = x^2$ reflected through the x-axis. We describe the general concept of an **x-axis reflection** as follows.

---

x-axis Reflection

The graph of $y = -f(x)$ is the graph of $y = f(x)$ reflected through the x-axis.

In Figure 8.20 the graph of $f(x) = -\sqrt{x}$ is obtained by reflecting the graph of $f(x) = \sqrt{x}$ through the x-axis. We sometimes refer to reflections as **mirror images**. Thus, in Figure 8.20, if we think of the x-axis as a mirror, the graphs of $f(x) = \sqrt{x}$ and $f(x) = -\sqrt{x}$ are mirror images of each other.

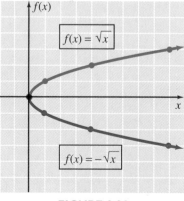

FIGURE 8.20

In Chapter 7 we did not consider a y-axis reflection of the basic parabola $f(x) = x^2$ because it is symmetric with respect to the y-axis. In other words, a y-axis reflection of $f(x) = x^2$ produces the same figure in the same location. However, at this time we will describe the general concept of a y-axis reflection.

---

y-axis Reflection

The graph of $y = f(-x)$ is the graph of $y = f(x)$ reflected through the y-axis.

Now suppose that we want to do a y-axis reflection of $f(x) = \sqrt{x}$. Since $f(x) = \sqrt{x}$ is defined for $x \geq 0$, the y-axis reflection $f(x) = \sqrt{-x}$ is defined for $-x \geq 0$, which is equivalent to $x \leq 0$. Figure 8.21 shows the y-axis reflection of $f(x) = \sqrt{x}$.

FIGURE 8.21

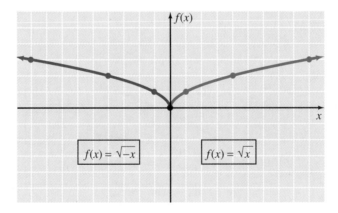

$$f(x) = \sqrt{-x} \qquad f(x) = \sqrt{x}$$

Vertical Stretching and Shrinking

Translations and reflections are called **rigid transformations** because the basic shape of the curve being transformed is not changed. In other words, only the positions of the graphs are changed. Now we want to consider some transformations that distort the shape of the original figure somewhat.

In Section 7.9 we graphed the equation $y = 2x^2$ by doubling the y-coordinates of the ordered pairs that satisfy the equation $y = x^2$. We obtained a parabola with its vertex at the origin, symmetric to the y-axis, but *narrower* than the basic parabola. Likewise, we graphed the equation $y = \frac{1}{2}x^2$ by halving the y-coordinates of the ordered pairs that satisfy $y = x^2$. In this case, we obtained a parabola with its vertex at the origin, symmetric to the y-axis, but *wider* than the basic parabola.

The concepts *narrower* and *wider* can be used to describe parabolas but cannot be used to accurately describe some other curves. Instead, we use the more general concepts of vertical stretching and shrinking.

Vertical Stretching and Shrinking

The graph of $y = cf(x)$ is obtained from the graph of $y = f(x)$ by multiplying the y-coordinates of $y = f(x)$ by c. If $c > 1$, the graph is said to be *stretched* by a factor of c, and if $0 < c < 1$, the graph is said to be *shrunk* by a factor of c.

In Figure 8.22 the graph of $f(x) = 2\sqrt{x}$ is obtained by doubling the y-coordinates of points on the graph of $f(x) = \sqrt{x}$. Likewise, in Figure 8.22 the graph of $f(x) = \frac{1}{2}\sqrt{x}$ is obtained by halving the y-coordinates of points on the graph of $f(x) = \sqrt{x}$.

FIGURE 8.22

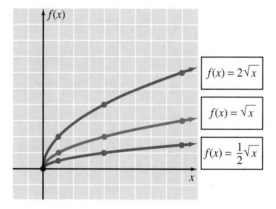

$f(x) = 2\sqrt{x}$

$f(x) = \sqrt{x}$

$f(x) = \frac{1}{2}\sqrt{x}$

Successive Transformations

Some curves are the result of performing more than one transformation on a basic curve. Let's consider the graph of a function that involves a stretching, a reflection, a horizontal translation, and a vertical translation of the basic absolute value function.

Example 1

Graph $f(x) = -2|x - 3| + 1$.

Solution

This is the basic absolute value curve stretched by a factor of two, reflected through the x-axis, shifted three units to the right, and shifted one unit upward. To sketch the graph we locate the point $(3, 1)$, and then determine a point on each of the rays. The graph is shown in Figure 8.23.

FIGURE 8.23

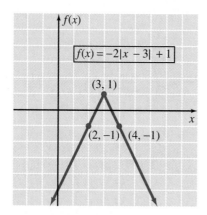

REMARK Note in Example 1 that we did not sketch the original basic curve $f(x) = |x|$ nor any of the intermediate transformations. However, it is helpful to mentally picture each transformation. This locates the point $(3, 1)$ and establishes the fact that the two rays point downward. Then a point on each ray determines the final graph. △

You also need to realize that changing the order of doing the transformations may produce an incorrect graph. In Example 1, performing the translations first followed by the stretching and x-axis reflection would produce an incorrect graph that has its vertex at $(3, -1)$ instead of $(3, 1)$. Unless parentheses indicate otherwise, stretchings, shrinkings, and x-axis reflections should be performed before translations.

Finally, let's use a graphing utility to give another illustration of the concept of stretching and shrinking a curve.

Example 2

If $f(x) = \sqrt{25 - x^2}$, sketch a graph of $y = 2(f(x))$ and $y = \frac{1}{2}(f(x))$.

Solution

If $y = f(x) = \sqrt{25 - x^2}$, then

$$y = 2(f(x))$$
$$= 2\sqrt{25 - x^2}$$

and

$$y = \frac{1}{2}(f(x))$$
$$= \frac{1}{2}\sqrt{25 - x^2}.$$

Graphing all three of these functions on the same set of axes produces Figure 8.24.

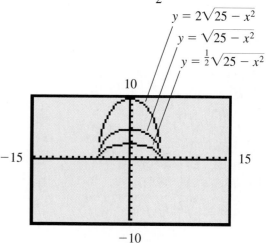

FIGURE 8.24 ▲

Problem Set 8.3

For Problems 1–34, graph each of the functions.

1. $f(x) = -x^3$

2. $f(x) = x^3 - 2$

3. $f(x) = -(x - 4)^2 + 2$

4. $f(x) = -2(x + 3)^2 - 4$

5. $f(x) = \dfrac{1}{x} - 2$

6. $f(x) = \dfrac{1}{x - 2}$

7. $f(x) = |x - 1| + 2$

8. $f(x) = -|x + 2|$

9. $f(x) = \dfrac{1}{2}|x|$

10. $f(x) = -2|x|$

11. $f(x) = -2\sqrt{x}$

12. $f(x) = 2\sqrt{x - 1}$

13. $f(x) = \sqrt{x + 2} - 3$

14. $f(x) = -\sqrt{x + 2} + 2$

15. $f(x) = \dfrac{2}{x - 1} + 3$

16. $f(x) = \dfrac{3}{x + 3} - 4$

17. $f(x) = \sqrt{2 - x}$

18. $f(x) = \sqrt{1 - x}$

19. $f(x) = -3(x - 2)^2 - 1$

20. $f(x) = (x + 5)^2 - 2$

21. $f(x) = 3(x - 2)^3 - 1$

22. $f(x) = -2(x + 1)^3 + 2$

23. $f(x) = 2x^3 + 3$

24. $f(x) = -2x^3 - 1$

25. $f(x) = -2\sqrt{x + 3} + 4$

26. $f(x) = -3\sqrt{x - 1} + 2$

27. $f(x) = \dfrac{-2}{x + 2} + 2$

28. $f(x) = \dfrac{-1}{x - 1} - 1$

29. $f(x) = \dfrac{x - 1}{x}$

30. $f(x) = \dfrac{x + 2}{x}$

31. $f(x) = -3|x + 4| + 3$

32. $f(x) = -2|x - 3| - 4$

33. $f(x) = 4|x| + 2$

34. $f(x) = -3|x| - 4$

35. The graph of $y = f(x)$ with a domain of $-2 \le x \le 2$ is shown in Figure 8.25.

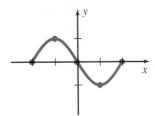

FIGURE 8.25

Sketch the graph of each of the following transformations of $y = f(x)$.

(a) $y = f(x) + 3$

(b) $y = f(x - 2)$

(c) $y = -f(x)$

(d) $y = f(x + 3) - 4$

36. Is the graph of $f(x) = x^2 + 2x + 4$ a y-axis reflection of $f(x) = x^2 - 2x + 4$? Defend your answer.

37. Is the graph of $f(x) = x^2 - 4x - 7$ an x-axis reflection of $f(x) = x^2 + 4x + 7$? Defend your answer.

38. Your friend claims that the graph of $f(x) = \dfrac{2x + 1}{x}$ is the graph of $f(x) = \dfrac{1}{x}$ shifted two units upward. How could you decide whether or not she is correct?

Graphics Calculator Activities

39. Use a graphics calculator to check your graphs for Problems 1–34.

40. For each of the following, answer the question based on your knowledge of transformations and then use a graphics calculator to check your answer.

(a) Is the graph of $f(x) = 2x^2 + 8x + 13$ a y-axis reflection of $f(x) = 2x^2 - 8x + 13$?

(b) Is the graph of $f(x) = 3x^2 - 12x + 16$ an x-axis reflection of $f(x) = -3x^2 + 12x - 16$?

(c) Is the graph of $f(x) = \sqrt{4 - x}$ a y-axis reflection of $f(x) = \sqrt{x + 4}$?

(d) Is the graph of $f(x) = \sqrt{3 - x}$ a y-axis reflection of $f(x) = \sqrt{x - 3}$?

(e) Is the graph of $f(x) = -x^3 + x + 1$ an y-axis reflection of $f(x) = x^3 - x + 1$?

(f) Is the graph of $f(x) = -(x - 2)^3$ an x-axis reflection of $f(x) = (x - 2)^3$?

(g) Is the graph of $f(x) = -x^3 - x^2 - x + 1$ an x-axis reflection of $f(x) = x^3 + x^2 + x - 1$?

(h) Is the graph of $f(x) = \dfrac{3x + 1}{x}$ a vertical translation of $f(x) = \dfrac{1}{x}$ two units upward?

(i) Is the graph of $f(x) = 2 + \dfrac{1}{x}$ a y-axis reflection of $f(x) = \dfrac{2x - 1}{x}$?

41. Are the graphs of the two functions $f(x) = \sqrt{x - 2}$ and $g(x) = \sqrt{2 - x}$ y-axis reflections of each other? Defend your answer.

42. Are the graphs of $f(x) = 2\sqrt{x}$ and $g(x) = \sqrt{2x}$ identical? Defend your answer.

43. Are the graphs of $f(x) = \sqrt{x + 4}$ and $g(x) = \sqrt{-x + 4}$ y-axis reflections of each other? Defend your answer.

44. Graph $f(x) = x^4 + x^3$. Now predict the graph for each of the following and check each prediction with your graphics calculator.

(a) $f(x) = x^4 + x^3 - 4$
(b) $f(x) = (x - 3)^4 + (x - 3)^3$
(c) $f(x) = -x^4 - x^3$
(d) $f(x) = x^4 - x^3$

45. Graph $f(x) = \sqrt[3]{x}$. Now predict the graph for each of the following and check each prediction with your graphics calculator.

(a) $f(x) = 5 + \sqrt[3]{x}$
(b) $f(x) = \sqrt[3]{x + 4}$
(c) $f(x) = -\sqrt[3]{x}$
(d) $f(x) = \sqrt[3]{x - 3} - 5$
(e) $f(x) = \sqrt[3]{-x}$

8.4 Combining Functions

In subsequent mathematics courses, it is common to encounter functions that are defined in terms of sums, differences, products, and quotients of simpler functions. For example, if $h(x) = x^2 + \sqrt{x - 1}$, then we may consider the function h as the sum of f and g, where $f(x) = x^2$ and $g(x) = \sqrt{x - 1}$. In general, *if f and g are functions and D is the intersection of their domains*, then the following definitions can be made.

Sum	$(f + g)(x) = f(x) + g(x)$
Difference	$(f - g)(x) = f(x) - g(x)$
Product	$(f \cdot g)(x) = f(x) \cdot g(x)$
Quotient	$\left(\dfrac{f}{g}\right)(x) = \dfrac{f(x)}{g(x)}, \quad g(x) \neq 0$

Example 1

If $f(x) = 3x - 1$ and $g(x) = x^2 - x - 2$, find (a) $(f + g)(x)$, (b) $(f - g)(x)$, (c) $(f \cdot g)(x)$, and (d) $(f/g)(x)$. Determine the domain of each.

Solutions

(a) $(f + g)(x) = f(x) + g(x) = (3x - 1) + (x^2 - x - 2) = x^2 + 2x - 3$

(b) $(f - g)(x) = f(x) - g(x)$
$$= (3x - 1) - (x^2 - x - 2)$$
$$= 3x - 1 - x^2 + x + 2$$
$$= -x^2 + 4x + 1$$

(c) $(f \cdot g)(x) = f(x) \cdot g(x)$
$$= (3x - 1)(x^2 - x - 2)$$
$$= 3x^3 - 3x^2 - 6x - x^2 + x + 2$$
$$= 3x^3 - 4x^2 - 5x + 2$$

(d) $\left(\dfrac{f}{g}\right)(x) = \dfrac{f(x)}{g(x)} = \dfrac{3x - 1}{x^2 - x - 2}$

The domain of both f and g is the set of all real numbers. Therefore, the domain of $f + g$, $f - g$, and $f \cdot g$ is the set of all real numbers. For f/g, the denominator $x^2 - x - 2$ cannot equal zero. Solving $x^2 - x - 2 = 0$ produces

$$(x - 2)(x + 1) = 0$$
$$x - 2 = 0 \quad \text{or} \quad x + 1 = 0$$
$$x = 2 \quad \text{or} \quad x = -1.$$

Therefore, the domain for f/g is the set of all real numbers except 2 and -1.

Composition of Functions

Besides adding, subtracting, multiplying, and dividing functions, there is another important operation called *composition*. The composition of two functions can be defined as follows.

DEFINITION 8.3

The **composition** of functions f and g is defined by

$$(f \circ g)(x) = f(g(x)),$$

for all x in the domain of g such that $g(x)$ is in the domain of f.

The left side, $(f \circ g)(x)$, of the equation in Definition 8.3 can be read as "the composition of f and g" and the right side, $f(g(x))$, can be read as "f of g of x." It may also be helpful for you to mentally picture Definition 8.3 as two function machines *hooked together* to produce another function (often called a **composite function**) as illustrated in Figure 8.26. Notice that what comes out of the function g is substituted into the function f. Thus, composition is sometimes called the substitution of functions.

FIGURE 8.26

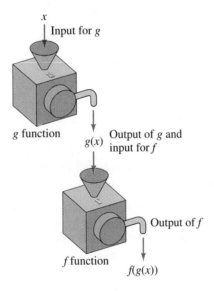

Figure 8.26 also vividly illustrates the fact that $f \circ g$ is defined *for all x in the domain of g such that g(x) is the domain of f*. In other words, what comes out of g must be capable of being fed into f. Let's consider some examples.

Example 2

If $f(x) = x^2$ and $g(x) = x - 3$, find $(f \circ g)(x)$ and determine its domain.

Solution

Apply Definition 8.3 to obtain

$$(f \circ g)(x) = f(g(x)) = f(x - 3) = (x - 3)^2.$$

Because g and f are both defined for all real numbers, so is $f \circ g$. ▲

Example 3

Solution

If $f(x) = \sqrt{x}$ and $g(x) = x - 4$, find $(f \circ g)(x)$ and determine its domain.

Apply Definition 8.3 to obtain

$$(f \circ g)(x) = f(g(x))$$
$$= f(x - 4)$$
$$= \sqrt{x - 4}.$$

The domain of g is all real numbers but the domain of f is only the nonnegative real numbers. Thus $g(x)$, which is $x - 4$, has to be nonnegative. So

$$x - 4 \geq 0$$
$$x \geq 4,$$

and the domain of $f \circ g$ is $D = \{x \mid x \geq 4\}$.   ▲

Definition 8.3, with f and g interchanged, defines the composition of g and f as $(g \circ f)(x) = g(f(x))$.

Example 4

Solution

If $f(x) = x^2$ and $g(x) = x - 3$, find $(g \circ f)(x)$ and determine its domain.

$$(g \circ f)(x) = g(f(x))$$
$$= g(x^2)$$
$$= x^2 - 3$$

Since f and g are both defined for all real numbers, the domain of $g \circ f$ is the set of all real numbers.   ▲

The results of Examples 2 and 4 demonstrate an important idea, namely, that the composition of functions is *not a commutative operation*. In other words, it is not true that $f \circ g = g \circ f$ for all functions f and g. However, as we will see in the next section, there is a special class of functions where $f \circ g = g \circ f$.

Example 5

Solution

If $f(x) = \sqrt{x}$ and $g(x) = 2x - 1$, find $(f \circ g)(x)$ and $(g \circ f)(x)$. Also determine the domain of each composite function.

$$(f \circ g)(x) = f(g(x))$$
$$= f(2x - 1)$$
$$= \sqrt{2x - 1}$$

The domain and range of g is the set of all real numbers, but the domain of f is all *nonnegative* real numbers. Therefore $g(x)$, which is $2x - 1$, must be nonnegative.

$$2x - 1 \geq 0$$

$$2x \geq 1$$

$$x \geq \frac{1}{2}$$

Thus the domain of $f \circ g$ is $D = \left\{ x \mid x \geq \frac{1}{2} \right\}$.

$$
\begin{aligned}
(g \circ f)(x) &= g(f(x)) \\
&= g(\sqrt{x}) \\
&= 2\sqrt{x} - 1
\end{aligned}
$$

The domain and range of f is the set of nonnegative real numbers. The domain of g is the set of all real numbers. Therefore, the domain of $g \circ f$ is $D = \{ x \mid x \geq 0 \}$. ▲

Example 6

If $f(x) = 2/(x - 1)$ and $g(x) = 1/x$, find $(f \circ g)(x)$ and $(g \circ f)(x)$. Determine the domain for each composite function.

Solution

$$
\begin{aligned}
(f \circ g)(x) &= f(g(x)) \\
&= f\left(\frac{1}{x}\right) \\
&= \frac{2}{\dfrac{1}{x} - 1} = \frac{2}{\dfrac{1 - x}{x}} \\
&= \frac{2x}{1 - x}
\end{aligned}
$$

The domain of g is all real numbers except zero, and the domain of f is all real numbers except one. Since $g(x)$, which is $1/x$, cannot equal 1,

$$\frac{1}{x} \neq 1$$

$$x \neq 1.$$

Therefore, the domain of $f \circ g$ is $D = \{ x \mid x \neq 0 \text{ and } x \neq 1 \}$.

$$
\begin{aligned}
(g \circ f)(x) &= g(f(x)) \\
&= g\left(\frac{2}{x - 1}\right) \\
&= \frac{1}{\dfrac{2}{x - 1}} \\
&= \frac{x - 1}{2}
\end{aligned}
$$

The domain of f is all real numbers except 1, and the domain of g is all real numbers except 0. Since $f(x)$, which is $2/(x - 1)$ will never equal 0, the domain of $g \circ f$ is $D = \{x \mid x \neq 1\}$.

Example 7

If $f(x) = 2x + 3$ and $g(x) = \sqrt{x - 1}$, determine each of the following.

    (a) $(f \circ g)(x)$
    (b) $(g \circ f)(x)$
    (c) $(f \circ g)(5)$
    (d) $(g \circ f)(7)$

Solution

    (a) $(f \circ g)(x) = f(g(x))$
$$= f(\sqrt{x - 1})$$
$$= 2\sqrt{x - 1} + 3, \qquad D = \{x \mid x \geq 1\}$$

    (b) $(g \circ f)(x) = g(f(x))$
$$= g(2x + 3)$$
$$= \sqrt{2x + 3 - 1}$$
$$= \sqrt{2x + 2}, \qquad D = \{x \mid x \geq -1\}$$

    (c) By using the composite function formed in part (a), we obtain
$$(f \circ g)(5) = 2\sqrt{5 - 1} + 3$$
$$= 2\sqrt{4} + 3$$
$$= 2(2) + 3$$
$$= 7.$$

    (d) By using the composite function formed in part (b) we obtain
$$(g \circ f)(7) = \sqrt{2(7) + 2}$$
$$= \sqrt{16} = 4.$$

A graphing utility can be used to find the graph of a composite function without actually forming the function algebraically. Let's see how this works.

Example 8

If $f(x) = x^3$ and $g(x) = x - 4$, use a graphing utility to obtain the graph of $y = (f \circ g)(x)$ and of $y = (g \circ f)(x)$.

Solution

To find the graph of $y = (f \circ g)(x)$ we can make the following assignments.

$$Y_1 = x - 4$$

$$Y_2 = (Y_1)^3$$

(Note that we have substituted Y_1 for x in $f(x)$ and assigned this expression to Y_2, much the same way as we would do it algebraically.) The graph of $y = (f \circ g)(x)$ is shown in Figure 8.27.

FIGURE 8.27

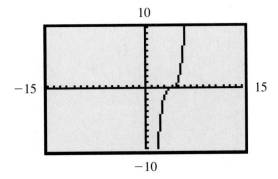

To find the graph of $y = (g \circ f)(x)$ we can make the following assignments.

$$Y_1 = x^3$$

$$Y_2 = Y_1 - 4$$

The graph of $y = (g \circ f)(x)$ is shown in Figure 8.28.

FIGURE 8.28

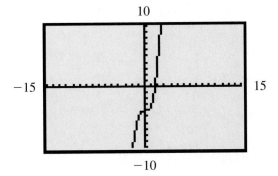

Take another look at Figures 8.27 and 8.28. Note that in Figure 8.27 the graph of $y = (f \circ g)(x)$ is the basic cubic curve $f(x) = x^3$ shifted 4 units to the right. Likewise, in Figure 8.28 the graph of $y = (g \circ f)(x)$ is the basic cubic curve shifted 4 units downward. These are examples of a more general concept of using composite functions to represent various geometric transformations.

Problem Set 8.4

For Problems 1–8, find $f + g, f - g, f \cdot g,$ and $\dfrac{f}{g}$.

1. $f(x) = 3x - 4, \quad g(x) = 5x + 2$

2. $f(x) = -6x - 1, \quad g(x) = -8x + 7$ —

3. $f(x) = x^2 - 6x + 4, \quad g(x) = -x - 1$

4. $f(x) = 2x^2 - 3x + 5, \quad g(x) = x^2 - 4$

5. $f(x) = x^2 - x - 1, \quad g(x) = x^2 + 4x - 5$

6. $f(x) = x^2 - 2x - 24, \quad g(x) = x^2 - x - 30$

7. $f(x) = \sqrt{x - 1}, \quad g(x) = \sqrt{x} \quad \sqrt{x - 1}$

8. $f(x) = \sqrt{x + 2}, \quad g(x) = \sqrt{3x - 1} \quad \sqrt{x + 2}$

For Problems 9–20, determine the indicated functional values.

9. If $f(x) = 9x - 2$ and $g(x) = -4x + 6$, find $(f \circ g)(-2)$ and $(g \circ f)(4)$.

10. If $f(x) = -2x - 6$ and $g(x) = 3x + 10$, find $(f \circ g)(5)$ and $(g \circ f)(-3)$

11. If $f(x) = 4x^2 - 1$ and $g(x) = 4x + 5$, find $(f \circ g)(1)$ and $(g \circ f)(4)$

12. If $f(x) = -5x + 2$ and $g(x) = -3x^2 + 4$, find $(f \circ g)(-2)$ and $(g \circ f)(-1)$.

13. If $f(x) = \dfrac{1}{x}$ and $g(x) = \dfrac{2}{x - 1}$, find $(f \circ g)(2)$ and $(g \circ f)(-1)$.

14. If $f(x) = \dfrac{2}{x - 1}$ and $g(x) = -\dfrac{3}{x}$, find $(f \circ g)(1)$ and $(g \circ f)(-1)$.

15. If $f(x) = \dfrac{1}{x - 2}$ and $g(x) = \dfrac{4}{x - 1}$, find $(f \circ g)(3)$ and $(g \circ f)(2)$.

16. If $f(x) = \sqrt{x + 6}$ and $g(x) = 3x - 1$, find $(f \circ g)(-2)$ and $(g \circ f)(-2)$.

17. If $f(x) = \sqrt{3x - 2}$ and $g(x) = -x + 4$, find $(f \circ g)(1)$ and $(g \circ f)(6)$. $\quad \sqrt{7}$ and 0

18. If $f(x) = -5x + 1$ and $g(x) = \sqrt{4x + 1}$, find $(f \circ g)(6)$ and $(g \circ f)(-1)$. —

19. If $f(x) = |4x - 5|$ and $g(x) = x^3$, find $(f \circ g)(-2)$ and $(g \circ f)(2)$.

20. If $f(x) = -x^3$ and $g(x) = |2x + 4|$, find $(f \circ g)(-1)$ and $(g \circ f)(-3)$. —

For Problems 21–38, determine $(f \circ g)(x)$ and $(g \circ f)(x)$ for each pair of functions. Also specify the domain of $(f \circ g)(x)$ and $(g \circ f)(x)$.

21. $f(x) = 3x$ and $g(x) = 5x - 1$

22. $f(x) = 4x - 3$ and $g(x) = -2x$

23. $f(x) = -2x + 1$ and $g(x) = 7x + 4$

24. $f(x) = 6x - 5$ and $g(x) = -x + 6$

25. $f(x) = 3x + 2$ and $g(x) = x^2 + 3$

26. $f(x) = -2x + 4$ and $g(x) = 2x^2 - 1$

27. $f(x) = 2x^2 - x + 2$ and $g(x) = -x + 3$

28. $f(x) = 3x^2 - 2x - 4$ and $g(x) = -2x + 1$

29. $f(x) = \dfrac{3}{x}$ and $g(x) = 4x - 9$

30. $f(x) = -\dfrac{2}{x}$ and $g(x) = -3x + 6$

31. $f(x) = \sqrt{x + 1}$ and $g(x) = 5x + 3$

32. $f(x) = 7x - 2$ and $g(x) = \sqrt{2x - 1}$

33. $f(x) = \dfrac{1}{x}$ and $g(x) = \dfrac{1}{x - 4}$

34. $f(x) = \dfrac{2}{x + 3}$ and $g(x) = -\dfrac{3}{x}$

35. $f(x) = \sqrt{x}$ and $g(x) = \dfrac{4}{x}$

36. $f(x) = \dfrac{2}{x}$ and $g(x) = |x|$

37. $f(x) = \dfrac{3}{2x}$ and $g(x) = \dfrac{1}{x + 1}$

38. $f(x) = \dfrac{4}{x - 2}$ and $g(x) = \dfrac{3}{4x}$

For Problems 39–46, show that $(f \circ g)(x) = x$ and $(g \circ f)(x) = x$ for each pair of functions.

39. $f(x) = 3x$ and $g(x) = \dfrac{1}{3}x$

40. $f(x) = -2x$ and $g(x) = -\dfrac{1}{2}x$

41. $f(x) = 4x + 2$ and $g(x) = \dfrac{x - 2}{4}$

42. $f(x) = 3x - 7$ and $g(x) = \dfrac{x + 7}{3}$

43. $f(x) = \dfrac{1}{2}x + \dfrac{3}{4}$ and $g(x) = \dfrac{4x - 3}{2}$

44. $f(x) = \dfrac{2}{3}x - \dfrac{1}{5}$ and $g(x) = \dfrac{3}{2}x + \dfrac{3}{10}$

45. $f(x) = -\dfrac{1}{4}x - \dfrac{1}{2}$ and $g(x) = -4x - 2$

46. $f(x) = -\dfrac{3}{4}x + \dfrac{1}{3}$ and $g(x) = -\dfrac{4}{3}x + \dfrac{4}{9}$

THOUGHTS INTO WORDS

47. Discuss whether or not addition, subtraction, multiplication, and division of functions are commutative operations.

48. How would you explain the concept of composition of functions to a friend who missed class the day it was discussed?

49. Explain why the composition of functions is not a commutative operation.

Graphics Calculator Activities

50. For each of the following (1) predict the general shape and location of the graph, and then (2) use your graphics calculator to graph the function to check your prediction. (Your knowledge of the graphs of the basic functions being added or subtracted should be helpful when making your predictions.)
(a) $f(x) = x^3 + x^2$　　(b) $f(x) = x^3 - x^2$
(c) $f(x) = x^2 - x^3$　　(d) $f(x) = |x| + \sqrt{x}$
(e) $f(x) = |x| - \sqrt{x}$　(f) $f(x) = \sqrt{x} - |x|$

51. For each of the following, use your graphics calculator to find the graph of $y = (f \circ g)(x)$ and of $y = (g \circ f)(x)$. Then algebraically find $(f \circ g)(x)$ and $(g \circ f)(x)$ to see if your results agree.
(a) $f(x) = x^2$ and $g(x) = x - 3$
(b) $f(x) = x^3$ and $g(x) = x + 4$
(c) $f(x) = x - 2$ and $g(x) = -x^3$
(d) $f(x) = x + 6$ and $g(x) = \sqrt{x}$
(e) $f(x) = \sqrt{x}$ and $g(x) = x - 5$

8.5   Inverse Functions

Graphically, the distinction between a relation and a function is easily recognizable. In Figure 8.29 we have sketched four graphs. Which of these are graphs of functions and which are graphs of relations that are not functions? Think in terms of "to each member of the domain there is assigned one and only one member of the range"; this is the basis for what is known as the **vertical line test for functions.** Because each value of x produces only one value of $f(x)$, any vertical line drawn through a **graph of a function must not intersect the graph in more than one point.** Therefore, parts (a) and (c) of Figure 8.29 are graphs of functions and parts (b) and (d) are graphs of relations that are not functions.

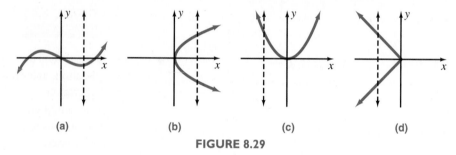

(a)   (b)   (c)   (d)

FIGURE 8.29

There is also a useful distinction we make between two basic types of functions. Consider the graphs of the two functions $f(x) = 2x - 1$ and $f(x) = x^2$ in Figure 8.30. In part (a) any *horizontal line* will intersect the graph in no more than one point. Therefore, every value of $f(x)$ has only one value of x associated with it. Any function that has the additional property of having only one value of x associated with each value of $f(x)$ is called a **one-to-one function.** The function $f(x) = x^2$ is not a one-to-one function because the horizontal line in part (b) of Figure 8.30 intersects the parabola in two points.

FIGURE 8.30

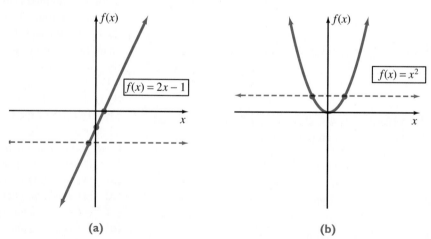

$f(x) = 2x - 1$

$f(x) = x^2$

(a)   (b)

The statement that for a function f to be a one-to-one function *every value of f(x) has only one value of x associated with it* can be equivalently stated as *if $f(x_1) = f(x_2)$ for x_1 and x_2 in the domain of f, then $x_1 = x_2$*. Let's use this last if-then statement to verify that $f(x) = 2x - 1$ is a one-to-one function. We start with the assumption that $f(x_1) = f(x_2)$.

$$2x_1 - 1 = 2x_2 - 1$$

$$2x_1 = 2x_2$$

$$x_1 = x_2$$

Thus, $f(x) = 2x - 1$ is a one-to-one function.

To show that $f(x) = x^2$ is not a one-to-one function we simply need to find two distinct real numbers in the domain of f that produce the same functional value. For example, $f(-2) = (-2)^2 = 4$ and $f(2) = 2^2 = 4$. Thus, $f(x) = x^2$ is not a one-to-one function.

In terms of ordered pairs, a one-to-one function does not contain any ordered pairs having the same second component. For example,

$$f = \{(1, 3), (2, 6), (4, 12)\}$$

is a one-to-one function, but

$$g = \{(1, 2), (2, 5), (-2, 5)\}$$

is not a one-to-one function because 5 is paired with both 2 and -2.

If the components of each ordered pair of a given one-to-one function are interchanged, the resulting function and the given function are called **inverses** of each other. Thus,

$$\{(1, 3), (2, 6), (4, 12)\} \quad \text{and} \quad \{(3, 1), (6, 2), (12, 4)\}$$

are **inverse functions.** The inverse of a function f is denoted by f^{-1} (read "f inverse" or "the inverse of f"). If (a, b) is an ordered pair of f, then (b, a) is an ordered pair of f^{-1}. The domain and range of f^{-1} are the range and domain, respectively, of f.

REMARK Do not confuse the -1 in f^{-1} with a negative exponent. The symbol f^{-1} does not mean $\dfrac{1}{f^1}$, but refers to the inverse function of function f. △

Graphically, two functions that are inverses of each other are mirror images with reference to the line $y = x$. This is due to the fact that ordered pairs (a, b) and (b, a) are mirror images with respect to the line $y = x$ as demonstrated in Figure 8.31.

FIGURE 8.31

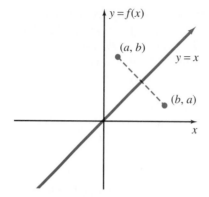

Therefore, if the graph of a function f is known, as in Figure 8.32(a), then the graph of f^{-1} can be determined by reflecting f across the line $y = x$ (Figure 8.32(b)).

FIGURE 8.32

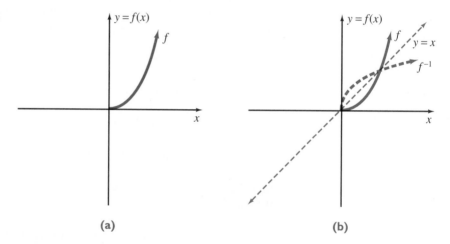

(a)   (b)

Another useful way of viewing inverse functions is in terms of composition. Basically, inverse functions *undo* each other and this can be more formally stated as follows. If f and g are inverses of each other, then

1. $(f \circ g)(x) = f(g(x)) = x$ for all x in domain of g; and

2. $(g \circ f)(x) = g(f(x)) = x$ for all x in domain of f.

As we will see in a moment, this relationship of inverse functions can be used to verify whether two functions are indeed inverses of each other.

Finding Inverse Functions

The idea of inverse functions *undoing each other* provides the basis for a rather informal approach to finding the inverse of a function. Consider the function

$$f(x) = 2x + 1.$$

To each x this function assigns *twice x plus 1*. To *undo* this function, we could *subtract 1 and divide by 2*. So, the inverse should be

$$f^{-1}(x) = \frac{x-1}{2}.$$

Now let's verify that f and f^{-1} are inverses of each other.

$$(f \circ f^{-1})(x) = f(f^{-1}(x)) = f\left(\frac{x-1}{2}\right) = 2\left(\frac{x-1}{2}\right) + 1 = x$$

and

$$(f^{-1} \circ f)(x) = f^{-1}(f(x)) = f^{-1}(2x+1) = \frac{2x+1-1}{2} = x$$

Thus, the inverse of f is given by

$$f^{-1}(x) = \frac{x-1}{2}.$$

Let's consider another example of finding an inverse function by the *undoing* process.

Example 1

Find the inverse of $f(x) = 3x - 5$.

Solution

To each x, the function f assigns *three times x minus 5*. To *undo* this we can *add 5 and then divide by 3*. So, the inverse should be

$$f^{-1}(x) = \frac{x+5}{3}.$$

To verify that f and f^{-1} are inverses we can show that

$$(f \circ f^{-1})(x) = f(f^{-1}(x)) = f\left(\frac{x+5}{3}\right) = 3\left(\frac{x+5}{3}\right) - 5 = x$$

and

$$(f^{-1} \circ f)(x) = f^{-1}(f(x)) = f^{-1}(3x-5) = \frac{3x-5+5}{3} = x.$$

Thus, f and f^{-1} are inverses and we can write

$$f^{-1}(x) = \frac{x+5}{3}.$$

▲

This informal approach may not work very well with more complex functions, but it does emphasize how inverse functions are related to each other. A more formal and systematic technique for finding the inverse of a function can be described as follows.

1. Replace the symbol $f(x)$ by y.
2. Interchange x and y.

3. Solve the equation for y in terms of x.

4. Replace y by the symbol $f^{-1}(x)$.

Now let's use two examples to illustrate this technique.

Example 2

Find the inverse of $f(x) = -3x + 11$.

Solution

Replace $f(x)$ by y and the given equation becomes

$$y = -3x + 11.$$

Interchange x and y to produce

$$x = -3y + 11.$$

Now, solve for y to obtain

$$x = -3y + 11$$

$$3y = -x + 11$$

$$y = \frac{-x + 11}{3}.$$

Finally, replacing y by $f^{-1}(x)$ we can express the inverse function as

$$f^{-1}(x) = \frac{-x + 11}{3}.$$

Example 3

Find the inverse of $f(x) = \frac{3}{2}x - \frac{1}{4}$.

Solution

Replace $f(x)$ by y and the given equation becomes

$$y = \frac{3}{2}x - \frac{1}{4}.$$

Interchange x and y to produce

$$x = \frac{3}{2}y - \frac{1}{4}.$$

Now, solve for y to obtain

$$x = \frac{3}{2}y - \frac{1}{4}$$

$$4x = 6y - 1$$

$$4x + 1 = 6y$$

$$\frac{4x + 1}{6} = y$$

$$\frac{2}{3}x + \frac{1}{6} = y.$$

Finally, replacing y by $f^{-1}(x)$ we can express the inverse function as

$$f^{-1}(x) = \frac{2}{3}x + \frac{1}{6}.$$ ▲

For both Examples 2 and 3 you should be able to show that

$$(f \circ f^{-1})(x) = x \quad \text{and} \quad (f^{-1} \circ f)(x) = x.$$

Does $f(x) = x^2 - 2$ have an inverse function? Sometimes a graph of the function helps to answer such a question. In Figure 8.33(a), it should be evident that f is not a one-to-one function and therefore cannot have an inverse. However, it should also be apparent from the graph that if we restrict the domain of f to be the nonnegative real numbers, then it is a one-to-one function and should have an inverse (Figure 8.33(b)). The next example illustrates how to find the inverse function.

FIGURE 8.33

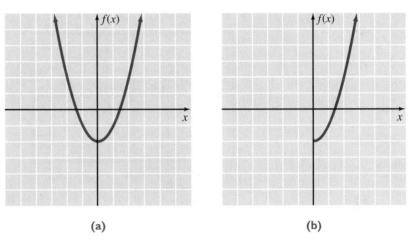

(a) (b)

Example 4

Solution

Find the inverse of $f(x) = x^2 - 2$, where $x \geq 0$.

Replace $f(x)$ by y, and the equation becomes

$$y = x^2 - 2, \quad x \geq 0.$$

Interchange x and y to produce

$$x = y^2 - 2, \quad y \geq 0.$$

Now let's solve for y, keeping in mind that y is to be nonnegative.

$$x = y^2 - 2$$
$$x + 2 = y^2$$
$$\sqrt{x + 2} = y, \quad \text{where } x \geq -2.$$

Finally, replace y by $f^{-1}(x)$ and the inverse function can be expressed

$$f^{-1}(x) = \sqrt{x + 2}, \quad x \geq -2.$$

The domain of f equals the range of f^{-1} (both are the nonnegative real numbers), and the range of f equals the domain of f^{-1} (both are the real numbers greater than or equal to -2). It can be shown that $(f \circ f^{-1})(x) = x$ and $(f^{-1} \circ f)(x) = x$. Again, we leave this for you to complete. ▲

As functions become more complex, a graphing utility can be used to help with the problems we have discussed in this section. For example, suppose that we want to know if the function $f(x) = \dfrac{3x + 1}{x - 4}$ is a one-to-one function and therefore has an inverse. Using a graphing utility, we can quickly get a sketch of the graph as shown in Figure 8.34. Then by applying the horizontal line test to the graph, we feel fairly certain that the function is one-to-one. (Later we will develop some concepts that will allow us to be absolutely certain of this conclusion.)

FIGURE 8.34

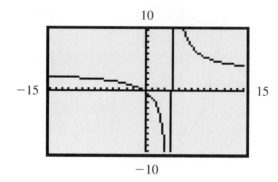

Problem Set 8.5

For Problems 1–8, identify each graph as (a) the graph of a function or (b) the graph of a relation that is not a function. Use the vertical line test.

1.

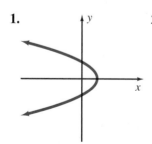

FIGURE 8.35

2.

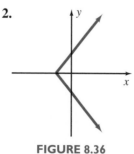

FIGURE 8.36

3.

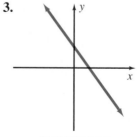

FIGURE 8.37

4.

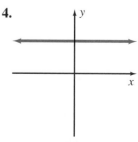

FIGURE 8.38

5.

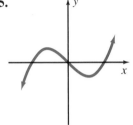

FIGURE 8.39

6.

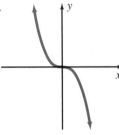

FIGURE 8.40

13.

FIGURE 8.47

14.

FIGURE 8.48

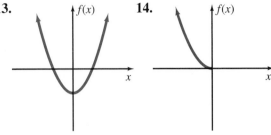

7.

FIGURE 8.41

8.

FIGURE 8.42

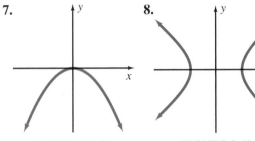

15.

FIGURE 8.49

16.

FIGURE 8.50

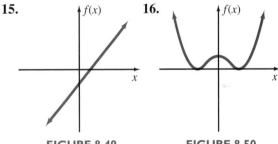

For Problems 9–16, identify each graph as (a) the graph of a one-to-one function or (b) the graph of a function that is not one-to-one. Use the horizontal line test.

9.

FIGURE 8.43

10.

FIGURE 8.44

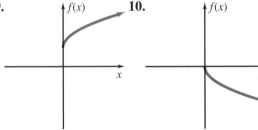

11.

FIGURE 8.45

12.

FIGURE 8.46

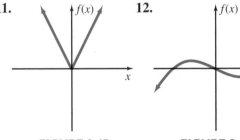

For Problems 17–24, determine whether the function is one-to-one without graphing the function.

17. $f(x) = 7x - 2$ **18.** $f(x) = -4x + 6$

19. $f(x) = x^4$ **20.** $f(x) = x^2 + 1$

21. $f(x) = |x|$ **22.** $f(x) = 2|x| + 3$

23. $f(x) = -2x$ **24.** $f(x) = -x$

For Problems 25–28 (a) list the domain and range of the given function, (b) form the inverse function, and (c) list the domain and range of the inverse function.

25. $f = \{(1, 3), (2, 6), (3, 11), (4, 18)\}$

26. $f = \{(0, -4), (1, -3), (4, -2)\}$

27. $f = \{(-2, -1), (-1, 1), (0, 5), (5, 10)\}$

28. $f = \{(-1, 1), (-2, 4), (1, 9), (2, 12)\}$

For Problems 29–38, find the inverse of the given function by using the "undoing process" and then verify that $(f \circ f^{-1})(x) = x$ and $(f^{-1} \circ f)(x) = x$.

29. $f(x) = 5x - 4$

30. $f(x) = 7x + 9$

31. $f(x) = -2x + 1$

32. $f(x) = -4x - 3$

33. $f(x) = \dfrac{4}{5}x$

34. $f(x) = -\dfrac{2}{3}x$

35. $f(x) = \dfrac{1}{2}x + 4$

36. $f(x) = \dfrac{3}{4}x - 2$

37. $f(x) = \dfrac{1}{3}x - \dfrac{2}{5}$

38. $f(x) = \dfrac{2}{5}x + \dfrac{1}{3}$

For Problems 39–48, find the inverse of the given function by using the process illustrated in Examples 2 and 3 of this section and then verify that $(f \circ f^{-1})(x) = x$ and $(f^{-1} \circ f)(x) = (x)$.

39. $f(x) = 9x + 4$

40. $f(x) = 8x - 5$

41. $f(x) = -5x - 4$

42. $f(x) = -6x + 2$

43. $f(x) = -\dfrac{2}{3}x + 7$

44. $f(x) = -\dfrac{3}{5}x + 1$

45. $f(x) = \dfrac{4}{3}x - \dfrac{1}{4}$

46. $f(x) = \dfrac{5}{2}x + \dfrac{2}{7}$

47. $f(x) = -\dfrac{3}{7}x - \dfrac{2}{3}$

48. $f(x) = -\dfrac{3}{5}x + \dfrac{3}{4}$

49. $f(x) = \sqrt{x}$ for $x \geq 0$

50. $f(x) = \dfrac{1}{x}$ for $x \neq 0$

51. $f(x) = x^2 + 4$ for $x \geq 0$

52. $f(x) = 1 + \dfrac{1}{x}$ for $x > 0$

For Problems 53–64 (a) find the inverse of the given function and (b) graph the given function and its inverse on the same set of axes.

53. $f(x) = 4x$

54. $f(x) = \dfrac{2}{5}x$

55. $f(x) = -\dfrac{1}{3}x$

56. $f(x) = -6x$

57. $f(x) = 3x - 3$

58. $f(x) = 2x + 2$

59. $f(x) = -2x - 4$

60. $f(x) = -3x + 9$

61. $f(x) = x^2, x \geq 0$

62. $f(x) = x^2 + 2, x \geq 0$

63. $f(x) = \sqrt{x - 1}$ for $x \geq 1$

64. $f(x) = \sqrt{x + 3}$ for $x \geq -3$

65. Does the function $f(x) = 4$ have an inverse? Explain your answer.

66. Explain why every nonconstant linear function has an inverse.

Further Investigations

67. The composition idea can also be used to find the inverse of a function. For example, to find the inverse of $f(x) = 5x + 3$, we could proceed as follows.

$$f(f^{-1}(x)) = 5(f^{-1}(x)) + 3 \quad \text{and} \quad f(f^{-1}(x)) = x$$

Therefore, equating the two expressions for $f(f^{-1}(x))$ we obtain

$$5(f^{-1}(x)) + 3 = x$$
$$5(f^{-1}(x)) = x - 3$$
$$f^{-1}(x) = \frac{x - 3}{5}.$$

Use this approach to find the inverse of each of the following functions.
(a) $f(x) = 2x + 1$
(b) $f(x) = 3x - 2$
(c) $f(x) = -4x + 5$
(d) $f(x) = -x + 1$
(e) $f(x) = 2x$
(f) $f(x) = -5x$

68. If $f(x) = 2x + 3$ and $g(x) = 3x - 5$, find
(a) $(f \circ g)^{-1}(x)$ **(b)** $(f^{-1} \circ g^{-1})(x)$
(c) $(g^{-1} \circ f^{-1})(x)$

Graphics Calculator Activities

69. For Problems 39–52, graph the given function, the inverse function that you found, and $f(x) = x$ on the same set of axes. In each case the given function and its inverse should produce graphs that are reflections of each other through the line $f(x) = x$.

70. Let's use a graphics calculator to show that $(f \circ g)(x) = x$ and $(g \circ f)(x) = x$ for two functions that we think are inverses of each other. Consider the following functions $f(x) = 3x + 4$ and $g(x) = \frac{x - 4}{3}$. We can make the following assignments.

$$f: \quad Y_1 = 3x + 4$$
$$g: \quad Y_2 = \frac{x - 4}{3}$$
$$f \circ g: \quad Y_3 = 3Y_2 + 4$$
$$g \circ f: \quad Y_4 = \frac{Y_1 - 4}{3}$$

Now we can graph Y_3 and Y_4 and show that they both produce the line $f(x) = x$.

Use this approach to check your answers for Problems 53–64.

71. Use the approach demonstrated in Problem 70 to show that $f(x) = x^2 - 2$, for $x \geq 0$, and $g(x) = \sqrt{x + 2}$, for $x \geq -2$, are inverses of each other.

8.6 Direct and Inverse Variations

"The distance a car travels at a fixed rate *varies directly* as the time." "At a constant temperature, the volume of an enclosed gas *varies inversely* as the pressure." Such statements illustrate two basic types of functional relationships, called **direct** and **inverse variation**, which are widely used, especially in the physical sciences. These relationships can be expressed by equations that specify functions. The purpose of this section is to investigate these special functions.

The statement *y varies directly as x* means

$$\boxed{y = kx}$$

where k is a nonzero constant, called the **constant of variation**. The phrase *y is directly proportional to x,* is also used to indicate direct variation; k is then referred to as the **constant of proportionality**.

REMARK Notice that the equation $y = kx$ defines a function; using function notation, it could be written as $f(x) = kx$. However, in this section it is more convenient to avoid the function notation and use variables that are meaningful in terms of the physical entities in the problem. △

Statements that indicate direct variation may also involve powers of x. For example, *y varies directly as the square of x* can be written as

$$y = kx^2.$$

In general, *y varies directly as the nth power of x(n > 0)* means

$$y = kx^n.$$

There are basically three types of problems in dealing with direct variation: (1) translating an English statement into an equation that expresses the direct variation; (2) finding the constant of variation from given values of the variables; and (3) finding additional values of the variables once the constant of variation has been determined. Let's consider an example of each of these types of problems.

Example 1

Translate the statement *the tension on a spring varies directly as the distance it is stretched* into an equation that uses k as the constant of variation.

Solution

Let t represent the tension and d the distance and the equation becomes $t = kd$.

▲

Example 2

If A varies directly as the square of s, and $A = 28$ when $s = 2$, find the constant of variation.

Solution

Since A varies directly as the square of s, we have

$$A = ks^2.$$

Substitute $A = 28$ and $s = 2$ to obtain

$$28 = k(2)^2.$$

Solve this equation for k to yield

$$28 = 4k$$
$$7 = k.$$

The constant of variation is 7.

▲

Example 3

If y is directly proportional to x and if $y = 6$ when $x = 9$, find the value of y when $x = 24$.

Solution

The statement y *is directly proportional to* x translates into $y = kx$. Let $y = 6$ and $x = 9$ and the constant of variation becomes

$$6 = k(9)$$

$$6 = 9k$$

$$\frac{6}{9} = k$$

$$\frac{2}{3} = k.$$

So, the specific equation is $y = \frac{2}{3}x$. Now, let $x = 24$ to obtain

$$y = \frac{2}{3}(24) = 16.$$

The required value of y is 16. ▲

Inverse Variation

The second basic type of variation, called **inverse variation**, is defined as follows. The statement y *varies inversely as* x means

$$y = \frac{k}{x}$$

where k is a nonzero constant and is again referred to as the constant of variation. The phrase y *is inversely proportional to* x is also used to express inverse variation. As with direct variation, statements that indicate inverse variation may involve powers of x. For example, y *varies inversely as the square of* x can be written as

$$y = \frac{k}{x^2}.$$

In general, y *varies inversely as the nth power of* $x(n > 0)$ means

$$y = \frac{k}{x^n}.$$

The following examples illustrate the three basic kinds of problems that we run across involving inverse variation.

Example 4

Translate the statement *the length of a rectangle of fixed area varies inversely as the width* into an equation using k as the constant of variation.

Solution

Let l represent the length and w the width and the equation is

$$l = \frac{k}{w}.$$

▲

Example 5

If y is inversely proportional to x and $y = 4$ when $x = 12$, find the constant of variation.

Solution

Since y is inversely proportional to x, we have

$$y = \frac{k}{x}.$$

Substitute $y = 4$ and $x = 12$ to obtain

$$4 = \frac{k}{12}.$$

Solve this equation for k to yield

$$k = 48.$$

The constant of variation is 48.   ▲

Example 6

Suppose the number of days it takes to complete a construction job varies inversely as the number of people assigned to the job. If it takes 7 people 8 days to do the job, how long would it take 14 people to complete the job?

Solution

Let d represent the number of days and p the number of people. The phrase *number of days ... varies inversely as the number of people* translates into

$$d = \frac{k}{p}.$$

Let $d = 8$ when $p = 7$ and the constant of variation becomes

$$8 = \frac{k}{7}$$

$$k = 56.$$

So, the specific equation is

$$d = \frac{56}{p}.$$

Now, let $p = 14$ to obtain

$$d = \frac{56}{14}$$

$$d = 4.$$

It should take 14 people 4 days to complete the job.   ▲

The terms *direct* and *inverse*, as applied to variation, refer to the relative behavior of the variables involved in the equation. That is to say, in direct variation $(y = kx)$ an assignment of *increasing absolute values* for x produces *increasing absolute values* for y. However in inverse variation $\left(y = \dfrac{k}{x}\right)$ an assignment of *increasing absolute values* for x produces *decreasing absolute values* for y.

Joint Variation

Variation may involve more than two variables. The following table illustrates some variation statements and their equivalent algebraic equations using k as the constant of variation.

Variation statement	Algebraic equation
1. y varies jointly as x and z	$y = kxz$
2. y varies jointly as x, z, and w	$y = kxzw$
3. V varies jointly as h and the square of r	$V = khr^2$
4. h varies directly as V and inversely as w	$h = \dfrac{kV}{w}$
5. y is directly proportional to x and inversely proportional to the square of z	$y = \dfrac{kx}{z^2}$
6. y varies jointly as w and z, and inversely as x	$y = \dfrac{kwz}{x}$

Statements 1, 2, and 3 illustrate the concept of **joint variation**. Statements 4 and 5 show that both direct and inverse variation may occur in the same problem. Statement 6 combines joint variation with inverse variation.

The two final examples involve some of these possible variation situations.

Example 7

The length of a rectangular box with a fixed height varies directly as the volume and inversely as the width. If the length is 12 centimeters when the volume is 960 cubic centimeters and the width is 8 centimeters, find the length when the volume is 700 centimeters and the width is 5 centimeters.

Solution

Use l for length, V for volume, and w for width and the phrase *length varies directly as the volume and inversely as the width* translates into

$$l = \frac{kV}{w}.$$

Substitute $l = 12$, $V = 960$, and $w = 8$ and the constant of variation becomes

$$12 = \frac{k(960)}{8}$$

$$12 = 120k$$

$$\frac{1}{10} = k.$$

So the specific equation is

$$l = \frac{\frac{1}{10}V}{w}$$

$$= \frac{V}{10w}.$$

Now, let $V = 700$ and $w = 5$ to obtain

$$l = \frac{700}{10(5)}$$

$$= \frac{700}{50}$$

$$= 14.$$

The length is 14 centimeters.　　　▲

Example 8

Suppose that y varies jointly as x and z, and inversely as w. If $y = 154$ when $x = 6$, $z = 11$, and $w = 3$, find the constant of variation.

Solution

The statement *y varies jointly as x and z, and inversely as w* translates into

$$y = \frac{kxz}{w}.$$

Substitute $y = 154$, $x = 6$, $z = 11$, and $w = 3$ to obtain

$$154 = \frac{k(6)(11)}{3}$$

$$154 = 22k$$

$$7 = k.$$

The constant of variation is 7.　　　▲

Problem Set 8.6

For Problems 1–10, translate each statement of variation into an equation using k as the constant of variation.

1. y varies inversely as the square of x.

2. y varies directly as the cube of x.

3. C varies directly as g and inversely as the cube of t.

4. V varies jointly as l and w.

5. The volume (V) of a sphere is directly proportional to the cube of its radius (r).

6. At a constant temperature, the volume (V) of a gas varies inversely as the pressure (P).

7. The surface area (S) of a cube varies directly as the square of the length of an edge (e).

8. The intensity of illumination (I) received from a source of light is inversely proportional to the square of the distance (d) from the source.

9. The volume (V) of a cone varies jointly as its height and the square of its radius.

10. The volume (V) of a gas varies directly as the absolute temperature (T) and inversely as the pressure (P).

For Problems 11–24, find the constant of variation for each of the stated conditions.

11. y varies directly as x, and $y = 8$ when $x = 12$.

12. y varies directly as x, and $y = 60$ when $x = 24$.

13. y varies directly as the square of x, and $y = -144$ when $x = 6$.

14. y varies directly as the cube of x, and $y = 48$ when $x = -2$.

15. V varies jointly as B and h, and $V = 96$ when $B = 24$ and $h = 12$.

16. A varies jointly as b and h, and $A = 72$ when $b = 16$ and $h = 9$.

17. y varies inversely as x, and $y = -4$ when $x = \frac{1}{2}$.

18. y varies inversely as x, and $y = -6$ when $x = \frac{4}{3}$.

19. r varies inversely as the square of t, and $r = \frac{1}{8}$ when $t = 4$.

20. r varies inversely as the cube of t, and $r = \frac{1}{16}$ when $t = 4$.

21. y varies directly as x and inversely as z, and $y = 45$ when $x = 18$ and $z = 2$.

22. y varies directly as x and inversely as z, and $y = 24$ when $x = 36$ and $z = 18$.

23. y is directly proportional to x and inversely proportional to the square of z, and $y = 81$ when $x = 36$ and $z = 2$.

24. y is directly proportional to the square of x and inversely proportional to the cube of z, and $y = 4\frac{1}{2}$ when $x = 6$ and $z = 4$.

Solve each of the following problems.

25. If y is directly proportional to x, and $y = 36$ when $x = 48$, find the value of y when $x = 12$.

26. If y is directly proportional to x, and $y = 42$ when $x = 28$, find the value of y when $x = 38$.

27. If y is inversely proportional to x, and $y = \frac{1}{9}$ when $x = 12$, find the value of y when $x = 8$.

28. If y is inversely proportional to x, and $y = \frac{1}{35}$ when $x = 14$, find the value of y when $x = 16$.

29. If A varies jointly as b and h, and $A = 60$ when $b = 12$ and $h = 10$, find A when $b = 16$ and $h = 14$.

30. If V varies jointly as B and h, and $V = 51$ when $B = 17$ and $h = 9$, find V when $B = 19$ and $h = 12$.

31. The volume of a gas at a constant temperature varies inversely as the pressure. What is the volume of a gas under a pressure of 25 pounds if the gas occupies 15 cubic centimeters under a pressure of 20 pounds?

32. The time required for a car to travel a certain distance varies inversely as the rate at which it travels. If it takes 4 hours at 50 miles per hour to travel the distance, how long will it take at 40 miles per hour?

33. The volume (V) of a gas varies directly as the temperature (T) and inversely as the pressure (P). If $V = 48$ when $T = 320$ and $P = 20$, find V when $T = 280$ and $P = 30$.

34. The distance that a freely falling body falls varies directly as the square of the time it falls. If a body falls 144 feet in 3 seconds, how far will it fall in 5 seconds?

35. The period (the time required for one complete oscillation) of a simple pendulum varies directly as the square root of its length. If a pendulum 12 feet long has a period of 4 seconds, find the period of a pendulum of length 3 feet.

36. The simple interest earned by a certain amount of money varies jointly as the rate of interest and the time (in years) that the money is invested. If $120 is earned for the money invested at 12% for 2 years, how much is earned if the money is invested at 14% for 3 years?

37. The electrical resistance of a wire varies directly as its length and inversely as the square of its diameter. If the resistance of 200 meters of wire having a diameter of $\frac{1}{2}$ centimeter is 1.5 ohms, find the resistance of 400 meters of wire with a diameter of $\frac{1}{4}$ centimeter.

38. The volume of a cylinder varies jointly as its altitude and the square of the radius of its base. If the volume of a cylinder is 1386 cubic centimeters when the radius of the base is 7 centimeters and its altitude is 9 centimeters, find the volume of a cylinder that has a base of radius 14 centimeters; the altitude of the cylinder is 5 centimeters.

39. The simple interest earned by a certain amount of money varies jointly as the rate of interest and the time (in years) that the money is invested.

(a) If some money invested at 11% for 2 years earns $385, how much would the same amount earn at 12% for 1 year?
(b) If some money invested at 12% for 3 years earns $819, how much would the same amount earn at 14% for 2 years?
(c) If some money invested at 14% for 4 years earns $1960, how much would the same amount earn at 15% for 2 years?

40. The period (the time required for one complete oscillation) of a simple pendulum varies directly as the square root of its length. If a pendulum 9 inches long has a period of 2.4 seconds, find the period of a pendulum of length 12 inches. Express answer to the nearest one-tenth of a second.

41. The volume of a cylinder varies jointly as its altitude and the square of the radius of its base. If the volume of a cylinder is 549.5 cubic meters when the radius of the base is 5 meters and its altitude is 7 meters, find the volume of a cylinder that has a base of radius 9 meters and an altitude of 14 meters.

42. If y is directly proportional to x and inversely proportional to the square of z, and if $y = .336$ when $x = 6$ and $z = 5$, find the constant of variation.

43. If y is inversely proportional to the square root of x, and $y = .08$ when $x = 225$, find y when $x = 625$.

THOUGHTS INTO WORDS

44. How would you explain the difference between direct variation and inverse variation?

45. Suppose that y varies directly as the square of x. Does doubling the value of x also double the value of y? Explain your answer.

46. Suppose that y varies inversely as x. Does doubling the value of x also double the value of y? Explain your answer.

SUMMARY

(8.1) A **relation** is a set of ordered pairs; a **function** is a relation in which no two ordered pairs have the same first component. The **domain** of a relation (or function) is the set of all first components and the **range** is the set of all second components.

Single symbols such as f, g, and h are commonly used to name functions. The symbol $f(x)$ represents the element in the range associated with x from the domain. Thus, if $f(x) = 3x + 7$, then $f(1) = 3(1) + 7 = 10$.

(8.2) Any function that can be written in the form

$$f(x) = ax + b,$$

where a and b are real numbers, is a **linear function**. The graph of a linear function is a straight line.

Any function that can be written in the form

$$f(x) = ax^2 + bx + c,$$

where a, b, and c are real numbers and $a \neq 0$, is a **quadratic function**. The graph of any quadratic function is a **parabola**, which can be drawn using either one of the following methods.

1. Express the function in the form $f(x) = a(x - h)^2 + k$ and use the values of a, h, and k to determine the parabola.

2. Express the function in the form $f(x) = ax^2 + bx + c$ and use the fact that the vertex is at

$$\left(-\frac{b}{2a}, f\left(-\frac{b}{2a} \right) \right)$$

and the axis of symmetry is

$$x = -\frac{b}{2a}.$$

We offer the following graphing suggestions.

1. Determine the domain of the function.

2. Determine any type of symmetry that the equation exhibits.

3. Find the intercepts.

4. Set up a table of ordered pairs that satisfy the equation.

5. Plot the points, connect them with a smooth curve, and reflect this portion of the curve according to any symmetry exhibited by the equation.

(8.3) Another important graphing technique is to be able to recognize equations of the transformations of basic curves. We have worked with the following transformations in this chapter.

Vertical Translation The graph of $y = f(x) + k$ is the graph of $y = f(x)$ shifted k units upward if $k > 0$ or shifted $|k|$ units downward if $k < 0$.

Horizontal Translation The graph of $y = f(x - h)$ is the graph of $y = f(x)$ shifted h units to the right if $h > 0$ or shifted $|h|$ units to the left if $h < 0$.

x-axis Reflection The graph of $y = -f(x)$ is the graph of $y = f(x)$ reflected through the x-axis.

y-axis Reflection The graph of $y = f(-x)$ is the graph of $y = f(x)$ reflected through the y-axis.

Vertical Stretching and Shrinking The graph of $y = cf(x)$ is obtained from the graph of $y = f(x)$ by multiplying the y-coordinates of $y = f(x)$ by c. If $c > 1$, the graph is said to be stretched by a factor of c, and if $0 < c < 1$, the graph is said to be shrunk by a factor of c.

(8.4) *Operations on Functions*

Sum of two functions $(f + g)(x) = f(x) + g(x)$

Difference of two functions $(f - g)(x) = f(x) - g(x)$

Product of two functions $(f \cdot g)(x) = f(x) \cdot g(x)$

Quotient of two functions $\left(\dfrac{f}{g}\right)(x) = \dfrac{f(x)}{g(x)}, \qquad g(x) \neq 0$

The **composition** of two functions f and g is defined by

$$(f \circ g)(x) = f(g(x))$$

for all x in the domain of g such that $g(x)$ is in the domain of f. Remember that the composition of functions is not a commutative operation.

(8.5) A **one-to-one function** is a function such that no two ordered pairs have the same second component.

If the components of each ordered pair of a given one-to-one function are interchanged, the resulting function and the given function are **inverses** of each other. The inverse of a function f is denoted by f^{-1}.

Graphically, two functions that are inverses of each other are mirror images with reference to the line $y = x$.

Two functions f and f^{-1} can be shown to be inverses of each other by verifying that

1. $(f^{-1} \circ f)(x) = x$ for all x in the domain of f.
2. $(f \circ f^{-1})(x) = x$ for all x in the domain of f^{-1}.

A technique for finding the inverse of a function can be described as follows.

1. Let $y = f(x)$.
2. Interchange x and y.
3. Solve the equation for y in terms of x.
4. $f^{-1}(x)$ is determined by the final equation.

(8.6) The equation $y = kx$ (k is a nonzero constant) defines a function called a **direct variation**. The equation $y = \dfrac{k}{x}$ defines a function called **inverse variation**. In both cases, k is called the **constant of variation**.

Chapter 8 Review Problem Set

For Problems 1–4, specify the domain of each function.

1. $f = \{(1, 3), (2, 5), (4, 9)\}$

2. $f(x) = \dfrac{4}{x - 5}$

3. $f(x) = \dfrac{3}{x^2 + 4x}$

4. $f(x) = \sqrt{x^2 - 25}$

5. If $f(x) = x^2 - 2x - 1$, find $f(2), f(-3)$, and $f(a)$.

6. If $f(x) = 2x^2 + x - 7$, find $\dfrac{f(a + h) - f(a)}{h}$.

For Problems 7–16, graph each of the functions.

7. $f(x) = 4$

8. $f(x) = -3x + 2$

9. $f(x) = x^2 + 2x + 2$

10. $f(x) = |x| + 4$

11. $f(x) = -|x - 2|$

12. $f(x) = \sqrt{x - 2} - 3$

13. $f(x) = \dfrac{1}{x^2}$

14. $f(x) = -\dfrac{1}{2}x^2$

15. $f(x) = -3x^2 + 6x - 2$

16. $f(x) = -\sqrt{x + 1} - 2$

17. Find the coordinates of the vertex and the equation of the line of symmetry for each of the following parabolas.

(a) $f(x) = x^2 + 10x - 3$

(b) $f(x) = -2x^2 - 14x + 9$

For Problems 18–20, determine $(f \circ g)(x)$ and $(g \circ f)(x)$ for each pair of functions.

18. $f(x) = 2x - 3$ and $g(x) = 3x - 4$

19. $f(x) = x - 4$ and $g(x) = x^2 - 2x + 3$

20. $f(x) = x^2 - 5$ and $g(x) = -2x + 5$

For Problems 21–23, find the inverse (f^{-1}) of the given function.

21. $f(x) = 6x - 1$

22. $f(x) = \frac{2}{3}x + 7$

23. $f(x) = -\frac{3}{5}x - \frac{2}{7}$

24. If y varies directly as x and inversely as z, and if $y = 21$ when $x = 14$ and $z = 6$, find the constant of variation.

25. If y varies jointly as x and the square root of z, and if $y = 60$ when $x = 2$ and $z = 9$, find y when $x = 3$ and $z = 16$.

26. The weight of a body above the surface of the earth varies inversely as the square of its distance from the center of the earth. Assuming the radius of the earth to be 4000 miles, how much would a man weigh 1000 miles above the earth's surface if he weighs 200 pounds on the surface?

27. Find two numbers whose sum is 40 and whose product is a maximum.

28. Find two numbers whose sum is 50 such that the square of one number plus six times the other number is a minimum.

29. Suppose that 50 students are able to raise $250 for a party by each student contributing $5. They figure that for each additional student they can find to contribute, the cost per student will decrease by a nickel. How many additional students do they need to find to maximize the amount they will have for a party?

30. The surface area of a cube varies directly as the square of the length of an edge. If the surface area of a cube having edges 8 inches long is 384 square inches, find the surface area of a cube having edges 10 inches long.

CHAPTER 8 TEST

1. Determine the domain of the function $f(x) = \dfrac{-3}{2x^2 + 7x - 4}$.

2. Determine the domain of the function $f(x) = \sqrt{5 - 3x}$.

3. If $f(x) = -\dfrac{1}{2}x + \dfrac{1}{3}$, find $f(-3)$.

4. If $f(x) = -x^2 - 6x + 3$, find $f(-2)$.

5. Find the vertex of the parabola $f(x) = -2x^2 - 24x - 69$.

6. If $f(x) = 3x^2 + 2x - 5$, find $\dfrac{f(a + h) - f(a)}{h}$.

7. If $f(x) = -3x + 4$ and $g(x) = 7x + 2$, find $(f \circ g)(x)$.

8. If $f(x) = 2x + 5$ and $g(x) = 2x^2 - x + 3$, find $(g \circ f)(x)$.

9. If $f(x) = \dfrac{3}{x - 2}$ and $g(x) = \dfrac{2}{x}$, find $(f \circ g)(x)$.

For Problems 10–12, find the inverse of the given function.

10. $f(x) = 5x - 9$

11. $f(x) = -3x - 6$

12. $f(x) = \dfrac{2}{3}x - \dfrac{3}{5}$

13. If y varies inversely as x and $y = \dfrac{1}{2}$ when $x = -8$, find the constant of variation.

14. If y varies jointly as x and z, and if $y = 18$ when $x = 8$ and $z = 9$, find y when $x = 5$ and $z = 12$.

15. Find two numbers whose sum is 60, such that the sum of the square of one number plus twelve times the other number is a minimum.

16. The simple interest earned by a certain amount of money varies jointly as the rate of interest and the time (in years) that the money is invested. If $140 is earned for the money invested at 7% for 5 years, how much is earned if the same amount is invested at 8% for 3 years?

For Problems 17–19, use the concepts of translation and/or reflection to describe how the second curve can be obtained from the first curve.

17. $f(x) = x^3$, $f(x) = (x - 6)^3 - 4$

(continued on next page)

CHAPTER 8 TEST *(continued)*

18. $f(x) = |x|, f(x) = -|x| + 8$

19. $f(x) = \sqrt{x}, \; f(x) = -\sqrt{x + 5} + 7$

For Problems 20–25, graph each of the functions.

20. $f(x) = -x - 1$

21. $f(x) = -2x^2 - 12x - 14$

22. $f(x) = 2\sqrt{x} - 2$

23. $f(x) = 3|x - 2| - 1$

24. $f(x) = -\dfrac{1}{x} + 3$

25. $f(x) = \sqrt{-x + 2}$

Polynomial and Rational Functions

Earlier in this text we solved linear and quadratic equations and graphed linear and quadratic functions. In this chapter we will expand our equation solving processes and graphing techniques to include more general polynomial equations and functions. Then our knowledge of polynomial functions will allow us to work with rational functions. The function concept will again serve as a unifying thread throughout the chapter. To facilitate our study in this chapter, we will first review the concept of dividing polynomials and we will introduce a special division technique called synthetic division.

▼9.1　Synthetic Division

In Section 4.5 we discussed the process of dividing polynomials by using the following format.

$$
\begin{array}{r}
x^2 - 2x + 4 \\
3x + 1\overline{)3x^3 - 5x^2 + 10x + 1} \\
\underline{3x^3 + x^2} \\
-6x^2 + 10x + 1 \\
\underline{-6x^2 - 2x} \\
12x + 1 \\
\underline{12x + 4} \\
-3
\end{array}
$$

We also suggested writing the final result as

$$
\frac{3x^3 - 5x^2 + 10x + 1}{3x + 1} = x^2 - 2x + 4 + \frac{-3}{3x + 1}.
$$

Multiplying both sides of this equation by $3x + 1$ produces

$$
3x^3 - 5x^2 + 10x + 1 = (3x + 1)(x^2 - 2x + 4) + (-3),
$$

which is of the familiar form,

$$
\text{Dividend} = (\text{Divisor})(\text{Quotient}) + \text{Remainder}.
$$

This result is commonly called the **Division Algorithm for Polynomials** and it can be stated in general terms as follows.

Division Algorithm for Polynomials

If $f(x)$ and $d(x)$ are polynomials and $d(x) \neq 0$, then there exist unique polynomials $q(x)$ and $r(x)$ such that

$$
f(x) = d(x)q(x) + r(x)
$$

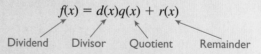

Dividend　　Divisor　　Quotient　　Remainder

where $r(x) = 0$ or the degree of $r(x)$ is less than the degree of $d(x)$.

If the divisor is of the form $x - c$, where c is a constant, then the typical long division algorithm can be conveniently simplified into a process called **synthetic division**. First, let's consider an example using the usual algorithm. Then, in a step-by-step fashion, we will list some shortcuts to use that will lead us into the synthetic division procedure. Consider the division problem $(2x^4 + x^3 - 17x^2 + 13x + 2) \div (x - 2)$.

$$
\begin{array}{r}
2x^3 + 5x^2 - 7x - 1 \\
x - 2\overline{)2x^4 + x^3 - 17x^2 + 13x + 2} \\
\underline{2x^4 - 4x^3} \\
5x^3 - 17x^2 \\
\underline{5x^3 - 10x^2} \\
-7x^2 + 13x \\
\underline{-7x^2 + 14x} \\
-x + 2 \\
-x + 2
\end{array}
$$

Notice that because the dividend $(2x^4 + x^3 - 17x^2 + 13x + 2)$ is written in descending powers of x, the quotient $(2x^3 + 5x^2 - 7x - 1)$ is also in descending powers of x. In other words, the numerical coefficients are the key. So let's rewrite the above problem in terms of its coefficients.

$$
\begin{array}{r}
2 \quad 5 \quad -7 \quad -1 \\
1 -2\overline{)2 \quad 1 \quad -17 \quad 13 \quad 2} \\
②\quad -4 \\
\underline{} \\
5 \quad ⟨-17⟩ \\
⑤ \quad -10 \\
\underline{} \\
-7 \quad 13 \\
⟨-7⟩ \quad 14 \\
\underline{} \\
-1 \quad ② \\
⟨-1⟩ \quad 2
\end{array}
$$

Now observe that the numbers circled are simply repetitions of the numbers directly above them in the format. Thus, the circled numbers could be omitted, and the format would be as follows. (Disregard the arrows for the moment.)

$$
\begin{array}{r}
2 \quad 5 \quad -7 \quad -1 \\
1 - 2\overline{)2 \quad 1 \quad -17 \quad 13 \quad 2} \\
-4 \quad \uparrow \quad \uparrow \quad \uparrow \\
5 \quad | \\
-10 \quad | \\
-7 \quad | \\
14 \quad | \\
-1 \quad 2
\end{array}
$$

Next, move some numbers up as indicated by the arrows and omit writing 1 as the coefficient of x in the divisor to yield the following more compact form.

$$
\begin{array}{r}
2 \quad 5 \quad -7 \quad -1 \\
-2\overline{)2 \quad 1 \quad -17 \quad 13 \quad 2} \\
-4 \quad -10 \quad 14 \quad 2 \\
\underline{} \\
5 \quad -7 \quad -1
\end{array}
$$

(1)
(2)
(3)
(4)

Notice that line (4) reveals all of the coefficients of the quotient [line (1)] except for the first coefficient, 2. Thus, we can omit line (1), begin line (4) with the first coefficient, and then use the following form.

$$
\begin{array}{r|rrrr}
-2) & 2 & 1 & -17 & 13 & 2 \\
& & -4 & -10 & 14 & 2 \\
\hline
& 2 & 5 & -7 & -1 & 0
\end{array}
$$

       (5)
(6)
(7)

Line (7) contains the coefficients of the quotient; the 0 indicates the remainder. Finally, changing the constant in the divisor to 2 (instead of -2), which will change the signs of the numbers in line (6), allows us to add the corresponding entries in lines (5) and (6) rather than subtract them. Thus, the final synthetic division form for this problem is as follows.

$$
\begin{array}{r|rrrr}
2) & 2 & 1 & -17 & 13 & 2 \\
& & 4 & 10 & -14 & -2 \\
\hline
& 2 & 5 & -7 & -1 & 0
\end{array}
$$

Now we will consider another problem and follow a step-by-step procedure for setting up and carrying out the synthetic division process. Suppose that we want to do the following division problem.

$$x + 4)\overline{2x^3 + 5x^2 - 13x - 2}$$

1. Write the coefficients of the dividend as follows.

$$)\overline{2 \quad 5 \quad -13 \quad -2}$$

2. In the divisor, use -4 instead of 4 so that later we can add rather than subtract.

$$-4)\overline{2 \quad 5 \quad -13 \quad -2}$$

3. Bring down the first coefficient of the dividend.

$$
\begin{array}{r|rrr}
-4) & 2 & 5 & -13 & -2 \\
\hline
& 2
\end{array}
$$

4. Multiply that first coefficient times the divisor, which yields $2(-4) = -8$. This result is added to the second coefficient of the dividend.

$$
\begin{array}{r|rrr}
-4) & 2 & 5 & -13 & -2 \\
& & -8 \\
\hline
& 2 & -3
\end{array}
$$

5. Multiply $(-3)(-4)$, which yields 12; this result is added to the third coefficient of the dividend.

$$
\begin{array}{r|rrr}
-4) & 2 & 5 & -13 & -2 \\
& & -8 & 12 \\
\hline
& 2 & -3 & -1
\end{array}
$$

6. Multiply $(-1)(-4)$, which yields 4; this result is added to the last term of the dividend.

$$
\begin{array}{r}
-4\overline{)25-13-2} \\
\phantom{-4)2}-8124 \\
\hline
2-3-12
\end{array}
$$

The last row indicates a quotient of $2x^2 - 3x - 1$ and a remainder of 2.

Let's consider three more examples showing only the final compact form for synthetic division.

Example 1

Solution

Find the quotient and remainder for $(2x^3 - 5x^2 + 6x + 4) \div (x - 2)$.

$$
\begin{array}{r}
2\overline{)2-564} \\
\phantom{2)2}4-28 \\
\hline
2-1412
\end{array}
$$

Therefore, the quotient is $2x^2 - x + 4$, and the remainder is 12.   ▲

Example 2

Solution

Find the quotient and remainder for $(4x^4 - 2x^3 + 6x - 1) \div (x - 1)$.

$$
\begin{array}{r}
1\overline{)4-206-1} \\
\phantom{1)4}4228 \\
\hline
42287
\end{array}
$$

Notice that a zero has been inserted as the coefficient of the missing x^2 term.

Thus, the quotient is $4x^3 + 2x^2 + 2x + 8$, and the remainder is 7.   ▲

Example 3

Solution

Find the quotient and remainder for $(x^3 + 8x^2 + 13x - 6) \div (x + 3)$.

$$
\begin{array}{r}
-3\overline{)1813-6} \\
\phantom{-3)1}-3-156 \\
\hline
15-20
\end{array}
$$

Thus, the quotient is $x^2 + 5x - 2$, and the remainder is 0.   ▲

In Example 3, because the remainder is 0, we can say that $x + 3$ is a factor of $x^3 + 8x^2 + 13x - 6$. We will use this idea a bit later when we solve polynomial equations.

Problem Set 9.1

Use synthetic division to determine the quotient and remainder for each of the following.

1. $(4x^2 - 5x - 6) \div (x - 2)$

2. $(5x^2 - 9x + 4) \div (x - 1)$

3. $(2x^2 - x - 21) \div (x + 3)$

4. $(3x^2 + 8x + 4) \div (x + 2)$

5. $(3x^2 - 16x + 17) \div (x - 4)$

6. $(6x^2 - 29x - 8) \div (x - 5)$

7. $(4x^2 + 19x - 32) \div (x + 6)$

8. $(7x^2 + 26x - 2) \div (x + 4)$

9. $(x^3 + 2x^2 - 7x + 4) \div (x - 1)$

10. $(2x^3 - 7x^2 + 2x + 3) \div (x - 3)$

11. $(3x^3 + 8x^2 - 8) \div (x + 2)$

12. $(4x^3 + 17x^2 + 75) \div (x + 5)$

13. $(5x^3 - 9x^2 - 3x - 2) \div (x - 2)$

14. $(x^3 - 6x^2 - 5x + 14) \div (x - 4)$

15. $(x^3 + 6x^2 - 8x + 1) \div (x + 7)$

16. $(2x^3 + 11x^2 - 5x + 1) \div (x + 6)$

17. $(-x^3 + 7x^2 - 14x + 6) \div (x - 3)$

18. $(-2x^3 - 3x^2 + 4x + 5) \div (x + 1)$

19. $(-3x^3 + x^2 + 2x + 2) \div (x + 1)$

20. $(-x^3 + 4x^2 + 31x + 2) \div (x - 8)$

21. $(3x^3 - 2x - 5) \div (x - 2)$

22. $(2x^3 - x - 4) \div (x + 3)$

23. $(2x^4 + x^3 + 3x^2 + 2x - 2) \div (x + 1)$

24. $(x^4 - 3x^3 - 6x^2 + 11x - 12) \div (x - 4)$

25. $(x^4 + 4x^3 - 7x - 1) \div (x - 3)$

26. $(3x^4 - x^3 + 2x^2 - 7x - 1) \div (x + 1)$

27. $(x^4 + 5x^3 - x^2 + 25) \div (x + 5)$

28. $(2x^4 + 3x^2 + 3) \div (x + 2)$

29. $(x^4 - 16) \div (x - 2)$

30. $(x^4 - 16) \div (x + 2)$

31. $(x^5 - 1) \div (x + 1)$

32. $(x^5 - 1) \div (x - 1)$

33. $(x^5 + 1) \div (x + 1)$

34. $(x^5 + 1) \div (x - 1)$

35. $(x^5 + 3x^4 - 5x^3 - 3x^2 + 3x - 4) \div (x + 4)$

36. $(2x^5 + 3x^4 - 4x^3 - x^2 + 5x - 2) \div (x + 2)$

37. $(4x^5 - 6x^4 + 2x^3 + 2x^2 - 5x + 2) \div (x - 1)$

38. $(3x^5 - 8x^4 + 5x^3 + 2x^2 - 9x + 4) \div (x - 2)$

39. $(9x^3 - 6x^2 + 3x - 4) \div \left(x - \dfrac{1}{3}\right)$

40. $(2x^3 + 3x^2 - 2x + 3) \div \left(x + \dfrac{1}{2}\right)$

41. $(3x^4 - 2x^3 + 5x^2 - x - 1) \div \left(x + \dfrac{1}{3}\right)$

42. $(4x^4 - 5x^2 + 1) \div \left(x - \dfrac{1}{2}\right)$

THOUGHTS INTO WORDS

43. How would you give a general description of what is accomplished with synthetic division to someone who had just completed an elementary algebra course?

44. Why is synthetic division restricted to situations where the divisor is of the form $x - c$?

9.2 Remainder and Factor Theorems

Let's consider the division algorithm (stated in the previous section) when the dividend, $f(x)$, is divided by a linear polynomial of the form $x - c$. Then the division algorithm

$$f(x) = d(x)q(x) + r(x)$$

Dividend   Divisor   Quotient   Remainder

becomes

$$f(x) = (x - c)q(x) + r(x).$$

Because the degree of the remainder, $r(x)$, must be less than the degree of the divisor, $x - c$, the remainder is a constant. Therefore, letting R represent the remainder, we have

$$f(x) = (x - c)q(x) + R.$$

If the functional value at c is found, we obtain

$$\begin{aligned} f(c) &= (c - c)q(c) + R \\ &= 0 \cdot q(c) + R \\ &= R. \end{aligned}$$

In other words, if a polynomial is divided by a linear polynomial of the form $x - c$, then the remainder is given by the value of the polynomial at c. Let's state this result more formally as the remainder theorem.

PROPERTY 9.1

Remainder Theorem

If the polynomial $f(x)$ is divided by $x - c$, then the remainder is equal to $f(c)$.

Example 1

If $f(x) = x^3 + 2x^2 - 5x - 1$, find $f(2)$ by (a) using synthetic division and the remainder theorem and (b) evaluating $f(2)$ directly.

Solution

(a)
```
2)1   2   -5   -1
        2    8    6
    1   4    3   ⑤  ←  R = f(2)
```

(b) $f(2) = 2^3 + 2(2)^2 - 5(2) - 1 = 8 + 8 - 10 - 1 = 5$ ▲

Example 2

If $f(x) = x^4 + 7x^3 + 8x^2 + 11x + 5$, find $f(-6)$ by (a) using synthetic division and the remainder theorem and (b) evaluating $f(-6)$ directly.

Solution

(a)
```
-6)1   7    8    11    5
       -6  -6   -12    6
    1   1    2   -1   ⑪  ←  R = f(-6)
```

(b) $f(-6) = (-6)^4 + 7(-6)^3 + 8(-6)^2 + 11(-6) + 5$
$= 1296 - 1512 + 288 - 66 + 5$
$= 11$ ▲

In Example 2, notice that the computations involved in finding $f(-6)$ using synthetic division and the remainder theorem are much easier than those required to evaluate $f(-6)$ directly. This is not always the case, but often using synthetic division is easier than evaluating $f(c)$ directly.

Example 3

Solution

Find the remainder when $x^3 + 3x^2 - 13x - 15$ is divided by $x + 1$.

Let $f(x) = x^3 + 3x^2 - 13x - 15$, write $x + 1$ as $x - (-1)$ and apply the remainder theorem.

$$f(-1) = (-1)^3 + 3(-1)^2 - 13(-1) - 15 = 0.$$

Thus, the remainder is 0. ▲

Example 3 illustrates an important aspect of the remainder theorem—the situation in which the remainder is *zero*. Thus, we can say that $x + 1$ is a factor of $x^3 + 3x^2 - 13x - 15$.

Factor Theorem

A general factor theorem can be formulated by considering the equation

$$f(x) = (x - c)q(x) + R.$$

If $x - c$ is a factor of $f(x)$, then the remainder R, which is also $f(c)$, must be zero. Conversely, if $R = f(c) = 0$, then $f(x) = (x - c)q(x)$; in other words, $x - c$ is a factor of $f(x)$. The factor theorem can be stated as follows.

PROPERTY 9.2

Factor Theorem

A polynomial $f(x)$ has a factor $x - c$ if and only if $f(c) = 0$.

Example 4

Solution

Is $x - 1$ a factor of $x^3 + 5x^2 + 2x - 8$?

Let $f(x) = x^3 + 5x^2 + 2x - 8$ and compute $f(1)$ to obtain

$$f(1) = 1^3 + 5(1)^2 + 2(1) - 8 = 0.$$

By the factor theorem, therefore, $x - 1$ is a factor of $f(x)$. ▲

Example 5

Solution

Is $x + 3$ a factor of $2x^3 + 5x^2 - 6x - 7$?

Use synthetic division to obtain the following.

$$
\begin{array}{r}
-3 \overline{)}\; 1 \quad\;\; 5 \quad -6 \quad -7 \\
\; -6 \quad\;\; 3 \quad\;\; 9 \\
\hline
\; 2 \quad -1 \quad -3 \quad\;\; ②
\end{array}
\quad \longleftarrow \quad R = f(-3)
$$

Since $R \neq 0$, we know that $x + 3$ is not a factor of the given polynomial. ▲

In Examples 4 and 5 we were concerned only with determining whether a linear polynomial of the form $x - c$ was a factor of another polynomial. For such problems, it is reasonable to compute $f(c)$ either directly or by synthetic division, whichever way seems easier for a particular problem. However, if more information is required, such as the complete factorization of the given polynomial, then the use of synthetic division becomes appropriate, as the next two examples illustrate.

Example 6

Show that $x - 1$ is a factor of $x^3 - 2x^2 - 11x + 12$, and find the other linear factors of the polynomial.

Solution

Let's use synthetic division to divide $x^3 - 2x^2 - 11x + 12$ by $x - 1$.

$$
\begin{array}{r|rrrr}
1) & 1 & -2 & -11 & 12 \\
   &   & 1 & -1 & -12 \\
\hline
   & 1 & -1 & -12 & 0
\end{array}
$$

The last line indicates a quotient of $x^2 - x - 12$ and a remainder of 0. The remainder of 0 means that $x - 1$ is a factor. Furthermore, we can write

$$x^3 - 2x^2 - 11x + 12 = (x - 1)(x^2 - x - 12).$$

The quadratic polynomial $x^2 - x - 12$ can be factored as $(x - 4)(x + 3)$ using our conventional factoring techniques. Thus, we obtain

$$x^3 - 2x^2 - 11x + 12 = (x - 1)(x - 4)(x + 3). \qquad \blacktriangle$$

Example 7

Show that $x + 4$ is a factor of $f(x) = x^3 - 5x^2 - 22x + 56$, and complete the factorization of $f(x)$.

Solution

Use synthetic division to divide $x^3 - 5x^2 - 22x + 56$ by $x + 4$ and obtain

$$
\begin{array}{r|rrrr}
-4) & 1 & -5 & -22 & 56 \\
    &   & -4 & 36 & -56 \\
\hline
    & 1 & -9 & 14 & 0.
\end{array}
$$

The last line indicates a quotient of $x^2 - 9x + 14$ and a remainder of 0. The remainder of 0 means that $x + 4$ is a factor. Furthermore, we can write

$$x^3 - 5x^2 - 22x + 56 = (x + 4)(x^2 - 9x + 14)$$

and then complete the factoring to obtain

$$f(x) = x^3 - 5x^2 - 22x + 56 = (x + 4)(x - 7)(x - 2). \qquad \blacktriangle$$

The factor theorem also plays a significant role in determining some general factorization ideas, as the last example of this section demonstrates.

Example 8

Verify that $x + 1$ is a factor of $x^n + 1$ for all odd positive integral values of n.

Solution

Let $f(x) = x^n + 1$ and compute $f(-1)$ to obtain

$$
\begin{aligned}
f(-1) &= (-1)^n + 1 \\
      &= -1 + 1 \qquad \text{Any odd power of } -1 \text{ is } -1. \\
      &= 0.
\end{aligned}
$$

Since $f(-1) = 0$, we know that $x + 1$ is a factor of $f(x)$. $\qquad \blacktriangle$

Problem Set 9.2

For Problems 1–10, find $f(c)$ by (a) evaluating $f(c)$ directly and (b) using synthetic division and the remainder theorem.

1. $f(x) = x^2 + 2x - 6$ and $c = 3$
2. $f(x) = x^2 - 7x + 4$ and $c = 2$
3. $f(x) = x^3 - 2x^2 + 3x - 1$ and $c = -1$
4. $f(x) = x^3 + 3x^2 - 4x - 7$ and $c = -2$
5. $f(x) = 2x^4 - x^3 - 3x^2 + 4x - 1$ and $c = 2$

6. $f(x) = 3x^4 - 4x^3 + 5x^2 - 7x + 6$ and $c = 1$

7. $f(n) = 6n^3 - 35n^2 + 8n - 10$ and $c = 6$

8. $f(n) = 8n^3 - 39n^2 - 7n - 1$ and $c = 5$
9. $f(n) = 2n^5 - 1$ and $c = -2$
10. $f(n) = 3n^4 - 2n^3 + 4n - 1$ and $c = 3$

For Problems 11–20, find $f(c)$ *either* by using synthetic division and the remainder theorem *or* by evaluating $f(c)$ directly.

11. $f(x) = 6x^5 - 3x^3 + 2$ and $c = -1$
12. $f(x) = -4x^4 + x^3 - 2x^2 - 5$ and $c = 2$

13. $f(x) = 2x^4 - 15x^3 - 9x^2 - 2x - 3$ and $c = 8$

14. $f(x) = x^4 - 8x^3 + 9x^2 - 15x + 2$ and $c = 7$

15. $f(n) = 4n^7 + 3$ and $c = 3$
16. $f(n) = -3n^6 - 2$ and $c = -3$
17. $f(n) = 3n^5 + 17n^4 - 4n^3 + 10n^2 - 15n + 13$ and $c = -6$
18. $f(n) = -2n^5 - 9n^4 + 7n^3 + 14n^2 + 19n - 38$ and $c = -5$
19. $f(x) = -4x^4 - 6x^2 + 7$ and $c = 4$
20. $f(x) = 3x^5 - 7x^3 - 6$ and $c = 5$

For Problems 21–34, use the factor theorem to help answer some questions about factors.

21. Is $x - 2$ a factor of $5x^2 - 17x + 14$?
22. Is $x + 1$ a factor of $3x^2 - 5x - 8$?
23. Is $x + 3$ a factor of $6x^2 + 13x - 14$?
24. Is $x - 5$ a factor of $8x^2 - 47x + 32$?
25. Is $x - 1$ a factor of $4x^3 - 13x^2 + 21x - 12$?
26. Is $x - 4$ a factor of $2x^3 - 11x^2 + 10x + 8$?
27. Is $x + 2$ a factor of $x^3 + 7x^2 + x - 18$?
28. Is $x + 3$ a factor of $x^3 + x^2 - 14x - 24$?
29. Is $x - 3$ a factor of $3x^3 - 5x^2 - 17x + 17$?
30. Is $x + 4$ a factor of $2x^3 + 9x^2 - 5x - 39$?
31. Is $x + 2$ a factor of $x^3 + 8$?
32. Is $x - 2$ a factor of $x^3 - 8$?
33. Is $x - 3$ a factor of $x^4 - 81$?
34. Is $x + 3$ a factor of $x^4 - 81$?

For Problems 35–44, use synthetic division to show that $g(x)$ is a factor of $f(x)$, and complete the factorization of $f(x)$.

35. $g(x) = x - 2$, $f(x) = x^3 - 6x^2 - 13x + 42$

36. $g(x) = x + 1$, $f(x) = x^3 + 6x^2 - 31x - 36$

37. $g(x) = x + 2$, $\quad f(x) = 12x^3 + 29x^2 + 8x - 4$

38. $g(x) = x - 3$, $f(x) = 6x^3 - 17x^2 - 5x + 6$

39. $g(x) = x + 1$, $f(x) = x^3 - 2x^2 - 7x - 4$

40. $g(x) = x - 5$, $f(x) = 2x^3 + x^2 - 61x + 30$

41. $g(x) = x - 6$, $\quad f(x) = x^5 - 6x^4 - 16x + 96$

42. $g(x) = x + 3$, $\quad f(x) = x^5 + 3x^4 - x - 3$

43. $g(x) = x + 5$, $\quad f(x) = 9x^3 + 21x^2 - 104x + 80$

44. $g(x) = x + 4$, $\quad f(x) = 4x^3 + 4x^2 - 39x + 36$

For Problems 45–48, find the value(s) of k that makes the second polynomial a factor of the first.

45. $k^2x^4 + 3kx^2 - 4; x - 1$

46. $x^3 - kx^2 + 5x + k; x - 2$

47. $kx^3 + 19x^2 + x - 6; x + 3$

48. $x^3 + 4x^2 - 11x + k; x + 2$

49. Argue that $f(x) = 3x^4 + 2x^2 + 5$ has no factor of the form $x - c$, where c is a real number.

50. Show that $x + 2$ is a factor of $x^{12} - 4096$.

51. Verify that $x + 1$ is a factor of $x^n - 1$ for all even positive integral values of n.

52. Verify that $x - 1$ is a factor of $x^n - 1$ for all positive integral values of n.

53. (a) Verify that $x - y$ is a factor of $x^n - y^n$ for all positive integral values of n.

(b) Verify that $x + y$ is a factor of $x^n - y^n$ for all even positive integral values of n.

(c) Verify that $x + y$ is a factor of $x^n + y^n$ for all odd positive integral values of n.

THOUGHTS INTO WORDS

54. State the remainder theorem in your own words.

55. Discuss some of the uses of the factor theorem.

Further Investigations

The remainder and factor theorems are true for any complex value of c. Therefore, for Problems 56–58, find $f(c)$ by (a) using synthetic division and the remainder theorem and (b) evaluating $f(c)$ directly.

56. $f(x) = x^3 - 5x^2 + 2x + 1$ and $c = i$

57. $f(x) = x^2 + 4x - 2$ and $c = 1 + i$

58. $f(x) = x^3 + 2x^2 + x - 2$ and $c = 2 - 3i$

59. Show that $x - 2i$ is a factor of $f(x) = x^4 + 6x^2 + 8$.

60. Show that $x + 3i$ is a factor of $f(x) = x^4 + 14x^2 + 45$.

61. Consider changing the form of the polynomial $f(x) = x^3 + 4x^2 - 3x + 2$ as follows.

$$f(x) = x^3 + 4x^2 - 3x + 2$$
$$= x(x^2 + 4x - 3) + 2$$
$$= x[x(x + 4) - 3] + 2.$$

The final form $f(x) = x[x(x + 4) - 3] + 2$ is called the **nested form** of the polynomial. It is particularly well suited for evaluating functional values of f either by hand or with a calculator. For each of the following, find the indicated functional values using the nested form of the given polynomial.

(a) $f(4), f(-5)$, and $f(7)$ for $f(x) = x^3 + 5x^2 - 2x + 1$

(b) $f(3), f(6)$, and $f(-7)$ for $f(x) = 2x^3 - 4x^2 - 3x + 2$

(c) $f(4)$, $f(5)$, and $f(-3)$ for $f(x) = -2x^3 + 5x^2 - 6x - 7$

(d) $f(5), f(6)$, and $f(-3)$ for $f(x) = x^4 + 3x^3 - 2x^2 + 5x - 1$

9.3 Polynomial Equations

We have solved a large variety of linear equations of the form $ax + b = 0$ and quadratic equations of the form $ax^2 + bx + c = 0$. Linear and quadratic equations are special cases of a general class of equations we refer to as **polynomial equations**. The equation

$$a_n x^n + a_{n-1} x^{n-1} + \cdots + a_1 x + a_0 = 0,$$

where the coefficients $a_0, a_1, \ldots, a_n$ are real numbers and n is a positive integer, is called a **polynomial equation of degree** n. The following are examples of polynomial equations.

$$\sqrt{2}x - 6 = 0 \qquad \text{Degree 1}$$

$$\frac{3}{4}x^2 - \frac{2}{3}x + 5 = 0 \qquad \text{Degree 2}$$

$$4x^3 - 3x^2 - 7x - 9 = 0 \qquad \text{Degree 3}$$

$$5x^4 - x + 6 = 0 \qquad \text{Degree 4}$$

REMARK The most general polynomial equation would allow complex numbers as coefficients. However, for our purposes in this text we will restrict the coefficients to real numbers. We often refer to such equations as **polynomial equations over the reals** △

In general, solving polynomial equations of degree greater than 2 can be very difficult and often requires mathematics beyond the scope of this text. However, there are some general properties pertaining to the solving of polynomial equations that you should be familiar with, and furthermore, there are certain types of polynomial equations that we can solve using the techniques available to us at this time.

Let's begin by listing some polynomial equations and corresponding solution sets that we have already encountered in this text.

Equation	Solution set
$3x + 4 = 7$	$\{1\}$
$x^2 + x - 6 = 0$	$\{-3, 2\}$
$2x^3 - 3x^2 - 2x + 3 = 0$	$\left\{-1, 1, \frac{3}{2}\right\}$
$x^4 - 16 = 0$	$\{-2, 2, -2i, 2i\}$

Notice that in each of the above examples the number of solutions corresponds to the degree of the equation. The first-degree equation has one solution, the second-degree equation has two solutions, the third-degree equation has three solutions, and the fourth-degree equation has four solutions. Now consider the equation

$$(x - 4)^2(x + 5)^3 = 0.$$

It can be written as

$$(x - 4)(x - 4)(x + 5)(x + 5)(x + 5) = 0,$$

which implies that

$$x - 4 = 0 \quad \text{or} \quad x - 4 = 0 \quad \text{or} \quad x + 5 = 0 \quad \text{or}$$
$$x + 5 = 0 \quad \text{or} \quad x + 5 = 0.$$

Therefore,

$$x = 4 \quad \text{or} \quad x = 4 \quad \text{or} \quad x = -5 \quad \text{or}$$
$$x = -5 \quad \text{or} \quad x = -5.$$

We state that the solution set of the original equation is $\{-5, 4\}$, but we also say that the equation has a solution of 4 with a **multiplicity of two**, and a solution of -5 with a **multiplicity of three**. Furthermore, notice that the sum of the multiplicities is 5, which agrees with the degree of the equation. The following general property can be stated.

PROPERTY 9.3

A polynomial equation of degree n has n solutions, where any solution of multiplicity p is counted p times.

Finding Rational Solutions

Although solving polynomial equations of a degree greater than 2 can, in general, be very difficult, **rational solutions of polynomial equations with integral coefficients** can be found using techniques presented in this chapter. The following property restricts the potential rational solutions of such equations.

PROPERTY 9.4

Rational Root Theorem

Consider the polynomial equation

$$a_n x^n + a_{n-1} x^{n-1} + \cdots + a_1 x + a_0 = 0,$$

where the coefficients $a_0, a_1, \ldots, a_n$ are *integers*. If the rational number $\dfrac{c}{d}$, reduced to lowest terms, is a solution of the equation, then c is a factor of the constant term a_0 and d is a factor of the leading coefficient a_n.

The "why" behind the rational root theorem is based on some simple factoring ideas, as indicated by the following outline of a proof for the theorem.

Outline of Proof   If $\dfrac{c}{d}$ is to be a solution, then

$$a_n\left(\frac{c}{d}\right)^n + a_{n-1}\left(\frac{c}{d}\right)^{n-1} + \cdots + a_1\left(\frac{c}{d}\right) + a_0 = 0.$$

Multiply both sides of this equation by d^n and add $-a_0 d^n$ to both sides to yield

$$a_n c^n + a_{n-1}c^{n-1}d + \cdots + a_1 cd^{n-1} = -a_0 d^n.$$

Because c is a factor of the left side of this equation, c must also be a factor of $-a_0 d^n$. Furthermore, because $\dfrac{c}{d}$ is in reduced form, c and d have no common factors other than -1 or 1. Thus, c is a factor of a_0. In the same way, from the equation

$$a_{n-1}c^{n-1}d + \cdots + a_1 cd^{n-1} + a_0 d^n = -a_n c^n$$

we can conclude that d is a factor of the left side and therefore d is also a factor of a_n.

The rational root theorem, synthetic division, the factor theorem, and some previous knowledge pertaining to solving linear and quadratic equations form a basis for finding rational solutions. Let's consider some examples.

Example 1

Find all rational solutions of $3x^3 + 8x^2 - 15x + 4 = 0$.

Solution

If $\dfrac{c}{d}$ is a rational solution, then c must be a factor of 4 and d must be a factor of 3. Therefore, the possible values for c and d are as follows.

For c: $\pm 1, \pm 2, \pm 4$
For d: $+1, \pm 3$

Thus, the possible values for $\dfrac{c}{d}$ are

$$\pm 1, \pm\frac{1}{3}, \pm 2, \pm\frac{2}{3}, \pm 4, \pm\frac{4}{3}.$$

Using synthetic division, we obtain

$$
\begin{array}{r|rrrr}
1 & 3 & 8 & -15 & 4 \\
  &   & 3 & 11 & -4 \\
\hline
  & 3 & 11 & -4 & 0
\end{array}
$$

which shows that $x - 1$ is a factor of the given polynomial; therefore, 1 is a rational solution of the equation. Furthermore, the synthetic division result also indicates that we can factor the given polynomial as follows.

$$3x^3 + 8x^2 - 15x + 4 = 0$$

$$(x - 1)(3x^2 + 11x - 4) = 0$$

The quadratic factor can be factored further using our previous techniques; we can proceed as follows.

$$(x - 1)(3x^2 + 11x - 4) = 0$$

$$(x - 1)(3x - 1)(x + 4) = 0$$

$$x - 1 = 0 \quad \text{or} \quad 3x - 1 = 0 \quad \text{or} \quad x + 4 = 0$$

$$x = 1 \quad \text{or} \quad x = \frac{1}{3} \quad \text{or} \quad x = -4$$

Thus, the entire solution set consists of rational numbers, which can be listed as $\left\{-4, \dfrac{1}{3}, 1\right\}$.   ▲

In Example 1, we were fortunate that a rational solution was the result of our first use of synthetic division. This does not happen frequently; we often need to conduct an organized search, as the next example demonstrates.

Example 2

Find all rational solutions of $3x^3 + 7x^2 - 22x - 8 = 0$.

Solution

If $\dfrac{c}{d}$ is a rational solution, then c must be a factor of -8 and d must be a factor of 3. Therefore, the possible values for c and d are as follows.

For c: $\pm 1, \pm 2, \pm 4, \pm 8$
For d: $\pm 1, \pm 3$

Thus, the possible values for $\dfrac{c}{d}$ are

$$\pm 1, \pm\frac{1}{3}, \pm 2, \pm\frac{2}{3}, \pm 4, \pm\frac{4}{3}, \pm 8, \pm\frac{8}{3}.$$

Let's begin our search for rational solutions; we will try the integers first.

```
1)3      7    -22    - 8
         3     10    -12
   ─────────────────────────
   3     10    -12   (-20)  ◄───── This remainder indicates that x − 1
                                   is not a factor and thus 1 is
                                   not a solution.
```

```
-1)3      7    -22    - 8
          - 3   - 4     26
   ─────────────────────────
   3      4    -26    (18)  ◄───── This remainder indicates that −1 is not a
                                   solution.
```

```
2)3      7    -22    - 8
         6     26      8
   ─────────────────────────
   3     13     4      0
```

Now we know that $x - 2$ is a factor; we can proceed as follows.

$$3x^3 + 7x^2 - 22x - 8 = 0$$
$$(x - 2)(3x^2 + 13x + 4) = 0$$
$$(x - 2)(3x + 1)(x + 4) = 0$$

$$x - 2 = 0 \quad \text{or} \quad 3x + 1 = 0 \quad \text{or} \quad x + 4 = 0$$
$$x = 2 \quad \text{or} \quad 3x = -1 \quad \text{or} \quad x = -4$$
$$x = 2 \quad \text{or} \quad x = -\frac{1}{3} \quad \text{or} \quad x = -4$$

The solution set is $\left\{-4, -\frac{1}{3}, 2\right\}$. ▲

In Examples 1 and 2 we were solving third-degree equations. Therefore, after finding one linear factor by synthetic division we were able to factor the remaining quadratic factor in the usual way. However, if the given equation is of degree 4 or more, we may need to find more than one linear factor by synthetic division, as the next example illustrates.

Example 3

Solve $x^4 - 6x^3 + 22x^2 - 30x + 13 = 0$.

Solution

The possible values for $\frac{c}{d}$ are as follows.

For $\frac{c}{d}$: $\quad \pm 1, \pm 13$

By synthetic division we find that

$$\begin{array}{r|rrrr} 1) & 1 & -6 & 22 & -30 & 13 \\ & & 1 & -5 & 17 & -13 \\ \hline & 1 & -5 & 17 & -13 & 0 \end{array}$$

which indicates that $x - 1$ is a factor of the given polynomial. The bottom line of the synthetic division indicates that the given polynomial can be factored as follows.

$$x^4 - 6x^3 + 22x^2 - 30x + 13 = 0$$
$$(x - 1)(x^3 - 5x^2 + 17x - 13) = 0$$

Therefore,

$$x - 1 = 0 \quad \text{or} \quad x^3 - 5x^2 + 17x - 13 = 0.$$

Now we can use the same approach to look for rational solutions of the expression $x^3 - 5x^2 + 17x - 13 = 0$. The possible values for $\frac{c}{d}$ are as follows.

For $\frac{c}{d}$: $\quad \pm 1, \pm 13$

By synthetic division we find that

$$
\begin{array}{r|rrrr}
1) & 1 & -5 & 17 & -13 \\
 & & 1 & -4 & 13 \\
\hline
 & 1 & -4 & 13 & 0
\end{array}
$$

which indicates that $x - 1$ is a factor of $x^3 - 5x^2 + 17x - 13$ and that the other factor is $x^2 - 4x + 13$.

Now we can solve the original equation as follows.

$$x^4 - 6x^3 + 22x^2 - 30x + 13 = 0$$

$$(x - 1)(x^3 - 5x^2 + 17x - 13) = 0$$

$$(x - 1)(x - 1)(x^2 - 4x + 13) = 0$$

$$x - 1 = 0 \quad \text{or} \quad x - 1 = 0 \quad \text{or} \quad x^2 - 4x + 13 = 0$$

$$x = 1 \quad \text{or} \quad x = 1 \quad \text{or} \quad x^2 - 4x + 13 = 0$$

Use the quadratic formula on $x^2 - 4x + 13 = 0$ to produce

$$x = \frac{4 \pm \sqrt{16 - 52}}{2} = \frac{4 \pm \sqrt{-36}}{2}$$

$$= \frac{4 \pm 6i}{2} = 2 \pm 3i.$$

Thus, the original equation has a rational solution of 1 with a multiplicity of 2 and has two complex solutions, $2 + 3i$ and $2 - 3i$. The solution set would be listed as $\{1, 2 \pm 3i\}$. ▲

Example 3 illustrates two general properties. First, notice that the coefficient of x^4 is 1, and thus the possible rational solutions must be integers. *In general, the possible rational solutions of $x^n + a_{n-1}, x^{n-1} + \cdots + a_1x + a_0 = 0$ are the integral factors of a_0.* Second, notice that the complex solutions of Example 3 are conjugates of each other. The following general property can be stated.

PROPERTY 9.5 Nonreal complex solutions of polynomial equations with real coefficients, if they exist, must occur in conjugate pairs.

REMARK The justification for Property 9.5 is based on some properties of conjugate complex numbers. We will not show the details of such a proof in this text. △

Each of Properties 9.3, 9.4, and 9.5 yields some information about the solutions of a polynomial equation. Before we state the final property of this section, which will give us some additional information, we need to consider two ideas.

First, in a polynomial that is arranged in descending powers of x, if two successive terms differ in sign, then there is said to be a **variation in sign**. (We disregard terms with zero coefficients when sign variations are counted.) For example, the polynomial

$$3x^3 - 2x^2 + 4x + 7$$

has *two* sign variations, whereas the polynomial

$$x^5 - 4x^3 + x - 5$$

has *three* variations.

Second, the solutions of

$$a_n(-x)^n + a_{n-1}(-x)^{n-1} + \cdots + a_1(-x) + a_0 = 0$$

are the opposites of the solutions of

$$a_n x^n + a_{n-1} x^{n-1} + \cdots + a_1 x + a_0 = 0.$$

In other words, if a new equation is formed by replacing x with $-x$ in a given equation, then the solutions of the newly formed equation are the opposites of the solutions of the given equation. For example, the solution set of $x^2 + 7x + 12 = 0$ is $\{-4, -3\}$, and the solution set of $(-x)^2 + 7(-x) + 12 = 0$, which simplifies to $x^2 - 7x + 12 = 0$, is $\{3, 4\}$.

Now we can state a property that can help us to determine the nature of the solutions of a polynomial equation without actually solving the equation.

PROPERTY 9.6

Descartes' Rule of Signs

Let $a_n x^n + a_{n-1} x^{n-1} + \cdots + a_1 x + a_0 = 0$ be a polynomial equation with real coefficients.

1. The number of *positive real solutions* of the given equation is either equal to the number of variations in sign of the polynomial or else less than the number of variations by a positive even integer.

2. The number of *negative real solutions* of the given equation is either equal to the number of variations in sign of the polynomial $a_n(-x)^n + a_{n-1}(-x)^{n-1} + \ldots + a_1(-x) + a_0$ or else less than the number of variations by a positive even integer.

Along with Properties 9.3 and 9.5, Property 9.6 allows us to acquire some information about the solutions of a polynomial equation without actually solving the equation. Let's consider some equations and see how much we know about their solutions without solving them.

1. $x^3 + 3x^2 + 5x + 4 = 0$

    (a) No variations of sign in $x^3 + 3x^2 + 5x + 4$ means that there are *no positive solutions*.

    (b) Replacing x with $-x$ in the given polynomial produces $(-x)^3 + 3(-x)^2 + 5(-x) + 4$, which simplifies to $-x^3 + 3x^2 - 5x + 4$ and contains three variations of sign; thus there are *three or one negative* solutions.

Conclusion   The given equation has three negative real solutions or else one negative real solution and two nonreal complex solutions.

2. $2x^4 + 3x^2 - x - 1 = 0$

    (a) There is one variation of the sign in the given polynomial; thus, the equation has *one positive solution*.

    (b) Replacing x with $-x$ produces $2(-x)^4 + 3(-x)^2 - (-x) - 1$, which simplifies to $2x^4 + 3x^2 + x - 1$ and contains one variation of sign. Thus, the given equation has *one negative solution*.

Conclusion   The given equation has one positive, one negative, and two nonreal complex solutions.

3. $3x^4 + 2x^2 + 5 = 0$

    (a) No variations of sign in the given polynomial means that there are *no positive solutions*.

    (b) Replacing x with $-x$ produces $3(-x)^4 + 2(-x)^2 + 5$, which simplifies to $3x^4 + 2x^2 + 5$ and contains no variations of sign. Thus, there are *no negative solutions*.

Conclusion   The given equation contains four nonreal complex solutions. These solutions will appear in conjugate pairs.

4. $2x^5 - 4x^3 + 2x - 5 = 0$

    (a) The fact that there are three variations of sign in the given polynomial implies that the *number of positive solutions is three or one*.

    (b) Replacing x with $-x$ produces $2(-x)^5 - 4(-x)^3 + 2(-x) - 5$, which simplifies to $-2x^5 + 4x^3 - 2x - 5$ and contains two variations of sign. Thus, the *number of negative solutions is two or zero*.

Conclusion   The given equation has either three positive and two negative solutions; three positive and two nonreal complex solutions; one positive, two negative, and two nonreal complex solutions; or one positive and four nonreal complex solutions.

It should be evident from the previous discussions that sometimes we can truly pinpoint the nature of the solutions of a polynomial equation. However, for some equations (such as the last example) the best we can do with the properties discussed in this section is to restrict the possibilities for the nature of the solutions. It might be helpful for you to review Examples 1, 2, and 3 of this section and show that the solution sets do satisfy Properties 9.3, 9.5, and 9.6.

A graphing utility can be very helpful when solving polynomial equations, especially if they are of degree greater than two. Even the search for possible rational solutions can be simplified by looking at a graph. To find the rational solutions of $3x^3 + 8x^2 - 15x + 4 = 0$ (Example 1), we could begin by graphing the equation $y = 3x^3 + 8x^2 - 15x + 4$. This graph is shown in Figure 9.1. From the graph it looks as if 1 and -4 are two of the x-intercepts and therefore solutions of the original equation. Let's check them in the equation.

FIGURE 9.1

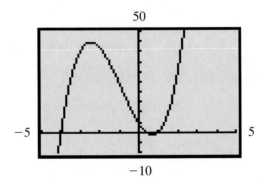

$$3(1)^3 + 8(1)^2 - 15(1) + 4 = 3 + 8 - 15 + 4 = 0$$

$$3(-4)^3 + 8(-4)^2 - 15(-4) + 4 = -192 + 128 + 60 + 4 = 0$$

Thus, $x - 1$ and $x + 4$ are factors of $3x^3 + 8x^2 - 15x + 4$ and the remaining factor could be found by division. We could then determine the solution set as we did in Example 1. Let's consider an example where we use a graphing utility to approximate the real number solutions of a polynomial equation.

Example 4

Find the real number solutions of the equation $x^4 - 2x^3 - 5 = 0$.

Solution

Let's use a graphing utility to get a sketch of the graph of $y = x^4 - 2x^3 - 5$ (Figure 9.2). From this graph we see that one x-intercept is between -1 and -2 and another between 2 and 3. We can use the zoom and trace features to approximate these values at -1.2 and 2.4, to the nearest tenth. Thus, the real solutions for the equation $x^4 - 2x^3 - 5 = 0$ are approximately -1.2 and 2.4. (The other two solutions must be conjugate complex numbers.)

FIGURE 9.2

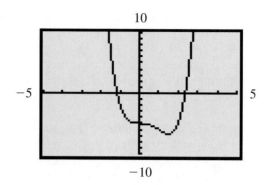

Problem Set 9.3

For Problems 1–20, use the rational root theorem and the factor theorem to help solve each equation. Be sure that the number of solutions for each equation agrees with Property 9.3, taking into account multiplicity of solutions.

1. $x^3 - 2x^2 - 11x + 12 = 0$
2. $x^3 + x^2 - 4x - 4 = 0$
3. $15x^3 + 14x^2 - 3x - 2 = 0$
4. $3x^3 + 13x^2 - 52x + 28 = 0$
5. $8x^3 - 2x^2 - 41x - 10 = 0$
6. $6x^3 + x^2 - 10x + 3 = 0$
7. $x^3 - x^2 - 8x + 12 = 0$
8. $x^3 - 2x^2 - 7x - 4 = 0$
9. $x^3 - 4x^2 + 8 = 0$
10. $x^3 - 10x - 2 = 0$
11. $x^4 + 4x^3 - x^2 - 16x - 12 = 0$
12. $x^4 - 4x^3 - 7x^2 + 34x - 24 = 0$
13. $x^4 + x^3 - 3x^2 - 17x - 30 = 0$
14. $x^4 - 3x^3 + 2x^2 + 2x - 4 = 0$
15. $x^3 - x^2 + x - 1 = 0$
16. $6x^4 - 13x^3 - 19x^2 + 12x = 0$
17. $2x^4 + 3x^3 - 11x^2 - 9x + 15 = 0$
18. $3x^4 - x^3 - 8x^2 + 2x + 4 = 0$
19. $4x^4 + 12x^3 + x^2 - 12x + 4 = 0$
20. $2x^5 - 5x^4 + x^3 + x^2 - x + 6 = 0$

For Problems 21–26, verify that the equations do not have any rational number solutions.

21. $x^4 + 3x - 2 = 0$
22. $x^4 - x^3 - 8x^2 - 3x + 1 = 0$
23. $3x^4 - 4x^3 - 10x^2 + 3x - 4 = 0$
24. $2x^4 - 3x^3 + 6x^2 - 24x + 5 = 0$
25. $x^5 + 2x^4 - 2x^3 + 5x^2 - 2x - 3 = 0$
26. $x^5 - 2x^4 + 3x^3 + 4x^2 + 7x - 1 = 0$

For Problems 27–30, solve each equation by first applying the multiplication property of equality to produce an equivalent equation with integral coefficients.

27. $\frac{1}{10}x^3 + \frac{1}{5}x^2 - \frac{1}{2}x - \frac{3}{5} = 0$

28. $\frac{1}{10}x^3 + \frac{1}{2}x^2 + \frac{1}{5}x - \frac{4}{5} = 0$

29. $x^3 - \frac{5}{6}x^2 - \frac{22}{3}x + \frac{5}{2} = 0$

30. $x^3 + \frac{9}{2}x^2 - x - 12 = 0$

For Problems 31–40, use Descartes' Rule of Signs (Property 9.6) to help list the possibilities for the nature of the solutions for each equation. *Do not* solve the equations.

31. $6x^2 + 7x - 20 = 0$

32. $8x^2 - 14x + 3 = 0$

33. $2x^3 + x - 3 = 0$

34. $4x^3 + 3x + 7 = 0$

35. $3x^3 - 2x^2 + 6x + 5 = 0$

36. $4x^3 + 5x^2 - 6x - 2 = 0$

37. $x^5 - 3x^4 + 5x^3 - x^2 + 2x - 1 = 0$

38. $2x^5 + 3x^3 - x + 1 = 0$

39. $x^5 + 32 = 0$

40. $2x^6 + 3x^4 - 2x^2 - 1 = 0$

THOUGHTS INTO WORDS

41. Explain what it means to say that the equation $(x + 3)^2 = 0$ has a solution of -3 with a multiplicity of two.

42. Describe how to use the rational root theorem to show that the equation $x^2 - 3 = 0$ has no rational solutions.

Further Investigations

43. Use the rational root theorem to argue that $\sqrt{2}$ is not a rational number. [*Hint*: The solutions of $x^2 - 2 = 0$ are $\pm \sqrt{2}$.]

44. Use the rational root theorem to argue that $\sqrt{12}$ is not a rational number.

45. Defend this statement: "Every polynomial equation of odd degree with real coefficients has at least one real number solution."

46. The following synthetic division shows that 2 is a solution of $x^4 + x^3 + x^2 - 9x - 10 = 0$.

$$
\begin{array}{r|rrrrr}
2 & 1 & 1 & 1 & -9 & -10 \\
  &   & 2 & 6 & 14 & 10 \\
\hline
  & 1 & 3 & 7 & 5 & 0 \leftarrow
\end{array}
$$

Notice that the new quotient row (indicated by the arrow) consists entirely of nonnegative numbers. This indicates that searching for solutions greater than 2 would be a waste of time, since larger divisors would continue to increase each of the numbers (except the one on the far left) in the new quotient row. (Try 3 as a divisor!) Thus, we say that 2 is an *upper bound* for the real number solutions of the given equation.

Now consider the following synthetic division, which shows that -1 is also a solution of $x^4 + x^3 + x^2 - 9x - 10 = 0$.

$$
\begin{array}{r|rrrrr}
-1 & 1 & 1 & 1 & -9 & -10 \\
   &   & -1 & 0 & -1 & 10 \\
\hline
   & 1 & 0 & 1 & -10 & 0 \leftarrow
\end{array}
$$

The new quotient row (indicated by the arrow) shows that there is no need to look for solutions less than -1, because any divisor less than -1 would increase the absolute value of each number (except the one on the far left) in the new quotient row. (Try -2 as a divisor!) Thus, we say that -1 is a *lower bound* for the real number solutions of the given equation.

The following general property can be stated.

If $a_n x^n + a_{n-1} x^{n-1} + \cdots + a_1 x + a_0 = 0$ is a polynomial equation with real coefficients, where $a_n > 0$, and if the polynomial is divided synthetically by $x - c$, then

1. If $c > 0$ and all numbers in the new quotient row of the synthetic division are nonnegative, c is an upper bound of the solutions of the given equation;

2. If $c < 0$ and the numbers in the new quotient row alternate in sign (with 0 considered either positive or negative, as needed), c is a lower bound of the solutions of the given equation.

Find the smallest positive integer and the largest negative integer that are upper and lower bounds, respectively, for the real number solutions of each of the following equations. Keep in mind that the integers that serve as bounds do not necessarily have to be solutions of the equation.

(a) $x^3 - 3x^2 + 25x - 75 = 0$

(b) $x^3 + x^2 - 4x - 4 = 0$

(c) $x^4 + 4x^3 - 7x^2 - 22x + 24 = 0$

(d) $3x^3 + 7x^2 - 22x - 8 = 0$

(e) $x^4 - 2x^3 - 9x^2 + 2x + 8 = 0$

Graphics Calculator Activities

47. Solve each of the following equations using a graphics calculator whenever it seems to be helpful. Express all irrational solutions in lowest radical form.

(a) $x^3 + 2x^2 - 14x - 40 = 0$
(b) $x^3 + x^2 - 7x + 65 = 0$
(c) $x^4 - 6x^3 - 6x^2 + 32x + 24 = 0$
(d) $x^4 + 3x^3 - 39x^2 + 11x + 24 = 0$

(e) $x^3 - 14x^2 + 26x - 24 = 0$
(f) $x^4 + 2x^3 - 3x^2 - 4x + 4 = 0$

48. Use a graphics calculator to help determine the nature of the solutions for each of the following equations.

(a) $2x^3 - 3x^2 - 3x + 2 = 0$

(b) $3x^3 + 7x^2 + 8x + 2 = 0$

(c) $2x^4 + 3x^2 + 1 = 0$
(d) $4x^5 - 8x^4 - 5x^3 + 10x^2 + x - 2 = 0$

(e) $x^4 - x^3 + 2x^2 - x - 1 = 0$

(f) $x^5 - x^4 + x^3 - x^2 + x - 3 = 0$

49. Find approximations, to the nearest hundredth, of the real number solutions of each of the following equations.

(a) $x^2 - 4x + 1 = 0$
(b) $3x^3 - 2x^2 + 12x - 8 = 0$
(c) $x^4 - 8x^3 + 14x^2 - 8x + 13 = 0$
(d) $x^4 + 6x^3 - 10x^2 - 22x + 161 = 0$

(e) $7x^5 - 5x^4 + 35x^3 - 25x^2 + 28x - 20 = 0$

9.4 Graphing Polynomial Functions

The terms that classify functions are analogous to the linear equation–quadratic equation–polynomial equation vocabulary. In Chapter 8 we defined a linear function in terms of the equation.

$$f(x) = ax + b$$

and a quadratic function in terms of the equation

$$f(x) = ax^2 + bx + c.$$

Both of these are special cases of a general class of functions called polynomial functions. Any function of the form

$$f(x) = a_nx^n + a_{n-1}x^{n-1} + \cdots + a_1x + a_0,$$

is called a **polynomial function of degree** n, where a_n is a nonzero real number, $a_{n-1}, \ldots, a_1, a_0$ are real numbers, and n is a nonnegative integer. The following are examples of polynomial functions.

$$f(x) = 5x^3 - 2x^2 + x - 4 \qquad \text{Degree 3}$$

$$f(x) = -2x^4 - 5x^3 + 3x^2 + 4x - 1 \qquad \text{Degree 4}$$

$$f(x) = 3x^5 + 2x^2 - 3 \qquad \text{Degree 5}$$

REMARK   Our previous work with polynomial equations is sometimes presented as "finding zeros of polynomial functions." The **solutions**, or **roots**, of a polynomial equation are also called the **zeros** of the polynomial function. For example, -2 and 2 are solutions of $x^2 - 4 = 0$, and they are zeros of $f(x) = x^2 - 4$. That is, $f(-2) = 0$ and $f(2) = 0$.   △

For a complete discussion of graphing polynomial functions, we would need some tools from calculus. However, the graphing techniques that we have discussed in this text will allow us to graph certain kinds of polynomial functions. For example, polynomial functions of the form

$$f(x) = ax^n$$

are quite easy to graph. We know from our previous work that if $n = 1$, then functions such as $f(x) = 2x$, $f(x) = -3x$, and $f(x) = \frac{1}{2}x$ are lines through the origin that have slopes of 2, -3, and $\frac{1}{2}$, respectively.

Furthermore, if $n = 2$, we know that the graphs of functions of the form $f(x) = ax^2$ are parabolas symmetrical with respect to the y-axis that have their vertices at the origin.

We have also previously graphed the special case of $f(x) = ax^n$, where $a = 1$ and $n = 3$—namely, the function $f(x) = x^3$. This graph is shown in Figure 9.3.

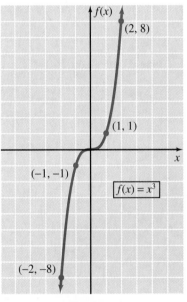

FIGURE 9.3

The graphs of functions of the form $f(x) = ax^3$ where $a \neq 1$ are slight variations of $f(x) = x^3$ and can be easily determined by plotting a few points. The graphs of $f(x) = \frac{1}{2}x^3$ and $f(x) = -x^3$ appear in Figure 9.4.

FIGURE 9.4

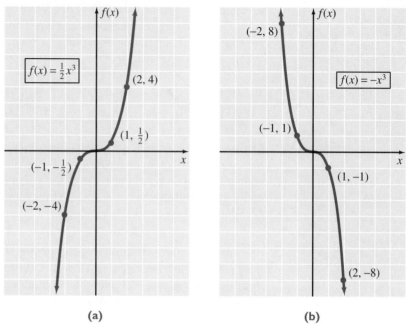

(a)   (b)

There are two general patterns that emerge from studying functions of the form $f(x) = x^n$. If n is odd and greater than 3, the graphs closely resemble Figure 9.3. The graph of $f(x) = x^5$ is shown in Figure 9.5. Notice that the curve "flattens out" a little more around the origin than it does in the graph of $f(x) = x^3$; it increases and decreases more rapidly because of the larger exponent. If n is even and greater than 2, the graphs of $f(x) = x^n$ are not parabolas. They resemble the basic parabola, but they are flatter at the bottom, and steeper on the sides. Figure 9.6 shows the graph of $f(x) = x^4$.

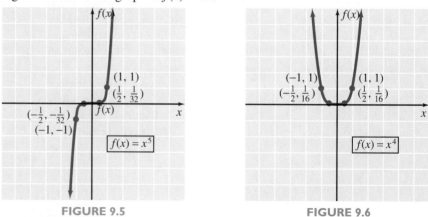

FIGURE 9.5   FIGURE 9.6

Graphs of functions of the form $f(x) = ax^n$, where n is an integer greater than 2 and $a \neq 1$, are variations of those shown in Figures 9.3 and 9.6. If n is odd, the curve is symmetrical about the origin. If n is even, the graph is symmetrical about the y-axis.

Transformations of these basic curves are easy to sketch. For example, in Figure 9.7 we translated the graph of $f(x) = x^3$ upward 2 units to produce the graph of $f(x) = x^3 + 2$. In Figure 9.8 we obtained the graph of $f(x) = (x - 1)^5$ by translating the graph of $f(x) = x^5$ one unit to the right. In Figure 9.9 we sketched the graph of $f(x) = -x^4$ as the x-axis reflection of $f(x) = x^4$.

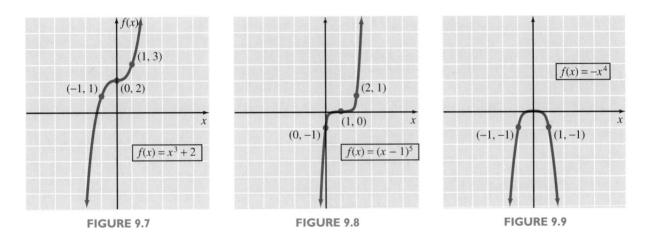

FIGURE 9.7   **FIGURE 9.8**   **FIGURE 9.9**

Graphing Polynomial Functions in Factored Form

As the degree of the polynomial increases, the graphs often become more complicated. We do know, however, that polynomial functions produce smooth continuous curves with a number of turning points, as illustrated in Figures 9.10 and 9.11. Some typical graphs of polynomial functions of odd degree are shown in Figure 9.10. As the graphs suggest, every polynomial function of odd degree has at least one *real zero*—that is, at least one real number c such that $f(c) = 0$. Geometrically, the zeros of the function are the x-intercepts of the graph. Figure 9.11 illustrates some possible graphs of polynomial functions of even degree.

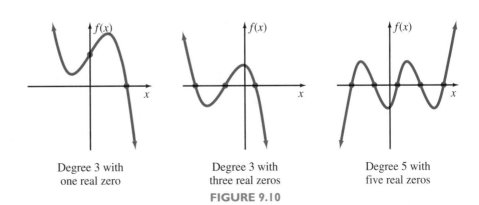

Degree 3 with   Degree 3 with   Degree 5 with
one real zero   three real zeros   five real zeros

FIGURE 9.10

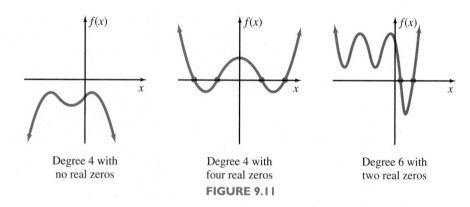

| Degree 4 with no real zeros | Degree 4 with four real zeros | Degree 6 with two real zeros |

FIGURE 9.11

The **turning points** are the places where the function either changes from increasing to decreasing or from decreasing to increasing. Using calculus we are able to verify that **a polynomial function of degree n has at most $n - 1$ turning points**. Now let's illustrate how we can use this information along with some other techniques to graph polynomial functions that are expressed in factored form.

Example 1

Graph $f(x) = (x + 2)(x - 1)(x - 3)$.

Solution

First let's find the x-intercepts (zeros of the function) by setting each factor equal to zero and solving for x.

$$x + 2 = 0 \qquad \text{or} \qquad x - 1 = 0 \qquad \text{or} \qquad x - 3 = 0$$
$$x = -2 \qquad \text{or} \qquad x = 1 \qquad \text{or} \qquad x = 3$$

Thus, the points $(-2, 0)$, $(1, 0)$, and $(3, 0)$ are on the graph. Second, the points associated with the x-intercepts divide the x-axis into four intervals as follows (Figure 9.12).

FIGURE 9.12

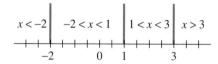

In each of these intervals, $f(x)$ is either always positive or always negative. That is to say, the graph is either above or below the x-axis. Selecting a test value for x in each of the intervals will determine whether x is positive or negative. Any additional points that are easily obtained improve the accuracy of the graph. The following table summarizes these results.

Interval	Test value	Sign of f(x)	Location of graph
$x < -2$	$f(-3) = -24$	Negative	Below x-axis
$-2 < x < 1$	$f(0) = 6$	Positive	Above x-axis
$1 < x < 3$	$f(2) = -4$	Negative	Below x-axis
$x > 3$	$f(4) = 18$	Positive	Above x-axis

Additional values: $f(-1) = 8$

Making use of the x-intercepts and the information in the table, we can sketch the graph in Figure 9.13. The points $(-3, -24)$ and $(4, 18)$ are not shown, but they are used to indicate a rapid decrease and increase of the curve in those regions.

FIGURE 9.13

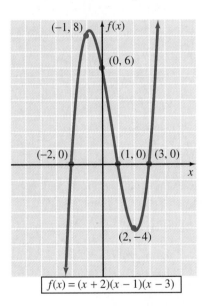

$f(x) = (x + 2)(x - 1)(x - 3)$

REMARK In Figure 9.13 turning points of the graph are indicated at $(2, -4)$ and $(-1, 8)$. Keep in mind that these are only approximations. Again, the tools of calculus are needed to find the exact turning points.

Example 2 Graph $f(x) = -x^4 + 3x^3 - 2x^2$.

Solution The polynomial can be factored as follows.

$$f(x) = -x^4 + 3x^3 - 2x^2$$
$$= -x^2(x^2 - 3x + 2)$$
$$= -x^2(x - 1)(x - 2)$$

Now we can find the x-intercepts.

$$-x^2 = 0 \quad \text{or} \quad x - 1 = 0 \quad \text{or} \quad x - 2 = 0$$

$$x = 0 \quad \text{or} \quad x = 1 \quad \text{or} \quad x = 2$$

The points $(0, 0)$, $(1, 0)$, and $(2, 0)$ are on the graph and divide the x-axis into four intervals as follows in Figure 9.14.

FIGURE 9.14

$$x < 0 \quad | \quad 0 < x < 1 \quad | \quad 1 < x < 2 \quad | \quad x > 2$$

In the following table we determine some points and summarize the sign behavior of $f(x)$.

Interval	Test value	Sign of f(x)	Location of graph
$x < 0$	$f(-1) = -6$	Negative	Below x-axis
$0 < x < 1$	$f\left(\frac{1}{2}\right) = -\frac{3}{16}$	Negative	Below x-axis
$1 < x < 2$	$f\left(\frac{3}{2}\right) = \frac{9}{16}$	Positive	Above x-axis
$x > 2$	$f(3) = -18$	Negative	Below x-axis

Making use of the table and the x-intercepts, we can determine the graph, as illustrated in Figure 9.15.

FIGURE 9.15

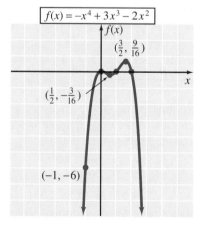

$$f(x) = -x^4 + 3x^3 - 2x^2$$

$\left(\frac{3}{2}, \frac{9}{16}\right)$

$\left(\frac{1}{2}, -\frac{3}{16}\right)$

$(-1, -6)$

Example 3

Graph $f(x) = x^3 + 3x^2 - 4$.

Solution

Use the rational root theorem, synthetic division, and the factor theorem, to factor the given polynomial as follows.

$$f(x) = x^3 + 3x^2 - 4$$
$$= (x - 1)(x^2 + 4x + 4)$$
$$= (x - 1)(x + 2)^2$$

Now we can find the x-intercepts.

$$x - 1 = 0 \quad \text{or} \quad (x + 2)^2 = 0$$
$$x = 1 \quad \text{or} \quad x = -2$$

The points $(-2, 0)$ and $(1, 0)$ are on the graph and divide the x-axis into three intervals as follows (Figure 9.16).

FIGURE 9.16

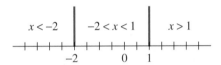

In the following table we determine some points and summarize the sign behavior of $f(x)$.

Interval	Test value	Sign of f(x)	Location of graph
$x < -2$	$f(-3) = -4$	Negative	Below x-axis
$-2 < x < 1$	$f(0) = -4$	Negative	Below x-axis
$x > 1$	$f(2) = 16$	Positive	Above x-axis

Additional values: $f(-1) = -2$
$f(-4) = -20$

As a result of the table and the x-intercepts, we can sketch the graph, as shown in Figure 9.17.

FIGURE 9.17

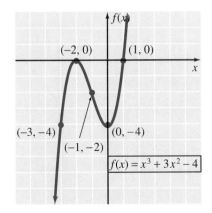

Finally, let's use a graphical approach to solve a problem involving a polynomial function.

Example 4

Suppose that we have a rectangular piece of cardboard that measures 20 inches by 14 inches. From each corner a square piece is cut out and then the flaps are turned up to form an open box (see Figure 9.18). Determine the length of a side of the square pieces to be cut out so that the volume of the box is as large as possible.

FIGURE 9.18

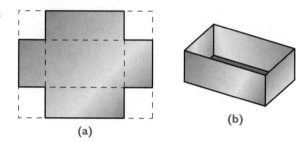

(a)

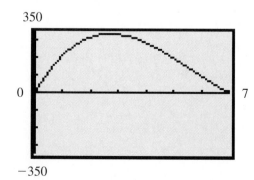

(b)

Solution

Let x represent the length of a side of the squares to be cut from each corner. Then $20 - 2x$ represents the length of the open box and $14 - 2x$ represents the width. The volume of a rectangular box is given by the formula $V = lwh$. So the volume of this box can be represented by $V = x(20 - 2x)(14 - 2x)$. Now let $y = V$ and graph the function $y = x(20 - 2x)(14 - 2x)$ as shown in Figure 9.19. For this problem we are interested only in the part of the graph between $x = 0$ and $x = 7$ because the length of a side of the squares has to be less than 7 inches for a box to be formed. Figure 9.20 gives us a view of that part of the graph. Now we can use the zoom and trace features to determine that as x equals approximately 2.7 the value of y is a maximum of approximately 339.0. Thus, square pieces of length approximately 2.7 inches on a side should be cut from each corner of the rectangular piece of cardboard. The open box formed will have a volume of approximately 339.0 cubic inches.

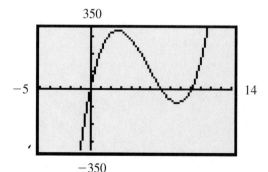

FIGURE 9.19

FIGURE 9.20

Problem Set 9.4

For Problems 1–22, graph each of the polynomial functions.

1. $f(x) = -(x - 3)^3$

2. $f(x) = (x - 2)^3 + 1$

3. $f(x) = (x + 1)^3$

4. $f(x) = x^3 - 3$

5. $f(x) = (x + 3)^4$

6. $f(x) = x^4 - 2$

7. $f(x) = -(x - 2)^4$

8. $f(x) = (x - 1)^5 + 2$

9. $f(x) = (x + 1)^4 + 3$

10. $f(x) = -x^5$

11. $f(x) = (x - 2)(x + 1)(x + 3)$

12. $f(x) = (x - 1)(x + 1)(x - 3)$

13. $f(x) = x(x + 2)(2 - x)$

14. $f(x) = (x + 4)(x + 1)(1 - x)$

15. $f(x) = -x^2(x - 1)(x + 1)$

16. $f(x) = -x(x + 3)(x - 2)$

17. $f(x) = (2x - 1)(x - 2)(x - 3)$

18. $f(x) = x(x - 2)^2(x - 1)$

19. $f(x) = (x - 2)(x - 1)(x + 1)(x + 2)$

20. $f(x) = (x - 1)^2(x + 2)$

21. $f(x) = x(x - 2)^2(x + 1)$

22. $f(x) = (x + 1)^2(x - 1)^2$

For Problems 23–34, graph each polynomial function by first factoring the given polynomial. You may need to use some factoring techniques from Chapter 3 as well as the rational root theorem and the factor theorem.

23. $f(x) = -x^3 - x^2 + 6x$

24. $f(x) = x^3 + x^2 - 2x$

25. $f(x) = x^4 - 5x^3 + 6x^2$

26. $f(x) = -x^4 - 3x^3 - 2x^2$

27. $f(x) = x^3 + 2x^2 - x - 2$

28. $f(x) = x^3 - x^2 - 4x + 4$

29. $f(x) = x^3 - 8x^2 + 19x - 12$

30. $f(x) = x^3 + 6x^2 + 11x + 6$

31. $f(x) = 2x^3 - 3x^2 - 3x + 2$

32. $f(x) = x^3 + 2x^2 - x - 2$

33. $f(x) = x^4 - 5x^2 + 4$

34. $f(x) = -x^4 + 5x^2 - 4$

For each of the following, (a) find the y-intercepts, (b) find the x-intercepts, and (c) find the intervals of x where $f(x) > 0$ and those where $f(x) < 0$. *Do not sketch the graphs.* (Problems 35–42)

35. $f(x) = (x + 3)(x - 6)(8 - x)$

36. $f(x) = (x - 5)(x + 4)(x - 3)$

37. $f(x) = (x + 3)^4(x - 1)^3$

38. $f(x) = (x - 4)^2(x + 3)^3$

39. $f(x) = x(x - 6)^2(x + 4)$

40. $f(x) = (x + 2)^2(x - 1)^3(x - 2)$

41. $f(x) = x^2(2 - x)(x + 3)$

42. $f(x) = (x + 2)^5(x - 4)^2$

THOUGHTS INTO WORDS

43. How would you defend the statement that the equation $2x^4 + 3x^3 + x^2 + 5 = 0$ has no positive solutions? Does it have any negative solutions? Defend your answer.

44. How do you know by inspection that the graph of $f(x) = (x + 1)^2(x - 2)^2$ in Figure 9.21 is incorrect?

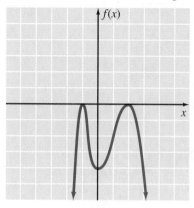

FIGURE 9.21

Further Investigations

45. A polynomial function with real coefficients is continuous everywhere; that is, its graph has no holes or breaks. This is the basis for the following property: "If $f(x)$ is a polynomial with real coefficients, and if $f(a)$ and $f(b)$ are of opposite sign, then there is at least one real zero between a and b." This property, along with our knowledge of polynomial functions, provides the basis for locating and approximating irrational solutions of a polynomial equation.

Consider the equation $x^3 + 2x - 4 = 0$. Applying Descartes' Rule of Signs, we can determine that this equation has one positive real solution and two nonreal complex solutions. (You may want to confirm this!) The rational root theorem indicates that the only possible rational solutions are 1, 2, and 4. Using a little more compact format for synthetic division, we obtain the following results when testing for 1 and 2 as possible solutions.

	1	0	2	−4
1	1	1	3	−1
2	1	2	6	8

Since $f(1) = -1$ and $f(2) = 8$, there must be an irrational solution between 1 and 2. Furthermore, since -1 is closer to 0 than is 8, our guess is that the solu-

tion is closer to 1 than to 2. Let's start looking at 1.0, 1.1, 1.2, etc., until we can place the solution between two numbers.

	1	0	2	−4
1.0	1	1	3	−1
1.1	1	1.1	3.21	−.469
1.2	1	1.2	3.44	.128

A calculator is very helpful at this time.

Since $f(1.1) = -.469$ and $f(1.2) = .128$, the irrational solution must be between 1.1 and 1.2. Furthermore, since .128 is closer to 0 than is $-.469$, our guess is that the solution is closer to 1.2 than to 1.1. Let's start looking at 1.15, 1.16, etc.

	1	0	2	−4
1.15	1	1.15	3.3225	−.179
1.16	1	1.16	3.3456	−.119
1.17	1	1.17	3.3689	−.058
1.18	1	1.18	3.3924	.003

Since $f(1.17) = -.058$ and $f(1.18) = .003$, the irrational solution must be between 1.17 and 1.18. Therefore, we can use 1.2 as a rational approximation to the nearest tenth.

For each of the following equations, (a) verify that the equation has exactly one irrational solution and (b) find an approximation, to the nearest tenth, of that solution.

(a) $x^3 + x - 6 = 0$
(b) $x^3 - 6x - 6 = 0$
(c) $x^3 - 27x - 60 = 0$
(d) $x^3 - x^2 - x - 1 = 0$
(e) $x^3 - 24x - 32 = 0$
(f) $x^3 - 5x^2 - 1 = 0$

Graphics Calculator Activities

46. Graph $f(x) = x^3$. Now predict the graphs for $f(x) = x^3 + 2$, $f(x) = -x^3 + 2$, and $f(x) = -x^3 - 2$. Graph these three functions on the same set of axes with the graph of $f(x) = x^3$.

47. Draw a rough sketch of the graphs of the functions $f(x) = x^3 - x^2$, $f(x) = -x^3 + x^2$, and $f(x) = -x^3 - x^2$. Now graph these three functions to check your sketches.

48. Graph $f(x) = x^4 + x^3 + x^2$. What should the graphs of $f(x) = x^4 - x^3 + x^2$ and $f(x) = -x^4 - x^3 - x^2$ look like? Graph them to see if you were right.

49. How should the graphs of $f(x) = x^3$, $f(x) = x^5$, and $f(x) = x^7$ compare? Graph these three functions on the same set of axes.

50. How should the graphs of $f(x) = x^2$, $f(x) = x^4$, and $f(x) = x^6$ compare? Graph these three functions on the same set of axes.

51. For each of the following functions, find the x-intercepts and find the intervals of x where $f(x) > 0$ and those where $f(x) < 0$.
(a) $f(x) = x^3 - 3x^2 - 6x + 8$

(b) $f(x) = x^3 - 8x^2 - x + 8$

(c) $f(x) = x^3 - 7x^2 + 16x - 12$

(d) $f(x) = x^3 - 19x^2 + 90x - 72$

(e) $f(x) = x^4 + 3x^3 - 3x^2 - 11x - 6$

(f) $f(x) = x^4 + 12x^2 - 64$

52. Find the coordinates of the turning points of each of the following graphs. Express x- and y-values to the nearest integer.
(a) $f(x) = 2x^3 - 3x^2 - 12x + 40$

(b) $f(x) = 2x^3 - 33x^2 + 60x + 1050$

(c) $f(x) = -2x^3 - 9x^2 + 24x + 100$

(d) $f(x) = x^4 - 4x^3 - 2x^2 + 12x + 3$

(e) $f(x) = x^3 - 30x^2 + 288x - 900$

(f) $f(x) = x^5 - 2x^4 - 3x^3 - 2x^2 + x - 1$

53. For each of the following functions, find the x-intercepts and find the turning points. Express your answers to the nearest tenth.
(a) $f(x) = x^3 + 2x^2 - 3x + 4$
(b) $f(x) = 42x^3 - x^2 - 246x - 35$

(c) $f(x) = x^4 - 4x^2 - 4$

54. A rectangular piece of cardboard is 13 inches long and 9 inches wide. From each corner a square piece is cut out and then the flaps are turned up to form an open box. Determine the length of a side of the square pieces so that the volume of the box is as large as possible.

55. A company determines that its weekly profit from manufacturing and selling x units of a certain item is given by $P(x) = -x^3 + 3x^2 + 2880x - 500$. What weekly production rate will maximize the profit?

▼ 9.5   Graphing Rational Functions

A function of the form

$$f(x) = \frac{p(x)}{q(x)}, \qquad q(x) \neq 0,$$

where $p(x)$ and $q(x)$ are polynomial functions, is called a **rational function**. The following are examples of rational functions.

$$f(x) = \frac{2}{x - 1}, \qquad f(x) = \frac{x}{x - 2},$$

$$f(x) = \frac{x^2}{x^2 - x - 6}, \qquad f(x) = \frac{x^3 - 8}{x + 4}$$

In each of these examples the domain of the rational function is the set of all real numbers except for those that make the denominator zero. For example, the domain of $f(x) = \dfrac{2}{x - 1}$ is the set of all real numbers except 1. As we will soon see, these exclusions from the domain are important numbers from a graphing standpoint; they represent breaks in an otherwise continuous curve.

Let's set the stage for graphing rational functions by considering in detail the function $f(x) = \dfrac{1}{x}$. First, note that at $x = 0$ the function is undefined. Second, let's consider a rather extensive table of values to find some number trends and to build a basis for defining the concept of an asymptote.

x	$f(x) = \dfrac{1}{x}$
1	1
2	.5
10	.1
100	.01
1000	.001

These values indicate that the value of $f(x)$ is positive and approaches zero from above as x gets larger and larger.

x	$f(x) = \dfrac{1}{x}$
.5	2
.1	10
.01	100
.001	1000
.0001	10000

These values indicate that $f(x)$ is positive and is getting larger and larger as x approaches zero from the right.

x	$f(x) = \dfrac{1}{x}$
−.5	−2
−.1	−10
−.01	−100
−.001	−1000
−.0001	−10000

These values indicate that $f(x)$ is negative and is getting smaller and smaller as x approaches zero from the left.

x	$f(x) = \dfrac{1}{x}$
−1	−1
−2	−.5
−10	−.1
−100	−.01
−1000	−.001

These values indicate that $f(x)$ is negative and approaches zero from below as x gets smaller and smaller.

Figure 9.22 shows a sketch of $f(x) = \dfrac{1}{x}$, which is drawn using a few points from this table and the patterns discussed. Notice that the graph approaches but does not touch either axis. We say that the y-axis (or $f(x)$-axis) is a **vertical asymptote** and the x-axis is a **horizontal asymptote**

FIGURE 9.22

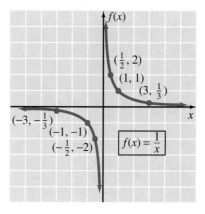

REMARK We know that the equation $f(x) = \dfrac{1}{x}$ exhibits origin symmetry because $f(-x) = -f(x)$. Thus, the graph in Figure 9.22 could have been drawn by first determining the part of the curve in the first quadrant and then reflecting that curve through the origin. △

Now let's define the concepts of vertical and horizontal asymptotes.

Vertical Asymptote

A line $x = a$ is a vertical asymptote for the graph of a function f if

1. $f(x)$ either increases or decreases without bound as x approaches a from the left as in Figure 9.23.

or

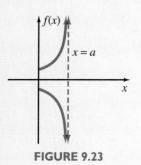

FIGURE 9.23

2. $f(x)$ either increases or decreases without bound as x approaches a from the right as in Figure 9.24.

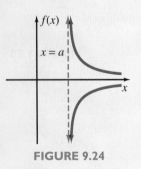

FIGURE 9.24

Horizontal Asymptote

A line $y = b$ [or $f(x) = b$] is a horizontal asymptote for the graph of a function f if

1. $f(x)$ aproaches b from above or below as x gets infinitely small as in Figure 9.25.

or

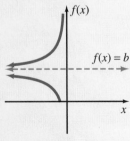

FIGURE 9.25

2. $f(x)$ aproaches b from above or below as x gets infinitely large as in Figure 9.26.

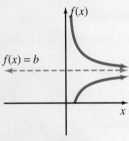

FIGURE 9.26

Following are some suggestions for graphing rational functions of the type we are considering in this section.

1. Check for y-axis and origin symmetry.
2. Find any vertical asymptote by setting the denominator equal to zero and solving for x.
3. Find any horizontal asymptote by studying the behavior of $f(x)$ as x gets infinitely large or as x gets infinitely small.
4. Study the behavior of the graph when it is close to the asymptotes.
5. Plot as many points as necessary to determine the shape of the graph. The number needed may be affected by whether or not the graph has any kind of symmetry.

Keep these suggestions in mind as you study the following examples.

Example 1

Graph $f(x) = \dfrac{-2}{x-1}$.

Solution

Since $x = 1$ makes the denominator zero, the line $x = 1$ is a vertical asymptote. We have indicated this with a dashed line in Figure 9.27. Now let's look for a horizontal asymptote by checking some large and some small values of x.

x	$f(x)$
10	$-\dfrac{2}{9}$
100	$-\dfrac{2}{99}$
1000	$-\dfrac{2}{999}$
-10	$\dfrac{2}{11}$
-100	$\dfrac{2}{101}$
-1000	$\dfrac{2}{1001}$

This portion of the table shows that as x gets very large, the value of $f(x)$ approaches zero from below.

This portion shows that as x gets very small, the value of $f(x)$ approaches zero from above.

Therefore, the x-axis is a horizontal asymptote.
Finally, let's check the behavior of the graph near the vertical asymptote.

x	f(x)
2	−2
1.5	−4
1.1	−20
1.01	−200
1.001	−2000

As x approaches 1 from the right side, the value of f(x) gets smaller and smaller.

0	2
.5	4
.9	20
.99	200
.999	2000

As x approaches 1 from the left side, the value of f(x) gets larger and larger.

The graph of $f(x) = \dfrac{-2}{x-1}$ is shown in Figure 9.27.

FIGURE 9.27

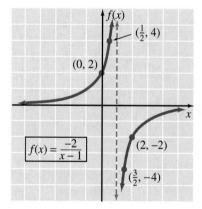

Example 2

Graph $f(x) = \dfrac{x}{x+2}$.

Solution

Since $x = -2$ makes the denominator zero, the line $x = -2$ is a vertical asymptote. To study the behavior of $f(x)$ as x gets very large or very small, let's change the form of the rational expression by dividing numerator and denominator by x.

$$f(x) = \frac{x}{x+2} = \frac{\dfrac{x}{x}}{\dfrac{x+2}{x}} = \frac{1}{\dfrac{x}{x}+\dfrac{2}{x}} = \frac{1}{1+\dfrac{2}{x}}$$

Now we can see that as x gets larger and larger, the value of $f(x)$ approaches 1 from below; and as x gets smaller and smaller, the value of $f(x)$ approaches 1 from above. (Perhaps you should check these claims by plugging in some values for x.) Thus, the line $f(x) = 1$ is a horizontal asymptote. Drawing the asymptotes (dashed lines) and plotting a few points allows us to complete the graph in Figure 9.28.

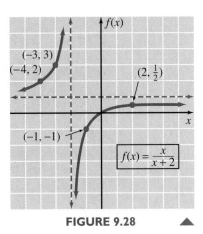

FIGURE 9.28 ▲

In the next two examples, pay special attention to the role of symmetry. It will allow us to direct our efforts toward quadrants I and IV and then to reflect those portions of the curve across the vertical axis to complete the graph.

Example 3

Graph $f(x) = \dfrac{2x^2}{x^2 + 4}$.

Solution

First, notice that $f(-x) = f(x)$; therefore this graph is symmetrical with respect to the vertical axis. Second, the denominator $x^2 + 4$ cannot equal zero for any real number value of x. Thus, there is no vertical asymptote. Third, dividing both numerator and denominator of the rational expression by x^2 produces

$$f(x) = \frac{2x^2}{x^2 + 4} = \frac{\dfrac{2x^2}{x^2}}{\dfrac{x^2 + 4}{x^2}} = \frac{2}{\dfrac{x^2}{x^2} + \dfrac{4}{x^2}}$$

$$= \frac{2}{1 + \dfrac{4}{x^2}}.$$

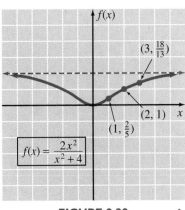

Now we can see that as x gets larger and larger, the value of $f(x)$ approaches 2 from below. Therefore, the line $f(x) = 2$ is a horizontal asymptote. So we can plot a few points using positive values for x, sketch this part of the curve, and then reflect across the $f(x)$-axis to obtain the complete graph, as shown in Figure 9.29.

FIGURE 9.29 ▲

Example 4

Graph $f(x) = \dfrac{3}{x^2 - 4}$.

Solution

First, notice that $f(-x) = f(x)$; therefore, this graph is symmetrical about the y-axis. So by setting the denominator equal to zero and solving for x, we obtain

$$x^2 - 4 = 0$$
$$x^2 = 4$$
$$x = \pm 2.$$

The lines $x = 2$ and $x = -2$ are vertical asymptotes. Next, we can see that $\dfrac{3}{x^2 - 4}$ approaches zero from above as x gets larger and larger. Finally, we can plot a few points using positive values for x (other than 2), sketch this part of the curve, and then reflect it across the $f(x)$-axis to obtain the complete graph. shown in Figure 9.30.

FIGURE 9.30

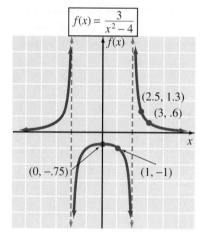

Now suppose that we are going to use a graphing utility to obtain a graph of the function $f(x) = \dfrac{4x^2}{x^4 - 16}$. Before we enter this function into a graphing utility, let's analyze what we know about the graph.

1. Since $f(0) = 0$, the origin is a point on the graph.
2. Since $f(-x) = f(x)$, the graph is symmetric with respect to the y-axis.
3. By setting the denominator equal to zero and solving for x, we can determine the vertical asymptotes.

$$x^4 - 16 = 0$$
$$(x^2 + 4)(x^2 - 4) = 0$$

$$x^2 + 4 = 0 \quad \text{or} \quad x^2 - 4 = 0$$
$$x^2 = -4 \quad \text{or} \quad x^2 = 4$$
$$x = \pm 2i \quad \text{or} \quad x = \pm 2$$

Remember that we are working with ordered pairs of real numbers. Thus, the lines $x = -2$ and $x = 2$ are vertical asymptotes.

4. Divide both the numerator and the denominator of the rational expression by x^4 to produce

$$\frac{4x^2}{x^4 - 16} = \frac{\dfrac{4x^2}{x^4}}{\dfrac{x^4 - 16}{x^4}} = \frac{\dfrac{4}{x^2}}{1 - \dfrac{16}{x^4}}.$$

From the last expression, we see that as $|x|$ gets larger and larger, the value of $f(x)$ approaches zero from above. Therefore, the x-axis is a horizontal asymptote.

Let's enter the function in a graphing utility and obtain a graph as shown in Figure 9.31. Note that the graph is consistent with all of the previous information that we determined before we used the graphing utility. In other words, our knowledge of graphing techniques enhances our use of a graphing utility.

FIGURE 9.31

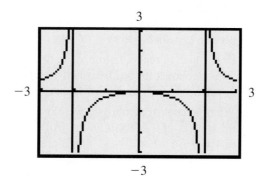

Back in Section 2.4 we solved problems of the following type: How much pure alcohol should be added to 6 liters of a 40% alcohol solution to raise it to a 60% alcohol solution? The answer of 3 liters can be found by solving the following equation, where x represents the amount of pure alcohol to be added.

$$\underset{\substack{\text{Pure alcohol} \\ \text{to start with}}}{\underset{\downarrow}{.40(6)}} + \underset{\substack{\text{Pure alcohol} \\ \text{added}}}{\underset{\downarrow}{x}} = \underset{\substack{\text{Pure alcohol in} \\ \text{final solution}}}{\underset{\downarrow}{.60(6 + x)}}$$

Now let's consider this problem in a more general setting. Again, x represents the amount of pure alcohol to be added and the rational expression $\dfrac{2.4 + x}{6 + x}$ represents the concentration of pure alcohol in the final solution. Let's graph the rational function $y = \dfrac{2.4 + x}{6 + x}$ as shown in Figure 9.32. For this particular problem x is nonnegative, so we are interested only in the part of the graph that is in the first quadrant. Change the boundaries of the viewing rectangle so that $0 \leq x \leq 15$ and $0 \leq y \leq 2$ to obtain Figure 9.33. Now we are ready to answer questions about this situation.

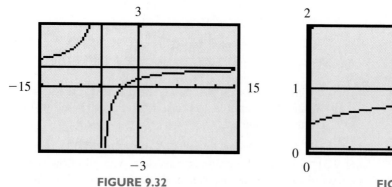

FIGURE 9.32

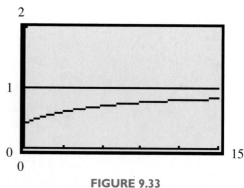

FIGURE 9.33

1. How much pure alcohol needs to be added to raise the 40% solution to a 60% alcohol solution? (*Answer*: Using the trace feature of the graphing utility we find that $y = .6$ when $x = 3$. Therefore, 3 liters of pure alcohol need to be added.)

2. How much pure alcohol needs to be added to raise the 40% solution to a 70% alcohol solution? (*Answer*: Using the trace feature we find that $y = .7$ when $x = 6$. Therefore, 6 liters of pure alcohol need to be added.)

3. What percent of alcohol do we have if we add 9 liters of pure alcohol to the 6 liters of a 40% solution? (*Answer*: Using the trace feature we find that $y = .76$ when $x = 9$. Therefore, adding 9 liters of pure alcohol will give us a 76% alcohol solution.)

Problem Set 9.5

Graph each of the following rational functions.

1. $f(x) = \dfrac{1}{x^2}$

2. $f(x) = \dfrac{-1}{x}$

3. $f(x) = \dfrac{-1}{x - 3}$

4. $f(x) = \dfrac{3}{x + 1}$

5. $f(x) = \dfrac{-3}{(x + 2)^2}$

6. $f(x) = \dfrac{2}{(x - 1)^2}$

7. $f(x) = \dfrac{2x}{x - 1}$

8. $f(x) = \dfrac{x}{x - 3}$

9. $f(x) = \dfrac{-x}{x + 1}$

10. $f(x) = \dfrac{-3x}{x + 2}$

11. $f(x) = \dfrac{-2}{x^2 - 4}$

12. $f(x) = \dfrac{1}{x^2 - 1}$

13. $f(x) = \dfrac{3}{(x + 2)(x - 4)}$

14. $f(x) = \dfrac{-2}{(x + 1)(x - 2)}$

15. $f(x) = \dfrac{-1}{x^2 + x - 6}$

16. $f(x) = \dfrac{2}{x^2 + x - 2}$

17. $f(x) = \dfrac{2x - 1}{x}$

18. $f(x) = \dfrac{x + 2}{x}$

19. $f(x) = \dfrac{4x^2}{x^2 + 1}$

20. $f(x) = \dfrac{4}{x^2 + 2}$

21. $f(x) = \dfrac{x^2 - 4}{x^2}$

22. $f(x) = \dfrac{2x^4}{x^4 + 1}$

Further Investigations

23. The rational function $f(x) = \dfrac{(x - 2)(x + 3)}{x - 2}$ has a domain of all real numbers except 2 and can be simplified to $f(x) = x + 3$. Thus, its graph is a straight line with a hole at (2, 5). Graph each of the following functions.

(a) $f(x) = \dfrac{(x + 4)(x - 1)}{x + 4}$

(b) $f(x) = \dfrac{x^2 - 5x + 6}{x - 2}$

(c) $f(x) = \dfrac{x - 1}{x^2 - 1}$

(d) $f(x) = \dfrac{x + 2}{x^2 + 6x + 8}$

24. Graph the function $f(x) = x + 2 + \dfrac{3}{x - 2}$. It may be necessary to plot a rather large number of points. Also defend the statement that $f(x) = x + 2$ is an **oblique asymptote**.

Graphics Calculator Activities

25. Use a graphics calculator to check your graphs for Problem 23. What feature of the graphs does not show up on the calculator?

26. Each of the following graphs is a transformation of $f(x) = 1/x$. First, predict the general shape and location of the graph and then check your prediction with a graphics calculator.

(a) $f(x) = \dfrac{1}{x} - 2$ **(b)** $f(x) = \dfrac{1}{x + 3}$

(c) $f(x) = -\dfrac{1}{x}$ **(d)** $f(x) = \dfrac{1}{x - 2} + 3$

(e) $f(x) = \dfrac{2x + 1}{x}$

27. Graph $f(x) = \dfrac{1}{x^2}$. How should the graph of $f(x) = \dfrac{1}{(x - 4)^2}$, $f(x) = \dfrac{1 + 3x^2}{x^2}$, and $f(x) = -\dfrac{1}{x^2}$ compare to the graph of $f(x) = \dfrac{1}{x^2}$? Graph the three functions on the same set of axes with the graph of $f(x) = \dfrac{1}{x^2}$.

28. Graph $f(x) = \dfrac{1}{x^3}$. How should the graphs of $f(x) = \dfrac{2x^3 + 1}{x^3}$, $f(x) = \dfrac{1}{(x + 2)^3}$, and $f(x) = -\dfrac{1}{x^3}$ compare to the graph of $f(x) = \dfrac{1}{x^3}$? Graph the three functions on the same set of axes with the graph of $f(x) = \dfrac{1}{x^3}$.

29. Use a graphics calculator to check your graphs for Problems 19–22.

30. Suppose that x ounces of pure acid have been added to 14 ounces of a 15% acid solution.

(a) Set up the rational expression that represents the concentration of pure acid in the final solution.
(b) Graph the rational function that displays the level of concentration.
(c) How many ounces of pure acid need to be added to the 14 ounces of a 15% solution to raise it to a 40.5% solution? Check your answer.
(d) How many ounces of pure acid need to be added to the 14 ounces of a 15% solution to raise it to a 50% solution? Check your answer.
(e) What percent of acid do we obtain if we add 12 ounces of pure acid to the 14 ounces of a 15% solution? Check your answer.

31. Solve the following problem both algebraically and graphically: One solution contains 50% alcohol and another solution contains 80% alcohol. How many liters of each solution should be mixed to produce 10.5 liters of a 70% alcohol solution? Check your answer.

32. Graph the function $f(x) = \dfrac{x^2}{x^2 - x - 2}$ and the horizontal asymptote $f(x) = 1$ on the same set of axes. Notice that the left branch of the curve intersects the asymptote and approaches it from the bottom. Find this point of intersection both algebraically and graphically.

33. Graph each of the following functions and for each one determine the point of intersection of some part of the curve and the horizontal asymptote.

(a) $f(x) = \dfrac{x}{x^2 - 4}$ **(b)** $f(x) = \dfrac{3x}{x^2 + 1}$

(c) $f(x) = \dfrac{x^2}{x^2 + x - 2}$ **(d)** $f(x) = \dfrac{2x^2}{x^2 - 2x - 8}$

(e) $f(x) = \dfrac{x}{x^2 + x - 6}$ **(f)** $f(x) = \dfrac{x^2}{x^2 - 4x + 3}$

34. Consider the function $f(x) = \dfrac{x^2 - 1}{x - 2}$. Since the degree of the numerator is greater than the degree of the denominator, we can change the form of the rational expression by division. Using synthetic division we obtain

$$\begin{array}{r} 2\overline{)1 \quad 0 \quad -1} \\ \underline{2 \quad 4}\cdot \\ 1 \quad 2 \quad 3 \end{array}$$

Therefore, the original function can be written as

$$f(x) = x + 2 + \frac{3}{x - 2}.$$

Now for very large values of $|x|$, the fraction $\dfrac{3}{x - 2}$ is close to zero. Thus, as $|x|$ gets larger and larger, the graph of $f(x) = x + 2 + \dfrac{3}{x - 2}$ gets closer and closer to the line $f(x) = x + 2$. We call this line an **oblique asymptote**.

For each of the following, first determine and graph any oblique asymptote; then on the same set of axes, graph the function.

(a) $f(x) = \dfrac{x^2 - 1}{x - 2}$

(b) $f(x) = \dfrac{x^2 + 1}{x + 2}$

(c) $f(x) = \dfrac{2x^2 + x + 1}{x + 1}$

(d) $f(x) = \dfrac{x^2 + 4}{x - 3}$

(e) $f(x) = \dfrac{3x^2 - x - 2}{x - 2}$

(f) $f(x) = \dfrac{4x^2 + x + 1}{x + 1}$

(g) $f(x) = \dfrac{x^3 + x^2 - x - 1}{x^2 + 2x + 3}$

(h) $f(x) = \dfrac{x^3 + 2x^2 + x - 3}{x^2 - 4}$

SUMMARY

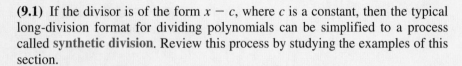

(9.1) If the divisor is of the form $x - c$, where c is a constant, then the typical long-division format for dividing polynomials can be simplified to a process called **synthetic division**. Review this process by studying the examples of this section.

The division algorithm for polynomials states that if $f(x)$ and $d(x)$ are polynomials and $d(x) \neq 0$, then there exist unique polynomials $q(x)$ and $r(x)$ such that

$$f(x) = d(x)q(x) + r(x),$$

where $r(x) = 0$ or the degree of $r(x)$ is less than the degree of $d(x)$.

(9.2) The remainder theorem states that if a polynomial $f(x)$ is divided by $x - c$, then the remainder is equal to $f(c)$. Thus, a polynomial can be evaluated for a given number either by direct substitution or by using synthetic division.

The factor theorem states that a polynomial $f(x)$ has a factor $x - c$ if and only if $f(c) = 0$.

(9.3) The following concepts and properties provide a basis for solving polynomial equations.

1. Synthetic division.
2. The factor theorem.
3. Property 9.3: A polynomial equation of degree n has n solutions, where any solution of multiplicity p is counted p times.
4. The rational root theorem: Consider the polynomial equation

   $$a_n x^n + a_{n-1} x^{n-1} + \cdots + a_1 x + a_0 = 0,$$

   where the *coefficients are integers*. If the rational number $\dfrac{c}{d}$, reduced to lowest terms, is a solution of the equation, then c is a factor of the constant term a_0 and d is a factor of the leading coefficient a_n.
5. Property 9.5: Nonreal complex solutions of polynomial equations with real coefficients, if they exist, must occur in conjugate pairs.
6. Descartes' Rule of Signs: Let $a_n x^n + a_{n-1} x^{n-1} + \cdots + a_1 x + a_0 = 0$ be a polynomial equation with real coefficients.

   (a) The number of *positive real solutions* is either equal to the number of sign variations or else less than the number of sign variations by a positive even integer.

SUMMARY

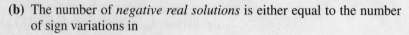

(b) The number of *negative real solutions* is either equal to the number of sign variations in

$$a_n(-x)^n + a_{n-1}(-x)^{n-1} + \cdots + a_1(-x) + a_0,$$

or else less than the number of sign variations by a positive even integer.

(9.4) The following steps may be used to graph a polynomial function that is expressed in factored form.

1. Find the x-intercepts, which are also called the **zeros of the polynomial**.
2. Use a test value in each of the intervals determined by the x-intercepts to find out whether the function is positive or negative over that interval.
3. Plot any additional points that are needed to determine the graph.

(9.5) The following steps may be used to graph a rational function.

1. Check for vertical axis and origin symmetry.
2. Find any vertical asymptotes by setting the denominator equal to zero and solving for x.
3. Find any horizontal asymptotes by studying the behavior of $f(x)$ as x gets very large or very small. This may require changing the form of the original rational expression.
4. Study the behavior of the graph when it is close to the asymptotic lines.
5. Plot as many points as necessary to determine the graph. The number needed may be affected by whether or not the graph has any kind of symmetry.

Chapter 9 Review Problem Set

For Problems 1–4, use synthetic division to determine the quotient and the remainder.

1. $(3x^3 - 4x^2 + 6x - 2) \div (x - 1)$

2. $(5x^3 + 7x^2 - 9x + 10) \div (x + 2)$

3. $(-2x^4 + x^3 - 2x^2 - x - 1) \div (x + 4)$

4. $(-3x^4 - 5x^2 + 9) \div (x - 3)$

For Problems 5–8, find $f(c)$ either by using synthetic division and the remainder theorem or by evaluating $f(c)$ directly.

5. $f(x) = 4x^5 - 3x^3 + x^2 - 1$ and $c = 1$

6. $f(x) = 4x^3 - 7x^2 + 6x - 8$ and $c = -3$

7. $f(x) = -x^4 + 9x^2 - x - 2$ and $c = -2$

8. $f(x) = x^4 - 9x^3 + 9x^2 - 10x + 16$ and $c = 8$

For Problems 9–12, use the factor theorem to help answer some questions about factors.

9. Is $x + 2$ a factor of $2x^3 + x^2 - 7x - 2$?

10. Is $x - 3$ a factor of $x^4 + 5x^3 - 7x^2 - x + 3$?

11. Is $x - 4$ a factor of $x^5 - 1024$?

12. Is $x + 1$ a factor of $x^5 + 1$?

For Problems 13–16, use the rational root theorem and the factor theorem to help solve each of the equations.

13. $x^3 - 3x^2 - 13x + 15 = 0$

14. $8x^3 + 26x^2 - 17x - 35 = 0$

15. $x^4 - 5x^3 + 34x^2 - 82x + 52 = 0$

16. $x^3 - 4x^2 - 10x + 4 = 0$

For Problems 17 and 18, use Descartes' Rule of Signs (Property 9.6) to help list the possibilities for the nature of the solutions. *Do not solve the equations.*

17. $4x^4 - 3x^3 + 2x^2 + x + 4 = 0$

18. $x^5 + 3x^3 + x + 7 = 0$

For Problems 19–22, graph each of the polynomial functions.

19. $f(x) = -(x - 2)^3 + 3$

20. $f(x) = (x + 3)(x - 1)(3 - x)$

21. $f(x) = x^4 - 4x^2$

22. $f(x) = x^3 - 4x^2 + x + 6$

For Problems 23 and 24, graph each of the rational functions. Be sure to identify the asymptotes.

23. $f(x) = \dfrac{2x}{x - 3}$

24. $f(x) = \dfrac{-3}{x^2 + 1}$

CHAPTER 9 TEST

1. Find the quotient and remainder when $3x^3 + 5x^2 - 14x - 6$ is divided by $x + 3$.

2. Find the quotient and remainder when $4x^4 - 7x^2 - x + 4$ is divided by $x - 2$.

3. If $f(x) = x^5 - 8x^4 + 9x^3 - 13x^2 - 9x - 10$, find $f(7)$.

4. If $f(x) = 3x^4 + 20x^3 - 6x^2 + 9x + 19$, find $f(-7)$.

5. If $f(x) = x^5 - 35x^3 - 32x + 15$, find $f(6)$.

6. Is $x - 5$ a factor of $3x^3 - 11x^2 - 22x - 20$?

7. Is $x + 2$ a factor of $5x^3 + 9x^2 - 9x - 17$?

8. Is $x + 3$ a factor of $x^4 - 16x^2 - 17x + 12$?

9. Is $x - 6$ a factor of $x^4 - 2x^2 + 3x - 12$?

For Problems 10–14, solve each equation.

10. $x^3 - 13x + 12 = 0$

11. $2x^3 + 5x^2 - 13x - 4 = 0$

12. $x^4 - 4x^3 - 5x^2 + 38x - 30 = 0$

13. $2x^3 + 3x^2 - 17x + 12 = 0$

14. $3x^3 - 7x^2 - 8x + 20 = 0$

15. Use Descartes' Rule of Signs to determine the nature of the roots of $5x^4 + 3x^3 - x^2 - 9 = 0$.

16. Find the x-intercepts of the graph of the function $f(x) = 3x^3 + 19x^2 - 14x$.

17. Find the equation of the vertical asymptote for the graph of the function $f(x) = \dfrac{5x}{x + 3}$.

18. Find the equation of the horizontal asymptote for the graph of the function $f(x) = \dfrac{5x^2}{x^2 - 4}$.

19. What type of symmetry does the equation $f(x) = \dfrac{x^2}{x^2 + 2}$ exhibit?

(continued on next page)

CHAPTER 9 TEST *(continued)*

20. What type of symmetry does the equation $f(x) = \dfrac{-3x}{x^2 + 1}$ exhibit?

For Problems 21–25, graph each of the functions.

21. $f(x) = (2 - x)(x - 1)(x + 1)$

22. $f(x) = -x(x - 3)(x + 2)$

23. $f(x) = \dfrac{-x}{x - 3}$

24. $f(x) = \dfrac{-2}{x^2 - 4}$

25. $f(x) = (x + 2)^2(x - 1)$

CHAPTER 10

Exponential and Logarithmic Functions

How long will it take $100 to triple itself if it is invested at 8% interest compounded continuously? **We can use the formula $A = Pe^{rt}$ to generate the equation $300 = 100e^{.08t}$, which can be solved for t using logarithms. It will take approximately 13.7 years for the money to triple itself.**

In this chapter we will (1) extend the meaning of an exponent, (2) work with some exponential functions, (3) consider the concept of a logarithm, (4) work with some logarithmic functions, and (5) use the concepts of exponents and logarithms to expand our problem solving skills. Your calculator will be a valuable tool throughout this chapter.

10.1 **Exponents and Exponential Functions**

In Chapter 1, the expression b^n was defined to mean n factors of b, where n is any positive integer and b is any real number. For example,

$$2^3 = 2 \cdot 2 \cdot 2 = 8, \qquad \left(\frac{1}{3}\right)^4 = \left(\frac{1}{3}\right)\left(\frac{1}{3}\right)\left(\frac{1}{3}\right)\left(\frac{1}{3}\right) = \frac{1}{81},$$

$$(-4)^2 = (-4)(-4) = 16, \qquad -(.5)^3 = -[(.5)(.5)(.5)] = -.125.$$

Then in Chapter 5, by defining $b^0 = 1$ and $b^{-n} = \dfrac{1}{b^n}$, where n is any positive integer and b is any nonzero real number, we extended the concept of an exponent to include all integers. Examples include

$$(.76)^0 = 1, \qquad 2^{-3} = \frac{1}{2^3} = \frac{1}{8},$$

$$\left(\frac{2}{3}\right)^{-2} = \frac{1}{\left(\frac{2}{3}\right)^2} = \frac{1}{\frac{4}{9}} = \frac{9}{4}, \qquad (.4)^{-1} = \frac{1}{(.4)^1} = \frac{1}{.4} = 2.5.$$

In Chapter 5, we also provided for the use of all rational numbers as exponents by defining

$$b^{m/n} = \sqrt[n]{b^m} = (\sqrt[n]{b})^m,$$

where n is a positive integer greater than 1 and b is a real number such that $\sqrt[n]{b}$ exists. Some examples are

$$27^{2/3} = (\sqrt[3]{27})^2 = 9, \qquad 16^{1/4} = \sqrt[4]{16^1} = 2,$$

$$\left(\frac{1}{9}\right)^{1/2} = \sqrt{\frac{1}{9}} = \frac{1}{3}, \qquad 32^{-1/5} = \frac{1}{32^{1/5}} = \frac{1}{\sqrt[5]{32}} = \frac{1}{2}.$$

Formally extending the concept of an exponent to include the use of irrational numbers requires some ideas from calculus and is therefore beyond the scope of this text. However, we can take a brief glimpse at the general idea involved. Consider the number $2^{\sqrt{3}}$. By using the nonterminating and nonrepeating decimal representation 1.73205 ... for $\sqrt{3}$, we can form the sequence of numbers 2^1, $2^{1.7}$, $2^{1.73}$, $2^{1.732}$, $2^{1.7320}$, $2^{1.73205}$, ... It should seem reasonable that each successive power gets closer to $2^{\sqrt{3}}$. This is precisely what happens if b^n, where n is irrational, is properly defined using the concept of a limit. Furthermore, this will ensure that an expression such as 2^x will yield exactly one value for each value of x.

So from now on we can use any real number as an exponent, and the basic properties we stated in Chapter 5 can be extended to include all real numbers as exponents. Let's restate those properties with the restriction that the bases a and b must be positive numbers so that we can avoid expressions such as $(-4)^{1/2}$, which do not represent real numbers.

PROPERTY 10.1

If a and b are positive real numbers and m and n are any real numbers, then

1. $b^n \cdot b^m = b^{n+m}$ Product of two powers

2. $(b^n)^m = b^{mn}$ Power of a power

3. $(ab)^n = a^n b^n$ Power of a product

4. $\left(\dfrac{a}{b}\right)^n = \dfrac{a^n}{b^n}$ Power of a quotient

5. $\dfrac{b^n}{b^m} = b^{n-m}$ Quotient of two powers

Another property that can be used to solve certain types of equations that involve exponents can be stated as follows.

PROPERTY 10.2

If $b > 0$, $b \neq 1$, and m and n are real numbers, then $b^n = b^m$ if and only if $n = m$.

The following examples illustrate the use of Property 10.2.

Example 1

Solve $2^x = 32$.

Solution

$2^x = 32$

$2^x = 2^5$ $32 = 2^5$

$x = 5.$ Property 10.2

The solution set is $\{5\}$.   ▲

Example 2

Solve $3^{2x} = \dfrac{1}{9}$.

Solution

$3^{2x} = \dfrac{1}{9} = \dfrac{1}{3^2}$

$3^{2x} = 3^{-2}$

$2x = -2$ Property 10.2

$x = -1.$

The solution set is $\{-1\}$.   ▲

Example 3

Solve $\left(\dfrac{1}{5}\right)^{x-4} = \dfrac{1}{125}$.

Solution

$$\left(\frac{1}{5}\right)^{x-4} = \frac{1}{125}$$

$$\left(\frac{1}{5}\right)^{x-4} = \left(\frac{1}{5}\right)^{3}$$

$$x - 4 = 3 \qquad \text{Property 10.2}$$

$$x = 7.$$

The solution set is $\{7\}$. ▲

Example 4

Solve $8^x = 32$.

Solution

$$8^x = 32$$

$$(2^3)^x = 2^5 \qquad 8 = 2^3$$

$$2^{3x} = 2^5$$

$$3x = 5 \qquad \text{Property 10.2}$$

$$x = \frac{5}{3}$$

The solution set is $\left\{\dfrac{5}{3}\right\}$. ▲

Example 5

Solve $(3^{x+1})(9^{x-2}) = 27$.

Solution

$$(3^{x+1})(9^{x-2}) = 27$$

$$(3^{x+1})(3^2)^{x-2} = 3^3$$

$$(3^{x+1})(3^{2x-4}) = 3^3$$

$$3^{3x-3} = 3^3$$

$$3x - 3 = 3 \qquad \text{Property 10.2}$$

$$3x = 6$$

$$x = 2$$

The solution set is $\{2\}$. ▲

Exponential Functions

If b is any positive number, then the expression b^x designates exactly one real number for every real value of x. Therefore, the equation $f(x) = b^x$ defines a function whose domain is the set of real numbers. Furthermore, if we include the additional restriction $b \neq 1$, then any equation of the form $f(x) = b^x$ describes a one-to-one function, which we call an **exponential function**. See the following definition.

DEFINITION 10.1

If $b > 0$ and $b \neq 1$, then the function f defined by

$$f(x) = b^x,$$

where x is any real number, is called the **exponential function with base b**

REMARK The function $f(x) = 1^x$ is a constant function and therefore it is not a one-to-one function. Remember from Chapter 8 that one-to-one functions have inverses; this becomes a key issue in a later section. △

Now let's consider graphing some exponential functions.

Example 6

Graph the function $f(x) = 2^x$.

Solution

Let's set up a table of values; keep in mind that the domain is the set of real numbers and that the equation $f(x) = 2^x$ exhibits no symmetry. Plot these points and connect them with a smooth curve to produce Figure 10.1.

x	2^x
-2	$\dfrac{1}{4}$
-1	$\dfrac{1}{2}$
0	1
1	2
2	4
3	8

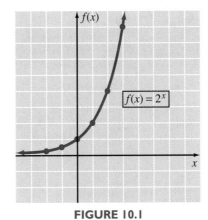

FIGURE 10.1 ▲

In the table for Example 6 we chose integral values for x to keep the computation simple. However, with the use of a calculator we could easily acquire functional values by using nonintegral exponents. Consider the following additional values for $f(x) = 2^x$.

$$f(.5) \approx 1.41, \qquad f(1.7) \approx 3.25,$$
$$f(-.5) \approx .71, \qquad f(-2.6) \approx .16$$

Use your calculator to check these results. Also notice that the points generated by these values do fit the graphs in Figure 10.1.

Example 7

Graph $f(x) = \left(\dfrac{1}{2}\right)^x$.

Solution

Again, let's set up a table of values, plot the points, and connect them with a smooth curve. The graph is shown in Figure 10.2.

x	$\left(\dfrac{1}{2}\right)^x$
-3	8
-2	4
-1	2
0	1
1	$\dfrac{1}{2}$
2	$\dfrac{1}{4}$
3	$\dfrac{1}{8}$

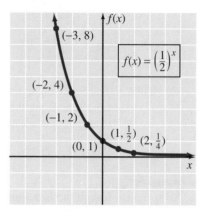

FIGURE 10.2

REMARK Since $\left(\dfrac{1}{2}\right)^x = \dfrac{1}{2^x} = 2^{-x}$, the graphs of $f(x) = 2^x$ and $f(x) = \left(\dfrac{1}{2}\right)^x$ are reflections of each other across the y-axis. Therefore, Figure 10.2 could have been drawn by reflecting Figure 10.1 across the y-axis. △

The graphs in Figures 10.1 and 10.2 illustrate a general behavior pattern of exponential functions. That is to say, if $b > 1$, then the graph of $f(x) = b^x$ goes up to the right and the function is called an **increasing function**. If $0 < b < 1$, then the graph of $f(x) = b^x$ goes down to the right and the function is called a **decreasing function**. These facts are illustrated in Figure 10.3. Notice that $b^0 = 1$ for any $b > 0$; thus, all graphs of $f(x) = b^x$ contain the point $(0, 1)$.

FIGURE 10.3

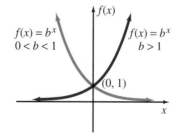

As you graph exponential functions, don't forget your previous graphing experiences.

1. The graph of $f(x) = 2^x - 4$ is the graph of $f(x) = 2^x$ *moved down four units.*

2. The graph of $f(x) = 2^{x+3}$ is the graph of $f(x) = 2^x$ *moved three units to the left.*

3. The graph of $f(x) = -2^x$ is the graph of $f(x) = 2^x$ *reflected across the x-axis.*

We used a graphics calculator to graph these four functions on the same set of axes as shown in Figure 10.4.

FIGURE 10.4

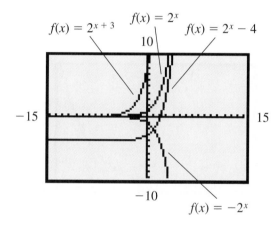

If you are faced with an exponential function that is not of the basic form $f(x) = b^x$ or a variation thereof, don't forget the graphing suggestions we offered in Chapters 7 and 8. Let's consider one such example.

Example 8

Graph $f(x) = 2^{-x^2}$.

Solution

Since $f(-x) = 2^{-(-x)^2} = 2^{-x^2} = f(x)$, we know that this curve is symmetrical with respect to the y-axis. Therefore, let's set up a table of values using nonnegative values for x. Plot these points, connect them with a smooth curve, and reflect this portion of the curve across the y-axis to produce the graph in Figure 10.5.

x	2^{-x^2}
0	1
$\frac{1}{2}$	.84
1	.5
$\frac{3}{2}$	.21
2	.06

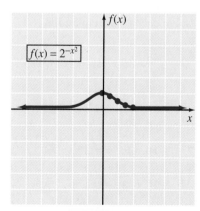

FIGURE 10.5

Finally, let's consider a problem in which a graphing utility gives us an approximate solution.

Example 9

Use a graphing utility to obtain a graph of $f(x) = 50(2^x)$ and find an approximate value for x when $f(x) = 15,000$.

Solution

First, we must find an appropriate viewing rectangle. Since $50(2^{10}) = 51,200$, let's set the boundaries so that $0 \le x \le 10$ and $0 \le y \le 50,000$ with a scale of 10,000 on the y-axis. (Certainly other boundaries could be used, but these will give us a graph that we can work with for this problem.) The graph of $f(x) = 50(2^x)$ is shown in Figure 10.6. Now we can use the trace and zoom-in features of the graphing utility to find that $x \approx 8.2$ at $y = 15,000$.

FIGURE 10.6

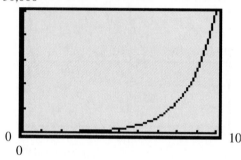

REMARK In Example 9 we used a graphical approach to solve the equation $50(2^x) = 15,000$. In Section 10.5 we will use an algebraic approach for solving that same kind of equation. △

Problem Set 10.1

For Problems 1–26, solve each of the equations.

1. $2^x = 64$

2. $3^x = 81$

3. $3^{2x} = 27$

4. $2^{2x} = 16$

5. $\left(\dfrac{1}{2}\right)^x = \dfrac{1}{128}$

6. $\left(\dfrac{1}{4}\right)^x = \dfrac{1}{256}$

7. $3^{-x} = \dfrac{1}{243}$

8. $3^{x+1} = 9$

9. $6^{3x-1} = 36$

10. $2^{2x+3} = 32$

11. $\left(\dfrac{3}{4}\right)^n = \dfrac{64}{27}$

12. $\left(\dfrac{2}{3}\right)^n = \dfrac{9}{4}$

13. $16^x = 64$

14. $4^x = 8$

15. $27^{4x} = 9^{x+1}$

16. $32^x = 16^{1-x}$

17. $9^{4x-2} = \dfrac{1}{81}$

18. $8^{3x+2} = \dfrac{1}{16}$

19. $10^x = .1$

20. $10^x = .0001$

21. $(2^{x+1})(2^x) = 64$

22. $(2^{2x-1})(2^{x+2}) = 32$

23. $(27)(3^x) = 9^x$

24. $(3^x)(3^{5x}) = 81$

25. $(4^x)(16^{3x-1}) = 8$

26. $(8^{2x})(4^{2x-1}) = 16$

For Problems 27–46, graph each of the exponential functions.

27. $f(x) = 3^x$

28. $f(x) = 4^x$

29. $f(x) = \left(\dfrac{1}{3}\right)^x$

30. $f(x) = \left(\dfrac{1}{4}\right)^x$

31. $f(x) = \left(\dfrac{3}{2}\right)^x$

32. $f(x) = \left(\dfrac{2}{3}\right)^x$

33. $f(x) = 2^x - 3$

34. $f(x) = 2^x + 1$

35. $f(x) = 2^{x+2}$

36. $f(x) = 2^{x-1}$

37. $f(x) = -2^x$

38. $f(x) = -3^x$

39. $f(x) = 2^{-x-2}$

40. $f(x) = 2^{-x+1}$

41. $f(x) = 2^{x^2}$

42. $f(x) = 2^x + 2^{-x}$

43. $f(x) = 2^{|x|}$

44. $f(x) = 3^{1-x^2}$

45. $f(x) = 2^x - 2^{-x}$

46. $f(x) = 2^{-|x|}$

Graphics Calculator Activities

50. Use a graphics calculator to check your graphs for Problems 27–46.

51. Graph $f(x) = 2^x$. Where should the graphs of $f(x) = 2^{x-5}$, $f(x) = 2^{x-7}$, and $f(x) = 2^{x+5}$ be located? Graph all three functions on the same set of axes with $f(x) = 2^x$.

52. Graph $f(x) = 3^x$. Where should the graphs of $f(x) = 3^x + 2$, $f(x) = 3^x - 3$, and $f(x) = 3^x - 7$ be located? Graph all three functions on the same set of axes with $f(x) = 3^x$.

53. Graph $f(x) = \left(\dfrac{1}{2}\right)^x$. Where should the graphs of $f(x) = -\left(\dfrac{1}{2}\right)^x$, $f(x) = \left(\dfrac{1}{2}\right)^{-x}$, and $f(x) = -\left(\dfrac{1}{2}\right)^{-x}$ be located? Graph all three functions on the same set of axes with $f(x) = \left(\dfrac{1}{2}\right)^x$.

54. Graph $f(x) = (1.5)^x$, $f(x) = (5.5)^x$, $f(x) = (.3)^x$, and $f(x) = (.7)^x$ on the same set of axes. Are these graphs consistent with Figure 10.3?

55. What is the solution for $3^x = 5$? Do you agree that it is between 1 and 2 since $3^1 = 3$ and $3^2 = 9$? Now graph $f(x) = 3^x - 5$ and use the zoom and trace features of your graphics calculator to find an approximation, to the nearest hundredth, for the x-intercept. You should get an answer of 1.46, to the nearest hundredth. Do you see that this is an approximation for the solution of $3^x = 5$? Try it; raise 3 to the 1.46 power.

Find an approximate solution, to the nearest hundredth, for each of the following equations by graphing the appropriate function and finding the x-intercept.

(a) $2^x = 19$ **(b)** $3^x = 50$ **(c)** $4^x = 47$

(d) $5^x = 120$ **(e)** $2^x = 1500$ **(f)** $3^{x-1} = 34$

THOUGHTS INTO WORDS

47. Explain how you would solve the equation

$(2^{x+1})(8^{2x-3}) = 64$.

48. Why is the base of an exponential function restricted to positive numbers not including one?

49. Explain how you would graph the function

$f(x) = -\left(\dfrac{1}{3}\right)^x$.

10.2 Applications of Exponential Functions

We can represent many real world situations exhibiting growth or decay with equations that describe exponential functions. For example, suppose an economist predicts an annual inflation rate of 5% per year for the next 10 years. This means that an item that presently costs $8 will cost $8(105\%) = 8(1.05) = \$8.40$ a year from now. The same item will cost $[8(105\%)](105\%) = 8(1.05)^2 = \$8.82$ in 2 years. In general, the equation

$$P = P_0(1.05)^t$$

yields the predicted price P of an item in t years if the present cost is P_0 and the annual inflation rate is 5%. Using this equation, we can look at some future prices based on the prediction of a 5% inflation rate.

1. A $3.27 container of cocoa mix will cost $\$3.27(1.05)^3 = \$3.79$ in 3 years.

2. A $4.07 jar of coffee will cost $\$4.07(1.05)^5 = \$5.19$ in 5 years.

3. A $9500 car will cost $\$9500(1.05)^7 = \$13{,}367$ (to the nearest dollar) in 7 years.

Compound Interest

Compound interest provides another illustration of exponential growth. Suppose that $500, called the **principal**, is invested at an interest rate of 8% *compounded annually*. The interest earned the first year is $500(.08) = \$40$, and this amount is added to the original $500 to form a new principal of $540 for the second year. The interest earned during the second year is $540(.08) = \$43.20$, and this amount is added to $540 to form a new principal of $583.20 for the third year. Each year a new principal is formed by reinvesting the interest earned during that year.

In general, suppose that a sum of money P (the principal) is invested at an interest rate of r percent compounded annually. The interest earned the first year is Pr, and the new principal for the second year is $P + Pr$, or $P(1 + r)$. Note that the new principal for the second year can be found by multiplying the original principal P by $(1 + r)$. In like fashion, the new principal for the third year can be found by multiplying the previous principal $P(1 + r)$ by $1 + r$, thus obtaining $P(1 + r)^2$. If this process is continued, *after t years the total amount of money accumulated, A, is given by*

$$\boxed{A = P(1 + r)^t.}$$

Consider the following examples of investments made at a certain rate of interest compounded annually.

1. $750 invested for 5 years at 9% compounded annually produces
$$A = \$750(1.09)^5 = \$1153.97.$$

2. $1000 invested for 10 years at 11% compounded annually produces
$$A = \$1000(1.11)^{10} = \$2839.42.$$

3. $5000 invested for 20 years at 12% compounded annually produces
$$A = \$5000(1.12)^{20} = \$48,231.47.$$

We can use the compound interest formula to determine what rate of interest is needed to accumulate a certain amount of money based on a given initial investment. The next example illustrates this idea.

Example I

What rate of interest is needed for an investment of $1000 to yield $4000 in 10 years if the interest is compounded annually?

Solution

Let's substitute $1000 for P, $4000 for A, and 10 years for t in the compound interest formula, and solve for r.

$$A = P(1 + r)^t$$

$$4000 = 1000(1 + r)^{10}$$

$$4 = (1 + r)^{10}$$

$$4^{.1} = [(1 + r)^{10}]^{.1} \qquad \text{Raise both sides to the .1 power.}$$

$$1.148698355 \approx 1 + r$$

$$.148698355 \approx r$$

$$r = 14.9\% \quad \text{to the nearest tenth of a percent}$$

Therefore, a rate of interest of approximately 14.9% is needed. (Perhaps you should check this answer.) ▲

If money invested at a certain rate of interest is compounded more than once a year, then the basic formula $A = P(1 + r)^t$ can be adjusted according to the number of compounding periods in a year. For example, for **semiannual compounding**, the formula becomes $A = P\left(1 + \dfrac{r}{2}\right)^{2t}$; for **quarterly compounding**, the formula becomes $A = P\left(1 + \dfrac{r}{4}\right)^{4t}$. In general, if n represents the number of compounding periods in a year, the formula becomes

$$\boxed{A = P\left(1 + \frac{r}{n}\right)^{nt}.}$$

The following examples illustrate the use of the formula.

1. $750 invested for 5 years at 9% compounded semiannually produces

$$A = \$750\left(1 + \frac{.09}{2}\right)^{2(5)} = \$750(1.045)^{10} = \$1164.73.$$

2. $1000 invested for 10 years at 11% compounded quarterly produces

$$A = \$1000\left(1 + \frac{.11}{4}\right)^{4(10)} = \$1000(1.0275)^{40} = \$2959.87.$$

3. $5000 invested for 20 years at 12% compounded monthly produces

$$A = \$5000\left(1 + \frac{.12}{12}\right)^{12(20)} = \$5000(1.01)^{240} = \$54,462.77.$$

You may find it interesting to compare these results with those we obtained earlier for annual compounding.

Exponential Decay

Suppose that the value of a car depreciates 15% per year for the first five years. Therefore, a car that costs $9500 will be worth $9500(100% − 15%) = $9500(85%) = $9500(.85) = $8075 in 1 year. In 2 years the value of the car will have depreciated $9500(.85)^2 = $6864 (to the nearest dollar). The equation

$$V = V_0(.85)^t$$

yields the value V of a car in t years if the initial cost is V_0 and the value depreciates 15% per year. Therefore, we can estimate some car values to the nearest dollar as follows.

1. A $6900 car will be worth $6900(.85)^3 = $4237 in 3 years.
2. A $10,900 car will be worth $10,900(.85)^4 = $5690 in 4 years.
3. A $13,000 car will be worth $13,000(.85)^5 = $5768 in 5 years.

Another example of exponential decay is associated with radioactive substances. The rate of decay can be described exponentially and is based on the half-life of a substance. The **half-life** of a radioactive substance is the amount of time that it takes for one-half of an initial amount of the substance to disappear as the result of decay. For example, suppose that we have 200 grams of a certain substance that has a half-life of 5 days. After 5 days, $200\left(\frac{1}{2}\right) = 100$ grams remain. After 10 days, $200\left(\frac{1}{2}\right)^2 = 50$ grams remain. After 15 days, $200\left(\frac{1}{2}\right)^3 = 25$ grams remain. In general, after t days, $200\left(\frac{1}{2}\right)^{\frac{t}{5}}$ grams remain.

The previous discussion leads to the following half-life formula. Suppose there is an initial amount, Q_0, of a radioactive substance with a half-life of h.

The amount of substance remaining, Q, after a time period of t, is given by the formula

$$Q = Q_0 \left(\frac{1}{2}\right)^{\frac{t}{h}}.$$

The units of measure for t and h must be the same.

Example 2

Barium-140 has a half-life of 13 days. If there are 500 milligrams of barium initially, how many milligrams remain after 26 days? After 100 days?

Solution

Using $Q_0 = 500$ and $h = 13$, the half-life formula becomes

$$Q = 500 \left(\frac{1}{2}\right)^{\frac{t}{13}}.$$

If $t = 26$, then

$$Q = 500 \left(\frac{1}{2}\right)^{\frac{26}{13}}$$

$$= 500 \left(\frac{1}{2}\right)^2$$

$$= 500 \left(\frac{1}{4}\right)$$

$$= 125.$$

So, 125 milligrams remain after 26 days. If $t = 100$, then

$$Q = 500 \left(\frac{1}{2}\right)^{\frac{100}{13}}$$

$$= 5000(.5)^{\frac{100}{13}}$$

$$= 2.4 \quad \text{to the nearest tenth of a milligram.}$$

So, approximately 2.4 milligrams remain after 100 days.   ▲

REMARK   Example 2 clearly illustrates that a calculator is useful at times, but unnecessary at other times. We solved the first part of the problem very easily without a calculator but it certainly was helpful for the second part of the problem.   △

Number e

An interesting situation occurs if we consider the compound interest formula for $P = \$1$, $r = 100\%$, and $t = 1$ year. The formula becomes $A = 1\left(1 + \frac{1}{n}\right)^n$. The following table shows some values, rounded to eight decimal places, of $\left(1 + \frac{1}{n}\right)^n$ for different values of n.

n	$\left(1 + \frac{1}{n}\right)^n$
1	2.00000000
10	2.59374246
100	2.70481383
1000	2.71692393
10,000	2.71814593
100,000	2.71826824
1,000,000	2.71828047
10,000,000	2.71828169
100,000,000	2.71828181
1,000,000,000	2.71828183

The table suggests that as n increases, the value of $\left(1 + \frac{1}{n}\right)^n$ gets closer and closer to some fixed number. This does happen and the fixed number is called e. To five decimal places, $e = 2.71828$.

The function defined by the equation $f(x) = e^x$ is the **natural exponential function**. It has a great many real world applications, some of which we will look at in a moment. First, however, let's get a picture of the natural exponential function. Since $2 < e < 3$, the graph of $f(x) = e^x$ must fall between the graphs of $f(x) = 2^x$ and $f(x) = 3^x$. To be more specific, let's use our calculator to determine a table of values. Use the $\boxed{e^x}$ key, and round the results to the nearest tenth to obtain the following table. Plot the points determined by this table and connect them with a smooth curve to produce Figure 10.7.

x	$f(x) = e^x$
0	1.0
1	2.7
2	7.4
-1	.4
-2	.1

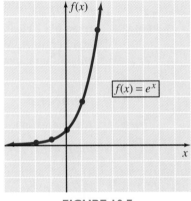

FIGURE 10.7

Back to Compound Interest

Let's return to the concept of compound interest. If the number of compounding periods in a year is increased indefinitely, we arrive at the concept of **compounding continuously**. Mathematically, we can accomplish this by applying the limit concept to the expression $P\left(1 + \dfrac{r}{n}\right)^{nt}$. We will not show the details here, but the following result is obtained. The formula

$$\boxed{A = Pe^{rt}}$$

yields the accumulated value, A, of a sum of money, P, that has been invested for t years at a rate of r percent compounded continuously. The following examples illustrate the use of the formula.

1. $750 invested for 5 years at 9% compounded continuously produces

$$A = 750e^{(.09)(5)} = 750e^{.45} = \$1176.23.$$

2. $1000 invested for 10 years at 11% compounded continuously produces

$$A = 1000e^{(.11)(10)} = 1000e^{1.1} = \$3004.17.$$

3. $5000 invested for 20 years at 12% compounded continuously produces

$$A = 5000e^{(.12)(20)} = 5000e^{2.4} = \$55,115.88.$$

Again, you may find it interesting to compare these results with those you obtained earlier when you were using a different number of compounding periods.

Is it better to invest at 6% compounded quarterly or at 5.75% compounded continuously? To answer such a question, we can use the concept of **effective yield** (sometimes called *effective annual rate of interest*). The effective yield of an investment is the simple interest rate that would yield the same amount in one year. Thus, for the 6%-compounded-quarterly investment, we can calculate the effective yield as follows.

$$P(1 + r) = P\left(1 + \frac{.06}{4}\right)^4$$

$$1 + r = \left(1 + \frac{.06}{4}\right)^4 \qquad \text{Multiply both sides by } \frac{1}{P}.$$

$$1 + r = (1.015)^4$$

$$r = (1.015)^4 - 1$$

$$r \approx .0613635506$$

$$r = 6.14\% \quad \text{to the nearest hundredth of a percent}$$

Likewise, for the 5.75%-compounded-continuously investment we can calculate the effective yield as follows.

$$P(1 + r) = Pe^{.0575}$$

$$1 + r = e^{.0575}$$

$$r = e^{.0575} - 1$$

$$r \approx .0591852707$$

$$r = 5.92\% \quad \text{to the nearest hundredth of a percent}$$

Therefore, comparing the two effective yields, we see that it is better to invest at 6% compounded quarterly than to invest at 5.75% compounded continuously.

Law of Exponential Growth

The ideas behind "compounded continuously" carry over to other growth situations. We use the law of exponential growth

$$\boxed{Q(t) = Q_0 e^{kt}}$$

as a mathematical model for numerous growth-and-decay applications. In this equation, $Q(t)$ represents the quantity of a given substance at any time t, Q_0 is the initial amount of the substance (when $t = 0$), and k is a constant that depends on the particular application. If $k < 0$, then $Q(t)$ decreases as t increases, and we refer to the model as the **law of decay**.

Let's consider some growth-and-decay applications.

Example 3

Suppose that in a certain culture the equation $Q(t) = 15000e^{.3t}$ expresses the number of bacteria present as a function of the time t, where t is expressed in hours. Find (a) the initial number of bacteria, and (b) the number of bacteria after 3 hours.

Solution

(a) The initial number of bacteria is produced when $t = 0$.

$$Q(0) = 15,000e^{.3(0)}$$
$$= 15,000e^0$$
$$= 15,000 \qquad e^0 = 1$$

(b) $Q(3) = 15,000e^{.3(3)}$
$$= 15,000e^{.9}$$
$$= 36,984 \quad \text{to the nearest whole number}$$

Therefore, there should be approximately 36,894 bacteria present after 3 hours.

▲

Example 4

Suppose the number of bacteria present in a certain culture after t minutes is given by the equation $Q(t) = Q_0 e^{.05t}$, where Q_0 represents the initial number of bacteria. If 5000 bacteria are present after 20 minutes, how many bacteria were present initially?

Solution

If 5000 bacteria are present after 20 minutes, then $Q(20) = 5000$.

$$5000 = Q_0 e^{.05(20)}$$

$$5000 = Q_0 e^1$$

$$\frac{5000}{e} = Q_0$$

$$1839 = Q_0 \quad \text{to the nearest whole number}$$

Therefore, there were approximately 1839 bacteria present initially. ▲

Example 5

The number of grams of a certain radioactive substance present after t seconds is given by the equation $Q(t) = 200e^{-.3t}$. How many grams remain after 7 seconds?

Solution

Use $Q(t) = 200e^{-.3t}$ to obtain

$$Q(7) = 200e^{(-.3)(7)}$$

$$= 200e^{-2.1}$$

$$= 24.5. \quad \text{to the nearest tenth}$$

Thus, approximately 24.5 grams remain after 7 seconds. ▲

Finally, let's consider two examples where we use a graphing utility to produce the graph.

Example 6

Suppose that $1000 was invested at 6.5% interest compounded continuously. How long would it take for the money to double itself?

Solution

Substitute $1000 for P and .065 for r in the formula $A = Pe^{rt}$ to produce $A = 1000e^{.065t}$. If we let $y = A$ and $x = t$, we can graph the equation $y = 1000e^{.065x}$. By letting $x = 20$, we obtain $y = 1000e^{.065(20)} = 1000e^{1.3} \approx 3670$. Therefore, let's set the boundaries of the viewing rectangle so that $0 \le x \le 20$ and $0 \le y \le 3700$ with a y-scale of 1000. Then we obtain the graph in Figure 10.8. Now we want to find the value of x so that $y = 2000$. (The money is to double itself.) Using the zoom and trace features of the graphing utility we can determine that an x-value of approximately 10.7 will produce a y-value of 2000. Thus, it will take approximately 10.7 years for the $1000 investment to double itself.

FIGURE 10.8   3700

0

0   20

▲

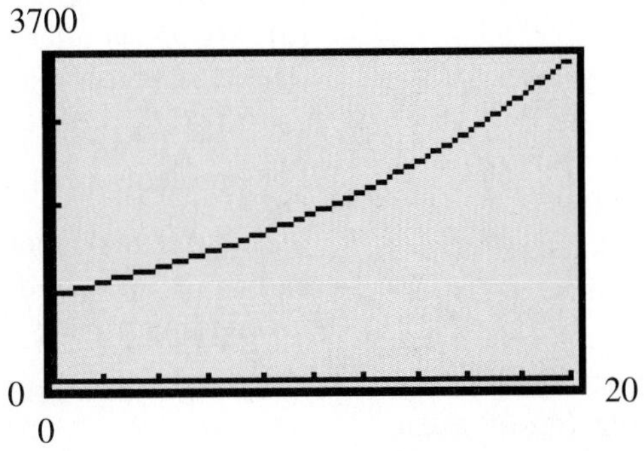

Example 7   Graph the function $y = \dfrac{1}{\sqrt{2\pi}} e^{-x^2/2}$ and find its maximum value.

Solution   If $x = 0$, then $y = \dfrac{1}{\sqrt{2\pi}} e^0 = \dfrac{1}{\sqrt{2\pi}} \approx .4$. So let's set the boundaries of the viewing rectangle so that $-5 \le x \le 5$ and $0 \le y \le 1$ with a y-scale of .1; the graph of the function is shown in Figure 10.9. From the graph, we see that the maximum value of the function occurs at $x = 0$, which we have already determined to be approximately .4.

FIGURE 10.9

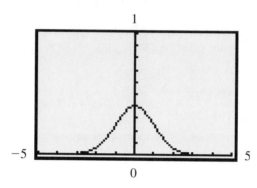

REMARK   The curve in Figure 10.9 is called the **normal distribution curve**. You may want to ask your instructor to explain what it means to assign grades based on the normal distribution curve.   △

Problem Set 10.2

1. Assuming that the rate of inflation is 4% per year, the equation $P = P_0(1.04)^t$ yields the predicted price P of an item in t years that presently costs P_0. Find the predicted price of each of the following items for the indicated years ahead.
(a) $.55 can of soup in 3 years
(b) $3.43 container of cocoa mix in 5 years
(c) $1.76 jar of coffee creamer in 4 years
(d) $.44 can of beans and bacon in 10 years
(e) $9000 car in 5 years (nearest dollar)
(f) $50,000 house in 8 years (nearest dollar)
(g) $500 TV set in 7 years (nearest dollar)

2. Suppose it is estimated that the value of a car depreciates 30% per year for the first 5 years. The equation $A = P_0(.7)^t$ yields the value (A) of a car

after t years if the original price is P_0. Find the value (to the nearest dollar) of each of the following priced cars after the indicated time.
(a) $9000 car after 4 years
(b) $5295 car after 2 years
(c) $6395 car after 5 years
(d) $15,595 car after 3 years

For Problems 3–14, use the formula $A = P\left(1 + \dfrac{r}{n}\right)^{nt}$ to find the total amount of money accumulated at the end of the indicated time period for each of the following investments.

3. $200 for 6 years at 6% compounded annually

4. $250 for 5 years at 7% compounded annually

5. $500 for 7 years at 8% compounded semiannually

6. $750 for 8 years at 8% compounded semiannually

7. $800 for 9 years at 9% compounded quarterly

8. $1200 for 10 years at 10% compounded quarterly

9. $1500 for 5 years at 12% compounded monthly

10. $2000 for 10 years at 9% compounded monthly

11. $5000 for 15 years at 8.5% compounded annually

12. $7500 for 20 years at 9.5% compounded semiannually

13. $8000 for 10 years at 10.5% compounded quarterly

14. $10,000 for 25 years at 9.25% compounded monthly

For Problems 15–23, use the formula $A = Pe^{rt}$ to find the total amount of money accumulated at the end of the indicated time period by compounding continuously.

15. $400 for 5 years at 7%

16. $500 for 7 years at 6%

17. $750 for 8 years at 8%

18. $1000 for 10 years at 9%

19. $2000 for 15 years at 10%

20. $5000 for 20 years at 11%

21. $7500 for 10 years at 8.5%

22. $10,000 for 25 years at 9.25%

23. $15,000 for 10 years at 7.75%

24. What rate of interest, to the nearest tenth of a percent, compounded annually is needed for an investment of $200 to grow to $350 in 5 years?

25. What rate of interest, to the nearest tenth of a percent, compounded quarterly is needed for an investment of $1500 to grow to $2700 in 10 years?

26. Find the effective yield, to the nearest tenth of a percent, of an investment at 7.5% compounded monthly.

27. Find the effective yield, to the nearest hundredth of a percent, of an investment at 7.75% compounded continuously.

28. What investment yields the greatest return: 7% compounded monthly or 6.85% compounded continuously?

29. What investment yields the greatest return: 8.25% compounded quarterly or 8.3% compounded semiannually?

30. Suppose that a certain radioactive substance has a half-life of 20 years. If there are presently 2500 milligrams of the substance, how much, to the nearest milligram, will remain after 40 years? After 50 years?

31. Strontium-90 has a half-life of 29 years. If there are 400 grams of strontium initially, how much, to the nearest gram, will remain after 87 years? After 100 years?

32. The half-life of radium is approximately 1600 years. If the present amount of radium in a certain location is 500 grams, how much will remain after 800 years? Express your answer to the nearest gram.

33. Suppose that in a certain culture, the equation $Q(t) = 1000e^{.4t}$ expresses the number of bacteria present as a function of the time t, where t is expressed in hours. How many bacteria are present at the end of 2 hours? 3 hours? 5 hours?

34. The number of bacteria present at a given time under certain conditions is given by the equation $Q = 5000e^{.05t}$, where t is expressed in minutes. How many bacteria are present at the end of 10 minutes? 30 minutes? 1 hour?

35. The number of bacteria present in a certain culture after t hours is given by the equation $Q = Q_0 e^{.3t}$, where Q_0 represents the initial number of bacteria. If 6640 bacteria are present after 4 hours, how many bacteria were present initially?

36. The number of grams Q of a certain radioactive substance present after t seconds is given by the equation $Q = 1500e^{-.4t}$. How many grams remain after 5 seconds? 10 seconds? 20 seconds?

37. The atmospheric pressure, measured in pounds per square inch, is a function of the altitude above sea level. The equation $P(a) = 14.7e^{-.21a}$, where a is the altitude measured in miles, can be used to approximate atmospheric pressure. Find the atmospheric pressure at each of the following locations.
(a) Mount McKinley in Alaska: altitude of 3.85 miles
(b) Denver, Colorado: the "mile-high" city

(c) Asheville, North Carolina: altitude of 1985 feet

(d) Phoenix, Arizona: altitude of 1090 feet

38. Suppose that the present population of a city is 75,000. Using the equation $P(t) = 75,000e^{.01t}$ to estimate future growth, estimate the population (a) 10 years from now, (b) 15 years from now, and (c) 25 years from now.

For Problems 39–44, graph each of the exponential functions.
39. $f(x) = e^x + 1$
40. $f(x) = e^x - 2$
41. $f(x) = 2e^x$
42. $f(x) = -e^x$
43. $f(x) = e^{2x}$
44. $f(x) = e^{-x}$

THOUGHTS INTO WORDS

45. Explain the difference between simple interest and compound interest.
46. Would it be better to invest $5000 at 6.25% interest compounded annually for 5 years or to invest $5000 at 6.25% interest compounded continuously for 5 years? Explain your answer.
47. How would you explain the concept of effective yield to someone who missed class when it was discussed?

48. How would you explain the half-life formula to someone who missed class when it was discussed?

Further Investigations

49. Complete the following chart that illustrates what happens to $1000 invested at various rates of interest for different lengths of time but always compounded continuously. Round your answers to the nearest dollar.

$1000 Compounded Continuously

	8%	10%	12%	14%
5 years				
10 years				
15 years				
20 years				
25 years				

50. Complete the following chart that illustrates what happens to $1000 invested at 12% for different lengths of time and different numbers of compounding periods. Round all of your answers to the nearest dollar.

$1000 at 12%

	1 year	5 years	10 years	20 years
Compounded annually				
Compounded semiannually				
Compounded quarterly				
Compounded monthly				
Compounded continuously				

51. Complete the following chart that illustrates what happens to $1000 in 10 years based on different rates of interest and different numbers of compounding periods. Round your answers to the nearest dollar.

$1000 for 10 years

	8%	10%	12%	14%
Compounded annually				
Compounded semiannually				
Compounded quarterly				
Compounded monthly				
Compounded continuously				

For Problems 52–56, graph each of the functions.

52. $f(x) = x(2^x)$

53. $f(x) = \dfrac{e^x + e^{-x}}{2}$

54. $f(x) = \dfrac{2}{e^x + e^{-x}}$

55. $f(x) = \dfrac{e^x - e^{-x}}{2}$

56. $f(x) = \dfrac{2}{e^x - e^{-x}}$

 **Graphics Calculator Activities**

57. Use a graphics calculator to check your graphs for Problems 52–56.

58. Graph $f(x) = 2^x$, $f(x) = e^x$, and $f(x) = 3^x$ on the same set of axes. Are these graphs consistent with the discussion prior to Figure 10.7?

59. Graph $f(x) = e^x$. Where should the graphs of $f(x) = e^{x-4}$, $f(x) = e^{x-6}$, and $f(x) = e^{x+5}$ be located? Graph all three functions on the same set of axes with $f(x) = e^x$.

60. Graph $f(x) = e^x$. Now predict the graphs for $f(x) = -e^x$, $f(x) = e^{-x}$, and $f(x) = -e^{-x}$. Graph all three functions on the same set of axes with $f(x) = e^x$.

61. How do you think the graphs of $f(x) = e^x$, $f(x) = e^{2x}$, and $f(x) = 2e^x$ will compare? Graph them on the same set of axes to see if you were correct.

62. Find an approximate solution, to the nearest hundredth, for each of the following equations by graphing the appropriate function and finding the x-intercept.

(a) $e^x = 7$ (b) $e^x = 21$ (c) $e^x = 53$
(d) $2e^x = 60$ (e) $e^{x+1} = 150$ (f) $e^{x-2} = 300$

63. Use a graphing approach to argue that it is better to invest money at 6% compounded quarterly than it is at 5.75% compounded continuously.

64. How long will it take $500 to be worth $1500 if it is invested at 7.5% interest compounded semiannually?

65. How long will it take $5000 to triple itself if it is invested at 6.75% interest compounded quarterly?

10.3 **Logarithms**

In Sections 10.1 and 10.2, we discussed exponential expressions of the form b^n, where b is any positive real number and n is any real number; we used exponential expressions of the form b^n to define exponential functions; and we used exponential functions to help solve problems. In the next three sections we will follow the same basic pattern with respect to a new concept—that of a logarithm. Let's begin with the following definition.

DEFINITION 10.2

If r is any positive real number, then the unique exponent t such that $b^t = r$ is called the **logarithm of r with base b** and is noted by $\log_b r$.

According to Definition 10.2, the logarithm of 16 base 2 is the exponent t such that $2^t = 16$; thus, we can write $\log_2 16 = 4$. Likewise, we can write $\log_{10} 1000 = 3$ because $10^3 = 1000$. In general, we can remember Definition 10.2 by the statement

$$\log_b r = t \quad \text{is equivalent to } b^t = r.$$

Therefore, we can easily switch back and forth between exponential and logarithmic forms of equations, as the next examples illustrate.

$$\log_2 8 = 3 \quad \text{is equivalent to } 2^3 = 8,$$
$$\log_{10} 100 = 2 \quad \text{is equivalent to } 10^2 = 100,$$
$$\log_3 81 = 4 \quad \text{is equivalent to } 3^4 = 81,$$
$$\log_{10} .001 = -3 \quad \text{is equivalent to } 10^{-3} = .001,$$
$$2^7 = 128 \quad \text{is equivalent to } \log_2 128 = 7,$$
$$5^3 = 125 \quad \text{is equivalent to } \log_5 125 = 3,$$
$$\left(\frac{1}{2}\right)^4 = \frac{1}{16} \quad \text{is equivalent to } \log_{1/2}\left(\frac{1}{16}\right) = 4,$$
$$10^{-2} = .01 \quad \text{is equivalent to } \log_{10} .01 = -2$$

Some logarithms can be determined by changing to exponential form and using the properties of exponents, as the next two examples illustrate.

Example 1

Evaluate $\log_{10} .0001$.

Solution

Let $\log_{10} .0001 = x$. Then by changing to exponential form, we have $10^x = .0001$, which can be solved as follows.

$$10^x = .0001$$
$$10^x = 10^{-4} \qquad .0001 = \frac{1}{10,000} = \frac{1}{10^4} = 10^{-4}$$
$$x = -4$$

Thus, we have $\log_{10} .0001 = -4$. ▲

Example 2

Evaluate $\log_9 \left(\frac{\sqrt[5]{27}}{3}\right)$.

Solution

Let $\log_9\left(\dfrac{\sqrt[5]{27}}{3}\right) = x$. Then by changing to exponential form, we have $9^x = \dfrac{\sqrt[5]{27}}{3}$, which can be solved as follows.

$$9^x = \frac{(27)^{1/5}}{3}$$

$$(3^2)^x = \frac{(3^3)^{1/5}}{3}$$

$$3^{2x} = \frac{3^{3/5}}{3}$$

$$3^{2x} = 3^{-2/5}$$

$$2x = -\frac{2}{5}$$

$$x = -\frac{1}{5}$$

Therefore, we have $\log_9\left(\dfrac{\sqrt[5]{27}}{3}\right) = -\dfrac{1}{5}$. ▲

Some equations that involve logarithms can also be solved by changing to exponential form and using our knowledge of exponents.

Example 3

Solve $\log_8 x = \dfrac{2}{3}$.

Solution

Changing $\log_8 x = \dfrac{2}{3}$ to exponential form, we obtain

$$8^{2/3} = x.$$

Therefore,

$$x = (\sqrt[3]{8})^2$$
$$= 2^2$$
$$= 4.$$

The solution set is $\{4\}$. ▲

Example 4

Solve $\log_b\left(\dfrac{27}{64}\right) = 3$.

Solution

Change $\log_b\left(\dfrac{27}{64}\right) = 3$ to exponential form to obtain

$$b^3 = \frac{27}{64}.$$

Therefore,

$$b = \sqrt[3]{\frac{27}{64}}$$

$$= \frac{3}{4}.$$

The solution set is $\left\{\frac{3}{4}\right\}$. ▲

Properties of Logarithms

There are some properties of logarithms that are a direct consequence of Definition 10.2 and the properties of exponents. For example, the following property is obtained by writing the exponential equation $b^1 = b$ and $b^0 = 1$ in logarithmic form.

PROPERTY 10.3

For $b > 0$ and $b \neq 1$,

$$\log_b b = 1 \qquad \text{and} \qquad \log_b 1 = 0.$$

Therefore, according to Property 10.3 we can write these equations.

$$\log_{10} 10 = 1, \qquad \log_4 4 = 1,$$
$$\log_{10} 1 = 0, \qquad \log_5 1 = 0$$

Also from Definition 10.2 we know that $\log_b r$ is the exponent t such that $b^t = r$. Therefore, raising b to the $\log_b r$ power must produce r. This fact is stated in Property 10.4.

PROPERTY 10.4

For $b > 0$, $b \neq 1$, and $r > 0$,

$$b^{\log_b r} = r.$$

Therefore, according to Property 10.4 we can write the following.

$$10^{\log_{10} 72} = 72 \qquad 3^{\log_3 85} = 85 \qquad e^{\log_e 7} = 7$$

Because a logarithm is by definition an exponent, it would seem reasonable to predict that some properties of logarithms correspond to the basic exponential properties. This is an accurate prediction; these properties provide a basis for computational work with logarithms. Let's state the first of these properties and show how we can use our knowledge of exponents to verify it.

PROPERTY 10.5

For positive numbers b, r, and s, where $b \neq 1$,

$$\log_b rs = \log_b r + \log_b s.$$

To verify Property 10.5, we can proceed as follows. Let $m = \log_b r$ and $n = \log_b s$. Change each of these equations to exponential form.

$$m = \log_b r \quad \text{becomes } r = b^m,$$
$$n = \log_b s \quad \text{becomes } s = b^n$$

Thus, the product rs becomes

$$rs = b^m \cdot b^n = b^{m+n}.$$

Now, by changing $rs = b^{m+n}$ back to logarithmic form, we obtain

$$\log_b rs = m + n.$$

Replace m with $\log_b r$ and replace n with $\log_b s$ to yield

$$\log_b rs = \log_b r + \log_b s.$$

The following two examples illustrate the use of Property 10.5.

Example 5

If $\log_2 5 = 2.3222$ and $\log_2 3 = 1.5850$, evaluate $\log_2 15$.

Solution

Because $15 = 5 \cdot 3$, we can apply Property 10.5 as follows.

$$\log_2 15 = \log_2 (5 \cdot 3)$$
$$= \log_2 5 + \log_2 3$$
$$= 2.3222 + 1.5850 = 3.9072 \qquad \blacktriangle$$

Example 6

Given that $\log_{10} 178 = 2.2504$ and $\log_{10} 89 = 1.9494$, evaluate $\log_{10}(178 \cdot 89)$.

Solution

$$\log_{10}(178 \cdot 89) = \log_{10} 178 + \log_{10} 89$$
$$= 2.2504 + 1.9494 = 4.1998 \qquad \blacktriangle$$

Since $\dfrac{b^m}{b^n} = b^{m-n}$, we would expect a corresponding property that pertains to logarithms. Property 10.6 is that property. We can verify it by using an approach similar to the one we used to verify Property 10.5. This verification is left for you to do as an exercise in the next problem set.

PROPERTY 10.6

For positive numbers b, r, and s, where $b \neq 1$,

$$\log_b \left(\frac{r}{s}\right) = \log_b r - \log_b s.$$

We can use Property 10.6 to change a division problem into an equivalent subtraction problem, as the next two examples illustrate.

Example 7

If $\log_5 36 = 2.2265$ and $\log_5 4 = .8614$, evaluate $\log_5 9$.

Solution

Since $9 = \dfrac{36}{4}$, we can use Property 10.6 as follows.

$$\log_5 9 = \log_5 \left(\frac{36}{4} \right)$$

$$= \log_5 36 - \log_5 4$$

$$= 2.2265 - .8614 = 1.3651$$ ▲

Example 8

Evaluate $\log_{10} \left(\dfrac{379}{86} \right)$ given that $\log_{10} 379 = 2.5786$ and $\log_{10} 86 = 1.9345$.

Solution

$$\log_{10} \left(\frac{379}{86} \right) = \log_{10} 379 - \log_{10} 86$$

$$= 2.5786 - 1.9345$$

$$= .6441$$ ▲

Another property of exponents states that $(b^n)^m = b^{mn}$. The corresponding property of logarithms is stated in Property 10.7. Again, we will leave the verification of this property as an exercise for you to do in the next set of problems.

PROPERTY 10.7

If r is a positive real number, b is a positive real number other than 1, and p is any real number, then

$$\log_b r^p = p(\log_b r).$$

We will use Property 10.7 in the next two examples.

Example 9

Evaluate $\log_2 22^{1/3}$ given that $\log_2 22 = 4.4598$.

Solution

$$\log_2 22^{1/3} = \frac{1}{3}\log_2 22 \qquad \text{Property 10.7}$$

$$= \frac{1}{3}(4.4598)$$

$$= 1.4866$$ ▲

Example 10

Evaluate $\log_{10}(8540)^{3/5}$ given that $\log_{10} 8540 = 3.9315$.

Solution

$$\log_{10}(8540)^{3/5} = \frac{3}{5}\log_{10} 8540$$

$$= \frac{3}{5}(3.9315)$$

$$= 2.3589 \qquad \blacktriangle$$

Used together, the properties of logarithms allow us to change the forms of various logarithmic expressions. For example, we can rewrite an expression such as $\log_b \sqrt{\dfrac{xy}{z}}$ in terms of sums and differences of simpler logarithmic quantities as follows.

$$\log_b \sqrt{\frac{xy}{z}} = \log_b \left(\frac{xy}{z}\right)^{1/2}$$

$$= \frac{1}{2}\log_b \left(\frac{xy}{z}\right) \qquad \text{Property 10.7}$$

$$= \frac{1}{2}(\log_b xy - \log_b z) \qquad \text{Property 10.6}$$

$$= \frac{1}{2}(\log_b x + \log_b y - \log_b z) \qquad \text{Property 10.5}$$

Sometimes we need to change from an indicated sum or difference of logarithmic quantities to an indicated product or quotient. This is especially helpful when solving certain kinds of equations that involve logarithms. Note in these next two examples how we can use the properties, along with the process of changing from logarithmic form to exponential form, to solve some equations.

Example 11

Solve $\log_{10} x + \log_{10}(x + 9) = 1$.

Solution

$$\log_{10} x + \log_{10}(x + 9) = 1$$

$$\log_{10}[x(x + 9)] = 1 \qquad \text{Property 10.5}$$

$$10^1 = x(x + 9) \qquad \text{Change to exponential form}$$

$$10 = x^2 + 9x$$

$$0 = x^2 + 9x - 10$$

$$0 = (x + 10)(x - 1)$$

$$x + 10 = 0 \qquad \text{or} \qquad x - 1 = 0$$

$$x = -10 \qquad \text{or} \qquad x = 1$$

Since logarithms are defined only for positive numbers, x and $x + 9$ have to be positive. Therefore, the solution of -10 must be discarded. The solution set is $\{1\}$.

▲

Example 12

Solution

Solve $\log_5(x + 4) - \log_5 x = 2$.

$$\log_5(x + 4) - \log_5 x = 2$$

$$\log_5\left(\frac{x + 4}{x}\right) = 2 \qquad \text{Property 10.6}$$

$$5^2 = \frac{x + 4}{x} \qquad \text{Change to exponential form}$$

$$25 = \frac{x + 4}{x}$$

$$25x = x + 4$$

$$24x = 4$$

$$x = \frac{4}{24} = \frac{1}{6}$$

The solution set is $\left\{\dfrac{1}{6}\right\}$.

▲

Problem Set 10.3

Write each of the following in logarithmic form. For example, $2^5 = 32$ becomes $\log_2 32 = 5$ in logarithmic form.

1. $2^7 = 128$
2. $3^3 = 27$
3. $5^3 = 125$
4. $2^6 = 64$
5. $10^3 = 1000$
6. $10^1 = 10$
7. $2^{-2} = \dfrac{1}{4}$
8. $3^{-4} = \dfrac{1}{81}$
9. $10^{-1} = .1$
10. $10^{-2} = .01$

Write each of the following in exponential form. For example, $\log_2 8 = 3$ becomes $2^3 = 8$ in exponential form.

11. $\log_3 81 = 4$
12. $\log_2 256 = 8$
13. $\log_4 64 = 3$
14. $\log_5 25 = 2$
15. $\log_{10} 10,000 = 4$
16. $\log_{10} 100,000 = 5$
17. $\log_2\left(\dfrac{1}{16}\right) = -4$
18. $\log_5\left(\dfrac{1}{125}\right) = -3$
19. $\log_{10} .001 = -3$
20. $\log_{10} .000001 = -6$

Evaluate each of the following.

21. $\log_2 16$

22. $\log_3 9$

23. $\log_3 81$

24. $\log_2 512$

25. $\log_6 216$

26. $\log_4 256$

27. $\log_7 \sqrt{7}$

28. $\log_2 \sqrt[3]{2}$

29. $\log_{10} 1$

30. $\log_{10} 10$

31. $\log_{10} .1$ —

32. $\log_{10} .0001$ —

33. $10^{\log_{10} 5}$

34. $10^{\log_{10} 14}$

35. $\log_2 \left(\dfrac{1}{32}\right)$ —

36. $\log_5 \left(\dfrac{1}{25}\right)$ —

37. $\log_5(\log_2 32)$

38. $\log_2(\log_4 16)$

39. $\log_{10}(\log_7 7)$

40. $\log_2(\log_5 5)$

Solve each of the following equations.

41. $\log_7 x = 2$

42. $\log_2 x = 5$

43. $\log_8 x = \dfrac{4}{3}$

44. $\log_{16} x = \dfrac{3}{2}$

45. $\log_9 x = \dfrac{3}{2}$

46. $\log_8 x = -\dfrac{2}{3}$

47. $\log_4 x = -\dfrac{3}{2}$

48. $\log_9 x = -\dfrac{5}{2}$

49. $\log_x 2 = \dfrac{1}{2}$

50. $\log_x 3 = \dfrac{1}{2}$

Given that $\log_2 5 = 2.3219$ and $\log_2 7 = 2.8074$, evaluate each of the following by using Properties 10.5–10.7.

51. $\log_2 35$

52. $\log_2 \left(\dfrac{7}{5}\right)$

53. $\log_2 125$

54. $\log_2 49$

55. $\log_2 \sqrt{7}$

56. $\log_2 \sqrt[3]{5}$

57. $\log_2 175$

58. $\log_2 56$

59. $\log_2 80$

Given that $\log_8 5 = .7740$ and $\log_8 11 = 1.1531$, evaluate each of the following using Properties 10.5–10.7.

60. $\log_8 55$

61. $\log_8 \left(\dfrac{5}{11}\right)$ —

62. $\log_8 25$

63. $\log_8 \sqrt{11}$

64. $\log_8 (5)^{\frac{2}{3}}$

65. $\log_8 88$

66. $\log_8 320$

67. $\log_8 \left(\dfrac{25}{11}\right)$

68. $\log_8 \left(\dfrac{121}{25}\right)$

Express each of the following as the sum or difference of simpler logarithmic quantities. Assume that all variables represent positive real numbers. For example,

$$\log_b \frac{x^3}{y^2} = \log_b x^3 - \log_b y^2$$
$$= 3 \log_b x - 2 \log_b y.$$

69. $\log_b xyz$

70. $\log_b 5x$

71. $\log_b \left(\dfrac{y}{z}\right)$

72. $\log_b \left(\dfrac{x^2}{y}\right)$

73. $\log_b y^3 z^4$

74. $\log_b x^2 y^3$

75. $\log_b \left(\dfrac{x^{1/2} y^{1/3}}{z^4}\right)$

76. $\log_b x^{2/3} y^{3/4}$

77. $\log_b \sqrt[3]{x^2 z}$

78. $\log_b \sqrt{xy}$

79. $\log_b \left(x \sqrt{\dfrac{x}{y}}\right)$

80. $\log_b \sqrt{\dfrac{x}{y}}$

Express each of the following as a single logarithm. (Assume that all variables represent positive real numbers.) For example,

$$3 \log_b x + 5 \log_b y = \log_b x^3 y^5.$$

81. $2 \log_b x - 4 \log_b y$

82. $\log_b x + \log_b y - \log_b z$

83. $\log_b x - (\log_b y - \log_b z)$

84. $(\log_b x - \log_b y) - \log_b z$

85. $2 \log_b x + 4 \log_b y - 3 \log_b z$

86. $\log_b x + \dfrac{1}{2} \log_b y$

87. $\dfrac{1}{2} \log_b x - \log_b x + 4 \log_b y$

88. $2 \log_b x + \dfrac{1}{2} \log_b(x - 1) - 4 \log_b(2x + 5)$

Solve each of the following equations.

89. $\log_3 x + \log_3 4 = 2$

90. $\log_7 5 + \log_7 x = 1$

91. $\log_{10} x + \log_{10}(x - 21) = 2$

92. $\log_{10} x + \log_{10}(x - 3) = 1$

93. $\log_2 x + \log_2(x - 3) = 2$

94. $\log_3 x + \log_3(x - 2) = 1$

95. $\log_{10}(2x - 1) - \log_{10}(x - 2) = 1$

96. $\log_{10}(9x - 2) = 1 + \log_{10}(x - 4)$

97. $\log_5(3x - 2) = 1 + \log_5(x - 4)$

98. $\log_6 x + \log_6(x + 5) = 2$

99. $\log_8(x + 7) + \log_8 x = 1$

100. $\log_6(x + 1) + \log_6(x - 4) = 2$

101. Verify Property 10.6

102. Verify Property 10.7

THOUGHTS INTO WORDS

103. Explain, without using Property 10.4, why $4^{\log_4 9}$ equals 9.

104. How would you explain the concept of a logarithm to someone who had just completed an elementary algebra course?

105. In the next section we will show that the logarithmic function $f(x) = \log_2 x$ is the inverse of the exponential function $f(x) = 2^x$. From that information how could you sketch a graph of $f(x) = \log_2 x$?

10.4 Logarithmic Functions

We can now use the concept of a logarithm to define a logarithmic function as follows.

DEFINITION 10.3

If $b > 0$ and $b \neq 1$, then the function defined by

$$f(x) = \log_b x,$$

where x is any positive real number, is called the **logarithmic function with base b**

We can obtain the graph of a specific logarithmic function in various ways. For example, the equation $y = \log_2 x$ can be changed to the exponential equation $2^y = x$, where we can determine a table of values. The next set of exercises asks you to use this approach to graph some logarithmic functions. We can also set up a table of values directly from the logarithmic equation and sketch the graph from the table. Example 1 illustrates this approach.

Example I

Solution

Graph $f(x) = \log_2 x$.

Let's choose some values for x where we can easily determine the corresponding values for $\log_2 x$. (Remember that logarithms are only defined for the positive real numbers.)

x	$f(x)$	
$\frac{1}{8}$	-3	$\text{Log}_2 \frac{1}{8} = -3$ because $2^{-3} = \frac{1}{2^3} = \frac{1}{8}$
$\frac{1}{4}$	-2	
$\frac{1}{2}$	-1	
1	0	$\text{Log}_2 1 = 0$ because $2^0 = 1$
2	1	
4	2	
8	3	

Plot these points and connect them with a smooth curve to produce Figure 10.10.

FIGURE 10.10

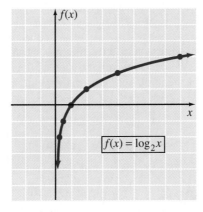

$$f(x) = \log_2 x$$

Now suppose that we consider two functions f and g as follows.

$f(x) = b^x$ Domain: all real numbers
Range: positive real numbers

$g(x) = \log_b x$ Domain: positive real numbers
Range: all real numbers

Furthermore, suppose that we consider the composition of f and g and the composition of g and f.

$$(f \circ g)(x) = f(g(x)) = f(\log_b x) = b^{\log_b x} = x$$

$$(g \circ f)(x) = g(f(x)) = g(b^x) = \log_b b^x = x \log_b b = x(1) = x$$

Because the domain of f is the range of g, the range of f is the domain of g, $f(g(x)) = x$, and $g(f(x)) = x$, the two functions f and g are *inverses of each other*.

Remember from Chapter 8 that the graph of a function and the graph of its inverse are reflections of each other through the line $y = x$. Thus, we can determine the graph of a logarithmic function by reflecting the graph of its inverse exponential function through the line $y = x$. We demonstrate this idea in Figure 10.11 where the graph of $y = 2^x$ has been reflected across the line $y = x$ to produce the graph of $y = \log_2 x$.

FIGURE 10.11

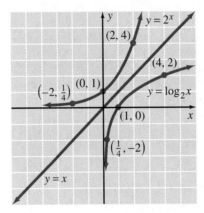

The general behavior patterns of exponential functions were illustrated back in Figure 10.3. We can now reflect each of these graphs through the line $y = x$ and observe the general behavior patterns of logarithmic functions, shown in Figure 10.12.

FIGURE 10.12

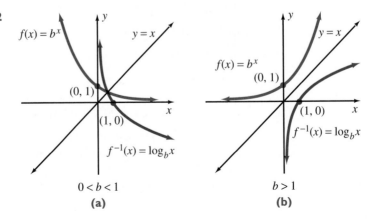

As you graph logarithmic functions, don't forget about transformations of basic curves.

1. The graph of $f(x) = 3 + \log_2 x$ is the graph of $f(x) = \log_2 x$ *moved up three units.* (Since $\log_2 x + 3$ is apt to be confused with $\log_2(x + 3)$, we commonly write $3 + \log_2 x$.)

2. The graph of $f(x) = \log_2(x - 4)$ is the graph of $f(x) = \log_2 x$ *moved four units to the right.*

3. The graph of $f(x) = -\log_2 x$ is the graph of $f(x) = \log_2 x$ *reflected across the x-axis.*

Common Logarithms—Base 10

The properties of logarithms we discussed in Section 10.3 are true for any valid base. However, since the Hindu-Arabic numeration system that we use is a base-10 system, logarithms to base 10 have historically been used for computational purposes. Base-10 logarithms are called **common logarithms**.

Originally, common logarithms were developed to aid in complicated numerical calculations that involve products, quotients, and powers of real numbers. Today they are seldom used for that purpose because the calculator and computer can much more effectively handle the messy computational problems. However, common logarithms do still occur in applications, so they deserve our attention.

REMARK In Appendix A we have included a short discussion relative to the computational aspects of common logarithms. You may find it interesting to at least browse through this material. It probably will enhance your appreciation of the calculator. △

As we know from earlier work, the definition of a logarithm provides the basis for evaluating $\log_{10} x$ for values of x that are integral powers of 10. Consider the following examples.

$\log_{10} 1000 = 3$ because $10^3 = 1000$,

$\log_{10} 100 = 2$ because $10^2 = 100$,

$\log_{10} 10 = 1$ because $10^1 = 10$,

$\log_{10} 1 = 0$ because $10^0 = 1$,

$\log_{10} .1 = -1$ because $10^{-1} = \frac{1}{10} = .1$,

$\log_{10} .01 = -2$ because $10^{-2} = \frac{1}{10^2} = .01$,

$\log_{10} .001 = -3$ because $10^{-3} = \frac{1}{10^3} = .001$

When working exclusively with base-10 logarithms, it is customary to omit writing the numeral 10 to designate the base. Thus, the expression $\log_{10} x$ is written as $\log x$ and a statement such as $\log_{10} 1000 = 3$ becomes $\log 1000 = 3$. We will follow this practice from now on in this chapter, but don't forget that the base is understood to be 10.

$$\boxed{\log_{10} x = \log x}$$

To find the common logarithm of a positive number that is not an integral power of 10, we can use an appropriately equipped calculator or a table such as the one that appears in Appendix A. A calculator equipped with a common logarithm function (ordinarily a key labeled $\boxed{\log}$ is used) gives us the following results rounded to four decimal places.

$\log 1.75 = .2430,$

$\log 23.8 = 1.3766,$ Be sure that you can use a calculator and obtain these results.

$\log 134 = 2.1271,$

$\log .192 = -.7167,$

$\log .0246 = -1.6091$

In order to use logarithms to solve problems we sometimes need to be able to determine a number when the logarithm of the number is known. That is to say, we may need to determine x if $\log x$ is known. Let's consider an example.

Example 2   Find x if $\log x = .2430$.

Solution   If $\log x = .2430$, then by changing to exponential form we have $10^{.2430} = x$. Therefore, use the $\boxed{10^x}$ key to find x.

$$x = 10^{.2430} \approx 1.749846689$$

Therefore, $x = 1.7498$ rounded to five significant digits.   ▲

Be sure that you can use your calculator and obtain the following results. We rounded the values for x to 5 significant digits.

If $\log x = .7629$, then $x = 10^{.7629} = 5.7930$.

If $\log x = 1.4825$, then $x = 10^{1.4825} = 30.374$.

If $\log x = 4.0214$, then $x = 10^{4.0214} = 10505$.

If $\log x = -1.5162$, then $x = 10^{-1.5162} = .030465$.

If $\log x = -3.8921$, then $x = 10^{-3.8921} = .00012820$.

The **common logarithmic function** is defined by the equation $f(x) = \log x$. It should now be a simple matter to set up a table of values and sketch the function.

We will have you do this in the next set of exercises. Remember that $f(x) = 10^x$ and $g(x) = \log x$ are inverses of each other. Therefore, we could also get the graph of $g(x) = \log x$ by reflecting the exponential curve $f(x) = 10^x$ across the line $y = x$.

Natural Logarithms—Base e

In many practical applications of logarithms, the number e (remember $e \approx 2.71828$) is used as a base. Logarithms with a base of e are called **natural logarithms** and the symbol $\ln x$ is commonly used instead of $\log_e x$.

$$\log_e x = \ln x$$

Natural logarithms can be found with an appropriately equipped calculator or with a table of natural logarithms. (A table of natural logarithms is provided in Appendix B.) A calculator with a natural logarithm function (ordinarily a key labeled $\boxed{\ln x}$), gives us the following results rounded to four decimal places.

$\ln 3.21 = 1.1663$

$\ln 47.28 = 3.8561$

$\ln 842 = 6.7358$

$\ln .21 = -1.5606$

$\ln .0046 = -5.3817$

$\ln 10 = 2.3026$

Be sure that you can use your calculator to obtain these results. Keep in mind the significance of a statement such as $\ln 3.21 = 1.1663$. By changing to exponential form we are claiming that e raised to the 1.1663 power is approximately 3.21. Using a calculator we obtain $e^{1.1663} = 3.210093293$.

Let's do a few more problems to find x when given $\ln x$. Be sure that you agree with these results.

If $\ln x = 2.4156$, then $x = e^{2.4156} = 11.196$.

If $\ln x = .9847$, then $x = e^{.9847} = 2.6770$.

If $\ln x = 4.1482$, then $x = e^{4.1482} = 63.320$.

If $\ln x = 1.7654$, then $x = e^{-1.7654} = .17112$.

The **natural logarithmic function** is defined by the equation $f(x) = \ln x$. It is the inverse of the natural exponential function $f(x) = e^x$. Thus, one way to graph $f(x) = \ln x$ is to reflect the graph of $f(x) = e^x$ across the line $y = x$. We will ask you to do this in the next set of problems.

In Figure 10.13 we used a graphing utility to sketch the graph of $f(x) = e^x$. Now, based on our previous work with transformations, we should be able to make the following statements.

FIGURE 10.13

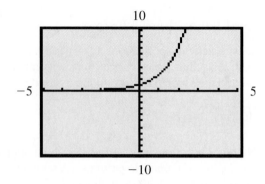

1. The graph of $f(x) = -e^x$ is the graph of $f(x) = e^x$ reflected through the x-axis.

2. The graph of $f(x) = e^{-x}$ is the graph of $f(x) = e^x$ reflected through the y-axis.

3. The graph of $f(x) = e^x + 4$ is the graph of $f(x) = e^x$ shifted upward 4 units.

4. The graph of $f(x) = e^{x+2}$ is the graph of $f(x) = e^x$ shifted 2 units to the left.

These statements are verified in Figure 10.14, which shows the result of graphing these four functions on the same set of axes using a graphing utility.

FIGURE 10.14

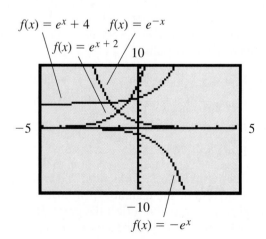

Problem Set 10.4

For Problems 1–10, use a calculator to find each **common logarithm**. Express answers to four decimal places.

1. log 7.24

2. log 2.05

3. log 52.23

4. log 825.8

5. log 3214.1

6. log 14,189

7. log .729

8. log .04376

9. log .00034

10. log .000069

For Problems 11–20, use your calculator to find x when given log x. Express answers to five significant digits.

11. log x = 2.6143

12. log x = 1.5263

13. log x = 4.9547

14. log x = 3.9335

15. log x = 1.9006

16. log x = .5517

17. log x = −1.3148

18. log x = −.1452

19. log x = −2.1928

20. log x = −2.6542

For Problems 21–30, use your calculator to find each **natural logarithm**. Express answers to four decimal places.

21. ln 5

22. ln 18

23. ln 32.6

24. ln 79.5

25. ln 430

26. ln 371.8

27. ln .46

28. ln .524

29. ln .0314

30. ln .008142

For Problems 31–40, use your calculator to find x when given ln x. Express answers to five significant digits.

31. ln x = .4721

32. ln x = .9413

33. ln x = 1.1425

34. ln x = 2.7619

35. ln x = 4.6873

36. ln x = 3.0259

37. ln x = −.7284

38. ln x = −1.6246

39. ln x = −3.3244

40. ln x = −2.3745

41. (a) Complete the following table and then graph $f(x) = \log x$. (Express the values for log x to the nearest tenth.)

x	.1	.5	1	2	4	8	10
log x							

(b) Complete the following table expressing values for 10^x to the nearest tenth.

x	−1	−.3	0	.3	.6	.9	1
10^x							

Then graph $f(x) = 10^x$ and reflect it across the line $y = x$ to produce the graph for $f(x) = \log x$.

42. (a) Complete the following table and then graph $f(x) = \ln x$. (Express the values for ln x to the nearest tenth.)

x	.1	.5	1	2	4	8	10
ln x							

(b) Complete the following table expressing values for e^x to the nearest tenth.

x	−2.3	−.7	0	.7	1.4	2.1	2.3
e^x							

Then graph $f(x) = e^x$ and reflect it across the line $y = x$ to produce the graph for $f(x) = \ln x$.

43. Graph $y = \log_{\frac{1}{2}} x$ by graphing $\left(\dfrac{1}{2}\right)^y = x$.

44. Graph $y = \log_2 x$ by graphing $2^y = x$.

45. Graph $f(x) = \log_3 x$ by reflecting the graph of $g(x) = 3^x$ across the line $y = x$.

46. Graph $f(x) = \log_4 x$ by reflecting the graph of $g(x) = 4^x$ across the line $y = x$.

For Problems 47–53, graph each of the functions. Remember that the graph of $f(x) = \log_2 x$ is given in Figure 10.10.

47. $f(x) = 3 + \log_2 x$

48. $f(x) = -2 + \log_2 x$

49. $f(x) = \log_2(x + 3)$

50. $f(x) = \log_2(x - 2)$

51. $f(x) = \log_2 2x$

52. $f(x) = -\log_2 x$

53. $f(x) = 2 \log_2 x$

For Problems 54–61, perform the following calculations and express answers to the nearest hundredth. (These calculations are in preparation for our work in the next section.)

54. $\dfrac{\log 7}{\log 3}$

55. $\dfrac{\ln 2}{\ln 7}$

56. $\dfrac{2 \ln 3}{\ln 8}$

57. $\dfrac{\ln 5}{2 \ln 3}$

58. $\dfrac{\ln 3}{.04}$

59. $\dfrac{\ln 2}{.03}$

60. $\dfrac{\log 2}{5 \log 1.02}$

61. $\dfrac{\log 5}{3 \log 1.07}$

THOUGHTS INTO WORDS

62. Why is the number one excluded from being a base of a logarithm?

63. How do we know that $\log_2 6$ is between 2 and 3?

Graphics Calculator Activities

64. Graph $f(x) = x$, $f(x) = e^x$, and $f(x) = \ln x$ on the same set of axes.

65. Graph $f(x) = x$, $f(x) = 10^x$, and $f(x) = \log x$ on the same set of axes.

66. Graph $f(x) = \ln x$. How should the graphs of $f(x) = 2 \ln x$, $f(x) = 4 \ln x$, and $f(x) = 6 \ln x$ compare to the graph of $f(x) = \ln x$? Graph the three functions on the same set of axes with $f(x) = \ln x$.

67. Graph $f(x) = \log x$. Now predict the graphs for $f(x) = 2 + \log x$, $f(x) = -2 + \log x$, and $f(x) = -6 + \log x$. Graph the three functions on the same set of axes with $f(x) = \log x$.

68. Graph $\ln x$. Now predict the graphs for $f(x) = \ln(x - 2)$, $f(x) = \ln(x - 6)$, and $f(x) = \ln(x + 4)$. Graph the three functions on the same set of axes with $f(x) = \ln x$.

69. For each of the following, (a) predict the general shape and location of the graph, and (b) use your graphics calculator to graph the function to check your prediction.

(a) $f(x) = \log x + \ln x$ **(b)** $f(x) = \log x - \ln x$

(c) $f(x) = \ln x - \log x$ **(d)** $f(x) = \ln x^2$

10.5 Exponential Equations, Logarithmic Equations, and Problem Solving

In Section 10.1 we solved exponential equations such as $3^x = 81$ by expressing both sides of the equation as a power of 3 and then applying the property "if $b^n = b^m$, then $n = m$." However, if we try this same approach with an equation such as $3^x = 5$, we face the difficulty of expressing 5 as a power of 3. We can solve this type of problem by using the properties of logarithms and the following property of equality.

PROPERTY 10.8   If $x > 0$, $y > 0$, $b > 0$, and $b \neq 1$, then $x = y$ if and only if $\log_b x = \log_b y$.

Property 10.8 is stated in terms of any valid base b; however, for most applications we use either common logarithms or natural logarithms. Let's consider some examples.

Example 1

Solve $3^x = 5$ to the nearest hundredth.

Solution

By using common logarithms we can proceed as follows.

$$3^x = 5$$

$$\log 3^x = \log 5 \qquad \text{Property 10.8}$$

$$x \log 3 = \log 5 \qquad \log r^p = p \log r$$

$$x = \frac{\log 5}{\log 3}$$

$$x = 1.46 \quad \text{to the nearest hundredth}$$

 Check   Since $3^{1.46} \approx 4.972754647$, we say that, to the nearest hundredth, the solution set for $3^x = 5$ is $\{1.46\}$.   ▲

Example 2

Solve $e^{x+1} = 5$ to the nearest hundredth.

Solution

Since base e is used in the exponential expression, let's use natural logarithms to help solve this equation.

$$e^{x+1} = 5$$

$$\ln e^{x+1} = \ln 5 \qquad\qquad\qquad \text{Property 10.8}$$

$$(x + 1) \ln e = \ln 5 \qquad\qquad\qquad \ln r^p = p \ln r$$

$$(x + 1)(1) = \ln 5 \qquad\qquad\qquad \ln e = 1$$

$$x = \ln 5 - 1$$

$$x = .61 \quad \text{to the nearest hundredth}$$

The solution set is $\{.61\}$. Check it!   ▲

Example 3

Solution

Solve $2^{3x-2} = 3^{2x+1}$ to the nearest hundredth.

$$2^{3x-2} = 3^{2x+1}$$

$$\log 2^{3x-2} = \log 3^{2x+1}$$

$$(3x - 2)\log 2 = (2x + 1)\log 3$$

$$3x \log 2 - 2 \log 2 = 2x \log 3 + \log 3$$

$$3x \log 2 - 2x \log 3 = \log 3 + 2 \log 2$$

$$x(3 \log 2 - 2 \log 3) = \log 3 + 2 \log 2$$

$$x = \frac{\log 3 + 2 \log 2}{3 \log 2 - 2 \log 3}.$$

$$x = -21.10 \quad \text{to the nearest hundredth}$$

The solution set is $\{-21.10\}$. Check it! ▲

Logarithmic Equations

In Example 11 of Section 10.3 we solved the logarithmic equation

$$\log_{10} x + \log_{10}(x + 9) = 1$$

by simplifying the left side of the equation to $\log_{10}[x(x + 9)]$ and then changing the equation to exponential form to complete the solution. Now, using Property 10.8, we can solve such a logarithmic equation another way and also expand our equation solving capabilities. Let's consider some examples.

Example 4

Solution

Solve $\log x + \log(x - 15) = 2$.

Since $\log 100 = 2$, the given equation becomes

$$\log x + \log(x - 15) = \log 100.$$

Now simplify the left side, apply Property 10.8, and proceed as follows.

$$\log(x)(x - 15) = \log 100$$

$$x(x - 15) = 100$$

$$x^2 - 15x - 100 = 0$$

$$(x - 20)(x + 5) = 0$$

$$x - 20 = 0 \qquad \text{or} \qquad x + 5 = 0$$

$$x = 20 \qquad \text{or} \qquad x = -5$$

The domain of a logarithmic function must contain only positive numbers, so x and $x - 15$ must be positive in this problem. Therefore, we discard the solution of -5; the solution set is $\{20\}$. ▲

Example 5

Solution

Solve $\ln(x + 2) = \ln(x - 4) + \ln 3$.

$$\ln(x + 2) = \ln(x - 4) + \ln 3$$
$$\ln(x + 2) = \ln[3(x - 4)]$$
$$x + 2 = 3(x - 4)$$
$$x + 2 = 3x - 12$$
$$14 = 2x$$
$$7 = x$$

The solution set is $\{7\}$. ▲

Example 6

Solution

Solve $\log_b(x + 2) + \log_b(2x - 1) = \log_b x$.

$$\log_b(x + 2) + \log_b(2x - 1) = \log_b x$$
$$\log_b[(x + 2)(2x - 1)] = \log_b x$$
$$(x + 2)(2x - 1) = x$$
$$2x^2 + 3x - 2 = x$$
$$2x^2 + 2x - 2 = 0$$
$$x^2 + x - 1 = 0$$

Using the quadratic formula we obtain

$$x = \frac{-1 \pm \sqrt{1 + 4}}{2}$$
$$= \frac{-1 \pm \sqrt{5}}{2}.$$

Since $x + 2$, $2x - 1$, and x have to be positive, we must discard the solution of $\frac{-1 - \sqrt{5}}{2}$; the solution set is $\left\{ \frac{-1 + \sqrt{5}}{2} \right\}$. ▲

Problem Solving

In Section 10.2 we used the compound interest formula

$$A = P\left(1 + \frac{r}{n}\right)^{nt}$$

to determine the amount of money (A) accumulated at the end of t years if P dollars is invested at rate of interest r compounded n times per year. Now let's use this formula to solve other types of problems that deal with compound interest.

Example 7

How long will it take for $500 to double itself if it is invested at 12% compounded quarterly?

Solution

"To double itself" means that the $500 must grow into $1000. Thus,

$$1000 = 500\left(1 + \frac{.12}{4}\right)^{4t}$$

$$= 500(1 + .03)^{4t}$$

$$= 500(1.03)^{4t}.$$

Multiplying both sides of $1000 = 500(1.03)^{4t}$ by $\frac{1}{500}$ yields

$$2 = (1.03)^{4t}.$$

Therefore,

$$\log 2 = \log(1.03)^{4t} \qquad \text{Property 10.8}$$

$$= 4t \log 1.03. \qquad \log r^p = p \log r$$

Now let's solve for t.

$$4t \log 1.03 = \log 2$$

$$t = \frac{\log 2}{4 \log 1.03}$$

$$t = 5.9 \quad \text{to the nearest tenth}$$

Therefore, we are claiming that $500 invested at 12% interest compounded quarterly will double itself in approximately 5.9 years.

 Check $500 invested at 12% compounded quarterly for 5.9 years will produce

$$A = \$500\left(1 + \frac{.12}{4}\right)^{4(5.9)}$$

$$= \$500(1.03)^{23.6}$$

$$= \$1004.45.$$ ▲

Example 8

Suppose that the number of bacteria present in a certain culture after t minutes is given by the equation $Q(t) = Q_0 e^{.04t}$, where Q_0 represents the initial number of bacteria. How long would it take for the bacteria count to grow from 500 to 2000?

Solution

Substituting into $Q(t) = Q_0 e^{.04t}$ and solving for t, we obtain the following.

$$2000 = 500e^{.04t}$$

$$4 = e^{.04t}$$

$$\ln 4 = \ln e^{.04t}$$

$$\ln 4 = .04t \ln e$$

$$\ln 4 = .04t \qquad \ln e = 1$$

$$\frac{\ln 4}{.04} = t$$

$$34.7 = t \quad \text{to the nearest tenth.}$$

It should take approximately 34.7 minutes.   ▲

Richter Numbers

Seismologists use the Richter scale to measure and report the magnitude of earthquakes. The equation

$$\boxed{R = \log \frac{I}{I_0}} \qquad \text{R is called a Richter number}$$

compares the intensity I of an earthquake to a minimal or reference intensity I_0. The reference intensity is the smallest earth movement that can be recorded on a seismograph. Suppose that the intensity of an earthquake was determined to be 50,000 times the reference intensity. In this case, $I = 50{,}000\ I_0$ and the Richter number would be calculated as follows.

$$R = \log \frac{50{,}000\ I_0}{I_0}$$

$$R = \log 50{,}000$$

$$R \approx 4.698970004$$

Thus, a Richter number of 4.7 would be reported. Let's consider two more examples that involve Richter numbers.

Example 9

An earthquake in San Francisco in 1989 was reported to have a Richter number of 6.9. How did its intensity compare to the reference intensity?

Solution

$$6.9 = \log \frac{I}{I_0}$$

$$10^{6.9} = \frac{I}{I_0}$$

$$I = (10^{6.9})(I_0)$$

$$I \approx 7{,}943{,}282\ I_0$$

So its intensity was a little less than 8 million times the reference intensity.   ▲

Example 10

An earthquake in Iran in 1990 had a Richter number of 7.7. Compare the intensity level of that earthquake to the one in San Francisco we refer to in Example 9.

Solution

From Example 9 we have $I = (10^{6.9})(I_0)$ for the earthquake in San Francisco. Then using a Richter number of 7.7 we obtain $I = (10^{7.7})(I_0)$ for the earthquake in Iran. Therefore, by comparison

$$\frac{(10^{7.7})(I_0)}{(10^{6.9})(I_0)} = 10^{7.7-6.9} = 10^{.8} \approx 6.3.$$

The earthquake in Iran was about 6 times as intense as the one in San Francisco. ▲

Logarithms with Base Other Than 10 or e

The basic approach whereby we apply Property 10.8 and use either common or natural logarithms can also be used to evaluate a logarithm to some base other than 10 or e. Consider the following example.

Example 11

Evaluate $\log_3 41$.

Solution

Let $x = \log_3 41$. Change to exponential form to obtain

$3^x = 41.$

Now we can apply Property 10.8 and proceed as follows.

$$\log 3^x = \log 41$$
$$x \log 3 = \log 41$$
$$x = \frac{\log 41}{\log 3}$$
$$x = 3.3802 \quad \text{rounded to four decimal places}$$

Therefore, we are claiming that 3 raised to the 3.3802 power will produce approximately 41. Check it! ▲

The method of Example 11 to evaluate $\log_a r$ produces the following formula, which we often refer to as the **change-of-base formula for logarithms**.

PROPERTY 10.9

If a, b, and r are positive numbers, with $a \neq 1$ and $b \neq 1$, then

$$\log_a r = \frac{\log_b r}{\log_b a}.$$

By using Property 10.9 we can easily determine a relationship between logarithms of different bases. For example, suppose that in Property 10.9 we let $a = 10$ and $b = e$.

$$\log_a r = \frac{\log_b r}{\log_b a}$$

becomes

$$\log_{10} r = \frac{\log_e r}{\log_e 10}$$

$$\log_e r = (\log_e 10)(\log_{10} r)$$

$$\log_e r = (2.3026)(\log_{10} r).$$

Thus, the natural logarithm of any positive number is approximately equal to the common logarithm of the number times 2.3026.

Now we can use a graphing utility to graph logarithmic functions such as $f(x) = \log_2 x$. Using the change-of-base formula, we can express this function as $f(x) = \dfrac{\log x}{\log 2}$ or as $f(x) = \dfrac{\ln x}{\ln 2}$. The graph of $f(x) = \log_2 x$ is shown in Figure 10.15.

FIGURE 10.15

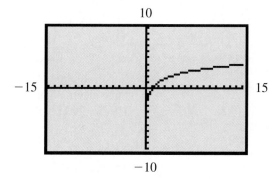

Finally, let's use a graphical approach to solve an equation that is cumbersome to solve with an algebraic approach.

Example 12

Solve the equation $(5^x - 5^{-x})/2 = 3$.

Solution

First, we need to recognize that the solutions for the equation $(5^x - 5^{-x})/2 = 3$ are the x-intercepts of the graph of the equation $y = (5^x - 5^{-x})/2 - 3$. So let's use a graphing utility to obtain the graph of this equation as shown in Figure 10.16. Use the zoom and trace features to determine that the graph crosses the x-axis at approximately 1.13. Thus, the solution set of the original expression is $\{1.13\}$.

FIGURE 10.16

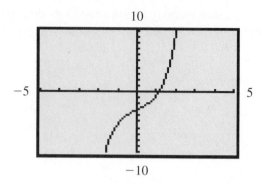

Problem Set 10.5

For Problems 1–20, solve each exponential equation and express approximate solutions to the nearest hundredth.

1. $3^x = 13$
2. $2^x = 21$
3. $4^n = 35$
4. $5^n = 75$
5. $2^x + 7 = 50$
6. $3^x - 6 = 25$
7. $3^{x-2} = 11$
8. $2^{x+1} = 7$
9. $5^{3t+1} = 9$
10. $7^{2t-1} = 35$
11. $e^x = 27$
12. $e^x = 86$
13. $e^{x-2} = 13.1$
14. $e^{x-1} = 8.2$
15. $3e^x - 1 = 17$
16. $2e^x = 12.4$
17. $5^{2x+1} = 7^{x+3}$
18. $3^{x-1} = 2^{x+3}$
19. $3^{2x+1} = 2^{3x+2}$
20. $5^{x-1} = 2^{2x+1}$

For Problems 21–32, solve each logarithmic equation and express irrational solutions in lowest radical form.

21. $\log x + \log(x + 21) = 2$
22. $\log x + \log(x + 3) = 1$
23. $\log(3x - 1) = 1 + \log(5x - 2)$
24. $\log(2x - 1) - \log(x - 3) = 1$
25. $\log(x + 1) = \log 3 - \log(2x - 1)$
26. $\log(x - 2) = 1 - \log(x + 3)$

27. $\log(x + 2) - \log(2x + 1) = \log x$
28. $\log(x + 1) - \log(x + 2) = \log \dfrac{1}{x}$
29. $\ln(2t + 5) = \ln 3 + \ln(t - 1)$
30. $\ln(3t - 4) - \ln(t + 1) = \ln 2$
31. $\log \sqrt{x} = \sqrt{\log x}$
32. $\log x^2 = (\log x)^2$

For Problems 33–42, approximate each logarithm to three decimal places. (Example 11 and/or Property 10.9 should be of some help.)

33. $\log_2 40$
34. $\log_2 93$
35. $\log_3 16$
36. $\log_3 37$
37. $\log_4 1.6$
38. $\log_4 3.2$
39. $\log_5 .26$
40. $\log_5 .047$
41. $\log_7 500$
42. $\log_8 750$

For Problems 43–54, solve each problem and express answers to the nearest tenth unless stated otherwise.

43. How long will it take \$750 to be worth \$1000 if it is invested at 12% interest compounded quarterly?

44. How long will it take \$1000 to double itself if it is invested at 9% interest compounded semiannually?

45. How long will it take $2000 to double itself if it is invested at 13% interest compounded continuously?

46. How long will it take $500 to triple itself if it is invested at 9% interest compounded continuously?

47. What rate of interest compounded continuously is needed for an investment of $500 to grow to $900 in 10 years?

48. What rate of interest compounded continuously is needed for an investment of $2500 to grow to $10,000 in 20 years?

49. For a certain strain of bacteria, the number of bacteria present after t hours is given by the equation $Q = Q_0 e^{.34t}$, where Q_0 represents the initial number of bacteria. How long will it take 400 bacteria to increase to 4000 bacteria?

50. A piece of machinery valued at $30,000 depreciates at a rate of 10% yearly. How long will it take for it to reach a value of $15,000?

51. The equation $P(a) = 14.7e^{-.21a}$, where a is the altitude above sea level measured in miles, yields the atmospheric pressure in pounds per square inch. If the atmospheric pressure at Cheyenne, Wyoming is approximately 11.53 pounds per square inch, find that city's altitude above sea level. Express your answer to the nearest hundred feet.

52. The number of grams of a certain radioactive substance present after t hours is given by the equation $Q = Q_0 e^{-.45t}$, where Q_0 represents the initial number of grams. How long would it take 2500 grams to be reduced to 1250 grams?

53. For a certain culture the equation $Q(t) = Q_0 e^{.4t}$, where Q_0 is an initial number of bacteria and t is time measured in hours, yields the number of bacteria as a function of time. How long will it take 500 bacteria to increase to 2000?

54. Suppose that the equation $P(t) = P_0 e^{.02t}$, where P_0 represents an initial population and t is the time in years, is used to predict population growth. How long would it take a city of 50,000 to double its population?

55. An earthquake in Los Angeles in 1971 had an intensity of approximately five million times the reference intensity. What was the Richter number associated with that earthquake?

56. An earthquake in San Francisco in 1906 was reported to have a Richter number of 8.3. How did its intensity compare to the reference intensity?

57. Calculate how many times more intense an earthquake with a Richter number of 7.3 is than an earthquake with a Richter number of 6.4.

58. Calculate how many times more intense an earthquake with a Richter number of 8.9 is than an earthquake with a Richter number of 6.2.

THOUGHTS INTO WORDS

59. Explain how to determine $\log_4 76$ without using Property 10.9.

60. Explain the concept of a Richter number.

61. Explain how you would solve the equation $2^x = 64$ and also how you would solve the equation $2^x = 53$.

62. How do logarithms with a base of 9 compare to logarithms with a base of 3? Explain how you reached this conclusion.

Further Investigations

63. Use the approach of Example 11 to develop Property 10.9.

64. Let $r = b$ in Property 10.9, and verify that $\log_a b = \dfrac{1}{\log_b a}$.

65. Solve the equation $\dfrac{5^x - 5^{-x}}{2} = 3$. Express your answer to the nearest hundredth.

66. Solve the equation $y = \dfrac{10^x + 10^{-x}}{2}$ for x in terms of y.

67. Solve the equation $y = \dfrac{e^x - e^{-x}}{2}$ for x in terms of y.

Graphics Calculator Activities

68. Check your answers for Problems 17–20 by graphing the appropriate function and finding the x-intercept.

69. Graph $f(x) = x$, $f(x) = 2^x$, and $f(x) = \log_2 x$ on the same set of axes.

70. Graph $f(x) = x$, $f(x) = (.5)^x$, and $f(x) = \log_{.5} x$ on the same set of axes.

71. Graph $f(x) = \log_2 x$. Now predict the graphs for $f(x) = \log_3 x$, $f(x) = \log_4 x$, and $f(x) = \log_8 x$. Graph these three functions on the same set of axes with $f(x) = \log_2 x$.

72. Graph $f(x) = \log_5 x$. Now predict the graphs for $f(x) = 2\log_5 x$, $f(x) = -4\log_5 x$, and $f(x) = \log_5(x + 4)$. Graph these three functions on the same set of axes with $f(x) = \log_5 x$.

73. Use both a graphical and an algebraic approach to solve the equation $\dfrac{2^x - 2^{-x}}{3} = 4$.

SUMMARY

(10.1) If a and b are positive real numbers and m and n are any real numbers, then

1. $b^n \cdot b^m = b^{n+m}$ Product of two powers

2. $(b^n)^m = b^{mn}$ Power of a power

3. $(ab)^n = a^n b^n$ Power of a product

4. $\left(\dfrac{a}{b}\right)^n = \dfrac{a^n}{b^n}$ Power of a quotient

5. $\dfrac{b^n}{b^m} = b^{n-m}$.   Quotient of two powers

If $b > 0$, $b \neq 1$, and m and n are real numbers, then $b^n = b^m$ if and only if $n = m$. A function defined by an equation of the form

$$f(x) = b^x, \qquad b > 0 \text{ and } b \neq 1$$

is called an **exponential function**.

(10.2) A general formula for any principal P compounded n times per year for any number t, at a rate r is

$$A = P\left(1 + \frac{r}{n}\right)^{nt}$$

where A represents the total amount of money accumulated at the end of the t years. The value of $\left(1 + \dfrac{1}{n}\right)^n$, as n gets infinitely large, approaches the number e, where e equals 2.71828 to five decimal places.

The formula

$$A = Pe^{rt}$$

yields the accumulated value A of a sum of money P that has been invested for t years at a rate of r percent **compounded continuously**.

The formula

$$Q = Q_0\left(\frac{1}{2}\right)^{\frac{t}{h}}$$

is referred to as the **half-life** formula.

The equation

$$Q(t) = Q_0 e^{kt}$$

is used as a mathematical model for many growth-and-decay applications.

(10.3) If r is any positive real number, then the unique exponent t such that $b^t = r$ is called the **logarithm of r with base b** and is denoted by $\log_b r$. For $b \geq 0$, $b \neq 1$, and $r > 0$,

1. $\log_b b = 1$
2. $\log_b 1 = 0$
3. $r = b^{\log_b r}$.

The following properties of logarithms are derived from the definition of a logarithm and the properties of exponents. For positive real numbers b, r, and s, where $b \neq 1$,

1. $\log_b rs = \log_b r + \log_b s$,
2. $\log_b\left(\dfrac{r}{s}\right) = \log_b r - \log_b s$, and
3. $\log_b r^p = p \log_b r$.   where p is any real number

(10.4) A function defined by an equation of the form

$$f(x) = \log_b x, \qquad b > 0 \text{ and } b \neq 1$$

is called a **logarithmic function**. The equation $y = \log_b x$ is equivalent to $x = b^y$. The two functions $f(x) = b^x$ and $g(x) = \log_b x$ are inverses of each other.

Logarithms with a base of 10 are called **common logarithms**. The expression $\log_{10} x$ is commonly written as $\log x$.

Many calculators are equipped with a common logarithm function. Often a key labeled $\boxed{\log}$ is used to find common logarithms.

Natural logarithms are logarithms that have a base of e, where e is an irrational number whose decimal approximation to eight digits is 2.7182818. Natural logarithms are denoted by $\log_e x$ or $\ln x$.

Many calculators are also equipped with a natural logarithm function. Often a key labeled $\boxed{\ln x}$ is used for this purpose.

(10.5) The properties of equality along with the properties of exponents and logarithms merge to help us solve a variety of exponential and logarithmic equations. These properties also help us solve the problems that deal with various applications, including compound interest and growth problems.

The formula

$$R = \log\frac{I}{I_0}$$

yields the Richter number associated with an earthquake.

The formula

$$\log_a r = \frac{\log_b r}{\log_b a}$$

is often called the **change-of-base formula**.

Chapter 10 Review Problem Set

Evaluate each of the following.

1. $8^{5/3}$

2. $-25^{3/2}$

3. $(-27)^{4/3}$

4. $\log_6 216$

5. $\log_7\left(\dfrac{1}{49}\right)$

6. $\log_2\sqrt[3]{2}$

7. $\log_2\left(\dfrac{\sqrt[4]{32}}{2}\right)$

8. $\log_{10} .00001$

9. $\ln e$

10. $7^{\log_7 12}$

Solve each of the following equations. Express approximate solutions to the nearest hundredth.

11. $\log_{10} 2 + \log_{10} x = 1$

12. $\log_3 x = -2$

13. $4^x = 128$

14. $3^t = 42$

15. $\log_2 x = 3$

16. $\left(\dfrac{1}{27}\right)^{3x} = 3^{2x-1}$

17. $2e^x = 14$

18. $2^{2x+1} = 3^{x+1}$

19. $\ln(x + 4) - \ln(x + 2) = \ln x$

20. $\log x + \log(x - 15) = 2$

21. $\log(\log x) = 2$

22. $\log(7x - 4) - \log(x - 1) = 1$

23. $\ln(2t - 1) = \ln 4 + \ln(t - 3)$

24. $64^{2t+1} = 8^{-t+2}$

For Problems 25–28, if $\log 3 = 0.4771$ and $\log 7 = .8451$, evaluate each of the following.

25. $\log\left(\dfrac{7}{3}\right)$

26. $\log 21$

27. $\log 27$

28. $\log 7^{2/3}$

29. Express each of the following as the sum or difference of simpler logarithmic quantities. Assume that all variables represent positive real numbers.

(a) $\log_b\left(\dfrac{x}{y^2}\right)$

(b) $\log_b \sqrt[4]{xy^2}$

(c) $\log_b\left(\dfrac{\sqrt{x}}{y^3}\right)$

30. Express each of the following as a single logarithm. Assume that all variables represent positive real numbers.

(a) $3 \log_b x + 2 \log_b y$

(b) $\dfrac{1}{2} \log_b y - 4 \log_b x$

(c) $\dfrac{1}{2}(\log_b x + \log_b y) - 2 \log_b z$

For Problems 31–34, approximate each of the logarithms to three decimal places.

31. $\log_2 3$

32. $\log_3 2$

33. $\log_4 191$

34. $\log_2 .23$

For Problems 35–42, graph each of the functions.

35. $f(x) = \left(\dfrac{3}{4}\right)^x$

36. $f(x) = 2^{x+2}$

37. $f(x) = e^{x-1}$

38. $f(x) = -1 + \log x$

39. $f(x) = 3^x - 3^{-x}$

40. $f(x) = e^{-x^2/2}$

41. $f(x) = \log_2(x - 3)$

42. $f(x) = 3 \log_3 x$

For Problems 43–45, use the compound interest formula $A + P\left(1 + \dfrac{r}{n}\right)^{nt}$ to find the total amount of money accumulated at the end of the indicated time period for each of the investments.

43. $750 for 10 years at 11% compounded quarterly

44. $1250 for 15 years at 9% compounded monthly

45. $2500 for 20 years at 9.5% compounded semi-annually

46. How long will it take $100 to double itself if it is invested at 14% interest compounded annually?

47. How long will it take $1000 to be worth $3500 if it is invested at 10.5% interest compounded quarterly?

48. What rate of interest (to the nearest tenth of a percent) compounded continuously is needed for an investment of $500 to grow to $1000 in 8 years?

49. Suppose that the present population of a city is 50,000. Use the equation $P(t) = P_0 e^{.02t}$ (where P_0 represents an initial population) to estimate future populations, and estimate the population of that city in 10 years, 15 years, and 20 years.

50. The number of bacteria present in a certain culture after t hours is given by the equation $Q = Q_0 e^{.29t}$, where Q_0 represents the initial number of bacteria. How long will it take 500 bacteria to increase to 2000 bacteria?

51. Suppose that a certain radioactive substance has a half-life of 40 days. If there are presently 750 grams of the substance, how much, to the nearest gram, will remain after 100 days?

52. An earthquake occurred in Mexico City in 1985 that had an intensity level about 125,000,000 times the reference intensity. Find the Richter number for that earthquake.

CHAPTER 10 TEST

For Problems 1–4, evaluate each expression.

1. $\log_3 \sqrt{3}$

2. $\log_2(\log_2 4)$

3. $-2 + \ln e^3$

4. $\log_2(.5)$

For Problems 5–10, solve each equation.

5. $4^x = \dfrac{1}{64}$

6. $9^x = \dfrac{1}{27}$

7. $2^{3x-1} = 128$

8. $\log_9 x = \dfrac{5}{2}$

9. $\log x + \log(x + 48) = 2$

10. $\ln x = \ln 2 + \ln(3x - 1)$

For Problems 11–14, given that $\log_3 4 = 1.2619$ and $\log_3 5 = 1.4650$, evaluate each of the following.

11. $\log_3 100$

12. $\log_3 1.25$

13. $\log_3 \sqrt{5}$

14. $\log_3(16 \cdot 25)$

15. Solve $e^x = 176$ to the nearest hundredth.

16. Solve $2^{x-2} = 314$ to the nearest hundredth.

17. Determine $\log_5 632$ to four decimal places.

18. Express $3\log_b x + 2 \log_b y + \log_b z$ as a single logarithm with a coefficient of one.

19. If \$3500 is invested at 7.5% interest compounded quarterly, how much money has accumulated at the end of 8 years?

20. How long will it take \$5000 to be worth \$12,500 if it is invested at 7% compounded annually? Express your answer to the nearest tenth of a year.

21. The number of bacteria present in a certain culture after t hours is given by $Q(t) = Q_0 e^{.23t}$, where Q_0 represents the initial number of bacteria. How long will it take 400 bacteria to increase to 2400 bacteria? Express your answer to the nearest tenth of an hour.

22. Suppose that a certain radioactive substance has a half-life of 50 years. If there are presently 7500 grams of the substance, how much will remain after 32 years? Express your answer to the nearest gram.

For Problems 23–25, graph each of the functions.

23. $f(x) = e^x - 2$

24. $f(x) = -3^{-x}$

25. $f(x) = \log_2(x - 2)$

Systems of Equations and Inequalities

A 10% salt solution is to be mixed with a 20% salt solution to produce 20 gallons of a 17.5% salt solution. How many gallons of the 10% solution and how many gallons of the 20% solution should be mixed? **The two equations $x + y = 20$ and $.10x + .20y = .175(20)$ algebraically represent the conditions of the problem; x represents the number of gallons of the 10% solution and y represents the number of gallons of the 20% solution. The two equations considered together form a** system of linear equations **and the problem can be solved by solving the system of equations.**

Throughout most of this chapter we consider systems of linear equations and their applications. We will discuss various techniques for solving systems of linear equations.

11.1 **Systems of Two Linear Equations in Two Variables**

In Chapter 7 we stated that any equation of the form $Ax + By = C$, where A, B, and C are real numbers (A and B not both zero) is a **linear equation** in the two variables x and y, and its graph is a straight line. Two linear equations in two variables considered together form a **system of two linear equations in two variables**, as illustrated by the following.

$$\begin{pmatrix} x + y = 6 \\ x - y = 2 \end{pmatrix}, \quad \begin{pmatrix} 3x + 2y = 1 \\ 5x - 2y = 23 \end{pmatrix}, \quad \begin{pmatrix} 4x - 5y = 21 \\ 3x + 7y = -38 \end{pmatrix}$$

To solve a system like the three above means to find all of the ordered pairs that satisfy both equations in the system. For example, if we graph the two equations $x + y = 6$ and $x - y = 2$ on the same set of axes, as in Figure 11.1, then the ordered pair associated with the point of intersection of the two lines is the **solution of the system**. Thus, we say that $\{(4, 2)\}$ is the solution set of the system

$$\begin{pmatrix} x + y = 6 \\ x - y = 2 \end{pmatrix}.$$

FIGURE 11.1

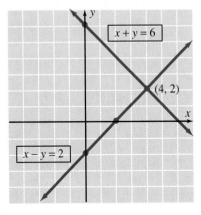

To check, we substitute 4 for x and 2 for y in the two equations.

$x + y$ becomes $4 + 2 = 6$ A true statement
$x - y$ becomes $4 - 2 = 2$ A true statement

Because the graph of a linear equation in two variables is a straight line, there are only three possible situations that can arise when a system of two linear equations in two variables is solved. We demonstrate these situations in Figure 11.2.

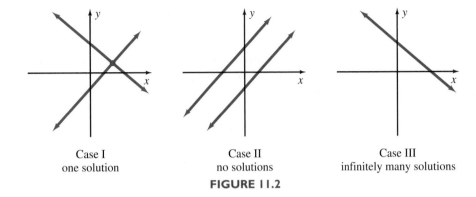

Case I
one solution

Case II
no solutions

Case III
infinitely many solutions

FIGURE 11.2

CASE I The graphs of the two equations are two lines that intersect at *one* point. There is *one solution*, and the system is called a **consistent system**.

CASE II The graphs of the two equations are parallel lines. There is *no solution*, and the system is called an **inconsistent system**.

CASE III The graphs of the two equations are the same line, and there are *infinitely many solutions* to the system. Any pair of real numbers that satisfies one of the equations will also satisfy the other equation, and we say that the equations are **dependent**.

Thus, when we solve a system of two linear equations in two variables, we know what to expect. The system will have *no* solutions, *one* ordered pair as a solution, or *infinitely many* ordered pairs as solutions.

Substitution Method

It should be evident that to solve systems of equations by graphing we must have accurate graphs. In fact, unless the solutions are integers, it is quite difficult to obtain exact solutions from a graph. Thus, we will consider some other methods for solving systems of equations.

The **substitution method**, which works quite well with systems of two linear equations in two unknowns, can be described as follows.

STEP 1 Solve one of the equations for one variable in terms of the other variable if neither equation is in such a form. (If possible, make a choice that will avoid fractions.)

STEP 2 Substitute the expression you obtained in step 1 into the other equation to produce an equation with one variable.

STEP 3 Solve the equation you obtained in step 2.

STEP 4 Use the solution you obtained in step 3, along with the expression obtained in step 1, to determine the solution of the system.

Now let's look at some examples illustrating the substitution method.

Example 1

Solve the system $\left(\begin{array}{l} x + y = 16 \\ y = x + 2 \end{array} \right)$.

Solution

Because the second equation states that y equals $x + 2$, we can substitute $x + 2$ for y in the first equation.

$$x + y = 16 \quad \underrightarrow{\text{Substitute } x + 2 \text{ for } y} \quad x + (x + 2) = 16$$

Now we have an equation with one variable that can be solved in the usual way.

$$x + (x + 2) = 16$$
$$2x + 2 = 16$$
$$2x = 14$$
$$x = 7$$

Substitute 7 for x in one of the two original equations (let's use the second one) to yield

$$y = 7 + 2 = 9.$$

To check, we can substitute 7 for x and 9 for y in both of the original equations.

$$7 + 9 = 16 \qquad \text{A true statement}$$
$$9 = 7 + 2 \qquad \text{A true statement}$$

The solution set is $\{(7, 9)\}$.   ▲

Example 2

Solve the system $\left(\begin{array}{l} x = 3y - 25 \\ 4x + 5y = 19 \end{array} \right)$.

Solution

In this case the first equation states that x equals $3y - 25$. Therefore, we can substitute $3y - 25$ for x in the second equation.

$$4x + 5y = 19 \quad \underrightarrow{\text{Substitute } 3y - 25 \text{ for } x} \quad 4(3y - 25) + 5y = 19$$

Solve this equation to yield

$$4(3y - 25) + 5y = 19$$
$$12y - 100 + 5y = 19$$
$$17y = 119$$
$$y = 7.$$

Substitute 7 for y in the first equation to produce

$$x = 3(7) - 25$$
$$= 21 - 25 = -4.$$

The solution set is $\{(-4, 7)\}$; check it.   ▲

Example 3

Solve the system $\begin{pmatrix} 3x - 7y = 2 \\ x + 4y = 1 \end{pmatrix}$.

Solution

Let's solve the second equation for x in terms of y.

$$x + 4y = 1$$
$$x = 1 - 4y$$

Now we can substitute $1 - 4y$ for x in the first equation.

$$3x - 7y = 2 \quad \underrightarrow{\text{Substitute } 1 - 4y \text{ for } x} \quad 3(1 - 4y) - 7y = 2$$

Let's solve the equation for y.

$$3(1 - 4y) - 7y = 2$$
$$3 - 12y - 7y = 2$$
$$-19y = -1$$
$$y = \frac{1}{19}$$

Finally, we can substitute $\frac{1}{19}$ for y in the equation $x = 1 - 4y$.

$$x = 1 - 4\left(\frac{1}{19}\right)$$
$$= 1 - \frac{4}{19}$$
$$= \frac{15}{19}$$

The solution set is $\left\{ \left(\frac{15}{19}, \frac{1}{19} \right) \right\}$.

Example 4

Solve the system $\begin{pmatrix} 5x - 6y = -4 \\ 3x + 2y = -8 \end{pmatrix}$.

Solution

Note that solving either equation for either variable will produce a fractional form. Let's solve the second equation for y in terms of x.

$$3x + 2y = -8$$
$$2y = -8 - 3x$$
$$y = \frac{-8 - 3x}{2}$$

Now we can substitute $\frac{-8 - 3x}{2}$ for y in the first equation.

$$5x - 6y = -4 \quad \underrightarrow{\text{Substitute } \frac{-8 - 3x}{2} \text{ for } y} \quad 5x - 6\left(\frac{-8 - 3x}{2}\right) = -4$$

Solve this equation to yield

$$5x - 6\left(\frac{-8 - 3x}{2}\right) = -4$$

$$5x - 3(-8 - 3x) = -4$$

$$5x + 24 + 9x = -4$$

$$14x = -28$$

$$x = -2.$$

Substitute -2 for x in $y = \dfrac{-8 - 3x}{2}$ to yield

$$y = \frac{-8 - 3(-2)}{2}$$

$$= \frac{-8 + 6}{2} = \frac{-2}{2} = -1.$$

The solution set is $\{(-2, -1)\}$.   ▲

Problem Solving

Many word problems that we solved earlier in this text using one variable and one equation can also be solved using a system of two linear equations in two variables. In fact, in many of these problems you may find it much more natural to use two variables. Let's consider some examples.

Problem 1

Anita invested some money at 8% and $400 more than that amount at 9%. The yearly interest from the two investments was $87. How much did Anita invest at each rate?

Solution

Let x represent the amount invested at 8% and let y represent the amount invested at 9%. The problem translates into the following system.

The amount invested at 9% was $400
more than 8%. →

The yearly interest from the two
investments was $87. ↗

$$\left(\begin{array}{l} y = x + 400 \\ .08x + .09y = 87 \end{array}\right)$$

From the first equation we can substitute $x + 400$ for y in the second equation and solve for x.

$$.08x + .09(x + 400) = 87$$

$$.08x + .09x + 36 = 87$$

$$.17x = 51$$

$$x = 300$$

Therefore, $300 is invested at 8% and $300 + $400 = $700 is invested at 9%.

▲

The two-variable expression $10t + u$ can be used to represent any two-digit number. The t represents the tens digit and the u represents the units digit. For example, if $t = 5$ and $u = 2$, then $10t + u$ becomes $10(5) + 2 = 52$. We use this general representation for a two-digit number in the next problem.

Problem 2

The tens digit of a two-digit number is 2 more than twice the units digit. The number with the digits reversed is 45 less than the original number. Find the original number.

Solution

Let u represent the units digit of the original number. Let t represent the tens digit of the original number. Then $10t + u$ represents the original number and $10u + t$ represents the number with the digits reversed. The problem translates into the following system.

$$\left(\begin{array}{c} t = 2u + 2 \\ 10u + t = 10t + u - 45 \end{array} \right)$$

The tens digit is 2 more than twice the units digit.

The number with the digits reversed is 45 less than the original number.

Simplify the second equation and the system becomes

$$\left(\begin{array}{c} t = 2u + 2 \\ -9t + 9u = -45 \end{array} \right).$$

From the first equation we can *substitute* $2u + 2$ for t in the second equation and solve.

$$-9t + 9u = -45$$
$$-9(2u + 2) + 9u = -45$$
$$-18u - 18 + 9u = -45$$
$$-9u = -27$$
$$u = 3$$

Substitute 3 for u in $t = 2u + 2$ to obtain

$$t = 2u + 2$$
$$= 2(3) + 2 = 8.$$

The tens digit is 8 and the units digit is 3, so the number is 83. (You should check to see if 83 satisfies the original conditions stated in the problem.) ▲

In our final example of this section, we will use a graphing utility to help solve a system of equations.

Example 5

Solve the system $\begin{pmatrix} 1.14x + 2.35y = -7.12 \\ 3.26x - 5.05y = 26.72 \end{pmatrix}$.

Solution

First, we need to solve each equation for y in terms of x. Thus, the system becomes

$$\begin{pmatrix} y = \dfrac{-7.12 - 1.14x}{2.35} \\ y = \dfrac{3.26x - 26.72}{5.05} \end{pmatrix}.$$

Now we can enter both of these equations into a graphing utility and obtain Figure 11.3.

In this figure it appears that the point of intersection is at approximately $x = 2$ and $y = -4$. By direct substitution into the given equations we can verify that the point of intersection is exactly $(2, -4)$.

FIGURE 11.3

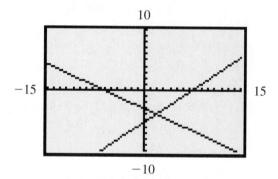

Problem Set 11.1

For Problems 1–10, use the graphing approach to determine whether each of the systems is *consistent*, *inconsistent*, or *dependent*. If the system is consistent, find the solution set from the graph and check it.

1. $\begin{pmatrix} x - y = 1 \\ 2x + y = 8 \end{pmatrix}$

2. $\begin{pmatrix} 3x + y = 0 \\ x - 2y = -7 \end{pmatrix}$

3. $\begin{pmatrix} 4x + 3y = -5 \\ 2x - 3y = -7 \end{pmatrix}$

4. $\begin{pmatrix} 2x - y = 9 \\ 4x - 2y = 11 \end{pmatrix}$

5. $\begin{pmatrix} \dfrac{1}{2}x + \dfrac{1}{4}y = 9 \\ 4x + 2y = 72 \end{pmatrix}$

6. $\begin{pmatrix} 5x + 2y = -9 \\ 4x - 3y = 2 \end{pmatrix}$

7. $\begin{pmatrix} \dfrac{1}{2}x - \dfrac{1}{3}y = 3 \\ x + 4y = -8 \end{pmatrix}$

8. $\begin{pmatrix} 4x - 9y = -60 \\ \frac{1}{3}x - \frac{3}{4}y = -5 \end{pmatrix}$

9. $\begin{pmatrix} x - \frac{1}{2}y = -4 \\ 8x - 4y = -1 \end{pmatrix}$

10. $\begin{pmatrix} 3x - 2y = 7 \\ 6x + 5y = -4 \end{pmatrix}$

For Problems 11–36, solve each system by using the substitution method.

11. $\begin{pmatrix} x + y = 20 \\ x = y - 4 \end{pmatrix}$

12. $\begin{pmatrix} x + y = 23 \\ y = x - 5 \end{pmatrix}$

13. $\begin{pmatrix} y = -3x - 18 \\ 5x - 2y = -8 \end{pmatrix}$

14. $\begin{pmatrix} 4x - 3y = 33 \\ x = -4y - 25 \end{pmatrix}$

15. $\begin{pmatrix} x = -3y \\ 7x - 2y = -69 \end{pmatrix}$

16. $\begin{pmatrix} 9x - 2y = -38 \\ y = -5x \end{pmatrix}$

17. $\begin{pmatrix} 2x + 3y = 11 \\ 3x - 2y = -3 \end{pmatrix}$

18. $\begin{pmatrix} 3x - 4y = -14 \\ 4x + 3y = 23 \end{pmatrix}$

19. $\begin{pmatrix} 3x - 4y = 9 \\ x = 4y - 1 \end{pmatrix}$

20. $\begin{pmatrix} y = 3x - 5 \\ 2x + 3y = 6 \end{pmatrix}$

21. $\begin{pmatrix} y = \frac{2}{5}x - 1 \\ 3x + 5y = 4 \end{pmatrix}$

22. $\begin{pmatrix} y = \frac{3}{4}x - 5 \\ 5x - 4y = 9 \end{pmatrix}$

23. $\begin{pmatrix} 7x - 3y = -2 \\ x = \frac{3}{4}y + 1 \end{pmatrix}$

24. $\begin{pmatrix} 5x - y = 9 \\ x = \frac{1}{2}y - 3 \end{pmatrix}$

25. $\begin{pmatrix} 2x + y = 12 \\ 3x - y = 13 \end{pmatrix}$

26. $\begin{pmatrix} -x + 4y = -22 \\ x - 7y = 34 \end{pmatrix}$

27. $\begin{pmatrix} 4x + 3y = -40 \\ 5x - y = -12 \end{pmatrix}$

28. $\begin{pmatrix} x - 5y = 33 \\ -4x + 7y = -41 \end{pmatrix}$

29. $\begin{pmatrix} .06x + .07y = 86 \\ y = x + 300 \end{pmatrix}$

30. $\begin{pmatrix} .07x + .08y = 100 \\ x = y - 500 \end{pmatrix}$

31. $\begin{pmatrix} 3x + 5y = 22 \\ 4x - 7y = -39 \end{pmatrix}$

32. $\begin{pmatrix} 2x - 3y = -16 \\ 6x + 7y = 16 \end{pmatrix}$

33. $\begin{pmatrix} 4x - 5y = 3 \\ 8x + 15y = -24 \end{pmatrix}$

34. $\begin{pmatrix} 2x + 3y = 3 \\ 4x - 9y = -4 \end{pmatrix}$

35. $\begin{pmatrix} x + y = 750 \\ .06x + .07y = 50 \end{pmatrix}$

36. $\begin{pmatrix} x + y = 300 \\ .09x + .08y = 25 \end{pmatrix}$

For Problems 37–50, solve each problem by setting up and solving an appropriate system of equations.

37. Doris invested some money at 7% and some money at 8%. She invested $6000 more at 8% than she did at 7%. Her total yearly interest from the two

investments was $780. How much did Doris invest at each rate?

38. Gus invested a total of $8000, part of it at 8% and the remainder at 9%. His yearly income from the two investments was $690. How much did he invest at each rate?

39. The sum of the digits of a two-digit number is 11. The tens digit is one more than four times the units digit. Find the number.

40. The units digit of a two-digit number is one less than three times the tens digit. If the sum of the digits is 11, find the number.

41. Find two numbers whose sum is 131 such that one number is 5 less than three times the other.

42. The difference of two numbers is 75. The larger number is 3 less than four times the smaller number. Find the numbers.

43. In a class of 50 students, the number of females is two more than five times the number of males. How many females are there in the class?

44. In a recent survey, one thousand registered voters were asked about their political preferences. The number of males in the survey was five less than one-half of the number of females. Find the number of males in the survey.

45. The perimeter of a rectangle is 94 inches. The length of the rectangle is 7 inches more than the width. Find the dimensions of the rectangle.

46. Two angles are supplementary and the measure of one of them is 20° less than three times the measure of the other angle. Find the measure of each angle.

47. A deposit slip listed $700 in cash to be deposited. There were 100 bills, some of them five-dollar bills and the remainder ten-dollar bills. How many bills of each denomination were deposited?

48. Cindy has 30 coins, consisting of dimes and quarters, which total $5.10. How many coins of each kind does she have?

49. The income from a student production was $10,000. The price of a student ticket was $3 and

nonstudent tickets were sold at $5 each. Three thousand tickets were sold. How many tickets of each kind were sold?

50. Sue bought 3 packages of cookies and 2 sacks of potato chips for $3.65. Later she bought 2 more packages of cookies and 5 additional sacks of potato chips for $4.23. Find the price of a package of cookies.

THOUGHTS INTO WORDS

51. Describe how to use the substitution method to solve a system of two linear equations in two variables.

52. Is it possible for a system of two linear equations in two variables to have exactly two solutions? Defend your answer.

53. Explain how you would solve the system $\begin{pmatrix} 2x + 5y = 5 \\ 5x - y = 9 \end{pmatrix}$ using the substitution method.

Graphics Calculator Activities

54. Use your graphics calculator to help determine whether each of the systems in Problems 1–10 is consistent, inconsistent, or dependent.

55. Use your graphics calculator to help determine the solution set for each of the following systems. Be sure to check your answers.

(a) $\begin{pmatrix} 3x - y = 30 \\ 5x - y = 46 \end{pmatrix}$ (b) $\begin{pmatrix} 1.2x + 3.4y = 25.4 \\ 3.7x - 2.3y = 14.4 \end{pmatrix}$

(c) $\begin{pmatrix} 1.98x + 2.49y = 13.92 \\ 1.19x + 3.45y = 16.18 \end{pmatrix}$ (d) $\begin{pmatrix} 2x - 3y = 10 \\ 3x + 5y = 53 \end{pmatrix}$

(e) $\begin{pmatrix} 4x - 7y = -49 \\ 6x + 9y = 219 \end{pmatrix}$

(f) $\begin{pmatrix} 3.7x - 2.9y = -14.3 \\ 1.6x + 4.7y = -30 \end{pmatrix}$

11.2 Elimination-by-Addition Method

We found in the previous section that the substitution method for solving a system of two equations and two unknowns works rather well. However, as the number of equations and unknowns increases, the substitution method becomes unwieldy. In this section we are going to introduce another method called the **elimination-by-addition** method. We introduce it here using systems of two linear equations in two unknowns and then expand its use to three linear equations in three unknowns in the next section.

The elimination-by-addition method involves the replacement of systems of equations with simpler equivalent systems until we obtain a system whereby we can easily extract the solutions. **Equivalent systems of equations are systems that have exactly the same solution set.** We can apply the following operations or transformations to a system of equations to produce an equivalent system.

1. Any two equations of the system can be interchanged.
2. Both sides of any equation of the system can be multiplied by any nonzero real number.
3. Any equation of the system can be replaced by the *sum* of that equation and a nonzero multiple of another equation.

Now let's see how to apply these operations to solve a system of two linear equations in two unknowns.

Example 1

Solve the system $\begin{pmatrix} 3x + 2y = 1 \\ 5x - 2y = 23 \end{pmatrix}$.

$\qquad$ (1)
$\qquad$ (2)

Solution

Let's replace equation (2) with an equation we form by multiplying equation (1) by 1 and then adding that result to equation (2).

$$\begin{pmatrix} 3x + 2y = 1 \\ 8x = 24 \end{pmatrix}$$

$\qquad$ (3)
$\qquad$ (4)

From equation (4) we can easily obtain the value of x.

$$8x = 24$$

$$x = 3$$

Then we can substitute 3 for x in equation (3).

$$3x + 2y = 1$$

$$3(3) + 2y = 1$$

$$2y = -8$$

$$y = -4$$

The solution set is $\{(3, -4)\}$. Check it!

Example 2

Solve the system $\left(\begin{array}{l} x + 5y = -2 \\ 3x - 4y = -25 \end{array} \right).$ **(1)**   **(2)**

Solution

Let's replace equation (2) with an equation we form by multiplying equation (1) by -3 and then adding that result to equation (2).

$$\left(\begin{array}{l} x + 5y = -2 \\ -19y = -19 \end{array} \right) \qquad \begin{array}{l} \textbf{(3)} \\ \textbf{(4)} \end{array}$$

From equation (4) we can obtain the value of y.

$$-19y = -19$$
$$y = 1$$

Now we can substitute 1 for y in equation (3).

$$x + 5y = -2$$
$$x + 5(1) = -2$$
$$x = -7$$

The solution set is $\{(-7, 1)\}$.   ▲

    Notice that our objective has been to produce an equivalent system of equations whereby one of the variables can be *eliminated* from one equation. We accomplish this by multiplying one equation of the system by an appropriate number and then *adding* that result to the other equation. We call the method **elimination-by-addition**. Let's look at another example.

Example 3

Solve the system $\left(\begin{array}{l} 2x + 5y = 4 \\ 5x - 7y = -29 \end{array} \right).$ **(1)**   **(2)**

Solution

Let's form an equivalent system where the second equation has no x-term. First, we can multiply equation (2) by -2.

$$\left(\begin{array}{l} 2x + 5y = 4 \\ -10x + 14y = 58 \end{array} \right) \qquad \begin{array}{l} \textbf{(3)} \\ \textbf{(4)} \end{array}$$

Now we can replace equation (4) with an equation that we form by multiplying equation (3) by 5 and then adding that result to equation (4).

$$\left(\begin{array}{l} 2x + 5y = 4 \\ 39y = 78 \end{array} \right) \qquad \begin{array}{l} \textbf{(5)} \\ \textbf{(6)} \end{array}$$

From equation (6) we can find the value of y.

$$39y = 78$$
$$y = 2$$

Now we can substitute 2 for y in equation (5).

$$2x + 5y = 4$$

$$2x + 5(2) = 4$$

$$2x = -6$$

$$x = -3$$

The solution set is $\{(-3, 2)\}$. ▲

Example 4 Solve the system $\begin{pmatrix} 3x - 2y = 5 \\ 2x + 7y = 9 \end{pmatrix}$. (1)
(2)

Solution We can start by multiplying equation (2) by -3.

$$\begin{pmatrix} 3x - 2y = 5 \\ -6x - 21y = -27 \end{pmatrix}$$ (3)
(4)

Now we can replace equation (4) with an equation we form by multiplying equation (3) by 2 and then adding that result to equation (4).

$$\begin{pmatrix} 3x - 2y = 5 \\ -25y = -17 \end{pmatrix}$$ (5)
(6)

From equation (6) we can find the value of y.

$$-25y = -17$$

$$y = \frac{17}{25}$$

Now we can substitute $\frac{17}{25}$ for y in equation (5).

$$3x - 2y = 5$$

$$3x - 2\left(\frac{17}{25}\right) = 5$$

$$3x - \frac{34}{25} = 5$$

$$3x = 5 + \frac{34}{25}$$

$$3x = \frac{125}{25} + \frac{34}{25}$$

$$3x = \frac{159}{25}$$

$$x = \left(\frac{159}{25}\right)\left(\frac{1}{3}\right) = \frac{53}{25}$$

The solution set is $\left\{\left(\frac{53}{25}, \frac{17}{25}\right)\right\}$. (Perhaps you should check this result!) ▲

Which Method to Use

Both the elimination-by-addition and the substitution methods can be used to obtain exact solutions for any system of two linear equations in two unknowns. Sometimes the issue is that of deciding which method to use on a particular system. As we have seen with the examples thus far in this section and those of the previous section, many systems lend themselves to one or the other method by the original format of the equations. Let's emphasize that point with some more examples.

Example 5

Solve the system $\begin{pmatrix} 4x - 3y = 4 \\ 10x + 9y = -1 \end{pmatrix}$.

(1)
(2)

Solution

Because changing the form of either equation in preparation for the substitution method would produce a fractional form, we are probably better off using the elimination-by-addition method.

Let's replace equation (2) with an equation we form by multiplying equation (1) by 3 and then adding that result to equation (2).

$$\begin{pmatrix} 4x - 3y = 4 \\ 22x \quad = 11 \end{pmatrix}$$

(3)
(4)

From equation (4) we can determine the value of x.

$$22x = 11$$

$$x = \frac{11}{22} = \frac{1}{2}$$

Now we can substitute $\frac{1}{2}$ for x in equation (3).

$$4x - 3y = 4$$

$$4\left(\frac{1}{2}\right) - 3y = 4$$

$$2 - 3y = 4$$

$$-3y = 2$$

$$y = -\frac{2}{3}$$

The solution set is $\left\{\left(\frac{1}{2}, -\frac{2}{3}\right)\right\}$.   ▲

Example 6

Solve the system $\begin{pmatrix} 6x + 5y = -3 \\ y = -2x - 7 \end{pmatrix}$.

(1)
(2)

Solution

Because the second equation is of the form y *equals*, let's use the substitution method. From the second equation we can substitute $-2x - 7$ for y in the first equation.

$$6x + 5y = -3 \quad \underrightarrow{\text{Substitute } -2x - 7 \text{ for } y} \quad 6x + 5(-2x - 7) = -3$$

Solving this equation yields

$$6x + 5(-2x - 7) = -3$$
$$6x - 10x - 35 = -3$$
$$-4x - 35 = -3$$
$$-4x = 32$$
$$x = -8.$$

Substitute -8 for x in the second equation to yield

$$y = -2(-8) - 7 = 16 - 7 = 9.$$

The solution set is $\{(-8, 9)\}$. ▲

Sometimes we need to simplify the equations of a system before we can decide which method to use for solving the system. Let's consider an example of that type.

Example 7

Solve the system $\left(\begin{array}{l} \dfrac{x - 2}{4} + \dfrac{y + 1}{3} = 2 \qquad (1) \\ \dfrac{x + 1}{7} + \dfrac{y - 3}{2} = \dfrac{1}{2} \end{array} \right)$. $\qquad (2)$

Solution

First, we need to simplify the two equations. Let's multiply both sides of equation (1) by 12 and simplify.

$$12\left(\frac{x - 2}{4} + \frac{y + 1}{3}\right) = 12(2)$$
$$3(x - 2) + 4(y + 1) = 24$$
$$3x - 6 + 4y + 4 = 24$$
$$3x + 4y - 2 = 24$$
$$3x + 4y = 26$$

Let's multiply both sides of equation (2) by 14.

$$14\left(\frac{x + 1}{7} + \frac{y - 3}{2}\right) = 14\left(\frac{1}{2}\right)$$
$$2(x + 1) + 7(y - 3) = 7$$
$$2x + 2 + 7y - 21 = 7$$
$$2x + 7y - 19 = 7$$
$$2x + 7y = 26$$

Now we have the following system to solve.

$$\begin{pmatrix} 3x + 4y = 26 \\ 2x + 7y = 26 \end{pmatrix}$$

(3)
(4)

Probably the easiest approach is to use the elimination-by-addition method. We can start by multiplying equation (4) by -3.

$$\begin{pmatrix} 3x + 4y = 26 \\ -6x - 21y = -78 \end{pmatrix}$$

(5)
(6)

Now we can replace equation (6) with an equation we form by multiplying equation (5) by 2 and then adding that result to equation (6).

$$\begin{pmatrix} 3x + 4y = 26 \\ -13y = -26 \end{pmatrix}$$

(7)
(8)

From equation (8) we can find the value of y.

$$-13y = -26$$
$$y = 2$$

Now we can substitute 2 for y in equation (7).

$$3x + 4y = 26$$
$$3x + 4(2) = 26$$
$$3x = 18$$
$$x = 6$$

The solution set is $\{(6, 2)\}$.   ▲

REMARK   Don't forget that to check a problem like Example 7 you must check the potential solutions back into the *original* equations.   △

In Section 11.1 we discussed the fact that you can tell whether a system of two linear equations in two unknowns has no solution, one solution, or infinitely many solutions by graphing the equations of the system. That is, the two lines may be parallel (no solution), or they may intersect in one point (one solution), or they may coincide (infinitely many solutions).

From a practical viewpoint, the systems that have one solution deserve most of our attention. However, we do need to be able to deal with the other situations since they do occur occasionally. Let's use two examples to illustrate the type of thing that happens when we hit a *no solution* or *infinitely many solutions* situation when using either the elimination-by-addition method or the substitution method.

Example 8

Solve the system $\begin{pmatrix} y = 3x - 1 \\ -9x + 3y = 4 \end{pmatrix}$.

(1)
(2)

Solution

Using the substitution method, we can proceed as follows.

$$-9x + 3y = 4 \quad \underrightarrow{\text{Substitute } 3x - 1 \text{ for } y} \quad -9x + 3(3x - 1) = 4$$

Solving this equation yields

$$-9x + 3(3x - 1) = 4$$
$$-9x + 9x - 3 = 4$$
$$-3 = 4.$$

The *false numerical statement*, $-3 = 4$, implies that the system has *no solution*. (You may want to graph the two lines to verify this conclusion!)   ▲

Example 9

Solve the system $\begin{pmatrix} 5x + y = 2 \\ 10x + 2y = 4 \end{pmatrix}$.   (1)
    (2)

Solution

Use the elimination-by-addition method and proceed as follows. Let's replace equation (2) with an equation we form by multiplying equation (1) by -2 and then adding that result to equation (2).

$$\begin{pmatrix} 5x + y = 2 \\ 0 + 0 = 0 \end{pmatrix}$$ (3)
    (4)

The *true numerical statement*, $0 + 0 = 0$, implies that the system has *infinitely many solutions*. Any ordered pair that satisfies one of the equations will also satisfy the other equation. Thus, we can express the solution set as

$$\{(x, y) \mid 5x + y = 2\}.$$ ▲

Problem Set 11.2

For Problems 1–16, use the elimination-by-addition method to solve each system.

1. $\begin{pmatrix} 2x + 3y = -1 \\ 5x - 3y = 29 \end{pmatrix}$

2. $\begin{pmatrix} 3x - 4y = -30 \\ 7x + 4y = 10 \end{pmatrix}$

3. $\begin{pmatrix} 6x - 7y = 15 \\ 6x + 5y = -21 \end{pmatrix}$

4. $\begin{pmatrix} 5x + 2y = -4 \\ 5x - 3y = 6 \end{pmatrix}$

5. $\begin{pmatrix} x - 2y = -12 \\ 2x + 9y = 2 \end{pmatrix}$

6. $\begin{pmatrix} x - 4y = 29 \\ 3x + 2y = -11 \end{pmatrix}$

7. $\begin{pmatrix} 4x + 7y = -16 \\ 6x - y = -24 \end{pmatrix}$

8. $\begin{pmatrix} 6x + 7y = 17 \\ 3x + y = -4 \end{pmatrix}$

9. $\begin{pmatrix} 3x - 2y = 5 \\ 2x + 5y = -3 \end{pmatrix}$

10. $\begin{pmatrix} 4x + 3y = -4 \\ 3x - 7y = 34 \end{pmatrix}$

11. $\begin{pmatrix} 7x - 2y = 4 \\ 7x - 2y = 9 \end{pmatrix}$

12. $\begin{pmatrix} 5x - y = 6 \\ 10x - 2y = 12 \end{pmatrix}$

13. $\begin{pmatrix} 5x + 4y = 1 \\ 3x - 2y = -1 \end{pmatrix}$

14. $\begin{pmatrix} 2x - 7y = -2 \\ 3x + y = 1 \end{pmatrix}$

15. $\begin{pmatrix} 8x - 3y = 13 \\ 4x + 9y = 3 \end{pmatrix}$

16. $\begin{pmatrix} 10x - 8y = -11 \\ 8x + 4y = -1 \end{pmatrix}$

For Problems 17–44, solve each system by using either the substitution or the elimination-by-addition method, whichever seems more appropriate.

17. $\begin{pmatrix} 5x + 3y = -7 \\ 7x - 3y = 55 \end{pmatrix}$

18. $\begin{pmatrix} 4x - 7y = 21 \\ -4x + 3y = -9 \end{pmatrix}$

19. $\begin{pmatrix} x = 5y + 7 \\ 4x + 9y = 28 \end{pmatrix}$

20. $\begin{pmatrix} 11x - 3y = -60 \\ y = -38 - 6x \end{pmatrix}$

21. $\begin{pmatrix} x = -6y + 79 \\ x = 4y - 41 \end{pmatrix}$

22. $\begin{pmatrix} y = 3x + 34 \\ y = -8x - 54 \end{pmatrix}$

23. $\begin{pmatrix} 4x - 3y = 2 \\ 5x - y = 3 \end{pmatrix}$

24. $\begin{pmatrix} 3x - y = 9 \\ 5x + 7y = 1 \end{pmatrix}$

25. $\begin{pmatrix} 5x - 2y = 1 \\ 10x - 4y = 7 \end{pmatrix}$

26. $\begin{pmatrix} 4x + 7y = 2 \\ 9x - 2y = 1 \end{pmatrix}$

27. $\begin{pmatrix} 3x - 2y = 7 \\ 5x + 7y = 1 \end{pmatrix}$

28. $\begin{pmatrix} 2x - 3y = 4 \\ y = \dfrac{2}{3}x - \dfrac{4}{3} \end{pmatrix}$

29. $\begin{pmatrix} -2x + 5y = -16 \\ x = \dfrac{3}{4}y + 1 \end{pmatrix}$

30. $\begin{pmatrix} y = \dfrac{2}{3}x - \dfrac{3}{4} \\ 2x + 3y = 11 \end{pmatrix}$

31. $\begin{pmatrix} y = \dfrac{2}{3}x - 4 \\ 5x - 3y = 9 \end{pmatrix}$

32. $\begin{pmatrix} 5x - 3y = 7 \\ x = \dfrac{3y}{4} - \dfrac{1}{3} \end{pmatrix}$

33. $\begin{pmatrix} \dfrac{x}{6} + \dfrac{y}{3} = 3 \\ \dfrac{5x}{2} - \dfrac{y}{6} = -17 \end{pmatrix}$

34. $\begin{pmatrix} \dfrac{3x}{4} - \dfrac{2y}{3} = 31 \\ \dfrac{7x}{5} + \dfrac{y}{4} = 22 \end{pmatrix}$

35. $\begin{pmatrix} -(x - 6) + 6(y + 1) = 58 \\ 3(x + 1) - 4(y - 2) = -15 \end{pmatrix}$

36. $\begin{pmatrix} -2(x + 2) + 4(y - 3) = -34 \\ 3(x + 4) - 5(y + 2) = 23 \end{pmatrix}$

37. $\begin{pmatrix} 5(x + 1) - (y + 3) = -6 \\ 2(x - 2) + 3(y - 1) = 0 \end{pmatrix}$

38. $\begin{pmatrix} 2(x - 1) - 3(y + 2) = 30 \\ 3(x + 2) + 2(y - 1) = -4 \end{pmatrix}$

39. $\begin{pmatrix} \dfrac{1}{2}x - \dfrac{1}{3}y = 12 \\ \dfrac{3}{4}x + \dfrac{2}{3}y = 4 \end{pmatrix}$

40. $\left(\begin{array}{l} \dfrac{2}{3}x + \dfrac{1}{5}y = 0 \\ \dfrac{3}{2}x - \dfrac{3}{10}y = -15 \end{array} \right)$

41. $\left(\begin{array}{l} \dfrac{2x}{3} - \dfrac{y}{2} = -\dfrac{5}{4} \\ \dfrac{x}{4} + \dfrac{5y}{6} = \dfrac{17}{16} \end{array} \right)$

42. $\left(\begin{array}{l} \dfrac{x}{2} + \dfrac{y}{3} = \dfrac{5}{72} \\ \dfrac{x}{4} + \dfrac{5y}{2} = -\dfrac{17}{48} \end{array} \right)$

43. $\left(\begin{array}{l} \dfrac{3x + y}{2} + \dfrac{x - 2y}{5} = 8 \\ \dfrac{x - y}{3} - \dfrac{x + y}{6} = \dfrac{10}{3} \end{array} \right)$

44. $\left(\begin{array}{l} \dfrac{x - y}{4} - \dfrac{2x - y}{3} = -\dfrac{1}{4} \\ \dfrac{2x + y}{3} + \dfrac{x + y}{2} = \dfrac{17}{6} \end{array} \right)$

For problems 45–57, solve each problem by setting up and solving an appropriate system of equations.

45. A 10% salt solution is to be mixed with a 20% salt solution to produce 20 gallons of a 17.5% salt solution. How many gallons of the 10% solution and how many gallons of the 20% solution will be needed?

46. A library buys a total of 35 books that cost $462. Some of the books cost $12 and the remainder cost $14 per book. How many books of each price did they buy?

47. The cost of 3 tennis balls and 2 golf balls is $7. The cost of 6 tennis balls and 3 golf balls is $12. Find the cost of 1 tennis ball and also the cost of 1 golf ball.

48. For moving purposes, the Hendersons bought 25 cardboard boxes for $97.50. There were two kinds of boxes; the large ones cost $7.50 per box and the small ones were $3 per box. How many boxes of each kind did they buy?

49. A motel rents double rooms for $42 per day and single rooms for $22 per day. If a total of 55 rooms was rented for $2010, how many of each kind were rented?

50. One solution contains 50% alcohol and another contains 80% alcohol. How many liters of each solution should be mixed to make 10.5 liters of a 70% solution?

51. Suppose that a fulcrum is placed so that weights of 40 pounds and 80 pounds are in balance. Furthermore, suppose that when 20 pounds is added to the 40-pound weight, the 80-pound weight must be moved $1\frac{1}{2}$ feet further from the fulcrum to obtain a balance. Find the original distance between the two weights.

52. If a certain two-digit number is divided by the sum of its digits, the quotient is 2. If the digits are reversed, the new number is 9 less than five times the original number. Find the original number.

53. If the numerator of a certain fraction is increased by 5 and the denominator is decreased by 1, the resulting fraction is $\frac{8}{3}$. However, if the numerator of the original fraction is doubled and the denominator is increased by 7, the resulting fraction is $\frac{6}{11}$. Find the original fraction.

54. A man bought 2 pounds of coffee and 1 pound of butter for a total of $7.75. A month later the prices had not changed (this makes it a fictitious problem) and he bought 3 pounds of coffee and 2 pounds of butter for $12.50. Find the price per pound of both the coffee and the butter.

55. Suppose that we have a rectangular-shaped book cover. If the width is increased by 2 centimeters and the length is decreased by 1 centimeter, the area is increased by 28 square centimeters. However, if the width is decreased by 1 centimeter and the length is increased by 2 centimeters, then the area is increased by 10 square centimeters. Find the dimensions of the book cover.

56. A blueprint indicates a master bedroom in the shape of a rectangle. If the width is increased by 2 feet and the length remains the same, then the area is increased by 36 square feet. However, if the width is increased by 1 foot and the length is increased by 2 feet, then the area is increased by 48 square feet. Find the dimensions of the room as indicated on the blueprint.

57. A fulcrum is placed so that weights of 60 pounds and 100 pounds are in balance. If 20 pounds are subtracted from the 100-pound weight, then the 60-pound weight must be moved 1 foot closer to the fulcrum to preserve the balance. Find the original distance between the 60-pound and 100-pound weights.

THOUGHTS INTO WORDS

58. Describe how to use the elimination-by-addition method to solve a system of two linear equations in two variables.

59. How do you decide whether to solve a system of linear equations in two variables by using the substitution method or the elimination-by-addition method?

Further Investigations

60. There is another way of telling whether a system of two linear equations in two unknowns is consistent, inconsistent, or dependent without taking the time to graph each equation. It can be shown that any system of the form

$$a_1x + b_1y = c_1$$
$$a_2x + b_2y = c_2$$

has one and only one solution if

$$\frac{a_1}{a_2} \neq \frac{b_1}{b_2} \qquad \text{Consistent}$$

has no solution if

$$\frac{a_1}{a_2} = \frac{b_1}{b_2} \neq \frac{c_1}{c_2} \qquad \text{Inconsistent}$$

and has infinitely many solutions if

$$\frac{a_1}{a_2} = \frac{b_1}{b_2} = \frac{c_1}{c_2} \qquad \text{Dependent}$$

Determine whether each of the following systems is consistent, inconsistent, or dependent.

(a) $\begin{pmatrix} 4x - 3y = 7 \\ 9x + 2y = 5 \end{pmatrix}$

(b) $\begin{pmatrix} 5x - y = 6 \\ 10x - 2y = 19 \end{pmatrix}$

(c) $\begin{pmatrix} 5x - 4y = 11 \\ 4x + 5y = 12 \end{pmatrix}$

(d) $\begin{pmatrix} x + 2y = 5 \\ x - 2y = 9 \end{pmatrix}$

(e) $\begin{pmatrix} x - 3y = 5 \\ 3x - 9y = 15 \end{pmatrix}$

(f) $\begin{pmatrix} 4x + 3y = 7 \\ 2x - y = 10 \end{pmatrix}$

(g) $\begin{pmatrix} 3x + 2y = 4 \\ y = -\frac{3}{2}x - 1 \end{pmatrix}$

(h) $\begin{pmatrix} y = \frac{4}{3}x - 2 \\ 4x - 3y = 6 \end{pmatrix}$

61. A system such as

$$\begin{pmatrix} \frac{3}{x} + \frac{2}{y} = 2 \\ \frac{2}{x} - \frac{3}{y} = \frac{1}{4} \end{pmatrix}$$

is not a system of linear equations but can be transformed into a linear system by changing variables. For example, substituting u for $\frac{1}{x}$ and v for $\frac{1}{y}$, the above system becomes

$$\begin{pmatrix} 3u + 2v = 2 \\ 2u - 3v = \frac{1}{4} \end{pmatrix}.$$

We can solve this "new" system by either elimination-by-addtion or substitution (we will leave the details for you) to produce $u = \dfrac{1}{2}$ and $v = \dfrac{1}{4}$. Therefore, since $u = \dfrac{1}{x}$ and $v = \dfrac{1}{y}$, we have

$$\frac{1}{x} = \frac{1}{2} \qquad \text{and} \qquad \frac{1}{y} = \frac{1}{4}.$$

Solving these equations yields

$$x = 2 \qquad \text{and} \qquad y = 4.$$

The solution set of the original system is $\{(2, 4)\}$.

Solve each of the following systems.

(a) $\left(\begin{array}{l} \dfrac{1}{x} + \dfrac{2}{y} = \dfrac{7}{12} \\ \dfrac{3}{x} - \dfrac{2}{y} = \dfrac{5}{12} \end{array} \right)$

(b) $\left(\begin{array}{l} \dfrac{2}{x} + \dfrac{3}{y} = \dfrac{19}{15} \\ -\dfrac{2}{x} + \dfrac{1}{y} = -\dfrac{7}{15} \end{array} \right)$

(c) $\left(\begin{array}{l} \dfrac{3}{x} - \dfrac{2}{y} = \dfrac{13}{6} \\ \dfrac{2}{x} + \dfrac{3}{y} = 0 \end{array} \right)$

(d) $\left(\begin{array}{l} \dfrac{4}{x} + \dfrac{1}{y} = 11 \\ \dfrac{3}{x} - \dfrac{5}{y} = -9 \end{array} \right)$

(e) $\left(\begin{array}{l} \dfrac{5}{x} - \dfrac{2}{y} = 23 \\ \dfrac{4}{x} + \dfrac{3}{y} = \dfrac{23}{2} \end{array} \right)$

(f) $\left(\begin{array}{l} \dfrac{2}{x} - \dfrac{7}{y} = \dfrac{9}{10} \\ \dfrac{5}{x} + \dfrac{4}{y} = -\dfrac{41}{20} \end{array} \right)$

62. Solve the following system for x and y.

$$\left(\begin{array}{l} a_1 x + b_1 y = c_1 \\ a_2 x + b_2 y = c_2 \end{array} \right)$$

Graphics Calculator Activities

63. Use a graphics calculator to check your answers for Problem 60.

64. Use a graphics calculator to check your answers for Problem 61.

11.3 Systems of Three Linear Equations in Three Variables

Consider a linear equation in three variables x, y, and z, such as $3x - 2y + z = 7$. Any **ordered triple** (x, y, z) that makes the equation a true numerical statement is said to be a solution of the equation. For example, the ordered triple $(2, 1, 3)$ is a solution because $3(2) - 2(1) + 3 = 7$. However, the ordered triple $(5, 2, 4)$ is not a solution because $3(5) - 2(2) + 4 \neq 7$. There are infinitely many solutions in the solution set.

REMARK   The idea of a **linear** equation is generalized to include equations of more than two variables. Thus, an equation such as $5x - 2y + 9z = 8$ is called a **linear equation in three variables**, the equation $5x - 7y + 2z - 11w = 1$ is called a **linear equation in four variables**, and so on. △

To *solve* a system of three linear equations in three variables, such as

$$\begin{pmatrix} 3x - y + 2z = 13 \\ 4x + 2y + 5z = 30 \\ 5x - 3y - z = 3 \end{pmatrix},$$

means to find all of the ordered triples that satisfy all three equations. In other words, the solution set of the system is the intersection of the solution sets of all three equations in the system.

The graph of a linear equation in three variables is a plane, not a line. In fact, graphing equations in three variables requires the use of a three-dimensional coordinate system. Thus, using a graphing approach to solve systems of three linear equations in three variables is not at all practical. However, a simple graphic analysis does provide us with some idea as to what we can expect as we begin solving such systems.

In general, because each linear equation in three variables produces a plane, a system of three such equations produces three planes. There are various ways in which three planes can be related. For example, they may be mutually parallel, or two of the planes may be parallel and the third intersects each of the two. (You may want to analyze all of the other possibilities for the three planes!) However, for our purposes at this time we need to realize that from a solution-set viewpoint, a system of three linear equations in three variables produces one of the following possibilities.

1. There is *one ordered triple* that satisfies all three equations. The three planes have a common point of intersection, as indicated in Figure 11.4.

FIGURE 11.4

2. There are *infinitely many ordered triples* in the solution set, all of which are coordinates of points on a line common to the planes. This can happen if the three planes have a common line of intersection, as in Figure 11.5(a), or if two of the planes coincide and the third plane intersects them, as in Figure 11.5(b).

FIGURE 11.5

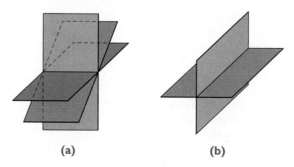

(a)   (b)

3. There are *infinitely many ordered triples* in the solution set, all of which are coordinates of points on a plane. This can happen if the three planes coincide, as illustrated in Figure 11.6.

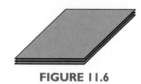

FIGURE 11.6

4. The *solution set is empty*; thus we write ∅. This can happen in various ways, as we see in Figure 11.7. Notice that in each situation there are no points common to all three planes.

FIGURE 11.7

(a) Three parallel planes

(b) Two planes coincide and the third one is parallel to the coinciding planes.

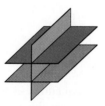

(c) Two planes are parallel and the third intersects them in parallel lines.

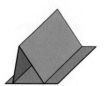

(d) No two planes are parallel, but two of them intersect in a line that is parallel to the third plane.

Now that we know what possibilities exist, let's consider finding the solution sets for some systems. Our approach will be the elimination-by-addition method, whereby systems are replaced with equivalent systems until a system is obtained where we can easily determine the solution set. Let's start with an example that allows us to determine the solution set without changing to another equivalent system.

Example 1

Solve the system $\begin{pmatrix} 2x - 3y + 5z = -5 \\ 2y - 3z = 4 \\ 4z = -8 \end{pmatrix}$.

$$(1)$$
$$(2)$$
$$(3)$$

Solution

From equation (3) we can find the value of z.

$$4z = -8$$

$$z = -2$$

Now we can substitute -2 for z in equation (2).

$$2y - 3z = 4$$

$$2y - 3(-2) = 4$$

$$2y + 6 = 4$$

$$2y = -2$$

$$y = -1$$

Finally, we can substitute -2 for z and -1 for y in equation (1).

$$2x - 3y + 5z = -5$$

$$2x - 3(-1) + 5(-2) = -5$$

$$2x + 3 - 10 = -5$$

$$2x - 7 = -5$$

$$2x = 2$$

$$x = 1$$

The solution set is $\{(1, -1, -2)\}$. ▲

Notice the format of the equations in the system of Example 1. The first equation contains all three variables, the second equation has only two variables, and the third equation has only one variable. This allowed us to solve the third equation and then use "back substitution" to find the values of the other variables. Let's consider another example where we have to make one replacement of an equivalent system.

Example 2

Solve the system
$$\begin{cases} 3x + 2y - 7z = -34 & (1) \\ y + 5z = 21 & (2) \\ 3y - 2z = -22 & (3) \end{cases}$$

Solution

Let's replace equation (3) with an equation we form by multiplying equation (2) by -3 and then adding that result to equation (3).

$$\begin{cases} 3x + 2y - 7z = -34 & (4) \\ y + 5z = 21 & (5) \\ -17z = -85 & (6) \end{cases}$$

From equation (6) we can find the value of z.

$$-17z = -85$$

$$z = 5$$

Now we can substitute 5 for z in equation (5).

$$y + 5z = 21$$

$$y + 5(5) = 21$$

$$y = -4$$

Finally, we can substitute 5 for z and -4 for y in equation (4).

$$3x + 2y - 7z = -34$$

$$3x + 2(-4) - 7(5) = -34$$

$$3x - 8 - 35 = -34$$

$$3x - 43 = -34$$

$$3x = 9$$

$$x = 3$$

The solution set is $\{(3, -4, 5)\}$. ▲

Now let's consider some examples where we have to make more than one replacement of equivalent systems.

Example 3

Solve the system $\begin{cases} x - y + 4z = -29 \\ 3x - 2y - z = -6 \\ 2x - 5y + 6z = -55 \end{cases}$.

    (1)
    (2)
    (3)

Solution

Let's replace equation (2) with an equation we form by multiplying equation (1) by -3 and then adding that result to equation (2). Let's also replace equation (3) with an equation we form by multiplying equation (1) by -2 and then adding that result to equation (3).

$$\begin{cases} x - y + 4z = -29 \\ y - 13z = 81 \\ -3y - 2z = 3 \end{cases}$$

    (4)
    (5)
    (6)

Now let's replace equation (6) with an equation we form by multiplying equation (5) by 3 and then adding that result to equation (6).

$$\begin{cases} x - y + 4z = -29 \\ y - 13z = 81 \\ -41z = 246 \end{cases}$$

    (7)
    (8)
    (9)

From equation (9) we can determine the value of z.

$$-41z = 246$$

$$z = -6$$

Now we can substitute -6 for z in equation (8).

$$y - 13z = 81$$

$$y - 13(-6) = 81$$

$$y + 78 = 81$$

$$y = 3$$

Finally, we can substitute -6 for z and 3 for y in equation (7).

$$x - y + 4z = -29$$

$$x - 3 + 4(-6) = -29$$

$$x - 3 - 24 = -29$$

$$x - 27 = -29$$

$$x = -2$$

The solution set is $\{(-2, 3, -6)\}$.

▲

Example 4

Solve the system $\begin{pmatrix} 3x - 4y + z = 14 \\ 5x + 3y - 2z = 27 \\ 7x - 9y + 4z = 31 \end{pmatrix}$.

(1)
(2)
(3)

Solution

Glancing at the coefficients in the system it appears that eliminating the z-terms from equations (2) and (3) would be easy to do. Let's replace equation (2) with an equation we form by multiplying equation (1) by 2 and then adding that result to equation (2). Also let's replace equation (3) with an equation we form by multiplying equation (1) by -4 and then adding that result to equation (3).

$$\begin{pmatrix} 3x - 4y + z = 14 \\ 11x - 5y = 55 \\ -5x + 7y = -25 \end{pmatrix}$$

(4)
(5)
(6)

Now let's eliminate the y-terms from equations (5) and (6). First, let's multiply equation (6) by 5.

$$\begin{pmatrix} 3x - 4y + z = 14 \\ 11x - 5y = 55 \\ -25x + 35y = -125 \end{pmatrix}$$

(7)
(8)
(9)

Now we can replace equation (9) with an equation we form by multiplying equation (8) by 7 and then adding that result to equation 9.

$$\begin{pmatrix} 3x - 4y + z = 14 \\ 11x - 5y = 55 \\ 52x = 260 \end{pmatrix}$$

(10)
(11)
(12)

From equation (12) we can determine the value of x.

$$52x = 260$$

$$x = 5$$

Now we can substitute 5 for x in equation (11).

$$11x - 5y = 55$$

$$11(5) - 5y = 55$$

$$-5y = 0$$

$$y = 0$$

Finally, we can substitute 5 for x and 0 for y in equation (10).

$$3x - 4y + z = 14$$

$$3(5) - 4(0) + z = 14$$

$$15 - 0 + z = 14$$

$$z = -1$$

The solution set is $\{(5, 0, -1)\}$.   ▲

Example 5

Solve the system $\begin{pmatrix} x - 2y + 3z = 1 \\ 3x - 5y - 2z = 4 \\ 2x - 4y + 6z = 7 \end{pmatrix}$.

(1)
(2)
(3)

Solution

A glance at the coefficients indicates that it should be easy to eliminate the x-terms from equations (2) and (3). We can replace equation (2) with an equation we form by multiplying equation (1) by -3 and then adding that result to equation (2). Likewise, we can replace equation (3) with an equation we form by multiplying equation (1) by -2 and then adding that result to equation (3).

$$\begin{pmatrix} x - 2y + 3z = 1 \\ y - 11z = 1 \\ 0 + 0 + 0 = 5 \end{pmatrix}$$

(4)
(5)
(6)

The false statement, $0 = 5$, indicates that the system is inconsistent and therefore the solution set is $\varnothing$. (If you were to graph this system, equations (1) and (3) would produce parallel planes, which is the situation depicted in Figure 11.7(c).)   ▲

Example 6

Solve the system $\begin{pmatrix} 2x - y + 4z = 1 \\ 3x + 2y - z = 5 \\ 5x - 6y + 17z = -1 \end{pmatrix}$.

(1)
(2)
(3)

Solution

A glance at the coefficients indicates that it is easy to eliminate the y-terms from equations (2) and (3). We can replace equation (2) with an equation we form by multiplying equation (1) by 2 and then adding that result to equation (2). Likewise, we can replace equation 3 with an equation we form by multiplying equation (1) by -6 and then adding that result to equation (3).

$\begin{pmatrix} 2x - y + 4z = 1 \\ 7x \quad + 7z = 7 \\ -7x \quad - 7z = -7 \end{pmatrix}$.

(4)
(5)
(6)

Now let's replace equation (6) with an equation we form by multiplying equation (5) by 1 and then adding that result to equation (6).

$\begin{pmatrix} 2x - y + 4z = 1 \\ 7x \quad + 7z = 7 \\ 0 + 0 = 0 \end{pmatrix}$.

(7)
(8)
(9)

The true numerical statement $0 + 0 = 0$, indicates that the system has *infinitely many solutions*. (The graph of this system is shown in Figure 11.5(a).) ▲

REMARK It can be shown that the solutions for the system in Example 6 are of the form $(t, 3 - 2t, 1 - t)$, where t is any real number. For example, if we let $t = 2$, then we get the ordered triple $(2, -1, -1)$ and this triple will satisfy all three of the original equations. For our purposes in this text we shall simply indicate that such a system has *infinitely many solutions*. △

▼ Problem Set 11.3

Solve each of the following systems. If the solution set is $\varnothing$ or if it contains infinitely many solutions, then so indicate.

1. $\begin{pmatrix} x + 2y - 3z = 2 \\ 3y - z = 13 \\ 3y + 5z = 25 \end{pmatrix}$

2. $\begin{pmatrix} 2x + 3y - 4z = -10 \\ 2y + 3z = 16 \\ 2y - 5z = -16 \end{pmatrix}$

3. $\begin{pmatrix} 3x + 2y - 2z = 14 \\ x \quad - 6z = 16 \\ 2x \quad + 5z = -2 \end{pmatrix}$

4. $\begin{pmatrix} 3x + 2y - z = -11 \\ 2x - 3y = -1 \\ 4x + 5y = -13 \end{pmatrix}$

5. $\begin{pmatrix} 2x - y + z = 0 \\ 3x - 2y + 4z = 11 \\ 5x + y - 6z = -32 \end{pmatrix}$

6. $\begin{pmatrix} x - 2y + 3z = 7 \\ 2x + y + 5z = 17 \\ 3x - 4y - 2z = 1 \end{pmatrix}$

7. $\begin{pmatrix} 4x - y + z = 5 \\ 3x + y + 2z = 4 \\ x - 2y - z = 1 \end{pmatrix}$

8. $\begin{pmatrix} 2x - y + 3z = -14 \\ 4x + 2y - z = 12 \\ 6x - 3y + 4z = -22 \end{pmatrix}$

9. $\begin{pmatrix} x - y + 2z = 4 \\ 2x - 2y + 4z = 7 \\ 3x - 3y + 6z = 1 \end{pmatrix}$

10. $\begin{pmatrix} x + y - z = 2 \\ 3x - 4y + 2z = 5 \\ 2x + 2y - 2z = 7 \end{pmatrix}$

11. $\begin{pmatrix} x - 2y + z = -4 \\ 2x + 4y - 3z = -1 \\ -3x - 6y + 7z = 4 \end{pmatrix}$

12. $\begin{pmatrix} 2x - y + 3z = 1 \\ 4x + 7y - z = 7 \\ x + 4y - 2z = 3 \end{pmatrix}$

13. $\begin{pmatrix} 3x - 2y + 4z = 6 \\ 9x + 4y - z = 0 \\ 6x - 8y - 3z = 3 \end{pmatrix}$

14. $\begin{pmatrix} 2x - y + 3z = 0 \\ 3x + 2y - 4z = 0 \\ 5x - 3y + 2z = 0 \end{pmatrix}$

15. $\begin{pmatrix} 3x - y + 4z = 9 \\ 3x + 2y - 8z = -12 \\ 9x + 5y - 12z = -23 \end{pmatrix}$

16. $\begin{pmatrix} 5x - 3y + z = 1 \\ 2x - 5y = -2 \\ 3x - 2y - 4z = -27 \end{pmatrix}$

17. $\begin{pmatrix} 4x - y + 3z = -12 \\ 2x + 3y - z = 8 \\ 6x + y + 2z = -8 \end{pmatrix}$

18. $\begin{pmatrix} x + 3y - 2z = 19 \\ 3x - y - z = 7 \\ -2x + 5y + z = 2 \end{pmatrix}$

19. $\begin{pmatrix} x + y + z = 1 \\ 2x - 3y + 6z = 1 \\ -x + y + z = 0 \end{pmatrix}$

20. $\begin{pmatrix} 3x + 2y - 2z = -2 \\ x - 3y + 4z = -13 \\ -2x + 5y + 6z = 29 \end{pmatrix}$

Solve each of the following problems by setting up and solving a system of three linear equations in three variables.

21. The sum of the digits of a three-digit number is 14. The number is 14 larger than twenty times the tens digit. The sum of the tens digit and the units digit is 12 larger than the hundreds digit. Find the number.

22. The sum of the digits of a three-digit number is 13. The sum of the hundreds digit and the tens digit is one less than the units digit. The sum of three times the hundreds digit and four times the units digit is 26 more than twice the tens digit. Find the number.

23. Two bottles of catsup, 2 jars of peanut butter, and 1 jar of pickles cost $4.20. Three bottles of catsup, 4 jars of peanut butter, and 2 jars of pickles cost $7.70. Four bottles of catsup, 3 jars of peanut butter, and 5 jars of pickles cost $9.80. Find the cost per bottle of catsup and per jar for peanut butter and pickles.

24. Five pounds of potatoes, 1 pound of onions, and 2 pounds of apples cost $1.26. Two pounds of potatoes, 3 pounds of onions, and 4 pounds of apples cost $1.88. Three pounds of potatoes, 4 pounds of onions, and 1 pound of apples cost $1.24. Find the price per pound for each item.

25. The sum of three numbers is 20. The sum of the first and third numbers is 2 more than twice the second number. The third number minus the first yields three times the second number. Find the numbers.

26. The sum of three numbers is 40. The third number is 10 less than the sum of the first two numbers. The second number is one larger than the first. Find the numbers.

27. The sum of the measures of the angles of a triangle is 180°. The largest angle is twice the smallest angle. The sum of the smallest and the largest angle is twice the other angle. Find the measure of each angle.

28. A box contains $2 in nickels, dimes, and quarters. There are 19 coins in all with twice as many nickels as dimes. How many coins of each kind are there?

29. Part of $3000 is invested at 12%, another part at 13%, and the remainder at 14%. The total yearly income from the three investments is $400. The sum of the amounts invested at 12% and 13% equals the amount invested at 14%. Determine how much is invested at each rate.

30. The perimeter of a triangle is 45 centimeters. The longest side is 4 centimeters less than twice the shortest side. The sum of the lengths of the shortest and longest sides is 7 centimeters less than three times the length of the remaining side. Find the lengths of all three sides of the triangle.

THOUGHTS INTO WORDS

31. Give a step-by-step description of how to solve the system

$$\left(\begin{array}{rcr} x - 2y + 3z &=& -23 \\ 5y - 2z &=& 32 \\ 4z &=& -24 \end{array} \right).$$

32. Describe how you would solve the system

$$\left(\begin{array}{rcr} x \quad\quad - 3z &=& 4 \\ 3x - 2y + 7z &=& -1 \\ 2x \quad\quad + z &=& 9 \end{array} \right).$$

11.4 Systems Involving Nonlinear Equations and Systems of Inequalities

Thus far in this chapter we have solved systems of linear equations. In this section, we will consider some systems in which at least one of the equations is nonlinear; we will also consider some systems of **linear inequalities**. Let's begin by considering a system of one linear equation and one second-degree equation.

Example 1

Solve the system $\left(\begin{array}{l} x^2 + y^2 = 17 \\ x + y = 5 \end{array} \right)$.

Solution

First, we will graph the system so that we can predict approximate solutions. From our previous graphing experiences in Chapters 7 and 8, we should recognize $x^2 + y^2 = 17$ as a circle and $x + y = 5$ as a straight line (Figure 11.8). The graph indicates that the solutions for this system should be two ordered pairs with positive components (the points of intersection occur in the first quadrant). In fact, we can guess that these solutions are $(1, 4)$ and $(4, 1)$ and then verify our guess by checking these solutions in the given equations.

FIGURE 11.8

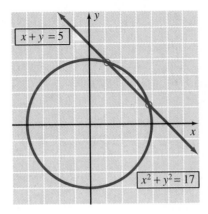

Let's also solve the system analytically using the substitution method as follows. First we change the form of $x + y = 5$ to $y = 5 - x$ and substitute $5 - x$ for y in the first equation.

$$x^2 + y^2 = 17$$
$$x^2 + (5 - x)^2 = 17$$
$$x^2 + 25 - 10x + x^2 = 17$$
$$2x^2 - 10x + 8 = 0$$
$$x^2 - 5x + 4 = 0$$
$$(x - 4)(x - 1) = 0$$
$$x - 4 = 0 \quad \text{or} \quad x - 1 = 0$$
$$x = 4 \quad \text{or} \quad x = 1$$

Substitute 4 for x and then 1 for x in the second equation of the system to produce the following.

$$x + y = 5 \qquad x + y = 5$$
$$4 + y = 5 \qquad 1 + y = 5$$
$$y = 1 \qquad y = 4$$

Therefore, the solution set is $\{(1, 4), (4, 1)\}$. ▲

Example 2

Solve the system $\begin{pmatrix} y = -x^2 + 1 \\ y = x^2 - 2 \end{pmatrix}$.

Solution

Again, let's get an idea of approximate solutions by graphing the system. Both equations produce parabolas, as indicated in Figure 11.9. From the graph we can predict two nonintegral ordered-pair solutions, one in the third quadrant and the other in the fourth quadrant.

FIGURE 11.9

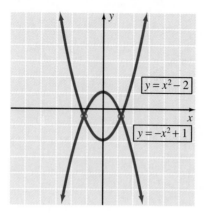

Substitute $-x^2 + 1$ for y in the second equation to obtain

$$y = x^2 - 2$$

$$-x^2 + 1 = x^2 - 2$$

$$3 = 2x^2$$

$$\frac{3}{2} = x^2$$

$$\pm\sqrt{\frac{3}{2}} = x$$

$$\pm\frac{\sqrt{6}}{2} = x. \qquad \sqrt{\frac{3}{2}} = \frac{\sqrt{3}}{\sqrt{2}} = \frac{\sqrt{3}}{\sqrt{2}}\frac{\sqrt{2}}{\sqrt{2}} = \frac{\sqrt{6}}{2}$$

Substitute $\dfrac{\sqrt{6}}{2}$ for x in the second equation to yield

$$y = x^2 - 2$$

$$y = \left(\frac{\sqrt{6}}{2}\right)^2 - 2$$

$$= \frac{6}{4} - 2 = -\frac{1}{2}.$$

Substitute $-\dfrac{\sqrt{6}}{2}$ for x in the second equation to yield

$$y = x^2 - 2$$

$$y = \left(-\frac{\sqrt{6}}{2}\right)^2 - 2$$

$$= \frac{6}{4} - 2 = -\frac{1}{2}.$$

The solution set is $\left\{\left(-\dfrac{\sqrt{6}}{2}, -\dfrac{1}{2}\right), \left(\dfrac{\sqrt{6}}{2}, -\dfrac{1}{2}\right)\right\}$. (You should check both of these

solutions in the original equations!) ▲

Example 3

Solve the system $\left(\begin{array}{r} x^2 + y^2 = 16 \\ -x^2 + y^2 = 4 \end{array}\right)$.

Solution

Graphing the system produces Figure 11.10, which indicates that there are four solutions to the system.

FIGURE 11.10

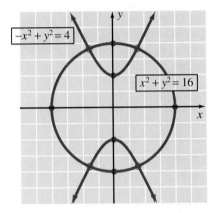

The elimination method works nicely. We can simply add the two equations, which eliminates the xs.

$$x^2 + y^2 = 16$$
$$\underline{-x^2 + y^2 = 4}$$
$$2y^2 = 20$$
$$y^2 = 10$$
$$y = \pm\sqrt{10}$$

Substitute $\sqrt{10}$ for y in the first equation to yield

$$x^2 + y^2 = 16$$
$$x^2 + (\sqrt{10})^2 = 16$$
$$x^2 + 10 = 16$$
$$x^2 = 6$$
$$x = \pm\sqrt{6}.$$

Thus, $(\sqrt{6}, \sqrt{10})$ and $(-\sqrt{6}, \sqrt{10})$ are solutions. Substitute $-\sqrt{10}$ for y in the first equation to yield

$$x^2 + y^2 = 16$$
$$x^2 + (-\sqrt{10})^2 = 16$$
$$x^2 + 10 = 16$$
$$x^2 = 6$$
$$x = \pm\sqrt{6}.$$

Thus, $(\sqrt{6}, -\sqrt{10})$ and $(-\sqrt{6}, -\sqrt{10})$ are solutions. The solution set is $\{(-\sqrt{6}, \sqrt{10}), (-\sqrt{6}, -\sqrt{10}), (\sqrt{6}, \sqrt{10}), (6, -\sqrt{10})\}$. ▲

REMARK   The graphing of systems has been given additional emphasis in this section. Perhaps you will need to return to Chapters 7 and 8 to review the various graphing techniques covered there. Be sure to reread Section 7.2; the ideas from that section will be used in a moment. △

Systems of Linear Inequalities

Finding solution sets for systems of **linear inequalities** relies heavily on the graphing approach. The solution set of a system of linear inequalities such as

$$\begin{pmatrix} x + y > 2 \\ x - y < 2 \end{pmatrix}$$

is the intersection of the solution sets of the individual inequalities. Figure 11.11(a) shows the solution set for $x + y > 2$, and Figure 11.11(b) shows the solution set for $x - y < 2$. Then the shaded region in Figure 11.11(c) represents the intersection of the two solution sets and therefore is the graph of the system. Remember that dashed lines are used to indicate that points on the lines are not included in the solution set.

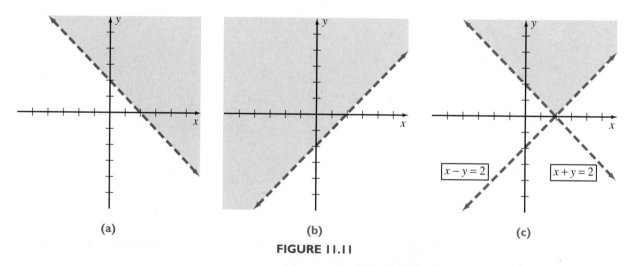

(a)   (b)   (c)

FIGURE 11.11

In the next examples only the final solution set for the systems is indicated.

Example 4

Solve the following system by graphing: $\begin{pmatrix} 2x - y \geq 4 \\ x + 2y < 2 \end{pmatrix}$.

Solution

The graph of $2x - y \geq 4$ consists of all points *on or below* the line $2x - y = 4$. The graph of $x + 2y < 2$ consists of all points *below* the line $x + 2y = 2$. The graph of the system is indicated by the shaded region in Figure 11.12. Notice that all points in the shaded region are on or below the line $2x - y = 4$ <u>and</u> below the line $x + 2y = 2$.

FIGURE 11.12

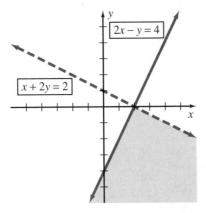

Example 5

Solve the following system by graphing: $\begin{pmatrix} x \leq 2 \\ y \geq -1 \end{pmatrix}$.

Solution

Remember that even though each inequality contains only one variable, we are working in a rectangular coordinate system that involves ordered pairs. That is to say, the system could be written as

$$\begin{pmatrix} x + 0(y) \leq 2 \\ 0(x) + y \geq -1 \end{pmatrix}.$$

The graph of this system is the shaded region in Figure 11.13. Notice that all points in the shaded region are on or to the left of the line $x = 2$ <u>and</u> on or above the line $y = -1$.

FIGURE 11.13

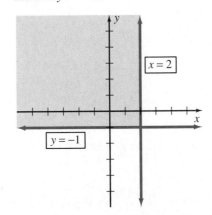

For the final example of this section, let's consider a system that contains four inequalities.

Example 6

Solve the following system by graphing: $\begin{pmatrix} x \geq 0 \\ y \geq 0 \\ 2x + 3y \leq 12 \\ 3x + y \leq 6 \end{pmatrix}$.

Solution

The solution set for the system is the intersection of the solution sets of the four inequalities. The shaded region in Figure 11.14 indicates the solution set for the system. Notice that all points in the shaded region are on or to the right of the y-axis, on or above the x-axis, on or below the line $2x + 3y = 12$, and on or below the line $3x + y = 6$.

FIGURE 11.14

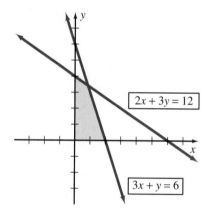

$2x + 3y = 12$

$3x + y = 6$

Problem Set 11.4

For each of the following systems, (a) graph the system so that approximate solutions can be predicted and (b) solve the system by the substitution or elimination method.

1. $\begin{pmatrix} y = (x + 2)^2 \\ y = -2x - 4 \end{pmatrix}$

2. $\begin{pmatrix} y = x^2 \\ y = x + 2 \end{pmatrix}$

3. $\begin{pmatrix} x^2 + y^2 = 13 \\ 3x + 2y = 0 \end{pmatrix}$

4. $\begin{pmatrix} x^2 + y^2 = 26 \\ x + y = 6 \end{pmatrix}$

5. $\begin{pmatrix} x + y = -8 \\ x^2 - y^2 = 16 \end{pmatrix}$

6. $\begin{pmatrix} x - y = 2 \\ x^2 - y^2 = 16 \end{pmatrix}$

7. $\begin{pmatrix} y = x^2 + 6x + 7 \\ 2x + y = -5 \end{pmatrix}$

8. $\begin{pmatrix} y = x^2 - 4x + 5 \\ -x + y = 1 \end{pmatrix}$

9. $\begin{pmatrix} xy = 4 \\ y = x \end{pmatrix}$

10. $\begin{pmatrix} y = x^2 + 2 \\ y = 2x^2 + 1 \end{pmatrix}$

11. $\begin{pmatrix} x^2 + 2y^2 = 8 \\ x^2 - y^2 = 1 \end{pmatrix}$

12. $\begin{pmatrix} x^2 - y^2 = 4 \\ x^2 + y^2 = 4 \end{pmatrix}$

13. $\begin{pmatrix} y = x^2 \\ y = x^2 - 4x + 4 \end{pmatrix}$

14. $\begin{pmatrix} y = -x^2 + 3 \\ y = x^2 + 1 \end{pmatrix}$

15. $\begin{pmatrix} y = x^2 + 2x - 1 \\ y = x^2 + 4x - 5 \end{pmatrix}$

16. $\begin{pmatrix} 2x^2 + y^2 = 8 \\ x^2 + y^2 = 4 \end{pmatrix}$

Indicate the solution set for each of the following systems of inequalities by graphing the system and shading the appropriate region.

17. $\begin{pmatrix} x - 3y < 6 \\ x + 2y \geq 4 \end{pmatrix}$

18. $\begin{pmatrix} 2x - y \leq 4 \\ 2x + y > 4 \end{pmatrix}$

19. $\begin{pmatrix} x + y < 4 \\ x - y > 2 \end{pmatrix}$

20. $\begin{pmatrix} x + y > 1 \\ x - y < 1 \end{pmatrix}$

21. $\begin{pmatrix} y > x \\ y > 2 \end{pmatrix}$

22. $\begin{pmatrix} 2x + y > 6 \\ 2x + y < 2 \end{pmatrix}$

23. $\begin{pmatrix} x \geq -1 \\ y < 4 \end{pmatrix}$

24. $\begin{pmatrix} x < 3 \\ y > 2 \end{pmatrix}$

25. $\begin{pmatrix} 2x - y < 4 \\ 2x - y > 0 \end{pmatrix}$

26. $\begin{pmatrix} x + y > 4 \\ x + y > 6 \end{pmatrix}$

27. $\begin{pmatrix} 3x - 2y < 6 \\ 2x - 3y < 6 \end{pmatrix}$

28. $\begin{pmatrix} 2x + 5y > 10 \\ 5x + 2y > 10 \end{pmatrix}$

29. $\begin{pmatrix} y < x + 1 \\ y \geq x \end{pmatrix}$

30. $\begin{pmatrix} y > x - 3 \\ y < x \end{pmatrix}$

31. $\begin{pmatrix} 3x - 4y \geq 0 \\ 2x + 3y \leq 0 \end{pmatrix}$

32. $\begin{pmatrix} 3x + 2y \leq 6 \\ 2x - 3y \geq 6 \end{pmatrix}$

33. $\begin{pmatrix} x \geq 0 \\ y \geq 0 \\ x - y \leq 5 \\ 4x + 7y \leq 28 \end{pmatrix}$

34. $\begin{pmatrix} x \geq 0 \\ y \geq 0 \\ x + y \leq 4 \\ 2x + y \leq 6 \end{pmatrix}$

35. $\begin{pmatrix} x \geq 0 \\ y \geq 0 \\ 3x + 5y \geq 15 \\ 5x + 3y \geq 15 \end{pmatrix}$

36. $\begin{pmatrix} x \geq 0 \\ y \geq 0 \\ 2x + y \leq 4 \\ 2x - 3y \leq 6 \end{pmatrix}$

For Problems 37–40, solve each problem by setting up and solving a system of equations.

37. The sum of the squares of two numbers is 34. The difference of the squares of the same two numbers is 16. Find the numbers.

38. The sum of the squares of two numbers is 13. One number is 1 larger than the other number. Find the numbers.

39. The area of a rectangular region is 54 square meters, and its perimeter is 30 meters. Find the length and the width of the rectangle.

40. A number is 1 larger than the square of another number. The sum of the two numbers is 7. Find the numbers.

THOUGHTS INTO WORDS

41. What happens if you try to graph the system
$\begin{pmatrix} x^2 + 4y^2 = & 16 \\ 2x^2 + 5y^2 = & -12 \end{pmatrix}$?

42. How do you know by inspection without graphing that the solution set of the system $\begin{pmatrix} 3x - 2y > 5 \\ 3x - 2y < 2 \end{pmatrix}$ is the null set?

Further Investigations

43. Figure 11.15 shows a graph of the system
$\begin{pmatrix} y = x^2 \\ y = -1 \end{pmatrix}$.

The graph indicates that the system has no solutions, but we must realize that we are graphing in the *real number plane*. Thus, the fact that the parabola and the line do not intersect simply indicates that the system has no *real number solutions*.

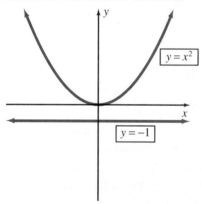

FIGURE 11.15

Using the substitution method, we can find some complex solutions as follows. Substitute x^2 for y in the second equation.

$$y = -1$$

becomes

$$x^2 = -1$$
$$x = \pm\sqrt{-1} = \pm i.$$

Substitute $\pm i$ for x in the first equation to yield the following.

$$y = x^2 \qquad y = x^2$$
$$y = i^2 \qquad y = (-i)^2$$
$$y = -1 \qquad y = i^2$$
$$\qquad\qquad y = -1$$

The solution set is $\{(i, -1), (-i, -1)\}$.

For each of the following systems, (a) graph the system to show that there are no real number solutions and (b) solve the system by substitution or elimination to find the complex solutions.

(a) $\begin{pmatrix} y = x^2 + 1 \\ y = -3 \end{pmatrix}$

(b) $\begin{pmatrix} y = -x^2 + 1 \\ y = 3 \end{pmatrix}$

(c) $\begin{pmatrix} y = x^2 \\ x - y = 4 \end{pmatrix}$

(d) $\begin{pmatrix} y = & x^2 + 1 \\ y = -x^2 \end{pmatrix}$

(e) $\begin{pmatrix} x^2 + y^2 = 1 \\ x + y = 2 \end{pmatrix}$

(f) $\begin{pmatrix} x^2 + y^2 = 2 \\ x^2 - y^2 = 6 \end{pmatrix}$

Graphics Calculator Activities

44. Use a graphics calculator to graph the systems in Problems 1–16 and check the reasonableness of your answers to those problems.

45. Use a graphics calculator to graph each of the systems in Problem 43 to demonstrate that they have no real number solutions.

46. Graph the system $\begin{pmatrix} y = x^2 + 2 \\ 6x - 4y = -5 \end{pmatrix}$ and use the trace and zoom features of your calculator to clearly demonstrate that this system has no real number solutions.

SUMMARY

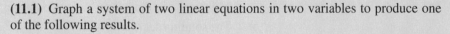

(11.1) Graph a system of two linear equations in two variables to produce one of the following results.

1. The graphs of the two equations are two intersecting lines, which indicates that there is **one unique solution** of the system. Such a system is called a **consistent system**.

2. The graphs of the two equations are two parallel lines that indicate that there is **no solution** for the system. Such a system is called an **inconsistent system**.

3. The graphs of the two equations are the same line, which indicates **infinitely many solutions** for the system. The equations are called **dependent** equations.

We can describe the **substitution method** of solving a system of equations as follows.

STEP 1 Solve one of the equations for one variable in terms of the other variable if neither equation is in such a form. (If possible make a choice that will avoid fractions.)

STEP 2 Substitute the expression you obtained in step 1 into the other equation to produce an equation with one variable.

STEP 3 Solve the equation you obtained in step 2.

STEP 4 Use the solution you obtained in step 3, along with the expression you obtained in step 1, to determine the solution of the system.

(11.2) The **elimination-by-addition method** involves the replacement of a system of equations with equivalent systems until a system is obtained whereby the solutions can be easily determined. The following operations or transformations can be performed on a system to produce an equivalent system.

1. Any two equations of the system can be interchanged.

2. Both sides of any equation of the system can be multiplied by any nonzero real number.

3. Any equation of the system can be replaced by the *sum* of that equation and a nonzero multiple of another equation.

(11.3) Solving a system of three linear equations in three variables produces one of the following results.

1. There is **one ordered triple** that satisfies all three equations.

2. There are **infinitely many ordered triples** in the solution set, all of which are coordinates of points on a line common to the planes.

3. There are **infinitely many ordered triples** in the solution set, all of which are coordinates on a common plane.

4. The solution set is **empty; it is** $\varnothing$

(11.4) The **substitution** and **elimination methods** can also be used to solve systems involving nonlinear equations.

The solution set of a system of **linear inequalities** is the intersection of the solution sets of the individual inequalities.

Chapter 11 Review Problem Set

In Problems 1–6, solve each of the systems using either the elimination or the substitution method.

1. $\begin{pmatrix} 3x - 2y = -6 \\ 2x + 5y = 34 \end{pmatrix}$

2. $\begin{pmatrix} x + 4y = 25 \\ y = -3x - 2 \end{pmatrix}$

3. $\begin{pmatrix} x = \dfrac{y + 9}{2} \\ 2x + 3y = 5 \end{pmatrix}$

4. $\begin{pmatrix} 9x + 5y = 12 \\ 7x + 4y = 10 \end{pmatrix}$

5. $\begin{pmatrix} \dfrac{1}{2}x + \dfrac{1}{3}y = -4 \\ \dfrac{3}{4}x - \dfrac{2}{3}y = 22 \end{pmatrix}$

6. $\begin{pmatrix} -2x + 5y = 7 \\ y = \dfrac{6x + 9}{15} \end{pmatrix}$

For Problems 7–11, solve each of the systems.

7. $\begin{pmatrix} x + y - z = -2 \\ 2x - 3y + 4z = 17 \\ -3x + 2y + 5z = -7 \end{pmatrix}$

8. $\begin{pmatrix} x - 2y + z = 7 \\ 2x - 5y + 2z = 17 \\ -3x + 7y + 5z = -32 \end{pmatrix}$

9. $\begin{pmatrix} x + 3y - 2z = -7 \\ 4x + 13y - 7z = -21 \\ 5x + 16y - 8z = -23 \end{pmatrix}$

10. $\begin{pmatrix} -x + 2y + 3z = 22 \\ 2x - 5y - 7z = -51 \\ -3x + 5y + 11z = 71 \end{pmatrix}$

11. $\begin{pmatrix} x - 3y - 4z = -1 \\ 2x + y - 2z = 3 \\ 5x - 8y - 14z = 0 \end{pmatrix}$

For Problems 12–15, graph the system and then solve it using either the elimination or the substitution method.

12. $\begin{pmatrix} y = 2x^2 - 1 \\ 2x + y = 3 \end{pmatrix}$

13. $\begin{pmatrix} x^2 + y^2 = 7 \\ x^2 - y^2 = 1 \end{pmatrix}$

14. $\begin{pmatrix} xy = -4 \\ x + y = 0 \end{pmatrix}$

15. $\begin{pmatrix} y = x^2 + 4x + 7 \\ y = -x^2 - 4x + 1 \end{pmatrix}$

For Problems 16–20, indicate the solution set of the system by graphing the system and shading the appropriate region.

16. $\begin{pmatrix} 3x + y > 6 \\ x - 2y \leq 4 \end{pmatrix}$

17. $\begin{pmatrix} 3x - 4y \geq 0 \\ 2x + 3y \leq 0 \end{pmatrix}$

18. $\begin{pmatrix} 3x - 2y < 6 \\ 2x - 3y < 6 \end{pmatrix}$

19. $\begin{pmatrix} x - 4y < 4 \\ 2x + y \geq 2 \end{pmatrix}$

20. $\begin{pmatrix} x \geq 0 \\ y \geq 0 \\ x + 2y \leq 4 \\ 2x - y \leq 4 \end{pmatrix}$

For Problems 21–27, solve each problem by setting up and solving a system of equations.

21. The sum of the squares of two numbers is 13. If one number is 1 larger than the other number, find the numbers.

22. The sum of the squares of two numbers is 34. The difference of the squares of the same two numbers is 16. Find the numbers.

23. A number is 1 larger than the square of another number. The sum of the two numbers is 7. Find the numbers.

24. The area of a rectangular region is 54 square meters and its perimeter is 30 meters. Find the length and width of the rectangle.

25. At a local confectionery, 7 pounds of cashews and 5 pounds of Spanish peanuts cost $88, and 3 pounds of cashews and 2 pounds of Spanish peanuts cost $37. Find the price per pound for cashews and for Spanish peanuts.

26. We bought 2 six-packs of soda and 4 pounds of candy for $12. The next day we bought 3 six-packs of soda and 2 pounds of candy for $9. Find the price of a six-pack of soda and also the price of a pound of candy.

27. A mail order company charges a fixed fee for shipping merchandise that weighs 1 pound or less, plus an additional fee for each pound over 1 pound. If the shipping charge for 5 pounds is $2.40 and for 12 pounds is $3.10, find the fixed fee and the additional fee.

CHAPTER 11 TEST

For Problems 1–4, refer to the following systems of equations.

I. $\begin{pmatrix} 3x - 2y = 4 \\ 9x - 6y = 12 \end{pmatrix}$ **II.** $\begin{pmatrix} 5x - y = 4 \\ 3x + 7y = 9 \end{pmatrix}$ **III.** $\begin{pmatrix} 2x - y = 4 \\ 2x - y = -6 \end{pmatrix}$

1. For which system are the graphs parallel lines?
2. For which system are the equations dependent?
3. For which system is the solution set $\varnothing$?
4. Which system is consistent?
5. Solve the system $\begin{pmatrix} 3x - 2y = -14 \\ 7x + 2y = -6 \end{pmatrix}$.

6. Solve the system $\begin{pmatrix} y = -3x + 8 \\ 4x - 5y = 17 \end{pmatrix}$.

7. Find the value of x in the solution for the system

$$\begin{pmatrix} \dfrac{3}{4}x - \dfrac{1}{2}y = -21 \\[2mm] \dfrac{2}{3}x + \dfrac{1}{6}y = -4 \end{pmatrix}.$$

8. Find the value of y in the solution for the system

$$\begin{pmatrix} y = -5x - 14 \\ y = 7x + 10 \end{pmatrix}.$$

9. Find the value of x in the solution for the system

$$\begin{pmatrix} 5x + 2y = -1 \\ 3x - 4y = 3 \end{pmatrix}.$$

10. Find the value of y in the solution for the system

$$\begin{pmatrix} 4x - y = 7 \\ 3x + 2y = 2 \end{pmatrix}.$$

11. Find the value of x in the solution for the system

$$\begin{pmatrix} 2x - y + 3z = -8 \\ 3x + y - 2z = -3 \\ 5x + y + 7z = -16 \end{pmatrix}.$$

12. Find the value of y in the solution for the system

$$\begin{pmatrix} x - 3y + z = -13 \\ 3x + 5y - z = 17 \\ 5x - 2y + 2z = -13 \end{pmatrix}.$$

(continued on next page)

CHAPTER II TEST *(continued)*

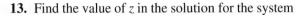

13. Find the value of z in the solution for the system

$$\begin{pmatrix} x + 2y - 3z = -7 \\ -x + y + 4z = 6 \\ x - 3y + 8z = 20 \end{pmatrix}.$$

14. Is $(1, -1, 4)$ a solution of the system

$$\begin{pmatrix} 2x - y + z = 7 \\ 3x - 2y + 2z = 13 \\ x - 4y + 5z = 17 \end{pmatrix}?$$

15. How many ordered pairs of real numbers are in the solution set for the system

$$\begin{pmatrix} y = 3x - 4 \\ 9x - 3y = 12 \end{pmatrix}?$$

16. How many ordered pairs of real numbers are in the solution set for the system

$$\begin{pmatrix} x^2 + 2y^2 = 8 \\ y = -x^2 + 1 \end{pmatrix}?$$

17. How many ordered pairs of real numbers are in the solution set for the system

$$\begin{pmatrix} y = x^2 - 4x \\ y = -2x^2 + 8x - 12 \end{pmatrix}?$$

18. How many ordered pairs of real numbers are in the solution set for the system

$$\begin{pmatrix} x^2 + y^2 = 16 \\ y^2 - x^2 = -1 \end{pmatrix}?$$

19. Find the values of x in the solutions for the system

$$\begin{pmatrix} y = x^2 \\ y = -x^2 + 8 \end{pmatrix}.$$

20. Find the values of y in the solutions for the system

$$\begin{pmatrix} x^2 + y^2 = 4 \\ x - 3y = 1 \end{pmatrix}.$$

21. The tens digit of a two-digit number is 1 more than twice the units digit. The number formed by reversing the digits is 27 smaller than the original number. Find the original number.

(continued on next page)

CHAPTER 11 TEST *(continued)*

22. One solution contains 30% alcohol and another solution is 70% alcohol. Some of each of the two solutions is mixed to produce 8 liters of a 40% solution. How many liters of the 70% solution should be used?

23. A box contains $7.25 in nickels, dimes, and quarters. There are 43 coins and the number of quarters is 1 more than three times the number of nickels. Find the number of quarters in the box.

24. Graph the solution set for the system

$$\begin{pmatrix} 2x - 3y < -6 \\ x + 2y > 4 \end{pmatrix}.$$

25. Graph the solution set for the system

$$\begin{pmatrix} y \geq 2x \\ x + y \leq 2 \end{pmatrix}.$$

C H A P T E R

Using Matrices and Determinants to Solve Linear Systems

In Chapter 11 we found that the techniques of elimination-by-addition and substitution worked effectively with two equations and two unknowns, but they started to become a bit cumbersome with three equations and three unknowns. In this chapter we will continue to work primarily with small systems, but we will consider some techniques that can be easily expanded to larger systems.

12.1   Matrix Approach to Solving Systems

A matrix is simply an array of numbers arranged in horizontal rows and vertical columns. For example, the matrix

$$2 \text{ rows} \longrightarrow \begin{bmatrix} 2 & 1 & -4 \\ 5 & -7 & 6 \end{bmatrix}$$

3 columns

has 2 rows and 3 columns, which we refer to as a 2×3 (read "two by three") matrix. Each number in a matrix is called an **element** of the matrix. Some additional examples of matrices (*matrices* is the plural of matrix) are as follows.

$$\begin{array}{cccc} 3 \times 2 & 2 \times 2 & 1 \times 2 & 4 \times 1 \\ \begin{bmatrix} 2 & 1 \\ 1 & 0 \\ \frac{1}{2} & 3 \end{bmatrix} & \begin{bmatrix} 17 & 16 \\ -14 & 12 \end{bmatrix} & [2 \quad 13] & \begin{bmatrix} 3 \\ -2 \\ 1 \\ 8 \end{bmatrix} \end{array}$$

In general, a matrix of m rows and n columns is called a matrix of **dimension** $m \times n$ or **order** $m \times n$.

With every system of linear equations we can associate a matrix that consists of the coefficients and constant terms. For example, with the system

$$\begin{pmatrix} a_1x + b_1y + c_1z = d_1 \\ a_2x + b_2y + c_2z = d_2 \\ a_3x + b_3y + c_3z = d_3 \end{pmatrix}$$

we can associate the matrix

$$\begin{bmatrix} a_1 & b_1 & c_1 & \vdots & d_1 \\ a_2 & b_2 & c_2 & \vdots & d_2 \\ a_3 & b_3 & c_3 & \vdots & d_3 \end{bmatrix}$$

which is commonly called the **augmented matrix** of the system. The dashed line separates the coefficients from the constant terms and reminds us that we are working with an augmented matrix.

Our previous work with systems of linear equations was based on the following properties.

1. Any two equations of a system may be interchanged. *Example*: Interchanging the two equations makes the system $\begin{pmatrix} 2x - 5y = 9 \\ x + 3y = 4 \end{pmatrix}$ equivalent to the system $\begin{pmatrix} x + 3y = 4 \\ 2x - 5y = 9 \end{pmatrix}$.

2. Any equation of the system may be multiplied by a nonzero constant. *Example*: Multiplying the top equation by -2 makes the system $\begin{pmatrix} x + 3y = 4 \\ 2x - 5y = 9 \end{pmatrix}$ equivalent to the system $\begin{pmatrix} -2x - 6y = -8 \\ 2x - 5y = 9 \end{pmatrix}$.

3. Any equation of the system can be replaced by adding a nonzero multiple of another equation to that equation. *Example*: Adding -2 times the first equation to the second equation makes the system $\begin{pmatrix} x + 3y = 4 \\ 2x - 5y = 9 \end{pmatrix}$ equivalent to the system $\begin{pmatrix} x + 3y = 4 \\ -11y = 1 \end{pmatrix}$.

Each of the properties geared to solving a system of equations produces a corresponding property of the augmented matrix of the system. For example, exchanging two equations of a system corresponds to exchanging two rows of the augmented matrix that represents the system. We usually refer to the properties of an augmented matrix as *elementary row operations* and we can state them as follows.

For any augmented matrix of a system of linear equations, the following **elementary row operations** will produce a matrix of an equivalent system.

1. Any two rows of the matrix can be interchanged.

2. Any row of the matrix can be multiplied by a nonzero real number.

3. Any row of the matrix can be replaced by adding a nonzero multiple of another row to that row.

Using the elementary row operations on an augmented matrix provides a basis for solving systems of linear equations. Study the following examples very carefully; keep in mind that the general scheme, called **Gaussian elimination**, is one of using the elementary row operations on a matrix to continue replacing a system of equations with equivalent systems until we obtain a system whereby we can easily determine the solutions.

Example I

Solve the system $\begin{pmatrix} x - 3y = -17 \\ 2x + 7y = 31 \end{pmatrix}$.

Solution

The augmented matrix of the system is

$$\begin{bmatrix} 1 & -3 & \vdots & -17 \\ 2 & 7 & \vdots & 31 \end{bmatrix}.$$

Let's multiply row 1 by -2 and then add that result to row 2 to produce a new row 2.

$$\begin{bmatrix} 1 & -3 & \vdots & -17 \\ 0 & 13 & \vdots & 65 \end{bmatrix}$$

The system represented by this matrix,

$$\begin{pmatrix} x - 3y = -17 \\ 13y = 65 \end{pmatrix},$$

is said to be in **triangular form**. We can easily solve the second equation for y and then substitute this value for y in the first equation in order to determine x.

$$13y = 65$$

$$y = 5$$

Substitute 5 for y in the first equation to produce

$$x - 3(5) = -17$$

$$x - 15 = -17$$

$$x = -2.$$

The solution set is $\{(-2, 5)\}$.　　　　　　　　　　　　　　　　▲

Example 2

Solve the system $\begin{pmatrix} 3x + 2y = 5 \\ 5x - y = -9 \end{pmatrix}$.

Solution

The augmented matrix of the system is

$$\begin{bmatrix} 3 & 2 & \vdots & 5 \\ 5 & -1 & \vdots & -9 \end{bmatrix}.$$

To get a 0 in the row 2-column 1 position, we multiply the first row by $-\dfrac{5}{3}$ and add this result to the second row to form a new second row.

$$\begin{bmatrix} 3 & 2 & \vdots & 5 \\ 0 & -\dfrac{13}{3} & \vdots & -\dfrac{52}{3} \end{bmatrix}$$

Now, from the second row we obtain

$$-\frac{13}{3}y = -\frac{52}{3}$$

$$y = 4.$$

Substitute 4 for y in the first equation to produce

$$3x + 2y = 5$$
$$3x + 2(4) = 5$$
$$3x = -3$$
$$x = -1.$$

The solution set is $\{(-1, 4)\}$. ▲

It should be evident from Examples 1 and 2 that the matrix approach does not provide us with much extra power for solving systems of two linear equations in two unknowns. However, as the systems become larger, the compactness of the matrix approach becomes more convenient.

Suppose that we had the following system to solve.

$$\begin{pmatrix} 5x - 3y - 2z = & -1 \\ 4y + 7z = & 3 \\ 4z = & -12 \end{pmatrix}$$

It would be easy to determine z from the last equation and then substitute that value in the second equation to determine y. The values for y and z could then be substituted in the first equation to determine x. In other words, the form of this system makes it convenient to solve. We say that the system is in **triangular form**; notice the location of zeros in its augmented matrix.

$$\begin{bmatrix} 5 & -3 & -2 & \vdots & -1 \\ 0 & 4 & 7 & \vdots & 3 \\ 0 & 0 & 4 & \vdots & -12 \end{bmatrix}$$

Example 3

Solve the system $\begin{pmatrix} x - 2y + z = & 16 \\ 2x - y - z = & 14 \\ 3x + 5y - 4z = & -10 \end{pmatrix}$.

Solution

The augmented matrix of the system is

$$\begin{bmatrix} 1 & -2 & 1 & \vdots & 16 \\ 2 & -1 & -1 & \vdots & 14 \\ 3 & 5 & -4 & \vdots & -10 \end{bmatrix}.$$

Let's multiply row 1 by -2 and add this result to row 2 to form a new row 2. Also, let's multiply row 1 by -3 and add this result to row 3 to produce a new row 3.

$$\begin{bmatrix} 1 & -2 & 1 & \vdots & 16 \\ 0 & 3 & -3 & \vdots & -18 \\ 0 & 11 & -7 & \vdots & -58 \end{bmatrix}$$

Now let's multiply row 2 by $\frac{1}{3}$.

$$\begin{bmatrix} 1 & -2 & 1 & | & 16 \\ 0 & 1 & -1 & | & -6 \\ 0 & 11 & -7 & | & -58 \end{bmatrix}$$

Now we can multiply row 2 by -11 and add to row 3 to produce a new row 3.

$$\begin{bmatrix} 1 & -2 & 1 & | & 16 \\ 0 & 1 & -1 & | & -6 \\ 0 & 0 & 4 & | & 8 \end{bmatrix}$$

The system represented by this matrix,

$$\begin{pmatrix} x - 2y + z = 16 \\ y - z = -6 \\ 4z = 8 \end{pmatrix},$$

is in triangular form. From the last equation we obtain

$$4z = 8$$
$$z = 2.$$

Substitute 2 for z in the second equation to produce

$$y - z = -6$$
$$y - 2 = -6$$
$$y = -4.$$

Finally, substituting 2 for z and -4 for y in the first equation produces

$$x - 2y + z = 16$$
$$x - 2(-4) + 2 = 16$$
$$x + 10 = 16$$
$$x = 6.$$

The solution set is $\{(6, -4, 2)\}$.

Example 4   Solve the system $\begin{pmatrix} 3x - 2y + 4z = -16 \\ 2x - y + 3z = -10 \\ 5x + 3y + 9z = -10 \end{pmatrix}$.

Solution

The augmented matrix of the system is

$$\begin{bmatrix} 3 & -2 & 4 & | & -16 \\ 2 & -1 & 3 & | & -10 \\ 5 & 3 & 9 & | & -10 \end{bmatrix}.$$

Let's begin by multiplying row 2 by -1 and then adding the result to row 1 to produce a new row 1.

$$\begin{bmatrix} 1 & -1 & 1 & | & -6 \\ 2 & -1 & 3 & | & -10 \\ 5 & 3 & 9 & | & -10 \end{bmatrix}$$

Notice that we now have a 1 at the top of the first column. This makes it easier to get zeros in the other two positions of the first column. Let's multiply row 1 by -2 and add the result to row 2 to produce a new row 2. Also, let's multiply row 1 by -5 and add the result to row 3 to produce a new row 3.

$$\begin{bmatrix} 1 & -1 & 1 & | & -6 \\ 0 & 1 & 1 & | & 2 \\ 0 & 8 & 4 & | & 20 \end{bmatrix}$$

Now we can multiply row 2 by -8 and add the result to row 3 to produce a new row 3.

$$\begin{bmatrix} 1 & -1 & 1 & | & -6 \\ 0 & 1 & 1 & | & 2 \\ 0 & 0 & -4 & | & 4 \end{bmatrix}$$

This matrix represents a system in triangular form. From the bottom row we obtain

$$-4z = 4$$
$$z = -1.$$

Substitute -1 for z in the equation represented by the second row to produce

$$y + z = 2$$
$$y - 1 = 2$$
$$y = 3.$$

Finally, substitute -1 for z and 3 for y in the equation represented by the top row to produce

$$x - y + z = -6$$
$$x - 3 - 1 = -6$$
$$x - 4 = -6$$
$$x = -2.$$

The solution set is $\{(-2, 3, -1)\}$. ▲

Problem Set 12.1

Use a matrix approach to solve each of the following systems.

1. $\begin{pmatrix} x + 3y = & 8 \\ 7x - 2y = & -13 \end{pmatrix}$

2. $\begin{pmatrix} x - 4y = & 12 \\ 2x + 5y = & -2 \end{pmatrix}$

3. $\begin{pmatrix} 5x + 2y = & -20 \\ x - 4y = & 18 \end{pmatrix}$

4. $\begin{pmatrix} 4x - 3y = & 33 \\ x + 5y = & -32 \end{pmatrix}$

5. $\begin{pmatrix} x - y = 6 \\ 3x + 2y = 7 \end{pmatrix}$

6. $\begin{pmatrix} x + 3y = 7 \\ 2x - 4y = 9 \end{pmatrix}$

7. $\begin{pmatrix} 3x + 7y = -45 \\ 5x - y = 1 \end{pmatrix}$

8. $\begin{pmatrix} 3x + 10y = -33 \\ 6x - 7y = 15 \end{pmatrix}$

9. $\begin{pmatrix} 3x + 2y = 17 \\ 11x - 4y = 85 \end{pmatrix}$

10. $\begin{pmatrix} 7x + y = -29 \\ 4x - 5y = -50 \end{pmatrix}$

11. $\begin{pmatrix} 2x + 9y = 98 \\ 6x - 5y = -26 \end{pmatrix}$

12. $\begin{pmatrix} 5x - 3y = -25 \\ 3x + 2y = 4 \end{pmatrix}$

13. $\begin{pmatrix} 9x - 2y = -6 \\ 4x + 5y = 15 \end{pmatrix}$

14. $\begin{pmatrix} 7x - 3y = 4 \\ 2x + y = 5 \end{pmatrix}$

15. $\begin{pmatrix} 2x - 9y + 4z = -27 \\ x - 5y - 2z = -11 \\ 5x + 2y - 3z = 19 \end{pmatrix}$

16. $\begin{pmatrix} x + 2y - 3z = -16 \\ 2x - y + 4z = 23 \\ 3x + 5y - 2z = -12 \end{pmatrix}$

17. $\begin{pmatrix} 2x + 3y - z = -3 \\ 5x - 2y - 4z = -1 \\ 3x - 8y + 3z = 17 \end{pmatrix}$

18. $\begin{pmatrix} 3x - y + 4z = 11 \\ -5x + 2y + z = 21 \\ 2x - 3y + 5z = 12 \end{pmatrix}$

19. $\begin{pmatrix} 2x + 5y - 2z = -8 \\ 3x - 11y + 4z = -5 \\ 4x + 3y - z = -13 \end{pmatrix}$

20. $\begin{pmatrix} 3x - 2y + 5z = -18 \\ 2x + 5y - 3z = 26 \\ 5x + y - 2z = 8 \end{pmatrix}$

21. $\begin{pmatrix} x + 5y - 2z = -5 \\ 3x - 2y - 3z = 24 \\ -2x - 9y + 5z = 3 \end{pmatrix}$

22. $\begin{pmatrix} 3x - y + z = 12 \\ 2x + 3y + 2z = 3 \\ 5x - 4y - z = 25 \end{pmatrix}$

23. $\begin{pmatrix} 2x - 3y - 4z = 1 \\ x - 2y - 5z = -1 \\ -3x + 5y + 2z = 2 \end{pmatrix}$

24. $\begin{pmatrix} x - 2y - 3z = 1 \\ -2x + 5y + 2z = 4 \\ 3x - 4y - 4z = 3 \end{pmatrix}$

25. $\begin{pmatrix} -x + y - 4z = -15 \\ -2x + 3y + 5z = 25 \\ 4x - 6y - z = -14 \end{pmatrix}$

26. $\begin{pmatrix} -x + 2y - z = 7 \\ 4x - y + z = 1 \\ -3x + 5y + 3z = 28 \end{pmatrix}$

27. $\begin{pmatrix} 4x - 3y + 2z = 17 \\ 2y + 5z = 24 \\ 3x - z = -9 \end{pmatrix}$

28. $\begin{pmatrix} 2x & + 3z = -7 \\ -y - 4z = 0 \\ 3x + 2y - 5z = 7 \end{pmatrix}$

29. $\begin{pmatrix} 4x - y - 2z = -7 \\ 5x + y + 3z = -20 \\ 2x - 3y - z = 9 \end{pmatrix}$

30. $\begin{pmatrix} 2x - y + 3z = 8 \\ 3x + 2y + z = -2 \\ 3x - y - 5z = -26 \end{pmatrix}$

33. $\begin{pmatrix} 3x_1 + x_2 - x_3 = 5 \\ x_1 - 2x_2 + 3x_3 = -5 \\ -4x_1 + 3x_2 + 5x_3 = -14 \end{pmatrix}$

34. $\begin{pmatrix} x_1 + 2x_2 - x_3 = 5 \\ 2x_1 - x_2 + 4x_3 = -1 \\ 3x_1 + 3x_2 - 2x_3 = 11 \end{pmatrix}$

35. $\begin{pmatrix} 5x_1 - x_2 + x_3 = 5 \\ 4x_1 + 2x_2 + x_3 = 7 \\ 3x_1 - 4x_2 - 2x_3 = -3 \end{pmatrix}$

36. $\begin{pmatrix} 2x_1 + 3x_2 + x_3 = 6 \\ 3x_1 - 4x_2 - x_3 = 10 \\ 5x_1 + 2x_2 - 3x_3 = 4 \end{pmatrix}$

THOUGHTS INTO WORDS

31. What is a matrix? What is an augmented matrix of a system of linear equations?

32. Describe how to use matrices to solve the system $\begin{pmatrix} x - 2y = 5 \\ 2x + 7y = 9 \end{pmatrix}$.

Further Investigations

We can also use subscripts to identify the different variables in an equation. For example, the equation $2x + 3y + 5z = 7$ can be written as $2x_1 + 3x_2 + 5x_3 = 7$. Solution sets are then expressed in terms of ordered triples of the form (x_1, x_2, x_3). Use a matrix approach to solve each of the following systems.

Graphics Calculator Activities

37. If your graphics calculator has the capability to manipulate matrices, this is a good time to become familiar with those operations. You may need to refer to your calculator manual for the specific instructions. To begin the familiarization process, load your calculator with the four augmented matrices in Examples 1–4 of this section. Then for each one, carry out the row operations as described in the text.

12.2 Reduced Echelon Form

In Example 3 of the previous section we took the augmented matrix of the system

$$\begin{pmatrix} x - 2y + z = 16 \\ 2x - y - z = 14 \\ 3x + 5y - 4z = -10 \end{pmatrix}$$

and changed it to the following triangular form

$$\begin{bmatrix} 1 & -2 & 1 & \vdots & 16 \\ 0 & 1 & -1 & \vdots & -6 \\ 0 & 0 & 4 & \vdots & 8 \end{bmatrix}.$$

From the equation represented by the bottom row of this matrix ($4z = 8$), we found the value for z, and then we used back-substitution to find the values of y and x. It is possible to continue working with that matrix until the solution set can be read directly from the matrix. Let's try this process.

Multiply row 2 by 2 and add the result to row 1 to form a new row 1. (We want a zero in the row 1–column 2 position.)

$$\begin{bmatrix} 1 & 0 & -1 & \vdots & 4 \\ 0 & 1 & -1 & \vdots & -6 \\ 0 & 0 & 4 & \vdots & 8 \end{bmatrix}.$$

Multiply row 3 by $\dfrac{1}{4}$.

$$\begin{bmatrix} 1 & 0 & -1 & \vdots & 4 \\ 0 & 1 & -1 & \vdots & -6 \\ 0 & 0 & 1 & \vdots & 2 \end{bmatrix}$$

Add row 3 to row 1 to form a new row 1. Also add row 3 to row 2 to form a new row 2. (We want zeros above the 1 in the third column.)

$$\begin{bmatrix} 1 & 0 & 0 & \vdots & 6 \\ 0 & 1 & 0 & \vdots & -4 \\ 0 & 0 & 1 & \vdots & 2 \end{bmatrix}$$

Now, reading directly from the matrix, we have $x = 6$, $y = -4$, and $z = 2$. In other words, the solution set is $\{(6, -4, 2)\}$. The matrix

$$\begin{bmatrix} 1 & 0 & 0 & \vdots & 6 \\ 0 & 1 & 0 & \vdots & -4 \\ 0 & 0 & 1 & \vdots & 2 \end{bmatrix}$$

is said to be in **reduced echelon form**. In general, a matrix is in reduced echelon form if the following conditions are satisfied.

1. The first (reading from left to right) nonzero entry of each row is 1.
2. The remaining entries in the *column* containing the leftmost 1 of a row are all zeros.
3. The leftmost 1 of any row is to the right of the leftmost 1 of the preceding rows.
4. Rows that contain only zeros are below rows that contain nonzero entries.

Each of the following matrices is in reduced echelon form.

$$\begin{bmatrix} 1 & 0 & | & 4 \\ 0 & 1 & | & -7 \end{bmatrix}, \qquad \begin{bmatrix} 1 & 0 & 0 & | & 9 \\ 0 & 1 & 0 & | & -10 \\ 0 & 0 & 1 & | & 14 \end{bmatrix},$$

$$\begin{bmatrix} 1 & 2 & | & -3 \\ 0 & 0 & | & 0 \end{bmatrix}, \qquad \begin{bmatrix} 1 & 0 & -2 & | & 5 \\ 0 & 1 & 4 & | & 7 \\ 0 & 0 & 0 & | & 0 \end{bmatrix}.$$

In contrast, the following matrices are not in reduced echelon form for the reason indicated below each matrix.

$$\begin{bmatrix} 1 & 0 & 0 & | & 11 \\ 0 & 3 & 0 & | & -1 \\ 0 & 0 & 1 & | & -2 \end{bmatrix}, \qquad \begin{bmatrix} 1 & 2 & -3 & | & 5 \\ 0 & 1 & 7 & | & 9 \\ 0 & 0 & 1 & | & -6 \end{bmatrix},$$
$$\text{Violates 1} \qquad\qquad\qquad \text{Violates 2}$$

$$\begin{bmatrix} 1 & 0 & 0 & | & 7 \\ 0 & 0 & 1 & | & -8 \\ 0 & 1 & 0 & | & 14 \end{bmatrix}, \qquad \begin{bmatrix} 1 & 0 & 0 & 0 & | & -1 \\ 0 & 0 & 0 & 0 & | & 0 \\ 0 & 0 & 1 & 0 & | & 7 \\ 0 & 0 & 0 & 0 & | & 0 \end{bmatrix}$$
$$\text{Violates 3} \qquad\qquad\qquad \text{Violates 4}$$

Once we have an augmented matrix in reduced echelon form, we can easily determine the solution set of the system. Furthermore, we can describe the procedure for changing a given augmented matrix to reduced echelon form in a very systematic way. For example, the augmented matrix of a system of three linear equations in three unknowns having a unique solution can be changed to reduced echelon form as follows.

Augmented matrix

$$\begin{bmatrix} * & * & * & * \\ * & * & * & * \\ * & * & * & * \end{bmatrix}$$

Get a 1 in the upper left-hand corner.

$$\longrightarrow \begin{bmatrix} 1 & * & * & * \\ * & * & * & * \\ * & * & * & * \end{bmatrix}$$

Get zeros in the first column beneath the 1.

$$\longrightarrow \begin{bmatrix} 1 & * & * & * \\ 0 & * & * & * \\ 0 & * & * & * \end{bmatrix}$$

Get a 1 in the row 2-column 1 position.

$$\longrightarrow \begin{bmatrix} 1 & * & * & * \\ 0 & 1 & * & * \\ 0 & * & * & * \end{bmatrix}$$

Get zeros above and below
the I in column 2.

$$\begin{bmatrix} 1 & 0 & * & * \\ 0 & 1 & * & * \\ 0 & 0 & * & * \end{bmatrix}$$

Get a I in the row
3-column I position.

$$\begin{bmatrix} 1 & 0 & * & * \\ 0 & 1 & * & * \\ 0 & 0 & 1 & * \end{bmatrix}$$

Get zeros above the I
in column 3.

$$\begin{bmatrix} 1 & 0 & 0 & * \\ 0 & 1 & 0 & * \\ 0 & 0 & 1 & * \end{bmatrix}.$$

We can identify inconsistent and dependent systems during the process of changing a matrix to reduced echelon form. We will look at some examples of such cases in a moment, but first let's consider some examples of systems that have a unique solution.

Example I

Solve the system $\left(\begin{array}{l} 5x + 7y = -13 \\ x - 3y = -7 \end{array} \right)$.

Solution

The augmented matrix

$$\begin{bmatrix} 5 & 7 & \vdots & -13 \\ 1 & -3 & \vdots & -7 \end{bmatrix}$$

does not have a 1 in the upper left-hand corner, but this problem can be remedied by exchanging the two rows.

$$\begin{bmatrix} 1 & -3 & \vdots & -7 \\ 5 & 7 & \vdots & -13 \end{bmatrix}$$

Now we can multiply row 1 by -5 and add the result to row 2 to form a new row 2.

$$\begin{bmatrix} 1 & -3 & \vdots & -7 \\ 0 & 22 & \vdots & 22 \end{bmatrix}$$

Next, let's multiply row 2 by $\dfrac{1}{22}$.

$$\begin{bmatrix} 1 & -3 & \vdots & -7 \\ 0 & 1 & \vdots & 1 \end{bmatrix}$$

Finally, we can multiply row 2 by 3 and add the result to row 1 to form a new row 1.

$$\begin{bmatrix} 1 & 0 & \vdots & -4 \\ 0 & 1 & \vdots & 1 \end{bmatrix}$$

The solution set of $\{(-4, 1)\}$ can be read directly from the last matrix. ▲

Example 2

Solve the system $\begin{pmatrix} x + 2y - 3z = 15 \\ -2x - 3y + z = -15 \\ 4x + 9y - 4z = 49 \end{pmatrix}$.

Solution

The augmented matrix of this system is

$$\begin{bmatrix} 1 & 2 & -3 & \vdots & 15 \\ -2 & -3 & 1 & \vdots & -15 \\ 4 & 9 & -4 & \vdots & 49 \end{bmatrix}.$$

Multiply row 1 by 2 and add this result to row 2 to produce a new row 2. Likewise, multiply row 1 by -4 and add this result to row 3 to produce a new row 3.

$$\begin{bmatrix} 1 & 2 & -3 & \vdots & 15 \\ 0 & 1 & -5 & \vdots & 15 \\ 0 & 1 & 8 & \vdots & -11 \end{bmatrix}$$

Now multiply row 2 by -2 and add to row 1 to produce a new row 1. Also, multiply row 2 by -1 and add the result to row 3 to produce a new row 3.

$$\begin{bmatrix} 1 & 0 & 7 & \vdots & -15 \\ 0 & 1 & -5 & \vdots & 15 \\ 0 & 0 & 13 & \vdots & -26 \end{bmatrix}$$

Now let's multiply row 3 by $\dfrac{1}{13}$.

$$\begin{bmatrix} 1 & 0 & 7 & \vdots & -15 \\ 0 & 1 & -5 & \vdots & 15 \\ 0 & 0 & 1 & \vdots & -2 \end{bmatrix}$$

Finally, we can multiply row 3 by -7 and add the result to row 1 to produce a new row 1, and multiply row 3 by 5 and add the result to row 2 for a new row 2.

$$\begin{bmatrix} 1 & 0 & 0 & \vdots & -1 \\ 0 & 1 & 0 & \vdots & 5 \\ 0 & 0 & 1 & \vdots & -2 \end{bmatrix}$$

From this last matrix, we can see that the solution set of the original system is $\{(-1, 5, -2)\}$. ▲

Example 3

Solve the system $\begin{pmatrix} 2x + 4y - 5z = 37 \\ x + 3y - 4z = 29 \\ 5x - y + 3z = -20 \end{pmatrix}$.

Solution

The augmented matrix

$$\begin{bmatrix} 2 & 4 & -5 & | & 37 \\ 1 & 3 & -4 & | & 29 \\ 5 & -1 & 3 & | & -20 \end{bmatrix}$$

does not have a 1 in the upper left-hand corner, but this problem can be remedied by exchanging rows 1 and 2.

$$\begin{bmatrix} 1 & 3 & -4 & | & 29 \\ 2 & 4 & -5 & | & 37 \\ 5 & -1 & 3 & | & -20 \end{bmatrix}$$

Now we can get zeros in the first column beneath the 1 by multiplying -2 times row 1 and adding the result to row 2 and multiplying -5 times row 1 and adding the result to row 3.

$$\begin{bmatrix} 1 & 3 & -4 & | & 29 \\ 0 & -2 & 3 & | & -21 \\ 0 & -16 & 23 & | & -165 \end{bmatrix}$$

Next, we can get a 1 as the first nonzero entry of the second row by multiplying the second row by $-\dfrac{1}{2}$.

$$\begin{bmatrix} 1 & 3 & -4 & | & 29 \\ 0 & 1 & -\dfrac{3}{2} & | & \dfrac{21}{2} \\ 0 & -16 & 23 & | & -165 \end{bmatrix}$$

Now we can get zeros above and below the 1 in the second column by multiplying -3 times row 2 and adding the result to row 1, and multiplying 16 times row 2 and adding the result to row 3.

$$\begin{bmatrix} 1 & 0 & \dfrac{1}{2} & | & -\dfrac{5}{2} \\ 0 & 1 & -\dfrac{3}{2} & | & \dfrac{21}{2} \\ 0 & 0 & -1 & | & 3 \end{bmatrix}$$

Next, we can get a 1 as the first nonzero entry of the third row by multiplying the third row by -1.

$$\begin{bmatrix} 1 & 0 & \dfrac{1}{2} & | & -\dfrac{5}{2} \\ 0 & 1 & -\dfrac{3}{2} & | & \dfrac{21}{2} \\ 0 & 0 & 1 & | & -3 \end{bmatrix}$$

Finally, we can get zeros above the 1 in the third column by multiplying $-\dfrac{1}{2}$ times row 3 and adding the result to row 1, and multiplying $\dfrac{3}{2}$ times row 3 and adding the result to row 2.

$$\begin{bmatrix} 1 & 0 & 0 & | & -1 \\ 0 & 1 & 0 & | & 6 \\ 0 & 0 & 1 & | & -3 \end{bmatrix}$$

From the matrix above, notice that the solution set of the original system is $\{(-1, 6, -3)\}$. ▲

Example 3 illustrates that even though the process of changing to reduced echelon form can be systematically described, it can involve some rather messy arithmetical calculations. For our purposes in this text we have chosen mostly examples and problems that involve systems where the messy calculations are minimized. This should allow you to concentrate on developing an understanding of the procedures being used. Furthermore, sometimes when attempting to change a matrix to reduced form it is more reasonable to stop at the triangular form. For example, consider the matrix

$$\begin{bmatrix} 1 & 3 & -4 & | & 29 \\ 0 & 1 & -\dfrac{3}{2} & | & \dfrac{21}{2} \\ 0 & -16 & 23 & | & -165 \end{bmatrix}$$

which is obtained about halfway through the solution of Example 3. At this step it seems evident that the calculations are getting a little messy. Therefore, instead of continuing toward the reduced echelon form, let's multiply row 2 by 16 and add the result to row 3 to produce a new row 3.

$$\begin{bmatrix} 1 & 3 & -4 & | & 29 \\ 0 & 1 & -\dfrac{3}{2} & | & \dfrac{21}{2} \\ 0 & 0 & -1 & | & 3 \end{bmatrix}$$

We now have a matrix that represents a system in triangular form. The last row determines the value of z, and then the values for y and x can be determined by back-substitution. For some problems it may be somewhat easier to stop at the triangular form rather than to continue on to reduced echelon form.

Finally, let's consider two examples to illustrate what happens when we use the matrix approach on inconsistent and dependent systems.

Example 4

Solve the system $\begin{pmatrix} x - 2y + 3z = & 3 \\ 5x - 9y + 4z = & 2 \\ 2x - 4y + 6z = & -1 \end{pmatrix}$.

Solution

The augmented matrix of the system is

$$\begin{bmatrix} 1 & -2 & 3 & \vdots & 3 \\ 5 & -9 & 4 & \vdots & 2 \\ 2 & -4 & 6 & \vdots & -1 \end{bmatrix}.$$

We can get zeros below the 1 in the first column by multiplying -5 times row 1 and adding the result to row 2; and multiplying -2 times row 1 and adding the result to row 3.

$$\begin{bmatrix} 1 & -2 & 3 & \vdots & 3 \\ 0 & 1 & -11 & \vdots & -13 \\ 0 & 0 & 0 & \vdots & -7 \end{bmatrix}$$

At this point we can stop, because the bottom row of the matrix represents the statement $0(x) + 0(y) + 0(z) = -7$, which is obviously a false statement for all values of x, y, and z. Thus, the original system is *inconsistent*; its solution set is $\varnothing$. ▲

Example 5

Solve the system $\begin{pmatrix} x + 2y + 2z = 9 \\ x + 3y - 4z = 5 \\ 2x + 5y - 2z = 14 \end{pmatrix}$.

Solution

The augmented matrix of the system is

$$\begin{bmatrix} 1 & 2 & 2 & \vdots & 9 \\ 1 & 3 & -4 & \vdots & 5 \\ 2 & 5 & -2 & \vdots & 14 \end{bmatrix}.$$

We can get zeros in the first column below the 1 in the upper left-hand corner by multiplying -1 times row 1 and adding the result to row 2; and multiplying -2 times row 1 and adding the result to row 3.

$$\begin{bmatrix} 1 & 2 & 2 & \vdots & 9 \\ 0 & 1 & -6 & \vdots & -4 \\ 0 & 1 & -6 & \vdots & -4 \end{bmatrix}$$

Now we can get zeros in the second column above and below the 1 in the second row by multiplying -2 times row 2 and adding the result to row 1; and multiplying -1 times row 2 and adding the result to row 3.

$$\begin{bmatrix} 1 & 0 & 14 & \vdots & 17 \\ 0 & 1 & -6 & \vdots & -4 \\ 0 & 0 & 0 & \vdots & 0 \end{bmatrix}$$

The bottom row of zeros represents the statement $0(x) + 0(y) + 0(z) = 0$, which is true for all values of x, y, and z. Thus, the original system is a *dependent system*, and its solution set consists of infinitely many ordered triples.

We can represent these ordered triples as follows. The second row of the previous matrix represents the statement $y - 6z = -4$, which can be written as $y = 6z - 4$. The top row represents the statement $x + 14z = 17$, which can be written as $x = -14z + 17$. Therefore, if we let $z = k$, where k is any real number, the solution set of infinitely many ordered triples can be represented by $\{(-14k + 17, 6k - 4, k)\}$. Specific solutions can be generated by letting k take on specific values. For example, if $k = 2$, then $6k - 4$ becomes $6(2) - 4 = 8$ and $-14k + 17$ becomes $-14(2) + 17 = -11$. Therefore, the ordered triple $(-11, 8, 2)$ is a member of the solution set. ▲

Problem Set 12.2

For Problems 1–10, indicate whether each matrix is in reduced echelon form.

1. $\begin{bmatrix} 1 & 2 & | & 8 \\ 0 & 0 & | & 0 \end{bmatrix}$

2. $\begin{bmatrix} 1 & 0 & | & -4 \\ 0 & 1 & | & 14 \end{bmatrix}$

3. $\begin{bmatrix} 1 & 0 & 3 & | & 8 \\ 0 & 1 & 2 & | & -6 \\ 0 & 0 & 0 & | & 0 \end{bmatrix}$

4. $\begin{bmatrix} 1 & 1 & 0 & | & -3 \\ 0 & 1 & 2 & | & 5 \\ 0 & 0 & 1 & | & 7 \end{bmatrix}$

5. $\begin{bmatrix} 1 & 0 & 0 & | & -7 \\ 0 & 1 & 0 & | & 0 \\ 0 & 0 & 1 & | & 9 \end{bmatrix}$

6. $\begin{bmatrix} 1 & 0 & 0 & | & 17 \\ 0 & 0 & 0 & | & 0 \\ 0 & 1 & 0 & | & -14 \end{bmatrix}$

7. $\begin{bmatrix} 1 & 0 & 0 & | & 5 \\ 0 & 3 & 0 & | & 8 \\ 0 & 0 & 1 & | & -11 \end{bmatrix}$

8. $\begin{bmatrix} 1 & 0 & 2 & | & 5 \\ 0 & 1 & 3 & | & 7 \\ 0 & 0 & 0 & | & 0 \end{bmatrix}$

9. $\begin{bmatrix} 1 & 0 & 0 & 0 & | & 2 \\ 0 & 0 & 1 & 0 & | & 4 \\ 0 & 1 & 0 & 0 & | & -3 \\ 0 & 0 & 0 & 1 & | & 9 \end{bmatrix}$

10. $\begin{bmatrix} 1 & 0 & 0 & 3 & | & 4 \\ 0 & 1 & 0 & 5 & | & -3 \\ 0 & 0 & 1 & -1 & | & 7 \\ 0 & 0 & 0 & 0 & | & 0 \end{bmatrix}$

For Problems 11–30, use a matrix approach as presented in this section to solve each of the systems.

11. $\begin{pmatrix} 2x + 7y = -55 \\ x - 4y = 25 \end{pmatrix}$

12. $\begin{pmatrix} 3x - 4y = 33 \\ x + 7y = -39 \end{pmatrix}$

13. $\begin{pmatrix} x + 5y = -18 \\ -2x + 3y = -16 \end{pmatrix}$

14. $\begin{pmatrix} x - 3y = 14 \\ 3x + 2y = -13 \end{pmatrix}$

15. $\begin{pmatrix} 3x + 9y = -1 \\ x + 3y = 10 \end{pmatrix}$

16. $\begin{pmatrix} 3x - 5y = 39 \\ 2x + 7y = -67 \end{pmatrix}$

17. $\begin{pmatrix} 2x - 3y = -12 \\ 3x + 2y = 8 \end{pmatrix}$

18. $\begin{pmatrix} x - 6y = -2 \\ 2x - 12y = 5 \end{pmatrix}$

19. $\begin{pmatrix} x + 3y - 4z = 13 \\ 2x + 7y - 3z = 11 \\ -2x - y + 2z = -8 \end{pmatrix}$

20. $\begin{pmatrix} x - 2y - 3z = -6 \\ 3x - 5y - z = 4 \\ 2x + y + 2z = 2 \end{pmatrix}$

21. $\begin{pmatrix} x - 4y + 3z = 16 \\ 2x + 3y - 4z = -22 \\ -3x + 11y - z = -36 \end{pmatrix}$

22. $\begin{pmatrix} x - 3y - z = 2 \\ 3x + y - 4z = -18 \\ -2x + 5y + 3z = 2 \end{pmatrix}$

23. $\begin{pmatrix} -3x + 2y + z = 17 \\ x - y + 5z = -2 \\ 4x - 5y - 3z = -36 \end{pmatrix}$

24. $\begin{pmatrix} -2x - 5y + 3z = 11 \\ x + 3y - 3z = -12 \\ 3x - 2y + 5z = 31 \end{pmatrix}$

25. $\begin{pmatrix} x + 2y - 5z = -1 \\ 2x + 3y - 2z = 2 \\ 3x + 5y - 7z = 4 \end{pmatrix}$

26. $\begin{pmatrix} x - y + 2z = 1 \\ -3x + 4y - z = 4 \\ -x + 2y + 3z = 6 \end{pmatrix}$

27. $\begin{pmatrix} 4x - 10y + 3z = -19 \\ 2x + 5y - z = -7 \\ x - 3y - 2z = -2 \end{pmatrix}$

28. $\begin{pmatrix} -2x + y + 5z = -5 \\ 3x + 8y - z = -34 \\ x + 2y + z = -12 \end{pmatrix}$

29. $\begin{pmatrix} 4x + 3y - z = 0 \\ 3x + 2y + 5z = 6 \\ 5x - y - 3z = 3 \end{pmatrix}$

30. $\begin{pmatrix} 2x + 3y - z = 7 \\ 3x + 4y + 5z = -2 \\ 5x + y + 3z = 13 \end{pmatrix}$

Each matrix in Problems 31–38 is the reduced echelon matrix for a system with variables x, y, z, and w. Find the solution set of each system.

31. $\begin{bmatrix} 1 & 0 & 0 & 0 & | & 0 \\ 0 & 1 & 0 & 0 & | & -5 \\ 0 & 0 & 1 & 0 & | & 0 \\ 0 & 0 & 0 & 1 & | & 4 \end{bmatrix}$

32. $\begin{bmatrix} 1 & 0 & 0 & 0 & | & -2 \\ 0 & 1 & 0 & 0 & | & 4 \\ 0 & 0 & 1 & 0 & | & -3 \\ 0 & 0 & 0 & 1 & | & 0 \end{bmatrix}$

33. $\begin{bmatrix} 1 & 0 & 0 & 0 & | & 2 \\ 0 & 1 & 0 & 2 & | & -3 \\ 0 & 0 & 1 & 3 & | & 4 \\ 0 & 0 & 0 & 0 & | & 0 \end{bmatrix}$

34. $\begin{bmatrix} 1 & 0 & 0 & 0 & | & -8 \\ 0 & 1 & 0 & 0 & | & 5 \\ 0 & 0 & 1 & 0 & | & -2 \\ 0 & 0 & 0 & 0 & | & 1 \end{bmatrix}$

35. $\begin{bmatrix} 1 & 3 & 0 & 0 & | & 0 \\ 0 & 0 & 1 & 0 & | & 0 \\ 0 & 0 & 0 & 0 & | & 1 \\ 0 & 0 & 0 & 0 & | & 0 \end{bmatrix}$

36. $\begin{bmatrix} 1 & 0 & 0 & 3 & | & 5 \\ 0 & 1 & 0 & 0 & | & -1 \\ 0 & 0 & 1 & 4 & | & 2 \\ 0 & 0 & 0 & 0 & | & 0 \end{bmatrix}$

37. $\begin{bmatrix} 1 & 0 & 0 & 0 & | & 7 \\ 0 & 1 & 0 & 0 & | & -3 \\ 0 & 0 & 1 & -2 & | & 5 \\ 0 & 0 & 0 & 0 & | & 0 \end{bmatrix}$

38. $\begin{bmatrix} 1 & 3 & 0 & 0 & | & 9 \\ 0 & 0 & 1 & 0 & | & 2 \\ 0 & 0 & 0 & 1 & | & -3 \\ 0 & 0 & 0 & 0 & | & 0 \end{bmatrix}$

THOUGHTS INTO WORDS

39. Suppose that your friend missed class the day that the topic *reduced echelon form* was discussed. How would you explain the concept to him and also explain the advantage of being able to change an augmented matrix to reduced echelon form?

40. Give a step-by-step description of how to change the matrix

$$\begin{bmatrix} 1 & -2 & -3 & | & -6 \\ 3 & 5 & -1 & | & 4 \\ 2 & 1 & 2 & | & 2 \end{bmatrix}$$

to reduced echelon form.

Further Investigations

For Problems 41–44, use a matrix approach to solve each system. Express the solutions as 4-tuples of the form (x_1, x_2, x_3, x_4).

41. $\begin{pmatrix} x_1 - 2x_2 + 2x_3 - x_4 = -2 \\ -3x_1 + 5x_2 - x_3 - 3x_4 = 2 \\ 2x_1 + 3x_2 + 3x_3 + 5x_4 = -9 \\ 4x_1 - x_2 - x_3 - 2x_4 = 8 \end{pmatrix}$

42. $\begin{pmatrix} x_1 - 3x_2 - 2x_3 + x_4 = -3 \\ -2x_1 + 7x_2 + x_3 - 2x_4 = -1 \\ 3x_1 - 7x_2 - 3x_3 + 3x_4 = -5 \\ 5x_1 + x_2 + 4x_3 - 2x_4 = 18 \end{pmatrix}$

43. $\begin{pmatrix} x_1 + 2x_2 - 3x_3 + x_4 = -2 \\ -2x_1 - 3x_2 + x_3 - x_4 = 5 \\ 4x_1 + 9x_2 - 2x_3 - 2x_4 = -28 \\ -5x_1 - 9x_2 + 2x_3 - 3x_4 = 14 \end{pmatrix}$

44. $\begin{pmatrix} x_1 + 3x_2 - x_3 + 2x_4 = -2 \\ 2x_1 + 7x_2 + 2x_3 - x_4 = 19 \\ -3x_1 - 8x_2 + 3x_3 + x_4 = -7 \\ 4x_1 + 11x_2 - 2x_3 - 3x_4 = 19 \end{pmatrix}$

For Problems 45–50, change the augmented matrix of each system to reduced echelon form and then indicate the solutions of the system.

45. $\begin{pmatrix} x + 3y - 2z = -1 \\ -2x - 5y + 7z = 4 \end{pmatrix}$

46. $\begin{pmatrix} x - 2y + 3z = 4 \\ 3x - 5y - z = 7 \end{pmatrix}$

47. $\begin{pmatrix} 3x + 6y - z = 9 \\ 2x - 3y + 4z = 1 \end{pmatrix}$

48. $\begin{pmatrix} 2x - 4y + 3z = 8 \\ 3x + 5y - z = 7 \end{pmatrix}$

49. $\begin{pmatrix} x + y - 2z = -1 \\ 3x + 3y - 6z = -3 \end{pmatrix}$

50. $\begin{pmatrix} x - 2y + 4z = 9 \\ 2x - 4y + 8z = 3 \end{pmatrix}$

Graphics Calculator Activities

51. Load the augmented matrices for Examples 1–3 of this section into your calculator. Then carry out the row operations as described in the text.

52. Use your calculator to change the augmented matrices for Problems 17–20 to reduced echelon form.

12.3   Determinants and Cramer's Rule

A **square matrix** is one that has the same number of rows as columns. Associated with each square matrix having real number entries is a real number called the **determinant** of the matrix. For a 2×2 matrix

$$\begin{bmatrix} a_1 & b_1 \\ a_2 & b_2 \end{bmatrix},$$

the determinant is written as

$$\begin{vmatrix} a_1 & b_1 \\ a_2 & b_2 \end{vmatrix}$$

and defined by

$$\begin{vmatrix} a_1 & b_1 \\ a_2 & b_2 \end{vmatrix} = a_1 b_2 - a_2 b_1.$$

(1)

Notice that a determinant is simply a number; the determinant notation used on the left side of equation (1) is a way of expressing the number on the right side. (Remember that a matrix has no such numerical value; it is simply a rectangular array of numbers.)

Example 1

Find the determinant of the matrix $\begin{bmatrix} 3 & -2 \\ 5 & 8 \end{bmatrix}$.

Solution

In this case, $a_1 = 3$, $b_1 = -2$, $a_2 = 5$, and $b_2 = 8$. Thus, we have

$$\begin{vmatrix} 3 & -2 \\ 5 & 8 \end{vmatrix} = 3(8) - 5(-2) = 24 + 10 = 34.$$ ▲

Finding the determinant of a square matrix is commonly called **evaluating the determinant**, and the matrix notation is sometimes omitted.

Example 2

Evaluate $\begin{vmatrix} -3 & 5 \\ 1 & 2 \end{vmatrix}$.

Solution

$$\begin{vmatrix} -3 & 5 \\ 1 & 2 \end{vmatrix} = -3(2) - 1(5) = -11.$$ ▲

Cramer's Rule

Determinants provide the basis for another method of solving linear systems. Consider the system

$$\begin{pmatrix} a_1x + b_1y = c_1 \\ a_2x + b_2y = c_2 \end{pmatrix}.$$

(1)
(2)

We will solve this system by using the elimination method. To solve for x, we can multiply equation (1) by b_2 and equation (2) by $-b_1$ and then add.

$$\begin{aligned} a_1b_2x + b_1b_2y &= c_1b_2 \\ -a_2b_1x - b_1b_2y &= -c_2b_1 \\ \hline a_1b_2x - a_2b_1x &= c_1b_2 - c_2b_1 \\ (a_1b_2 - a_2b_1)x &= c_1b_2 - c_2b_1 \\ x &= \frac{c_1b_2 - c_2b_1}{a_1b_2 - a_2b_1} \quad \text{If } a_1b_2 - a_2b_1 \neq 0 \end{aligned}$$

To solve for y, we can multiply equation (1) by $-a_2$ and equation (2) by a_1 and add.

$$-a_1a_2x - a_2b_1y = -a_2c_1$$
$$\underline{a_1a_2x + a_1b_2y = a_1c_2}$$
$$a_1b_2y - a_2b_1y = a_1c_2 - a_2c_1$$
$$(a_1b_2 - a_2b_1)y = a_1c_2 - a_2c_1$$
$$y = \frac{a_1c_2 - a_2c_1}{a_1b_2 - a_2b_1} \qquad \text{If } a_1b_2 - a_2b_1 \neq 0$$

The solutions for x and y can be expressed in determinant form as follows.

$$x = \frac{c_1b_2 - c_2b_1}{a_1b_2 - a_2b_1} = \frac{\begin{vmatrix} c_1 & b_1 \\ c_2 & b_2 \end{vmatrix}}{\begin{vmatrix} a_1 & b_1 \\ a_2 & b_2 \end{vmatrix}},$$

$$y = \frac{a_1c_2 - a_2c_1}{a_1b_2 - a_2b_1} = \frac{\begin{vmatrix} a_1 & c_1 \\ a_2 & c_2 \end{vmatrix}}{\begin{vmatrix} a_1 & b_1 \\ a_2 & b_2 \end{vmatrix}}$$

For convenience, we will denote the three determinants in the solution as

$$\begin{vmatrix} a_1 & b_1 \\ a_2 & b_2 \end{vmatrix} = D, \qquad \begin{vmatrix} c_1 & b_1 \\ c_2 & b_2 \end{vmatrix} = D_x, \qquad \begin{vmatrix} a_1 & c_1 \\ a_2 & c_2 \end{vmatrix} = D_y.$$

Notice that the elements of D are the coefficients of the variables in the given system. In D_x, the elements are obtained by replacing the coefficients of x by the respective constants. In D_y, the coefficients of y are replaced by the respective constants.

This method of using determinants to solve a system of two linear equations in two variables, called **Cramer's rule**, can be stated as follows.

Cramer's Rule

Given the system

$$\begin{pmatrix} a_1x + b_1y = c_1 \\ a_2x + b_2y = c_2 \end{pmatrix} \quad \text{with } a_1b_2 - a_2b_1 \neq 0$$

then

$$x = \frac{\begin{vmatrix} c_1 & b_1 \\ c_2 & b_2 \end{vmatrix}}{\begin{vmatrix} a_1 & b_1 \\ a_2 & b_2 \end{vmatrix}} = \frac{D_x}{D} \quad \text{and} \quad y = \frac{\begin{vmatrix} a_1 & c_1 \\ a_2 & c_2 \end{vmatrix}}{\begin{vmatrix} a_1 & b_1 \\ a_2 & b_2 \end{vmatrix}} = \frac{D_y}{D}.$$

Let's illustrate the use of Cramer's rule to solve some systems.

Example 3

Solve the system $\left(\begin{array}{l} x + 2y = 11 \\ 2x - y = 2 \end{array} \right)$.

Solution

Let's find D, D_x, and D_y.

$$D = \begin{vmatrix} 1 & 2 \\ 2 & -1 \end{vmatrix} = -1 - 4 = -5,$$

$$D_x = \begin{vmatrix} 11 & 2 \\ 2 & -1 \end{vmatrix} = -11 - 4 = -15,$$

$$D_y = \begin{vmatrix} 1 & 11 \\ 2 & 2 \end{vmatrix} = 2 - 22 = -20$$

Thus, we have

$$x = \frac{D_x}{D} = \frac{-15}{-5} = 3,$$

$$y = \frac{D_y}{D} = \frac{-20}{-5} = 4.$$

The solution set is $\{(3, 4)\}$. This can be verified, as always, by substituting back into the original equations.   ▲

REMARK   Notice that Cramer's rule has a restriction in that $a_1 b_2 - a_2 b_1 \neq 0$; that is, $D \neq 0$. Thus, it is a good idea to find D first. If $D = 0$, Cramer's rule does not apply and you can revert to one of the other methods to determine whether the solution set is empty or has infinitely many solutions.   △

Example 4

Solve the system $\left(\begin{array}{l} 2x - 3y = -8 \\ 3x + 5y = 7 \end{array} \right)$.

Solution

$$D = \begin{vmatrix} 2 & -3 \\ 3 & 5 \end{vmatrix} = 10 - (-9) = 19,$$

$$D_x = \begin{vmatrix} -8 & -3 \\ 7 & 5 \end{vmatrix} = -40 - (-21) = -19,$$

$$D_y = \begin{vmatrix} 2 & -8 \\ 3 & 7 \end{vmatrix} = 14 - (-24) = 38$$

Thus, we obtain

$$x = \frac{D_x}{D} = \frac{-19}{19} = -1,$$

$$y = \frac{D_y}{D} = \frac{38}{19} = 2.$$

The solution set is $\{(-1, 2)\}$. ▲

Example 5

Solve the system $\begin{pmatrix} y = -2x - 2 \\ 4x - 5y = 17 \end{pmatrix}$.

Solution

First, we must change the form of the first equation so that the system fits the form given in Cramer's rule. The equation $y = -2x - 2$ can be written as $2x + y = -2$. The system now becomes

$$\begin{pmatrix} 2x + y = -2 \\ 4x - 5y = 17 \end{pmatrix},$$

and we can proceed as before.

$$D = \begin{vmatrix} 2 & 1 \\ 4 & -5 \end{vmatrix} = -10 - 4 = -14,$$

$$D_x = \begin{vmatrix} -2 & 1 \\ 17 & -5 \end{vmatrix} = 10 - 17 = -7,$$

$$D_y = \begin{vmatrix} 2 & -2 \\ 4 & 17 \end{vmatrix} = 34 - (-8) = 42$$

Thus, the solutions are

$$x = \frac{D_x}{D} = \frac{-7}{-14} = \frac{1}{2},$$

$$y = \frac{D_y}{D} = \frac{42}{-14} = -3.$$

The solution set is $\left\{ \left(\frac{1}{2}, -3 \right) \right\}$. ▲

Example 6

Solve the system $\begin{pmatrix} \frac{1}{2}x + \frac{2}{3}y = -4 \\ \frac{1}{4}x - \frac{3}{2}y = 20 \end{pmatrix}$.

Solution

$$D = \begin{vmatrix} \dfrac{1}{2} & \dfrac{2}{3} \\[2ex] \dfrac{1}{4} & -\dfrac{3}{2} \end{vmatrix} = \frac{1}{2}\left(-\frac{3}{2}\right) - \frac{1}{4}\left(\frac{2}{3}\right)$$

$$= -\frac{3}{4} - \frac{1}{6} = -\frac{11}{12},$$

$$D_x = \begin{vmatrix} -4 & \dfrac{2}{3} \\[2ex] 20 & -\dfrac{3}{2} \end{vmatrix} = -4\left(-\frac{3}{2}\right) - 20\left(\frac{2}{3}\right)$$

$$= 6 - \frac{40}{3} = -\frac{22}{3},$$

$$D_y = \begin{vmatrix} \dfrac{1}{2} & -4 \\[2ex] \dfrac{1}{4} & 20 \end{vmatrix} = \frac{1}{2}(20) - \frac{1}{4}(-4) = 11$$

Thus, the solutions are

$$x = \frac{D_x}{D} = \frac{-\dfrac{22}{3}}{-\dfrac{11}{12}} = \left(-\frac{22}{3}\right)\left(-\frac{12}{11}\right) = 8,$$

$$y = \frac{D_y}{D} = \frac{11}{-\dfrac{11}{12}} = (11)\left(-\frac{12}{11}\right) = -12.$$

The solution set is $\{(8, -12)\}$.

▲

Problem Set 12.3

For Problems 1–20, evaluate each determinant.

1. $\begin{vmatrix} 6 & 2 \\ 4 & 3 \end{vmatrix}$ **2.** $\begin{vmatrix} 7 & 6 \\ 2 & 5 \end{vmatrix}$ **7.** $\begin{vmatrix} 8 & -3 \\ 6 & 4 \end{vmatrix}$ **8.** $\begin{vmatrix} 5 & 9 \\ -3 & 6 \end{vmatrix}$

3. $\begin{vmatrix} 4 & 7 \\ 8 & 2 \end{vmatrix}$ **4.** $\begin{vmatrix} 3 & 9 \\ 6 & 4 \end{vmatrix}$ **9.** $\begin{vmatrix} -3 & 2 \\ 5 & -6 \end{vmatrix}$ **10.** $\begin{vmatrix} -2 & 4 \\ 9 & -7 \end{vmatrix}$

5. $\begin{vmatrix} -3 & 2 \\ 7 & 5 \end{vmatrix}$ **6.** $\begin{vmatrix} 5 & 1 \\ 8 & -4 \end{vmatrix}$ **11.** $\begin{vmatrix} 3 & -3 \\ -6 & 8 \end{vmatrix}$ **12.** $\begin{vmatrix} 6 & -5 \\ -8 & 12 \end{vmatrix}$

13. $\begin{vmatrix} -7 & -2 \\ -2 & 4 \end{vmatrix}$

14. $\begin{vmatrix} 6 & -1 \\ -8 & -3 \end{vmatrix}$

15. $\begin{vmatrix} -2 & -3 \\ -4 & -5 \end{vmatrix}$

16. $\begin{vmatrix} -9 & -7 \\ -6 & -4 \end{vmatrix}$

17. $\begin{vmatrix} \dfrac{1}{4} & -2 \\ \dfrac{3}{2} & 8 \end{vmatrix}$

18. $\begin{vmatrix} -\dfrac{2}{3} & 10 \\ -\dfrac{1}{2} & 6 \end{vmatrix}$

19. $\begin{vmatrix} \dfrac{3}{2} & -\dfrac{1}{2} \\ \dfrac{1}{2} & -\dfrac{2}{5} \end{vmatrix}$

20. $\begin{vmatrix} -\dfrac{1}{4} & \dfrac{1}{3} \\ \dfrac{3}{4} & \dfrac{2}{3} \end{vmatrix}$

For Problems 21–40, use Cramer's rule to help find the solution set for each system of equations.

21. $\begin{pmatrix} 2x + y = 14 \\ 3x - y = 1 \end{pmatrix}$

22. $\begin{pmatrix} 4x - y = 11 \\ 2x + 3y = 23 \end{pmatrix}$

23. $\begin{pmatrix} -x + 3y = 17 \\ 4x - 5y = -33 \end{pmatrix}$

24. $\begin{pmatrix} 5x + 2y = -15 \\ 7x - 3y = 37 \end{pmatrix}$

25. $\begin{pmatrix} 9x + 5y = -8 \\ 7x - 4y = -22 \end{pmatrix}$

26. $\begin{pmatrix} 8x - 11y = 3 \\ -x + 4y = -3 \end{pmatrix}$

27. $\begin{pmatrix} x + 5y = 4 \\ 3x + 15y = -1 \end{pmatrix}$

28. $\begin{pmatrix} 4x - 7y = 0 \\ 7x + 2y = 0 \end{pmatrix}$

29. $\begin{pmatrix} 6x - y = 0 \\ 5x + 4y = 29 \end{pmatrix}$

30. $\begin{pmatrix} 3x - 4y = 2 \\ 9x - 12y = 6 \end{pmatrix}$

31. $\begin{pmatrix} -4x + 3y = 3 \\ 4x - 6y = -5 \end{pmatrix}$

32. $\begin{pmatrix} x - 2y = -1 \\ x = -6y + 5 \end{pmatrix}$

33. $\begin{pmatrix} 6x - 5y = 1 \\ 4x + 7y = 2 \end{pmatrix}$

34. $\begin{pmatrix} y = 3x + 5 \\ y = 6x + 6 \end{pmatrix}$

35. $\begin{pmatrix} 7x + 2y = -1 \\ y = -x + 2 \end{pmatrix}$

36. $\begin{pmatrix} 9x - y = -2 \\ y = 4 - 8x \end{pmatrix}$

37. $\begin{pmatrix} -\dfrac{2}{3}x + \dfrac{1}{2}y = -7 \\ \dfrac{1}{3}x - \dfrac{3}{2}y = 6 \end{pmatrix}$

38. $\begin{pmatrix} \dfrac{1}{2}x + \dfrac{2}{3}y = -6 \\ \dfrac{1}{4}x - \dfrac{1}{3}y = -1 \end{pmatrix}$

39. $\begin{pmatrix} x + \dfrac{2}{3}y = -6 \\ -\dfrac{1}{4}x + 3y = -8 \end{pmatrix}$

40. $\begin{pmatrix} 3x - \dfrac{1}{2}y = 6 \\ -2x + \dfrac{1}{3}y = -4 \end{pmatrix}$

For Problems 41–47, solve each problem by using a system of two equations and two variables.

41. At a local confectionery, 7 pounds of cashews and 5 pounds of Spanish peanuts cost $88, and 3 pounds of cashews and 2 pounds of Spanish peanuts cost $37. Find the price per pound for cashews and for Spanish peanuts.

42. We bought 2 cartons of pop and 4 pounds of candy for $12. The next day we bought 3 cartons of pop and 2 pounds of candy for $9. Find the price of a carton of pop and the price of a pound of candy.

43. A mail-order company charges a fixed fee for shipping merchandise that weighs 1 pound or less, plus an additional fee for each pound over 1 pound. If the shipping charge for 5 pounds is $2.40 and for 12 pounds is $3.10, find the fixed fee and the additional fee.

44. The sum of two numbers is 23, and the difference of the two numbers is 5. Find the numbers.

45. The sum of two numbers is 19. The larger number is 1 larger than twice the smaller number. Find the numbers.

46. A two-digit number is 4 times the sum of its digits. If the digits are interchanged, the new number is 36 larger than the original number. Find the original number.

47. A woman invested a sum of money at 8% and another amount at 9%. The total yearly interest from both investments is $101. If she interchanged her investments, the total interest would be $103. How much does she have invested at each rate?

THOUGHTS INTO WORDS

48. Explain the difference between a matrix and a determinant.

49. Give a step-by-step description of how you would solve the system $\begin{pmatrix} 3x - 2y = 7 \\ 5x + 9y = 14 \end{pmatrix}$ using determinants.

Further Investigations

50. Verify each of the following. The variables represent real numbers.

(a) $\begin{vmatrix} a & b \\ a & b \end{vmatrix} = 0$ **(b)** $\begin{vmatrix} a & a \\ b & b \end{vmatrix} = 0$

(c) $\begin{vmatrix} a & b \\ c & d \end{vmatrix} = -\begin{vmatrix} b & a \\ d & c \end{vmatrix}$

(d) $\begin{vmatrix} a & b \\ c & d \end{vmatrix} = -\begin{vmatrix} c & d \\ a & b \end{vmatrix}$

(e) $k\begin{vmatrix} a & b \\ c & d \end{vmatrix} = \begin{vmatrix} ka & b \\ kc & d \end{vmatrix}$

(f) $k\begin{vmatrix} a & b \\ c & d \end{vmatrix} = \begin{vmatrix} ka & kb \\ c & d \end{vmatrix}$

Graphics Calculator Activities

51. Use the determinant function of your graphics calculator and check your answers in Problems 1–20.

52. Make up two or three examples for each part of Problem 50 and evaluate the determinants using your graphics calculator.

12.4   3 × 3 Determinants and Cramer's Rule

In this section we will extend the concept of a determinant to include 3 × 3 determinants, and then we will also extend the use of determinants to solve systems of three linear equations in three variables.

For a 3 × 3 matrix

$$\begin{bmatrix} a_1 & b_1 & c_1 \\ a_2 & b_2 & c_2 \\ a_3 & b_3 & c_3 \end{bmatrix},$$

the determinant is written as

$$\begin{vmatrix} a_1 & b_1 & c_1 \\ a_2 & b_2 & c_2 \\ a_3 & b_3 & c_3 \end{vmatrix}$$

and is defined by

$$\begin{vmatrix} a_1 & b_1 & c_1 \\ a_2 & b_2 & c_2 \\ a_3 & b_3 & c_3 \end{vmatrix} = a_1b_2c_3 + b_1c_2a_3 + c_1a_2b_3 - a_3b_2c_1 - b_3c_2a_1 - c_3a_2b_1. \quad (1)$$

It is evident that the definition given by equation (1) is a bit too complicated to be very useful in practice. Fortunately, a method called **expansion of a determinant by minors** can be used to calculate such a determinant. The **minor** of an element in a determinant is the determinant that remains after the row and column in which the element appears are deleted. For example, consider the determinant of equation (1).

The minor of a_1 is $\begin{vmatrix} b_2 & c_2 \\ b_3 & c_3 \end{vmatrix}$,

the minor of a_2 is $\begin{vmatrix} b_1 & c_1 \\ b_3 & c_3 \end{vmatrix}$,

the minor of a_3 is $\begin{vmatrix} b_1 & c_1 \\ b_2 & c_2 \end{vmatrix}$, etc.

Now, let's consider the terms of the right side of equation (1) in pairs, and show the tie-up with minors.

$$a_1b_2c_3 - b_3c_2a_1 = a_1(\underbrace{b_2c_3 - b_3c_2})$$

$$\downarrow$$

$$\begin{vmatrix} b_2 & c_2 \\ b_3 & c_3 \end{vmatrix}$$

$$c_1a_2b_3 - c_3a_2b_1 = -(c_3a_2b_1 - c_1a_2b_3)$$
$$= -a_2\underbrace{(b_1c_3 - b_3c_1)}$$

$$\begin{vmatrix} b_1 & c_1 \\ b_3 & c_3 \end{vmatrix}$$

$$b_1c_2a_3 - a_3b_2c_1 = a_3\underbrace{(b_1c_2 - b_2c_1)}$$

$$\begin{vmatrix} b_1 & c_1 \\ b_2 & c_2 \end{vmatrix}$$

Therefore, we have

$$\begin{vmatrix} a_1 & b_1 & c_1 \\ a_2 & b_2 & c_2 \\ a_3 & b_3 & c_3 \end{vmatrix} = a_1 \begin{vmatrix} b_2 & c_2 \\ b_3 & c_3 \end{vmatrix} - a_2 \begin{vmatrix} b_1 & c_1 \\ b_3 & c_3 \end{vmatrix} + a_3 \begin{vmatrix} b_1 & c_1 \\ b_2 & c_2 \end{vmatrix}.$$

This process is called the **expansion of the determinant by minors about the first column.**

Example I

Evaluate $\begin{vmatrix} 1 & 2 & -1 \\ 3 & 1 & 2 \\ 2 & 4 & 3 \end{vmatrix}$ by expanding by minors about the first column.

Solution

$$\begin{vmatrix} 1 & 2 & -1 \\ 3 & 1 & 2 \\ 2 & 4 & 3 \end{vmatrix} = 1 \begin{vmatrix} 1 & 2 \\ 4 & 3 \end{vmatrix} - 3 \begin{vmatrix} 2 & -1 \\ 4 & 3 \end{vmatrix} + 2 \begin{vmatrix} 2 & -1 \\ 1 & 2 \end{vmatrix}$$

$$= 1(3 - 8) - 3[6 - (-4)] + 2[4 - (-1)]$$
$$= 1(-5) - 3(10) + 2(5)$$
$$= -5 - 30 + 10$$
$$= -25 \qquad \blacktriangle$$

It is possible to expand a determinant by minors about *any row* or *any column*. The following sign array is very useful for determining the signs of the terms in the expansion.

$$\begin{array}{ccc} + & - & + \\ - & + & - \\ + & - & + \end{array}$$

For example, let's expand the determinant in Example 1 by minors about the *second row*. The second row in the sign array is $-$ $+$ $-$. Therefore,

$$\begin{vmatrix} 1 & 2 & -1 \\ 3 & 1 & 2 \\ 2 & 4 & 3 \end{vmatrix} = -3\begin{vmatrix} 2 & -1 \\ 4 & 3 \end{vmatrix} + 1\begin{vmatrix} 1 & -1 \\ 2 & 3 \end{vmatrix} - 2\begin{vmatrix} 1 & 2 \\ 2 & 4 \end{vmatrix}$$

$$= -3[6 - (-4)] + 1[3 - (-2)] - 2(4 - 4)$$
$$= -3(10) + 1(5) - 2(0)$$
$$= -25. \qquad \blacktriangle$$

Your decision as to which row or column to use for expanding a particular determinant by minors may depend on the numbers involved in the determinant. A row or column with one or more zeros is frequently a good choice, as the next example illustrates.

Example 2

Evaluate $\begin{vmatrix} 3 & -1 & 4 \\ 5 & 2 & 0 \\ -2 & 6 & 0 \end{vmatrix}$.

Solution

Since the third column has two zeros, we will expand about it.

$$\begin{vmatrix} 3 & -1 & 4 \\ 5 & 2 & 0 \\ -2 & 6 & 0 \end{vmatrix} = 4\begin{vmatrix} 5 & 2 \\ -2 & 6 \end{vmatrix} - 0\begin{vmatrix} 3 & -1 \\ -2 & 6 \end{vmatrix} + 0\begin{vmatrix} 3 & -1 \\ 5 & 2 \end{vmatrix}$$

$$= 4[30 - (-4)] - 0 + 0$$
$$= 136$$

(Because of the zeros, there is no need to evaluate the last two minors.) $\qquad \blacktriangle$

REMARK   The *expansion-by-minors method* can be expanded to determinants of size 4×4, 5×5, and so on. However, it should be obvious that it becomes increasingly more tedious with bigger determinants. Fortunately, the computer handles the calculation of such determinants by using a different technique. $\qquad \triangle$

Cramer's Rule Expanded

Without showing all of the details, we will simply state that Cramer's rule also applies to solving systems of three linear equations in three variables. It can be stated as follows.

Cramer's Rule Expanded

Given the system

$$\begin{cases} a_1x + b_1y + c_1z = d_1 \\ a_2x + b_2y + c_2z = d_2 \\ a_3x + b_3y + c_3z = d_3 \end{cases}$$

with

$$D = \begin{vmatrix} a_1 & b_1 & c_1 \\ a_2 & b_2 & c_2 \\ a_3 & b_3 & c_3 \end{vmatrix} \neq 0, \qquad D_x = \begin{vmatrix} d_1 & b_1 & c_1 \\ d_2 & b_2 & c_2 \\ d_3 & b_3 & c_3 \end{vmatrix},$$

$$D_y = \begin{vmatrix} a_1 & d_1 & c_1 \\ a_2 & d_2 & c_2 \\ a_3 & d_3 & c_3 \end{vmatrix}, \qquad D_z = \begin{vmatrix} a_1 & b_1 & d_1 \\ a_2 & b_2 & d_2 \\ a_3 & b_3 & d_3 \end{vmatrix},$$

then

$$x = \frac{D_x}{D}, \qquad y = \frac{D_y}{D}, \qquad \text{and} \qquad z = \frac{D_z}{D}.$$

Notice that the elements of D are the coefficients of the variables in the given system. Then D_x, D_y, and D_z are formed by replacing the elements in the x, y, or z column, respectively, by the constants of the system d_1, d_2, and d_3. Again, note the restriction $D \neq 0$. As before, if $D = 0$, then Cramer's rule does not apply and you can use the elimination method to determine whether the system has *no solution* or *infinitely many solutions*, so calculate D first.

Example 3

Use Cramer's rule to solve the system $\begin{cases} x - 2y + z = -4 \\ 2x + y - z = 5 \\ 3x + 2y + 4z = 3 \end{cases}$.

Solution

To find D, we expand about row 1.

$$D = \begin{vmatrix} 1 & -2 & 1 \\ 2 & 1 & -1 \\ 3 & 2 & 4 \end{vmatrix} = 1\begin{vmatrix} 1 & -1 \\ 2 & 4 \end{vmatrix} - (-2)\begin{vmatrix} 2 & -1 \\ 3 & 4 \end{vmatrix} + 1\begin{vmatrix} 2 & 1 \\ 3 & 2 \end{vmatrix}$$

$$= 1[4 - (-2)] + 2[8 - (-3)] + 1(4 - 3)$$

$$= 1(6) + 2(11) + 1(1) = 29$$

To find D_x, we expand about column 3.

$$D_x = \begin{vmatrix} -4 & -2 & 1 \\ 5 & 1 & -1 \\ 3 & 2 & 4 \end{vmatrix} = 1 \begin{vmatrix} 5 & 1 \\ 3 & 2 \end{vmatrix} - (-1) \begin{vmatrix} -4 & -2 \\ 3 & 2 \end{vmatrix} + 4 \begin{vmatrix} -4 & -2 \\ 5 & 1 \end{vmatrix}$$

$$= 1(10 - 3) + 1[-8 - (-6)] + 4[-4 - (-10)]$$

$$= 1(7) + 1(-2) + 4(6) = 29$$

To find D_y, we expand about row 1.

$$D_y = \begin{vmatrix} 1 & -4 & 1 \\ 2 & 5 & -1 \\ 3 & 3 & 4 \end{vmatrix} = 1 \begin{vmatrix} 5 & -1 \\ 3 & 4 \end{vmatrix} - (-4) \begin{vmatrix} 2 & -1 \\ 3 & 4 \end{vmatrix} + 1 \begin{vmatrix} 2 & 5 \\ 3 & 3 \end{vmatrix}$$

$$= 1[20 - (-3)] + 4[8 - (-3)] + 1(6 - 15)$$

$$= 1(23) + 4(11) + 1(-9) = 58$$

To find D_z, we expand about column 1.

$$D_z = \begin{vmatrix} 1 & -2 & -4 \\ 2 & 1 & 5 \\ 3 & 2 & 3 \end{vmatrix} = 1 \begin{vmatrix} 1 & 5 \\ 2 & 3 \end{vmatrix} - 2 \begin{vmatrix} -2 & -4 \\ 2 & 3 \end{vmatrix} + 3 \begin{vmatrix} -2 & -4 \\ 1 & 5 \end{vmatrix}$$

$$= 1(3 - 10) - 2[-6 - (-8)] + 3[-10 - (-4)]$$

$$= 1(-7) - 2(2) + 3(-6)$$

$$= -29$$

Thus,

$$x = \frac{D_x}{D} = \frac{29}{29} = 1,$$

$$y = \frac{D_y}{D} = \frac{58}{29} = 2,$$

$$z = \frac{D_z}{D} = \frac{-29}{29} = -1.$$

The solution set is $\{(1, 2, -1)\}$. (Be sure to check it!)

Example 4

Use Cramer's rule to solve the system $\begin{pmatrix} 2x - y + 3z = -17 \\ 3y + z = 5 \\ x - 2y - z = -3 \end{pmatrix}$.

Solution

To find D, we expand about column 1.

$$D = \begin{vmatrix} 2 & -1 & 3 \\ 0 & 3 & 1 \\ 1 & -2 & -1 \end{vmatrix} = 2\begin{vmatrix} 3 & 1 \\ -2 & -1 \end{vmatrix} - 0\begin{vmatrix} -1 & 3 \\ -2 & -1 \end{vmatrix} + 1\begin{vmatrix} -1 & 3 \\ 3 & 1 \end{vmatrix}$$

$$= 2[-3 - (-2)] - 0 + 1(-1 - 9)$$

$$= 2(-1) - 0 - 10 = -12$$

To find D_x, we expand about column 3.

$$D_x = \begin{vmatrix} -17 & -1 & 3 \\ 5 & 3 & 1 \\ -3 & -2 & -1 \end{vmatrix}$$

$$= 3\begin{vmatrix} 5 & 3 \\ -3 & -2 \end{vmatrix} - 1\begin{vmatrix} -17 & -1 \\ -3 & -2 \end{vmatrix} + (-1)\begin{vmatrix} -17 & -1 \\ 5 & 3 \end{vmatrix}$$

$$= 3[-10 - (-9)] - 1(34 - 3) - 1[-51 - (-5)]$$

$$= 3(-1) - 1(31) - 1(-46) = 12$$

To find D_y, we expand about column 1.

$$D_y = \begin{vmatrix} 2 & -17 & 3 \\ 0 & 5 & 1 \\ 1 & -3 & -1 \end{vmatrix}$$

$$= 2\begin{vmatrix} 5 & 1 \\ -3 & -1 \end{vmatrix} - 0\begin{vmatrix} -17 & 3 \\ -3 & -1 \end{vmatrix} + 1\begin{vmatrix} -17 & 3 \\ 5 & 1 \end{vmatrix}$$

$$= 2[-5 - (-3)] - 0 + 1(-17 - 15)$$

$$= 2(-2) - 0 + 1(-32) = -36$$

To find D_z, we expand about column 1.

$$D_z = \begin{vmatrix} 2 & -1 & -17 \\ 0 & 3 & 5 \\ 1 & -2 & -3 \end{vmatrix} = 2\begin{vmatrix} 3 & 5 \\ -2 & -3 \end{vmatrix} - 0\begin{vmatrix} -1 & -17 \\ -2 & -3 \end{vmatrix} + 1\begin{vmatrix} -1 & -17 \\ 3 & 5 \end{vmatrix}$$

$$= 2[-9 - (-10)] - 0 + 1[-5 - (-51)]$$

$$= 2(1) - 0 + 1(46) = 48$$

Thus,

$$x = \frac{D_x}{D} = \frac{12}{-12} = -1,$$

$$y = \frac{D_y}{D} = \frac{-36}{-12} = 3,$$

$$z = \frac{D_z}{D} = \frac{48}{-12} = -4.$$

The solution set is $\{(-1, 3, -4)\}$.

Problem Set 12.4

Use *expansion by minors* to evaluate each of the following determinants.

1. $\begin{vmatrix} 2 & 7 & 5 \\ 1 & -1 & 1 \\ -4 & 3 & 2 \end{vmatrix}$

2. $\begin{vmatrix} 2 & 4 & 1 \\ -1 & 5 & 1 \\ -3 & 6 & 2 \end{vmatrix}$

3. $\begin{vmatrix} 3 & -2 & 1 \\ 2 & 1 & 4 \\ -1 & 3 & 5 \end{vmatrix}$

4. $\begin{vmatrix} 1 & -1 & 2 \\ 2 & 1 & 3 \\ -1 & -2 & 1 \end{vmatrix}$

5. $\begin{vmatrix} -3 & -2 & 1 \\ 5 & 0 & 6 \\ 2 & 1 & -4 \end{vmatrix}$

6. $\begin{vmatrix} -5 & 1 & -1 \\ 3 & 4 & 2 \\ 0 & 2 & -3 \end{vmatrix}$

7. $\begin{vmatrix} 3 & -4 & -2 \\ 5 & -2 & 1 \\ 1 & 0 & 0 \end{vmatrix}$

8. $\begin{vmatrix} -6 & 5 & 3 \\ 2 & 0 & -1 \\ 4 & 0 & 7 \end{vmatrix}$

9. $\begin{vmatrix} 4 & -2 & 7 \\ 1 & -1 & 6 \\ 3 & 5 & -2 \end{vmatrix}$

10. $\begin{vmatrix} -5 & 2 & 6 \\ 1 & -1 & 3 \\ 4 & -2 & -4 \end{vmatrix}$

Use Cramer's rule to find the solution set for each of the following systems.

11. $\begin{pmatrix} 2x - y + 3z = -10 \\ x + 2y - 3z = 2 \\ 3x - 2y + 5z = -16 \end{pmatrix}$

12. $\begin{pmatrix} -x + y - z = 1 \\ 2x + 3y - 4z = 10 \\ -3x - y + z = -5 \end{pmatrix}$

13. $\begin{pmatrix} x - y + 2z = -8 \\ 2x + 3y - 4z = 18 \\ -x + 2y - z = 7 \end{pmatrix}$

14. $\begin{pmatrix} x - 2y + z = 3 \\ 3x + 2y + z = -3 \\ 2x - 3y - 3z = -5 \end{pmatrix}$

15. $\begin{pmatrix} 3x - 2y - 3z = -5 \\ x + 2y + 3z = -3 \\ -x + 4y - 6z = 8 \end{pmatrix}$

16. $\begin{pmatrix} 2x - 3y + 3z = -3 \\ -2x + 5y - 3z = 5 \\ 3x - y + 6z = -1 \end{pmatrix}$

17. $\begin{pmatrix} -x + y + z = -1 \\ x - 2y + 5z = -4 \\ 3x + 4y - 6z = -1 \end{pmatrix}$

18. $\begin{pmatrix} x - 2y + 3z = 1 \\ 2x + y + z = 4 \\ 4x - 3y + 7z = 6 \end{pmatrix}$

19. $\begin{pmatrix} x - y + 2z = 4 \\ 3x - 2y + 4z = 6 \\ 2x - 2y + 4z = -1 \end{pmatrix}$

20. $\begin{pmatrix} -x - 2y + z = 8 \\ 3x + y - z = 5 \\ 5x - y + 4z = 33 \end{pmatrix}$

21. $\begin{pmatrix} 2x - y + 3z = -5 \\ 3x + 4y - 2z = -25 \\ -x + z = 6 \end{pmatrix}$

22. $\begin{pmatrix} 3x - 2y + z = 11 \\ 5x + 3y = 17 \\ x + y - 2z = 6 \end{pmatrix}$

23. $\begin{pmatrix} 2y - z = 10 \\ 3x + 4y = 6 \\ x - y + z = -9 \end{pmatrix}$

24. $\begin{pmatrix} 6x - 5y + 2z = 7 \\ 2x + 3y - 4z = -21 \\ 2y + 3z = 10 \end{pmatrix}$

25. $\begin{pmatrix} -2x + 5y - 3z = -1 \\ 2x - 7y + 3z = 1 \\ 4x - y - 6z = -6 \end{pmatrix}$

26. $\begin{pmatrix} 7x - 2y + 3z = -4 \\ 5x + 2y - 3z = 4 \\ -3x - 6y + 12z = -13 \end{pmatrix}$

27. $\begin{pmatrix} -x - y + 5z = 4 \\ x + y - 7z = -6 \\ 2x + 3y + 4z = 13 \end{pmatrix}$

28. $\begin{pmatrix} x + 7y - z = -1 \\ -x - 9y + z = 3 \\ 3x + 4y - 6z = 5 \end{pmatrix}$

29. $\begin{pmatrix} 5x - y + 2z = 10 \\ 7x + 2y - 2z = -4 \\ -3x - y + 4z = 1 \end{pmatrix}$

30. $\begin{pmatrix} 4x - y - 3z = -12 \\ 5x + y + 6z = 4 \\ 6x - y - 3z = -14 \end{pmatrix}$

THOUGHTS INTO WORDS

31. How would you explain the process of evaluating 3×3 determinants to a friend who missed class the day it was discussed?

32. Explain how to use determinants to solve the system

$$\begin{pmatrix} x - 2y + z = 1 \\ 2x - y - z = 5 \\ 5x + 3y + 4z = -6 \end{pmatrix}.$$

Further Investigations

33. Evaluate the following determinant by expanding about the *second column*.

$$\begin{vmatrix} a & e & a \\ b & f & b \\ c & g & c \end{vmatrix}$$

Make a conjecture about determinants that contain two identical columns.

34. Show that $\begin{vmatrix} 1 & -1 & 2 \\ 2 & 3 & -1 \\ -1 & 2 & 4 \end{vmatrix} = - \begin{vmatrix} -1 & 1 & 2 \\ 3 & 2 & -1 \\ 2 & -1 & 4 \end{vmatrix}.$

Make a conjecture about the result of interchanging two columns of a determinant.

35. (a) Show that $\begin{vmatrix} 2 & 1 & 2 \\ 4 & -1 & -2 \\ 6 & 3 & 1 \end{vmatrix} = 2 \begin{vmatrix} 1 & 1 & 2 \\ 2 & -1 & -2 \\ 3 & 3 & 1 \end{vmatrix}.$

Make a conjecture about the result of factoring a common factor from each element of a column in a determinant.

(b) Use your conjecture from part (a) to help evaluate the following determinant.

$$\begin{vmatrix} 2 & 4 & -1 \\ -3 & -4 & -2 \\ 5 & 4 & 3 \end{vmatrix}$$

36. We can describe another technique for evaluating 3×3 determinants as follows. First, let's write the given determinant with its first two columns repeated on the right.

$$\begin{vmatrix} a_1 & b_1 & c_1 \\ a_2 & b_2 & c_2 \\ a_3 & b_3 & c_3 \end{vmatrix} \begin{matrix} a_1 & b_1 \\ a_2 & b_2 \\ a_3 & b_3 \end{matrix}$$

Then we can add the three products shown with $+$, and subtract the three products shown with $-$.

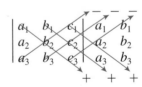

(a) Be sure that the previous description will produce equation (1) on page 651.

(b) Use this technique to do Problems 1–10.

 Graphics Calculator Activities

37. Use your graphics calculator to check your answers for Problems 1–10.

38. Let matrix

$$A = \begin{bmatrix} 2 & 5 & 7 & 9 \\ -4 & 6 & 2 & 4 \\ 6 & 9 & 12 & 3 \\ 5 & 4 & -2 & 8 \end{bmatrix}.$$

Form matrix B by interchanging rows 1 and 3 of matrix A. Now use your calculator to show that $|B| = -|A|$.

39. Let matrix

$$A = \begin{bmatrix} 2 & 1 & 7 & 6 & 8 \\ 3 & -2 & 4 & 5 & -1 \\ 6 & 7 & 9 & 12 & 13 \\ -4 & -7 & 6 & 2 & 1 \\ 9 & 8 & 12 & 14 & 17 \end{bmatrix}.$$

Form matrix B by multiplying each element of the second row of A by 3. Now use your calculator to show that $|B| = 3|A|$.

40. Let matrix

$$A = \begin{bmatrix} 4 & 3 & 2 & 1 & 5 & -3 \\ 5 & 2 & 7 & 8 & 6 & 3 \\ 0 & 9 & 1 & 4 & 7 & 2 \\ 4 & 3 & 2 & 1 & 5 & -3 \\ -4 & -6 & 7 & 12 & 11 & 9 \\ 5 & 8 & 6 & -3 & 2 & -1 \end{bmatrix}.$$

Use your calculator to show that $|A| = 0$.

41. Use determinants and your calculator to solve each of the following systems. Be sure to check your answers.

(a) $\begin{cases} 4x - 3y + z = 10 \\ 8x + 5y - 2z = -6 \\ -12x - 2y + 3z = -2 \end{cases}$

(b) $\begin{cases} 2x + y - z + w = -4 \\ x + 2y + 2z - 3w = 6 \\ 3x - y - z + 2w = 0 \\ 2x + 3y + z + 4w = -5 \end{cases}$

(c) $\begin{cases} x - 2y + z - 3w = 4 \\ 2x + 3y - z - 2w = -4 \\ 3x - 4y + 2z - 4w = 12 \\ 2x - y - 3z + 2w = -2 \end{cases}$

(d) $\begin{cases} 1.98x + 2.49y + 3.45z = 80.10 \\ 2.15x + 3.20y + 4.19z = 97.16 \\ 1.49x + 4.49y + 2.79z = 83.92 \end{cases}$

SUMMARY

(12.1) A rectangular array of numbers such as

$$\begin{bmatrix} 1 & 3 & 7 \\ 2 & 8 & -4 \end{bmatrix}$$

is called a **matrix**. With every system of linear equations we can associate a matrix that consists of the coefficients and constant terms. For example, with the system

$$\begin{pmatrix} a_1x + b_1y = c_1 \\ a_2x + b_2y = c_2 \end{pmatrix}$$

we can associate the matrix

$$\left[\begin{array}{cc|c} a_1 & b_1 & c_1 \\ a_2 & b_2 & c_2 \end{array}\right],$$

which is called the **augmented matrix** of the system.

Systems of linear equations can be solved by applying the following **elementary row operations** to the augmented matrices.

1. Any two rows of an augmented matrix can be interchanged.
2. Any row can be multiplied by a nonzero constant.
3. Any row can be replaced by adding a nonzero multiple of another row to that row.

A system of linear equations can be solved by changing the augmented matrix to a **triangular form** and then using back-substitution. See Examples 1–4 of this section for a review of this process.

(12.2) The form

$$\left[\begin{array}{ccc|c} 1 & 0 & 0 & * \\ 0 & 1 & 0 & * \\ 0 & 0 & 1 & * \end{array}\right]$$

is called the **reduced echelon form** for a system of three linear equations in three variables. The solution set of a system is obvious once the augmented matrix has been put in reduced echelon form.

(12.3) A rectangular array of numbers is called a **matrix**. A **square matrix** has the same number of rows as columns. For a 2×2 matrix

$$\begin{bmatrix} a_1 & b_1 \\ a_2 & b_2 \end{bmatrix},$$

the **determinant** of the matrix is written as

$$\begin{vmatrix} a_1 & b_1 \\ a_2 & b_2 \end{vmatrix}$$

and is defined by

$$\begin{vmatrix} a_1 & b_1 \\ a_2 & b_2 \end{vmatrix} = a_1b_2 - a_2b_1.$$

Cramer's rule for solving a system of two linear equations in two variables is stated as follows: Given the system

$$\begin{pmatrix} a_1x + b_1y = c_1 \\ a_2x + b_2y = c_2 \end{pmatrix},$$

with

$$D = \begin{vmatrix} a_1 & b_1 \\ a_2 & b_2 \end{vmatrix} \neq 0, \qquad D_x = \begin{vmatrix} c_1 & b_1 \\ c_2 & b_2 \end{vmatrix}, \qquad D_y = \begin{vmatrix} a_1 & c_1 \\ a_2 & c_2 \end{vmatrix},$$

then

$$x = \frac{D_x}{D} \qquad \text{and} \qquad y = \frac{D_y}{D}.$$

(12.4) A 3×3 determinant is defined by

$$\begin{vmatrix} a_1 & b_1 & c_1 \\ a_2 & b_2 & c_2 \\ a_3 & b_3 & c_3 \end{vmatrix} = a_1b_2c_3 + b_1c_2a_3 + c_1a_2b_3 - a_3b_2c_1 - b_3c_2a_1 - c_3a_2b_1.$$

The **minor** of an element in a determinant is the determinant that remains after the row and column in which the element appears are deleted. A determinant can be evaluated by **expansion by minors** of the elements of any row or any column.

Cramer's rule for solving a system of three linear equations in three variables is stated as follows: Given the system

$$\begin{pmatrix} a_1x + b_1y + c_1z = d_1 \\ a_2x + b_2y + c_2z = d_2 \\ a_3x + b_3y + c_3z = d_3 \end{pmatrix},$$

with

$$D = \begin{vmatrix} a_1 & b_1 & c_1 \\ a_2 & b_2 & c_2 \\ a_3 & b_3 & c_3 \end{vmatrix} \neq 0, \qquad D_x = \begin{vmatrix} d_1 & b_1 & c_1 \\ d_2 & b_2 & c_2 \\ d_3 & b_3 & c_3 \end{vmatrix},$$

$$D_y = \begin{vmatrix} a_1 & d_1 & c_1 \\ a_2 & d_2 & c_2 \\ a_3 & d_3 & c_3 \end{vmatrix}, \qquad D_z = \begin{vmatrix} a_1 & b_1 & d_1 \\ a_2 & b_2 & d_2 \\ a_3 & b_3 & d_3 \end{vmatrix},$$

then

$$x = \frac{D_x}{D}, \qquad y = \frac{D_y}{D}, \qquad \text{and} \qquad z = \frac{D_z}{D}.$$

Chapter 12 Review Problem Set

For Problems 1–10, use a matrix approach to solve each of the systems.

1. $\begin{pmatrix} 2x + 5y = 2 \\ 4x - 3y = 30 \end{pmatrix}$

2. $\begin{pmatrix} 6x + 5y = -21 \\ x = 4y + 11 \end{pmatrix}$

3. $\begin{pmatrix} 4x - 3y = 34 \\ 3x + 2y = 0 \end{pmatrix}$

4. $\begin{pmatrix} \dfrac{1}{2}x - \dfrac{2}{3}y = 1 \\ \dfrac{3}{4}x + \dfrac{1}{6}y = -1 \end{pmatrix}$

5. $\begin{pmatrix} x + y - 2z = -7 \\ 3x + 4y + z = 6 \\ 5x - y - 3z = -1 \end{pmatrix}$

6. $\begin{pmatrix} x - 2y - z = 11 \\ 2x - 3y + 5z = 4 \\ -3x + 5y - 3z = -17 \end{pmatrix}$

7. $\begin{pmatrix} 3x - 4y - 3z = 10 \\ -2x + 9y - 2z = 14 \\ x - 5y + z = -8 \end{pmatrix}$

8. $\begin{pmatrix} 2x - y + 3z = -19 \\ 3x + 2y - 4z = 21 \\ 5x - 4y - z = -8 \end{pmatrix}$

9. $\begin{pmatrix} 3x - 2y - 5z = 2 \\ -4x + 3y + 11z = 3 \\ 2x - y + z = -1 \end{pmatrix}$

10. $\begin{pmatrix} 3x + 2y - 4z = 4 \\ 5x + 3y - z = 2 \\ 4x - 2y + 3z = 11 \end{pmatrix}$

For Problems 11–17, evaluate each of the determinants.

11. $\begin{vmatrix} -2 & 6 \\ 3 & 8 \end{vmatrix}$

12. $\begin{vmatrix} 5 & -4 \\ 7 & -3 \end{vmatrix}$

13. $\begin{vmatrix} 5 & -3 \\ -4 & -2 \end{vmatrix}$

14. $\begin{vmatrix} 4 & -1 & -3 \\ 2 & 1 & 4 \\ -3 & 2 & 2 \end{vmatrix}$

15. $\begin{vmatrix} 2 & 3 & -1 \\ 3 & 4 & -5 \\ 6 & 4 & 2 \end{vmatrix}$

16. $\begin{vmatrix} 3 & -2 & 4 \\ 1 & 0 & 6 \\ 3 & -3 & 5 \end{vmatrix}$

17. $\begin{vmatrix} 5 & 4 & 3 \\ 2 & -7 & 0 \\ 3 & -2 & 0 \end{vmatrix}$

For Problems 18–29, use Cramer's rule to solve each of the systems.

18. $\begin{pmatrix} 7x - 2y = -53 \\ x + 5y = 40 \end{pmatrix}$

19. $\begin{pmatrix} 2x - 3y = 12 \\ 3x + 5y = -20 \end{pmatrix}$

20. $\begin{pmatrix} y = 3x - 16 \\ 5x + 7y = -34 \end{pmatrix}$

21. $\begin{pmatrix} \dfrac{3}{4}x - \dfrac{1}{2}y = -15 \\ \dfrac{2}{3}x + \dfrac{1}{4}y = -5 \end{pmatrix}$

22. $\begin{pmatrix} 7x - 3y = -49 \\ y = \dfrac{3}{5}x - 1 \end{pmatrix}$

23. $\begin{pmatrix} .2x + .3y = 2.6 \\ .5x + .1y = 1.4 \end{pmatrix}$

24. $\begin{pmatrix} -x - y + z = -3 \\ 3x + 2y - 4z = 12 \\ 5x + y + 2z = 5 \end{pmatrix}$

25. $\begin{pmatrix} 3x + y - z = -6 \\ 3x + 2y + 3z = 9 \\ 6x - 3y + 2z = 9 \end{pmatrix}$

26. $\begin{pmatrix} x - 2y + z = -7 \\ 2x - 3y + 4z = -14 \\ -3x + y - 2z = 10 \end{pmatrix}$

27. $\begin{pmatrix} -2x - 7y + z = 9 \\ x + 3y - 4z = -11 \\ 4x + 5y - 3z = -11 \end{pmatrix}$

28. $\begin{pmatrix} 2x - 3y - 3z = 25 \\ 3x + y + 2z = -5 \\ 5x - 2y - 4z = 32 \end{pmatrix}$

29. $\begin{pmatrix} 3x - y + z = -10 \\ 6x - 2y + 5z = -35 \\ 7x + 3y - 4z = 19 \end{pmatrix}$

For Problems 30–33, solve each problem by setting up and solving a system of linear equations.

30. The sum of the digits of a two-digit number is 9. If the digits are reversed, the newly formed number is 45 less than the original number. Find the original number.

31. Sara invested $2500, part at 10% and the rest at 12% yearly interest. The income on the 12% investment is $102 more than the income on the 10% investment. How much money did Sara invest at each rate?

32. A box contains $17.70 in nickels, dimes, and quarters. The number of dimes is 8 less than twice the number of nickels. The number of quarters is 2 more than the sum of the number of nickels and dimes. How many coins of each kind are there in the box?

33. The measure of the largest angle of a triangle is 10° more than four times that of the smallest angle. The sum of the smallest and the largest angles is three times the measure of the other angle. Find the measure of each angle of the triangle.

CHAPTER 12 TEST

For Problems 1–7, use a matrix approach we discussed in Sections 12.1 and 12.2.

1. Find the value of x in the solution for the system
$$\begin{pmatrix} 2x - 3y = -7 \\ 3x + 4y = -2 \end{pmatrix}.$$

2. Find the value of y in the solution for the system
$$\begin{pmatrix} -x + 4y = -19 \\ 3x - 5y = 29 \end{pmatrix}.$$

3. Find the value of x in the solution for the system
$$\begin{pmatrix} 2x + 7y = -33 \\ 3x - 5y = 28 \end{pmatrix}.$$

4. Find the value of y in the solution for the system
$$\begin{pmatrix} 3x - y = -5 \\ 5x + y = -3 \end{pmatrix}.$$

5. Find the value of x in the solution for the system
$$\begin{pmatrix} x + y + z = -1 \\ -2x - y + 2z = 3 \\ 3x + 4y - z = -2 \end{pmatrix}.$$

6. Find the value of y in the solution for the system
$$\begin{pmatrix} x - 2y + 2z = -6 \\ 2x - 3y + z = -8 \\ -3x - y + 5z = -10 \end{pmatrix}.$$

7. Find the value of z in the solution for the system
$$\begin{pmatrix} x + 2y + z = 5 \\ 3x + 5y - 2z = -15 \\ -2x - y - z = -4 \end{pmatrix}.$$

8. Suppose that the augmented matrix of a system of three linear equations in the three variables x, y, and z can be changed to the matrix
$$\begin{bmatrix} 1 & 1 & -4 & | & 3 \\ 0 & 1 & 4 & | & 5 \\ 0 & 0 & 3 & | & 6 \end{bmatrix}.$$

Find the value of x in the solution for the system.

(continued on next page)

CHAPTER 12 TEST (continued)

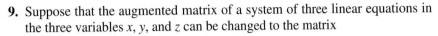

9. Suppose that the augmented matrix of a system of three linear equations in the three variables x, y, and z can be changed to the matrix

$$\begin{bmatrix} 1 & 2 & -3 & | & 4 \\ 0 & 1 & 2 & | & 5 \\ 0 & 0 & 2 & | & -8 \end{bmatrix}.$$

Find the value of y in the solution for the system.

10. How many ordered triples are there in the solution set for the system

$$\begin{pmatrix} x + 3y - z = 5 \\ 2x - y - z = 7 \\ 5x + 8y - 4z = 22 \end{pmatrix}?$$

11. How many ordered triples are there in the solution set for the system

$$\begin{pmatrix} 3x - y - 2z = 1 \\ 4x + 2y + z = 5 \\ 6x - 2y - 4z = 9 \end{pmatrix}?$$

12. Evaluate $\begin{vmatrix} -2 & 4 \\ -5 & 6 \end{vmatrix}$.

13. Evaluate $\begin{vmatrix} 3 & -2 \\ -5 & 4 \end{vmatrix}$.

14. Evaluate $\begin{vmatrix} -3 & -4 \\ -2 & -5 \end{vmatrix}$.

15. Evaluate $\begin{vmatrix} \dfrac{1}{2} & \dfrac{1}{3} \\ \dfrac{3}{4} & -\dfrac{2}{3} \end{vmatrix}$.

16. Evaluate $\begin{vmatrix} -1 & 2 & 1 \\ 3 & 1 & -2 \\ 2 & -1 & 1 \end{vmatrix}$.

17. Evaluate $\begin{vmatrix} 2 & 4 & -5 \\ -4 & 3 & 0 \\ -2 & 6 & 1 \end{vmatrix}$.

18. Evaluate $\begin{vmatrix} 2 & -3 & 4 \\ 5 & 0 & -1 \\ -6 & 0 & -2 \end{vmatrix}$.

(continued on next page)

CHAPTER 12 TEST *(continued)*

For Problems 19–25, use Cramer's rule to find the indicated values.

19. Find the values of x in the solution for the system

$$\begin{pmatrix} 5x - 2y = -41 \\ 3x + 4y = -9 \end{pmatrix}.$$

20. Find the value of y in the solution for the system

$$\begin{pmatrix} 5x - 7y = -33 \\ 3x + 5y = 17 \end{pmatrix}.$$

21. Find the value of x in the solution for the system

$$\begin{pmatrix} \dfrac{3}{2}x - \dfrac{1}{3}y = -22 \\ \dfrac{2}{3}x + \dfrac{1}{4}y = -5 \end{pmatrix}.$$

22. Find the value of y in the solution for the system

$$\begin{pmatrix} 3x - y = -1 \\ 2x + 5y = 1 \end{pmatrix}.$$

23. Find the value of x in the solution for the system

$$\begin{pmatrix} x - 4y + z = 12 \\ -2x + 3y - z = -11 \\ 5x - 3y + 2z = 17 \end{pmatrix}.$$

24. Find the value of y in the solution for the system

$$\begin{pmatrix} 3x - y + 2z = 4 \\ 3y + 4z = -3 \\ x + 2y + 5z = -1 \end{pmatrix}.$$

25. Find the value of z in the solution for the system

$$\begin{pmatrix} 3x - 2z = -6 \\ x - 3y + 5z = -11 \\ 2y + 3z = 6 \end{pmatrix}.$$

Cumulative Review Problem Set

For Problems 1–5, evaluate each algebraic expression for the given values of the variables.

1. $-5(x - 1) - 3(2x + 4) + 3(3x - 1)$ for $x = -2$

2. $\dfrac{14a^3 b^2}{7a^2 b}$ for $a = -1$ and $b = 4$

3. $\dfrac{2}{n} - \dfrac{3}{2n} + \dfrac{5}{3n}$ for $n = 4$

4. $4\sqrt{2x - y} + 5\sqrt{3x + y}$ for $x = 16$ and $y = 16$

5. $\dfrac{3}{x - 2} - \dfrac{5}{x + 3}$ for $x = 3$

For Problems 6–15, perform the indicated operations and express answers in simplified form.

6. $(-5\sqrt{6})(3\sqrt{12})$

7. $(2\sqrt{x} - 3)(\sqrt{x} + 4)$

8. $(3\sqrt{2} - \sqrt{6})(\sqrt{2} + 4\sqrt{6})$

9. $(2x - 1)(x^2 + 6x - 4)$

10. $\dfrac{x^2 - x}{x + 5} \cdot \dfrac{x^2 + 5x + 4}{x^4 - x^2}$

11. $\dfrac{16x^2 y}{24xy^3} \div \dfrac{9xy}{8x^2 y^2}$

12. $\dfrac{x + 3}{10} + \dfrac{2x + 1}{15} - \dfrac{x - 2}{18}$

13. $\dfrac{7}{12ab} - \dfrac{11}{15a^2}$

14. $\dfrac{8}{x^2 - 4x} + \dfrac{2}{x}$

15. $(8x^3 - 6x^2 - 15x + 4) \div (4x - 1)$

For Problems 16–19, simplify each of the complex fractions.

16. $\dfrac{\dfrac{5}{x^2} - \dfrac{3}{x}}{\dfrac{1}{y} + \dfrac{2}{y^2}}$

17. $\dfrac{\dfrac{2}{x} - 3}{\dfrac{3}{7} + 4}$

18. $\dfrac{2 - \dfrac{1}{n - 2}}{3 + \dfrac{4}{n + 3}}$

19. $\dfrac{3a}{2 - \dfrac{1}{a}} - 1$

For Problems 20–25, factor each of the algebraic expressions completely.

20. $20x^2 + 7x - 6$

21. $16x^3 + 54$

22. $4x^4 - 25x^2 + 36$

23. $12x^3 - 52x^2 - 40x$

24. $xy - 6x + 3y - 18$

25. $10 + 9x - 9x^2$

For Problems 26–35, evaluate each of the numerical expressions.

26. $\left(\dfrac{2}{3}\right)^{-4}$

27. $\dfrac{3}{\left(\dfrac{4}{3}\right)^{-1}}$

28. $\sqrt[3]{-\dfrac{27}{64}}$

29. $-\sqrt{.09}$

30. $(27)^{-\frac{4}{3}}$

31. $4^0 + 4^{-1} + 4^{-2}$

32. $\left(\dfrac{3^{-1}}{2^{-3}}\right)^{-2}$

33. $(2^{-3} - 3^{-2})^{-1}$

34. $\log_2 64$

35. $\log_3 \left(\dfrac{1}{9}\right)$

For Problems 36–38, find the indicated products and quotients; express final answers with positive integral exponents only.

36. $(-3x^{-1}y^2)(4x^{-2}y^{-3})$

37. $\dfrac{48x^{-4}y^2}{6xy}$

38. $\left(\dfrac{27a^{-4}b^{-3}}{-3a^{-1}b^{-4}}\right)^{-1}$

For Problems 39–46, express each radical expression in simplest radical form.

39. 80

40. $-2\sqrt{54}$

41. $\sqrt{\dfrac{75}{81}}$

42. $\dfrac{4\sqrt{6}}{3\sqrt{8}}$

43. $\sqrt[3]{56}$

44. $\dfrac{\sqrt[3]{3}}{\sqrt[3]{4}}$

45. $4\sqrt{52x^3y^2}$

46. $\sqrt{\dfrac{2x}{3y}}$

For Problems 47–49, use the distributive property to help simplify each of the following.

47. $-3\sqrt{24} + 6\sqrt{54} - \sqrt{6}$

48. $\dfrac{\sqrt{8}}{3} - \dfrac{3\sqrt{18}}{4} - \dfrac{5\sqrt{50}}{2}$

49. $8\sqrt[3]{3} - 6\sqrt[3]{24} - 4\sqrt[3]{81}$

For Problems 50 and 51, rationalize the denominator and simplify.

50. $\dfrac{\sqrt{3}}{\sqrt{6} - 2\sqrt{2}}$

51. $\dfrac{3\sqrt{5} - \sqrt{3}}{2\sqrt{3} + \sqrt{7}}$

For Problems 52–54, use scientific notation to help perform the indicated operations.

52. $\dfrac{(.00016)(300)(.028)}{.064}$

53. $\dfrac{.00072}{.0000024}$

54. $.00000009$

For Problems 55–58, find each of the indicated products or quotients and express answers in standard form.

55. $(5 - 2i)(4 + 6i)$

56. $(-3 - i)(5 - 2i)$

57. $\dfrac{5}{4i}$

58. $\dfrac{-1 + 6i}{7 - 2i}$

59. Find the slope of the line determined by the points $(2, -3)$ and $(-1, 7)$.

60. Find the slope of the line determined by the equation $4x - 7y = 9$.

61. Find the length of the line segment whose endpoints are $(4, 5)$ and $(-2, 1)$.

62. Write the equation of the line that contains the points $(3, -1)$ and $(7, 4)$.

63. Write the equation of the line that is perpendicular to the line $3x - 4y = 6$ and contains the point $(-3, -2)$.

64. Find the center and the length of a radius of the circle $x^2 + 4x + y^2 - 12y + 31 = 0$.

65. Find the coordinates of the vertex of the parabola $y = x^2 + 10x + 21$.

66. Find the length of the major axis of the ellipse $x^2 + 4y^2 = 16$.

For Problems 67–72, graph each of the equations.

67. $-x + 2y = -4$

68. $x^2 + y^2 = 9$

69. $x^2 - y^2 = 9$

70. $x^2 + 2y^2 = 8$

71. $y = -3x$

72. $x^2y = -4$

For Problems 73–82, graph each of the functions.

73. $f(x) = -2x - 4$

74. $f(x) = -2x^2 - 2$

75. $f(x) = x^2 - 2x - 2$

76. $f(x) = \sqrt{x + 1} + 2$

77. $f(x) = 2x^2 + 8x + 9$

78. $f(x) = -|x - 2| + 1$

79. $f(x) = 2^x + 2$

80. $f(x) = \log_2(x - 2)$

81. $f(x) = -x(x + 1)(x - 2)$

82. $f(x) = \dfrac{-x}{x + 2}$

83. If $f(x) = x - 3$ and $g(x) = 2x^2 - x - 1$, find $(g \circ f)(x)$ and $(f \circ g)(x)$.

84. Find the inverse (f^{-1}) of $f(x) = 3x - 7$.

85. Find the inverse of $f(x) = -\dfrac{1}{2}x + \dfrac{2}{3}$.

86. Find the constant of variation if y varies directly as x, and $y = 2$ when $x = -\dfrac{2}{3}$.

87. If y is inversely proportional to the square of x, and $y = 4$ when $x = 3$, find y when $x = 6$.

88. The volume of gas at a constant temperature varies inversely as the pressure. What is the volume of a gas under a pressure of 25 pounds if the gas occupies 15 cubic centimeters under a pressure of 20 pounds?

For Problems 89 and 90, evaluate each of the determinants.

89. $\begin{vmatrix} -2 & 4 \\ 7 & 6 \end{vmatrix}$

90. $\begin{vmatrix} 1 & -2 & -1 \\ 2 & 1 & 3 \\ -1 & -3 & 4 \end{vmatrix}$

For Problems 91–118, solve each equation.

91. $3(2x - 1) - 2(5x + 1) = 4(3x + 4)$

92. $n + \dfrac{3n - 1}{9} - 4 = \dfrac{3n + 1}{3}$

93. $.92 + .9(x - .3) = 2x - 5.95$

94. $|4x - 1| = 11$

95. $3x^2 = 7x$

96. $x^3 - 36x = 0$

97. $30x^2 + 13x - 10 = 0$

98. $8x^3 + 12x^2 - 36x = 0$

99. $x^4 + 8x^2 - 9 = 0$

100. $(n + 4)(n - 6) = 11$

101. $2 - \dfrac{3x}{x - 4} = \dfrac{14}{x + 7}$

102. $\dfrac{2n}{6n^2 + 7n - 3} - \dfrac{n - 3}{3n^2 + 11n - 4} = \dfrac{5}{2n^2 + 11n + 12}$

103. $\sqrt{3y} - y = -6$

104. $\sqrt{x + 19} - \sqrt{x + 28} = -1$

105. $(3x - 1)^2 = 45$

106. $(2x + 5)^2 = -32$

107. $2x^2 - 3x + 4 = 0$

108. $3n^2 - 6n + 2 = 0$

109. $\dfrac{5}{n - 3} - \dfrac{3}{n + 3} = 1$

110. $12x^4 - 19x^2 + 5 = 0$

111. $2x^2 + 5x + 5 = 0$

112. $x^3 - 4x^2 - 25x + 28 = 0$

113. $6x^3 - 19x^2 + 9x + 10 = 0$

114. $16^x = 64$

115. $\log_3 x = 4$

116. $\log_{10} x + \log_{10} 25 = 2$

117. $\ln(3x - 4) - \ln(x + 1) = \ln 2$

118. $27^{4x} = 9^{x+1}$

For Problems 119–128, solve each inequality.

119. $-5(y - 1) + 3 > 3y - 4 - 4y$

120. $.06x + .08(250 - x) \geq 19$

121. $|5x - 2| > 13$

122. $|6x + 2| < 8$

123. $\dfrac{x - 2}{5} - \dfrac{3x - 1}{4} \leq \dfrac{3}{10}$

124. $(x - 2)(x + 4) \leq 0$

125. $(3x - 1)(x - 4) > 0$

126. $x(x + 5) < 24$

127. $\dfrac{x - 3}{x - 7} \geq 0$

128. $\dfrac{2x}{x + 3} > 4$

For Problems 129–133, solve each of the systems of equations.

129. $\begin{pmatrix} 4x - 3y = 18 \\ 3x - 2y = 15 \end{pmatrix}$

130. $\begin{pmatrix} y = \dfrac{2}{5}x - 1 \\ 3x + 5y = 4 \end{pmatrix}$

131. $\begin{pmatrix} \dfrac{x}{2} - \dfrac{y}{3} = 1 \\ \dfrac{2x}{5} + \dfrac{y}{2} = 2 \end{pmatrix}$

132. $\begin{pmatrix} 4x - y + 3z = -12 \\ 2x + 3y - z = 8 \\ 6x + y + 2z = -8 \end{pmatrix}$

133. $\begin{pmatrix} x - y + 5z = -10 \\ 5x + 2y - 3z = 6 \\ -3x + 2y - z = 12 \end{pmatrix}$

For Problems 134–148, set up an equation, an inequality, or a system of equations to help solve each problem.

134. Find three consecutive odd integers whose sum is 57.

135. Eric has a collection of 63 coins consisting of nickels, dimes, and quarters. The number of dimes is 6 more than the number of nickels and the number of quarters is one more than twice the number of nickels. How many coins of each kind are in the collection?

136. One of two supplementary angles is 4° more than one-third of the other angle. Find the measure of each of the angles.

137. If a ring costs a jeweler $300, at what price should it be sold to make a profit of 50% on the selling price?

138. Beth invested a certain amount of money at 8% and $300 more than that amount at 9%. Her total yearly interest was $316. How much did she invest at each rate?

139. Two trains leave the same depot at the same time, one traveling east and the other traveling west. At the end of $4\frac{1}{2}$ hours they are 639 miles apart. If the rate of the train traveling east is 10 miles per hour faster than the other train, find their rates.

140. A 10-quart radiator contains a 50% solution of antifreeze. How much needs to be drained out and replaced with pure antifreeze to obtain a 70% antifreeze solution?

141. Sam shot rounds of 70, 73, and 76 on the first three days of a golf tournament. What must he shoot on the fourth day of the tournament to average 72 or less for the four days?

142. The cube of a number equals nine times the same number. Find the number.

143. A strip of uniform width is to be cut off of both sides and both ends of a sheet of paper that is 8 inches by 14 inches to reduce the size of the paper to an area of 72 square inches. Find the width of the strip.

144. A sum of $2450 is to be divided between two people in the ratio of 3 to 4. How much does each person receive?

145. Working together Sue and Dean can complete a task in $1\frac{1}{5}$ hours. Dean can do the task by himself in 2 hours. How long would it take Sue to complete the task by herself?

146. Dudley bought a number of shares of stock for $300. A month later he sold all but 10 shares at a profit of $5 per share and regained his original investment of $300. How many shares did he originally buy and at what price per share?

147. The units digit of a two-digit number is one more than twice the tens digit. The sum of the digits is 10. Find the number.

148. The sum of the two smallest angles of a triangle is 40° less than the other angle. The sum of the smallest and largest angle is twice the other angle. Find the measures of the three angles of the triangle.

Sequences and Series

Suppose that Math University had an enrollment of 7500 students in 1990 and each year thereafter through 1994 the enrollment increased by 500 students. The numbers

7500,   8000,   8500,   9000,   9500

represent the enrollment figures for the years 1990 through 1994. This list of numbers, where there is a constant difference of 500 between any two successive numbers in the list, is called an arithmetic sequence.

Suppose a woman's present yearly salary is $30,000 and she expects a 10% raise for each of the next four years. The numbers

30,000,   33,000,   36,300,   39,930,   43,923

represent her present salary and her salary for each of the next four years. This list of numbers, where each number after the first one is 1.1 times the previous number in the list, is called a geometric sequence. Arithmetic and geometric sequences are the focus of our attention in this chapter.

13.1 **Arithmetic Sequences**

An **infinite sequence** is a function whose domain is the set of positive integers. For example, consider the function defined by the equation

$$f(n) = 3n + 2,$$

where the domain is the set of positive integers. Let's substitute the numbers of the domain, in order, starting with 1. The resulting ordered pairs can be listed as

$$(1, 5), (2, 8), (3, 11), (4, 14), (5, 17),$$

and so on. Since we have agreed to use the domain of positive integers, in order, starting with 1, there is no need to use ordered pairs. We can simply express the infinite sequence as

$$5, 8, 11, 14, 17, \ldots$$

Frequently, the letter a is used to represent sequential functions, and the functional value at n is written as a_n (read "a sub n") instead of $a(n)$. The sequence is then expressed as

$$a_1, a_2, a_3, a_4, \ldots,$$

where a_1 is the *first term*, a_2 is the *second term*, a_3 is the *third term*, and so on. The expression a_n, which defines the sequence, is called the **general term** of the sequence. Knowing the general term of a sequence allows us to find as many terms of the sequence as needed and also to find any specific terms. Consider the following examples.

Example 1

Find the first five terms of each of the following sequences.

(*a*) $a_n = n^2 + 1$ (*b*) $a_n = 2^n$.

Solution

(*a*) We find the first five terms by replacing n with 1, 2, 3, 4, and 5.

$$a_1 = 1^2 + 1 = 2,$$
$$a_2 = 2^2 + 1 = 5,$$
$$a_3 = 3^2 + 1 = 10,$$
$$a_4 = 4^2 + 1 = 17,$$
$$a_5 = 5^2 + 1 = 26$$

Thus, the first five terms are 2, 5, 10, 17, and 26.

(*b*) $a_1 = 2^1 = 2,$
$$a_2 = 2^2 = 4,$$
$$a_3 = 2^3 = 8,$$

$$a_4 = 2^4 = 16,$$

$$a_5 = 2^5 = 32$$

The first five terms are 2, 4, 8, 16, and 32. ▲

Example 2

Solution

Find the 12th and 25th terms of the sequence $a_n = 5n - 1$.

$$a_{12} = 5(12) - 1 = 59$$

$$a_{25} = 5(25) - 1 = 124$$

The 12th term is 59, and the 25th term is 124. ▲

An **arithmetic sequence** (also called an arithmetic progression) is a sequence where there is a common difference between successive terms. The following are examples of arithmetic sequences.

1. 1, 4, 7, 10, 13, . . .
2. 6, 11, 16, 21, 26, . . .
3. 14, 25, 36, 47, 58, . . .
4. 4, 2, 0, −2, −4, . . .
5. −1, −7, −13, −19, −25, . . .

The common difference in the first sequence is 3. That is to say, $4 - 1 = 3$, $7 - 4 = 3$, $10 - 7 = 3$, $13 - 10 = 3$, and so on. The common differences for the second, third, fourth, and fifth sequences are 5, 11, −2, and −6, respectively. We sometimes state that arithmetic sequences exhibit *constant growth*. This is an accurate description if we keep in mind that the *growth* may be in a negative direction, as illustrated by the fourth and fifth sequences above.

In a more general setting we say that the sequence

$$a_1, a_2, a_3, a_4, \ldots, a_n, \ldots$$

is an arithmetic sequence if and only if there is a real number d such that

$$a_{k+1} - a_k = d \tag{1}$$

for every positive integer. We call the number d the **common difference**.

From equation (1) we see that $a_{k+1} = a_k + d$. In other words, we can generate an arithmetic sequence that has a common difference of d by starting with a first term of a_1 and then simply adding d to each successive term as follows.

First term	a_1	
Second term	$a_1 + d$	
Third term	$a_1 + 2d$	$(a_1 + d) + d = a_1 + 2d$
Fourth term	$a_1 + 3d$	$(a_1 + 2d) + d = a_1 + 3d$
$\vdots$		
nth term	$a_1 + (n-1)d$	

Thus, the general term of an arithmetic sequence is given by

$$\boxed{a_n = a_1 + (n - 1)d,}$$

where a_1 is the first term and d is the common difference. This general-term formula provides the basis for doing a variety of problems that involve arithmetic sequences.

Example 3

Find the general term for each of the following arithmetic sequences:

(**a**) $1, 5, 9, 13, \ldots$　　　(**b**) $5, 2, -1, -4, \ldots$

Solution

(**a**)　The common difference is 4, and the first term is 1. Substitute these values into $a_n = a_1 + (n - 1)d$ and simplify to obtain

$$\begin{aligned} a_n &= a_1 + (n - 1)d \\ &= 1 + (n - 1)4 \\ &= 1 + 4n - 4 \\ &= 4n - 3. \end{aligned}$$

(Perhaps you should verify that the general term $a_n = 4n - 3$ does produce the sequence $1, 5, 9, 13, \ldots$.)

(**b**)　Since the first term is 5 and the common difference is -3, we obtain

$$\begin{aligned} a_n &= a_1 + (n - 1)d \\ &= 5 + (n - 1)(-3) \\ &= 5 - 3n + 3 \\ &= -3n + 8. \end{aligned}$$

▲

Example 4

Find the 50th term of the arithmetic sequence $2, 6, 10, 14, \ldots$.

Solution

Certainly we could simply continue to write the terms of the given sequence until we reached the 50th term. However, let's use the general-term formula, $a_n = a_1 + (n - 1)d$, to find the 50th term without determining all of the other terms.

$$\begin{aligned} a_{50} &= 2 + (50 - 1)4 \\ &= 2 + 49(4) = 2 + 196 = 198. \end{aligned}$$

▲

Example 5

Find the first term of the arithmetic sequence where the 3rd term is 13 and the 10th term is 62.

Solution

Using $a_n = a_1 + (n - 1)d$ with $a_3 = 13$ (the 3rd term is 13) and $a_{10} = 62$ (the 10th term is 62), we have

$$13 = a_1 + (3 - 1)d = a_1 + 2d$$

$$62 = a_1 + (10 - 1)d = a_1 + 9d.$$

Solve the system of equations

$$\begin{pmatrix} a_1 + 2d = 13 \\ a_1 + 9d = 62 \end{pmatrix}$$

to yield $a_1 = -1$. Thus, the first term is -1. ▲

REMARK Perhaps you can think of another way to solve the problem in Example 5 without using a system of equations. [*Hint*: How many *differences* are there between the 3rd and 10th terms of an arithmetic sequence?] △

We commonly use phrases such as "the set of odd whole numbers," "the set of even whole numbers," and "the set of positive multiples of 5" in mathematical literature to refer to various subsets of the whole numbers. Though no specific ordering of the numbers is implied by these phrases, most of us would probably react with a natural ordering. For example, if we were asked to list the set of *odd whole numbers*, our answer probably would be 1, 3, 5, 7,. . . . So ordered, we can think of the set of odd whole numbers as an arithmetic sequence. Therefore, we can formulate a general representation for the set of odd whole numbers by using the general-term formula. Thus,

$$\begin{aligned} a_n &= a_1 + (n - 1)d \\ &= 1 + (n - 1)2 \\ &= 1 + 2n - 2 \\ &= 2n - 1. \end{aligned}$$

The final example of this section illustrates the use of an arithmetic sequence to solve a problem that deals with the *constant growth* of a man's yearly salary.

Example 6

A man started to work in 1970 at an annual salary of $9500. He received a $700 raise each year. How much was his annual salary in 1991?

Solution

The following arithmetic sequence represents the annual salary beginning in 1970.

9500, 10200, 10900, 11600, . . .

The general term of this sequence is

$$\begin{aligned} a_n &= a_1 + (n - 1)d \\ &= 9500 + (n - 1)700 \\ &= 9500 + 700n - 700 \\ &= 700n + 8800. \end{aligned}$$

The man's 1991 salary is the 22nd term of this sequence. Thus,

$$a_{22} = 700(22) + 8800 = 24200.$$

His salary in 1991 was $24,200. ▲

Problem Set 13.1

For Problems 1–14, write the first five terms of each sequence having the indicated general term.

1. $a_n = 3n - 4$
2. $a_n = 2n + 3$
3. $a_n = -2n + 5$
4. $a_n = -3n - 2$
5. $a_n = n^2 - 2$
6. $a_n = n^2 + 3$
7. $a_n = -n^2 + 1$
8. $a_n = -n^2 - 2$
9. $a_n = 2n^2 - 3$
10. $a_n = 3n^2 + 2$
11. $a_n = 2^{n-2}$
12. $a_n = 3^{n+1}$
13. $a_n = -2(3)^{n-2}$
14. $a_n = -3(2)^{n-3}$

15. Find the 8th and 12th terms of the sequence where $a_n = n^2 - n - 2$.
16. Find the 10th and 15th terms of the sequence where $a_n = -n^2 - 2n + 3$.
17. Find the 7th and 8th terms of the sequence where $a_n = (-2)^{n-2}$.
18. Find the 6th and 7th terms of the sequence where $a_n = -(3)^{n-3}$.

For Problems 19–28, find the general term (nth term) for each of the arithmetic sequences.

19. $1, 3, 5, 7, 9, \ldots$
20. $2, 4, 6, 8, 10, \ldots$
21. $-2, 2, 6, 10, 14, \ldots$
22. $-3, 2, 7, 12, 17, \ldots$
23. $5, 3, 1, -1, -3, \ldots$
24. $2, -1, -4, -7, -10, \ldots$
25. $-7, -10, -13, -16, -19, \ldots$
26. $-3, -5, -7, -9, -11, \ldots$
27. $1, \dfrac{3}{2}, 2, \dfrac{5}{2}, 3, \ldots$
28. $\dfrac{3}{2}, 3, \dfrac{9}{2}, 6, \dfrac{15}{2}, \ldots$

For Problems 29–34, find the indicated term of the arithmetic sequence.

29. The 10th term of $7, 10, 13, 16, \ldots$
30. The 12th term of $9, 11, 13, 15, \ldots$
31. The 20th term of $2, 6, 10, 14, \ldots$
32. The 50th term of $-1, 4, 9, 14, \ldots$
33. The 75th term of $-7, -9, -11, -13, \ldots$
34. The 100th term of $-7, -10, -13, -16, \ldots$

For Problems 35–40, find the number of terms of each of the finite arithmetic sequences.

35. $1, 3, 5, 7, \ldots, 211$
36. $2, 4, 6, 8, \ldots, 312$
37. $10, 13, 16, 19, \ldots, 157$
38. $9, 13, 17, 21, \ldots, 849$
39. $-7, -9, -11, -13, \ldots, -345$
40. $-4, -7, -10, -13, \ldots, -331$

41. If the 6th term of an arithmetic sequence is 24 and the 10th term is 44, find the first term.
42. If the 5th term of an arithmetic sequence is 26 and the 12th term is 75, find the first term.
43. If the 4th term of an arithmetic sequence is -9 and the 9th term is -29, find the 5th term.
44. If the 6th term of an arithmetic sequence is -4 and the 14th term is -20, find the 10th term.
45. In the arithmetic sequence $.97, 1.00, 1.03, 1.06, \ldots$, which term is 5.02?
46. In the arithmetic sequence $1, 1.2, 1.4, 1.6, \ldots$, which term is 35.4?

For Problems 47–50, set up an arithmetic sequence and use $a_n = a_1 + (n-1)d$ to solve each problem.

47. A woman started to work in 1975 at an annual salary of $12,500. She received a $900 raise each year. How much was her annual salary in 1992?

48. Math University had an enrollment of 8500 students in 1976. Each year the enrollment has increased by 350 students. What was the enrollment in 1990?

49. Suppose you are offered a job starting at $900 a month with a guaranteed increase of $30 a month every 6 months for the next 5 years. What will your monthly salary be for the last six months of the 5th year of your employment?

50. Between 1976 and 1990, a person invested $500 at 14% simple interest at the beginning of each year. By the end of 1990, how much interest had been earned by the $500 that was invested at the beginning of 1982?

THOUGHTS INTO WORDS

51. How would you explain the concept of an arithmetic sequence to someone taking an elementary algebra course?

52. Is the sequence whose nth term is $a_n = 2^n$ an arithmetic sequence? Defend your answer.

Further Investigations

The general term of a sequence can consist of one expression for certain values of n and another expression (or expressions) for other values of n. That is to say, we can give a **multiple description** of the sequence. For example,

$$a_n = \begin{cases} 2n + 3 & \text{for } n \text{ odd} \\ 3n - 2 & \text{for } n \text{ even} \end{cases}$$

means that we use $a_n = 2n + 3$ for $n = 1, 3, 5, 7, \ldots$, and we use $a_n = 3n - 2$ for $n = 2, 4, 6, 8, \ldots$. The first six terms of this sequence are 5, 4, 9, 10, 13, and 16.

For Problems 53–56, write the first six terms of each sequence.

53. $a_n = \begin{cases} 2n + 1 & \text{for } n \text{ odd} \\ 2n - 1 & \text{for } n \text{ even} \end{cases}$

54. $a_n = \begin{cases} \dfrac{1}{n} & \text{for } n \text{ odd} \\ n^2 & \text{for } n \text{ even} \end{cases}$

55. $a_n = \begin{cases} 3n + 1 & \text{for } n \leq 3 \\ 4n - 3 & \text{for } n > 3 \end{cases}$

56. $a_n = \begin{cases} 5n - 1 & \text{for } n \text{ a multiple of } 3 \\ 2n & \text{otherwise} \end{cases}$

The multiple description approach can also be used to give a **recursive description** for a sequence. A sequence is said to be described recursively if the first n terms are stated and then each succeeding term is defined as a function of one or more of the preceding terms. For example,

$$\begin{cases} a_1 = 2 \\ a_n = 2a_{n-1} & \text{for } n \geq 2 \end{cases}$$

means that the first term is 2 and each succeeding term is 2 times the previous term. Thus, the first six terms are 2, 4, 8, 16, 32, and 64.

For Problems 57–62, write the first six terms of each sequence.

57. $\begin{cases} a_1 = 4 \\ a_n = 3a_{n-1} & \text{for } n \geq 2 \end{cases}$

58. $\begin{cases} a_1 = 3 \\ a_n = a_{n-1} + 2 & \text{for } n \geq 2 \end{cases}$

59. $\begin{cases} a_1 = 1 \\ a_2 = 1 \\ a_n = a_{n-2} + a_{n-1} & \text{for } n \geq 3 \end{cases}$

60. $\begin{cases} a_1 = 2 \\ a_2 = 3 \\ a_n = 2a_{n-2} + 3a_{n-1} & \text{for } n \geq 3 \end{cases}$

61. $\begin{cases} a_1 = 3 \\ a_2 = 1 \\ a_n = (a_{n-1} - a_{n-2})^2 & \text{for } n \geq 3 \end{cases}$

62. $\begin{cases} a_1 = 1 \\ a_2 = 2 \\ a_3 = 3 \\ a_n = a_{n-1} + a_{n-2} + a_{n-3} & \text{for } n \geq 4 \end{cases}$

13.2 Arithmetic Series

Let's begin this section with a problem to solve. Study the solution very carefully.

Problem 1

Find the sum of the first 100 positive integers.

Solution

We are asked to find the sum of $1 + 2 + 3 + 4 + \cdots + 100$. Rather than use a calculator, let's find the sum in the following way.

$$
\begin{array}{c}
1 + 2 + 3 + 4 + \cdots + 100 \\
100 + 99 + 98 + 97 + \cdots + 1 \\
\hline
101 + 101 + 101 + 101 + \cdots + 101
\end{array}
$$

$$
\frac{\overset{50}{\cancel{100}}(101)}{\underset{2}{\cancel{2}}} = 5050.
$$

Note that we simply wrote the indicated sum *forward and backward* and then added the results. So doing produces 100 sums of 101, but half of them are "repeats." For example, $(100 + 1)$ and $(1 + 100)$ are both counted in this process. Thus, we divided the product $(100)(101)$ by 2, which yielded the final result of 5050. ▲

Now let's look at some terminology that applies to problems such as the one above. The indicated sum of a sequence is called a **series**. Associated with the finite sequence

$$a_1, a_2, a_3, \ldots, a_n$$

is a **finite series**

$$a_1 + a_2 + a_3 + \cdots + a_n.$$

Likewise, from the infinite sequence

$$a_1, a_2, a_3, \ldots$$

we can form the **infinite series**

$$a_1 + a_2 + a_3 + \cdots.$$

In this section we will direct your attention to working with **arithmetic series** that is, the indicated sums of arithmetic sequences.

Problem 1 above could have been stated as "find the sum of the first 100 terms of the arithmetic series that has a general term (nth term) of $a_n = n$." In fact, the *forward-backward approach* used to solve that problem can be applied to the general arithmetic series

$$a_1 + a_2 + a_3 + \cdots + a_n$$

to produce a formula for finding the sum of the first n terms of any arithmetic series. Use S_n to represent the sum of the first n terms and proceed as follows.

$$S_n = a_1 + (a_1 + d) + (a_1 + 2d) + \cdots + (a_n - 2d) + (a_n - d) + a_n.$$

Now we write this sum in reverse as

$$S_n = a_n + (a_n - d) + (a_n - 2d) + \cdots + (a_1 + 2d) + (a_1 + d) + a_1.$$

Add the two equations to produce

$$2S_n = (a_1 + a_n) + (a_1 + a_n) + (a_1 + a_n) + \cdots + (a_1 + a_n) + (a_1 + a_n)$$
$$+ (a_1 + a_n).$$

That is, we have n sums of $(a_1 + a_n)$, so

$$2S_n = n(a_1 + a_n),$$

from which we obtain

$$\boxed{S_n = \frac{n(a_1 + a_n)}{2}.}$$

Using the nth-term formula $a_n = a_1 + (n - 1)d$ along with the sum formula $S_n = \frac{n(a_1 + a_n)}{2}$, we can solve a variety of problems that involve arithmetic series.

Example 1

Find the sum of the first 50 terms of the series

$$2 + 5 + 8 + 11 + \cdots.$$

Solution

Using $a_n = a_1 + (n - 1)d$, we find the 50th term to be

$$a_{50} = 2 + 49(3) = 149.$$

Then using $S_n = \frac{n(a_1 + a_n)}{2}$, we obtain

$$S_{50} = \frac{50(2 + 149)}{2} = 3775. \qquad \blacktriangle$$

Example 2

Find the sum of all odd numbers between 7 and 433, inclusive.

Solution

We need to find the sum $7 + 9 + 11 + \cdots + 433$. To use $S_n = \frac{n(a_1 + a_n)}{2}$, we need the number of terms, n. Perhaps we could figure this out without a formula (try it), but suppose we use the nth-term formula.

$$a_n = a_1 + (n - 1)d$$
$$433 = 7 + (n - 1)2$$

$$433 = 7 + 2n - 2$$
$$433 = 2n + 5$$
$$428 = 2n$$
$$214 = n$$

Then, use $n = 214$ in the sum formula to yield

$$S_{214} = \frac{214(7 + 433)}{2} = 47080.$$ ▲

Example 3

Find the sum of the first 75 terms of the series that has a general term of

$$a_n = -5n + 9.$$

Solution

Using $a_n = -5n + 9$, we can generate the series as follows.

$$a_1 = -5(1) + 9 = 4,$$
$$a_2 = -5(2) + 9 = -1,$$
$$a_3 = -5(3) + 9 = -6,$$
$$\vdots$$
$$a_{75} = -5(75) + 9 = -366$$

Thus, we have the series

$$4 + (-1) + (-6) + \cdots + (-366).$$

Using the sum formula, we obtain

$$S_{75} = \frac{75[4 + (-366)]}{2} = -13575.$$ ▲

Example 4

Kera is saving quarters. She saves 1 quarter the first day, 2 quarters the second day, 3 quarters the third day, and so on. How much money will she have saved in 30 days?

Solution

The total number of quarters will be the sum of the series

$$1 + 2 + 3 + \cdots + 30.$$

Using the sum formula yields

$$S_{30} = \frac{30(1 + 30)}{2} = 465.$$

So Kera will have saved $(465)(\$.25) = \$116.25$ at the end of 30 days.   ▲

The sum formula, $S_n = \dfrac{n(a_1 + a_n)}{2}$, was developed using the *forward-backward technique* that we used to solve Problem 1 of this section. Now that you know the sum formula, you have two choices. You can either memorize the formula and use it as it applies or disregard the formula and use the forward-backward technique. However, if you choose to use the sum formula and then some day your memory fails you and you forget it, you can still use the forward-backward approach. In other words, once you understand the development of a formula you can solve some problems even though you have forgotten the formula itself.

Problem Set 13.2

For Problems 1–10, find the sum of the indicated number of terms of the given series.

1. First 50 terms of $2 + 4 + 6 + 8 + \cdots$

2. First 45 terms of $1 + 3 + 5 + 7 + \cdots$

3. First 60 terms of $3 + 8 + 13 + 18 + \cdots$

4. First 80 terms of $2 + 6 + 10 + 14 + \cdots$

5. First 65 terms of $(-1) + (-3) + (-5) + (-7) + \cdots$

6. First 100 terms of $(-1) + (-4) + (-7) + (-10) + \cdots$

7. First 40 terms of $\dfrac{1}{2} + 1 + \dfrac{3}{2} + 2 + \cdots$

8. First 50 terms of $1 + \dfrac{5}{2} + 4 + \dfrac{11}{2} + \cdots$

9. First 75 terms of $7 + 10 + 13 + 16 + \cdots$

10. First 90 terms of $(-8) + (-1) + 6 + 13 + \cdots$

For Problems 11–16, find the sum of each finite arithmetic series.

11. $4 + 8 + 12 + 16 + \cdots + 212$

12. $7 + 9 + 11 + 13 + \cdots + 179$

13. $(-4) + (-1) + 2 + 5 + \cdots + 173$

14. $5 + 10 + 15 + 20 + \cdots + 495$

15. $2.5 + 3.0 + 3.5 + 4.0 + \cdots + 18.5$

16. $1 + (-6) + (-13) + (-20) + \cdots + (-202)$

For Problems 17–22, find the sum of the indicated number of terms of the series with the given nth term.

17. First 50 terms of series with $a_n = 3n - 1$

18. First 150 terms of series with $a_n = 2n - 7$

19. First 125 terms of series with $a_n = 5n + 1$

20. First 75 terms of series with $a_n = 4n + 3$

21. First 65 terms of series with $a_n = -4n - 1$

22. First 90 terms of series with $a_n = -3n - 2$

For Problems 23–34, use arithmetic sequences and series to help solve the problem.

23. Find the sum of the first 350 positive even whole numbers.

24. Find the sum of the first 200 odd whole numbers.

25. Find the sum of all odd whole numbers between 15 and 397, inclusive.

26. Find the sum of all even whole numbers between 14 and 286, inclusive.

27. An auditorium has 20 seats in the front row, 24 seats in the second row, 28 seats in the third row, and so on, for 15 rows. How many seats are there in the last row? How many seats are there in the auditorium?

28. A pile of wood has 15 logs in the bottom row, 14 logs in the next-to-the-bottom row, and so on, with one less log in each row until the top row, which consists of 1 log. How many logs are there in the pile?

29. A raffle is organized so that the amount paid for each ticket is determined by a number on the ticket. The tickets are numbered with the consecutive odd whole numbers, 1, 3, 5, 7, ... Each participant pays as many cents as the number on the ticket drawn. How much money will the raffle take in if 1000 tickets are sold?

30. A woman invests $700 at 13% simple interest at the beginning of each year for a period of 15 years. Find the total accumulated value of all the investments at the end of the 15-year period.

31. A man started to work in 1970 at an annual salary of $18,500. He received a $1500 raise each year through 1982. What were his total earnings for the 13-year period?

32. A well-driller charges $9.00 per foot for the first 10 feet, $9.10 per foot for the next 10 feet, $9.20 per foot for the next 10 feet, and so on; he continues to increase the price by $0.10 per foot for succeeding intervals of 10 feet. How much would he charge to drill a well with a depth of 150 feet?

33. A display in a grocery store has cans stacked with 25 cans in the bottom row, 23 cans in the second row from the bottom, 21 cans in the third row from the bottom, and so on until there is only 1 can in the top row. How many cans are there in the display?

34. Suppose that a person starts on the first day of August and saves a dime the first day, $.20 the second day, $.30 the third day, and continues to save $.10 more per day than the previous day. How much could be saved in the 31 days of August?

THOUGHTS INTO WORDS

35. Explain how to find the sum $1 + 2 + 3 + \cdots + 150$ without using the sum formula.

36. Explain how you would find the sum of the first 150 terms of the arithmetic series that has a general term of $a_n = -3n + 7$.

Further Investigations

37. A series where the general term is known can be expressed in a convenient and compact form using the symbol $\sum$ along with the general term expression. For example, consider the finite arithmetic series

$$1 + 3 + 5 + 7 + 9 + 11,$$

where the general term is $a_n = 2n - 1$. This series can be expressed in *summation notation* as

$$\sum_{i=1}^{6}(2i - 1)$$

where the letter i is used as the *index of summation*. The individual terms of the series can be generated by successively replacing i in the expression $(2i - 1)$ with the numbers 1, 2, 3, 4, 5, and 6. Thus, the first term is $2(1) - 1 = 1$, the second term is $2(2) - 1 = 3$, the third term is $2(3) - 1 = 5$, and so on. Write out the terms and find the sum of each of the following series.

(a) $\displaystyle\sum_{i=1}^{3}(5i + 2)$

(b) $\displaystyle\sum_{i=1}^{4}(6i - 7)$

(c) $\displaystyle\sum_{i=1}^{6}(-2i - 1)$

(d) $\displaystyle\sum_{i=1}^{5}(-3i + 4)$

(e) $\displaystyle\sum_{i=1}^{5}3i$

(f) $\displaystyle\sum_{i=1}^{6} - 4i$

38. Write each of the following in summation nota-
tion. For example, since $3 + 8 + 13 + 18 + 23 + 28$ is an arithmetic series, the general term formula $a_n = a_1 + (n - 1)d$ yields

$$a_n = 3 + (n - 1)5$$
$$= 3 + 5n - 5$$
$$= 5n - 2.$$

Now using i as an index of summation, we can write

$$\sum_{i=1}^{6}(5i - 2).$$

(a) $2 + 5 + 8 + 11 + 14$

(b) $8 + 15 + 22 + 29 + 36 + 43$

(c) $1 + (-1) + (-3) + (-5) + (-7)$

(d) $(-5) + (-9) + (-13) + (-17) + (-21) + (-25) + (-29)$

13.3 Geometric Sequences and Series

A **geometric sequence** or **geometric progression** is a sequence in which each term after the first is obtained by multiplying the preceding term by a common multiplier. The common multiplier is called the **common ratio** of the sequence. The following geometric sequences have common ratios of 2, 3, $\frac{1}{2}$, and -4, respectively.

$$1, 2, 4, 8, 16, \ldots \qquad 3, 9, 27, 81, 243, \ldots$$
$$8, 4, 2, 1, \frac{1}{2}, \ldots \qquad 1, -4, 16, -64, 256, \ldots$$

We can find the common ratio of a geometric sequence by dividing a term (other than the first term) by the preceding term.

In a more general setting we say that the sequence

$$a_1, a_2, a_3, a_4, \ldots, a_n, \ldots$$

is a geometric sequence if and only if there is a nonzero real number r such that

$$a_{k+1} = ra_k \tag{1}$$

for every positive integer k. The nonzero real number r is called the **common ratio**.

Equation (1) can be used to generate a general geometric sequence that has a_1 as a first term and r as a common ratio. We can proceed as follows.

First term	a_1	
Second term	$a_1 r$	
Third term	$a_1 r^2$	$(a_1 r)(r) = a_1 r^2$
Fourth term	$a_1 r^3$	$(a_1 r^2)(r) = a_1 r^3$
$\vdots$		
nth term	$a_1 r^{n-1}$	

Thus, the general term of a geometric sequence is given by

$$\boxed{a_n = a_1 r^{n-1},}$$

where a_1 is the first term and r is the common ratio.

Example 1

Solution

Find the general term for the geometric sequence 2, 4, 8, 16, ...

Using $a_n = a_1 r^{n-1}$, we obtain

$$a_n = 2(2)^{n-1} \qquad r = \frac{4}{2} = \frac{8}{4} = \frac{16}{8} = 2$$
$$= 2^n. \qquad\qquad 2^1 (2)^{n-1} = 2^{1+n-1} = 2^n$$

▲

Example 2

Solution

Find the 10th term of the geometric sequence 9, 3, 1, ...

Using $a_n = a_1 r^{n-1}$, we can find the 10th term as follows.

$$a_{10} = 9\left(\frac{1}{3}\right)^{10-1}$$

$$= 9\left(\frac{1}{3}\right)^9$$

$$= 9\left(\frac{1}{19683}\right)$$

$$= \frac{1}{2187}$$

▲

A **geometric series** is the indicated sum of a geometric sequence. The following are examples of geometric series.

$$1 + 2 + 4 + 8 + \cdots,$$

$$3 + 9 + 27 + 81 + \cdots,$$

$$8 + 4 + 2 + 1 + \cdots,$$

$$1 + (-4) + 16 + (-64) + \cdots$$

Before we develop a general formula for finding the sum of a geometric series, let's consider a specific example.

Example 3

Solution

Find the sum of $1 + 2 + 4 + 8 + \cdots + 512$.

Let S represent the sum and we can proceed as follows.

$$S = 1 + 2 + 4 + 8 + \cdots + 512, \tag{2}$$

$$2S = \qquad 2 + 4 + 8 + \cdots + 512 + 1024 \tag{3}$$

Equation (3) is the result of multiplying both sides of equation (2) by 2. Subtract equation (2) from equation (3) to yield

$$S = 1024 - 1$$
$$= 1023.$$ ▲

Now let's consider the general geometric series

$$a_1 + a_1r + a_1r^2 + \cdots + a_1r^{n-1}.$$

By applying a procedure similar to the one we used in Example 3, we can develop a formula for finding the sum of the first n terms of any geometric series. Let S_n represent the sum of the first n terms. Thus,

$$S_n = a_1 + a_1r + a_1r^2 + \cdots + a_1r^{n-1}. \tag{4}$$

Multiply both sides of equation (4) by the common ratio r to produce

$$rS_n = a_1r + a_1r^2 + a_1r^3 + \cdots + a_1r^{n-1} + a_1r^n. \tag{5}$$

Subtract equation (4) from equation (5) to yield

$$rS_n - S_n = a_1r^n - a_1.$$

Apply the distributive property on the left side and then solve for S_n to obtain

$$S_n(r - 1) = a_1r^n - a_1$$

$$S_n = \frac{a_1r^n - a_1}{r - 1}, \qquad r \neq 1.$$

Therefore, the sum of the first n terms of a geometric series that has a first term of a_1 and a common ratio of r is given by

$$\boxed{S_n = \frac{a_1r^n - a_1}{r - 1}, \qquad r \neq 1.}$$

Example 4

Find the sum of the first 7 terms of the geometric series $2 + 6 + 18 + \cdots$.

Solution

Use the sum formula to obtain

$$S_7 = \frac{2(3)^7 - 2}{3 - 1}$$

$$= \frac{2(3^7 - 1)}{2}$$

$$= 3^7 - 1$$

$$= 2187 - 1$$

$$= 2186.$$ ▲

If the common ratio of a geometric series is less than 1, it may be more convenient to change the form of the sum formula. That is, we can change the fraction $\dfrac{a_1 r^n - a_1}{r - 1}$ to $\dfrac{a_1 - a_1 r^n}{1 - r}$ by multiplying both the numerator and the denominator by -1. Thus, using $S_n = \dfrac{a_1 - a_1 r^n}{1 - r}$ when $r < 1$ can sometimes allow us to avoid unnecessary work with negative numbers, as the next example demonstrates.

Example 5

Find the sum of the geometric series $1 + \dfrac{1}{2} + \dfrac{1}{4} + \cdots + \dfrac{1}{256}$.

Solution A

To use the sum formula, we need to know the number of terms, which can be found by simply counting them or by applying the nth-term formula, as follows.

$$a_n = a_1 r^{n-1}$$

$$\frac{1}{256} = 1\left(\frac{1}{2}\right)^{n-1}$$

$$\left(\frac{1}{2}\right)^8 = \left(\frac{1}{2}\right)^{n-1}$$

$$8 = n - 1 \qquad \text{Remember that ``if } b^n = b^m, \text{ then } n = m.\text{''}$$

$$9 = n.$$

Using $n = 9$, $a_1 = 1$, and $r = \dfrac{1}{2}$ in the form of the sum formula

$$S_n = \frac{a_1 - a_1 r^n}{1 - r},$$

we obtain

$$S_9 = \frac{1 - 1\left(\dfrac{1}{2}\right)^9}{1 - \dfrac{1}{2}}$$

$$= \frac{1 - \dfrac{1}{512}}{\dfrac{1}{2}}$$

$$= \frac{\dfrac{511}{512}}{\dfrac{1}{2}}$$

$$= \left(\frac{511}{512}\right)\left(\frac{2}{1}\right) = \frac{511}{256} = 1\frac{255}{256}.$$

You should realize that a problem such as Example 5 can be done *without* using the sum formula; we can simply apply the general technique used to develop the formula. Solution B illustrates this approach.

Solution B

Let S represent the desired sum. Thus,

$$S = 1 + \frac{1}{2} + \frac{1}{4} + \cdots + \frac{1}{256}.$$

Multiply both sides by $\frac{1}{2}$ (the common ratio):

$$\frac{1}{2}S = \frac{1}{2} + \frac{1}{4} + \cdots + \frac{1}{256} + \frac{1}{512}.$$

Subtract the second equation from the first equation to produce

$$\frac{1}{2}S = 1 - \frac{1}{512}$$

$$\frac{1}{2}S = \frac{511}{512}$$

$$S = \frac{511}{256} = 1\frac{255}{256}. \quad \blacktriangle$$

Example 6

Suppose your employer agrees to pay you a penny for your first day's wages and then double your pay on each succeeding day. How much will you earn on the 15th day? What will your total earnings for the first 15 days be?

Solution

The terms of the geometric series $1 + 2 + 4 + 8 + \cdots$ depict your daily wages, and the sum of the first 15 terms is your total earnings for the 15 days. We can use the formula $a_n = a_1 r^{n-1}$ to find the 15th day's wages.

$$a_{15} = (1)(2)^{14} = 16384.$$

Since the terms of the series are expressed in cents, your wages for the 15th day would be $163.84. Now, using the sum formula, we can find your total earnings as follows.

$$S_n = \frac{a_1 r^n - a_1}{r - 1}$$

$$S_{15} = \frac{1(2)^{15} - 1}{1}$$

$$= 32768 - 1 = 32767.$$

Thus, for the 15 days you would earn a total of $327.67. $\quad \blacktriangle$

Problem Set 13.3

For Problems 1–12, find the general term (nth term) of each geometric sequence.

1. $1, 3, 9, 27, \ldots$
2. $1, 2, 4, 8, \ldots$
3. $2, 8, 32, 128, \ldots$
4. $3, 9, 27, 81, \ldots$
5. $1, \dfrac{1}{3}, \dfrac{1}{9}, \dfrac{1}{27}, \ldots$
6. $\dfrac{1}{2}, \dfrac{1}{4}, \dfrac{1}{8}, \dfrac{1}{16}, \ldots$
7. $.2, .04, .008, .0016, \ldots$
8. $1, .3, .09, .027, \ldots$
9. $9, 6, 4, \dfrac{8}{3}, \ldots$
10. $6, 2, \dfrac{2}{3}, \dfrac{2}{9}, \ldots$
11. $1, -4, 16, -64, \ldots$
12. $1, -2, 4, -8, \ldots$

For Problems 13–18, find the indicated term of the geometric sequence.

13. 12th term of $\dfrac{1}{9}, \dfrac{1}{3}, 1, 3, \ldots$
14. 9th term of $2, 4, 8, 16, \ldots$
15. 10th term of $1, -2, 4, -8, \ldots$
16. 8th term of $\dfrac{1}{2}, \dfrac{1}{8}, \dfrac{1}{32}, \dfrac{1}{128}, \ldots$
17. 9th term of $-1, -\dfrac{3}{2}, -\dfrac{9}{4}, -\dfrac{27}{8}, \ldots$
18. 11th term of $1, \dfrac{2}{3}, \dfrac{4}{9}, \dfrac{8}{27}, \ldots$

For Problems 19–24, find the sum of the indicated number of terms of each geometric series.

19. First 10 terms of $\dfrac{1}{2} + \dfrac{3}{2} + \dfrac{9}{2} + \dfrac{27}{2} + \cdots$
20. First 9 terms of $1 + 2 + 4 + 8 + \cdots$

21. First 9 terms of $-2 + 6 + (-18) + 54 + \cdots$
22. First 10 terms of $-4 + 8 + (-16) + 32 + \cdots$
23. First 7 terms of $1 + 3 + 9 + 27 + \cdots$
24. First 8 terms of $4 + 2 + 1 + \dfrac{1}{2} + \cdots$

For Problems 25–30, find the sum of the indicated number of terms of the geometric series with the given nth term.

25. First 9 terms of series where $a_n = 2^{n-1}$.
26. First 8 terms of series where $a_n = 3^n$
27. First 8 terms of series where $a_n = 2(3)^n$
28. First 10 terms of series where $a_n = \dfrac{1}{2^{n-4}}$
29. First 12 terms of series where $a_n = (-2)^n$
30. First 9 terms of series where $a_n = (-3)^{n-1}$

For Problems 31–36, find the sum of each finite geometric series.

31. $1 + 3 + 9 + \cdots + 729$
32. $2 + 8 + 32 + \cdots + 2048$
33. $1 + \dfrac{1}{2} + \dfrac{1}{4} + \cdots + \dfrac{1}{1024}$
34. $1 + (-2) + 4 + \cdots + (-128)$
35. $8 + 4 + 2 + \cdots + \dfrac{1}{32}$
36. $2 + 6 + 18 + \cdots + 4374$

For Problems 37–48, use geometric sequences and series to help solve the problem.

37. Find the common ratio of a geometric sequence if the 2nd term is $\dfrac{1}{6}$ and the 5th term is $\dfrac{1}{48}$.
38. Find the first term of a geometric sequence if the 5th term is $\dfrac{32}{3}$ and the common ratio is 2.

39. Find the sum of the first 16 terms of the geometric series where $a_n = (-1)^n$. Also find the sum of the first 19 terms.

40. A fungus culture growing under controlled conditions doubles in size each day. How many units will the culture contain after 7 days if it originally contained 5 units?

41. A tank contains 16,000 liters of water. Each day one-half of the water in the tank is removed and not replaced. How much water remains in the tank at the end of the 7th day?

42. Suppose that you save 25 cents the first day of a week, 50 cents the second day, $1 the third day, and continue to double your savings each day. How much will you save on the 7th day? What will be your total savings for the week?

43. Suppose you save a nickel the first day of a month, a dime the second day, 20 cents the third day, and continue to double your savings each day. How much will you save on the 12th day of the month? What will your total savings be for the first 12 days?

44. Suppose an element has a half-life of 3 hours. This means that if n grams of it exist at a specific time, then only $\frac{1}{2}n$ grams remain 3 hours later. If at a particular moment we have 40 grams of the element, how much of it remains 24 hours later?

45. A rubber ball is dropped from a height of 486 meters and each time it rebounds one-third of the height from which it last fell. How far has the ball traveled by the time it strikes the ground for the 7th time?

46. A pump is attached to a container for the purpose of creating a vacuum. For each stroke of the pump, one-fourth of the air remaining in the container is removed.
(a) Form a geometric sequence where each term represents the fractional part of the air *that still remains* in the container after each stroke. Then use this sequence to find out how much of the air remains after 6 strokes.
(b) Form a geometric sequence where each term represents the fractional part of the air *being*

removed from the container on each stroke of the pump. Then use this sequence (or the associated *series*) to find out how much of the air remains after 6 strokes.

47. If you pay $9500 for a car and its value depreciates 10% per year, how much will it be worth in 5 years?

48. Suppose that you could get a job that pays only a penny for the first day of employment, but doubles your wages each succeeding day. How much would you be earning on the 31st day of your employment?

THOUGHTS INTO WORDS

49. Explain the difference between an arithmetic sequence and a geometric sequence.

50. Consider the sequence whose nth term is $a_n = 3n$. How can you determine if this is an arithmetic sequence or a geometric sequence?

Further Investigations

For Problems 51–56, use your calculator to help find the indicated term of each geometric sequence.

51. 20th term of 2, 4, 8, 16, . . .

52. 15th term of 3, 9, 27, 81, . . .

53. 12th term of $\frac{2}{3}, \frac{4}{9}, \frac{8}{27}, \frac{16}{81}, \ldots$

54. 10th term of $-\frac{3}{4}, \frac{9}{16}, -\frac{27}{64}, \frac{81}{256}, \ldots$

55. 11th term of $-\frac{3}{2}, \frac{9}{4}, -\frac{27}{8}, \frac{81}{16}, \ldots$

56. 6th term of the sequence for which $a_n = (.1)^n$

57. In Problem 37 of Problem Set 13.2, we introduced the summation notation, which can also be used with geometric series. For example, the series

$$2 + 4 + 8 + 16 + 32$$

can be expressed as

$$\sum_{i=1}^{5} 2^i.$$

Write out the terms and find the sum of each of the following.

(a) $\displaystyle\sum_{i=1}^{6} 2^i$

(b) $\displaystyle\sum_{i=1}^{5} 3^i$

(c) $\displaystyle\sum_{i=1}^{5} 2^{i-1}$

(d) $\displaystyle\sum_{i=1}^{6} \left(\frac{1}{2}\right)^{i+1}$

(e) $\displaystyle\sum_{i=1}^{4} \left(\frac{2}{3}\right)^i$

(f) $\displaystyle\sum_{i=1}^{5} \left(-\frac{3}{4}\right)^i$

13.4 Infinite Geometric Series

In Section 13.3 we used the formula

$$S_n = \frac{a_1 - a_1 r^n}{1 - r}, \qquad r \neq 1 \tag{1}$$

to find the sum of the first n terms of a geometric series. By using the property $\dfrac{a - b}{c} = \dfrac{a}{c} - \dfrac{b}{c}$, we can express the right side of equation (1) in terms of two fractions as follows.

$$S_n = \frac{a_1 - a_1 r^n}{1 - r} = \frac{a_1}{1 - r} - \frac{a_1 r^n}{1 - r}, \qquad r \neq 1 \tag{2}$$

Now let's examine the behavior of r^n for $|r| < 1$, that is, for $-1 < r < 1$. For example, suppose that $r = \dfrac{1}{3}$; then

$$r^2 = \left(\frac{1}{3}\right)^2 = \frac{1}{9}, \qquad r^3 = \left(\frac{1}{3}\right)^3 = \frac{1}{27},$$

$$r^4 = \left(\frac{1}{3}\right)^4 = \frac{1}{81}, \qquad r^5 = \left(\frac{1}{3}\right)^5 = \frac{1}{243},$$

and so on. We can make $\left(\dfrac{1}{3}\right)^n$ as close to zero as we please by taking sufficiently large values for n. In general, for values of r such that $|r| < 1$, the expression r^n will approach zero as n increases. Therefore, in equation (2), the fraction $\dfrac{a_1 r^n}{1 - r}$ will approach zero as n increases, and we say that the *sum of an infinite geometric series* is given by

$$\boxed{S_\infty = \frac{a_1}{1 - r}, \qquad |r| < 1.}$$

Example 1

Find the sum of the infinite geometric series

$$1 + \frac{2}{3} + \frac{4}{9} + \frac{8}{27} + \cdots.$$

Solution

Since $a_1 = 1$ and $r = \dfrac{2}{3}$,

$$S_\infty = \frac{1}{1 - \dfrac{2}{3}} = \frac{1}{\dfrac{1}{3}} = 3.$$ ▲

In Example 1, $S_\infty = 3$ means that as we add more and more terms, the sum approaches 3 as follows.

First term: $\quad 1$

Sum of first 2 terms: $\quad 1 + \dfrac{2}{3} = 1\dfrac{2}{3}$

Sum of first 3 terms: $\quad 1 + \dfrac{2}{3} + \dfrac{4}{9} = 2\dfrac{1}{9}$

Sum of first 4 terms: $\quad 1 + \dfrac{2}{3} + \dfrac{4}{9} + \dfrac{8}{27} = 2\dfrac{11}{27}$

Sum of first 5 terms: $\quad 1 + \dfrac{2}{3} + \dfrac{4}{9} + \dfrac{8}{27} + \dfrac{16}{81} = 2\dfrac{49}{81}$

etc.

Example 2

Find the sum of the infinite geometric series

$$\frac{1}{2} - \frac{1}{4} + \frac{1}{8} - \frac{1}{6} + \cdots .$$

Solution

Since $a_1 = \dfrac{1}{2}$ and $r = -\dfrac{1}{2}$, we obtain

$$S_\infty = \frac{\dfrac{1}{2}}{1 - \left(-\dfrac{1}{2}\right)} = \frac{\dfrac{1}{2}}{\dfrac{3}{2}} = \frac{1}{3}.$$ ▲

If $|r| > 1$, the absolute value of r^n increases without bound as n increases. Consider the following two examples and notice the unbounded growth of the absolute value of r^n.

Let r = 2

$r^2 = 2^2 = 4$

$r^3 = 2^3 = 8$

$r^4 = 2^4 = 16$

$r^5 = 2^5 = 32$

etc.

Let r = −3

$r^2 = (-3)^2 = 9$

$r^3 = (-3)^3 = -27 \qquad |-27| = 27$

$r^4 = (-3)^4 = 81$

$r^5 = (-3)^5 = -243 \qquad |-243| = 243$

etc.

If $r = 1$, then $S_n = na_1$, and as n increases without bound, $|S_n|$ also increases without bound. If $r = -1$, then S_n will either be a_1 or 0. Therefore, we say that the sum of any infinite geometric sequence where $|r| \geq 1$ *does not exist*.

Repeating Decimals as Infinite Geometric Series

In Section 1.1, we stated that a rational number is a number that has either a terminating or a repeating decimal representation. For example,

$$.23, \qquad .147, \qquad .\overline{3}, \qquad .\overline{14}, \qquad \text{and} \qquad .5\overline{81}$$

are examples of rational numbers. (Remember that $.\overline{3}$ means $.333\ldots$.) Our knowledge of place value provides the basis for changing terminating decimals such as .23 and .147 to $\dfrac{a}{b}$ form, where a and b are integers, $b \neq 0$.

$$.23 = \frac{23}{100},$$

$$.147 = \frac{147}{1000}$$

However, changing repeating decimals to $\dfrac{a}{b}$ form requires a different technique, and our work with infinite geometric series provides the basis for one such approach. Consider the following examples.

Example 3

Change $.\overline{3}$ to $\dfrac{a}{b}$ form, where a and b are integers, $b \neq 0$.

Solution

We can write the repeating decimal $.\overline{3}$ as the infinite geometric series

$$.3 + .03 + .003 + .0003 + \cdots,$$

with $a_1 = .3$ and $r = .1$. Therefore, we can use the sum formula and obtain

$$S_\infty = \frac{a_1}{1 - r} = \frac{.3}{1 - .1} = \frac{.3}{.9} = \frac{3}{9} = \frac{1}{3}.$$

So, $.\overline{3} = \dfrac{1}{3}$.

Example 4

Change $.\overline{14}$ to $\dfrac{a}{b}$ form, where a and b are integers, $b \neq 0$.

Solution

We can write the repeating decimal $.\overline{14}$ as the infinite geometric series

$$.14 + .0014 + .000014 + \cdots,$$

with $a_1 = .14$ and $r = .01$. The sum formula produces

$$S_\infty = \frac{.14}{1 - 0.01} = \frac{.14}{.99} = \frac{14}{99}.$$

Thus, $.\overline{14} = \frac{14}{99}.$ ▲

If the repeating block of digits does not begin immediately after the decimal point, we can make a slight adjustment, as the final example illustrates.

Example 5

Change $.5\overline{81}$ to $\frac{a}{b}$ form, where a and b are integers, $b \neq 0$.

Solution

We can write the repeating decimal $.5\overline{81}$ as

$$(.5 + .081 + .00081 + .0000081 + \cdots),$$

where

$$.081 + .00081 + .0000081 + \cdots$$

is an infinite geometric series with $a_1 = .081$ and $r = .01$. Thus,

$$S_\infty = \frac{.081}{1 - .01} = \frac{.081}{.99} = \frac{81}{990} = \frac{9}{110}.$$

Therefore,

$$.5\overline{81} = .5 + \frac{9}{110}$$

$$= \frac{5}{10} + \frac{9}{110}$$

$$= \frac{55}{110} + \frac{9}{110}$$

$$= \frac{64}{110} = \frac{32}{55}.$$ ▲

Problem Set 13.4

For Problems 1–20, find the sum of the infinite geometric series. If the series has no sum, so state.

1. $1 + \frac{3}{4} + \frac{9}{16} + \frac{27}{64} + \cdots$

2. $\frac{2}{3} + \frac{2}{9} + \frac{2}{27} + \frac{2}{81} + \cdots$

3. $\frac{1}{2} + \frac{1}{4} + \frac{1}{8} + \frac{1}{16} + \cdots$

4. $1 + \frac{1}{2} + \frac{1}{4} + \frac{1}{8} + \cdots$

5. $\frac{2}{3} + \frac{4}{9} + \frac{8}{27} + \frac{16}{81} + \cdots$

6. $\dfrac{1}{3} + \dfrac{1}{9} + \dfrac{1}{27} + \dfrac{1}{81} + \cdots$

7. $1 - \dfrac{1}{2} + \dfrac{1}{4} - \dfrac{1}{8} + \cdots$

8. $1 + 2 + 4 + 8 + \cdots$

9. $6 + 2 + \dfrac{2}{3} + \dfrac{2}{9} + \cdots$

10. $4 + (-2) + 1 + \left(-\dfrac{1}{2}\right) + \cdots$

11. $2 + (-6) + 18 + (-54) + \cdots$

12. $4 + 2 + 1 + \dfrac{1}{2} + \cdots$

13. $1 + \left(-\dfrac{3}{4}\right) + \dfrac{9}{16} + \left(-\dfrac{27}{64}\right) + \cdots$

14. $9 - 3 + 1 - \dfrac{1}{3} + \cdots$

15. $8 - 4 + 2 - 1 + \cdots$

16. $5 + 3 + \dfrac{9}{5} + \dfrac{27}{25} + \cdots$

17. $1 + \dfrac{3}{2} + \dfrac{9}{4} + \dfrac{27}{8} + \cdots$

18. $1 - \dfrac{4}{3} + \dfrac{16}{9} - \dfrac{64}{27} + \cdots$

19. $27 + 9 + 3 + 1 + \cdots$

20. $9 + 3 + 1 + \dfrac{1}{3} + \cdots$

For Problems 21–34, change each repeating decimal to $\dfrac{a}{b}$ form, where a and b are integers, $b \neq 0$. Express $\dfrac{a}{b}$ in reduced form.

21. $.\overline{4}$ **22.** $.\overline{7}$

23. $.\overline{47}$ **24.** $.\overline{23}$

25. $.\overline{45}$ **26.** $.\overline{72}$

27. $.\overline{427}$ **28.** $.\overline{129}$

29. $.4\overline{6}$ **30.** $.8\overline{6}$

31. $2.\overline{18}$ **32.** $2.9\overline{6}$

33. $.4\overline{27}$ **34.** $.2\overline{36}$

THOUGHTS INTO WORDS

35. What does it mean to say that the sum of the infinite geometric series $1 + \dfrac{1}{2} + \dfrac{1}{4} + \dfrac{1}{8} + \cdots$ is 2?

36. What do we mean when we state that the infinite geometric sequence $1 + 2 + 4 + 8 + \cdots$ has no sum?

37. Why don't we discuss the sum of an infinite arithmetic series?

13.5 Binomial Expansions

In Chapter 3, we used the pattern $(x + y)^2 = x^2 + 2xy + y^2$ to square binomials and the pattern $(x + y)^3 = x^3 + 3x^2y + 3xy^2 + y^3$ to cube binomials. At this time, we simply want to extend those ideas to arrive at a pattern that will allow us to write the expansion of $(x + y)^n$, where n is *any* positive integer. Let's begin by looking at some specific expansions, which can be verified by direct multiplication.

$$(x + y)^1 = x + y$$

$$(x + y)^2 = x^2 + 2xy + y^2$$

$$(x + y)^3 = x^3 + 3x^2y + 3xy^2 + y^3$$

$$(x + y)^4 = x^4 + 4x^3y + 6x^2y^2 + 4xy^3 + y^4$$

$$(x + y)^5 = x^5 + 5x^4y + 10x^3y^2 + 10x^2y^3 + 5xy^4 + y^5$$

First, note the patterns of the exponents for x and y on a term-by-term basis. The exponents of x begin with the exponent of the binomial, and term by term they decrease by 1, until the last term has x^0, which is 1. The exponents of y begin with $0(y^0 = 1)$, and term by term they increase by 1, until the last term contains y to the power of the original binomial. In other words, the variables in the expansion of $(x + y)^n$ have the following pattern:

$$x^n, \qquad x^{n-1}y, \qquad x^{n-2}y^2, \qquad x^{n-3}y^3, \qquad \ldots, \qquad xy^{n-1}, \qquad y^n.$$

Notice that the sum of the exponents of x and y for each term is n.

Next, let's arrange the **coefficients** in the following triangular formation, which yields an easy-to-remember pattern.

$$
\begin{array}{ccccccccccc}
 & & & & 1 & & 1 & & & & \\
 & & & 1 & & 2 & & 1 & & & \\
 & & 1 & & 3 & & 3 & & 1 & & \\
 & 1 & & 4 & & 6 & & 4 & & 1 & \\
1 & & 5 & & 10 & & 10 & & 5 & & 1 \\
\end{array}
$$

The number of the row of the formation contains the coefficients of the expansion of $(x + y)$ to that power. For example, the fifth row contains 1, 5, 10, 10, 5, 1, which are the coefficients of the terms of the expansion of $(x + y)^5$. Furthermore, each row can be formed from the previous row as follows.

1. Start and end each row with 1.

2. All other entries result from adding the two numbers in the row immediately above, one number to the left and one number to the right.

Thus, from row 5 we can form row 6 as follows.

Row 5:   1   5   10   10   5   1

     add   add   add   add   add

Row 6:   1   6   15   20   15   6   1   **(1)**

Using the row-6 coefficients and our previous observations relative to the exponents, we can write out the expansion for $(x + y)^6$.

$$(x + y)^6 = x^6 + 6x^5y + 15x^4y^2 + 20x^3y^3 + 15x^2y^4 + 6xy^5 + y^6$$

REMARK   We often refer to the triangular formation of numbers that we have been discussing as **Pascal's triangle** in honor of Blaise Pascal, the seventeenth-century mathematician to whom the discovery of this pattern is attributed.   △

Although Pascal's triangle will work for any positive integral power of a binomial, it does become somewhat impractical for large powers. So we need another technique for determining the coefficients. For this purpose, it is convenient to make the following notational agreements. $n!$ (read "n factorial") means $n(n - 1)(n - 2) \cdots 1$, where n is any positive integer. For example,

3! means $3 \cdot 2 \cdot 1 = 6$,

5! means $5 \cdot 4 \cdot 3 \cdot 2 \cdot 1 = 120$.

We also agree that $0! = 1$. (Note that both $0!$ and $1!$ equal 1.)

Let us now use the factorial notation to state the expansion of the general case $(x + y)^n$, where n is any positive integer.

$$(x + y)^n = x^n + nx^{n-1}y + \frac{n(n - 1)}{2!}x^{n-2}y^2 + \frac{n(n - 1)(n - 2)}{3!}x^{n-3}y^3 + \cdots + y^n.$$

The binomial expansion for the general case may look a little confusing, but actually it is quite easy to apply once you try it a few times on some specific examples. Remember the decreasing pattern for the exponents of x and the increasing pattern for the exponents of y. Furthermore, note the following pattern of the coefficients.

$$1, \quad n, \quad \frac{n(n - 1)}{2!}, \quad \frac{n(n - 1)(n - 2)}{3!}, \quad \text{etc.}$$

Keep these ideas in mind as you study the following examples.

Example 1

Expand $(x + y)^7$.

Solution

$$(x + y)^7 = x^7 + 7x^6y + \frac{7 \cdot 6}{2!}x^5y^2 + \frac{7 \cdot 6 \cdot 5}{3!}x^4y^3 + \frac{7 \cdot 6 \cdot 5 \cdot 4}{4!}x^3y^4 + \frac{7 \cdot 6 \cdot 5 \cdot 4 \cdot 3}{5!}x^2y^5$$
$$+ \frac{7 \cdot 6 \cdot 5 \cdot 4 \cdot 3 \cdot 2}{6!}xy^6 + y^7$$

$$= x^7 + 7x^6y + 21x^5y^2 + 35x^4y^3 + 35x^3y^4 + 21x^2y^5 + 7xy^6 + y^7 \qquad \blacktriangle$$

Example 2

Expand $(x - y)^5$.

Solution

We shall treat $(x - y)^5$ as $[x + (-y)]^5$.

$$[x + (-y)]^5 = x^5 + 5x^4(-y) + \frac{5 \cdot 4}{2!}x^3(-y)^2 + \frac{5 \cdot 4 \cdot 3}{3!}x^2(-y)^3 + \frac{5 \cdot 4 \cdot 3 \cdot 2}{4!}x(-y)^4 + (-y)^5$$

$$= x^5 - 5x^4y + 10x^3y^2 - 10x^2y^3 + 5xy^4 - y^5 \qquad \blacktriangle$$

Example 3

Expand and simplify $(2a + 3b)^4$.

Solution

Let $x = 2a$ and $y = 3b$.

$$(2a + 3b)^4 = (2a)^4 + 4(2a)^3(3b) + \frac{4 \cdot 3}{2!}(2a)^2(3b)^2 + \frac{4 \cdot 3 \cdot 2}{3!}(2a)(3b)^3 + (3b)^4$$

$$= 16a^4 + 96a^3b + 216a^2b^2 + 216ab^3 + 81b^4. \qquad \blacktriangle$$

Finding Specific Terms

Sometimes it is convenient to find a specific term of a binomial expansion without writing out the entire expansion. For example, suppose that we need the 6th term of the expansion $(x + y)^{12}$. We could proceed as follows.

The 6th term will contain y^5. Note that in the general expansion the *exponent of y is always 1 less than the number of the term.* Since the sum of the exponents for x and y must be 12 (the exponent of the binomial), the 6th term will also contain x^7. Again looking back at the general binomial expansion, note that each *denominator of a coefficient* is of the form $r!$, where the value of r agrees with the exponent of y for each term. Thus, if we have y^5, the denominator of the coefficient is $5!$. In the general expansion, each *numerator of a coefficient* contains r factors, where the first factor is the exponent of the binomial and each succeeding factor is 1 less than the preceding one. Thus, the 6th term of $(x + y)^{12}$ is $\dfrac{12 \cdot 11 \cdot 10 \cdot 9 \cdot 8}{5!} x^7 y^5$, which simplifies to $792x^7y^5$.

Example 4

Find the 4th term of $(3a + 2b)^7$.

Solution

The 4th term will contain $(2b)^3$ and therefore $(3a)^4$. The coefficient is $\dfrac{7 \cdot 6 \cdot 5}{3!}$. Therefore, the 4th term is $\dfrac{7 \cdot 6 \cdot 5}{3!}(3a)^4(2b)^3$, which simplifies to $22,680a^4b^3$.

▲

Problem Set 13.5

For Problems 1–6, use Pascal's triangle to help expand each of the following.

1. $(x + y)^8$

2. $(x + y)^7$

3. $(3x + y)^4$
4. $(x + 2y)^4$
5. $(x - y)^5$
6. $(x - y)^4$

For Problems 7–20, expand and simplify.

7. $(x + y)^{10}$

8. $(x + y)^9$

9. $(2x + y)^6$

10. $(x + 3y)^5$

11. $(x - 3y)^5$

12. $(2x - y)^6$

13. $(3a - 2b)^5$

14. $(2a - 3b)^4$

15. $(x + y^3)^6$

16. $(x^2 + y)^5$
17. $(x + 2)^7$

18. $(x + 3)^6$

19. $(x - 3)^4$
20. $(x - 1)^9$

For Problems 21–24, write the first four terms of the expansion.

21. $(x + y)^{15}$

22. $(x + y)^{12}$

23. $(a - 2b)^{13}$

24. $(x - y)^{20}$

For Problems 25–30, find the indicated term of the expansion.

25. 7th term of $(x + y)^{11}$

26. 4th term of $(x + y)^{8}$

27. 4th term of $(x - 2y)^{6}$

28. 5th term of $(x - y)^{9}$

29. 3rd term of $(2x - 5y)^{5}$

30. 6th term of $(3a + b)^{7}$

THOUGHTS INTO WORDS

31. How would you explain binomial expansions to an elementary algebra student?

32. Explain how to find the fifth term of the expansion of $(2x + 3y)^{9}$ without writing out the entire expansion.

SUMMARY

(13.1) An **infinite sequence** is a function whose domain is the set of positive integers. Frequently, a general infinite sequence is expressed as

$$a_1, a_2, a_3, \ldots, a_n, \ldots,$$

where a_1 is the first term, a_2 the second term, and so on, and a_n represents the general, or nth, term.

An **arithmetic sequence** is a sequence where there is a common difference between successive terms.

The general term of an arithmetic sequence is given by

$$a_n = a_1 + (n - 1)d,$$

where a_1 is the first term and d is the common difference.

(13.2) The indicated sum of a sequence is called a **series**. The sum of the first n terms of an arithmetic series is given by

$$S_n = \frac{n(a_1 + a_n)}{2}.$$

(13.3) A **geometric sequence** is a sequence in which each term after the first is obtained by multiplying the preceding term by a common multiplier. The common multiplier is called the **common ratio** of the sequence.

The general term of a geometric sequence is given by

$$a_n = a_1 r^{n-1},$$

where a_1 is the first term and r is the common ratio.

The sum of the first n terms of a geometric series is given by

$$S_n = \frac{a_1 r^n - a_1}{r - 1}, \qquad r \neq 1.$$

(13.4) The sum of an infinite geometric series is given by

$$S_\infty = \frac{a_1}{1 - r}, \qquad |r| < 1.$$

Any infinite geometric series where $|r| \geq 1$ has no sum.

(13.5) The expansion of $(x + y)^n$, where n is a positive integer, is given by

$$(x + y)^n = x^n + nx^{n-1}y + \frac{n(n-1)}{2!}x^{n-2}y^2 + \frac{n(n-1)(n-2)}{3!}x^{n-3}y^3 + \cdots + y^n.$$

Chapter 13 Review Problem Set

For Problems 1–10, find the general term (nth term) for each of the following sequences. These problems contain a mixture of arithmetic and geometric sequences.

1. $3, 9, 15, 21, \ldots$

2. $\dfrac{1}{3}, 1, 3, 9, \ldots$

3. $10, 20, 40, 80, \ldots$

4. $5, 2, -1, -4, \ldots$

5. $-5, -3, -1, 1, \ldots$

6. $9, 3, 1, \dfrac{1}{3}, \ldots$

7. $-1, 2, -4, 8, \ldots$

8. $12, 15, 18, 21, \ldots$

9. $\dfrac{2}{3}, 1, \dfrac{4}{3}, \dfrac{5}{3}, \ldots$

10. $1, 4, 16, 64, \ldots$

For Problems 11–16, find the indicated term of each of the sequences.

11. The 19th term of $1, 5, 9, 13, \ldots$

12. The 28th term of $-2, 2, 6, 10, \ldots$

13. The 9th term of $8, 4, 2, 1, \ldots$

14. The 8th term of $\dfrac{243}{32}, \dfrac{81}{16}, \dfrac{27}{8}, \dfrac{9}{4}, \ldots$

15. The 34th term of $7, 4, 1, -2, \ldots$

16. The 10th term of $-32, 16, -8, 4, \ldots$

17. If the 5th term of an arithmetic sequence is -19 and the 8th term is -34, find the common difference of the sequence.

18. If the 8th term of an arithmetic sequence is 37 and the 13th term is 57, find the 20th term.

19. Find the first term of a geometric sequence if the 3rd term is 5 and the 6th term is 135.

20. Find the common ratio of a geometric sequence if the 2nd term is $\frac{1}{2}$ and the 6th term is 8.

21. Find the sum of the first 9 terms of the sequence $81, 27, 9, 3, \ldots$.

22. Find the sum of the first 70 terms of the sequence $-3, 0, 3, 6, \ldots$.

23. Find the sum of the first 75 terms of the sequence $5, 1, -3, -7, \ldots$.

24. Find the sum of the first 10 terms of the sequence for which $a_n = 2^{5-n}$.

25. Find the sum of the first 95 terms of the sequence for which $a_n = 7n + 1$.

26. Find the sum $5 + 7 + 9 + \cdots + 137$.

27. Find the sum $64 + 16 + 4 + \cdots + \frac{1}{64}$.

28. Find the sum of all even numbers between 8 and 384, inclusive.

29. Find the sum of all multiples of 3 between 27 and 276, inclusive.

30. Find the sum of the infinite geometric sequence $64, 16, 4, 1, \ldots$.

31. Change $.\overline{36}$ to reduced $\frac{a}{b}$ form, where a and b are integers, $b \neq 0$.

32. Change $.4\overline{5}$ to reduced $\frac{a}{b}$ form, where a and b are integers, $b \neq 0$.

Solve Problems 33–37 by using your knowledge of arithmetic and geometric sequences.

33. Suppose that at the beginning of the year your savings account contains $3750. If you withdraw $250 per month from the account, how much will it contain at the end of the year?

34. Sonya has decided to start saving dimes. She plans to save 1 dime the first day of April, 2 dimes the second day, 3 dimes the third day, 4 dimes the fourth day, and so on, for the 30 days of April. How much money will she save in April?

35. Nancy has decided to start saving dimes. She plans to save 1 dime the first day of April, 2 dimes the second day, 4 dimes the third day, 8 dimes the fourth day, and so on, for the first 15 days of April. How much will she save in 15 days?

36. A tank contains 61,440 gallons of water. Each day one-fourth of the water is to be drained out. How much will remain in the tank at the end of 6 days?

37. An object falling from rest in a vacuum falls 16 feet the first second, 48 feet the second second, 80 feet the third second, 112 feet the fourth second, and so on. How far will the object fall in 15 seconds?

38. Expand $(2x + y)^7$.

39. Expand $(x - 3y)^4$.

40. Find the 5th term of the expansion of $(x + 2y)^8$.

CHAPTER 13 TEST

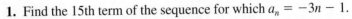

1. Find the 15th term of the sequence for which $a_n = -3n - 1$.

2. Find the 5th term of the sequence for which $a_n = 3(2)^{n-1}$.

3. Find the general term of the sequence $-3, 1, 5, 9, \ldots$.

4. Find the general term of the sequence $5, \dfrac{5}{2}, \dfrac{5}{4}, \dfrac{5}{8}, \ldots$.

5. Find the general term of the sequence $6, 3, 0, -3, \ldots$.

6. Find the 7th term of the sequence $8, 12, 18, 27, \ldots$.

7. Find the 75th term of the sequence $1, 4, 7, 10, \ldots$.

8. Find the number of terms in the sequence $7, 11, 15, \ldots, 243$.

9. If the 4th term of an arithmetic sequence is 13 and the 7th term is 22, find the 15th term.

10. Find the sum of the first 40 terms of the series $1 + 4 + 7 + 10 + \ldots$.

11. Find the sum of the first 8 terms of the series $3 + 6 + 12 + 24 + \ldots$.

12. Find the sum of the series $3 + 1 + (-1) + \ldots + (-55)$.

13. Find the sum of the series $3 + 9 + 27 + \ldots + 2187$.

14. Find the sum of the first 45 terms of the series for which $a_n = 7n - 2$.

15. Find the sum of the first 10 terms of the series for which $a_n = 3(2)^n$.

16. Find the sum of the first 150 positive even whole numbers.

17. Find the sum of the odd numbers between 11 and 193, inclusive.

18. A woman invests \$350 at 12% simple interest at the beginning of each year for a period of 10 years. Find the total accumulated value of all the investments at the end of the 10-year period.

19. Suppose that you save a dime the first day of a month, \$.20 the second day, \$.40 the third day, and continue to double your savings per day for 15 days. Find the total amount that you will save at the end of 15 days.

20. Find the sum of the infinite geometric series $9 + 3 + 1 + \dfrac{1}{3} + \ldots$.

21. Change the repeating decimal $.\overline{37}$ to $\dfrac{a}{b}$ form, where a and b are integers and $b \neq 0$.

22. Change the repeating decimal $.2\overline{6}$ to $\dfrac{a}{b}$ form, where a and b are integers and $b \neq 0$.

23. Expand $(x - 3y)^5$.

24. Write the first three terms of the expansion of $(2x + y)^7$.

25. Find the 5th term of the expansion of $(a + b)^{12}$.

Counting Techniques and Probability

When you are dealt a five-card hand from an ordinary deck of 52 playing cards, there is 1 chance out of 54,145 that you will be dealt four aces. The weatherman predicts a 40% chance of severe thunderstorms by late afternoon. The odds in favor of the Cubs winning the pennant are 2 to 3. In a box that contains 50 light bulbs, 45 are good and 5 are burned out; if 2 bulbs are chosen at random, the probability of getting at least 1 good bulb is $\frac{243}{245}$.

Historically, many basic probability concepts developed as a result of studying games of chance. However, in recent years probability applications have been surfacing at a phenomenal rate in a large variety of fields, such as physics, biology, psychology, economics, insurance, military science, manufacturing, and politics. It is the purpose of this chapter to introduce some counting techniques and then to apply those ideas to some basic concepts of probability.

14.1 Fundamental Principle of Counting

One very useful counting strategy is the **fundamental principle of counting**. After looking at some examples of the property, we will state the property and then use it to solve a variety of counting-type problems.

Example I

A woman has 4 skirts and 5 blouses. Assuming that each blouse can be worn with each skirt, how many different skirt-blouse outfits does the woman have?

Solution

For each of the 4 skirts, the woman has a choice of 5 blouses. Therefore, she has $4(5) = 20$ different skirt-blouse outfits from which to choose.   ▲

Example 2

Eric is shopping for a new bicycle. He has 2 different models (five-speed or ten-speed) and 4 different colors (red, white, blue, or silver) from which to choose. How many different choices does he have?

Solution

Eric's different choices can be "counted" through a **tree diagram** as follows.

Models	Colors	Choices
Five-speed •	Red	Five-speed red
	White	Five-speed white
	Blue	Five-speed blue
	Silver	Five-speed silver
Ten-speed •	Red	Ten-speed red
	White	Ten-speed white
	Blue	Ten-speed blue
	Silver	Ten-speed silver

For each of the 2 model choices, there are 4 choices of color. So altogether Eric has $2 \cdot 4 = 8$ choices.   ▲

The two previous examples motivate the following general principle.

Fundamental Principle of Counting

If one task can be accomplished in x different ways, and following this task, a second task can be accomplished in y different ways, then the first task followed by the second task can be accomplished in $x \cdot y$ different ways. (This counting principle can be extended to any finite number of tasks.)

In order to apply the fundamental principle of counting, it is often helpful to systematically analyze a problem in terms of the tasks to be accomplished. Let's consider some examples that illustrate this idea.

Example 3

How many numbers made up of 3 different digits each can be formed by choosing from the digits 1, 2, 3, 4, 5, and 6?

Solution

Let's analyze this problem in terms of three tasks as follows.

1. Choose the hundred's digit, for which there are 6 choices.
2. Now choose the ten's digit, for which there are only 5 choices, since 1 digit was used in the hundred's place.
3. Now choose the unit's digit, for which there are only 4 choices, since 2 digits have been used for the other places.

Therefore, task 1 followed by task 2 followed by task 3 can be accomplished in $(6)(5)(4) = 120$ ways. In other words, there are 120 numbers made up of 3 different digits that can be formed by choosing from the 6 given digits. ▲

To ensure that you understand the principle, answer the following questions; look back over the solution for Example 3 when necessary.

1. Could the problem be solved by choosing the unit's digit first, then the ten's digit, and finally the hundred's digit?
2. How many 3-digit numbers could be formed from 1, 2, 3, 4, 5, and 6 if it were not required that each number have three *different* digits? (Your answer should be 216.)
3. Suppose that the available digits were 0, 1, 2, 3, 4, and 5. Now how many numbers made up of 3 different digits each could be formed, assuming that 0 cannot be in the hundred's place? (Your answer should be 100.)
4. How many *even* numbers made up of 3 different digits each could be formed by choosing from the digits 1, 2, 3, 4, 5, and 6? (Your answer should be 60.)

Example 4

Employee ID numbers at a certain factory consist of one capital letter followed by a 3-digit number that contains no repeat digits. (For example, A-014 is an ID number.) How many such ID numbers can be formed? How many could be formed if repeat digits were allowed?

Solution

Again, let's analyze in terms of tasks to be completed.

1. Choose the letter part of the ID number; there are 26 choices.
2. Choose the first digit of the three-digit number; there are 10 choices.
3. Choose the second digit; there are 9 choices.
4. Choose the third digit; there are 8 choices.

Therefore, if we apply the fundamental principle of counting, we obtain $(26)(10)(9)(8) = 18,720$ possible ID numbers. If repeat digits were allowed, then there would be $(26)(10)(10)(10) = 26,000$ possible ID numbers. ▲

Example 5

How many ways can Al, Barb, Chad, Dan, and Edna be seated in a row of 5 seats so that Al and Barb are seated side by side?

Solution

We can analyze this problem in terms of the following tasks.

1. Choose 2 adjacent seats for Al and Barb. Figure 14.1 should help you to visualize that there are 4 choices for the 2 adjacent seats.

FIGURE 14.1

2. Determine the number of ways that Al and Barb can be seated. Since Al can be seated on the left and Barb on the right, or vice versa, there are 2 ways to seat Al and Barb for each of the 4 choices of the 2 adjacent seats.

3. The remaining 3 people must be seated in the remaining 3 seats. This can be done in $(3)(2)(1) = 6$ different ways.

Therefore, by the fundamental principle of counting, task 1 followed by task 2 followed by task 3 can be done in $(4)(2)(6) = 48$ ways.   ▲

Suppose that in Example 5 we wanted the number of ways in which the 5 people could be seated so that Al and Barb were *not* side by side. We could determine this number either by (1) analyzing and counting the number of "not side-by-side positions" for Al and Barb or (2) subtracting the side-by-side arrangements determined in Example 5 from the total number of ways that 5 people can be seated in 5 seats. Do this problem both ways and see if you come up with the answer of 72 ways.

As you apply the fundamental principle of counting, you may find that for certain problems simply thinking about an appropriate tree diagram is helpful, even though the size of the problem prohibits writing out the diagram in detail. Consider the following example.

Example 6

Suppose that the undergraduate students in three departments—geography, history, and psychology—are to be classified according to sex and year in school. How many categories are needed?

Solution

We can symbolically represent the various classifications as follows.

M: Male	1. Freshman	G: Geography
F: Female	2. Sophomore	H: History
	3. Junior	P: Psychology
	4. Senior	

We can mentally picture a tree diagram in which each of the two sex classifications branches into four year-in-school classifications, which in turn branch into three department classifications. Thus, we have

(2)(4)(3) = 24 different categories. ▲

Another technique that works on certain problems involves what some people call the "back door" approach. Suppose we know that our classroom contains 50 seats. To determine the number of students present on a given day, it may be easier to count the number of empty seats and subtract from 50, rather than to count the number of students in attendance. We suggest this back-door approach as one way to determine the "not side-by-side" seating arrangements back in the discussion following Example 5, and the next example further demonstrates this approach.

Example 7

When a pair of dice is rolled, in how many ways can a sum greater than 4 be obtained?

Solution

For purposes of clarification, let's use a red die and a white die. (It is not necessary to use different colored dice, but it does help us analyze the different possible outcomes.) It should be intuitively obvious that there are more ways of getting a sum greater than 4 than there are ways of getting a sum of 4 or less. Therefore, let's determine the possibilities for getting a sum of 4 or less, and then subtract from the total number of possible outcomes when a pair of dice is rolled.

First, we can simply list and count the ways of getting a sum of 4 or less.

Red die	White die
1	1
1	2
1	3
2	1
2	2
3	1

There are 6 ways of getting a sum of 4 or less

Second, since there are 6 possible outcomes on the red die and 6 possible outcomes on the white die, there is a total of (6)(6) = 36 possible outcomes when a pair of dice is rolled. Therefore, subtracting the number of ways of getting a sum of 4 or less from the total number of possible outcomes, we obtain 36 − 6 = 30 ways of getting a sum greater than 4. ▲

Problem Set 14.1

1. If a woman has 2 skirts and 10 blouses, how many different skirt-blouse combinations does she have?

2. If a man has 8 shirts, 5 pairs of slacks, and 3 pairs of shoes, how many different shirt-slack-shoe combinations does he have?

3. In how many ways can 4 people be seated in a row of 4 seats?

4. How many numbers of 2 different digits can be formed by choosing from the digits 1, 2, 3, 4, 5, 6, and 7?

5. How many even numbers of 3 different digits can be formed by choosing from the digits 2, 3, 4, 5, 6, 7, 8, and 9?

6. How many odd numbers of 4 different digits can be formed by choosing from the digits 1, 2, 3, 4, 5, 6, 7, and 8?

7. Suppose that the students at a certain university are to be classified according to college (College of Applied Science, College of Arts and Sciences, College of Business, College of Education, College of Fine Arts, College of Health and Physical Education), sex (female, male), and year in school (1, 2, 3, 4). How many categories are possible?

8. A medical researcher classifies subjects according to sex (female, male), smoking habits, (smoker, nonsmoker), and weight (below average, average, above average). How many different combined classifications are used?

9. A pollster classifies voters according to sex (female, male), party affiliation (Democrat, Republican, Independent) and family income (below $10,000, $10,000–$19,999, $20,000–$29,999, $30,000–$39,999, $40,000–$49,999, $50,000 and above). How many combined classifications does the pollster use?

10. A couple is planning to have 4 children. How many different arrangements of boys and girls are possible? (For example, *BBBG* would indicate that the first 3 children were boys and the last was a girl.)

11. In how many ways can 3 officers (president, secretary, and treasurer) be selected from a club that has 20 members?

12. In how many ways can 3 officers (president, secretary, and treasurer) be selected from a club with 15 female and 10 male members so that the president is female and the secretary and the treasurer are males?

13. A disc jockey has 6 tunes to be played once each in a half-hour program. How many different orderings of these tunes are possible?

14. State officials have decided to have their state's automobile license plates consist of 2 letters followed by 4 digits. They do not want to repeat any letters or digits in any license number. How many different license numbers do they have available?

15. In how many ways can 6 people be seated in a row of 6 seats?

16. In how many ways can Al, Bob, Carlos, Don, Vinay, and Fern be seated in a row of 6 seats if Al and Bob are to be seated side by side?

17. In how many ways can Amy, Bob, Juanita, Dan, and Elmer be seated in a row of 5 seats so that neither Amy nor Bob occupies an end seat?

18. In how many ways can Al, Bob, Carlos, Don, Vinay, and Fern be seated in a row of 6 seats if Al and Bob are not to be seated side by side? [*Hint*: Al and Bob will either be seated side by side or not be seated side by side.]

19. In how many ways can Al, Bob, Carlos, Dawn, and Vinay be seated in a row of 5 chairs if Al is to be seated in the middle chair?

20. In how many ways can 3 letters be dropped into 5 mail boxes?

21. In how many ways can 5 letters be dropped into 3 mail boxes?

22. In how many ways can 4 letters be dropped into 6 mail boxes so that no 2 letters are dropped into the same box?

23. In how many ways can 6 letters be dropped into 4 mail boxes so that no 2 letters are dropped into the same box?

24. If 5 coins are tossed, in how many ways can they fall?

25. If 3 dice are tossed, in how many ways can they fall?

26. In how many ways can a sum less than 10 be obtained when a pair of dice is tossed?

27. In how many ways can a sum greater than 5 be obtained when a pair of dice is tossed?

28. In how many ways can a sum greater than 4 be obtained when 3 dice are tossed?

29. How many numbers greater than 400, where no number contains repeated digits, can be formed from the digits 2, 3, 4, and 5? [*Hint*: Consider both 3-digit and 4-digit numbers.]

30. How many numbers greater than 5000, where no number contains repeated digits, can be formed by choosing from the digits 1, 2, 3, 4, 5, and 6?

31. In how many ways can 4 boys and 3 girls be seated in a row of 7 seats so that boys and girls occupy alternate seats?

32. In how many ways can 3 different mathematics books and 4 different history books be exhibited on a shelf so that all of the books in a subject area are side by side?

33. In how many ways can a true-false test of 10 questions be answered?

34. How many even numbers greater than 3000, where no number contains repeated digits, can be formed from the digits 1, 2, 3, and 4?

35. How many odd numbers greater than 40,000, where no number contains repeated digits, can be formed from the digits 1, 2, 3, 4, and 5?

36. In how many ways can Al, Bob, Carol, Dan, Vinay, Faye, and George be seated in a row of 7 seats so that Al, Bob, and Carol occupy consecutive seats in some order?

37. The license plates for a certain state consist of 2 letters followed by a 4-digit number such that the first digit of the number is not zero. An example would be PK-2446.

(a) How many different license plates can be produced?

(b) How many different plates would not have the same letter repeated?

(c) How many plates would not have any digits repeated in the number part of the plate?

(d) How many plates would not have the same letter repeated and would not have any digits repeated in the number part of the plate?

THOUGHTS INTO WORDS

38. How would you explain the fundamental principle of counting to a friend who missed class the day it was discussed?

39. Give two or three simple illustrations of the fundamental principle of counting.

40. Explain how you solved Problem 29.

14.2   Permutations and Combinations

As we develop the material in this section, **factorial notation** will become very useful. We use the notation $n!$ (read "n factorial") with positive integers as follows.

$1! = 1$

$2! = 2 \cdot 1 = 2$

$3! = 3 \cdot 2 \cdot 1 = 6$

$4! = 4 \cdot 3 \cdot 2 \cdot 1 = 24$

Notice that the factorial notation refers to an *indicated product.*

In general, we write

$$n! = n(n-1)(n-2) \cdot \cdots \cdot 3 \cdot 2 \cdot 1.$$

Now, as a lead-in to the first concept of this section, let's consider a counting problem that closely resembles some from the previous section.

Example 1

In how many ways can the 3 letters A, B, and C be arranged in a row?

Solution A

Certainly one approach to the problem is to simply list and count the arrangements.

$ABC, \quad ACB, \quad BAC, \quad BCA, \quad CAB, \quad CBA$

There are 6 arrangements of the 3 letters.

Solution B

Another approach, which can be generalized for more difficult problems, uses the fundamental principle of counting. Since there are 3 choices for the first letter of an arrangement, 2 choices for the second letter, and 1 choice for the third letter, there are $(3)(2)(1) = 6$ different arrangements. ▲

Ordered arrangements are called **permutations**. In general, a permutation of a set of n elements is an ordered arrangement of the n elements, and we will use the symbol $P(n, n)$ to note the number of these permutations. For example, from Example 1 we know that $P(3, 3) = 6$. Furthermore, by using the same basic approach as in Solution B of Example 1, we can obtain the following results.

$P(1, 1) = \quad\quad 1 \quad\quad = 1!$

$P(2, 2) = \quad\quad 2 \cdot 1 \quad = 2!$

$P(4, 4) = \quad 4 \cdot 3 \cdot 2 \cdot 1 \quad = 4!$

$P(5, 5) = 5 \cdot 4 \cdot 3 \cdot 2 \cdot 1 = 5!$

The following formula becomes evident.

$$\boxed{P(n, n) = n!}$$

Now suppose that we are interested in the number of 2-letter permutations that can be formed by choosing from the 4 letters A, B, C, and D. (Some of these permutations are AB, BA, AC, CA, BC, and CB.) In other words, we want to find the number of 2-element permutations that can be formed from a set of 4 elements. We denote this number by $P(4, 2)$. To find $P(4, 2)$, we can reason as follows: First, choose any one of the 4 letters to occupy the first position in the permutation, and then choose any one of the 3 remaining letters for the second position. Therefore, by the fundamental principle of counting, we have $4(3) = 12$ different 2-letter permutations; that is, $P(4, 2) = 12$. We can determine the following list using a similar line of reasoning. (Make sure that you agree with each of these.)

$$P(4, 3) = 4 \cdot 3 \cdot 2 = 24,$$
$$P(5, 2) = 5 \cdot 4 = 20,$$
$$P(6, 4) = 6 \cdot 5 \cdot 4 \cdot 3 = 360,$$
$$P(7, 3) = 7 \cdot 6 \cdot 5 = 210$$

In general, we say that the *number of r-element permutations that can be formed from a set of n elements is given by*

$$P(n, r) = n\underbrace{(n - 1)(n - 2) \cdots}_{r \text{ factors}}.$$

Notice that the indicated product for $P(n, r)$ begins with n, thereafter each factor is 1 less than the previous one, and there is a total of r factors. Following are some examples.

$$P(6, 2) = 6 \cdot 5 = 30,$$
$$P(8, 3) = 8 \cdot 7 \cdot 6 = 336,$$
$$P(9, 4) = 9 \cdot 8 \cdot 7 \cdot 6 = 3024$$

Let's consider two examples that illustrate the use of $P(n, n)$ and $P(n, r)$.

Example 2

In how many ways can 5 students be seated in a row of 5 seats?

Solution

We are looking for the number of 5-element permutations that can be formed from a set of 5 elements. Thus, we can apply $P(n, n) = n!$.

$$P(5, 5) = 5! = 5 \cdot 4 \cdot 3 \cdot 2 \cdot 1 = 120 \qquad \blacktriangle$$

Example 3

Suppose that 7 people enter a swimming race. In how many ways can the first-, second-, and third-place ribbons be awarded?

Solution

We are looking for the number of 3-element permutations that can be formed from a set of 7 elements. Therefore, using the formula for $P(n, r)$, we obtain

$$P(7, 3) = 7 \cdot 6 \cdot 5 = 210. \qquad \blacktriangle$$

It should be evident that both Example 2 and Example 3 could have been solved by applying the fundamental principle of counting. In fact, it should be noted that the formulas for $P(n, n)$ and $P(n, r)$ do not really give us much additional problem solving power. However, as we will see in a moment, they do provide the basis for developing a formula that is very useful as a problem solving tool.

Permutations Involving Nondistinguishable Objects

Suppose that we have 2 identical *H*s and 1 *T* in an arrangement such as *HTH*. Obviously, if we switched the 2 identical *H*s, the newly formed arrangement, *HTH*, would not be distinguishable from the original. In other words, if some of the *n* elements are identical, there are fewer distinguishable permutations than there would be of *n* distinctly different elements.

To see the effect of identical elements on the number of distinguishable permutations, let's look at some specific examples.

2 identical *H*s:　　　　1 permutation—namely, *HH*

2 different letters:　　　2! permutations

Therefore, having 2 different letters affects the number of permutations by a *factor of* 2!.

3 identical *H*s:　　　　1 permutation—namely, *HHH*

3 different letters:　　　3! permutations

Therefore, having 3 different letters affects the number of permutations by a *factor of* 3!.

4 identical *H*s:　　　　1 permutation–namely, *HHHH*

4 different letters:　　　4! permutations

Therefore, having 4 different letters affects the number of permutations by a *factor of* 4!.

Now let's solve a specific problem.

Example 4

How many distinguishable permutations can be formed from 3 identical *H*s and 2 identical *T*s?

Solution

If we had 5 distinctly different letters, we could form 5! permutations. But the 3 identical *H*s affect the number of distinguishable permutations by a factor of 3!, and the 2 identical *T*s affect the number of permutations by a factor of 2!. Therefore we must divide 5! by 3! and 2!. Thus, we obtain

$$\frac{5!}{(3!)(2!)} = \frac{5 \cdot \overset{2}{\cancel{4}} \cdot \cancel{3} \cdot \cancel{2} \cdot 1}{\cancel{3} \cdot \cancel{2} \cdot 1 \cdot \cancel{2} \cdot 1} = 10$$

distinguishable permutations of 3*H*s and 2*T*s.

▲

In Example 4, the type of reasoning we used leads us to the following general counting technique: If there are n elements to be arranged, where there are r_1 of one kind, r_2 of another kind, r_3 of another kind, ... r_k of a kth kind, then the total number of distinguishable permutations is given by the expression

$$\frac{n!}{(r_1!)(r_2!)(r_3!)\cdots(r_k!)}.$$

Example 5

How many different 11-letter permutations can be formed from the 11 letters of the word *MISSISSIPPI*?

Solution

Since there are 4 *I*s, 4 *S*s, and 2 *P*s, we can form

$$\frac{11!}{(4!)(4!)(2!)} = \frac{11\cdot 10\cdot 9\cdot 8\cdot 7\cdot 6\cdot 5\cdot 4\cdot 3\cdot 2\cdot 1}{4\cdot 3\cdot 2\cdot 1\cdot 4\cdot 3\cdot 2\cdot 1\cdot 2\cdot 1} = 34{,}650$$

distinguishable permutations.　　　　　　　　　　　　　　　▲

Combinations (Subsets)

Permutations are ordered arrangements; however, *order* is not always a consideration. For example, suppose that we want to determine the number of 3-person committees that can be formed from the 5 people Al, Barb, Carol, Dawn, and Eric. Certainly the committee that consists of Al, Barb, and Eric is the same as the committee that consists of Barb, Eric, and Al. In other words, the order in which we choose or list the members is not to be considered. Therefore, we are really dealing with subsets; that is, we are looking for the number of 3-element subsets that can be formed from a set of 5 elements. Traditionally, in this context, subsets are called **combinations**. So, stated another way, we are looking for the number of combinations of 5 things taken 3 at a time. In general, r-element subsets taken from a set of n elements are called *combinations of n things taken r at a time*. The symbol $C(n, r)$ will denote the number of these combinations.

Now let's restate the previous committee problem and look at a detailed solution that can be generalized to handle a variety of problems dealing with combinations.

Example 6

How many 3-person committees can be formed from the five people Al, Barb, Carol, Dawn, and Eric?

Solution

Let's use the set $\{A, B, C, D, E\}$ to represent the 5 people. Consider one possible 3-person committee (subset), such as $\{A, B, C\}$. There are 3! permutations of these 3 letters. Now take another committee, such as $\{A, B, D\}$. There are also 3! permutations of these 3 letters. If we were to continue this process with all of the 3-letter subsets that can be formed from the 5 letters, we would obtain all possible 3-letter permutations of the 5 letters. That is, we would obtain $P(5, 3)$. Therefore, if we let $C(5, 3)$ represent the number of 3-element subsets, then

$$[C(5, 3)](3!) = P(5, 3).$$

Solve this equation for $C(5, 3)$ to yield

$$C(5, 3) = \frac{P(5, 3)}{3!} = \frac{5 \cdot 4 \cdot 3}{3 \cdot 2 \cdot 1} = 10.$$

So, there are 10 3-person committees that can be formed from the 5 people. ▲

In general, $C(n, r)$ times $r!$ yields $P(n, r)$. Thus,

$$[C(n, r)](r!) = P(n, r),$$

and solving this equation for $C(n, r)$ produces

$$\boxed{C(n, r) = \frac{P(n, r)}{r!}.}$$

In other words, we can find the number of combinations of n things taken r at a time by dividing the number of permutations of n things taken r at a time by $r!$. The following examples illustrate this process.

$$C(7, 3) = \frac{P(7, 3)}{3!} = \frac{7 \cdot 6 \cdot 5}{3 \cdot 2 \cdot 1} = 35,$$

$$C(9, 2) = \frac{P(9, 2)}{2!} = \frac{9 \cdot 8}{2 \cdot 1} = 36,$$

$$C(10, 4) = \frac{P(10, 4)}{4!} = \frac{10 \cdot 9 \cdot 8 \cdot 7}{4 \cdot 3 \cdot 2 \cdot 1} = 210$$

Example 7

How many different 5-card hands can be dealt from a deck of 52 playing cards?

Solution

Since the order in which the cards are dealt is not an issue, we are working with a combination (subset) problem. Thus, using the formula for $C(n, r)$, we obtain

$$C(52, 5) = \frac{P(52, 5)}{5!} = \frac{52 \cdot 51 \cdot 50 \cdot 49 \cdot 48}{5 \cdot 4 \cdot 3 \cdot 2 \cdot 1} = 2,598,960.$$

There are 2,598,960 different 5-card hands that can be dealt from a deck of 52 playing cards. ▲

Some counting problems can be solved by using the fundamental principle of counting along with the combination formula, as the next example demonstrates.

Example 8

How many committees consisting of 3 women and 2 men can be formed from a group of 5 women and 4 men?

Solution

Let's think of this problem in terms of two tasks.

  1. Choose a subset of 3 women from the 5 women. This can be done in

$$C(5, 3) = \frac{P(5, 3)}{3!} = \frac{5 \cdot 4 \cdot 3}{3 \cdot 2 \cdot 1} = 10 \text{ ways.}$$

  2. Choose a subset of 2 men from the 4 men. This can be done in

$$C(4, 2) = \frac{P(4, 2)}{2!} = \frac{4 \cdot 3}{2 \cdot 1} = 6 \text{ ways.}$$

Task 1 followed by Task 2 can be done in $(10)(6) = 60$ ways. Therefore, there are 60 committees consisting of 3 women and 2 men that can be formed. ▲

Sometimes it takes a little thought to decide whether permutations or combinations should be used. Remember that if order is to be considered, permutations should be used, but if order does not matter, then combinations should be used. It may be helpful to think of combinations as subsets.

Example 9

A small accounting firm has 12 computer programmers. Three of these people are to be promoted to systems analysts. In how many ways can the 3 people to be promoted be selected?

Solution

Let's call the people $A, B, C, D, E, F, G, H, I, J, K$, and L. Suppose A, B, and C are chosen to be promoted. Is this any different than choosing B, C, and A? Obviously not. Therefore order does not matter, so we are considering combinations. Specifically, we need to find the number of combinations of 12 people taken 3 at a time. Thus, we obtain

$$C(12, 3) = \frac{P(12, 3)}{3!} = \frac{12 \cdot 11 \cdot 10}{3 \cdot 2 \cdot 1} = 220$$

different ways that the 3 people to be promoted can be chosen. ▲

Example 10

A club is to elect 3 officers—president, secretary, and treasurer—from a group of 6 people who are each willing to serve in any office. How many different ways can the officers be chosen?

Solution

Let's call the people A, B, C, D, E, and F. Is electing A as president, B as secretary, and C as treasurer different from electing B as president, C as secretary, and A as treasurer? Obviously it is, and therefore we are working with permutations. Thus we obtain

$$P(6, 3) = 6 \cdot 5 \cdot 4 = 120$$

different ways of filling the offices. ▲

Problem Set 14.2

In Problems 1–12, evaluate the expression.

1. $P(5, 3)$ **2.** $P(8, 2)$

3. $P(6, 4)$ **4.** $P(9, 3)$

5. $C(7, 2)$ **6.** $C(8, 5)$

7. $C(10, 5)$ **8.** $C(12, 4)$

9. $C(15, 2)$ **10.** $P(5, 5)$

11. $C(5, 5)$ **12.** $C(11, 1)$

13. How many permutations of the 4 letters A, B, C, and D can be formed by using all letters in each permutation?

14. In how many ways can 6 students be seated in a row of 6 seats?

15. How many 3-person committees can be formed from a group of 9 people?

16. How many 2-card hands can be dealt from a deck of 52 playing cards?

17. How many 3-letter permutations can be formed from the first 8 letters of the alphabet if
(a) repetitions are not allowed?
(b) repetitions are allowed?

18. In a 7-team baseball league, in how many ways can the top 3 positions in the final standings be filled?

19. In how many ways can the manager of a baseball team arrange his batting order if he wants his 4 best hitters in the top 4 positions?

20. In a baseball league of 9 teams, how many games are needed to complete the schedule if each team must play 12 games with every other team?

21. How many committees consisting of 4 women and 4 men can be chosen from a group of 7 women and 8 men?

22. How many 3-element subsets containing 1 vowel and 2 consonants can be formed from the set {a, b, c, d, e, f, g, h, i}?

23. There are 5 associate professors being considered for promotion to the rank of full professor, but only 3 will be promoted. How many different ways are there of selecting the 3 to be promoted?

24. From the digits 1, 2, 3, 4, 5, 6, 7, 8, and 9, how many numbers of 4 different digits each can be formed if each number consists of 2 odd and 2 even digits?

25. How many 3-element subsets containing the letter A can be formed from the set {A, B, C, D, E, F}?

26. How many 4-person committees can be chosen from a group of 5 women and 3 men so that each committee contains at least 1 man?

27. How many different 7-letter permutations can be formed from 4 identical Hs and 3 identical Ts?

28. How many different 8-letter permutations can be formed from 6 identical Hs and 2 identical Ts?

29. How many different 9-letter permutations can be formed from 3 identical As, 4 identical Bs, and 2 identical Cs?

30. How many different 10-letter permutations can be formed from 5 identical As, 4 identical Bs, and a C?

31. How many different 7-letter permutations can be formed from the 7 letters of the word ALGEBRA?

32. How many different 11-letter permutations can be formed from the 11 letters of the word MATHEMATICS?

33. In how many ways can $x^4 y^2$ be written without using exponents? [*Hint*: One way is *xxxxyy*.]

34. In how many ways can $x^3 y^4 z^3$ be written without using exponents?

35. In how many ways can 10 basketball players be divided into 2 teams of 5 players each?

36. In how many ways can 10 basketball players be divided into 2 teams of 5 players so that the 2 best players are on opposite teams?

37. A box contains 9 good light bulbs and 4 defective bulbs. How many samples of 3 bulbs would contain 1 defective bulb? How many samples of 3 bulbs would contain *at least* 1 defective bulb?

38. How many 5-person committees consisting of 2 juniors and 3 seniors can be formed from a group of 6 juniors and 8 seniors?

39. In how many ways can 6 people be divided into 2 groups so that there are 4 in one group and 2 in the other group? How many ways could 6 people be divided into groups of 3 each?

40. How many 5-element subsets containing both A and B can be formed from the set $\{A, B, C, D, E, F, G, H\}$?

41. How many 4-element subsets containing A or B, but not both A and B, can be formed from the set $\{A, B, C, D, E, F, G\}$?

42. How many different 5-person committees can be selected from a group of 9 people if 2 of those people refuse to serve together on a committee?

43. How many different line segments are determined by 5 points? By 6 points? By 7 points? By n points?

44. (a) How many 5-card hands consisting of 2 kings and 3 aces can be dealt from a deck of 52 playing cards?

(b) How many 5-card hands consisting of 3 kings and 2 aces can be dealt from a deck of 52 playing cards?

(c) How many 5-card hands consisting of 3 cards of one face value and 2 cards of another face value can be dealt from a deck of 52 playing cards?

THOUGHTS INTO WORDS

45. Explain the difference between a permutation and a combination. Give an illustration of each one to help with your explanation.

46. Your friend is having difficulty distinguishing between permutations and combinations in problem solving situations. What might you do to help her?

Further Investigations

47. In how many ways can 6 people be seated at a circular table? [*Hint*: Moving each person one place to the right (or left) does not create a new seating arrangement.]

48. The meaning of $P(8, 3)$ can be expressed completely in factorial notation as follows.

$$P(8, 3) = \frac{P(8, 3) \cdot 5!}{5!} = \frac{(8 \cdot 7 \cdot 6)(5 \cdot 4 \cdot 3 \cdot 2 \cdot 1)}{5!}$$

$$= \frac{8!}{5!}$$

Express each of the following in terms of factorial notation.
(a) $P(7, 3)$ **(b)** $P(9, 2)$ **(c)** $P(10, 7)$
(d) $P(n, r)$, where $r \leq n$ and $0!$ is defined to be 1

49. Sometimes the formula $C(n, r) = \dfrac{n!}{r!(n - r)!}$ is used to find the number of combinations of n things taken r at a time. Using the result from Problem 48(d), develop this formula.

50. Compute $C(7, 3)$ and $C(7, 4)$. Compute $C(8, 2)$ and $C(8, 6)$. Compute $C(9, 8)$ and $C(9, 1)$. Now argue that $C(n, r) = C(n, n - r)$ for $r \leq n$.

 **Graphics Calculator Activities**

Before you do Problems 51–56, be sure that you can use your calculator to compute the number of permutations and combinations. Your calculator may possess a special sequence of keys for such computations. You may need to refer to your calculator manual for this information.

51. Use your calculator to check your answers for Problems 1–12.

52. How many different 5-card hands can be dealt from a deck of 52 playing cards?

53. How many different 7-card hands can be dealt from a deck of 52 playing cards?

54. How many different 5-person committees can be formed from a group of 50 people?

55. How many different juries consisting of 11 people can be chosen from a group of 30 people?

56. How many 7-person committees consisting of 3 juniors and 4 seniors can be formed from 45 juniors and 53 seniors?

14.3   Probability

To introduce some terminology and symbolism, we will consider a simple experiment of tossing a regular 6-sided die. There are 6 possible outcomes to this experiment; either 1, 2, 3, 4, 5, or 6 will be on the side of the die facing up. The set of all possible outcomes of a given experiment is called a **sample space**, and the individual elements of the sample space are called **sample points**. (In this text we will be working only with finite sample spaces.) We will use S (sometimes with subscripts for identification purposes) to refer to a particular sample space of an experiment; the number of sample points will be denoted by $n(S)$. Thus, $S = \{1, 2, 3, 4, 5, 6\}$, where $n(S) = 6$, can be used for the experiment of tossing a die.

Now suppose that we are interested in some of the various possible outcomes of the die-tossing experiment. For example, we might be interested in the event "an even number comes up." In this case we are satisfied if 2, 4, or 6 appears on the upface of the die; therefore, the event "an even number comes up" is the subset $E = \{2, 4, 6\}$, where $n(E) = 3$. Or we might be interested in the event "a multiple of 3 comes up." This would be the subset $F = \{3, 6\}$, where $n(F) = 2$.

In general, any subset of a sample space is called an **event** or **event space**. If the event consists of exactly one element of the sample space, then it is called a **simple event**. Any nonempty event that is not simple is called a **compound event**. A compound event can be represented as the union of simple events.

It is now possible to give a very simple definition for **probability** the way we will use it in this text.

DEFINITION 14.1

The **probability** of an event E in an experiment in which all of the possible outcomes in the sample space S are equally likely to occur is defined by

$$P(E) = \frac{n(E)}{n(S)},$$

where $n(E)$ denotes the number of elements in the event E and $n(S)$ denotes the number of elements in the sample space S.

Many probability problems can be solved by applying Definition 14.1. We just need to be able to determine the number of elements in the sample space and the number of elements in the event space. For example, returning to the die-tossing experiment, the probability of getting an even number with 1 toss of a die is given by

$$P(E) = \frac{n(E)}{n(S)} = \frac{3}{6} = \frac{1}{2}.$$

Let's consider two examples where the number of elements in both the sample space and the event space is easy to determine.

Example 1

Toss a coin. Find the probability that a head (H) turns up.

Solution

Let the sample space be $S = \{H, T\}$, where $n(S) = 2$. The event of turning up a head is the subset $E = \{H\}$, where $n(E) = 1$. Therefore, we say that the probability of getting a head with 1 flip of a coin is given by

$$P(E) = \frac{n(E)}{n(S)} = \frac{1}{2}.$$
▲

Example 2

What is the probability that at least 1 head will turn up if 2 coins are tossed?

Solution

For purposes of clarification, let the coins be a penny and a nickel. The possible outcomes of this experiment are (1) a head on both coins, (2) a head on the penny and a tail on the nickel, (3) a tail on the penny and a head on the nickel, and (4) a tail on both coins. Using ordered-pair symbolism in which the first entry of a pair records the result of the penny and the second entry records the result of the nickel, we have a sample space of

$$S = \{(H, H), (H, T), (T, H), (T, T)\},$$

where $n(S) = 4$. Let E be the event of getting at least 1 head. Thus,

$$E = \{(H, H), (H, T), (T, H)\}$$

and $n(E) = 3$. Therefore, the probability of getting at least 1 head with one toss of 2 coins is

$$P(E) = \frac{n(E)}{n(S)} = \frac{3}{4}.$$
▲

As you might expect, the counting techniques we discussed in the first two sections of this chapter can frequently be used to help solve probability problems.

Example 3

Find the probability of getting 3 heads and 1 tail if 4 coins are tossed.

Solution

The sample space consists of the possible outcomes for tossing 4 coins. Since there are 2 things that can happen on each coin, by the fundamental principle of counting there are $2 \cdot 2 \cdot 2 \cdot 2 = 16$ possible outcomes for tossing 4 coins. So we know that $n(S) = 16$ without taking the time to list all of the elements. The event of getting 3 heads and 1 tail is the subset

$$E = \{(H, H, H, T), (H, H, T, H), (H, T, H, H), (T, H, H, H)\},$$

where $n(E) = 4$. Therefore, the requested probability is

$$P(E) = \frac{n(E)}{n(S)} = \frac{4}{16} = \frac{1}{4}.$$
▲

Example 4

Al, Pablo, Chad, Dawn, Eve, and Francis are randomly seated in a row of 6 chairs. What is the probability that Al and Pablo are seated in the end seats?

Solution

The sample space consists of all possible ways of seating 6 people in 6 chairs—in other words, the permutations of 6 things taken 6 at a time. Thus,

$$n(S) = P(6, 6) = 6! = 6 \cdot 5 \cdot 4 \cdot 3 \cdot 2 \cdot 1 = 720.$$

The event space consists of all possible ways of seating the 6 people so that Al and Pablo occupy end seats. The number of these possibilities can be counted as follows.

1. Put Al and Pablo in the end seats. This can be done in 2 ways, since Al can be on the left end and Pablo on the right end, or vice versa.

2. Put the other 4 people in the remaining 4 seats. This can be done in $4! = 4 \cdot 3 \cdot 2 \cdot 1 = 24$ different ways.

Therefore, task 1 followed by task 2 can be done in $(2)(24) = 48$ different ways; so $n(E) = 48$. Thus, the requested probability is

$$P(E) = \frac{n(E)}{n(S)} = \frac{48}{720} = \frac{1}{15}. \qquad \blacktriangle$$

Notice that in Example 3 we used the fundamental principle of counting to determine the number of elements in the sample space without actually listing all of the elements. For the event space, we listed the elements and counted them in the usual way. In Example 4 we used the permutation formula $P(n, n) = n!$ to determine the number of elements in the sample space, and then we used the fundamental principle of counting to help determine the number of elements in the event space. There are no definite rules as to when to list the elements and when to apply some sort of counting technique. In general, if you do not see a counting pattern for a particular problem, you should begin the listing process. If a counting pattern then emerges as you are listing the elements, use the pattern at that time.

The combination (subset) formula, $C(n, r) = \dfrac{P(n, r)}{r!}$, developed in Section 14.2, is also a very useful tool for solving certain kinds of probability problems. The next three examples illustrate some problems involving combinations.

Example 5

A committee of 3 people is randomly selected from a group composed of Swati, Barb, Chad, Dee, and Eric. What is the probability that Swati is on the committee?

Solution

The sample space S consists of all possible 3-person committees that can be formed from the 5 people. Therefore,

$$n(S) = C(5, 3) = \frac{P(5, 3)}{3!} = \frac{5 \cdot 4 \cdot 3}{3 \cdot 2 \cdot 1} = 10.$$

The event space E consists of all of the 3-person committees that have Swati as a member. Each of these committees contains Swati and 2 other people chosen from the 4 remaining people. Thus, the number of these committees is given by $C(4, 2)$. So we obtain

$$n(E) = C(4, 2) = \frac{P(4, 2)}{2!} = \frac{4 \cdot 3}{2 \cdot 1} = 6.$$

The requested probability is

$$P(E) = \frac{n(E)}{n(S)} = \frac{6}{10} = \frac{3}{5}.$$

Example 6

From a group of 5 seniors and 4 juniors, a committee of 4 is chosen at random. Find the probability that the committee will contain 2 seniors and 2 juniors.

Solution

The sample space S consists of all possible 4-person committees that can be formed from the 9 people. Thus,

$$n(S) = C(9, 4) = \frac{P(9, 4)}{4!} = \frac{9 \cdot 8 \cdot 7 \cdot 6}{4 \cdot 3 \cdot 2 \cdot 1} = 126.$$

The event space E consists of all of those 4-person committees that contain 2 seniors and 2 juniors. They can be counted as follows.

1. Choose 2 seniors from the 5 available seniors in $C(5, 2) = 10$ ways.
2. Choose 2 juniors from the 4 available juniors in $C(4, 2) = 6$ ways.

Therefore, there are $10 \cdot 6 = 60$ committees that consist of 2 seniors and 2 juniors. The requested probability is

$$P(E) = \frac{n(E)}{n(S)} = \frac{60}{126} = \frac{10}{21}.$$

Example 7

Find the probability of getting 2 heads and 6 tails if 8 coins are tossed.

Solution

Since two things can happen to each coin, the total number of possible outcomes, $n(S)$, is $2^8 = 256$. We can select the 2 coins that must fall heads up in $C(8, 2) = 28$ ways. For each of these ways, there is only 1 way to select the other 6 coins that must fall tails up. Therefore, there are $28 \cdot 1 = 28$ ways of getting 2 heads and 6 tails; so $n(E) = 28$. The requested probability is

$$P(E) = \frac{n(E)}{n(S)} = \frac{28}{256} = \frac{7}{64}.$$

Problem Set 14.3

For Problems 1–4, find the probability of each of the following events if 2 coins are tossed.

1. Getting 1 head and 1 tail
2. Getting 2 tails
3. Getting at least 1 tail
4. Getting no tails

For Problems 5–8, find the probability of each of the following events if 3 coins are tossed.

5. Getting 3 heads
6. Getting 2 heads and 1 tail
7. Getting at least 1 head
8. Getting exactly 1 tail

For Problems 9–12, find the probability of each of the following events if 4 coins are tossed.

9. Getting 4 heads
10. Getting 3 heads and 1 tail
11. Getting 2 heads and 2 tails
12. Getting at least 1 head

For Problems 13–16, find the probability of each of the following events if 1 die is tossed.

13. Getting a multiple of 3
14. Getting a prime number
15. Getting an even number
16. Getting a multiple of 7

For Problems 17–22, find the probability of each of the following events if 2 dice are tossed.

17. Getting a sum of 6
18. Getting a sum of 11
19. Getting a sum less than 5
20. Getting a 5 on exactly 1 die
21. Getting a 4 on at least 1 die
22. Getting a sum greater than 4

For Problems 23–26, find the probability of each of the following events if 1 card is drawn from a standard deck of 52 playing cards.

23. A heart is drawn.
24. A king is drawn.
25. A spade or a diamond is drawn.
26. A red jack is drawn.

For Problems 27–30, find the probability of each of the following events if 25 slips of paper numbered 1 to 25, inclusive, are put in a hat and then one is drawn out at random.

27. The slip with the 5 on it is drawn.
28. A slip with an even number on it is drawn.
29. A slip with a prime number on it is drawn.
30. A slip with a multiple of 6 on it is drawn.

For Problems 31–34, find the probability of each of the following events if a committee of 2 boys is chosen at random from a group composed of the 5 boys, Al, Bill, Carl, Dan, and Elmer.

31. Dan is on the committee.
32. Dan and Elmer are both on the committee.
33. Bill and Carl are not both on the committee.
34. Either Dan or Elmer, but not both, is on the committee.

For Problems 35–38, find the probability of each of the following events if a 5-person committee is selected at random from a group composed of the 8 people, Ramsey, Barb, Chad, Devon, Eric, Fern, George, and Harriet.

35. Ramsey and Barb are both on the committee.
36. George is not on the committee.
37. Either Chad or Devon, but not both, is on the committee.
38. Neither Ramsey nor Barb is on the committee.

For Problems 39–41, suppose that a box of 10 items from a manufacturing process is known to contain 2 defective and 8 nondefective items. Find the probability of each of the following events if a sample of 3 is selected at random.

39. The sample contains all nondefective items.

40. The sample contains 1 defective and 2 nondefective items.

41. The sample contains 2 defective and 1 nondefective items.

42. A building has 5 doors. Find the probability that 2 persons entering the building at random will choose the same door.

43. Bill, Carol, and Erin are seated at random in a row of 3 seats. Find the probability that Bill and Carol are seated side by side.

44. April, Terrell, Carl, and Denise are to be seated at random in a row of 4 chairs. What is the probability that April and Terrell will occupy the end seats?

45. A committee of 4 girls is to be chosen at random from a group composed of the 5 girls, Alice, Becky, Candy, Dee, and Elaine. Find the probability that Elaine will not be on the committee.

46. Three boys and 2 girls are randomly seated in a row of 5 seats. What is the probability that the boys and girls are in alternate seats?

47. Four different mathematics books and 5 different history books are randomly placed on a shelf. What is the probability that all the books on a subject are side by side?

48. Each of 3 letters is to be mailed in any of 5 different mail boxes. What is the probability that all will be mailed in the same mail box?

49. Randomly form a 4-digit number by using the digits 2, 3, 4, and 6 once each. What is the probability that the number formed will be greater than 4000?

50. Randomly select 1 of the 120 permutations of the letters a, b, c, d, and e. Find the probability that in the permutation chosen the letter a precedes the letter b (that is, a is to the left of b).

51. From a group of 6 women and 5 men, a committee of 4 is chosen at random. Find the probability that the committee contains 2 women and 2 men.

52. From a group of 4 women and 5 men, a committee of 3 is chosen at random. Find the probability that the committee contains at least 1 man.

53. Al, Bob, Carl, Dan, Ed, Frank, Gino, Harry, Jerry, and Mike are randomly divided into 2 5-man teams for a basketball game. What is the probability that Al, Bob, and Carl are on the same team?

54. Find the probability of getting 4 heads and 3 tails if 7 coins are tossed.

55. Find the probability of getting 3 heads and 6 tails if 9 coins are tossed.

56. Find the probability of getting 4 heads if 6 coins are tossed.

57. Find the probability of not getting more than 3 heads if 5 coins are tossed.

58. Each arrangement of the 11 letters of the word MISSISSIPPI is put on a slip of paper and placed in a hat. One slip is drawn at random from the hat. Find the probability that the slip contains an arrangement of the letters with the 4 Ss at the very beginning of the arrangement.

59. Each arrangement of the 7 letters of the word OSMOSIS is put on a slip of paper and placed in a hat. One slip is drawn at random from the hat. Find the probability that the slip contains an arrangement of the letters with an O at the beginning and an O at the end.

60. Consider all possible arrangements of 3 identical Hs and 3 identical Ts. Suppose that 1 of these arrangements is selected at random. What is the probability that the arrangement selected has the 3 Hs in consecutive positions?

THOUGHTS INTO WORDS

61. Explain the concepts of sample space and event space.

62. Why must probability answers fall in the range between 0 and 1, inclusive? Give an example of a situation for which the probability is zero. Also give an example for which the probability is one.

Further Investigations

In Example 7 of Section 14.2 we found that there are 2,598,960 different 5-card hands that can be dealt from a deck of 52 playing cards. Therefore, probabilities for certain kinds of 5-card poker hands can be calculated using 2,598,960 as the number of elements in the sample space.

For Problems 63–71, determine the probability of being dealt a 5-card hand of the type indicated. Express your answers in reduced fractional form.

63. Straight flush (5 cards in sequence and of the same suit; aces are used as both low and high cards— A2345 and 10JQKA are both acceptable)

64. Four of a kind (4 cards of the same face value, such as 4 kings)

65. Full house (3 cards of the same face value and 2 cards of another face value)

66. Flush (5 cards of the same suit but not in sequence)

67. Straight (5 cards in sequence but not all of the same suit)

68. Three of a kind (exactly 3 cards of the same face value and 2 other cards of different face values)

69. Two pairs

70. One pair

71. No pairs

14.4 Some Properties of Probability and Tree Diagrams

In this section we will investigate some basic properties that are helpful in the computation of probabilities. The first property may state the obvious, but it still needs to be emphasized.

PROPERTY 14.1

For all events E,

$$0 \le P(E) \le 1.$$

Property 14.1 merely states that probabilities must fall in the range from 0 to 1, inclusive. This should seem reasonable, since $P(E) = \dfrac{n(E)}{n(S)}$ and E is a subset of S. The next two examples illustrate circumstances where $P(E) = 0$ and $P(E) = 1$.

Example 1

What is the probability of getting a 7 when a regular 6-sided die is tossed?

Solution

The sample space is $S = \{1, 2, 3, 4, 5, 6\}$, and the event space is $E = \emptyset$. Since $n(E) = 0$, we have $P(E) = \dfrac{n(E)}{n(S)} = \dfrac{0}{6} = 0.$

Example 2

What is the probability of getting a head or a tail when a coin is tossed?

Solution

The sample space is $S = \{H, T\}$, and the event space is $E = \{H, T\}$. Therefore, $n(S) = n(E) = 2$ and $P(E) = \dfrac{n(E)}{n(S)} = \dfrac{2}{2} = 1$. ▲

An event that has a probability of 1 is sometimes called **certain success**, and an event with a probability of 0 is called **certain failure**. It should be noted that Property 14.1 also serves as a check for the *reasonableness* of answers. In other words, when computing probabilities we know that our answer must fall in the range from 0 to 1, inclusive. Any other probability answer is simply not reasonable.

Complementary events are complementary sets where S, the sample space, serves as the universal set. The following examples illustrate this idea.

Sample space	Event space	Complementary event space
$S = \{1, 2, 3, 4, 5, 6\}$	$E = \{1, 2\}$	$E' = \{3, 4, 5, 6\}$
$S = \{H, T\}$	$E = \{T\}$	$E' = \{H\}$
$S = \{2, 3, 4, \ldots, 12\}$	$E = \{2, 3, 4\}$	$E' = \{5, 6, 7, \ldots, 12\}$
$S = \{1, 2, 3, \ldots, 25\}$	$E = \{3, 4, 5, \ldots, 25\}$	$E' = \{1, 2\}$

In each row note that E' (the complement of E) consists of all elements of S that *are not* in E. Thus, E and E' are called *complementary events*. Also note that for each row, the statement $P(E) + P(E') = 1$ can be made. For example, in the first row $P(E) = \dfrac{2}{6} = \dfrac{1}{3}$ and $P(E') = \dfrac{4}{6} = \dfrac{2}{3}$. Likewise, in the last row $P(E) = \dfrac{23}{25}$ and $P(E') = \dfrac{2}{25}$. The following general property can be stated.

PROPERTY 14.2

If E is any event of a sample space S and if E' is the complementary event, then

$$P(E) + P(E') = 1.$$

From a computational viewpoint, Property 14.2 provides us with a double-barreled approach to some probability problems. That is to say, if we are able to compute either $P(E)$ or $P(E')$, then the other one can be determined by subtracting from 1. For example, suppose that for a particular problem we can find that $P(E) = \dfrac{3}{13}$. Then we immediately know that $P(E') = 1 - P(E) = 1 - \dfrac{3}{13} = \dfrac{10}{13}$. The following examples further illustrate the use of Property 14.2.

Example 3

Find the probability of getting a sum greater than 3 if two dice are tossed.

Solution

Let S be the familiar sample space of ordered pairs for this problem, where $n(S) = 36$. Let E be the event of obtaining a sum greater than 3. Then E', the event of obtaining a sum less than or equal to 3, is $[(1, 1), (1, 2), (2, 1)]$. Thus, $P(E') = \dfrac{n(E')}{n(S)} = \dfrac{3}{36} = \dfrac{1}{12}$, and thus

$$P(E) = 1 - P(E') = 1 - \frac{1}{12} = \frac{11}{12}.$$ ▲

Example 4

Find the probability of getting at least 1 head if three coins are tossed.

Solution

The sample space S consists of all possible outcomes for tossing 3 coins. Using the fundamental principle of counting, we know that there are $(2)(2)(2) = 8$ outcomes, so $n(S) = 8$. Let E be the event of getting at least 1 head; then E' is the complementary event of not getting at least 1 head. The set E' is easy to list— namely, $E' = [(T, T, T)]$. Thus, $n(E') = 1$ and $P(E') = \dfrac{1}{8}$. From this $P(E)$ can be determined to be

$$P(E) = 1 - P(E') = 1 - \frac{1}{8} = \frac{7}{8}.$$ ▲

Example 5

From a group of 5 women and 4 men a 3-person committee is chosen at random. Find the probability that the committee contains at least 1 woman.

Solution

Let the sample space S be the set of all possible 3-person committees that can be formed from 9 people. There are $C(9, 3) = 84$ such committees; thus $n(S) = 84$. Let E be the event that the committee contains at least 1 woman. Then E' is the complementary event that the committee contains all men. Thus, E' consists of all 3-man committees that can be chosen from 4 men. There are $C(4, 3) = 4$ such committees; thus, $n(E') = 4$. Therefore,

$$P(E') = \frac{n(E')}{n(S)} = \frac{4}{84} = \frac{1}{21},$$

which determines $P(E)$ to be

$$P(E) = 1 - P(E') = 1 - \frac{1}{21} = \frac{20}{21}.$$ ▲

Tree Diagrams

Tree diagrams provide a convenient way of analyzing certain kinds of probability problems involving *drawing with replacement* and *drawing without replacement*. Suppose that a marble is randomly drawn from a bag that contains 2 red and 3 white marbles. Since there are 5 marbles, where 2 are red, the probability of drawing a red marble is $\frac{2}{5}$. Likewise, there are 3 white marbles; so the probability of drawing a white marble is $\frac{3}{5}$. These probabilities can be recorded on a tree diagram as in Figure 14.2.

FIGURE 14.2

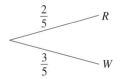

Now suppose that after we draw that first marble at random and record its color, we replace the marble in the bag before we draw a second marble and record its color. (This is called **drawing with replacement**.) The tree diagram in Figure 14.3 shows this two-step experiment. Each path along the tree corresponds to a possible outcome of this experiment. For example, the *RW* path means that we obtain a red marble on the first draw and a white marble on the second draw.

FIGURE 14.3

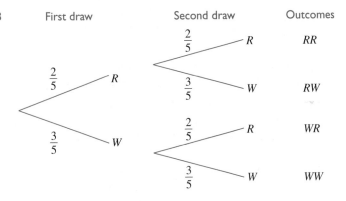

Probabilities for each of the outcomes of the two-step experiment can be calculated from the tree diagram. Let's analyze the *RW* outcome. [We will use $P(RW)$ to symbolize the probability of the *RW* outcome.]

The probability of drawing a red marble on the first draw is $\frac{2}{5}$, and the probability of drawing a white marble on the second draw is $\frac{3}{5}$. This means that we expect to draw a red marble on the first draw $\frac{2}{5}$ of the time and to draw a white marble on the second draw $\frac{3}{5}$ *of the times that we obtained a red marble on the first draw*. Thus, $P(RW) = \frac{2}{5}$ of $\frac{3}{5} = \frac{2}{5} \cdot \frac{3}{5} = \frac{6}{25}$.

In other words, the probability for each outcome of the two-step experiment can be found from the tree diagram by following the path to each of the outcomes and then *multiplying the probabilities along the paths*. Thus, for the other three outcomes in Figure 14.3, we obtain

$$P(RR) = \frac{2}{5} \cdot \frac{2}{5} = \frac{4}{25}, \qquad P(WR) = \frac{3}{5} \cdot \frac{2}{5} = \frac{6}{25},$$

$$P(WW) = \frac{3}{5} \cdot \frac{3}{5} = \frac{9}{25}.$$

Notice that $P(RR) + P(RW) + P(WR) + P(WW) = \frac{4}{25} + \frac{6}{25} + \frac{6}{25} + \frac{9}{25} = 1$.

From the tree diagram in Figure 14.3 we can also answer some other probability questions relative to this experiment. For example, suppose that we want to know the probability of drawing a red marble and a white marble. The two outcomes *RW* and *WR* both contain a red and a white marble. Since $P(RW) = \frac{6}{25}$ and $P(WR) = \frac{6}{25}$, the probability that *RW* or *WR* will occur is the *sum* of the probabilities. Thus, $P(RW \text{ or } WR) = \frac{6}{25} + \frac{6}{25} = \frac{12}{25}$.

Tree diagrams also provide a convenient way of analyzing **drawing without replacement**. For example, suppose that we randomly draw a marble from a bag containing 5 red and 2 white marbles and record its color. Then, *without replacing* the marble, we draw another marble and record its color. The tree diagram in Figure 14.4 shows this two-step experiment.

FIGURE 14.4

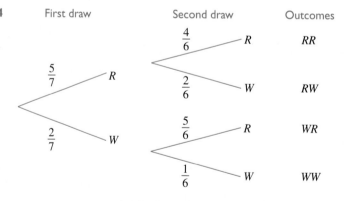

Notice that the denominators of the probabilities on the second draw are all 6. Since after the first draw the marble is *not* replaced, there are only 6 marbles remaining for the second draw. We can calculate probabilities for the various outcomes from the tree diagram as we did before.

$$P(RR) = \frac{5}{7} \cdot \frac{4}{6} = \frac{20}{42} = \frac{10}{21},$$

$$P(WR) = \frac{2}{7} \cdot \frac{5}{6} = \frac{10}{42} = \frac{5}{21},$$

$$P(RW) = \frac{5}{7} \cdot \frac{2}{6} = \frac{10}{42} = \frac{5}{21},$$

$$P(WW) = \frac{2}{7} \cdot \frac{1}{6} = \frac{2}{42} = \frac{1}{21}$$

Notice again that $P(RR) + P(RW) + P(WR) + P(WW) = \frac{10}{21} + \frac{5}{21} + \frac{5}{21} + \frac{1}{21} = \frac{21}{21} = 1.$

Problem Set 14.4

For Problems 1–11, compute the probability by using $P(E) = \frac{n(E)}{n(S)}$. For some of these problems, the property $P(E) + P(E') = 1$ will be helpful.

1. Find the probability of each of the following events if 2 dice are tossed.
(a) Getting a sum of 6
(b) Getting a sum greater than 2
(c) Getting a sum less than 8
(d) Getting a sum greater than 1

2. Find the probability of each of the following events if 3 dice are tossed.
(a) Getting a sum of 3
(b) Getting a sum greater than 4
(c) Getting a sum less than 17
(d) Getting a sum greater than 18

3. What is the probability of not getting a double when a pair of dice is tossed?

4. Find the probability of each of the following events if 4 coins are tossed.
(a) Getting 4 heads
(b) Getting 3 heads and 1 tail
(c) Getting at least 1 tail

5. Find the probability of each of the following events if 5 coins are tossed.
(a) Getting 5 tails
(b) Getting 4 heads and 1 tail
(c) Getting at least 1 tail

6. The probability that a certain horse will win the Kentucky Derby is $\frac{1}{20}$. What is the probability that it will lose the race?

7. One card is randomly drawn from a deck of 52 playing cards. What is the probability that it is not an ace?

8. Find the probability of getting at least 2 heads if 6 coins are tossed.

9. A subset of 2 letters is chosen at random from the set $\{a, b, c, d, e, f, g, h, i\}$. Find the probability that the subset will contain at least 1 vowel.

10. From a group of 4 men and 3 women, a 2-person committee is chosen at random. Find the probability that the committee contains at least 1 man.

11. From a group of 7 women and 5 men, a 3-person committee is chosen at random. Find the probability that the committee contains at least 1 man.

For Problems 12–18, use tree diagrams to help analyze the probability situations.

12. A bag contains 3 red and 2 white marbles. One marble is drawn and its color recorded. The marble is then replaced in the bag, and a second marble is drawn and its color recorded. Find the probability of each of the following.
(a) The first marble drawn is red, and the second marble drawn is white.

(b) The first marble drawn is white, and the second marble drawn is red.

(c) Both marbles drawn are white.

(d) At least 1 marble drawn is red.

(e) A red and a white marble are drawn.

13. A bag contains 5 blue and 6 gold marbles. Two marbles are drawn in succession, without replacement. Find the probability of each of the following.

(a) Both marbles drawn are blue.

(b) One marble drawn is blue, and the other is gold.

(c) At least 1 marble drawn is gold.

(d) Both marbles drawn are gold.

14. A bag contains 1 red and 2 white marbles. Two marbles are drawn in succession without replacement. Find the probability of each of the following.

(a) One marble drawn is red, and the other is white.

(b) Both marbles drawn are white.

(c) Both marbles drawn are red.

15. A bag contains 5 red and 12 white marbles. Two marbles are drawn in succession without replacement. Find the probability of each of the following.

(a) Both marbles drawn are red.

(b) Both marbles drawn are white.

(c) A red and a white marble are drawn.

(d) At least 1 marble drawn is red.

16. A bag contains 2 red, 3 white, and 4 blue marbles. Two marbles are drawn in succession without replacement. Find the probability of each of the following.

(a) Both marbles drawn are white.

(b) One marble drawn is white, and the other is blue.

(c) Both marbles drawn are blue.

(d) At least 1 red marble is drawn.

(e) No white marbles are drawn.

17. Two boxes with red and white marbles are shown. A marble is drawn at random from Box 1, then a marble is drawn from Box 2, and the colors are recorded in order. Find the probability of each of the following.

Box 1 — 3 red, 4 white   Box 2 — 2 red, 1 white

(a) Both marbles drawn are white.

(b) Both marbles drawn are red.

(c) A red and a white marble are drawn.

18. Three boxes containing red and white marbles are shown below. Randomly draw a marble from Box 1 and put it in Box 2. Then draw a marble from Box 2 and put it in Box 3. Then draw a marble from Box 3. What is the probability that the last marble drawn, from Box 3, is red? What is the probability that it is white?

Box 1 — 2 red, 2 white   Box 2 — 3 red, 1 white   Box 3 — 3 white

THOUGHTS INTO WORDS

19. If the probability of some event happening is .4, what is the probability of the event not happening? Explain your answer.

20. Explain, to a friend, the use of tree diagrams for certain probability problems.

Further Investigations

21. The term *odds* is sometimes used to express a probability statement. For example, we might say that "the odds in favor of the Cubs winning the pennant are 5 to 1" or "the odds against the White Sox winning the pennant are 50 to 1." "Odds in favor" and "odds against" for equally likely outcomes can be defined as follows.

$$\text{Odds in favor} = \frac{\text{Number of favorable outcomes}}{\text{Number of unfavorable outcomes}},$$

$$\text{Odds against} = \frac{\text{Number of unfavorable outcomes}}{\text{Number of favorable outcomes}}$$

We have used the fraction form to state the definitions for odds; however, in practice the "to" vocabulary is commonly used. Thus, the odds in favor of rolling a 4 with 1 roll of a die are usually stated as 1 to 5 instead of $\frac{1}{5}$. The odds against rolling a 4 are stated as 5 to 1.

The "odds in favor of" statement about the Cubs means that there are 5 favorable outcomes compared to 1 unfavorable, or a total of 6 possible outcomes. So the "5 to 1 in favor of" statement also means that the probability of the Cubs winning the pennant is $\frac{5}{6}$. Likewise, the "50 to 1 against" statement about the White Sox means that the probability that the White Sox will not win the pennant is $\frac{50}{51}$.

(a) What are the odds in favor of getting 2 heads with a toss of 2 coins?

(b) What are the odds against getting 2 heads and 1 tail with a toss of 3 coins?

(c) What are the odds against rolling a double with a pair of dice?

(d) Suppose that $P(E) = \frac{4}{7}$ for some event E. What are the odds against Es happening?

(e) Suppose that $P(E) = \frac{5}{9}$ for some event E. What are the odds in favor of Es happening?

(f) If the odds against an event's occurring are 5 to 2, find the probability that the event will occur.

22. In Problem 63 of Problem Set 14.3 you should have determined that the probability of being dealt a straight flush in a 5-card poker hand is $\frac{1}{64{,}974}$. Using the concept of odds (defined in Problem 21) we can say that the odds against being dealt a straight flush in a 5-card poker hand are 64,973 to 1.

Determine the odds against being dealt each of the following 5-card poker hands. (You may want to refer back to Problems 64–71 of Problem Set 14.3.)

(a) Four of a kind
(b) Full house
(c) Flush (not a straight flush)
(d) Straight (not a straight flush)
(e) Three of a kind
(f) Two pairs
(g) One pair
(h) No pair

SUMMARY

(14.1) The fundamental principle of counting can be stated as "if one task can be accomplished in x different ways, and following this task, a second task can be accomplished in y different ways, then the first task followed by the second task can be accomplished in $x \cdot y$ different ways." (This counting principle can be extended to any finite number of tasks.)

In order to apply the fundamental principle of counting, it is often helpful to analyze the problem in terms of the tasks to be accomplished.

Simply thinking about an appropriate tree diagram, without writing out the diagram in full, can also be helpful.

(14.2) The number of permutations of n things taken n at a time is given by

$$P(n, n) = n!.$$

The number of **permutations** of n things taken r at a time is given by

$$P(n, r) = \underbrace{n(n - 1)(n - 2). \dots}_{r \text{ factors}}$$

The number of **combinations** (subsets) of n things taken r at a time is given by

$$C(n, r) = \frac{P(n, r)}{r!}.$$

[Remember: If order matters, use permutations, but if order does not matter, use combinations.]

It is often very helpful to think of combinations as subsets.

(14.3) The set of all possible outcomes of a given experiment is called the **sample space**, and the individual elements of the sample space are called **sample points**. Any subset of a sample space is called an **event** or **event space**.

The probability of an event E in an experiment where all of the possible outcomes in the sample space S are equally likely to occur is defined by

$$P(E) = \frac{n(E)}{n(S)}.$$

(14.4) For all events E, $0 \leq P(E) \leq 1$.

If E is any event of a sample space S and E' is the complement of E, then

$$P(E) + P(E') = 1.$$

Tree diagrams provide a convenient way of analyzing probability problems that deal with **drawing with replacement** and **drawing without replacement**.

Chapter 14 Review Problem Set

Problems 1–14 are counting-type problems.

1. How many different arrangements of the letters A, B, C, D, E, and F can be made?

2. How many different 9-letter arrangements can be formed from the 9 letters of the word APPARATUS?

3. How many odd numbers of 3 different digits each can be formed by choosing from the digits 1, 2, 3, 5, 7, 8, and 9?

4. In how many ways can Arlene, Brent, Cindy, Dave, Ernie, Frank, and Gladys be seated in a row of 7 seats so that Arlene and Cindy are side by side?

5. In how many ways can a committee of 3 people be chosen from a group of 6 people?

6. How many committees consisting of 3 men and 2 women can be formed from a group of 7 men and 6 women?

7. How many different 5-card hands consisting of all hearts can be formed from a deck of 52 playing cards?

8. How many numbers greater than 500, where no number contains repeated digits, can be formed by choosing from the digits 2, 3, 4, 5, and 6?

9. How many 3-person committees can be formed from a group of 4 men and 5 women so that each committee contains at least 1 man?

10. How many different 4-person committees can be formed from a group of 8 people if a certain pair of people refuse to serve together on a committee?

11. How many 4-element subsets containing A or B but not both A and B can be formed from the set $\{A, B, C, D, E, F, G, H\}$?

12. How many different 6-letter permutations can be formed from 4 identical Hs and 2 identical Ts?

13. How many 4-person committees consisting of 2 seniors, 1 sophomore, and 1 junior can be formed from a group of 3 seniors, 4 juniors, and 5 sophomores?

14. In a baseball league of 6 teams, how many games are needed to complete a schedule if each team must play 8 games with each other team?

Problems 15–32 pose some probability questions.

15. Find the probability of getting 2 heads and 1 tail if 3 coins are tossed.

16. Find the probability of getting 3 heads and 2 tails if 5 coins are tossed.

17. What is the probability of getting a sum of 8 with 1 roll of a pair of dice?

18. What is the probability of getting a sum more than 5 with 1 roll of a pair of dice?

19. Amy, Brenda, Chuck, Dave, and Elmer are randomly seated in a row of 5 seats. Find the probability that Amy and Chuck are not seated side by side.

20. Four girls and 3 boys are randomly seated in a row of 7 seats. Find the probability that the girls and boys are seated in alternate seats.

21. Find the probability of getting at least 2 heads if 6 coins are tossed.

22. Two cards are randomly chosen from a deck of 52 playing cards. What is the probability that 2 jacks are drawn?

23. Each arrangement of the 6 letters of the word CYCLIC is put on a slip of paper and placed in a hat. One slip is drawn at random. Find the probability that the slip contains an arrangement with the Y at the beginning.

24. A committee of 3 is randomly chosen from a group of 1 man and 6 women. What is the probability that the man is not on the committee chosen?

25. A 4-person committee is selected at random from among the 8 people Alice, Bob, Carl, Dee, Edna, Fred, Gina, and Hilda. Find the probability that Alice or Bob, but not both, is on the committee.

26. From a group of 5 men and 4 women, a committee of 3 is chosen at random. Find the probability that the committee contains 2 men and 1 woman.

27. From a group of 6 men and 7 women, a committee of 4 is chosen at random. Find the probability that the committee contains at least 1 woman.

28. A bag contains 5 red and 8 white marbles. Two marbles are drawn in succession with replacement. Find the probability that at least 1 red marble is drawn.

29. A bag contains 4 red, 5 white, and 3 blue marbles. Two marbles are drawn in succession with replacement. Find the probability that 1 red and 1 blue marble are drawn.

30. A bag contains 4 red and 7 blue marbles. Two marbles are drawn in succession without replacement. Find the probability of drawing 1 red and 1 blue marble.

31. A bag contains 3 red, 2 white, and 2 blue marbles. Two marbles are drawn in succession without replacement. Find the probability of drawing at least 1 red marble.

32. Each of 3 letters is to be mailed in any of 4 different mail boxes. What is the probability that all 3 letters will be mailed in the same mail box?

CHAPTER 14 TEST

1. In how many ways can Ivan, Barb, Carol, and Aaron be seated in a row of 4 seats so that Ivan occupies an end seat?

2. How many numbers of 3 different digits each can be formed from the digits 1, 2, 3, 4, 5, and 6?

3. How many even numbers of 4 different digits each can be formed by choosing from the digits 1, 2, 3, 5, 7, 8, and 9?

4. In how many ways can 3 letters be mailed in 4 mailboxes?

5. In how many ways can a sum of 6 be obtained when tossing a pair of dice?

6. In how many ways can Alice, Barb, Chad, Dawn, and Ed be seated in a row of 5 seats so that either Alice or Bob occupies the middle seat?

7. In how many ways can a sum greater than 5 be obtained when tossing a pair of dice?

8. Evaluate $P(7, 4)$.

9. Evaluate $C(8, 6)$.

10. How many 4-element subsets containing A or B but not both A and B can be formed from the set $\{A, B, C, D, E, F, G\}$?

11. In a baseball league of 10 teams, how many games are needed to complete the schedule if each team plays each other team 6 games?

12. How many committees consisting of 4 men and 3 women can be formed from a group of 7 men and 5 women?

13. What is the probability of rolling a sum of 10 with a pair of dice?

14. What is the probability of rolling a sum less than 9 with a pair of dice?

15. Six coins are tossed. Find the probability of getting 5 heads and 1 tail.

16. Five coins are tossed. Find the probability of getting at least one head.

17. All possible numbers of 3 different digits each are formed from the digits 1, 2, 3, 4, 5, and 6. If one number is then chosen at random, find the probability that it is greater than 200.

18. A 4-person committee is selected at random from Ines, Barb, Chad, Dick, Edna, Fern, and Gerry. What is the probability that neither Ines nor Barb is on the committee?

19. Alice, Phyllis, Chad, Carlos, Ed, and Fred are randomly seated in a row of 6 seats. Find the probability that neither Alice nor Phyllis will occupy an end seat.

(continued on next page)

CHAPTER 14 TEST *(continued)*

20. From a group of 5 men and 5 women, a 3-person committee is chosen at random. Find the probability that the committee contains 2 women and 1 man.

21. From a group of 3 men and 5 women, a 2-person committee is selected at random. Find the probability that the committee contains at least 1 man.

22. A box of 12 items is known to contain 1 defective and 11 nondefective items. If a sample of 3 items is selected at random, what is the probability that all three items are nondefective?

23. A bag contains 7 white and 12 green marbles. Two marbles are drawn in succession, with replacement. Find the probability that one marble of each color is drawn.

24. A bag contains 3 white, 5 green, and 7 blue marbles. Two marbles are drawn, without replacement. Find the probability that 2 green marbles are drawn.

25. A bag contains 6 red and 7 blue marbles. Two marbles are drawn in succession, without replacement. Find the probability that at least one of the marbles is red.

Appendixes

A *Common Logarithms*

B *Natural Logarithms*

 Common Logarithms

TABLE OF COMMON LOGARITHMS

N	0	1	2	3	4	5	6	7	8	9
1.0	.0000	.0043	.0086	.0128	0.170	0.212	.0253	.0294	.0334	.0374
1.1	.0414	.0453	.0492	.0531	.0569	.0607	.0645	.0682	.0719	.0755
1.2	.0792	.0828	.0864	.0899	.0934	.0969	.1004	.1038	.1072	.1106
1.3	.1139	.1173	.1206	.1239	.1271	.1303	.1335	.1367	.1399	.1430
1.4	.1461	.1492	.1523	.1553	.1584	.1614	.1644	.1673	.1703	.1732
1.5	.1761	.1790	.1818	.1847	.1875	.1903	.1931	.1959	.1987	.2014
1.6	.2041	.2068	.2095	.2122	.2148	.2175	.2201	.2227	.2253	.2279
1.7	.2304	.2330	.2355	.2380	.2405	.2430	.2455	.2480	.2504	.2529
1.8	.2553	.2577	.2601	.2625	.2648	.2672	.2695	.2718	.2742	.2765
1.9	.2788	.2810	.2833	.2856	.2878	.2900	.2923	.2945	.2967	.2989

TABLE OF COMMON LOGARITHMS (CONTINUED)

N	0	1	2	3	4	5	6	7	8	9
2.0	.3010	.3032	.3054	.3075	.3096	.3118	.3139	.3160	.3181	.3201
2.1	.3222	.3243	.3263	.3284	.3304	.3324	.3345	.3365	.3385	.3404
2.2	.3424	.3444	.3464	.3483	.3502	.3522	.3541	.3560	.3579	.3598
2.3	.3617	.3636	.3655	.3674	.3692	.3711	.3729	.3747	.3766	.3784
2.4	.3802	.3820	.3838	.3856	.3874	.3892	.3909	.3927	.3945	.3962
2.5	.3979	.3997	.4014	.4031	.4048	.4065	.4082	.4099	.4116	.4133
2.6	.4150	.4166	.4183	.4200	.4216	.4232	.4249	.4265	.4281	.4298
2.7	.4314	.4330	.4346	.4362	.4378	.4393	.4409	.4425	.4440	.4456
2.8	.4472	.4487	.4502	.4518	.4533	.4548	.4564	.4579	.4594	.4609
2.9	.4624	.4639	.4654	.4669	.4683	.4698	.4713	.4728	.4742	.4757
3.0	.4771	.4786	.4800	.4814	.4829	.4843	.4857	.4871	.4886	.4900
3.1	.4914	.4928	.4942	.4955	.4969	.4983	.4997	.5011	.5024	.5038
3.2	.5051	.5065	.5079	.5092	.5105	.5119	.5132	.5145	.5159	.5172
3.3	.5185	.5198	.5211	.5224	.5237	.5250	.5263	.5276	.5289	.5302
3.4	.5315	.5328	.5340	.5353	.5366	.5378	.5391	.5403	.5416	.5428
3.5	.5441	.5453	.5465	.5478	.5490	.5502	.5514	.5527	.5539	.5551
3.6	.5563	.5575	.5587	.5599	.5611	.5623	.5635	.5647	.5658	.5670
3.7	.5682	.5694	.5705	.5717	.5729	.5740	.5752	.5763	.5775	.5786
3.8	.5798	.5809	.5821	.5832	.5843	.5855	.5866	.5877	.5888	.5899
3.9	.5911	.5922	.5933	.5944	.5955	.5966	.5977	.5988	.5999	.6010
4.0	.6021	.6031	.6042	.6053	.6064	.6075	.6085	.6096	.6107	.6117
4.1	.6128	.6138	.6149	.6160	.6170	.6180	.6191	.6201	.6212	.6222
4.2	.6232	.6243	.6253	.6263	.6274	.6284	.6294	.6304	.6314	.6325
4.3	.6335	.6345	.6355	.6365	.6375	.6385	.6395	.6405	.6415	.6425
4.4	.6435	.6444	.6454	.6464	.6474	.6484	.6493	.6503	.6513	.6522
4.5	.6532	.6542	.6551	.6561	.6571	.6580	.6590	.6599	.6609	.6618
4.6	.6628	.6637	.6646	.6656	.6665	.6675	.6684	.6693	.6702	.6712
4.7	.6721	.6730	.6739	.6749	.6758	.6767	.6776	.6785	.6794	.6803
4.8	.6812	.6821	.6830	.6839	.6848	.6857	.6866	.6875	.6884	.6893
4.9	.6902	.6911	.6920	.6928	.6937	.6946	.6955	.6964	.6972	.6981
5.0	.6990	.6998	.7007	.7016	.7024	.7033	.7042	.7050	.7059	.7067
5.1	.7076	.7084	.7093	.7101	.7110	.7118	.7126	.7135	.7143	.7152
5.2	.7160	.7168	.7177	.7185	.7193	.7202	.7210	.7218	.7226	.7235
5.3	.7243	.7251	.7259	.7267	.7275	.7284	.7292	.7300	.7308	.7316
5.4	.7324	.7332	.7340	.7348	.7356	.7364	.7372	.7380	.7388	.7396

TABLE OF COMMON LOGARITHMS (CONTINUED)

N	0	1	2	3	4	5	6	7	8	9
5.5	.7404	.7412	.7419	.7427	.7435	.7443	.7451	.7459	.7466	.7474
5.6	.7482	.7490	.7497	.7505	.7513	.7520	.7528	.7536	.7543	.7551
5.7	.7559	.7566	.7574	.7582	.7589	.7597	.7604	.7612	.7619	.7627
5.8	.7634	.7642	.7649	.7657	.7664	.7672	.7679	.7686	.7694	.7701
5.9	.7709	.7716	.7723	.7731	.7738	.7745	.7752	.7760	.7767	.7774
6.0	.7782	.7789	.7796	.7803	.7810	.7818	.7825	.7832	.7839	.7846
6.1	.7853	.7860	.7868	.7875	.7882	.7889	.7896	.7903	.7910	.7917
6.2	.7924	.7931	.7938	.7945	.7952	.7959	.7966	.7973	.7980	.7987
6.3	.7993	.8000	.8007	.8014	.8021	.8028	.8035	.8041	.8048	.8055
6.4	.8062	.8069	.8075	.8082	.8089	.8096	.8102	.8109	.8116	.8122
6.5	.8129	.8136	.8142	.8149	.8156	.8162	.8169	.8176	.8182	.8189
6.6	.8195	.8202	.8209	.8215	.8222	.8228	.8235	.8241	.8248	.8254
6.7	.8261	.8267	.8274	.8280	.8287	.8293	.8299	.8306	.8312	.8319
6.8	.8325	.8331	.8338	.8344	.8351	.8357	.8363	.8370	.8376	.8382
6.9	.8388	.8395	.8401	.8407	.8414	.8420	.8426	.8432	.8439	.8445
7.0	.8451	.8457	.8463	.8470	.8476	.8482	.8488	.8494	.8500	.8506
7.1	.8513	.8519	.8525	.8531	.8537	.8543	.8549	.8555	.8561	.8567
7.2	.8573	.8579	.8585	.8591	.8597	.8603	.8609	.8615	.8621	.8627
7.3	.8633	.8639	.8645	.8651	.8657	.8663	.8669	.8675	.8681	.8686
7.4	.8692	.8698	.8704	.8710	.8716	.8722	.8727	.8733	.8739	.8745
7.5	.8751	.8756	.8762	.8768	.8774	.8779	.8785	.8791	.8797	.8802
7.6	.8808	.8814	.8820	.8825	.8831	.8837	.8842	.8848	.8854	.8859
7.7	.8865	.8871	.8876	.8882	.8887	.8893	.8899	.8904	.8910	.8915
7.8	.8921	.8927	.8932	.8938	.8943	.8949	.8954	.8960	.8965	.8971
7.9	.8976	.8982	.8987	.8993	.8998	.9004	.9009	.9015	.9020	.9025
8.0	.9031	.9036	.9042	.9047	.9053	.9058	.9063	.9069	.9074	.9079
8.1	.9085	.9090	.9096	.9101	.9106	.9112	.9117	.9122	.9128	.9133
8.2	.9138	.9143	.9149	.9154	.9159	.9165	.9170	.9175	.9180	.9186
8.3	.9191	.9196	.9201	.9206	.9212	.9217	.9222	.9227	.9232	.9238
8.4	.9243	.9248	.9253	.9258	.9263	.9269	.9274	.9279	.9284	.9289
8.5	.9294	.9299	.9304	.9309	.9315	.9320	.9325	.9330	.9335	.9340
8.6	.9345	.9350	.9355	.9360	.9365	.9370	.9375	.9380	.9385	.9390
8.7	.9395	.9400	.9405	.9410	.9415	.9420	.9425	.9430	.9435	.9440
8.8	.9445	.9450	.9455	.9460	.9465	.9469	.9474	.9479	.9484	.9489
8.9	.9494	.9499	.9504	.9509	.9513	.9518	.9523	.9528	.9533	.9538

TABLE OF COMMON LOGARITHMS (CONTINUED)

N	0	1	2	3	4	5	6	7	8	9
9.0	.9542	.9547	.9552	.9557	.9562	.9566	.9571	.9576	.9581	.9586
9.1	.9590	.9595	.9600	.9605	.9609	.9614	.9619	.9624	.9628	.9633
9.2	.9638	.9643	.9647	.9652	.9657	.9661	.9666	.9671	.9675	.9680
9.3	.9685	.9689	.9694	.9699	.9703	.9708	.9713	.9717	.9722	.9727
9.4	.9731	.9736	.9741	.9745	.9750	.9754	.9759	.9763	.9768	.9773
9.5	.9777	.9782	.9786	.9791	.9795	.9800	.9805	.9809	.9814	.9818
9.6	.9823	.9827	.9832	.9836	.9841	.9845	.9850	.9854	.9859	.9863
9.7	.9868	.9872	.9877	.9881	.9886	.9890	.9894	.9899	.9903	.9908
9.8	.9912	.9917	.9921	.9926	.9930	.9934	.9939	.9943	.9948	.9952
9.9	.9956	.9961	.9965	.9969	.9974	.9978	.9983	.9987	.9991	.9996

Using a table to find a common logarithm is relatively easy but it does require a little more effort than pushing a few keys as you do with a calculator. Let's consider the table of common logarithms. Each number in the column headed n represents the first two significant digits of a number between 1 and 10, and each of the column headings 0 through 9 represents the third significant digit. To find the logarithm of a number such as 1.75, we look at the intersection of the row that contains 1.7 and the column headed 5. Thus, we obtain

$$\log 1.75 = .2430.$$

Similarly, we can find that

$$\log 2.09 = .3201 \quad \text{and} \quad \log 2.40 = .3802;$$

keep in mind that these values are also rounded to four decimal places.

Now suppose that we want to use the table to find the logarithm of a positive number greater than 10 or less than 1. To accomplish this we represent the number in scientific notation and then apply the property $\log rs = \log r + \log s$. For example, to find $\log 134$ we can proceed as follows.

$$\log 134 = \log(1.34 \cdot 10^2)$$
$$= \log 1.34 + \log 10^2$$
$$= .1271 + 2 = 2.1271$$

By inspection we know that the common logarithm of 10^2 is 2 (the exponent), and the common logarithm of 1.34 can be found in the table.

The decimal part (.1271) of the logarithm 2.1271 is called the **mantissa**, and the integral part, (2), is called the **characteristic**. Thus, we can find the characteristic of a common logarithm by inspection (since it is the exponent of 10 when the number is written in scientific notation) and the mantissa we can get from a table. Let's consider two more examples.

$$\log 23.8 = \log(2.38 \cdot 10^1)$$
$$= \log 2.38 + \log 10^1$$
$$= .3766 + 1$$

↑ From the table ↗ Exponent of 10

$$= 1.3766$$

$$\log .192 = \log(1.92 \cdot 10^{-1})$$
$$= \log 1.92 + \log 10^{-1}$$
$$= .2833 + (-1)$$

↑ From the table ↑ Exponent of 10

$$= .2833 + (-1)$$

Notice that in the last example we expressed the logarithm of .192 as .2833 + (−1); we did not add .2833 and −1. This is normal procedure when using a table of common logarithms because the mantissas given in the table are positive numbers. However, you should recognize that adding .2833 and −1 produces −.7167, which agrees with the result obtained earlier with a calculator.

We can also use the table to find a number when given the common logarithm of the number. That is to say, given log x we can determine x from the table. Traditionally, x is referred to as the **antilogarithm** (abbreviated **antilog**) of log x. Let's consider some examples.

Example 1

Determine antilog 1.3365.

Solution

Finding an antilogarithm simply reverses the process used before for finding a logarithm. Thus, antilog 1.3365 means that 1 is the characteristic and .3365 the mantissa. We look for .3365 in the body of the common logarithm table and we find that it is located at the intersection of the 2.1-row and the 7-column. Therefore, the antilogarithm is

$$2.17 \cdot 10^1 = 21.7.$$ ▲

Example 2

Determine antilog (.1523 + (−2)).

Solution

The mantissa, .1523, is located at the intersection of the 1.4-row and 2-column. The characteristic is −2 and therefore the antilogarithm is

$$1.42 \cdot 10^{-2} = .0142.$$ ▲

Example 3

Determine antilog -2.6038.

Solution

The mantissas given in a table are *positive* numbers. Thus, we need to express -2.6038 in terms of a positive mantissa and this can be done by adding and subtracting 3 as follows.

$$(-2.6038 + 3) - 3 = .3962 + (-3)$$

Now we can look for .3962 and find it at the intersection of the 2.4-row and 9-column. Therefore, the antilogarithm is

$$2.49 \cdot 10^{-3} = .00249.$$ ▲

Linear Interpolation

Now suppose that we want to determine log 2.774 from the table. Because the table contains only logarithms of numbers with, at most, three significant digits, we have a problem. However, by a process called **linear interpolation** we can expand the capabilities of the table to include numbers with four significant digits.

First, let's consider a geometric basis of linear interpolation and then we will use a systematic procedure for carrying out the necessary calculations. A portion of the graph of $y = \log x$, with the curvature exaggerated to help illustrate the principle involved, is shown in Figure A.1. The line segment that joins points P and Q is used to approximate the curve from P to Q. The actual value of log 2.744 is the ordinate of the point C, that is, the length of $\overline{AC}$. This cannot be determined from the table. Instead we will use the ordinate of point B (the length of $\overline{AB}$) as an approximation for log 2.744.

FIGURE A.I

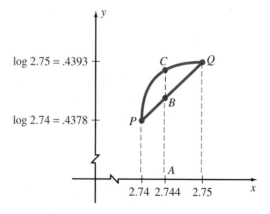

Consider Figure A.2 where line segments $\overline{DB}$ and $\overline{EQ}$ are drawn perpendicular to $\overline{PE}$. The right triangles formed $\triangle PDB$ and $\triangle PEQ$, are similar and therefore the lengths of their corresponding sides are proportional. Thus, we can write

FIGURE A.2

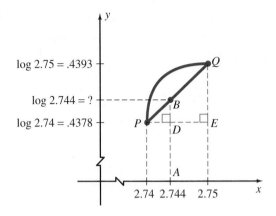

$$\frac{PD}{PE} = \frac{DB}{EQ}.$$ (1)

From Figure A.2 we see that

$$PD = 2.744 - 2.74 = .004$$

$$PE = 2.75 - 2.74 = .01$$

$$EQ = .4393 - .4378 = .0015.$$

Therefore, the proportion (1) becomes

$$\frac{.004}{.01} = \frac{DB}{.0015}.$$

Solving this proportion for DB yields

$$DB = .0006.$$

Since $AB = AD + DB$, we have

$$AB = .4378 + .0006 = .4384.$$

Thus, we obtain log 2.744 = .4384.

Now let's suggest an abbreviated format for carrying out the calculations necessary to find log 2.744.

$$
4\left\{\begin{array}{c} 2.740 \\ 2.744 \end{array}\right. \\
\left.\begin{array}{c} \\ \end{array}\right\}10 \\
2.750\Big)
\qquad
k\left\{\begin{array}{c} .4378 \\ ? \end{array}\right. \\
\left.\begin{array}{c} \\ \end{array}\right\}.0015 \\
.4393\Big)
$$

x   **log x**

Notice that we have used 4 and 10 for the differences for values of x instead of .004 and .01 because the ratio $\dfrac{.004}{.01}$ equals $\dfrac{4}{10}$. Setting up a proportion and solving for k yields

$$\frac{4}{10} = \frac{k}{.0015}$$

$$10k = 4(.0015) = .0060$$

$$k = .0006.$$

Thus, $\log 2.744 = .4378 + .0006 = .4384$.

Let's do another example to make sure of the process.

Example 4

Solution

Find $\log 617.6$.

$$\log 617.6 = \log(6.176 \cdot 10^2)$$
$$= \log 6.176 + \log 10^2$$

Thus, the characteristic is 2 and we can approximate the mantissa by using interpolation from the table as follows.

$$\frac{6}{10} = \frac{k}{.0007}$$

$$10k = 6(.0007) = .0042$$

$$k = .00042 \approx .0004$$

Therefore, $\log 6.176 = .7903 + .0004 = .7907$ and we can complete the solution for $\log 617.6$ as follows.

$$\log 617.6 = \log(6.176 \cdot 10^2)$$
$$= \log 6.176 + \log 10^2$$
$$= .7907 + 2$$
$$= 2.7907$$

The process of linear interpolation can also be used to approximate an antilogarithm when the mantissa is in between two values in the table. The following example illustrates this procedure.

Example 5

Find antilog 1.6157.

Solution

From the table we see that the mantissa, .6157, is between .6149 and .6160. We can carry out the interpolation as follows.

$$h \left\{ \begin{matrix} 4.120 \\ \\ ? \\ 4.130 \end{matrix} \right\} .010 \qquad 8 \left\{ \begin{matrix} .6149 \\ \\ .6157 \\ .6160 \end{matrix} \right\} 11 \qquad \frac{.0008}{.0011} = \frac{8}{11}$$

$$\frac{h}{.010} = \frac{8}{11}$$

$$h = 8(.010) = .080$$

$$h = \frac{1}{11}(.080) = .007 \quad \text{to the nearest thousandth}$$

Thus, antilog .6157 = 4.120 + .007 = 4.127.
Therefore,

$$\begin{aligned} \text{antilog } 1.6157 &= \text{antilog}(.6157 + 1) \\ &= 4.127 \cdot 10^1 \\ &= 41.27. \end{aligned}$$

▲

Computation with Common Logarithms

Let's first restate the basic properties of logarithms in terms of **common logarithms**. (Remember that we are writing $\log x$ instead of $\log_{10} x$.)

If x and y are positive real numbers, then

1. $\log xy = \log x + \log y$

2. $\log \dfrac{x}{y} = \log x - \log y$

3. $\log x^p = p \log x$. p is any real number.

The following two properties of equality that pertain to logarithms will also be used.

4. If $x = y$ (x and y are positive), then $\log x = \log y$.

5. If $\log x = \log y$, then $x = y$.

Example 6

Find the product $(49.1)(876)$.

Solution

Let $N = (49.1)(876)$; by Property 4,

$$\log N = \log(49.1)(876).$$

By Property 1,

$$\log N = \log 49.1 + \log 876.$$

From the table we find that $\log 49.1 = 1.6911$ and that $\log 876 = 2.9425$. Thus,

$$\begin{aligned} \log N &= 1.6911 + 2.9425 \\ &= 4.6336. \end{aligned}$$

Therefore,

$$N = \text{antilog } 4.6336.$$

By using linear interpolation, we can determine antilog .6336 to four significant digits. Thus, we obtain

$$\begin{aligned} N &= \text{antilog}(.6336 + 4) \\ &= 4.301 \cdot 10^4 \\ &= 43{,}010. \end{aligned}$$

✔ **Check**   By using a calculator we obtain

$$N = (49.1)(876) = 43{,}011.6.$$ ▲

Example 7

Find the quotient $\dfrac{942}{64.8}$.

Solution

Let $N = \dfrac{94.2}{64.8}$. Therefore,

$$\begin{aligned} \log N &= \log \frac{942}{64.8} \\ &= \log 942 - \log 64.8 \\ &= 2.9741 - 1.8116 \qquad \log \frac{x}{y} = \log x - \log y \\ &= 1.1625. \qquad\qquad\quad \text{from the table} \end{aligned}$$

Therefore,

$$\begin{aligned} N &= \text{antilog } 1.1625 \\ &= \text{antilog}(.01625 + 1) \\ &= 1.454 \cdot 10^1 \\ &= 14.54. \end{aligned}$$

✔ **Check**   By using a calculator we obtain

$$N = \frac{942}{64.8} = 14.537037.$$ ▲

Example 8

Evaluate $\dfrac{(571.4)(8.236)}{71.68}$.

Solution

Let $N = \dfrac{(571.4)(8.236)}{71.68}$. Therefore,

$$\log N = \log \dfrac{(571.4)(8.236)}{71.68}$$
$$= \log 571.4 + \log 8.236 - \log 71.68$$
$$= 2.7569 + .9157 - 1.8554 = 1.8172.$$

Therefore,

$$N = \text{antilog } 1.8172$$
$$= \text{antilog}(.8172 + 1)$$
$$= 6.564 \cdot 10^1 = 65.64.$$

✔ **Check** By using a calculator we obtain

$$N = \dfrac{(571.4)(8.236)}{71.68} = 65.653605.$$ ▲

Example 9

Evaluate $\sqrt[3]{3770}$.

Solution

Let $N = \sqrt[3]{3770} = (3770)^{\frac{1}{3}}$. Therefore,

$$\log N = \log(3770)^{\frac{1}{3}}$$
$$= \frac{1}{3}\log 3770 \qquad \log x^p = p \log x$$
$$= \frac{1}{3}(3.5763)$$
$$= 1.1921.$$

Therefore,

$$N = \text{antilog } 1.1921$$
$$= \text{antilog}(.1921 + 1)$$
$$= 1.556 \cdot 10^1 = 15.56.$$

✔ **Check** By using a calculator we obtain
$$N = \sqrt[3]{3370} = 15.563733.$$ ▲

When using tables of logarithms, it is sometimes necessary to change the form of writing a logarithm so that the decimal part (mantissa) is positive. The next example illustrates this idea.

Example 10

Find the quotient $\dfrac{1.73}{5.08}$.

Solution

Let $N = \dfrac{1.73}{5.08}$. Therefore,

$$\log N = \log \frac{1.73}{5.08}$$

$$= \log 1.73 - \log 5.08$$

$$= .2380 - .7059$$

$$= -.4679.$$

Now by adding 1 and subtracting 1, which changes the form but not the value, we obtain

$$\log N = -.4679 + 1 - 1$$

$$= .5321 - 1$$

$$= .5321 + (-1).$$

Therefore,

$$N = \text{antilog}(.5321 + (-1))$$

$$= 3.405 \cdot 10^{-1}$$

$$= .3405.$$

 Check   By using a calculator we obtain

$$N = \frac{1.73}{5.08} = .34055118. \qquad \blacktriangle$$

Sometimes it is also necessary to change the form of a logarithm so that a subsequent calculation will produce an **integer for the characteristic part of the logarithm**. Let's consider an example to illustrate this idea.

Example 11

Solution

Evaluate $\sqrt[4]{.0767}$.

Let $N = \sqrt[4]{.0767} = (.0767)^{\frac{1}{4}}$. Therefore,

$$\log N = \log(.0767)^{\frac{1}{4}} = \frac{1}{4} \log .0767$$

$$= \frac{1}{4}(.8848 + (-2))$$

$$= \frac{1}{4}(-2 + .8848).$$

At this stage we recognize that applying the distributive property will produce a nonintegral characteristic, namely, $-\dfrac{1}{2}$. Therefore, let's add 4 and subtract 4 inside the parentheses, which will change the form as follows.

$$\log N = \frac{1}{4}(-2 + .8848 + 4 - 4)$$

$$= \frac{1}{4}(4 - 2 + .8848 - 4)$$

$$= \frac{1}{4}(2.8848 - 4)$$

Now applying the distributive property we obtain

$$\log N = \frac{1}{4}(2.8848) - \frac{1}{4}(4)$$

$$= .7212 - 1 = .7212 + (-1).$$

Therefore

$$N = \text{antilog}(.7212 + (-1))$$

$$= 5.262 \cdot 10^{-1} = .5262.$$

Check   By using a calculator we obtain

$$N = \sqrt[4]{.0767} = .5262816.$$

Practice Exercises

Use the table of common logarithms and linear interpolation to find each of the following common logarithms.

1. log 4.327
2. log 27.43
3. log 128.9
4. log 3526
5. log .8761
6. log .07692
7. log .005186
8. log .0002558

Use the table of common logarithms and linear interpolation to find each of the following antilogarithms to four significant digits.

9. antilog .4690
10. antilog 1.7971
11. antilog 2.1925
12. antilog 3.7225
13. antilog(.5026 + (−1))
14. antilog(.9397 + (−2))

Use common logarithms and linear interpolation to help evaluate each of the following. Express your answers with four significant digits. Check your answers by using a calculator.

15. $(294)(71.2)$
16. $(192.6)(4.017)$
17. $\dfrac{23.4}{4.07}$
18. $\dfrac{718.5}{8.248}$
19. $(17.3)^5$
20. $(48.02)^3$
21. $\dfrac{(108)(76.2)}{13.4}$
22. $\dfrac{(126.3)(24.32)}{8.019}$
23. $\sqrt[5]{.821}$
24. $\sqrt[4]{645.3}$
25. $(79.3)^{\frac{3}{5}}$
26. $(176.8)^{\frac{3}{4}}$
27. $\sqrt{\dfrac{(7.05)(18.7)}{.521}}$
28. $\sqrt[3]{\dfrac{(41.3)(.271)}{8.05}}$

B Natural Logarithms

The following table contains the natural logarithms for numbers between .1 and 10, inclusive, at intervals of .1. Be sure that you agree with the following values taken directly from the table.

$$\ln 1.6 = .4700$$

$$\ln .5 = -.6931$$

$$\ln 4.8 = 1.5686$$

$$\ln 9.2 = 2.2192$$

TABLE OF NATURAL LOGARITHMS

n	$\ln n$	n	$\ln n$	n	$\ln n$	n	$\ln n$
.1	−2.3026	2.6	.9555	5.1	1.6292	7.6	2.0281
.2	−1.6094	2.7	.9933	5.2	1.6487	7.7	2.0412
.3	−1.2040	2.8	1.0296	5.3	1.6677	7.8	2.0541
.4	−.9163	2.9	1.0647	5.4	1.6864	7.9	2.0669
.5	−.6931	3.0	1.0986	5.5	1.7047	8.0	2.0794
.6	−.5108	3.1	1.1314	5.6	1.7228	8.1	2.0919
.7	−.3567	3.2	1.1632	5.7	1.7405	8.2	2.1041
.8	−.2231	3.3	1.1939	5.8	1.7579	8.3	2.1163
.9	−.1054	3.4	1.2238	5.9	1.7750	8.4	2.1282
1.0	.0000	3.5	1.2528	6.0	1.7918	8.5	2.1401
1.1	.0953	3.6	1.2809	6.1	1.8083	8.6	2.1518
1.2	.1823	3.7	1.3083	6.2	1.8245	8.7	2.1633
1.3	.2624	3.8	1.3350	6.3	1.8405	8.8	2.1748
1.4	.3365	3.9	1.3610	6.4	1.8563	8.9	2.1861
1.5	.4055	4.0	1.3863	6.5	1.8718	9.0	2.1972
1.6	.4700	4.1	1.4110	6.6	1.8871	9.1	2.2083
1.7	.5306	4.2	1.4351	6.7	1.9021	9.2	2.2192
1.8	.5878	4.3	1.4586	6.8	1.9169	9.3	2.2300
1.9	.6419	4.4	1.4816	6.9	1.9315	9.4	2.2407
2.0	.6931	4.5	1.5041	7.0	1.9459	9.5	2.2513
2.1	.7419	4.6	1.5261	7.1	1.9601	9.6	2.2618
2.2	.7885	4.7	1.5476	7.2	1.9741	9.7	2.2721
2.3	.8329	4.8	1.5686	7.3	1.9879	9.8	2.2824
2.4	.8755	4.9	1.5892	7.4	2.0015	9.9	2.2925
2.5	.9163	5.0	1.6094	7.5	2.0149	10	2.3026

When using a table, the natural logarithm of a positive number less than .1 or greater than 10 can be approximated by using the property $\ln rs = \ln r + \ln s$ as follows.

$$\ln 190 = \ln(1.9 \cdot 10^2)$$
$$= \ln 1.9 + \ln 10^2$$
$$= \ln 1.9 + 2 \ln 10$$
$$= .6419 + 2(2.3026)$$

From the table   From the table

$$= 5.2471.$$

$$\ln .0084 = \ln(8.4 \cdot 10^{-3})$$
$$= \ln 8.4 + \ln 10^{-3}$$
$$= \ln 8.4 + (-3)(\ln 10)$$
$$= 2.1282 - 3(2.3026)$$

From the table   From the table

$$= 2.1282 - 6.9078 = -4.7796.$$

Answers to Odd-Numbered Problems and All Chapter Review, Chapter Test, and Cumulative Review Problems

CHAPTER 1

Problem Set 1.1 (page 11)

1. True   **3.** False   **5.** True   **7.** False
9. True   **11.** 0 and 14

13. $0, 14, \frac{2}{3}, -\frac{11}{14}, 2.34, 3.2\overline{1}, 6\frac{7}{8}, -19$, and -2.6

15. 0 and 14   **17.** All of them   **19.** $\not\subset$ **21.** $\subseteq$
23. $\not\subseteq$ **25.** $\subseteq$ **27.** $\not\subseteq$ **29.** $\subseteq$ **31.** $\not\subseteq$
33. $[1, 2]$ **35.** $[0, 1, 2, 3, 4, 5]$ **37.** $[..., -1, 0, 1, 2]$
39. $\varnothing$ **41.** $[0, 1, 2, 3, 4]$ **43.** -6 **45.** 2
47. $3x + 1$ **49.** $5x$ **51.** 26   **53.** 84
55. 23   **57.** 65   **59.** 60   **61.** 33   **63.** 1320
65. 20   **67.** 119   **69.** 18   **71.** 4   **73.** 31

Problem Set 1.2 (page 20)

1. -7 **3.** -19 **5.** -22 **7.** -7 **9.** 108
11. -70 **13.** 14   **15.** -7 **17.** 28   **19.** 33
21. -204 **23.** -6 **25.** 0   **27.** Undefined
29. -60 **31.** -17 **33.** 72   **35.** -360
37. -126 **39.** -8 **41.** -12 **43.** -24
45. 15   **47.** 15   **49.** -17 **51.** 21   **53.** 5
55. 14   **57.** 26   **59.** 6   **61.** 25   **63.** 78
65. -10 **67.** 5   **69.** -5 **71.** -620
73. -4 **75.** -12.6 **77.** 6.6   **79.** -4.8
81. 7.35   **83.** -38.88 **85.** .2   **87.** -14
89. -6.5 **91.** -5.87 **93.** -6.78 **99.** $-7°F$
101. A profit of $225

Problem Set 1.3 (page 28)

1. Associative property of addition
3. Commutative property of addition
5. Additive inverse property
7. Multiplication property of negative one
9. Commutative property of multiplication
11. Distributive property
13. Associative property of multiplication

15. 18   **17.** 2   **19.** -1300 **21.** 1700
23. -47 **25.** 3200   **27.** -19 **29.** -41
31. -17 **33.** -39 **35.** 24   **37.** 20
39. 55   **41.** 16   **43.** 64   **45.** -216
47. -14 **49.** -8 **51.** -16 **53.** 6
57. 2187   **59.** -2048 **61.** $-15,625$
63. 3.9525416

Problem Set 1.4 (page 36)

1. $4x$ **3.** $-a^2$ **5.** $-6n$ **7.** $-5x + 2y$
9. $6a^2 + 5b^2$ **11.** $21x - 13$ **13.** $-2a^2b - ab^2$
15. $8x + 21$ **17.** $-5a + 2$ **19.** $-5n^2 + 11$
21. $-7x^2 + 32$ **23.** $22x - 3$ **25.** $-14x - 7$
27. $-10n^2 + 4$ **29.** $4x - 30y$ **31.** $-13x - 31$
33. $-21x - 9$ **35.** -17 **37.** 12   **39.** 4
41. 3   **43.** -38 **45.** -14 **47.** -58
49. 104   **51.** 5   **53.** 4   **55.** -54 **57.** 25
59. 221.6   **61.** 1092.4   **63.** 1420.5

65. $n + 12$ **67.** $n - 5$ **69.** $50n$ **71.** $\frac{1}{2}n - 4$

73. $\frac{n}{8}$ **75.** $2n - 9$ **77.** $10(n - 6)$

79. $n + 20$ **81.** $2t - 3$ **83.** $n + 47$ **85.** $8y$

87. $25m$ **89.** $\frac{c}{25}$ **91.** $n + 2$ **93.** $\frac{c}{5}$

95. $12d$ **97.** $3y + f$ **99.** $5280m$

Chapter 1 Review Problem Set (page 40)

1. (a) 67   **(b)** $0, -8$, and 67   **(c)** 0 and 67

(d) $0, \frac{3}{4}, -\frac{5}{6}, 8\frac{1}{3}, -8, .34, .2\overline{3}, 67$, and $\frac{9}{7}$

(e) $\sqrt{2}$ and $-\sqrt{3}$

2. Associative property for addition
3. Substitution property for equality
4. Multiplication property of negative one
5. Distributive property
6. Associative property for multiplication
7. Commutative property for addition

8. Distributive property
9. Multiplicative inverse property
10. Symmetric property of equality
11. -6 **12.** -6 **13.** -10 **14.** -15
15. 20   **16.** 49   **17.** -56 **18.** -24 **19.** 6
20. 4   **21.** -1000 **22.** 8   **23.** $-4a^2 - 5b^2$
24. $3x - 2$ **25.** $5ab^2 + 3a^2b$ **26.** $6x + 31$
27. $10n^2 - 17$ **28.** $-13a + 4$ **29.** $-2n + 2$
30. $-7x - 29y$ **31.** $-7a - 9$ **32.** $-9x^2 + 7$
33. -31 **34.** -5 **35.** -55 **36.** 144
37. -16 **38.** -44 **39.** 53   **40.** 53   **41.** 6
42. -221 **43.** $4 + 2n$ **44.** $3n - 50$
45. $\frac{2}{3}n - 6$ **46.** $10(n - 14)$ **47.** $5n + 8$
48. $\frac{n}{n-3}$ **49.** $5(n + 2) - 3$ **50.** $\frac{3}{4}(n + 12)$
51. $37 - n$ **52.** $\frac{w}{60}$ **53.** $2y - 7$ **54.** $n + 3$
55. $p + 5n + 25q$ **56.** $\frac{i}{48}$ **57.** $24f + 72y$
58. $10d$ **59.** $12f + i$ **60.** $25 - c$

Chapter 1 Test (page 43)

1. Symmetric property   **2.** Distributive property
3. -3 **4.** -23 **5.** -15 **6.** 11   **7.** 47
8. -94 **9.** -4 **10.** 960   **11.** -32
12. $-x^2 - 8x - 2$ **13.** $-19n - 20$ **14.** 27
15. -33 **16.** 32   **17.** 77   **18.** 6   **19.** -21
20. -5 **21.** $6n - 30$ **22.** $3n + 4$ **23.** $\frac{72}{n}$
24. $5n + 10d + 25q$ **25.** $6x + 2y$

CHAPTER 2

Problem Set 2.1 (page 53)

1. $\{4\}$ **3.** $\{-3\}$ **5.** $\{-14\}$ **7.** $\{6\}$ **9.** $\left\{\frac{19}{3}\right\}$
11. $\{1\}$ **13.** $\left\{-\frac{10}{3}\right\}$ **15.** $\{4\}$ **17.** $\left\{-\frac{13}{3}\right\}$
19. $\{3\}$ **21.** $\{8\}$ **23.** $\{-9\}$ **25.** $\{-3\}$
27. $\{0\}$ **29.** $\left\{-\frac{7}{2}\right\}$ **31.** $\{-2\}$ **33.** $\left\{-\frac{5}{3}\right\}$
35. $\left\{\frac{33}{2}\right\}$ **37.** $\{-35\}$ **39.** $\left\{\frac{1}{2}\right\}$ **41.** $\left\{\frac{1}{6}\right\}$
43. $\{5\}$ **45.** $\{-1\}$ **47.** $\left\{-\frac{21}{16}\right\}$ **49.** $\left\{\frac{12}{7}\right\}$
51. 14   **53.** 13, 14, and 15   **55.** 9, 11, and 13
57. 14 and 81   **59.** $11 per hour

61. 30 pennies, 50 nickels, and 70 dimes   **63.** $300
65. 20 three-bedroom, 70 two-bedroom, and 140 one-
bedroom   **71. (a)** $\varnothing$ **(c)** $\{0\}$ **(e)** $\varnothing$

Problem Set 2.2 (page 61)

1. $\{12\}$ **3.** $\left\{-\frac{3}{5}\right\}$ **5.** $\{3\}$ **7.** $\{-2\}$
9. $\{-36\}$ **11.** $\left\{\frac{20}{9}\right\}$ **13.** $\{3\}$ **15.** $\{3\}$
17. $\{-2\}$ **19.** $\left\{\frac{8}{5}\right\}$ **21.** $\{-3\}$ **23.** $\left\{\frac{48}{17}\right\}$
25. $\left\{\frac{103}{6}\right\}$ **27.** $\{3\}$ **29.** $\left\{\frac{40}{3}\right\}$ **31.** $\left\{-\frac{20}{7}\right\}$
33. $\left\{\frac{24}{5}\right\}$ **35.** $\{-10\}$ **37.** $\left\{-\frac{25}{4}\right\}$ **39.** $\{0\}$
41. 18   **43.** 16 inches long and 5 inches wide
45. 14, 15, and 16   **47.** 8 feet
49. Angie is 22 and her mother is 42.
51. 80, 90, and 94   **53.** $48°$ and $132°$ **55.** $78°$

Problem Set 2.3 (page 67)

1. $\{20\}$ **3.** $\{50\}$ **5.** $\{40\}$ **7.** $\{12\}$
9. $\{6\}$ **11.** $\{400\}$ **13.** $\{400\}$ **15.** $\{38\}$
17. $\{6\}$ **19.** $\{3000\}$ **21.** $\{3000\}$ **23.** $\{400\}$
25. $\{14\}$ **27.** $\{15\}$ **29.** $90   **31.** $54.40
33. $48   **35.** $400   **37.** 65%   **39.** $32,500
41. $3000 at 10% and $4500 at 11%   **43.** $1000
45. 8 pennies, 15 nickels, and 18 dimes
47. 15 dimes, 45 quarters, and 10 half-dollars
53. $\{7.5\}$ **55.** $\{-4775\}$ **57.** $\{8.7\}$ **59.** $\{17.1\}$
61. $\{13.5\}$

Problem Set 2.4 (page 76)

1. $120   **3.** 3 years   **5.** 6%   **7.** $800
9. $1600   **11.** 8%   **13.** $200
15. 6 feet; 14 feet; 10 feet; 20 feet; 7 feet; 2 feet
17. $h = \frac{V}{B}$ **19.** $h = \frac{V}{\pi r^2}$ **21.** $r = \frac{C}{2\pi}$
23. $C = \frac{100M}{I}$ **25.** $C = \frac{5}{9}(F - 32)$ or $C = \frac{5F - 160}{9}$
27. $x = \frac{y - b}{m}$ **29.** $x = \frac{y - y_1 + mx_1}{m}$
31. $x = \frac{ab + bc}{b - a}$ **33.** $x = a + bc$
35. $x = \frac{3b - 6a}{2}$ **37.** $x = \frac{5y + 7}{2}$
39. $y = -7x - 4$ **41.** $x = \frac{6y + 4}{3}$

43. $x = \dfrac{cy - ac - b^2}{b}$ **45.** $y = \dfrac{x - a + 1}{a - 3}$

47. 22 meters long and 6 meters wide   **49.** $11\frac{1}{9}$ years

51. $11\frac{1}{9}$ years   **53.** 4 hours   **55.** 3 hours

57. 40 miles   **59.** 25 milliliters

61. 15 quarts of 30% solution and 5 quarts of 70% solution

67. \$596.25   **69.** 1.5 years   **71.** 14.5%

73. \$1850

Problem Set 2.5 (page 85)

1. $(1, \infty)$

3. $[-1, \infty)$

5. $(-\infty, -2)$

7. $(-\infty, 2]$

9. $x < 4$ **11.** $x \le -7$ **13.** $x > 8$ **15.** $x \ge -7$

17. $(1, \infty)$

19. $(-\infty, -4]$

21. $(-\infty, -2]$

23. $(-\infty, 2)$

25. $(-1, \infty)$

27. $[-1, \infty)$

29. $(-2, \infty)$

31. $(-2, \infty)$

33. $(-\infty, -2)$

35. $[-3, \infty)$

37. $(0, \infty)$

39. $[4, \infty)$

41. $\left(\dfrac{7}{2}, \infty\right)$ **43.** $\left(\dfrac{12}{5}, \infty\right)$ **45.** $\left(-\infty, -\dfrac{5}{2}\right]$

47. $\left[\dfrac{5}{12}, \infty\right)$ **49.** $(-6, \infty)$ **51.** $(-5, \infty)$

53. $\left(-\infty, \dfrac{5}{3}\right]$ **55.** $(-36, \infty)$ **57.** $\left(-\infty, -\dfrac{8}{17}\right]$

59. $\left(-\dfrac{11}{2}, \infty\right)$ **61.** $(23, \infty)$ **63.** $(-\infty, 3)$

65. $\left(-\infty, -\dfrac{1}{7}\right]$ **67.** $(-22, \infty)$ **69.** $\left(-\infty, \dfrac{6}{5}\right)$

Problem Set 2.6 (page 93)

1. $(4, \infty)$ **3.** $\left(-\infty, \dfrac{23}{3}\right)$ **5.** $[5, \infty)$

7. $[-9, \infty)$ **9.** $\left(-\infty, -\dfrac{37}{3}\right]$ **11.** $\left(-\infty, -\dfrac{19}{6}\right)$

13. $(-\infty, 50]$ **15.** $(300, \infty)$ **17.** $[4, \infty)$

19. $(-1, 2)$

21. $(-1, 2]$

23. $(-\infty, -1) \cup (2, \infty)$

25. $(-\infty, 1] \cup (3, \infty)$

27. $(0, \infty)$

29. $\varnothing$

31. $(-\infty, \infty)$

33. $(-1, \infty)$

35. $(1, 3)$

37. $(-\infty, -5) \cup (1, \infty)$

39. $[3, \infty)$

41. $\left(\dfrac{1}{3}, \dfrac{2}{5}\right)$

43. $(-\infty, -1) \cup \left(-\dfrac{1}{3}, \infty\right)$

45. $(-2, 2)$ **47.** $[-5, 4]$ **49.** $\left(-\dfrac{1}{2}, \dfrac{3}{2}\right)$

51. $\left(-\dfrac{1}{4}, \dfrac{11}{4}\right)$ **53.** $[-11, 13]$ **55.** $(-1, 5)$

57. More than 10%   **59.** 5 feet and 10 inches or better

61. 168 or better   **63.** 77 or less

65. $163° \leq C \leq 218°$ **67.** $6.3 \leq M \leq 11.25$

Problem Set 2.7 (page 100)

1. $(-5, 5)$

3. $[-2, 2]$

5. $(-\infty, -2) \cup (2, \infty)$

7. $(-1, 3)$

9. $[-6, 2]$

11. $(-\infty, -3) \cup (-1, \infty)$

13. $(-\infty, 1] \cup [5, \infty)$

15. $[-7, 9]$ **17.** $(-\infty, -4) \cup (8, \infty)$ **19.** $(-8, 2)$

21. $[-1, 5]$ **23.** $[-4, 5]$ **25.** $\left(-\infty, -\dfrac{7}{2}\right] \cup \left[\dfrac{5}{2}, \infty\right)$

27. $\left\{-5, \dfrac{7}{3}\right\}$ **29.** $[-1, 5]$ **31.** $(-\infty, -2) \cup (6, \infty)$

33. $\left(-\dfrac{1}{2}, \dfrac{3}{2}\right)$ **35.** $\left[-5, \dfrac{7}{5}\right]$ **37.** $\left\{\dfrac{1}{12}, \dfrac{17}{12}\right\}$

39. $[-3, 10]$ **41.** $(-5, 11)$ **43.** $\left(-\infty, -\dfrac{3}{2}\right) \cup \left(\dfrac{1}{2}, \infty\right)$

45. $[0, 3]$ **47.** $(-\infty, -14] \cup [0, \infty)$ **49.** $[-2, 3]$

51. $\varnothing$ **53.** $(-\infty, \infty)$ **55.** $\left\{\dfrac{2}{5}\right\}$ **57.** $\varnothing$

59. $\varnothing$ **65.** $\left\{-2, -\dfrac{4}{3}\right\}$ **67.** $\{-2\}$ **69.** $\{0\}$

Chapter 2 Review Problem Set (page 103)

1. $\{18\}$ **2.** $\{-14\}$ **3.** $\{0\}$ **4.** $\left\{\dfrac{1}{2}\right\}$

5. $\{10\}$ **6.** $\left\{\dfrac{7}{3}\right\}$ **7.** $\left\{\dfrac{28}{17}\right\}$ **8.** $\left\{-\dfrac{1}{38}\right\}$

9. $\left\{\dfrac{27}{17}\right\}$ **10.** $\left\{-\dfrac{10}{3}, 4\right\}$ **11.** $\{50\}$ **12.** $\left\{-\dfrac{39}{2}\right\}$

13. $\{200\}$ **14.** $\{-8\}$ **15.** $\left\{-\dfrac{7}{2}, \dfrac{1}{2}\right\}$

16. $x = \dfrac{2b + 2}{a}$ **17.** $x = \dfrac{c}{a - b}$

18. $x = \dfrac{pb - ma}{m - p}$ **19.** $x = \dfrac{11 + 7y}{5}$

20. $x = \dfrac{by + b + ac}{c}$ **21.** $s = \dfrac{A - \pi r^2}{\pi r}$

22. $b_2 = \dfrac{2A - hb_1}{h}$ **23.** $R_1 = \dfrac{RR_2}{R_2 - R}$

24. $R = \dfrac{R_1 R_2}{R_1 + R_2}$ **25.** $[-5, \infty)$ **26.** $(4, \infty)$

27. $\left(-\dfrac{7}{3}, \infty\right)$ **28.** $\left[\dfrac{17}{2}, \infty\right)$ **29.** $\left(-\infty, \dfrac{1}{3}\right)$

30. $\left(\dfrac{5}{11}, \infty\right)$ **31.** $[6, \infty)$ **32.** $(-\infty, 100]$

33. $(-5, 6)$ **34.** $\left(-\infty, -\dfrac{11}{3}\right) \cup (3, \infty)$

35. $(-\infty, -17)$　　**36.** $\left(-\infty, -\dfrac{15}{4}\right)$

37. ⟵———————( )———————⟶
　　　　　　−1　1

38. ⟵——]——————(———————⟶
　　　　−3　　　　2

39. ⟵——————————(———————⟶
　　　　　　　　　3

40. ⟵——————————————————⟶

41. ⟵—————) ——(———————⟶
　　　　　−2　　1

42. ⟵—————(——) ———————⟶
　　　　　−2　　1

43. ⟵——————————(——]———⟶
　　　　　　　　$\frac{1}{2}$　3

44. $\varnothing$　　**45.** The length is 15 and the width is 7 meters.
46. $200 at 7% and $300 at 8%　　**47.** 88 or better
48. 4, 5, and 6　　**49.** $10.50 per hour
50. 20 nickels, 50 dimes, and 75 quarters　　**51.** $80°$
52. $45.60　　**53.** 30 or more　　**54.** 55 miles per hour

55. Sonya for $3\frac{1}{4}$ hours and Rita for $4\frac{1}{2}$ hours

56. $6\frac{1}{4}$ cups

Chapter 2 Test (page 106)

1. $\{-3\}$　　**2.** $\{5\}$　　**3.** $\left\{\dfrac{1}{2}\right\}$　　**4.** $\left\{\dfrac{16}{5}\right\}$

5. $\left\{-\dfrac{14}{5}\right\}$　　**6.** $\{-1\}$　　**7.** $\left\{-\dfrac{3}{2}, 3\right\}$　　**8.** $\{3\}$

9. $\left\{\dfrac{31}{3}\right\}$　　**10.** $\{650\}$　　**11.** $y = \dfrac{8x - 24}{9}$

12. $h = \dfrac{S - 2\pi r^2}{2\pi r}$　　**13.** $(-2, \infty)$　　**14.** $[-4, \infty)$

15. $(-\infty, -35]$　　**16.** $(-\infty, 10)$　　**17.** $(3, \infty)$

18. $(-\infty, 200]$　　**19.** $\left(-1, \dfrac{7}{3}\right)$

20. $\left(-\infty, -\dfrac{11}{4}\right] \cup \left[\dfrac{1}{4}, \infty\right)$　　**21.** $72

22. 19 centimeters　　**23.** $\dfrac{2}{3}$ of a cup　　**24.** 97 or better

25. $70°$

Problem Set 3.1 (page 114)

1. 2　　**3.** 3　　**5.** 2　　**7.** 6　　**9.** 0
11. $10x - 3$　　**13.** $-11t + 5$　　**15.** $-x^2 + 2x - 2$
17. $17a^2b^2 - 5ab$　　**19.** $-9x + 7$　　**21.** $-2x + 6$
23. $10a + 7$　　**25.** $4x^2 + 10x + 6$
27. $-6a^2 + 12a + 14$　　**29.** $3x^3 + x^2 + 13x - 11$
31. $7x + 8$　　**33.** $-3x - 16$　　**35.** $2x^2 - 2x - 8$
37. $-3x^3 + 5x^2 - 2x + 9$　　**39.** $5x^2 - 4x + 11$
41. $-6x^2 + 9x + 7$　　**43.** $-2x^2 + 9x + 4$
45. $-10n^2 + n + 9$　　**47.** $8x - 2$　　**49.** $8x - 14$
51. $-9x^2 - 12x + 4$　　**53.** $10x^2 + 13x - 18$
55. $-n^2 - 4n - 4$　　**57.** $-x + 6$　　**59.** $6x^2 - 4$
61. $-7n^2 + n + 6$　　**63.** $t^2 - 4t + 8$
65. $4n^2 - n - 12$　　**67.** $-4x - 2y$
69. $-x^3 - x^2 + 3x$　　**71. (a)** $8x + 4$　　**(c)** $12x + 6$
73. $8\pi h + 32\pi$　　**(a)** 226.1　　**(c)** 452.2

Problem Set 3.2 (page 121)

1. $36x^4$　　**3.** $-12x^5$　　**5.** $4a^3b^4$　　**7.** $-3x^3y^2z^6$
9. $-30xy^4$　　**11.** $27a^4b^5$　　**13.** $-m^3n^3$
15. $\dfrac{3}{10}x^3y^6$　　**17.** $-\dfrac{3}{20}a^3b^4$　　**19.** $-\dfrac{1}{6}x^3y^4$
21. $30x^6$　　**23.** $-18x^9$　　**25.** $-3x^6y^6$　　**27.** $-24y^9$
29. $-56a^4b^2$　　**31.** $-18a^3b^3$　　**33.** $-10x^7y^7$
35. $50x^5y^2$　　**37.** $27x^3y^6$　　**39.** $-32x^{10}y^5$
41. $x^{16}y^{20}$　　**43.** $a^6b^{12}c^{18}$　　**45.** $64a^{12}b^{18}$
47. $81x^2y^8$　　**49.** $81a^4b^{12}$　　**51.** $-16a^4b^4$
53. $-x^6y^{12}z^{18}$　　**55.** $-125a^6b^6c^3$　　**57.** $-x^7y^{28}z^{14}$
59. $3x^3y^3$　　**61.** $-5x^3y^2$　　**63.** $9bc^2$
65. $-18xyz^4$　　**67.** $-a^2b^3c^2$　　**69.** 9　　**71.** $-b^2$
73. $-18x^3$　　**75.** $6x^{3n}$　　**77.** a^{5n+3}　　**79.** x^{4n}
81. a^{5n+1}　　**83.** $-10x^{2n}$　　**85.** $12a^{n+4}$　　**87.** $6x^{3n+2}$
89. $12x^{n+2}$　　**91.** $22x^2; 6x^3$　　**93.** $\pi r^2 - 36\pi$

Problem Set 3.3 (page 128)

1. $10x^2y^3 + 6x^3y^4$　　**3.** $-12a^3b^3 + 15a^5b$
5. $24a^4b^5 - 16a^4b^6 + 32a^5b^6$
7. $-6x^3y^3 - 3x^4y^4 + x^5y^2$　　**9.** $ax + ay + 2bx + 2by$
11. $ac + 4ad - 3bc - 12bd$　　**13.** $x^2 + 16x + 60$
15. $y^2 + 6y - 55$　　**17.** $n^2 - 5n - 14$　　**19.** $x^2 - 36$
21. $x^2 - 12x + 36$　　**23.** $x^2 - 14x + 48$
25. $x^3 - 4x^2 + x + 6$　　**27.** $x^3 - x^2 - 9x + 9$
29. $t^2 + 18t + 81$　　**31.** $y^2 - 14y + 49$
33. $4x^2 + 33x + 35$　　**35.** $9y^2 - 1$
37. $14x^2 + 3x - 2$　　**39.** $5 + 3t - 2t^2$
41. $9t^2 + 42t + 49$　　**43.** $4 - 25x^2$

45. $49x^2 - 56x + 16$ **47.** $18x^2 - 39x - 70$

49. $2x^2 + xy - 15y^2$ **51.** $25x^2 - 4a^2$

53. $t^3 - 14t - 15$ **55.** $x^3 + x^2 - 24x + 16$

57. $2x^3 + 9x^2 + 2x - 30$ **59.** $12x^3 - 7x^2 + 25x - 6$

61. $x^4 + 5x^3 + 11x^2 + 11x + 4$

63. $2x^4 - x^3 - 12x^2 + 5x + 4$

65. $x^3 + 6x^2 + 12x + 8$ **67.** $x^3 - 12x^2 + 48x - 64$

69. $8x^3 + 36x^2 + 54x + 27$

71. $64x^3 - 48x^2 + 12x - 1$

73. $125x^3 + 150x^2 + 60x + 8$ **75.** $x^{2n} - 16$

77. $x^{2a} + 4x^a - 12$ **79.** $6x^{2n} + x^n - 35$

81. $x^{4a} - 10x^{2a} + 21$ **83.** $4x^{2n} + 20x^n + 25$

87. $2x^2 + 6$ **89.** $4x^3 - 64x^2 + 256x; 256 - 4x^2$

93. (a) $a^6 + 6a^5b + 15a^4b^2 + 20a^3b^3 + 15a^2b^4 + 6ab^5 + b^6$

(c) $a^8 + 8a^7b + 28a^6b^2 + 56a^5b^3 + 70a^4b^4 + 56a^3b^5 + 28a^2b^6 + 8ab^7 + b^8$

Problem Set 3.4 (page 137)

1. Composite   **3.** Prime   **5.** Composite

7. Composite   **9.** Prime   **11.** $2 \cdot 2 \cdot 7$

13. $2 \cdot 2 \cdot 11$ **15.** $2 \cdot 2 \cdot 2 \cdot 7$ **17.** $2 \cdot 2 \cdot 2 \cdot 3 \cdot 3$

19. $3 \cdot 29$ **21.** $3(2x + y)$ **23.** $2x(3x + 7)$

25. $4y(7y - 1)$ **27.** $5x(4y - 3)$ **29.** $x^2(7x + 10)$

31. $9ab(2a + 3b)$ **33.** $3x^3y^3(4y - 13x)$

35. $4x^2(2x^2 + 3x - 6)$ **37.** $x(5 + 7x + 9x^3)$

39. $5xy^2(3xy + 4 + 7x^2y^2)$ **41.** $(y + 2)(x + 3)$

43. $(2a + b)(3x - 2y)$ **45.** $(x + 2)(x + 5)$

47. $(a + 4)(x + y)$ **49.** $(a - 2b)(x + y)$

51. $(a - b)(3x - y)$ **53.** $(a + 1)(2x + y)$

55. $(a - 1)(x^2 + 2)$ **57.** $(a + b)(2c + 3d)$

59. $(a + b)(x - y)$ **61.** $(x + 9)(x + 6)$

63. $(x + 4)(2x + 1)$ **65.** $\{-7, 0\}$ **67.** $\{0, 1\}$

69. $\{0, 5\}$ **71.** $\left\{-\frac{1}{2}, 0\right\}$ **73.** $\left\{-\frac{7}{3}, 0\right\}$

75. $\left\{0, \frac{5}{4}\right\}$ **77.** $\left\{0, \frac{1}{4}\right\}$ **79.** $\{-12, 0\}$

81. $\left\{0, \frac{3a}{5b}\right\}$ **83.** $\left\{-\frac{3a}{2b}, 0\right\}$ **85.** $\{a, -2b\}$

87. 0 or 7   **89.** 6 units   **91.** $\frac{4}{\pi}$ units

93. The square is 100 feet by 100 feet and the rectangle is 50 feet by 100 feet.

95. 6 units   **101.** $x^a(2x^a - 3)$ **103.** $y^{2m}(y^m + 5)$

105. $x^{4a}(2x^{2a} - 3x^a + 7)$

Problem Set 3.5 (page 144)

1. $(x + 1)(x - 1)$ **3.** $(4x + 5)(4x - 5)$

5. $(3x + 5y)(3x - 5y)$ **7.** $(5xy + 6)(5xy - 6)$

9. $(2x + y^2)(2x - y^2)$ **11.** $(1 + 12n)(1 - 12n)$

13. $(x + 2 + y)(x + 2 - y)$

15. $(2x + y + 1)(2x - y - 1)$

17. $(3a + 2b + 3)(3a - 2b - 3)$

19. $-5(2x + 9)$ **21.** $9(x + 2)(x - 2)$

23. $5(x^2 + 1)$ **25.** $8(y + 2)(y - 2)$

27. $ab(a + 3)(a - 3)$ **29.** Not factorable

31. $(n + 3)(n - 3)(n^2 + 9)$ **33.** $3x(x^2 + 9)$

35. $4xy(x + 4y)(x - 4y)$ **37.** $6x(1 + x)(1 - x)$

39. $(1 + xy)(1 - xy)(1 + x^2y^2)$ **41.** $4(x + 4y)(x - 4y)$

43. $3(x + 2)(x - 2)(x^2 + 4)$ **45.** $(a - 4)(a^2 + 4a + 16)$

47. $(x + 1)(x^2 - x + 1)$

49. $(3x + 4y)(9x^2 - 12xy + 16y^2)$

51. $(1 - 3a)(1 + 3a + 9a^2)$

53. $(xy - 1)(x^2y^2 + xy + 1)$

55. $(x + y)(x - y)(x^4 + x^2y^2 + y^4)$ **57.** $\{-5, 5\}$

59. $\left\{-\frac{7}{3}, \frac{7}{3}\right\}$ **61.** $\{-2, 2\}$ **63.** $\{-1, 0, 1\}$

65. $\{-2, 2\}$ **67.** $\{-3, 3\}$ **69.** $\{0\}$

71. $-3, 0,$ or 3 **73.** 4 and 8

75. 10 meters long and 5 meters wide   **77.** 6 inches

79. 8 yards

Problem Set 3.6 (page 152)

1. $(x + 5)(x + 4)$ **3.** $(x - 4)(x - 7)$

5. $(a + 9)(a - 4)$ **7.** $(y + 6)(y + 14)$

9. $(x - 7)(x + 2)$ **11.** Not factorable

13. $(6 - x)(1 + x)$ **15.** $(x + 3y)(x + 12y)$

17. $(a - 8b)(a + 7b)$ **19.** $(3x + 1)(5x + 6)$

21. $(4x - 3)(3x + 2)$ **23.** $(2a + 3)(4a - 9)$

25. $(n - 4)(12n + 5)$ **27.** Not factorable

29. $(2n - 7)(10n + 3)$ **31.** $(8x - 5)(2x + 9)$

33. $(1 - 6x)(6 + 7x)$ **35.** $(5y + 9)(4y - 1)$

37. $(12n + 5)(2n - 1)$ **39.** $(5n + 3)(n + 6)$

41. $(x + 10)(x + 15)$ **43.** $(n - 16)(n - 20)$

45. $(t + 15)(t - 12)$ **47.** $(t^2 - 3)(t^2 - 2)$

49. $(2x^2 - 1)(5x^2 + 4)$ **51.** $(x + 1)(x - 1)(x^2 - 8)$

53. $(3n + 1)(3n - 1)(2n^2 + 3)$

55. $(x + 1)(x - 1)(x + 4)(x - 4)$

57. $2(t + 2)(t - 2)$ **59.** $(4x + 5y)(3x - 2y)$

61. $3n(2n + 5)(3n - 1)$ **63.** $(n - 12)(n - 5)$

65. $(6a - 1)^2$ **67.** $6(x^2 + 9)$ **69.** Not factorable

71. $(x + y - 7)(x - y + 7)$

73. $(1 + 4x^2)(1 + 2x)(1 - 2x)$ **75.** $(4n + 9)(3n + 8)$

77. $n(n + 7)(n - 7)$ **79.** $(x - 8)(x + 1)$

81. $3x(x - 3)(x^2 + 3x + 9)$ **83.** $(x^2 + 3)^2$

85. $(x + 3)(x - 3)(x^2 + 4)$ **87.** $(6w - 7)(3w + 5)$

89. Not factorable   **91.** $2n(n^2 + 7n - 10)$

93. $(2x + 1)(y + 3)$ **99.** $(x^a + 3)(x^a + 7)$

101. $(2x^a + 5)^2$ **103.** $(5x^n - 1)(4x^n + 5)$

105. $(x-4)(x-2)$ **107.** $(3x-11)(3x+2)$
109. $(3x+4)(5x+9)$

Problem Set 3.7 (page 159)

1. $\{-3,-1\}$ **3.** $\{-12,-6\}$ **5.** $\{4,9\}$
7. $\{-6,2\}$ **9.** $\{-1,5\}$ **11.** $\{-13,-12\}$
13. $\left\{-5,\frac{1}{3}\right\}$ **15.** $\left\{-\frac{7}{2},-\frac{2}{3}\right\}$ **17.** $\{0,4\}$
19. $\left\{\frac{1}{6},2\right\}$ **21.** $\{-6,0,6\}$ **23.** $\{-4,6\}$
25. $\{-2,2\}$ **27.** $\{-11,4\}$ **29.** $\{-5,5\}$
31. $\left\{-\frac{5}{3},-\frac{3}{5}\right\}$ **33.** $\left\{-\frac{1}{8},6\right\}$ **35.** $\left\{\frac{3}{7},\frac{5}{4}\right\}$
37. $\left\{-\frac{2}{7},\frac{4}{5}\right\}$ **39.** $\left\{-7,\frac{2}{3}\right\}$ **41.** $\{-20,18\}$
43. $\left\{-2,-\frac{1}{3},\frac{1}{3},2\right\}$ **45.** $\{-4,4\}$ **47.** $\{-1,1\}$
49. $\left\{-\frac{5}{2},-\frac{4}{3},0\right\}$ **51.** $\left\{-1,\frac{5}{3}\right\}$
53. $\left\{-\frac{3}{2},\frac{1}{2}\right\}$ **55.** 8 and 9 or -9 and -8
57. 7 and 15   **59.** 10 inches by 6 inches
61. -7 and -6 or 6 and 7
63. 4 centimeters by 4 centimeters and 6 centimeters by
8 centimeters   **65.** 3, 4, and 5 units
67. 9 inches and 12 inches
69. An altitude of 4 inches and a side 14 inches long.

Chapter 3 Review Problem Set (page 164)

1. $5x-3$ **2.** $3x^2+12x-2$ **3.** $12x^2-x+5$
4. $-20x^5y^7$ **5.** $-6a^5b^5$ **6.** $15a^4-10a^3-5a^2$
7. $24x^2+2xy-15y^2$ **8.** $3x^3+7x^2-21x-4$
9. $256x^8y^{12}$ **10.** $9x^2-12xy+4y^2$
11. $-8x^6y^9z^3$ **12.** $-13x^2y$ **13.** $2x+y-2$
14. $x^4+x^3-18x^2-x+35$ **15.** $21+26x-15x^2$
16. $-12a^5b^7$ **17.** $-8a^7b^3$ **18.** $7x^2+19x-36$
19. $6x^3-11x^2-7x+2$ **20.** $6x^{4n}$
21. $4x^2+20xy+25y^2$ **22.** $x^3-6x^2+12x-8$
23. $8x^3+60x^2+150x+125$ **24.** $(x+7)(x-4)$
25. $2(t+3)(t-3)$ **26.** Not factorable
27. $(4n-1)(3n-1)$ **28.** $x^2(x^2+1)(x+1)(x-1)$
29. $x(x-12)(x+6)$ **30.** $2a^2b(3a+2b-c)$
31. $(x-y+1)(x+y-1)$ **32.** $4(2x^2+3)$
33. $(4x+7)(3x-5)$ **34.** $(4n-5)^2$
35. $4n(n-2)$ **36.** $3w(w^2+6w-8)$
37. $(5x+2y)(4x-y)$ **38.** $16a(a-4)$
39. $3x(x+1)(x-6)$ **40.** $(n+8)(n-16)$
41. $(t+5)(t-5)(t^2+3)$ **42.** $(5x-3)(7x+2)$
43. $(3-x)(5-3x)$ **44.** $(4n-3)(16n^2+12n+9)$

45. $2(2x+5)(4x^2-10x+25)$ **46.** $\{-3,3\}$
47. $\{-6,1\}$ **48.** $\left\{\frac{2}{7}\right\}$ **49.** $\left\{-\frac{2}{5},\frac{1}{3}\right\}$
50. $\left\{-\frac{1}{3},3\right\}$ **51.** $\{-3,0,3\}$ **52.** $\{-1,0,1\}$
53. $\{-7,9\}$ **54.** $\left\{-\frac{4}{7},\frac{2}{7}\right\}$ **55.** $\left\{-\frac{4}{5},\frac{5}{6}\right\}$
56. $\{-2,2\}$ **57.** $\left\{\frac{5}{3}\right\}$ **58.** $\{-8,6\}$
59. $\left\{-5,\frac{2}{7}\right\}$ **60.** $\{-8,5\}$ **61.** $\{-12,1\}$
62. $\varnothing$ **63.** $\left\{-5,\frac{6}{5}\right\}$ **64.** $\{0,1,8\}$
65. $\left\{-10,\frac{1}{4}\right\}$ **66.** 8, 9, and 10 or $-1, 0,$ and 1
67. -6 and 8   **68.** 13 and 15
69. 12 miles and 16 miles   **70.** 4 meters by 12 meters
71. 9 rows and 16 chairs per row
72. The side is 13 feet long and the altitude is 6 feet.
73. 3 feet
74. 5 centimeters by 5 centimeters and 8 centimeters by 8
centimeters   **75.** 6 inches

Chapter 3 Test (page 166)

1. $2x-11$ **2.** $-48x^4y^4$ **3.** $-27x^6y^{12}$
4. $20x^2+17x-63$ **5.** $6n^2-13n+6$
6. $x^3-12x^2y+48xy^2-64y^3$
7. $2x^3+11x^2-11x-30$ **8.** $-14x^3y$
9. $(6x-5)(x+4)$ **10.** $3(2x+1)(2x-1)$
11. $(4+t)(16-4t+t^2)$ **12.** $2x(3-2x)(5+4x)$
13. $(x-y)(x+4)$ **14.** $(3n+8)(8n-3)$
15. $\{-12,4\}$ **16.** $\left\{0,\frac{1}{4}\right\}$ **17.** $\left\{\frac{3}{2}\right\}$ **18.** $\{-4,-1\}$
19. $\{-9,0,2\}$ **20.** $\left\{-\frac{3}{7},\frac{4}{5}\right\}$ **21.** $\left\{-\frac{1}{3},2\right\}$
22. $\{-2,2\}$ **23.** 9 inches   **24.** 15 rows
25. 8 feet

Cumulative Review Problem Set (page 167)

1. 4   **2.** -19 **3.** 9   **4.** 21   **5.** -78
6. -33 **7.** -43 **8.** -11 **9.** -39 **10.** 57
11. $2x-11$ **12.** $36a^2b^6$ **13.** $30x^2-37x-7$
14. $-2x^2-11x-12$ **15.** $-64a^6b^9$
16. $5x^3-6x^2-20x+24$ **17.** x^3-4x^2-x+12
18. $2x^4-x^3-2x^2-19x-28$ **19.** $7(x+1)(x-1)$
20. $(2a-b)^2$ **21.** $(3x+7)(x-8)$
22. $(1-x)(1+x+x^2)$ **23.** $(y-5)(x+2)$
24. $3(x-4)^2$ **25.** $(4n^2+3)(n+1)(n-1)$
26. $4x(2x+3)(4x^2-6x+9)$ **27.** $4(x^2+9)$

28. $(5x - 6)(4x + 5)$ **29.** $x = \dfrac{2y + 6}{5}$

30. $y = \dfrac{12 - 3x}{4}$ **31.** $h = \dfrac{V - 2\pi r^2}{2\pi r}$

32. $R_1 = \dfrac{RR_2}{R_2 - R}$ **33.** $[-6, 3]$ **34.** $\left\{-\dfrac{7}{3}, \dfrac{2}{5}\right\}$

35. $[-4]$ **36.** $[-9, 2]$ **37.** $[-1, 1]$ **38.** $[-10]$

39. $[15]$ **40.** $\left\{-\dfrac{25}{4}\right\}$ **41.** $\left\{-\dfrac{5}{3}, 3\right\}$ **42.** $[-1, 5]$

43. $[400]$ **44.** $[-4, 10]$ **45.** $[-4, 0, 4]$

46. $[-2, 3]$ **47.** $\left\{-\dfrac{8}{3}, \dfrac{1}{4}\right\}$ **48.** $[-6, 5]$

49. $[-5, 0, 2]$ **50.** $\left\{\dfrac{17}{12}\right\}$ **51.** $(-22, \infty)$

52. $(23, \infty)$ **53.** $(-\infty, -3) \cup (4, \infty)$

54. $\left(-7, \dfrac{7}{3}\right)$ **55.** $(300, \infty)$ **56.** $\left(-\infty, \dfrac{7}{8}\right]$

57. $\left[\dfrac{5}{2}, \infty\right)$ **58.** $\left(-\infty, \dfrac{32}{31}\right)$ **59.** 7, 9, and 11

60. 8 nickels, 15 dimes, and 25 quarters   **61.** 12 and 34
62. $62°$ and $118°$ **63.** \$400 at 8% and \$600 at 9%
64. 35 pennies, 40 nickels, and 70 dimes
65. 1 hour and 40 minutes   **66.** 25 milliliters
67. 154 or better   **68.** Better than 88
69. 4 inches   **70.** 7 meters by 14 meters
71. 8 rows and 12 chairs per row
72. 9 feet, 12 feet, and 15 feet

CHAPTER 4

Problem Set 4.1 (page 175)

1. $\dfrac{3}{4}$ **3.** $\dfrac{5}{6}$ **5.** $-\dfrac{2}{5}$ **7.** $\dfrac{2}{7}$ **9.** $\dfrac{2x}{7}$

11. $\dfrac{2a}{5b}$ **13.** $-\dfrac{y}{4x}$ **15.** $-\dfrac{9c}{13d}$ **17.** $\dfrac{5x^2}{3y^3}$

19. $\dfrac{x - 2}{x}$ **21.** $\dfrac{3x + 2}{2x - 1}$ **23.** $\dfrac{a + 5}{a - 9}$

25. $\dfrac{n - 3}{5n - 1}$ **27.** $\dfrac{5x^2 + 7}{10x}$ **29.** $\dfrac{3x + 5}{4x + 1}$

31. $\dfrac{3x}{x^2 + 4x + 16}$ **33.** $\dfrac{x + 6}{3x - 1}$ **35.** $\dfrac{x(2x + 7)}{y(x + 9)}$

37. $\dfrac{y + 4}{5y - 2}$ **39.** $\dfrac{3x(x - 1)}{x^2 + 1}$ **41.** $\dfrac{2(x + 3y)}{3x(3x + y)}$

43. $\dfrac{3n - 4}{7n + 2}$ **45.** $\dfrac{4 - x}{5 + 3x}$ **47.** $\dfrac{9x^2 + 3x + 1}{2(x + 2)}$

49. $\dfrac{-2(x-1)}{x + 1}$ **51.** $\dfrac{y + b}{y + c}$ **53.** $\dfrac{x + 2y}{2x + y}$

55. $\dfrac{x + 1}{x - 6}$ **57.** $\dfrac{2s + 5}{3s + 1}$ **59.** -1 **61.** $-n - 7$

63. $-\dfrac{2}{x + 1}$ **65.** -2 **67.** $-\dfrac{n + 3}{n + 5}$

Problem Set 4.2 (page 181)

1. $\dfrac{1}{10}$ **3.** $-\dfrac{4}{15}$ **5.** $\dfrac{3}{16}$ **7.** $-\dfrac{5}{6}$ **9.** $-\dfrac{4}{3}$

11. $\dfrac{10}{11}$ **13.** $-\dfrac{5x^3}{12y^2}$ **15.** $\dfrac{2a^3}{3b}$ **17.** $\dfrac{3x^3}{4}$

19. $\dfrac{25x^3}{108y^2}$ **21.** $\dfrac{ac^2}{2b^2}$ **23.** $\dfrac{3x}{4y}$ **25.** $\dfrac{3(x^2 + 4)}{5y(x + 8)}$

27. $\dfrac{5(a + 3)}{a(a - 2)}$ **29.** $\dfrac{3}{2}$ **31.** $\dfrac{3xy}{4(x + 6)}$

33. $\dfrac{5(x - 2y)}{7y}$ **35.** $\dfrac{5 + n}{3 - n}$ **37.** $\dfrac{x^2 + 1}{x^2 - 10}$

39. $\dfrac{6x + 5}{3x + 4}$ **41.** $\dfrac{2t^2 + 5}{2(t^2 + 1)(t + 1)}$ **43.** $\dfrac{t(t + 6)}{4t + 5}$

45. $\dfrac{n + 3}{n(n - 2)}$ **47.** $\dfrac{25x^3y^3}{4(x + 1)}$ **49.** $\dfrac{2(a - 2b)}{a(3a - 2b)}$

Problem Set 4.3 (page 188)

1. $\dfrac{13}{12}$ **3.** $\dfrac{11}{40}$ **5.** $\dfrac{19}{20}$ **7.** $\dfrac{49}{75}$ **9.** $\dfrac{17}{30}$

11. $-\dfrac{11}{84}$ **13.** $\dfrac{2x + 4}{x - 1}$ **15.** 4 **17.** $\dfrac{7y - 10}{7y}$

19. $\dfrac{5x + 3}{6}$ **21.** $\dfrac{12a + 1}{12}$ **23.** $\dfrac{n + 14}{18}$

25. $-\dfrac{11}{15}$ **27.** $\dfrac{3x - 25}{30}$ **29.** $\dfrac{43}{40x}$

31. $\dfrac{20y - 77x}{28xy}$ **33.** $\dfrac{16y + 15x - 12xy}{12xy}$

35. $\dfrac{21 + 22x}{30x^2}$ **37.** $\dfrac{10n - 21}{7n^2}$ **39.** $\dfrac{45 - 6n + 20n^2}{15n^2}$

41. $\dfrac{11x - 10}{6x^2}$ **43.** $\dfrac{42t + 43}{35t^3}$ **45.** $\dfrac{20b^2 - 33a^3}{96a^2b}$

47. $\dfrac{14 - 24y^3 + 45xy}{18xy^3}$ **49.** $\dfrac{2x^2 + 3x - 3}{x(x - 1)}$

51. $\dfrac{a^2 - a - 8}{a(a + 4)}$ **53.** $\dfrac{-41n - 55}{(4n + 5)(3n + 5)}$

55. $\dfrac{-3x + 17}{(x + 4)(7x - 1)}$ **57.** $\dfrac{-x + 74}{(3x - 5)(2x + 7)}$

59. $\dfrac{38x + 13}{(3x - 2)(4x + 5)}$ **61.** $\dfrac{5x + 5}{2x + 5}$ **63.** $\dfrac{x + 15}{x - 5}$

65. $\dfrac{-2x - 4}{2x + 1}$ **67.** (a) -1 (c) 0

Problem Set 4.4 (page 196)

1. $\dfrac{7x + 20}{x(x + 4)}$ **3.** $\dfrac{-x - 3}{x(x + 7)}$ **5.** $\dfrac{6x - 5}{(x + 1)(x - 1)}$

7. $\dfrac{1}{a + 1}$ **9.** $\dfrac{5n + 15}{4(n + 5)(n - 5)}$ **11.** $\dfrac{x^2 + 60}{x(x + 6)}$

13. $\dfrac{11x + 13}{(x + 2)(x + 7)(2x + 1)}$

15. $\dfrac{-3a + 1}{(a - 5)(a + 2)(a + 9)}$

17. $\dfrac{9a^2 + 17a + 1}{(5a + 1)(4a - 3)(3a + 4)}$

19. $\dfrac{3x^2 + 20x - 111}{(x^2 + 3)(x + 7)(x - 3)}$

21. $\dfrac{-7y - 14}{(y + 8)(y - 2)}$ **23.** $\dfrac{-2x^2 - 4x + 3}{(x + 2)(x - 2)}$

25. $\dfrac{2x^2 + 14x - 19}{(x + 10)(x - 2)}$ **27.** $\dfrac{2n + 1}{n - 6}$

29. $\dfrac{2x^2 - 32x + 16}{(x + 1)(2x - 1)(3x - 2)}$ **31.** $\dfrac{1}{(n^2 + 1)(n + 1)}$

33. $\dfrac{-16x}{(5x - 2)(x - 1)}$ **35.** $\dfrac{t + 1}{t - 2}$ **37.** $\dfrac{2}{11}$

39. $-\dfrac{7}{27}$ **41.** $\dfrac{x}{4}$ **43.** $\dfrac{3y - 2x}{4x - 7}$

45. $\dfrac{6ab^2 - 5a^2}{12b^2 + 2a^2 b}$ **47.** $\dfrac{2y - 3xy}{3x + 4xy}$ **49.** $\dfrac{3n + 14}{5n + 19}$

51. $\dfrac{5n - 17}{4n - 13}$ **53.** $\dfrac{-x + 5y - 10}{3y - 10}$

55. $\dfrac{-x + 15}{-2x - 1}$ **57.** $\dfrac{3a^2 - 2a + 1}{2a - 1}$

59. $\dfrac{-x^2 + 6x - 4}{3x - 2}$

Problem Set 4.5 (page 202)

1. $3x^3 + 6x^2$ **3.** $-6x^4 + 9x^6$ **5.** $3a^2 - 5a - 8$

7. $-13x^2 + 17x - 28$ **9.** $-3xy + 4x^2 y - 8xy^2$

11. $x - 13$ **13.** $x + 20$ **15.** $2x + 1 - \dfrac{3}{x - 1}$

17. $5x - 1$ **19.** $3x^2 - 2x - 7$ **21.** $x^2 + 5x - 6$

23. $4x^2 + 7x + 12 + \dfrac{30}{x - 2}$ **25.** $x^3 - 4x^2 - 5x + 3$

27. $x^2 + 5x + 25$ **29.** $x^2 - x + 1 + \dfrac{63}{x + 1}$

31. $2x^2 - 4x + 7 - \dfrac{20}{x + 2}$ **33.** $4a - 4b$

35. $4x + 7 + \dfrac{23x - 6}{x^2 - 3x}$ **37.** $8y - 9 + \dfrac{8y + 5}{y^2 + y}$

39. $2x - 1$ **41.** $x - 3$ **43.** $5a - 8 + \dfrac{42a - 41}{a^2 + 3a - 4}$

45. $2n^2 + 3n - 4$ **47.** $x^4 + x^3 + x^2 + x + 1$

49. $x^3 - x^2 + x - 1$ **51.** $3x^2 + x + 1 + \dfrac{7}{x^2 - 1}$

Problem Set 4.6 (page 209)

1. $\{2\}$ **3.** $\{-3\}$ **5.** $\{6\}$ **7.** $\left\{-\dfrac{85}{18}\right\}$ **9.** $\left\{\dfrac{7}{10}\right\}$

11. $\{5\}$ **13.** $\{58\}$ **15.** $\left\{\dfrac{1}{4}, 4\right\}$ **17.** $\left\{-\dfrac{2}{5}, 5\right\}$

19. $\{-16\}$ **21.** $\left\{-\dfrac{13}{3}\right\}$ **23.** $\{-3, 1\}$

25. $\left\{-\dfrac{5}{2}\right\}$ **27.** $\{-51\}$ **29.** $\left\{-\dfrac{5}{3}, 4\right\}$ **31.** $\varnothing$

33. $\left\{-\dfrac{11}{8}, 2\right\}$ **35.** $\{-29, 0\}$ **37.** $\{-9, 3\}$

39. $\left\{-2, \dfrac{23}{8}\right\}$ **41.** $\left\{\dfrac{11}{23}\right\}$ **43.** $\left\{3, \dfrac{7}{2}\right\}$

45. \$750 and \$1000   **47.** 48° and 72°   **49.** $\dfrac{2}{7}$ or $\dfrac{7}{2}$

51. \$1080   **53.** \$69 for Tammy and \$51.75 for Laura
55. 8 and 82   **57.** 14 feet and 6 feet
59. 690 females and 460 males

Problem Set 4.7 (page 218)

1. $\{-21\}$ **3.** $\{-1, 2\}$ **5.** $\{2\}$ **7.** $\left\{\dfrac{37}{15}\right\}$

9. $\{-1\}$ **11.** $\{-1\}$ **13.** $\left\{0, \dfrac{13}{2}\right\}$ **15.** $\left\{-2, \dfrac{19}{2}\right\}$

17. $\{-2\}$ **19.** $\left\{-\dfrac{1}{5}\right\}$ **21.** $\varnothing$ **23.** $\left\{\dfrac{7}{2}\right\}$

25. $\{-3\}$ **27.** $\left\{-\dfrac{7}{9}\right\}$ **29.** $\left\{-\dfrac{7}{6}\right\}$

31. $x = \dfrac{18y - 4}{15}$ **33.** $y = \dfrac{-5x + 22}{2}$

35. $M = \dfrac{IC}{100}$ **37.** $R = \dfrac{ST}{S + T}$

39. $y = \dfrac{bx - x - 3b + a}{a - 3}$ **41.** $y = \dfrac{ab - bx}{a}$

43. $y = \dfrac{-2x - 9}{3}$

45. 50 miles per hour for Dave and 54 miles per hour for Kent   **47.** 60 minutes
49. 60 words per minute for Connie and 40 words per minute for Katie
51. Plane B could travel at 400 miles per hour for 5 hours and plane A at 350 miles per hour for 4 hours or plane B could travel at 250 miles per hour for 8 hours and plane A at 200 miles per hour for 7 hours.
53. 60 minutes for Nancy and 120 minutes for Amy

55. 3 hours

57. 16 miles per hour on the way out and 12 miles per hour on the way back or 12 miles per hour out and 8 miles per hour back

Chapter 4 Review Problem Set (page 222)

1. $\dfrac{2y}{3x^2}$ **2.** $\dfrac{a-3}{a}$ **3.** $\dfrac{n-5}{n-1}$ **4.** $\dfrac{x^2+1}{x}$

5. $\dfrac{2x+1}{3}$ **6.** $\dfrac{x^2-10}{2x^2+1}$ **7.** $\dfrac{3}{22}$ **8.** $\dfrac{18y+20x}{48y-9x}$

9. $\dfrac{3x+2}{3x-2}$ **10.** $\dfrac{x-1}{2x-1}$ **11.** $\dfrac{2x}{7y^2}$ **12.** $3b$

13. $\dfrac{n(n+5)}{n-1}$ **14.** $\dfrac{x(x-3y)}{x^2+9y^2}$ **15.** $\dfrac{23x-6}{20}$

16. $\dfrac{57-2n}{18n}$ **17.** $\dfrac{3x^2-2x-14}{x(x+7)}$ **18.** $\dfrac{2}{x-5}$

19. $\dfrac{5n-21}{(n-9)(n+4)(n-1)}$ **20.** $\dfrac{6y-23}{(2y+3)(y-6)}$

21. $6x-1$ **22.** $3x^2-7x+22-\dfrac{90}{x+4}$ **23.** $\left\{\dfrac{4}{13}\right\}$

24. $\left\{\dfrac{3}{16}\right\}$ **25.** $\varnothing$ **26.** $\{-17\}$ **27.** $\left\{\dfrac{2}{7},\dfrac{7}{2}\right\}$

28. $\{22\}$ **29.** $\left\{-\dfrac{6}{7},3\right\}$ **30.** $\left\{\dfrac{3}{4},\dfrac{5}{2}\right\}$ **31.** $\left\{\dfrac{9}{7}\right\}$

32. $\left\{-\dfrac{5}{4}\right\}$ **33.** $y=\dfrac{3x+27}{4}$ **34.** $y=\dfrac{bx-ab}{a}$

35. \$525 and \$875

36. 20 minutes for Don and 30 minutes for Dan

37. 50 miles per hour and 55 miles per hour

38. 9 hours   **39.** 80 hours   **40.** 13 miles per hour

Chapter 4 Test (page 224)

1. $\dfrac{13y^2}{24x}$ **2.** $\dfrac{3x-1}{x(x-6)}$ **3.** $\dfrac{2n-3}{n+4}$ **4.** $-\dfrac{2x}{x+1}$

5. $\dfrac{3y^2}{8}$ **6.** $\dfrac{a-b}{4(2a+b)}$ **7.** $\dfrac{x+4}{5x-1}$ **8.** $\dfrac{13x+7}{12}$

9. $\dfrac{3x}{2}$ **10.** $\dfrac{10n-26}{15n}$ **11.** $\dfrac{3x^2+2x-12}{x(x-6)}$

12. $\dfrac{11-2x}{x(x-1)}$ **13.** $\dfrac{13n+46}{(2n+5)(n-2)(n+7)}$

14. $3x^2-2x-1$ **15.** $\dfrac{18-2x}{8+9x}$ **16.** $y=\dfrac{4x+20}{3}$

17. $\{1\}$ **18.** $\left\{\dfrac{1}{10}\right\}$ **19.** $\{-35\}$ **20.** $\{-1,5\}$

21. $\left\{\dfrac{5}{3}\right\}$ **22.** $\left\{-\dfrac{9}{13}\right\}$ **23.** $\dfrac{27}{72}$

24. 1 hour   **25.** 15 miles per hour

CHAPTER 5

Problem Set 5.1 (page 233)

1. $\dfrac{1}{27}$ **3.** $-\dfrac{1}{100}$ **5.** 81   **7.** -27 **9.** -8

11. 1   **13.** $\dfrac{9}{49}$ **15.** 16   **17.** $\dfrac{1}{1000}$

19. $\dfrac{1}{1000}$ **21.** 27   **23.** $\dfrac{1}{125}$ **25.** $\dfrac{9}{8}$

27. $\dfrac{256}{25}$ **29.** $\dfrac{2}{25}$ **31.** $\dfrac{81}{4}$ **33.** 81

35. $\dfrac{1}{10,000}$ **37.** $\dfrac{13}{36}$ **39.** $\dfrac{1}{2}$ **41.** $\dfrac{72}{17}$

43. $\dfrac{1}{x^6}$ **45.** $\dfrac{1}{a^3}$ **47.** $\dfrac{1}{a^8}$ **49.** $\dfrac{y^6}{x^2}$ **51.** $\dfrac{c^8}{a^4b^{12}}$

53. $\dfrac{y^{12}}{8x^9}$ **55.** $\dfrac{x^3}{y^{12}}$ **57.** $\dfrac{4a^4}{9b^2}$ **59.** $\dfrac{1}{x^2}$

61. a^5b^2 **63.** $\dfrac{6y^3}{x}$ **65.** $7b^2$ **67.** $\dfrac{7x}{y^2}$

69. $-\dfrac{12b^3}{a}$ **71.** $\dfrac{x^5y^5}{5}$ **73.** $\dfrac{b^{20}}{81}$ **75.** $\dfrac{x+1}{x^3}$

77. $\dfrac{y-x^3}{x^3y}$ **79.** $\dfrac{3b+4a^2}{a^2b}$ **81.** $\dfrac{1-x^2y}{xy^2}$

83. $\dfrac{2x-3}{x^2}$

Problem Set 5.2 (page 243)

1. 8   **3.** -10 **5.** 3   **7.** -4 **9.** 3

11. $\dfrac{4}{5}$ **13.** $-\dfrac{6}{7}$ **15.** $\dfrac{1}{2}$ **17.** $\dfrac{3}{4}$ **19.** 8

21. $3\sqrt{3}$ **23.** $4\sqrt{2}$ **25.** $4\sqrt{5}$ **27.** $4\sqrt{10}$

29. $12\sqrt{2}$ **31.** $-12\sqrt{5}$ **33.** $2\sqrt{3}$ **35.** $3\sqrt{6}$

37. $-\dfrac{5}{3}\sqrt{7}$ **39.** $\dfrac{\sqrt{19}}{2}$ **41.** $\dfrac{3\sqrt{3}}{4}$ **43.** $\dfrac{5\sqrt{3}}{9}$

45. $\dfrac{\sqrt{14}}{7}$ **47.** $\dfrac{\sqrt{6}}{3}$ **49.** $\dfrac{\sqrt{15}}{6}$ **51.** $\dfrac{\sqrt{66}}{12}$

53. $\dfrac{\sqrt{6}}{3}$ **55.** $\sqrt{5}$ **57.** $\dfrac{2\sqrt{21}}{7}$ **59.** $-\dfrac{8\sqrt{15}}{5}$

61. $\dfrac{\sqrt{6}}{4}$ **63.** $-\dfrac{12}{25}$ **65.** $2\sqrt[3]{2}$ **67.** $6\sqrt[3]{3}$

69. $\dfrac{2\sqrt[3]{3}}{3}$ **71.** $\dfrac{3\sqrt[3]{2}}{2}$ **73.** $\dfrac{\sqrt[3]{12}}{2}$

75. 42 miles per hour; 49 miles per hour; 65 miles per hour

77. 107 square centimeters   **79.** 140 square inches

85. (a) 1.414   **(c)** 12.490   **(e)** 57.000

(g) .374   **(i)** .930

Problem Set 5.3 (page 249)

1. $13\sqrt{2}$ **3.** $54\sqrt{3}$ **5.** $-30\sqrt{2}$ **7.** $-\sqrt{5}$

9. $-21\sqrt{6}$ **11.** $-\dfrac{7\sqrt{7}}{12}$ **13.** $\dfrac{37\sqrt{10}}{10}$

15. $\dfrac{41\sqrt{2}}{20}$ **17.** $-9\sqrt[3]{3}$ **19.** $10\sqrt[3]{2}$ **21.** $4\sqrt{2x}$

23. $5x\sqrt{3}$ **25.** $2x\sqrt{5y}$ **27.** $8xy^3\sqrt{xy}$

29. $3a^2b\sqrt{6b}$ **31.** $3x^3y^4\sqrt{7}$ **33.** $4a\sqrt{10a}$

35. $\dfrac{8y}{3}\sqrt{6xy}$ **37.** $\dfrac{\sqrt{10xy}}{5y}$ **39.** $\dfrac{\sqrt{15}}{6x^2}$

41. $\dfrac{5\sqrt{2y}}{6y}$ **43.** $\dfrac{\sqrt{14xy}}{4y^3}$ **45.** $\dfrac{3y\sqrt{2xy}}{4x}$

47. $\dfrac{2\sqrt{42ab}}{7b^2}$ **49.** $2\sqrt[3]{3y}$ **51.** $2x\sqrt[3]{2x}$

53. $2x^2y^2\sqrt[3]{7y^2}$ **55.** $\dfrac{\sqrt[3]{21x}}{3x}$ **57.** $\dfrac{\sqrt[3]{12x^2y}}{4x^2}$

59. $\dfrac{\sqrt[3]{4x^2y^2}}{xy^2}$ **61.** $2\sqrt{2x+3y}$ **63.** $4.\sqrt{x+3y}$

65. $33\sqrt{x}$ **67.** $-30\sqrt{2x}$ **69.** $7\sqrt{3n}$

71. $-40\sqrt{ab}$ **73.** $-7x\sqrt{2x}$

Problem Set 5.4 (page 255)

1. $6\sqrt{2}$ **3.** $18\sqrt{2}$ **5.** $-24\sqrt{10}$ **7.** $24\sqrt{6}$

9. 120 **11.** 24 **13.** $56\sqrt[3]{3}$ **15.** $\sqrt{6}+\sqrt{10}$

17. $6\sqrt{10}-3\sqrt{35}$ **19.** $24\sqrt{3}-60\sqrt{2}$

21. $-40-32\sqrt{15}$ **23.** $15\sqrt{2x}+3\sqrt{xy}$

25. $5xy-6x\sqrt{y}$ **27.** $2\sqrt{10xy}+2y\sqrt{15y}$

29. $-25\sqrt{6}$ **31.** $-25-3\sqrt{3}$ **33.** $23-9\sqrt{5}$

35. $6\sqrt{35}+3\sqrt{10}-4\sqrt{21}-2\sqrt{6}$

37. $8\sqrt{3}-36\sqrt{2}+6\sqrt{10}-18\sqrt{15}$

39. $11+13\sqrt{30}$ **41.** $141-51\sqrt{6}$ **43.** -10

45. -8 **47.** $2x-3y$ **49.** $10\sqrt[3]{12}+2\sqrt[3]{18}$

51. $12-36\sqrt[3]{2}$ **53.** $\dfrac{\sqrt{7}-1}{3}$ **55.** $\dfrac{-3\sqrt{2}-15}{23}$

57. $\dfrac{\sqrt{7}-\sqrt{2}}{5}$ **59.** $\dfrac{2\sqrt{5}+\sqrt{6}}{7}$

61. $\dfrac{\sqrt{15}-2\sqrt{3}}{2}$ **63.** $\dfrac{6\sqrt{7}+4\sqrt{6}}{13}$

65. $\sqrt{3}-\sqrt{2}$ **67.** $\dfrac{2\sqrt{x}-8}{x-16}$ **69.** $\dfrac{x+5\sqrt{x}}{x-25}$

71. $\dfrac{x-8\sqrt{x}+12}{x-36}$ **73.** $\dfrac{x-2\sqrt{xy}}{x-4y}$

75. $\dfrac{6\sqrt{xy}+9y}{4x-9y}$

Problem Set 5.5 (page 261)

1. $\{20\}$ **3.** $\varnothing$ **5.** $\left\{\dfrac{25}{4}\right\}$ **7.** $\left\{\dfrac{4}{9}\right\}$ **9.** $\{5\}$

11. $\left\{\dfrac{39}{4}\right\}$ **13.** $\varnothing$ **15.** $\{1\}$ **17.** $\left\{\dfrac{3}{2}\right\}$

19. $\{3\}$ **21.** $\left\{\dfrac{61}{25}\right\}$ **23.** $\{-3,3\}$ **25.** $\{-9,-4\}$

27. $\{0\}$ **29.** $\{3\}$ **31.** $\{4\}$ **33.** $\{-4,-3\}$

35. $\{12\}$ **37.** $\{25\}$ **39.** $\{29\}$ **41.** $\{-15\}$

43. $\left\{-\dfrac{1}{3}\right\}$ **45.** $\{-3\}$ **47.** $\{0\}$ **49.** $\{5\}$

51. $\{2,6\}$ **53.** 56 feet; 106 feet; 148 feet

55. 3.2 feet; 5.1 feet; 7.3 feet

Problem Set 5.6 (page 267)

1. 9 **3.** 3 **5.** -2 **7.** -5 **9.** $\dfrac{1}{6}$

11. 3 **13.** 8 **15.** 81 **17.** -1

19. -32 **21.** $\dfrac{81}{16}$ **23.** 4 **25.** $\dfrac{1}{128}$

27. -125 **29.** 625 **31.** $\sqrt[3]{x^4}$ **33.** $3\sqrt{x}$

35. $\sqrt[3]{2y}$ **37.** $\sqrt{2x-3y}$ **39.** $\sqrt[3]{(2a-3b)^2}$

41. $\sqrt[4]{x^2y}$ **43.** $-3\sqrt[5]{xy^2}$ **45.** $5^{\frac{1}{2}}y^{\frac{1}{2}}$ **47.** $3y^{\frac{1}{2}}$

49. $x^{\frac{1}{3}}y^{\frac{2}{3}}$ **51.** $a^{\frac{1}{2}}b^{\frac{3}{4}}$ **53.** $(2x-y)^{\frac{3}{5}}$ **55.** $5xy^{\frac{1}{2}}$

57. $-(x+y)^{\frac{1}{3}}$ **59.** $12x^{\frac{13}{20}}$ **61.** $y^{\frac{5}{12}}$ **63.** $\dfrac{4}{x^{\frac{1}{10}}}$

65. $16xy^2$ **67.** $2x^2y$ **69.** $4x^{\frac{4}{15}}$

71. $\dfrac{4}{b^{\frac{5}{12}}}$ **73.** $\dfrac{36x^{\frac{4}{5}}}{49y^{\frac{4}{3}}}$ **75.** $\dfrac{y^{\frac{3}{2}}}{x}$ **77.** $4x^{\frac{1}{6}}$

79. $\dfrac{16}{a^{\frac{11}{10}}}$ **81.** $\sqrt[6]{243}$ **83.** $\sqrt[4]{216}$ **85.** $\sqrt[12]{3}$

87. $\sqrt{2}$ **89.** $\sqrt[4]{3}$

93. (a) 12   **(c)** 7   **(e)** 11

95. (a) 1024   **(c)** 512   **(e)** 49

Problem Set 5.7 (page 272)

1. $(8.9)(10)^1$ **3.** $(4.29)(10)^3$ **5.** $(6.12)(10)^6$

7. $(4)(10)^7$ **9.** $(3.764)(10)^2$ **11.** $(3.47)(10)^{-1}$

13. $(2.14)(10)^{-2}$ **15.** $(5)(10)^{-5}$ **17.** $(1.94)(10)^{-9}$

19. 23   **21.** 4190   **23.** 500,000,000

25. 31,400,000,000   **27.** .43   **29.** .000914

31. .00000005123   **33.** .000000074   **35.** .77

37. 300,000,000,000   **39.** .000000004   **41.** 1000

43. 1000   **45.** 3000   **47.** 20   **49.** 27,000,000

53. (a) 7000 **(c)** 120 **(e)** 30
55. (a) $(4.385)(10)^{14}$ **(c)** $(2.322)(10)^{17}$
(e) $(3.052)(10)^{12}$

Chapter 5 Review Problem Set (page 276)

1. $\dfrac{1}{64}$ **2.** $\dfrac{9}{4}$ **3.** 3 **4.** -2 **5.** $\dfrac{2}{3}$ **6.** 32

7. 1 **8.** $\dfrac{4}{9}$ **9.** -64 **10.** 32 **11.** 1

12. 27 **13.** $3\sqrt{6}$ **14.** $4x\sqrt{3xy}$ **15.** $2\sqrt{2}$

16. $\dfrac{\sqrt{15x}}{6x^2}$ **17.** $2\sqrt[3]{7}$ **18.** $\dfrac{\sqrt[3]{6}}{3}$ **19.** $\dfrac{3\sqrt{5}}{5}$

20. $\dfrac{x\sqrt{21x}}{7}$ **21.** $3xy^2\sqrt[3]{4xy^2}$ **22.** $\dfrac{15\sqrt{6}}{4}$

23. $2y\sqrt{5xy}$ **24.** $2\sqrt{x}$ **25.** $24\sqrt{10}$ **26.** 60

27. $24\sqrt{3}-6\sqrt{14}$ **28.** $x-2\sqrt{x}-15$ **29.** 17

30. $12-8\sqrt{3}$ **31.** $6a-5\sqrt{ab}-4b$ **32.** 70

33. $\dfrac{2(\sqrt{7}+1)}{3}$ **34.** $\dfrac{2\sqrt{6}-\sqrt{15}}{3}$

35. $\dfrac{3\sqrt{5}-2\sqrt{3}}{11}$ **36.** $\dfrac{6\sqrt{3}+3\sqrt{5}}{7}$ **37.** $\dfrac{x^6}{y^8}$

38. $\dfrac{27a^3b^{12}}{8}$ **39.** $20x^{\frac{7}{10}}$ **40.** $7a^{\frac{5}{12}}$ **41.** $\dfrac{y^{\frac{4}{3}}}{x}$

42. $\dfrac{x^{12}}{9}$ **43.** $\sqrt{5}$ **44.** $5\sqrt[3]{3}$ **45.** $\dfrac{29\sqrt{6}}{5}$

46. $-15\sqrt{3x}$ **47.** $\dfrac{y+x^2}{x^2y}$ **48.** $\dfrac{b-2a}{a^2b}$

49. $\left\{\dfrac{19}{7}\right\}$ **50.** $\{4\}$ **51.** $\{8\}$ **52.** $\varnothing$ **53.** $\{14\}$

54. $\{-10, 1\}$ **55.** $\{2\}$ **56.** $\{8\}$ **57.** .000000006
58. 36,000,000,000 **59.** 6 **60.** .15
61. .000028 **62.** .002 **63.** .002
64. 8,000,000,000

Chapter 5 Test (page 278)

1. $\dfrac{1}{32}$ **2.** -32 **3.** $\dfrac{81}{16}$ **4.** $\dfrac{1}{4}$ **5.** $3\sqrt{7}$

6. $3\sqrt[3]{4}$ **7.** $2x^2y\sqrt{13y}$ **8.** $\dfrac{5\sqrt{6}}{6}$ **9.** $\dfrac{\sqrt{42x}}{12x^2}$

10. $72\sqrt{2}$ **11.** $-5\sqrt{6}$ **12.** $-38\sqrt{2}$

13. $\dfrac{3\sqrt{6}+3}{10}$ **14.** $\dfrac{9x^2y^2}{4}$ **15.** $-\dfrac{12}{a^{\frac{3}{10}}}$

16. $\dfrac{y^3+x}{xy^3}$ **17.** $-12x^{\frac{1}{4}}$ **18.** 33 **19.** 600

20. .003 **21.** $\left[\dfrac{8}{3}\right]$ **22.** $[2]$ **23.** $[4]$ **24.** $[5]$
25. $[4, 6]$

Cumulative Review Problem Set (page 279)

1. $\dfrac{64}{15}$ **2.** $\dfrac{11}{3}$ **3.** $\dfrac{1}{6}$ **4.** $-\dfrac{44}{5}$ **5.** -7

6. $-24a^4b^5$ **7.** $2x^3+5x^2-7x-12$ **8.** $\dfrac{3x^2y^2}{8}$

9. $\dfrac{a(a+1)}{2a-1}$ **10.** $\dfrac{-x+14}{18}$ **11.** $\dfrac{5x+19}{x(x+3)}$

12. $\dfrac{2}{n+8}$ **13.** $\dfrac{x-14}{(5x-2)(x+1)(x-4)}$

14. y^2-5y+6 **15.** x^2-3x-2 **16.** $20+7\sqrt{10}$

17. $2x-2\sqrt{xy}-12y$ **18.** $-\dfrac{3}{8}$ **19.** $-\dfrac{2}{3}$

20. .2 **21.** $\dfrac{1}{2}$ **22.** $\dfrac{13}{9}$ **23.** -27 **24.** $\dfrac{16}{9}$

25. $\dfrac{8}{27}$ **26.** $3x(x+3)(x^2-3x+9)$

27. $(6x-5)(x+4)$ **28.** $(4+7x)(3-2x)$
29. $(3x+2)(3x-2)(x^2+8)$ **30.** $(2x-y)(a-b)$
31. $(3x-2y)(9x^2+6xy+4y^2)$ **32.** $\left[-\dfrac{12}{7}\right]$

33. $[150]$ **34.** $[25]$ **35.** $[0]$ **36.** $[-2, 2]$

37. $[-7]$ **38.** $\left[-6, \dfrac{4}{3}\right]$ **39.** $\left[\dfrac{5}{4}\right]$ **40.** $[3]$

41. $\left[\dfrac{4}{5}, 1\right]$ **42.** $\left[-\dfrac{10}{3}, 4\right]$ **43.** $\left[\dfrac{1}{4}, \dfrac{2}{3}\right]$

44. $\left[-\dfrac{3}{2}, 3\right]$ **45.** $\left[\dfrac{1}{5}\right]$ **46.** $\left[\dfrac{5}{7}\right]$ **47.** $[-2, 2]$

48. $[0]$ **49.** $[-6, 19]$ **50.** $(-\infty, -2]$

51. $\left(-\infty, \dfrac{19}{5}\right)$ **52.** $\left(\dfrac{1}{4}, \infty\right)$ **53.** $(-2, 3)$

54. $\left(-\infty, -\dfrac{13}{3}\right) \cup (3, \infty)$ **55.** $(-\infty, 29]$ **56.** 6 liters

57. \$900 and \$1350 **58.** 12 inches by 17 inches
59. 5 hours **60.** 7 golf balls **61.** 12 minutes

CHAPTER 6

Problem Set 6.1 (page 289)

1. False **3.** True **5.** True **7.** True
9. $10+8i$ **11.** $-6+10i$ **13.** $-2-5i$
15. $-12+5i$ **17.** $-1-23i$ **19.** $-4-5i$
21. $1+3i$ **23.** $\dfrac{5}{3}-\dfrac{5}{12}i$ **25.** $-\dfrac{17}{9}+\dfrac{23}{30}i$

27. $9i$ **29.** $i\sqrt{14}$ **31.** $\dfrac{4}{5}i$ **33.** $3i\sqrt{2}$

35. $5i\sqrt{3}$ **37.** $6i\sqrt{7}$ **39.** $-8i\sqrt{5}$

41. $36i\sqrt{10}$ **43.** -8 **45.** $-\sqrt{15}$ **47.** $-3\sqrt{6}$

49. $-5\sqrt{3}$ **51.** $-3\sqrt{6}$ **53.** $4i\sqrt{3}$ **55.** $\dfrac{5}{2}$

57. $2\sqrt{2}$ **59.** $2i$ **61.** $-20 + 0i$ **63.** $42 + 0i$

65. $15 + 6i$ **67.** $-42 + 12i$ **69.** $7 + 22i$

71. $40 - 20i$ **73.** $-3 - 28i$ **75.** $-3 - 15i$

77. $-9 + 40i$ **79.** $-12 + 16i$ **81.** $85 + 0i$

83. $5 + 0i$ **85.** $\dfrac{3}{5} + \dfrac{3}{10}i$ **87.** $\dfrac{5}{17} - \dfrac{3}{17}i$

89. $2 + \dfrac{2}{3}i$ **91.** $0 - \dfrac{2}{7}i$ **93.** $\dfrac{22}{25} - \dfrac{4}{25}i$

95. $-\dfrac{18}{41} + \dfrac{39}{41}i$ **97.** $\dfrac{9}{2} - \dfrac{5}{2}i$ **99.** $\dfrac{4}{13} - \dfrac{1}{26}i$

Problem Set 6.2 (page 297)

1. $\{0, 9\}$ **3.** $\{-3, 0\}$ **5.** $\{-4, 0\}$ **7.** $\left\{0, \dfrac{9}{5}\right\}$

9. $\{-6, 5\}$ **11.** $\{7, 12\}$ **13.** $\left\{-8, -\dfrac{3}{2}\right\}$

15. $\left\{-\dfrac{7}{3}, \dfrac{2}{5}\right\}$ **17.** $\left\{\dfrac{3}{5}\right\}$ **19.** $\left\{-\dfrac{3}{2}, \dfrac{7}{3}\right\}$ **21.** $\{1, 4\}$

23. $\{8\}$ **25.** $\{12\}$ **27.** $\{0, 5k\}$ **29.** $\{0, 16k^2\}$

31. $\{5k, 7k\}$ **33.** $\left\{\dfrac{k}{2}, -3k\right\}$ **35.** $\{\pm 1\}$ **37.** $\{\pm 6i\}$

39. $\{\pm\sqrt{14}\}$ **41.** $\{\pm 2\sqrt{7}\}$ **43.** $\{\pm 3\sqrt{2}\}$

45. $\left\{\pm\dfrac{\sqrt{14}}{2}\right\}$ **47.** $\left\{\pm\dfrac{2\sqrt{3}}{3}\right\}$ **49.** $\left\{\pm\dfrac{2i\sqrt{30}}{5}\right\}$

51. $\left\{\pm\dfrac{\sqrt{6}}{2}\right\}$ **53.** $\{-1, 5\}$ **55.** $\{-8, 2\}$

57. $\{-6 \pm 2i\}$ **59.** $\{1, 2\}$ **61.** $\{4 \pm \sqrt{5}\}$

63. $\{-5 \pm 2\sqrt{3}\}$ **65.** $\left\{\dfrac{2 \pm 3i\sqrt{3}}{3}\right\}$ **67.** $\{-12, -2\}$

69. $\left\{\dfrac{2 \pm \sqrt{10}}{5}\right\}$ **71.** $2\sqrt{13}$ centimeters

73. $4\sqrt{5}$ inches   **75.** 8 yards   **77.** $6\sqrt{2}$ inches

79. $a = b = 4\sqrt{2}$ meters

81. $b = 3\sqrt{3}$ inches and $c = 6$ inches

83. $a = 7$ centimeters and $b = 7\sqrt{3}$ centimeters

85. $a = \dfrac{10\sqrt{3}}{3}$ feet and $c = \dfrac{20\sqrt{3}}{3}$ feet

87. 17.9 feet   **89.** 38 meters

91. 53 meters   **95.** 10.8 centimeters

Problem Set 6.3 (page 303)

1. $\{-6, 10\}$ **3.** $\{4, 10\}$ **5.** $\{-5, 10\}$ **7.** $\{-8, 1\}$

9. $\left\{-\dfrac{5}{2}, 3\right\}$ **11.** $\left\{-3, \dfrac{2}{3}\right\}$ **13.** $\{-16, 10\}$

15. $\{-2 \pm \sqrt{6}\}$ **17.** $\{-3 \pm 2\sqrt{3}\}$

19. $\{5 \pm \sqrt{26}\}$ **21.** $\{4 \pm i\}$ **23.** $\{-6 \pm 3\sqrt{3}\}$

25. $\{-1 \pm i\sqrt{5}\}$ **27.** $\left\{\dfrac{-3 \pm \sqrt{17}}{2}\right\}$

29. $\left\{\dfrac{-5 \pm \sqrt{21}}{2}\right\}$ **31.** $\left\{\dfrac{7 \pm \sqrt{37}}{2}\right\}$

33. $\left\{\dfrac{-2 \pm \sqrt{10}}{2}\right\}$ **35.** $\left\{\dfrac{3 \pm i\sqrt{6}}{3}\right\}$

37. $\left\{\dfrac{-5 \pm \sqrt{37}}{6}\right\}$ **39.** $\{-12, 4\}$ **41.** $\left\{\dfrac{4 \pm \sqrt{10}}{2}\right\}$

43. $\left\{-\dfrac{9}{2}, \dfrac{1}{3}\right\}$ **45.** $\{-3, 8\}$ **47.** $\{3 \pm 2\sqrt{3}\}$

49. $\left\{\dfrac{3 \pm i\sqrt{3}}{3}\right\}$ **51.** $\{-20, 12\}$ **53.** $\left\{-6, -\dfrac{11}{3}\right\}$

55. $\left\{-\dfrac{7}{3}, -\dfrac{3}{2}\right\}$ **57.** $\{-6 \pm 2\sqrt{10}\}$ **59.** $\left\{\dfrac{1}{4}, \dfrac{1}{3}\right\}$

61. $\left\{\dfrac{-b \pm \sqrt{b^2 - 4ac}}{2a}\right\}$ **65.** $x = \dfrac{a\sqrt{b^2 - y^2}}{b}$

67. $r = \dfrac{\sqrt{A\pi}}{\pi}$ **69.** $\{2a, 3a\}$ **71.** $\left\{\dfrac{a}{2}, -\dfrac{2a}{3}\right\}$

73. $\left\{\dfrac{2b}{3}\right\}$

Problem Set 6.4 (page 311)

1. Two real solutions; $\{-7, 3\}$ **3.** One real solution; $\left\{\dfrac{1}{3}\right\}$

5. Two complex solutions; $\left\{\dfrac{7 \pm i\sqrt{3}}{2}\right\}$

7. Two real solutions; $\left\{-\dfrac{4}{3}, \dfrac{1}{5}\right\}$

9. Two real solutions; $\left\{\dfrac{-2 \pm \sqrt{10}}{3}\right\}$

11. $\{-1 \pm \sqrt{2}\}$ **13.** $\left\{\dfrac{-5 \pm \sqrt{37}}{2}\right\}$ **15.** $\{4 \pm 2\sqrt{5}\}$

17. $\left\{\dfrac{-5 \pm i\sqrt{7}}{2}\right\}$ **19.** $\{8, 10\}$ **21.** $\left\{\dfrac{9 \pm \sqrt{61}}{2}\right\}$

23. $\left\{\dfrac{-1 \pm \sqrt{33}}{4}\right\}$ **25.** $\left\{\dfrac{-1 \pm i\sqrt{3}}{4}\right\}$ **27.** $\left\{\dfrac{4 \pm \sqrt{10}}{3}\right\}$

29. $\left\{-1, \dfrac{5}{2}\right\}$ **31.** $\left\{-5, -\dfrac{4}{3}\right\}$ **33.** $\left\{\dfrac{5}{6}\right\}$

35. $\left\{\dfrac{1 \pm \sqrt{13}}{4}\right\}$ **37.** $\left\{0, \dfrac{13}{5}\right\}$ **39.** $\left\{\pm\dfrac{\sqrt{15}}{3}\right\}$

41. $\left\{\dfrac{-1 \pm \sqrt{73}}{12}\right\}$ **43.** $\{-18, -14\}$ **45.** $\left\{\dfrac{11}{4}, \dfrac{10}{3}\right\}$

47. $\left\{\dfrac{2 \pm i\sqrt{2}}{2}\right\}$ **49.** $\left\{\dfrac{1 \pm \sqrt{7}}{6}\right\}$ **55.** $\{-1.381, 17.381\}$

57. $\{-13.426, 3.426\}$ **59.** $\{-.347, -8.653\}$

61. $\{.119, 1.681\}$ **63.** $\{-.708, 4.708\}$

65. $k = 4$ or $k = -4$

Problem Set 6.5 (page 319)

1. $\{2 \pm \sqrt{10}\}$ **3.** $\left\{-9, \dfrac{4}{3}\right\}$ **5.** $\{9 \pm 3\sqrt{10}\}$

7. $\left\{\dfrac{3 \pm i\sqrt{23}}{4}\right\}$ **9.** $\{-15, -9\}$ **11.** $\{-8, 1\}$

13. $\left\{\dfrac{2 \pm i\sqrt{10}}{2}\right\}$ **15.** $\{9 \pm \sqrt{66}\}$ **17.** $\left\{-\dfrac{5}{4}, \dfrac{2}{5}\right\}$

19. $\left\{\dfrac{-1 \pm \sqrt{2}}{2}\right\}$ **21.** $\left\{\dfrac{3}{4}, 4\right\}$ **23.** $\left\{\dfrac{11 \pm \sqrt{109}}{2}\right\}$

25. $\left\{\dfrac{3}{7}, 4\right\}$ **27.** $\left\{\dfrac{7 \pm \sqrt{129}}{10}\right\}$ **29.** $\left\{-\dfrac{10}{7}, 3\right\}$

31. $\{1 \pm \sqrt{34}\}$ **33.** $\{\pm\sqrt{6}, \pm 2\sqrt{3}\}$

35. $\left\{\pm 3, \pm\dfrac{2\sqrt{6}}{3}\right\}$ **37.** $\left\{\pm\dfrac{i\sqrt{15}}{3}, \pm 2i\right\}$

39. $\left\{\pm\dfrac{\sqrt{14}}{2}, \pm\dfrac{2\sqrt{3}}{3}\right\}$ **41.** 8 and 9   **43.** 9 and 12

45. $5 + \sqrt{3}$ and $5 - \sqrt{3}$ **47.** 3 and 6

49. 9 inches and 12 inches   **51.** 1 meter

53. 8 inches by 14 inches   **55.** 20 miles per hour for Lorraine and 25 miles per hour for Charlotte or 45 miles per hour for Lorraine and 50 miles per hour for Charlotte

57. 55 miles per hour

59. 6 hours for Tom and 8 hours for Terry   **61.** 30 hours

63. 8 people   **65.** 40 shares at \$20 per share

67. 50 numbers   **69.** 9%   **75.** $\{1, 9\}$

77. $\{-27, 8\}$ **79.** $\left\{-\dfrac{1}{6}, \dfrac{1}{2}\right\}$

Problem Set 6.6 (page 326)

1. $(-\infty, -2) \cup (1, \infty)$

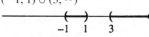

3. $(-4, -1)$

5. $\left(-\infty, -\dfrac{7}{3}\right] \cup \left[\dfrac{1}{2}, \infty\right)$

7. $\left[-2, \dfrac{3}{4}\right]$

9. $(-1, 1) \cup (3, \infty)$

11. $(-\infty, -2] \cup [0, 4]$

13. $(-\infty, -1) \cup (2, \infty)$

15. $(-2, 3)$

17. $(-\infty, 0) \cup \left[\dfrac{1}{2}, \infty\right)$

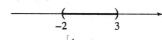

19. $(-\infty, 1) \cup [2, \infty)$

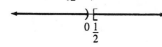

21. $(-7, 5)$ **23.** $(-\infty, 4) \cup (7, \infty)$ **25.** $\left[-5, \dfrac{2}{3}\right]$

27. $\left(-\infty, -\dfrac{5}{2}\right] \cup \left[-\dfrac{1}{4}, \infty\right)$ **29.** $\left(-\infty, -\dfrac{4}{5}\right) \cup (8, \infty)$

31. $(-\infty, \infty)$ **33.** $\left\{-\dfrac{5}{2}\right\}$ **35.** $(-1, 3) \cup (3, \infty)$

37. $(-6, -3)$ **39.** $(-\infty, 5) \cup [9, \infty)$

41. $\left(-\infty, \dfrac{4}{3}\right) \cup (3, \infty)$ **43.** $(-4, 6]$ **45.** $(-\infty, 2)$

Chapter 6 Review Problem Set (page 330)

1. $2 - 2i$ **2.** $-3 - i$ **3.** $30 + 15i$ **4.** $86 - 2i$

5. $-32 + 4i$ **6.** 25   **7.** $\dfrac{9}{20} + \dfrac{13}{20}i$

8. $-\dfrac{3}{29} + \dfrac{7}{29}i$ **9.** Two equal real solutions

10. Two nonreal complex solutions

11. Two unequal real solutions

12. Two unequal real solutions   **13.** $\{0, 17\}$

14. $\{-4, 8\}$ **15.** $\left\{\dfrac{1 \pm 8i}{2}\right\}$ **16.** $\{-3, 7\}$

17. $\{-1 + \sqrt{10}\}$ **18.** $\{3 \pm 5i\}$ **19.** $\{25\}$

20. $\left[-4, \dfrac{2}{3}\right]$　**21.** $[-10, 20]$　**22.** $\left\{\dfrac{-1 \pm \sqrt{61}}{6}\right\}$

23. $\left\{\dfrac{1 \pm i\sqrt{11}}{2}\right\}$　**24.** $\left\{\dfrac{5 \pm i\sqrt{23}}{4}\right\}$

25. $\left\{\dfrac{-2 \pm \sqrt{14}}{2}\right\}$　**26.** $[-9, 4]$　**27.** $\{-2 \pm i\sqrt{5}\}$

28. $[-6, 12]$　**29.** $\{1 \pm \sqrt{10}\}$　**30.** $\left\{\pm\dfrac{\sqrt{14}}{2}, \pm 2\sqrt{2}\right\}$

31. $\left\{\dfrac{-3 \pm \sqrt{97}}{2}\right\}$　**32.** $(-\infty, -5) \cup (2, \infty)$

33. $\left[-\dfrac{7}{2}, 3\right]$　**34.** $(-\infty, -6) \cup [4, \infty)$　**35.** $\left(-\dfrac{5}{2}, -1\right)$

36. $3 + \sqrt{7}$ and $3 - \sqrt{7}$

37. 20 shares at \$15 per share

38. 45 miles per hour and 52 miles per hour　**39.** 8 units

40. 8 and 10　**41.** 7 inches by 12 inches

42. 4 hours for Janet and 6 hours for Billy　**43.** 10 meters

25. The lengths of the three sides are $\sqrt{40}$, $\sqrt{200}$, and $\sqrt{160}$. Since $(\sqrt{40})^2 + (\sqrt{160})^2 = (\sqrt{200})^2$, it is a right triangle.

27. The distance between $(2, -1)$ and $(-2, -4)$ equals the distance between $(2, -1)$ and $(6, 2)$ which is 5 units.

29. $(15 + \sqrt{97})$ units.

31. Each side is 13 units long and the sum of the squares of two adjacent sides equal the square of the length of a diagonal.

33. $\dfrac{2}{3}$　**35.** $-\dfrac{3}{2}$　**37.** $\dfrac{4}{3}$　**39.** 0

41. $\dfrac{d - b}{c - a}$ or $\dfrac{b - d}{a - c}$　**43.** 1　**45.** Undefined

47. 7　**49.** 9　**51–57.** Answers will vary.

59. (a) 105.6 feet　**61.** 1.0 feet

67. (a) 7　**(c)** $\dfrac{1}{3}$　**(e)** -7

68. (a) (3.5)　**(c)** $(2, 5)$　**(e)** $\left(\dfrac{17}{8}, -7\right)$

Chapter 6 Test (page 332)

1. $39 - 2i$　**2.** $-\dfrac{6}{25} - \dfrac{17}{25}i$　**3.** $[0, 7]$

4. $[-1, 7]$　**5.** $[-6, 3]$　**6.** $\{1 - \sqrt{2}, 1 + \sqrt{2}\}$

7. $\left\{\dfrac{1 - 2i}{5}, \dfrac{1 + 2i}{5}\right\}$　**8.** $[-16, -14]$

9. $\left\{\dfrac{1 - 6i}{3}, \dfrac{1 + 6i}{3}\right\}$　**10.** $\left[-\dfrac{7}{4}, \dfrac{6}{5}\right]$　**11.** $\left[-3, \dfrac{19}{6}\right]$

12. $\left[-\dfrac{10}{3}, 4\right]$　**13.** $\{-2, 2, -4i, 4i\}$　**14.** $\left[-\dfrac{3}{4}, 1\right]$

15. $\left\{\dfrac{1 - \sqrt{10}}{3}, \dfrac{1 + \sqrt{10}}{3}\right\}$　**16.** Two equal real solutions

17. Two nonreal complex solutions　**18.** $[-6, 9]$

19. $(-\infty, -2) \cup \left(\dfrac{1}{3}, \infty\right)$　**20.** $[-10, -6]$

21. 20.8 feet　**22.** 29 meters　**23.** 150 shares

24. $6\dfrac{1}{2}$ inches　**25.** $3 + \sqrt{5}$

CHAPTER 7

Problem Set 7.1 (page 343)

1. 15　**3.** -5　**5.** -9　**7.** 11　**9.** 10

11. 5　**13.** 10　**15.** $\sqrt{26}$　**17.** $2\sqrt{13}$

19. 20　**21.** $4\sqrt{5}$　**23.** $3\sqrt{2}$

Problem Set 7.2 (page 354)

1.
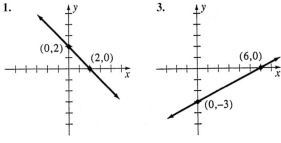

3.

5.
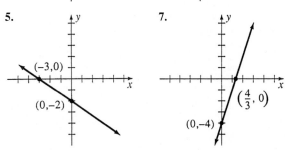

7.

9.
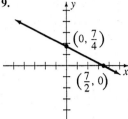

11.
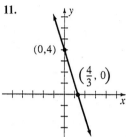

29.
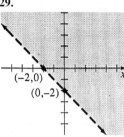

31.
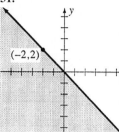

13.
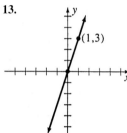

15.
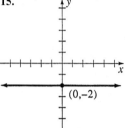

33.
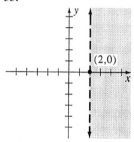

35.
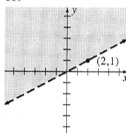

17.
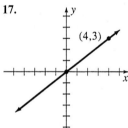

19.
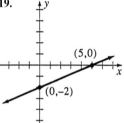

37. $-\dfrac{4}{5}$　　**39.** 3　　**41.** $-\dfrac{2}{3}$　　**43.** $\dfrac{3}{4}$　　**45.** -3

47. $\dfrac{1}{2}$

21.
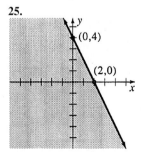

23.
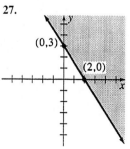

53.
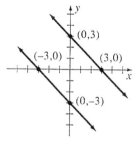

55.
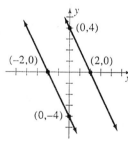

25.

27.

57.
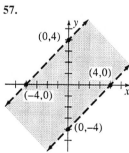

59.

61.

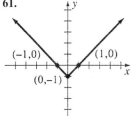

63.

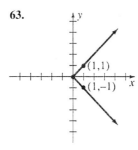

(e)

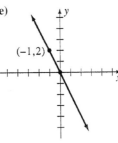

(g)

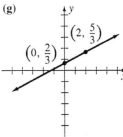

65.

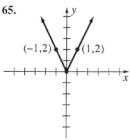

67. (a) $x - 4y = -3$ **(c)** $2x + 3y = 9$
69. (a) $2x + 3y = 26$ **(c)** $5x + 3y = -11$

Problem Set 7.3 (page 364)

1. $x - 2y = -5$ **3.** $5x + 6y = 32$ **5.** $3x - y = 1$
7. $3x + 2y = -10$ **9.** $x - y = -2$
11. $3x + 5y = 4$ **13.** $7x - 5y = -34$ **15.** $y = -4$
17. $2x - y = -6$ **19.** $y = \frac{2}{7}x + 3$ **21.** $y = 3x - 2$
23. $y = -\frac{1}{4}x - 5$ **25.** $y = -\frac{1}{2}x + \frac{2}{3}$
27. $4x - 3y = -12$ **29.** $x = 3$ **31.** $y = 7$
33. Parallel   **35.** Perpendicular
37. Intersecting lines that are not perpendicular
39. Parallel
41. Intersecting lines that are not perpendicular
43. Perpendicular   **45.** $5x - y = 6$
47. $3x + 2y = 0$ **49.** $x - 2y = 6$
51. $x - 3y = -1$ **53.** $y = \frac{1}{5}x + \frac{28}{5}$
55. $y = -\frac{5}{4}x - \frac{11}{2}$
57. (a)

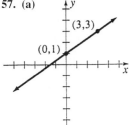

(c)

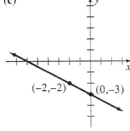

Problem Set 7.4 (page 376)

1.

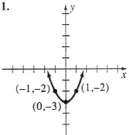

3.

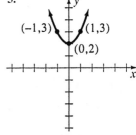

5.

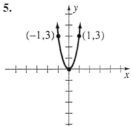

7.

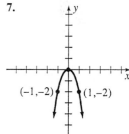

9.

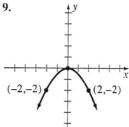

11.

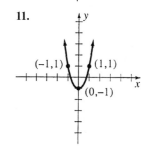

13.

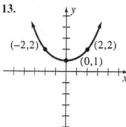

15.

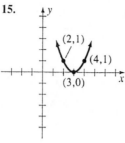

Problem Set 7.5 (page 384)

1.

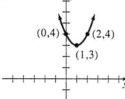

3.

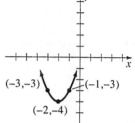

17.

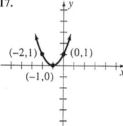

19.

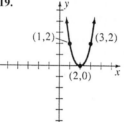

5.

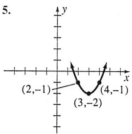

7.

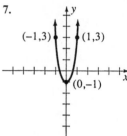

21.

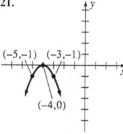

23.

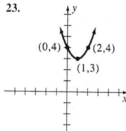

9.

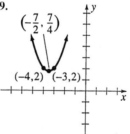

11.

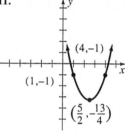

25.

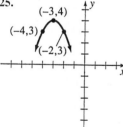

27.

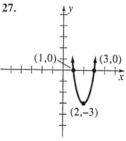

13.

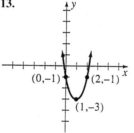

15.

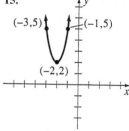

29.

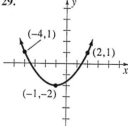

31.

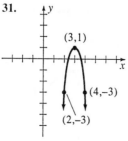

17.

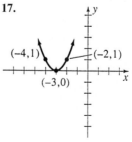

19.

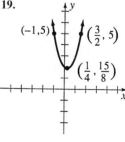

21.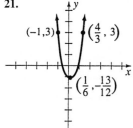

23.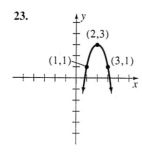

73.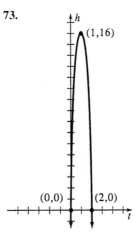

25.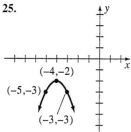

27.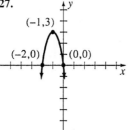

29.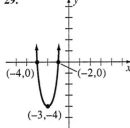

75. (a) 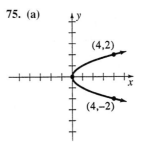   **(c)**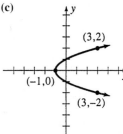

31. 3 and 5; $(4, -1)$ **33.** 6 and 8; $(7, -2)$
35. 4 and 6; $(5, 1)$ **37.** $7 + \sqrt{5}$ and $7 - \sqrt{5}$; $(7, -5)$
39. No x-intercepts; $\left(\dfrac{9}{2}, -\dfrac{3}{4}\right)$

41. $\dfrac{1 + \sqrt{5}}{2}$ and $\dfrac{1 - \sqrt{5}}{2}$; $\left(\dfrac{1}{2}, 5\right)$

43. $x^2 + y^2 - 6x - 10y + 30 = 0$
45. $x^2 + y^2 + 4x - 2y - 4 = 0$ **47.** $x^2 + y^2 = 3$
49. $x^2 + y^2 + 4x + 10y + 17 = 0$
51. $x^2 + y^2 + 6y - 27 = 0$ **53.** $(1, 4), r = 3$
55. $(3, -2), r = 6$ **57.** $(0, 0), r = 2\sqrt{6}$
59. $(-2, -6), r = 5$ **61.** $(-2, 0), r = \sqrt{7}$
63. $(-3, 4), r = 5$ **65.** $x^2 + y^2 + 10y = 0$
67. $x^2 + y^2 + 8x + 6y = 0$

(e) 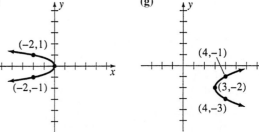   **(g)**

79. (a) $3 + \sqrt{6}$ and $3 - \sqrt{6}$; $(3, -6)$
(c) $4 + \sqrt{13}$ and $4 - \sqrt{13}$; $(4, 13)$
(e) No x-intercepts; $\left(\dfrac{1}{4}, \dfrac{1}{8}\right)$

Problem Set 7.6 (page 394)

1.

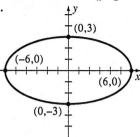

3.

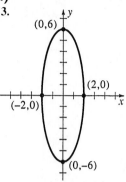

17.

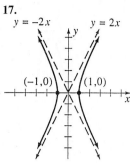

19.

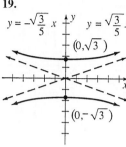

5.

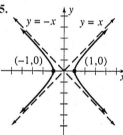

7.

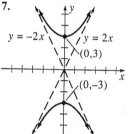

21. (a)

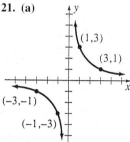

(c)

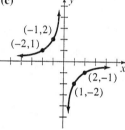

9.

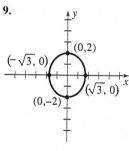

11.

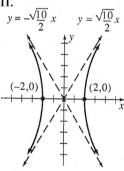

23. (a) Origin

(c)

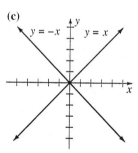

13.

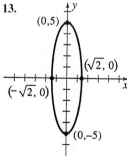

15.

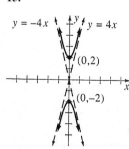

25.

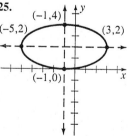

27.

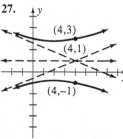

29.

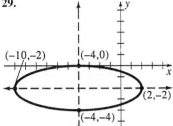

(−10,−2)   (−4,0)   (2,−2)   (−4,−4)

29.

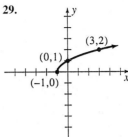

(3,2)   (0,1)   (−1,0)

31.

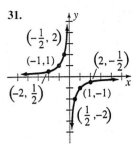

$\left(-\frac{1}{2}, 2\right)$ (−1,1) $\left(2,-\frac{1}{2}\right)$ $\left(-2, \frac{1}{2}\right)$ (1,−1) $\left(\frac{1}{2}, -2\right)$

31.

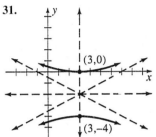

(3,0)   (3,−4)

33.

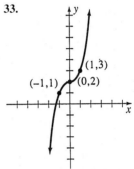

(1,3)   (−1,1)   (0,2)

35.

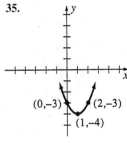

(0,−3)   (2,−3)   (1,−4)

Problem Set 7.7 (page 402)
1. $(3, -5); (-3, 5); (-3, -5)$
3. $(-2, 4); (2, -4); (2, 4)$ **5.** $(0, 1); (0, -1); (0, 1)$
7. *x*-axis   **9.** *y*-axis   **11.** *x*-axis   **13.** Origin
15. Origin   **17.** *x*-axis, *y*-axis, and origin
19. *y*-axis

37.

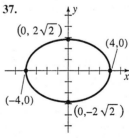

$\left(0, 2\sqrt{2}\right)$ (4,0)   (−4,0) $\left(0, -2\sqrt{2}\right)$

39.

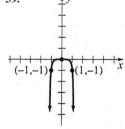

(−1,−1)   (1,−1)

21.

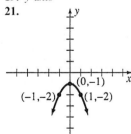

(0,−1)   (−1,−2)   (1,−2)

23.

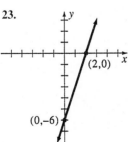

(2,0)   (0,−6)

41.

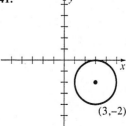

(3,−2)

43.

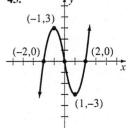

(−1,3)   (−2,0)   (2,0)   (1,−3)

25.

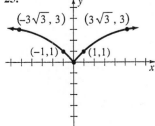

$\left(-3\sqrt{3}, 3\right)$ $\left(3\sqrt{3}, 3\right)$ (−1,1)   (1,1)

27.

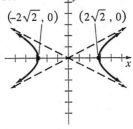

$\left(-2\sqrt{2}, 0\right)$ $\left(2\sqrt{2}, 0\right)$

45.

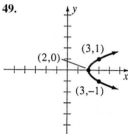

47.

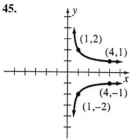

13.

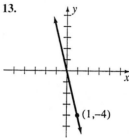

14.

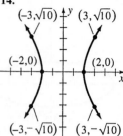

49.

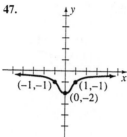

15.

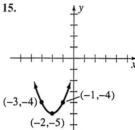

16.

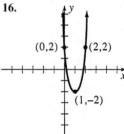

17.

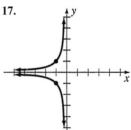

18.

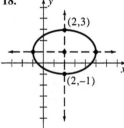

Chapter 7 Review Problem Set (page 406)

1. (a) $\frac{6}{5}$ (b) $-\frac{2}{3}$ **2.** (a) -4 (b) $\frac{2}{7}$

3. $5, 10$, and $\sqrt{97}$ **4.** $7x + 4y = 1$

5. $3x + 7y = 28$ **6.** $2x - 3y = 16$

7. $x - 2y = -8$ **8.** $2x - 3y = 14$

9.

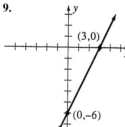

10.

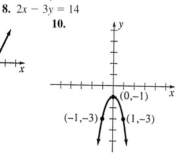

19.

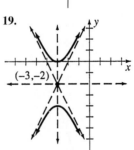

20.

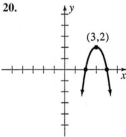

11.

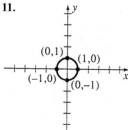

12.

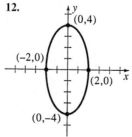

21.

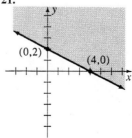

22.

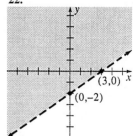

23. $(-3, 4), r = 3$ **24.** $(-4, -6)$

25. $y = \dfrac{3}{2}x$ and $y = -\dfrac{3}{2}x$ **26.** 6

24.

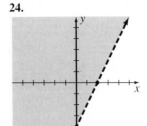

25.

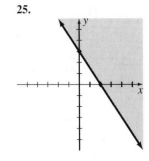

Chapter 7 Test (page 407)

1. $-\dfrac{6}{5}$ **2.** $\dfrac{3}{7}$ **3.** $\sqrt{58}$ **4.** $3x + 2y = 2$

5. $y = -\dfrac{1}{6}x + \dfrac{4}{3}$ **6.** $5x + 2y = -18$

7. $6x + y = 31$ **8.** $(2, -5)$ **9.** 6 units

10. $(-5, -2)$ **11.** $(3, -4)$ **12.** 8 units

13. $y = \pm\dfrac{4}{3}x$ **14.** Hyperbola **15.** Parabola

16. Straight line **17.** Circle

18.

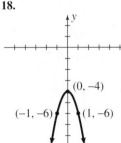

19.

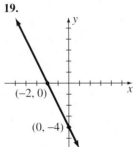

20.

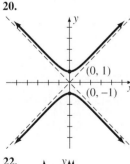

21.

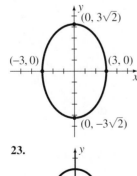

22.

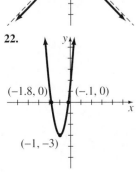

23.

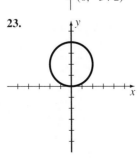

Cumulative Review Problem Set (page 408)

1. $4y\sqrt{2xy}$ **2.** $4ab^2\sqrt{3a}$ **3.** $\dfrac{\sqrt{6}}{2}$ **4.** $-\dfrac{\sqrt{10}}{2}$

5. $2x\sqrt[3]{6x^2}$ **6.** $\dfrac{9\sqrt{3}}{2}$ **7.** $\dfrac{6\sqrt{5}+3}{19}$

8. $-3\sqrt{2} + 2\sqrt{6}$ **9.** $24\sqrt{3}$ **10.** $x - \sqrt{x} - 6$

11. $-4 - 3\sqrt{15}$ **12.** $2x^3 - 3x^2 + 8x - 7$

13. $\dfrac{3xy}{4(x+6)}$ **14.** $\dfrac{25x^3y^3}{4(x+1)}$ **15.** $\dfrac{25x^2-2}{30x}$

16. $\dfrac{4x+43}{(x-9)(x+6)}$ **17.** $4x - 3$

18. $4x + 7 + \dfrac{23x-6}{x^2-3x}$ **19.** 90,000 **20.** 400

21. .003 **22.** $17 - 7i$ **23.** $27 + 34i$

24. $-32 - 24i$ **25.** $0 - \dfrac{2}{5}i$ **26.** $\dfrac{1}{10} + \dfrac{7}{10}i$

27. $-\dfrac{15}{34} + \dfrac{4}{17}i$ **28.** $-\dfrac{1}{2}$ **29.** -2 **30.** $\dfrac{25}{4}$

31. 2 **32.** $\dfrac{15}{8}$ **33.** $(3x + 2y)^2$

34. $(3x - 4y)(9x^2 + 12xy + 16y^2)$

35. $(2x + 3)(2x - 3)(x + 1)(x - 1)$

36. $3x(x - 12)(x + 2)$ **37.** $(x - 2y - 2)(x - 2y + 2)$

38. $\left\{\dfrac{5}{4}\right\}$ **39.** $\{-2\sqrt{2}, 0, 2\sqrt{2}\}$

40. $\{-2 + \sqrt{3}, -2 - \sqrt{3}\}$

41. $\{-2 + 2i\sqrt{6}, -2 - 2i\sqrt{6}\}$ **42.** $\left\{-\dfrac{2}{5}, 6\right\}$

43. $\{2, -1 + i\sqrt{3}, -1 - i\sqrt{3}\}$ **44.** $\left\{-4, \dfrac{7}{2}\right\}$

45. $\left\{-\dfrac{3}{4}, 6\right\}$ **46.** $\{3\}$ **47.** $\{3 - \sqrt{10}, 3 + \sqrt{10}\}$

48. $\{13.5\}$ **49.** $\{1 - i\sqrt{3}, 1 + i\sqrt{3}\}$ **50.** $\left\{-6, \dfrac{1}{4}\right\}$

51. $\{-12, 16\}$ **52.** $\left\{\dfrac{72}{13}\right\}$ **53.** $\{-2, 1\}$ **54.** $\{9\}$

55. $\{5\}$ **56.** $\{-3i, 3i, -2, 2\}$ **57.** $\{-5, 2\}$

58. $y = \dfrac{3x-10}{5}$ **59.** $y = \dfrac{3x-2}{4}$ **60.** $b = \dfrac{fa}{a-f}$

61. $T = \dfrac{NC - NV}{C}$

62. $(-\infty, -2) \cup (1, \infty)$

63. $\left(-\dfrac{7}{3}, -1\right)$

64. $(-\infty, -4) \cup (2, \infty)$

65. $\left(-1, \dfrac{3}{2}\right)$

66. $\left(-\infty, \dfrac{17}{2}\right]$

67. $(-\infty, 3] \cup (5, \infty)$

68. $\left(-\infty, -\dfrac{5}{2}\right) \cup \left(\dfrac{1}{3}, \infty\right)$

69. $(-6, -3)$

70. $(-4, 6]$

71. -7 and 5 **72.** 3 centimeters and 6 centimeters

73. $\dfrac{1}{2}$ inch   **74.** 4 hours for Cindy and 6 hours for Kent

CHAPTER 8

Problem Set 8.1 (page 417)

1. $D = \{1, 2, 3, 4\}$, $R = \{5, 8, 11, 14\}$ It is a function.

3. $D = \{0, 1\}$, $R = \{-2\sqrt{6}, -5, 5, 2\sqrt{6}\}$ It is a function.

5. $D = \{1, 2, 3, 4, 5\}$, $R = \{2, 5, 10, 17, 26\}$ It is a function.

7. $D = \{$all reals$\}$, $R = \{$all reals$\}$ It is a function.

9. $D = \{$all reals$\}$, $R = \{y \,|\, y \geq 0\}$ It is not a function.

11. $\{$All reals$\}$ **13.** $\{x \,|\, x \neq 1\}$ **15.** $\left\{x \,\middle|\, x \neq \dfrac{3}{4}\right\}$

17. $\{x \,|\, x \neq -1 \text{ and } x \neq 4\}$ **19.** $\{x \,|\, x \neq -8 \text{ and } x \neq 5\}$

21. $\{x \,|\, x \neq -6 \text{ and } x \neq 0\}$ **23.** $\{$All reals$\}$

25. $\{t \,|\, t \neq -2 \text{ and } t \neq 2\}$ **27.** $\{x \,|\, x \geq -4\}$

29. $\left\{s \,\middle|\, s \geq \dfrac{5}{4}\right\}$ **31.** $\{x \,|\, x \leq -4 \text{ or } x \geq 4\}$

33. $\{x \,|\, x \leq -3 \text{ or } x \geq 6\}$ **35.** $\{x \,|\, -1 \leq x \leq 1\}$

37. $f(0) = -2, f(2) = 8, f(-1) = -7, f(-4) = -22$

39. $f(-2) = -\dfrac{7}{4}, f(0) = -\dfrac{3}{4}, f\left(\dfrac{1}{2}\right) = -\dfrac{1}{2}, f\left(\dfrac{2}{3}\right) = -\dfrac{5}{12}$

41. $g(-1) = 0$; $g(2) = -9$; $g(-3) = 26$; $g(4) = 5$

43. $h(-2) = -2$; $h(-3) = -11$; $h(4) = -32$; $h(5) = -51$

45. $f(3) = \sqrt{7}$; $f(4) = 3$; $f(10) = \sqrt{21}$; $f(12) = 5$

47. $f(1) = -1$; $f(-1) = -2$; $f(3) = -\dfrac{2}{3}$; $f(-6) = \dfrac{4}{3}$

49. $f(-2) = 27$; $f(3) = 42$; $g(-4) = -37$; $g(6) = -17$

51. $f(-2) = 5$; $f(3) = 8$; $g(-4) = -3$; $g(5) = -4$

53. -3 **55.** $-2a - h$ **57.** $4a - 1 + 2h$

59. $-8a - 7 - 4h$

61. $h(1) = 48$; $h(2) = 64$; $h(3) = 48$; $h(4) = 0$

63. $C(75) = \$74$; $C(150) = \$98$; $C(225) = \$122$; $C(650) = \$258$

65. $I(.11) = 55$; $I(.12) = 60$; $I(.135) = 67.5$; $I(.15) = 75$

Problem Set 8.2 (page 430)

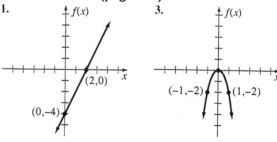

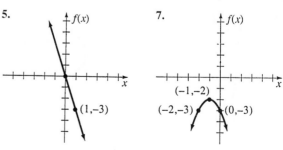

9.

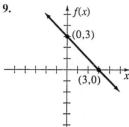

(0,3)

(3,0)

11.

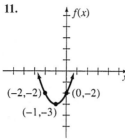

(−2,−2)   (0,−2)

(−1,−3)

29.

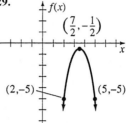

$\left(\frac{7}{2}, -\frac{1}{2}\right)$

(2,−5)   (5,−5)

13.

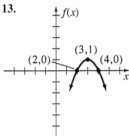

(3,1)

(2,0)   (4,0)

15.

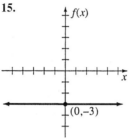

(0,−3)

31. 80   **33.** 5 and 25

35. 60 meters by 60 meters

37. 1100 subscribers at $13.75 per month

39.

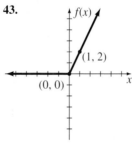

$(0, \sqrt{3})$ (1, 2)

(−3, 0)

41.

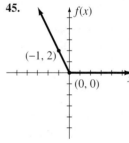

$\left(-1, \frac{1}{3}\right)$ (3, 3)

(4, 2)

(0, 0)

(1, −1)

17.

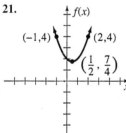

(4,4)   (6,4)

(5,2)

19.

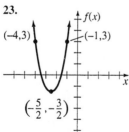

(1,3)

(2,0)

43.

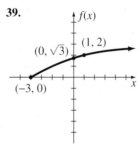

(1, 2)

(0, 0)

45.

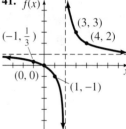

(−1, 2)

(0, 0)

21.

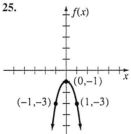

(−1,4)   (2,4)

$\left(\frac{1}{2}, \frac{7}{4}\right)$

23.

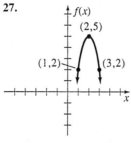

(−4,3)   (−1,3)

$\left(-\frac{5}{2}, -\frac{3}{2}\right)$

25.

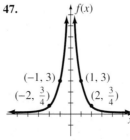

(0,−1)

(−1,−3)   (1,−3)

27.

(2,5)

(1,2)   (3,2)

47.

(−1, 3)   (1, 3)

$(-2, \frac{3}{4})$ $(2, \frac{3}{4})$

Problem Set 8.3 (page 438)

1.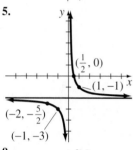

3.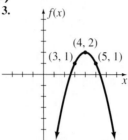

5.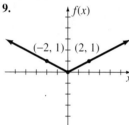

7.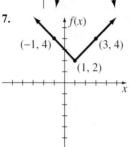

9.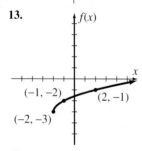

11.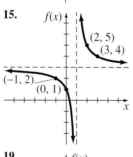

13.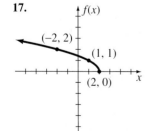

15.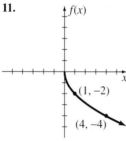

17.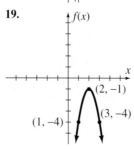

19.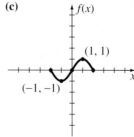

21.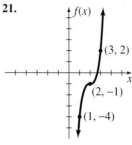

23.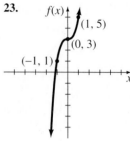

25.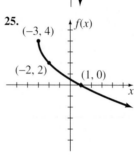

27.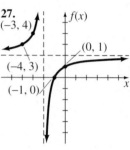

29.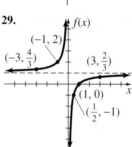

31.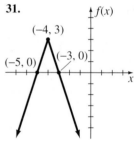

33.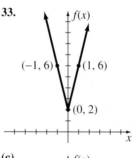

35. (a)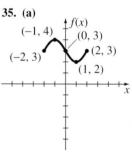

(c)

Problem Set 8.4 (page 446)

1. $8x - 2; -2x - 6; 15x^2 - 14x - 8; \dfrac{3x - 4}{5x + 2}$

3. $x^2 - 7x + 3; x^2 - 5x + 5; -x^3 + 5x^2 + 2x - 4;$
$\dfrac{x^2 - 6x + 4}{-x - 1}$

5. $2x^2 + 3x - 6; -5x + 4; x^4 + 3x^3 - 10x^2 + x + 5;$
$\dfrac{x^2 - x - 1}{x^2 + 4x - 5}$

7. $\sqrt{x - 1} + \sqrt{x}; \sqrt{x - 1} - \sqrt{x}; \sqrt{x^2 - x}; \dfrac{\sqrt{x(x - 1)}}{x}$

9. 124 and -130　　**11.** 323 and 257　　**13.** $\dfrac{1}{2}$ and -1

15. Undefined and undefined　　**17.** $\sqrt{7}$ and 0

19. 37 and 27

21. $(f \circ g)(x) = 15x - 3, D = \{\text{all reals}\}$
$(g \circ f)(x) = 15x - 1, D = \{\text{all reals}\}$

23. $(f \circ g)(x) = -14x - 7, D = \{\text{all reals}\}$
$(g \circ f)(x) = -14x + 11, D = \{\text{all reals}\}$

25. $(f \circ g)(x) = 3x^2 + 11, D = \{\text{all reals}\}$
$(g \circ f)(x) = 9x^2 + 12x + 7, D = \{\text{all reals}\}$

27. $(f \circ g)(x) = 2x^2 - 11x + 17, D = \{\text{all reals}\}$
$(g \circ f)(x) = -2x^2 + x + 1, D = \{\text{all reals}\}$

29. $(f \circ g)(x) = \dfrac{3}{4x - 9}, D = \left\{x \mid x \neq \dfrac{9}{4}\right\}$

$(g \circ f)(x) = \dfrac{12 - 9x}{x}, D = \{x \mid x \neq 0\}$

31. $(f \circ g)(x) = \sqrt{5x + 4}, D = \left\{x \mid x \geq -\dfrac{4}{5}\right\}$

$(g \circ f)(x) = 5\sqrt{x + 1} + 3, D = \{x \mid x \geq -1\}$

33. $(f \circ g)(x) = x - 4, D = \{x \mid x \neq 4\}$

$(g \circ f)(x) = \dfrac{x}{1 - 4x}, D = \left\{x \mid x \neq 0 \text{ and } x \neq \dfrac{1}{4}\right\}$

35. $(f \circ g)(x) = \dfrac{2\sqrt{x}}{x}, D = \{x \mid x > 0\}$

$(g \circ f)(x) = \dfrac{4\sqrt{x}}{x}, D = \{x \mid x > 0\}$

37. $(f \circ g)(x) = \dfrac{3x + 3}{2}, D = \{x \mid x \neq -1\}$

$(g \circ f)(x) = \dfrac{2x}{2x + 3}, D = \left\{x \mid x \neq 0 \text{ and } x \neq -\dfrac{3}{2}\right\}$

Problem Set 8.5 (page 454)

1. Not a function　　**3.** Function　　**5.** Function

7. Function　　**9.** One-to-one function

11. Not a one-to-one function

13. Not a one-to-one function　　**15.** One-to-one function

17. Yes　　**19.** No　　**21.** No　　**23.** Yes

25. Domain of f: $\{1, 2, 3, 4\}$
Range of f: $\{3, 6, 11, 18\}$
f^{-1}: $\{(3, 1), (6, 2), (11, 3), (18, 4)\}$
Domain of f^{-1}: $\{3, 6, 11, 18\}$
Range of f^{-1}: $\{1, 2, 3, 4\}$

27. Domain of f: $\{-2, -1, 0, 5\}$
Range of f: $\{-1, 1, 5, 10\}$
f^{-1}: $\{(-1, -2), (1, -1), (5, 0), (10, 5)\}$
Domain of f^{-1}: $\{-1, 1, 5, 10\}$
Range of f^{-1}: $\{-2, -1, 0, 5\}$

29. $f^{-1}(x) = \dfrac{x + 4}{5}$　　**31.** $f^{-1}(x) = \dfrac{1 - x}{2}$

33. $f^{-1}(x) = \dfrac{5}{4}x$　　**35.** $f^{-1}(x) = 2x - 8$

37. $f^{-1}(x) = \dfrac{15x + 6}{5}$　　**39.** $f^{-1}(x) = \dfrac{x - 4}{9}$

41. $f^{-1}(x) = \dfrac{-x - 4}{5}$　　**43.** $f^{-1}(x) = \dfrac{-3x + 21}{2}$

45. $f^{-1}(x) = \dfrac{3}{4}x + \dfrac{3}{16}$　　**47.** $f^{-1}(x) = -\dfrac{7}{3}x - \dfrac{14}{9}$

49. $f^{-1}(x) = x^2$ for $x \geq 0$

51. $f^{-1}(x) = \sqrt{x - 4}$ for $x \geq 4$

53. $f^{-1}(x) = \dfrac{1}{4}x$　　**55.** $f^{-1}(x) = -3x$

57. $f^{-1}(x) = \dfrac{x + 3}{3}$　　**59.** $f^{-1}(x) = \dfrac{-x - 4}{2}$

61. $f^{-1}(x) = \sqrt{x}, x \geq 0$　　**63.** $f^{-1}(x) = x^2 + 1$ for $x \geq 0$

67. (a) $f^{-1}(x) = \dfrac{x - 1}{2}$　　**(c)** $f^{-1}(x) = \dfrac{-x + 5}{4}$

(e) $f^{-1}(x) = \dfrac{1}{2}x$

Problem Set 8.6 (page 463)

1. $y = \dfrac{k}{x^2}$　　**3.** $C = \dfrac{kg}{t^3}$　　**5.** $V = kr^3$　　**7.** $S = ke^2$

9. $V = khr^2$　　**11.** $\dfrac{2}{3}$　　**13.** -4　　**15.** $\dfrac{1}{3}$

17. -2　　**19.** 2　　**21.** 5　　**23.** 9　　**25.** 9

27. $\dfrac{1}{6}$　　**29.** 112　　**31.** 12 cubic centimeters

33. 28　　**35.** 2 seconds　　**37.** 12 ohms

39. (a) \$210　　**(c)** \$1050

41. 3560.76 cubic meters　　**43.** .048

Chapter 8 Review Problem Set (page 467)

1. $D = \{1, 2, 4\}$　　**2.** $D = \{x \mid x \neq 5\}$

3. $D = \{x \mid x \neq 0 \text{ and } x \neq -4\}$

4. $D = \{x \mid x \geq 5 \text{ or } x \leq -5\}$

5. $f(2) = -1, f(-3) = 14; f(a) = a^2 - 2a - 1$

6. $4a + 2h + 1$

7.

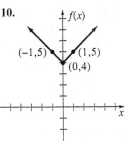

8.

9.

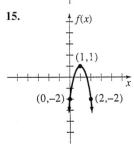

10.

11.

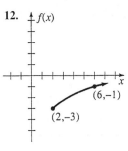

12.

13.

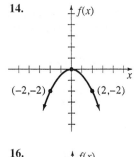

14.

15.

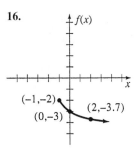

16.

17. (a) $(-5, -28); x = -5$ **(b)** $\left(-\dfrac{7}{2}, \dfrac{67}{2}\right); x = -\dfrac{7}{2}$

18. $(f \circ g)(x) = 6x - 11$ and $(g \circ f)(x) = 6x - 13$

19. $(f \circ g)(x) = x^2 - 2x - 1$ and $(g \circ f)(x) = x^2 - 10x + 27$

20. $(f \circ g)(x) = 4x^2 - 20x + 20$ and $(g \circ f)(x) = -2x^2 + 15$

21. $f^{-1}(x) = \dfrac{x+1}{6}$ **22.** $f^{-1}(x) = \dfrac{3x - 21}{2}$

23. $f^{-1}(x) = \dfrac{-35x - 10}{21}$ **24.** $k = 9$ **25.** $y = 120$

26. 128 pounds   **27.** 20 and 20   **28.** 3 and 47

29. 25 students   **30.** 600 square inches

Chapter 8 Test (page 469)

1. $\left\{x \mid x \neq -4 \text{ and } x \neq \dfrac{1}{2}\right\}$ **2.** $\left\{x \mid x \leq \dfrac{5}{3}\right\}$ **3.** $\dfrac{11}{6}$

4. 11   **5.** $(-6, 3)$ **6.** $6a + 3h + 2$

7. $(f \circ g)(x) = -21x - 2$

8. $(g \circ f)(x) = 8x^2 + 38x + 48$ **9.** $(f \circ g)(x) = \dfrac{3x}{2 - 2x}$

10. $f^{-1}(x) = \dfrac{x+9}{5}$ **11.** $f^{-1}(x) = \dfrac{-x - 6}{3}$

12. $f^{-1}(x) = \dfrac{15x + 9}{10}$ **13.** -4 **14.** 15

15. 6 and 54   **16.** \$96

17. The graph of $f(x) = (x - 6)^3 - 4$ is the graph of $f(x) = x^3$ translated 6 units to the right and 4 units downward.

18. The graph of $f(x) = -|x| + 8$ is the graph of $f(x) = |x|$ reflected across the x-axis and translated 8 units upward.

19. The graph of $f(x) = -\sqrt{x + 5} + 7$ is the graph of $f(x) = \sqrt{x}$ reflected across the x-axis and then translated 5 units to the left and 7 units upward.

20.

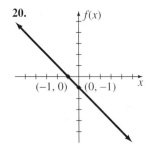

21.

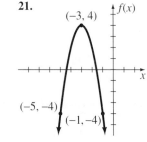

22.

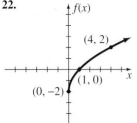

23.

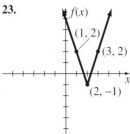

24.

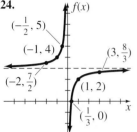

25.

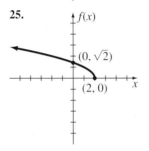

CHAPTER 9

Problem Set 9.1 (page 478)

1. $Q: 4x + 3; R: 0$ **3.** $Q: 2x - 7; R: 0$
5. $Q: 3x - 4; R: 1$ **7.** $Q: 4x - 5; R: -2$
9. $Q: x^2 + 3x - 4; R: 0$ **11.** $Q: 3x^2 + 2x - 4; R: 0$
13. $Q: 5x^2 + x - 1; R: -4$ **15.** $Q: x^2 - x - 1; R: 8$
17. $Q: -x^2 + 4x - 2; R: 0$
19. $Q: -3x^2 + 4x - 2; R: 4$
21. $Q: 3x^2 + 6x + 10; R: 15$
23. $Q: 2x^3 - x^2 + 4x - 2; R: 0$
25. $Q: x^3 + 7x^2 + 21x + 56; R: 167$
27. $Q: x^3 - x + 5; R: 0$
29. $Q: x^3 + 2x^2 + 4x + 8; R: 0$
31. $Q: x^4 - x^3 + x^2 - x + 1; R: -2$
33. $Q: x^4 - x^3 + x^2 - x + 1; R: 0$
35. $Q: x^4 - x^3 - x^2 + x - 1; R: 0$
37. $Q: 4x^4 - 2x^3 + 2x - 3; R: -1$
39. $Q: 9x^2 - 3x + 2; R: -\dfrac{10}{3}$
41. $Q: 3x^3 - 3x^2 + 6x - 3; R: 0$

Problem Set 9.2 (page 483)

1. $f(3) = 9$ **3.** $f(-1) = -7$ **5.** $f(2) = 19$
7. $f(6) = 74$ **9.** $f(-2) = -65$ **11.** $f(-1) = -1$
13. $f(8) = -83$ **15.** $f(3) = 8751$ **17.** $f(-6) = 31$
19. $f(4) = -1113$ **21.** Yes   **23.** No   **25.** Yes

27. Yes   **29.** No   **31.** Yes   **33.** Yes
35. $f(x) = (x - 2)(x + 3)(x - 7)$
37. $f(x) = (x + 2)(4x - 1)(3x + 2)$
39. $f(x) = (x + 1)^2(x - 4)$
41. $f(x) = (x - 6)(x + 2)(x - 2)(x^2 + 4)$
43. $f(x) = (x + 5)(3x - 4)^2$ **45.** $k = 1$ or $k = -4$
47. $k = 6$ **49.** $f(c) > 0$ for all values of c
51. Let $f(x) = x^n - 1$. Since $(-1)^n = 1$ for all even positive integral values of n, then $f(-1) = 0$ and $x - (-1) = x + 1$ is a factor
53. (a) Let $f(x) = x^n - y^n$. Therefore, $f(y) = y^n - y^n = 0$ and $x - y$ is a factor of $f(x)$.   (c) Let $f(x) = x^n + y^n$. Therefore, $f(-y) = (-y)^n + y^n = -y^n + y^n = 0$ when n is odd, and $x - (-y) = x + y$ is a factor of $f(x)$.
57. $f(1 + i) = 2 + 6i$
61. (a) $f(4) = 137; f(-5) = 11; f(7) = 575$
(c) $f(4) = -79; f(5) = -162; f(-3) = 110$

Problem Set 9.3 (page 494)

1. $\{-3, 1, 4\}$ **3.** $\left\{-1, -\dfrac{1}{3}, \dfrac{2}{5}\right\}$ **5.** $\left\{-2, -\dfrac{1}{4}, \dfrac{5}{2}\right\}$
7. $\{-3, 2\}$ **9.** $\{2, 1 \pm \sqrt{5}\}$ **11.** $\{-3, -2, -1, 2\}$
13. $\{-2, 3, -1 \pm 2i\}$ **15.** $\{1, \pm i\}$
17. $\left\{-\dfrac{5}{2}, 1, \pm\sqrt{3}\right\}$ **19.** $\left\{-2, \dfrac{1}{2}\right\}$ **27.** $\{-3, -1, 2\}$
29. $\left\{-\dfrac{5}{2}, \dfrac{1}{3}, 3\right\}$

31. 1 positive and 1 negative solution
33. 1 positive and 2 nonreal complex solutions
35. 1 negative and 2 positive *or* 1 negative and 2 nonreal complex solutions
37. 5 positive *or* 3 positive and 2 nonreal complex solutions *or* 1 positive and 4 nonreal complex solutions
39. 1 negative and 4 nonreal complex solutions
47. (a) $\{4, -3 \pm i\}$ (c) $\{-2, 6, 1 \pm \sqrt{3}\}$
(e) $\{12, 1 \pm i\}$
49. (a) .27 and 3.73   (c) 2.27 and 5.73   (e) .71

Problem Set 9.4 (page 505)

1.

3.

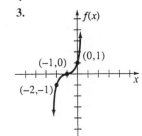

5.
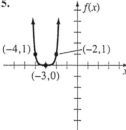

7.
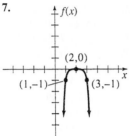

21.
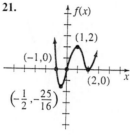

23.
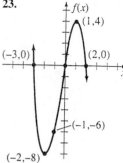

9.
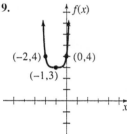

11.
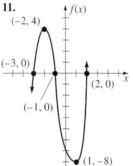

25.
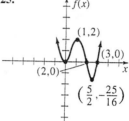

27.
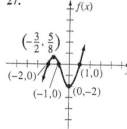

29.
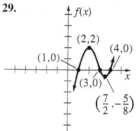

31.
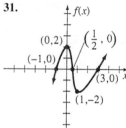

13.
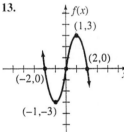

15.
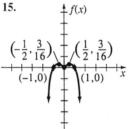

33.
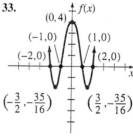

17.
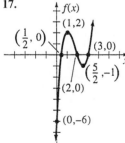

19.
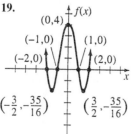

35. (a) -144 **(c)** $f(x) > 0$ for $\{x \mid x < -3$ or $6 < x < 8\}$;
$f(x) < 0$ for $\{x \mid -3 < x < 6$ or $x > 8\}$

37. (a) -81 **(c)** $f(x) > 0$ for $\{x \mid x > 1\}$; $f(x) < 0$ for
$\{x \mid x < -3$ or $-3 < x < 1\}$

39. (a) 0 **(c)** $f(x) > 0$ for $\{x \mid x < -4$ or $0 < x < 6$ or
$x > 6\}$; $f(x) < 0$ for $\{x \mid -4 < x < 0\}$

41. (a) 0 **(c)** $f(x) > 0$ for $\{x \mid -3 < x < 0$ or 0
$< x < 2\}$; $f(x) < 0$ for $\{x \mid x < -3$ or $x > 2\}$

45. **(a)** 1.6   **(c)** 4.8   **(e)** −1.5

51. **(a)** −2, 1, and 4: $f(x) > 0$ for $(-2, 1) \cup (4, \infty)$ and $f(x) < 0$ for $(-\infty, -2) \cup (1, 4)$

(c) 2 and 3; $f(x) > 0$ for $(3, \infty)$ and $f(x) < 0$ for $(2, 3) \cup (-\infty, 2)$

(e) −3, −1, and 2; $f(x) > 0$ for $(-\infty, -3) \cup (2, \infty)$ and $f(x) < 0$ for $(-3, -1) \cup (-1, 2)$

53. **(a)** −3.3; (.5, 3.1), (−1.9, 10.1)

(c) −2.2, 2.2; (−1.4, −8.0), (.0, −4.0), (1.4, 8.0)

55. 32 units

11.

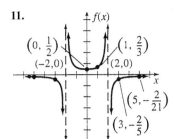

13.

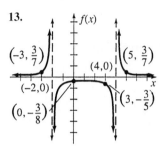

Problem Set 9.5 (page 516)

1.

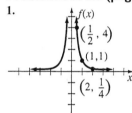

3.

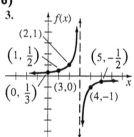

15.

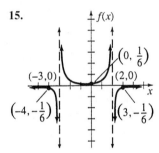

5.

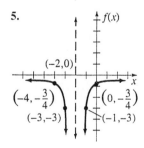

7.

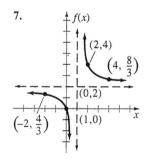

17.

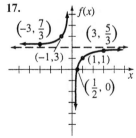

19.

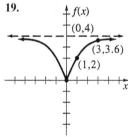

9.

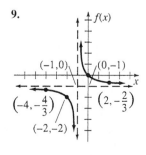

21.

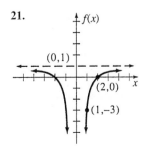

23. (a)

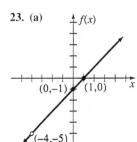

(c)

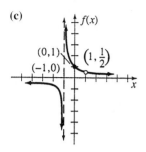

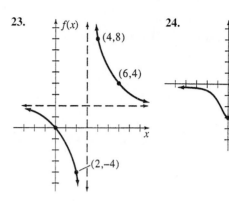

Chapter 9 Review Problem Set (page 520)

1. $Q: 3x^2 - x + 5; R: 3$ **2.** $Q: 5x^2 - 3x - 3; R: 16$
3. $Q: -2x^3 + 9x^2 - 38x + 151; R: -605$
4. $Q: -3x^3 - 9x^2 - 32x - 96; R: -279$ **5.** $f(1) = 1$
6. $f(-3) = -197$ **7.** $f(-2) = 20$ **8.** $f(8) = 0$
9. Yes   **10.** No   **11.** Yes   **12.** Yes
13. $\{-3, 1, 5\}$ **14.** $\left\{-\dfrac{7}{2}, -1, \dfrac{5}{4}\right\}$ **15.** $\{1, 2, 1 \pm 5i\}$
16. $\{-2, 3 \pm \sqrt{7}\}$
17. 2 positive solutions and 2 negative solutions *or* 2 positive solutions and 2 nonreal complex solutions *or* 2 negative solutions and 2 nonreal complex solutions *or* 4 nonreal complex solutions
18. 1 negative solution and 4 nonreal complex solutions
19.

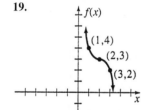

20.

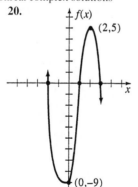

21.

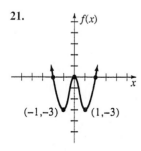

22.

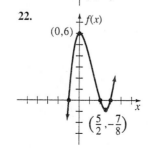

Chapter 9 Test (page 522)

1. $Q: 3x^2 - 4x - 2; R: 0$
2. $Q: 4x^3 + 8x^2 + 9x + 17; R: 38$ **3.** -24 **4.** 5
5. 39   **6.** No   **7.** No   **8.** Yes   **9.** No
10. $\{-4, 1, 3\}$ **11.** $\left\{-4, \dfrac{3 \pm \sqrt{17}}{4}\right\}$
12. $\{-3, 1, 3 \pm i\}$
13. $\left\{-4, 1, \dfrac{3}{2}\right\}$ **14.** $\left\{-\dfrac{5}{3}, 2\right\}$
15. One positive, one negative, and two nonreal complex numbers
16. $-7, 0,$ and $\dfrac{2}{3}$ **17.** $x = -3$
18. $f(x) = 5$ or $y = 5$ **19.** y-axis symmetry
20. Origin
21.

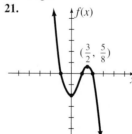

22.

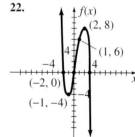

23.

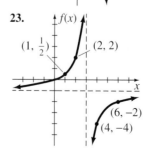

24.

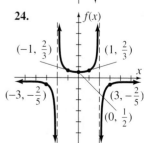

25.

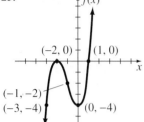

CHAPTER 10

Problem Set 10.1 (page 532)

1. $\{6\}$ **3.** $\left\{\frac{3}{2}\right\}$ **5.** $\{7\}$ **7.** $\{5\}$ **9.** $\{1\}$

11. $\{-3\}$ **13.** $\left\{\frac{3}{2}\right\}$ **15.** $\left\{\frac{1}{5}\right\}$ **17.** $\{0\}$

19. $\{-1\}$ **21.** $\left\{\frac{5}{2}\right\}$ **23.** $\{3\}$ **25.** $\left\{\frac{1}{2}\right\}$

27.

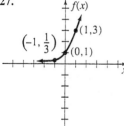

29.

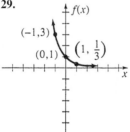

31.

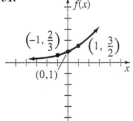

33.

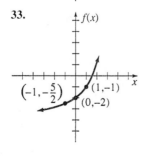

35.

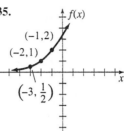

37.

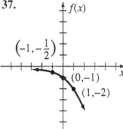

39.

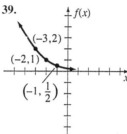

41.

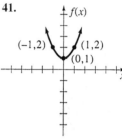

43.

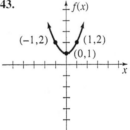

45.

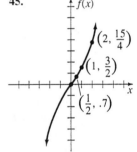

Problem Set 10.2 (page 542)

1. (a) \$.62   **(c)** \$2.06   **(e)** \$10,950   **(g)** \$658
3. \$283.70   **5.** \$865.84   **7.** \$1782.25
9. \$2725.05   **11.** \$16,998.71   **13.** \$22,553.65
15. \$567.63   **17.** \$1422.36   **19.** \$8963.38
21. \$17,547.35   **23.** \$32,558.88   **25.** 5.9%
27. 8.06%   **29.** 8.25% compounded quarterly
31. 50 grams; 37 grams   **33.** 2226; 3320; 7389
35. 2000   **37. (a)** 6.5 pounds per square inch
(c) 13.6 pounds per square inch

39.

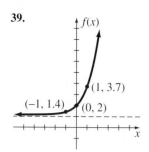

41.

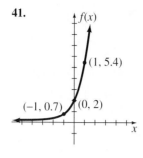

43.

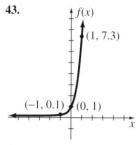

21. 4   **23.** 4   **25.** 3   **27.** $\frac{1}{2}$ **29.** 0

31. -1 **33.** 5   **35.** -5 **37.** 1   **39.** 0

41. $\{49\}$ **43.** $\{16\}$ **45.** $\{27\}$ **47.** $\left\{\frac{1}{8}\right\}$

49. $\{4\}$ **51.** 5.1293   **53.** 6.9657   **55.** 1.4037

57. 7.4512   **59.** 6.3219   **61.** $-.3791$

63. .5766   **65.** 2.1531   **67.** .3949

69. $\log_b x + \log_b y + \log_b z$ **71.** $\log_b y - \log_b z$

73. $3\log_b y + 4\log_b z$ **75.** $\frac{1}{2}\log_b x + \frac{1}{3}\log_b y - 4\log_b z$

77. $\frac{2}{3}\log_b x + \frac{1}{3}\log_b z$ **79.** $\frac{3}{2}\log_b x - \frac{1}{2}\log_b y$

81. $\log_b\left(\dfrac{x^2}{y^4}\right)$ **83.** $\log_b\left(\dfrac{xz}{y}\right)$ **85.** $\log_b\left(\dfrac{x^2 y^4}{z^3}\right)$

87. $\log_b\left(\dfrac{y^4\sqrt{x}}{x}\right)$ **89.** $\left\{\dfrac{9}{4}\right\}$ **91.** $\{25\}$

93. $\{4\}$ **95.** $\left\{\dfrac{19}{8}\right\}$ **97.** $\{9\}$ **99.** $\{1\}$

Problem Set 10.4 (page 561)

1. .8597   **3.** 1.7179   **5.** 3.5071   **7.** $-.1373$

9. -3.4685 **11.** 411.43   **13.** 90,095

15. 79.543   **17.** .048440   **19.** .0064150

21. 1.6094   **23.** 3.4843   **25.** 6.0638

27. $-.7765$ **29.** -3.4609 **31.** 1.6034

33. 3.1346   **35.** 108.56   **37.** .48268

39. .035994

49.

	8%	10%	12%	14%
5 yrs	$1492	1649	1822	2014
10 yrs	2226	2718	3320	4055
15 yrs	3320	4482	6050	8166
20 yrs	4953	7389	11023	16445
25 yrs	7389	12182	20086	33115

51.

	8%	10%	12%	14%
Compounded annually	$2159	2594	3106	3707
Compounded semiannually	2191	2653	3207	3870
Compounded quarterly	2208	2685	3262	3959
Compounded monthly	2220	2707	3300	4022
Compounded continuously	2226	2718	3320	4055

53.

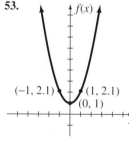

55.

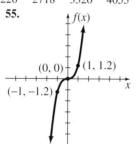

41.

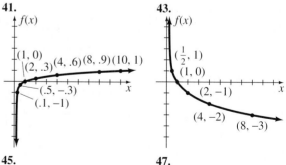

43.

45.

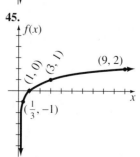

47.

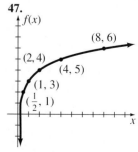

Problem Set 10.3 (page 552)

1. $\log_2 128 = 7$ **3.** $\log_5 125 = 3$

5. $\log_{10} 1000 = 3$ **7.** $\log_2\left(\dfrac{1}{4}\right) = -2$

9. $\log_{10} .1 = -1$ **11.** $3^4 = 81$ **13.** $4^3 = 64$

15. $10^4 = 10,000$ **17.** $2^{-4} = \dfrac{1}{16}$ **19.** $10^{-3} = .001$

49.

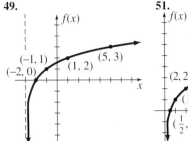

51.

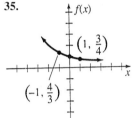

53.

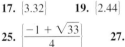

55. .36   **57.** .73   **59.** 23.10   **61.** 7.93

Problem Set 10.5 (page 570)

1. ⌈2.33⌉   **3.** ⌈2.56⌉   **5.** ⌈5.43⌉   **7.** ⌈4.18⌉
9. ⌈.12⌉   **11.** ⌈3.30⌉   **13.** ⌈4.57⌉   **15.** ⌈1.79⌉

17. ⌈3.32⌉   **19.** ⌈2.44⌉   **21.** ⌈4⌉   **23.** $\left\lceil \dfrac{19}{47} \right\rceil$

25. $\left\lceil \dfrac{-1 + \sqrt{33}}{4} \right\rceil$ **27.** ⌈1⌉   **29.** ⌈8⌉

31. ⌈1,10000⌉   **33.** 5.322   **35.** 2.524   **37.** .339
39. −.837   **41.** 3.194   **43.** 2.4 years
45. 5.3 years   **47.** 5.9%   **49.** 6.8 hours
51. 6100 feet   **53.** 3.5 hours   **55.** 6.7
57. Approximately 8 times   **65.** ⌈1.13⌉
67. $x = \ln(y + \sqrt{y^2 + 1})$

Chapter 10 Review Problem Set (page 574)

1. 32   **2.** −125   **3.** 81   **4.** 3   **5.** −2

6. $\dfrac{1}{3}$ **7.** $\dfrac{1}{4}$ **8.** −5   **9.** 1   **10.** 12

11. ⌈5⌉   **12.** $\left\lceil \dfrac{1}{9} \right\rceil$ **13.** $\left\lceil \dfrac{7}{2} \right\rceil$ **14.** ⌈3.40⌉   **15.** ⌈8⌉

16. $\left\lceil \dfrac{1}{11} \right\rceil$ **17.** ⌈1.95⌉   **18.** ⌈1.41⌉   **19.** ⌈1.56⌉

20. ⌈20⌉   **21.** ⌈10^{100}⌉   **22.** ⌈2⌉   **23.** $\left\lceil \dfrac{11}{2} \right\rceil$

24. ⌈0⌉   **25.** .3680   **26.** 1.3222   **27.** 1.4313
28. .5634

29. (a) $\log_b x - 2 \log_b y$ (b) $\dfrac{1}{4} \log_b x + \dfrac{1}{2} \log_b y$

(c) $\dfrac{1}{2} \log_b x - 3 \log_b y$

30. (a) $\log_b x^3 y^2$ (b) $\log_b\left(\dfrac{\sqrt{y}}{x^4}\right)$ (c) $\log_b\left(\dfrac{\sqrt{xy}}{z^2}\right)$

31. 1.58   **32.** .63   **33.** 3.79   **34.** −2.12
35.

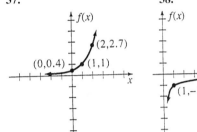

36.

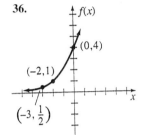

37.

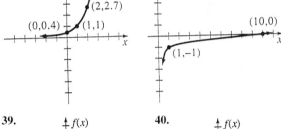

38.

39.

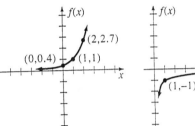

40.

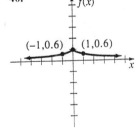

41.

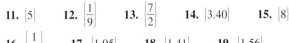

42.

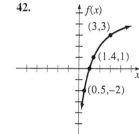

43. $2219.91   **44.** $4797.55   **45.** $15,999.31
46. Approximately 5.3 years
47. Approximately 12.1 years   **48.** Approximately 8.7%
49. 61,070; 67,493; 74,591

50. Approximately 4.8 hours   **51.** 133 grams
52. 8.1

Chapter 10 Test (page 576)

1. $\frac{1}{2}$ **2.** 1   **3.** 1   **4.** -1 **5.** $\{-3\}$

6. $\left\{-\frac{3}{2}\right\}$ **7.** $\left\{\frac{8}{3}\right\}$ **8.** $\{243\}$ **9.** $\{2\}$ **10.** $\left\{\frac{2}{5}\right\}$

11. 4.1919   **12.** .2031   **13.** .7325   **14.** 5.4538
15. $\{5.17\}$ **16.** $\{10.29\}$ **17.** 4.0069

18. $\log_b\!\left(\frac{x^3y^2}{z}\right)$ **19.** \$6342.08   **20.** 13.5 years

21. 7.8 hours   **22.** 4813 grams
23.

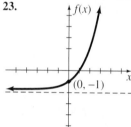

24

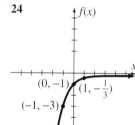

25.

CHAPTER 11

Problem Set 11.1 (page 586)

1. $\{(3, 2)\}$ **3.** $\{(2, 1)\}$ **5.** Dependent
7. $\{(4, -3)\}$ **9.** Inconsistent   **11.** $\{(8, 12)\}$
13. $\{(-4, -6)\}$ **15.** $\{(-9, 3)\}$ **17.** $\{(1, 3)\}$
19. $\left\{\left(5, \frac{3}{2}\right)\right\}$ **21.** $\left\{\left(\frac{9}{5}, -\frac{7}{25}\right)\right\}$ **23.** $\{(-2, -4)\}$
25. $\{(5, 2)\}$ **27.** $\{(-4, -8)\}$ **29.** $\{(500, 800)\}$
31. $\{(-1, 5)\}$ **33.** $\left\{\left(-\frac{3}{4}, -\frac{6}{5}\right)\right\}$ **35.** $\{(250, 500)\}$

37. \$2000 at 7% and \$8000 at 8%   **39.** 92
41. 34 and 97   **43.** 42 females

45. 20 inches by 27 inches
47. 60 five-dollar bills and 40 ten-dollar bills
49. 2500 student tickets and 500 nonstudent tickets

Problem Set 11.2 (page 595)

1. $\{(4, -3)\}$ **3.** $\{(-1, -3)\}$ **5.** $\{(-8, 2)\}$
7. $\{(-4, 0)\}$ **9.** $\{(1, -1)\}$ **11.** Inconsistent
13. $\left\{\left(-\frac{1}{11}, \frac{4}{11}\right)\right\}$ **15.** $\left\{\left(\frac{3}{2}, -\frac{1}{3}\right)\right\}$ **17.** $\{(4, -9)\}$
19. $\{(7, 0)\}$ **21.** $\{(7, 12)\}$ **23.** $\left\{\left(\frac{7}{11}, \frac{2}{11}\right)\right\}$
25. Inconsistent   **27.** $\left\{\left(\frac{51}{31}, -\frac{32}{31}\right)\right\}$ **29.** $\{(-2, -4)\}$
31. $\left\{\left(-1, -\frac{14}{3}\right)\right\}$ **33.** $\{(-6, 12)\}$ **35.** $\{(2, 8)\}$
37. $\{(-1, 3)\}$ **39.** $\{(16, -12)\}$ **41.** $\left\{\left(-\frac{3}{4}, \frac{3}{2}\right)\right\}$
43. $\{(5, -5)\}$
45. 5 gallons of 10% solution and 15 gallons of 20% solution
47. \$1 for a tennis ball and \$2 for a golf ball
49. 40 double rooms and 15 single rooms   **51.** 9 feet
53. $\frac{3}{4}$ **55.** 18 centimeters by 24 centimeters
57. 8 feet

61. **(a)** $\{(4, 6)\}$ **(c)** $\{(2, -3)\}$ **(e)** $\left\{\left(\frac{1}{4}, -\frac{2}{3}\right)\right\}$

Problem Set 11.3 (page 606)

1. $\{(-2, 5, 2)\}$ **3.** $\{(4, -1, -2)\}$ **5.** $\{(-1, 3, 5)\}$
7. Infinitely many solutions   **9.** $\varnothing$ **11.** $\left\{\left(-2, \frac{3}{2}, 1\right)\right\}$
13. $\left\{\left(\frac{1}{3}, -\frac{1}{2}, 1\right)\right\}$ **15.** $\left\{\left(\frac{2}{3}, -4, \frac{3}{4}\right)\right\}$ **17.** $\{(-2, 4, 0)\}$
19. $\left\{\left(\frac{1}{2}, \frac{1}{3}, \frac{1}{6}\right)\right\}$ **21.** 194
23. \$.70 per bottle for catsup, \$1 per jar of peanut butter, and
\$.80 per jar of pickles   **25.** -2, 6, and 16
27. 40°, 60°, and 80°
29. \$500 at 12%, \$1000 at 13%, and \$1500 at 14%

Problem Set 11.4 (page 614)

1. $\{(-2, 0), (-4, 4)\}$ **3.** $\{(2, -3), (-2, 3)\}$
5. $\{(-5, -3)\}$ **7.** $\{(-6, 7), (-2, -1)\}$
9. $\{(2, 2), (-2, -2)\}$
11. $\left\{\left(\frac{\sqrt{30}}{3}, \frac{\sqrt{21}}{3}\right), \left(\frac{\sqrt{30}}{3}, -\frac{\sqrt{21}}{3}\right), \left(-\frac{\sqrt{30}}{3}, \frac{\sqrt{21}}{3}\right),\right.$
$\left.\left(-\frac{\sqrt{30}}{3}, -\frac{\sqrt{21}}{3}\right)\right\}$ **13.** $\{(1, 1)\}$ **15.** $\{(-3, 2)\}$

17.
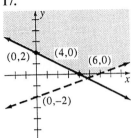

19.
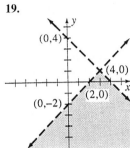

33.
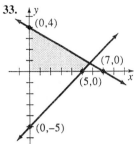

35.
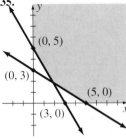

37. 5 and 3 or 5 and -3 or -5 and 3 or -5 and -3

39. 6 meters by 9 meters

43. (a) $\left|(2i, -3), (-2i, -3)\right|$

(c) $\left|\left(\dfrac{1 + i\sqrt{15}}{2}, \dfrac{-7 + i\sqrt{15}}{2}\right), \left(\dfrac{1 - i\sqrt{15}}{2}, \dfrac{-7 - i\sqrt{15}}{2}\right)\right|$

(e) $\left|\left(\dfrac{2 + i\sqrt{2}}{2}, \dfrac{2 - i\sqrt{2}}{2}\right), \left(\dfrac{2 - i\sqrt{2}}{2}, \dfrac{2 + i\sqrt{2}}{2}\right)\right|$

21.
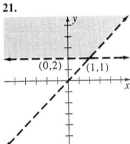

23.
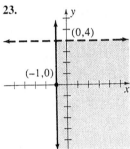

Chapter 11 Review Problem Set (page 618)

1. $\left|(2, 6)\right|$ **2.** $\left|(-3, 7)\right|$ **3.** $\left|(4, -1)\right|$

4. $\left|(-2, 6)\right|$ **5.** $\left|(8, -24)\right|$ **6.** $\varnothing$

7. $\left|(2, -3, 1)\right|$ **8.** $\left|(2, -3, -1)\right|$ **9.** $\left|(-3, 2, 5)\right|$

10. $\left|(-4, 3, 4)\right|$ **11.** Infinitely many

12. $\left|(1, 1), (-2, 7)\right|$

13. $\left|(2, \sqrt{3}), (2, -\sqrt{3}), (-2, \sqrt{3}), (-2, -\sqrt{3})\right|$

14. $\left|(2, -2), (-2, 2)\right|$ **15.** $\left|(-1, 4), (-3, 4)\right|$

25.
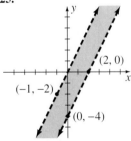

27.
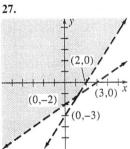

16.
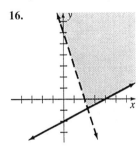

17.
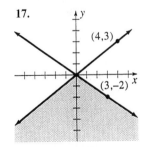

29.
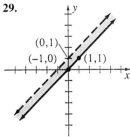

31.
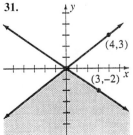

18.
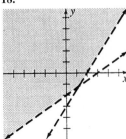

19.
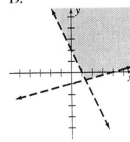

20.

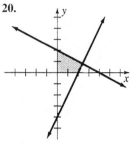

21. 2 and 3 or -3 and -2
22. 5 and 3, or 5 and -3, or -5 and 3, or -5 and -3
23. 2 and 5 or -3 and 10   **24.** 6 meters by 9 meters
25. $9 per pound for cashews and $5 per pound for Spanish peanuts
26. $1.50 for a carton of pop and $2.25 for a pound of candy
27. The fixed fee is $2 and the additional fee is $.10 per pound.

Chapter 11 Test (page 620)

1. III   **2.** I   **3.** III   **4.** II   **5.** $\{(-2, 4)\}$
6. $\{(3, -1)\}$ **7.** $x = -12$ **8.** $y = -4$
9. $x = \dfrac{1}{13}$ **10.** $y = -\dfrac{13}{11}$ **11.** $x = -2$
12. $y = 4$ **13.** $z = 2$ **14.** No
15. Infinitely many   **16.** Two   **17.** One
18. Four   **19.** ± 2 **20.** $\dfrac{-3 \pm \sqrt{39}}{10}$ **21.** 52
22. 2 liters   **23.** 22 quarters

24.

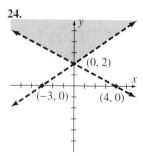

25.

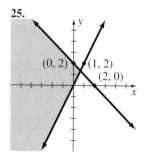

CHAPTER 12

Problem Set 12.1 (page 632)

1. $\{(-1, 3)\}$ **3.** $\{(-2, -5)\}$ **5.** $\left\{\left(\dfrac{19}{5}, -\dfrac{11}{5}\right)\right\}$
7. $\{(-1, -6)\}$ **9.** $\{(7, -2)\}$ **11.** $\{(4, 10)\}$
13. $\{(0, 3)\}$ **15.** $\{(2, 3, -1)\}$ **17.** $\{(1, -1, 2)\}$
19. $\{(-3, 0, 1)\}$ **21.** $\{(2, -3, -4)\}$
23. $\left\{\left(7, \dfrac{33}{7}, -\dfrac{2}{7}\right)\right\}$ **25.** $\{(2, 3, 4)\}$ **27.** $\{(-1, -3, 6)\}$
29. $\{(-3, -5, 0)\}$ **33.** $\{(1, 0, -2)\}$ **35.** $\{(1, 1, 1)\}$

Problem Set 12.2 (page 641)

1. Yes   **3.** Yes   **5.** Yes   **7.** No   **9.** No
11. $\{(-3, -7)\}$ **13.** $\{(2, -4)\}$ **15.** $\varnothing$
17. $\{(0, 4)\}$ **19.** $\{(1, 0, -3)\}$ **21.** $\{(-3, -4, 1)\}$
23. $\{(-2, 5, 1)\}$ **25.** $\varnothing$ **27.** $\{(-4, 0, -1)\}$
29. $\{(1, -1, 1)\}$ **31.** $\{(0, -5, 0, 4)\}$
33. $\{(2, -2k - 3, -3k + 4, k)\}$ **35.** $\varnothing$
37. $\{(7, -3, 2k + 5, k)\}$ **41.** $\{(1, 0, -2, -1)\}$
43. $\{(-1, -2, 0, 3)\}$ **45.** $\{(11k - 7, -3k + 2, k)\}$
47. $\left\{\left(-k + \dfrac{11}{7}, \dfrac{2}{3}k + \dfrac{5}{7}, k\right)\right\}$
49. $\{(x, y, z) \mid x + y - 2z = -1\}$

Problem Set 12.3 (page 648)

1. 10   **3.** -48 **5.** -29 **7.** 50   **9.** 8
11. 6   **13.** -32 **15.** -2 **17.** 5   **19.** $-\dfrac{7}{20}$
21. $\{(3, 8)\}$ **23.** $\{(-2, 5)\}$ **25.** $\{(-2, 2)\}$
27. $\varnothing$ **29.** $\{(1, 6)\}$ **31.** $\left\{\left(-\dfrac{1}{4}, \dfrac{2}{3}\right)\right\}$
33. $\left\{\left(\dfrac{17}{62}, \dfrac{4}{31}\right)\right\}$ **35.** $\{(-1, 3)\}$ **37.** $\{(9, -2)\}$
39. $\{(-4, -3)\}$
41. $9 per pound for cashews and $5 per pound for Spanish peanuts
43. The fixed fee is $2 and the additional fee is $.10 per pound.   **45.** 6 and 13
47. $700 at 8% and $500 at 9%

Problem Set 12.4 (page 657)

1. -57 **3.** 14   **5.** -41 **7.** -8 **9.** -96
11. $\{(-3, 1, -1)\}$ **13.** $\{(0, 2, -3)\}$
15. $\left\{\left(-2, \dfrac{1}{2}, -\dfrac{2}{3}\right)\right\}$ **17.** $\{(-1, -1, -1)\}$ **19.** $\varnothing$
21. $\{(-5, -2, 1)\}$ **23.** $\{(-2, 3, -4)\}$

25. $\left\{\left(-\dfrac{1}{2}, 0, \dfrac{2}{3}\right)\right\}$　　**27.** $\{(-6, 7, 1)\}$

29. $\left\{\left(1, -6, -\dfrac{1}{2}\right)\right\}$　　**33.** 0　　**35. (b)** -20

Chapter 12 Review Problem Set (page 662)

1. $\{(6, -2)\}$　　**2.** $\{(-1, -3)\}$　　**3.** $\{(4, -6)\}$

4. $\left\{\left(-\dfrac{6}{7}, -\dfrac{15}{7}\right)\right\}$　　**5.** $\{(2, -1, 4)\}$　　**6.** $\{(1, -4, -2)\}$

7. $\{(4, 2, -2)\}$　　**8.** $\{(-1, 2, -5)\}$　　**9.** $\varnothing$

10. $\{(2, -3, -1)\}$　　**11.** -34　　**12.** 13　　**13.** -22

14. -29　　**15.** -40　　**16.** 16　　**17.** 51

18. $\{(-5, 9)\}$　　**19.** $\{(0, -4)\}$　　**20.** $\{(3, -7)\}$

21. $\{(-12, 12)\}$　　**22.** $\{(-10, -7)\}$　　**23.** $\{(4, 6)\}$

24. $\{(2, -1, -2)\}$　　**25.** $\left\{\left(-\dfrac{1}{3}, -1, 4\right)\right\}$

26. $\{(-2, 2, -1)\}$　　**27.** $\{(0, -1, 2)\}$

28. $\{(2, -3, -4)\}$　　**29.** $\{(-1, 2, -5)\}$　　**30.** 72

31. \$900 at 10% and \$1600 at 12%

32. 20 nickels, 32 dimes, and 54 quarters

33. $25°$, $45°$, and $110°$

Chapter 12 Test (page 664)

1. $x = -2$　　**2.** $y = -4$　　**3.** $x = 1$　　**4.** $y = 2$

5. $x = -2$　　**6.** $y = 4$　　**7.** $z = 6$　　**8.** $x = 14$

9. $y = 13$　　**10.** Infinitely many　　**11.** None

12. 8　　**13.** 2　　**14.** 7　　**15.** $-\dfrac{7}{12}$　　**16.** -18

17. 112　　**18.** -48　　**19.** $x = -7$　　**20.** $y = 4$

21. $x = -12$　　**22.** $y = \dfrac{5}{17}$　　**23.** $x = 1$

24. $y = -1$　　**25.** $z = 0$

Cumulative Review Problem Set (page 667)

1. -6　　**2.** -8　　**3.** $\dfrac{13}{24}$　　**4.** 24　　**5.** $\dfrac{13}{6}$

6. $-90\sqrt{2}$　　**7.** $2x + 5\sqrt{x} - 12$　　**8.** $-18 + 22\sqrt{3}$

9. $2x^3 + 11x^2 - 14x + 4$　　**10.** $\dfrac{x + 4}{x(x + 5)}$

11. $\dfrac{16x^2}{27y}$　　**12.** $\dfrac{16x + 43}{90}$　　**13.** $\dfrac{35a - 44b}{60a^2b}$

14. $\dfrac{2}{x - 4}$　　**15.** $2x^2 - x - 4$　　**16.** $\dfrac{5y^2 - 3xy^2}{x^2y + 2x^2}$

17. $\dfrac{2y - 3xy}{3x + 4xy}$　　**18.** $\dfrac{(2n - 5)(n + 3)}{(n - 2)(3n + 13)}$

19. $\dfrac{3a^2 - 2a + 1}{2a - 1}$　　**20.** $(5x - 2)(4x + 3)$

21. $2(2x + 3)(4x^2 - 6x + 9)$

22. $(2x + 3)(2x - 3)(x + 2)(x - 2)$

23. $4x(3x + 2)(x - 5)$　　**24.** $(y - 6)(x + 3)$

25. $(5 - 3x)(2 + 3x)$　　**26.** $\dfrac{81}{16}$　　**27.** 4　　**28.** $-\dfrac{3}{4}$

29. $-.3$　　**30.** $\dfrac{1}{81}$　　**31.** $\dfrac{21}{16}$　　**32.** $\dfrac{9}{64}$　　**33.** 72

34. 6　　**35.** -2　　**36.** $\dfrac{-12}{x^3 y}$　　**37.** $\dfrac{8y}{x^5}$

38. $-\dfrac{a^3}{9b}$　　**39.** $4\sqrt{5}$　　**40.** $-6\sqrt{6}$　　**41.** $\dfrac{5\sqrt{3}}{9}$

42. $\dfrac{2\sqrt{3}}{3}$　　**43.** $2\sqrt[3]{7}$　　**44.** $\dfrac{\sqrt[3]{6}}{2}$　　**45.** $8xy\sqrt{13x}$

46. $\dfrac{\sqrt{6xy}}{3y}$　　**47.** $11\sqrt{6}$

48. $-\dfrac{169\sqrt{2}}{12}$

49. $-16\sqrt[3]{3}$　　**50.** $\dfrac{-3\sqrt{2} - 2\sqrt{6}}{2}$

51. $\dfrac{6\sqrt{15} - 3\sqrt{35} - 6 + \sqrt{21}}{5}$　　**52.** $.021$

53. 300　　**54.** $.0003$　　**55.** $32 + 22i$

56. $-17 + i$　　**57.** $0 - \dfrac{5}{4}i$　　**58.** $-\dfrac{19}{53} + \dfrac{40}{53}i$

59. $-\dfrac{10}{3}$　　**60.** $\dfrac{4}{7}$　　**61.** $2\sqrt{13}$

62. $5x - 4y = 19$　　**63.** $4x + 3y = -18$

64. $(-2, 6)$ and $r = 3$　　**65.** $(-5, -4)$　　**66.** 8 units

67.

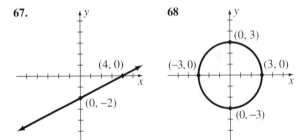

68

69.

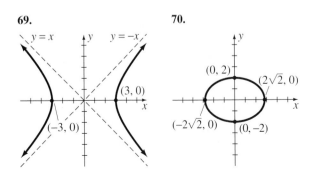

70.

71.
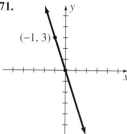

72.
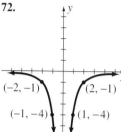

73.
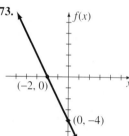

74.
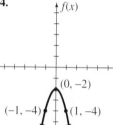

75.
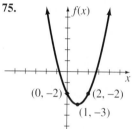

76.
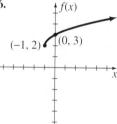

77.
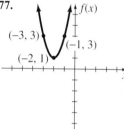

78.
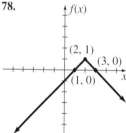

79.
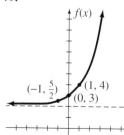

80.
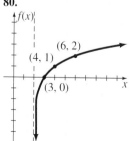

81.
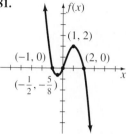

82
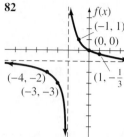

83. $(g \circ f)(x) = 2x^2 - 13x + 20; (f \circ g)(x) = 2x^2 - x - 4$

84. $f^{-1}(x) = \dfrac{x + 7}{3}$ **85.** $f^{-1}(x) = -2x + \dfrac{4}{3}$

86. $k = -3$ **87.** $y = 1$ **88.** 12 cubic centimeters

89. -40 **90.** 40   **91.** $\left\{-\dfrac{21}{16}\right\}$ **92.** $\left\{\dfrac{40}{3}\right\}$

93. $\{6\}$ **94.** $\left\{-\dfrac{5}{2}, 3\right\}$ **95.** $\left\{0, \dfrac{7}{3}\right\}$ **96.** $\{-6, 0, 6\}$

97. $\left\{-\dfrac{5}{6}, \dfrac{2}{5}\right\}$ **98.** $\left\{-3, 0, \dfrac{3}{2}\right\}$ **99.** $\{\pm 1, \pm 3i\}$

100. $\{-5, 7\}$ **101.** $\{-29, 0\}$ **102.** $\left\{\dfrac{7}{2}\right\}$

103. $\{12\}$ **104.** $\{-3\}$ **105.** $\left\{\dfrac{1 \pm 3\sqrt{5}}{3}\right\}$

106. $\left\{\dfrac{-5 \pm 4i\sqrt{2}}{2}\right\}$ **107.** $\left\{\dfrac{3 \pm i\sqrt{23}}{4}\right\}$

108. $\left\{\dfrac{3 \pm \sqrt{3}}{3}\right\}$ **109.** $\{1 \pm \sqrt{34}\}$

110. $\left\{\pm\dfrac{\sqrt{5}}{2}, \pm\dfrac{\sqrt{3}}{3}\right\}$ **111.** $\left\{\dfrac{-5 \pm i\sqrt{15}}{4}\right\}$

112. $\{-4, 1, 7\}$ **113.** $\left\{-\dfrac{1}{2}, \dfrac{5}{3}, 2\right\}$ **114.** $\left\{\dfrac{3}{2}\right\}$

115. $\{81\}$ **116.** $\{4\}$ **117.** $\{6\}$ **118.** $\left\{\dfrac{1}{5}\right\}$

119. $(-\infty, 3)$ **120.** $(-\infty, 50)$

121. $\left(-\infty, -\dfrac{11}{5}\right) \cup (3, \infty)$ **122.** $\left(-\dfrac{5}{3}, 1\right)$

123. $\left[-\dfrac{9}{11}, \infty\right)$ **124.** $[-4, 2]$

125. $\left(-\infty, \dfrac{1}{3}\right) \cup (4, \infty)$ **126.** $(-8, 3)$

127. $(-\infty, 3] \cup (7, \infty)$ **128.** $(-6, -3)$

129. $\{(9, 6)\}$ **130.** $\left\{\left(\dfrac{9}{5}, -\dfrac{7}{25}\right)\right\}$ **131.** $\left\{\left(\dfrac{70}{23}, \dfrac{36}{23}\right)\right\}$

132. $\{(-2, 4, 0)\}$

133. $\{(-1, 4, -1)\}$ **134.** 17, 19, and 21

135. 14 nickels, 20 dimes, and 29 quarters

136. $48°$ and $132°$ **137.** \$600

138. \$1700 at 8% and \$2000 at 9%

139. 66 miles per hour and 76 miles per hour

140. 4 quarts   **141.** 69 or less   **142.** $-3, 0$, or 3
143. 1 inch strip   **144.** \$1050 and \$1400
145. 3 hours   **146.** 30 shares at \$10 per share
147. 37   **148.** $10°, 60°$, and $110°$

CHAPTER 13

Problem Set 13.1 (page 678)

1. $-1, 2, 5, 8, 11$ **3.** $3, 1, -1, -3, -5$
5. $-1, 2, 7, 14, 23$ **7.** $0, -3, -8, -15, -24$
9. $-1, 5, 15, 29, 47$ **11.** $\frac{1}{2}, 1, 2, 4, 8$
13. $-\frac{2}{9}, -\frac{2}{3}, -2, -6, -18$
15. $a_8 = 54$ and $a_{12} = 130$ **17.** $a_7 = -32$ and $a_8 = 64$
19. $a_n = 2n - 1$ **21.** $a_n = 4n - 6$
23. $a_n = -2n + 7$ **25.** $a_n = -3n - 4$
27. $a_n = \frac{1}{2}(n + 1)$ **29.** 34   **31.** 78   **33.** -155
35. 106   **37.** 50   **39.** 170   **41.** -1
43. -13 **45.** 136   **47.** \$27,800   **49.** \$1170
53. $3, 3, 7, 7, 11, 11$ **55.** $4, 7, 10, 13, 17, 21$
57. $4, 12, 36, 108, 324, 972$ **59.** $1, 1, 2, 3, 5, 8$
61. $3, 1, 4, 9, 25, 256$

Problem Set 13.2 (page 683)

1. 2550   **3.** 9030   **5.** -4225 **7.** 410
9. 8850   **11.** 5724   **13.** 5070   **15.** 346.5
17. 3775   **19.** 39,500   **21.** -8645
23. 122,850   **25.** 39,552   **27.** 76 seats; 720 seats
29. \$10,000   **31.** \$357,500   **33.** 169
37. (a) $7 + 12 + 17; 36$
(c) $(-3) + (-5) + (-7) + (-9) + (-11) + (-13); -48$
(e) $3 + 6 + 9 + 12 + 15; 45$

Problem Set 13.3 (page 690)

1. $a_n = 3^{n-1}$ **3.** $a_n = 2(4)^{n-1}$
5. $a_n = \left(\frac{1}{3}\right)^{n-1}$ or $a_n = \frac{1}{3^{n-1}}$
7. $a_n = .2(.2)^{n-1}$ or $a_n = (.2)^n$
9. $a_n = 9\left(\frac{2}{3}\right)^{n-1}$ or $a_n = (3^{-n+3})(2)^{n-1}$ **11.** $a_n = (-4)^{n-1}$
13. 19,683   **15.** -512 **17.** $-\frac{6561}{256}$
19. 14,762   **21.** -9842 **23.** 1093   **25.** 511
27. 19,680   **29.** 2730   **31.** 1093   **33.** $1\frac{1023}{1024}$

35. $15\frac{31}{32}$ **37.** $\frac{1}{2}$ **39.** $0; -1$ **41.** 125 liters
43. \$102.40; \$204.75   **45.** $971.\overline{3}$ meters
47. \$5609.66   **51.** 1,048,576   **53.** $\frac{4096}{531,441}$
55. $-\frac{177,147}{2048}$ **57. (a)** 126   **(c)** 31   **(e)** $1\frac{49}{81}$

Problem Set 13.4 (page 695)

1. 4   **3.** 1   **5.** 2   **7.** $\frac{2}{3}$ **9.** 9
11. No sum   **13.** $\frac{4}{7}$ **15.** $\frac{16}{3}$ **17.** No sum
19. $\frac{81}{2}$ **21.** $\frac{4}{9}$ **23.** $\frac{47}{99}$ **25.** $\frac{5}{11}$ **27.** $\frac{427}{999}$
29. $\frac{7}{15}$ **31.** $\frac{24}{11}$ **33.** $\frac{47}{110}$

Problem Set 13.5 (page 699)

1. $x^8 + 8x^7y + 28x^6y^2 + 56x^5y^3 + 70x^4y^4 + 56x^3y^5 + 28x^2y^6 + 8xy^7 + y^8$
3. $81x^4 + 108x^3y + 54x^2y^2 + 12xy^3 + y^4$
5. $x^5 - 5x^4y + 10x^3y^2 - 10x^2y^3 + 5xy^4 - y^5$
7. $x^{10} + 10x^9y + 45x^8y^2 + 120x^7y^3 + 210x^6y^4 + 252x^5y^5 + 210x^4y^6 + 120x^3y^7 + 45x^2y^8 + 10xy^9 + y^{10}$
9. $64x^6 + 192x^5y + 240x^4y^2 + 160x^3y^3 + 60x^2y^4 + 12xy^5 + y^6$
11. $x^5 - 15x^4y + 90x^3y^2 - 270x^2y^3 + 405xy^4 - 243y^5$
13. $243a^5 - 810a^4b + 1080a^3b^2 - 720a^2b^3 + 240ab^4 - 32b^5$
15. $x^6 + 6x^5y^3 + 15x^4y^6 + 20x^3y^9 + 15x^2y^{12} + 6xy^{15} + y^{18}$
17. $x^7 + 14x^6 + 84x^5 + 280x^4 + 560x^3 + 672x^2 + 448x + 128$
19. $x^4 - 12x^3 + 54x^2 - 108x + 81$
21. $x^{15} + 15x^{14}y + 105x^{13}y^2 + 455x^{12}y^3$
23. $a^{13} - 26a^{12}b + 312a^{11}b^2 - 2288a^{10}b^3$
25. $462x^5y^6$ **27.** $-160x^3y^3$ **29.** $2000x^3y^2$

Chapter 13 Review Problem Set (page 701)

1. $a_n = 6n - 3$ **2.** $a_n = 3^{n-2}$ **3.** $a_n = 5 \cdot 2^n$
4. $a_n = -3n + 8$ **5.** $a_n = 2n - 7$ **6.** $a_n = 3^{3-n}$
7. $a_n = -(-2)^{n-1}$ **8.** $a_n = 3n + 9$ **9.** $a_n = \frac{n+1}{3}$
10. $a_n = 4^{n-1}$ **11.** 73   **12.** 106   **13.** $\frac{1}{32}$
14. $\frac{4}{9}$ **15.** -92 **16.** $\frac{1}{16}$ **17.** -5 **18.** 85
19. $\frac{5}{9}$ **20.** 2 or -2 **21.** $121\frac{40}{81}$ **22.** 7035

23. $-10,725$　　**24.** $31\frac{31}{32}$　　**25.** $32,015$　　**26.** 4757

27. $85\frac{21}{64}$　　**28.** $37,044$　　**29.** $12,726$　　**30.** $85\frac{1}{3}$

31. $\frac{4}{11}$　　**32.** $\frac{41}{90}$　　**33.** \$750　　**34.** \$46.50

35. \$3276.70　　**36.** $10,935$ gallons　　**37.** 3600 feet

38. $128x^7 + 448x^6y + 672x^5y^2 + 560x^4y^3 + 280x^3y^4 + 84x^2y^5 + 14xy^6 + y^7$

39. $x^4 - 12x^3y + 54x^2y^2 - 108xy^3 + 81y^4$

40. $1120x^4y^4$

Chapter 13 Test (page 703)

1. -46　　**2.** 48　　**3.** $a_n = 4n - 7$

4. $a_n = 5(2)^{1-n}$　　**5.** $a_n = -3n + 9$

6. $\frac{729}{8}$ or $91\frac{1}{8}$　　**7.** 223　　**8.** 60 terms　　**9.** 46

10. 2380　　**11.** 765　　**12.** -780　　**13.** 3279

14. 7155　　**15.** 6138　　**16.** $22,650$　　**17.** 9384

18. \$5810　　**19.** \$3276.70　　**20.** $13\frac{1}{2}$　　**21.** $\frac{37}{99}$

22. $\frac{4}{15}$

23. $x^5 - 15x^4y + 90x^3y^2 - 270x^2y^3 + 405xy^4 - 243y^5$

24. $128x^7 + 448x^6y + 672x^5y^2$　　**25.** $495a^8b^4$

CHAPTER 14

Problem Set 14.1 (page 710)

1. 20　　**3.** 24　　**5.** 168　　**7.** 48　　**9.** 36

11. 6840　　**13.** 720　　**15.** 720　　**17.** 36

19. 24　　**21.** 243　　**23.** Impossible　　**25.** 216

27. 26　　**29.** 36　　**31.** 144　　**33.** 1024

35. 30　　**37. (a)** $6,084,000$　　**(c)** $3,066,336$

Problem Set 14.2 (page 718)

1. 60　　**3.** 360　　**5.** 21　　**7.** 252　　**9.** 105

11. 1　　**13.** 24　　**15.** 84　　**17. (a)** 336

19. 2880　　**21.** 2450　　**23.** 10　　**25.** 10

27. 35　　**29.** 1260　　**31.** 2520　　**33.** 15

35. 126　　**37.** $144; 202$　　**39.** $15; 10$　　**41.** 20

43. $10; 15; 21; \frac{n(n-1)}{2}$　　**47.** 120

53. $133,784,560$　　**55.** $54,627,300$

Problem Set 14.3 (page 724)

1. $\frac{1}{2}$　　**3.** $\frac{3}{4}$　　**5.** $\frac{1}{8}$　　**7.** $\frac{7}{8}$　　**9.** $\frac{1}{16}$　　**11.** $\frac{3}{8}$

13. $\frac{1}{3}$　　**15.** $\frac{1}{2}$　　**17.** $\frac{5}{36}$　　**19.** $\frac{1}{6}$　　**21.** $\frac{11}{36}$

23. $\frac{1}{4}$　　**25.** $\frac{1}{2}$　　**27.** $\frac{1}{25}$　　**29.** $\frac{9}{25}$　　**31.** $\frac{2}{5}$

33. $\frac{9}{10}$　　**35.** $\frac{5}{14}$　　**37.** $\frac{15}{28}$　　**39.** $\frac{7}{15}$　　**41.** $\frac{1}{15}$

43. $\frac{2}{3}$　　**45.** $\frac{1}{5}$　　**47.** $\frac{1}{63}$　　**49.** $\frac{1}{2}$　　**51.** $\frac{5}{11}$

53. $\frac{1}{6}$　　**55.** $\frac{21}{128}$　　**57.** $\frac{13}{16}$　　**59.** $\frac{1}{21}$

63. $\frac{40}{2,598,960} = \frac{1}{64,974}$　　**65.** $\frac{3744}{2,598,960} = \frac{6}{4165}$

67. $\frac{10,200}{2,598,960} = \frac{5}{1274}$　　**69.** $\frac{123,552}{2,598,960} = \frac{198}{4165}$

71. $\frac{1,302,540}{2,598,960} = \frac{1277}{2548}$

Problem Set 14.4 (page 731)

1. (a) $\frac{5}{36}$　　**(c)** $\frac{7}{12}$　　**3.** $\frac{5}{6}$　　**5. (a)** $\frac{1}{32}$　　**(c)** $\frac{31}{32}$

7. $\frac{12}{13}$　　**9.** $\frac{7}{12}$　　**11.** $\frac{37}{44}$　　**13. (a)** $\frac{2}{11}$　　**(c)** $\frac{9}{11}$

15. (a) $\frac{5}{68}$　　**(c)** $\frac{15}{34}$　　**17. (a)** $\frac{4}{21}$　　**(c)** $\frac{11}{21}$

21. (a) 1 to 3　　**(c)** 5 to 1　　**(e)** 5 to 4

Chapter 14 Review Problem Set (page 735)

1. 720　　**2.** $30,240$　　**3.** 150　　**4.** 1440　　**5.** 20

6. 525　　**7.** 1287　　**8.** 264　　**9.** 74　　**10.** 55

11. 40　　**12.** 15　　**13.** 60　　**14.** 120　　**15.** $\frac{3}{8}$

16. $\frac{5}{16}$　　**17.** $\frac{5}{36}$　　**18.** $\frac{13}{18}$　　**19.** $\frac{3}{5}$　　**20.** $\frac{1}{35}$

21. $\frac{57}{64}$　　**22.** $\frac{1}{221}$　　**23.** $\frac{1}{6}$　　**24.** $\frac{4}{7}$　　**25.** $\frac{4}{7}$

26. $\frac{10}{21}$　　**27.** $\frac{140}{143}$　　**28.** $\frac{105}{169}$　　**29.** $\frac{1}{6}$　　**30.** $\frac{28}{55}$

31. $\frac{5}{7}$　　**32.** $\frac{1}{16}$

Chapter 14 Test (page 737)

1. 12　　**2.** 120　　**3.** 240　　**4.** 64　　**5.** 5

6. 48　　**7.** 26　　**8.** 840　　**9.** 28　　**10.** 20

11. 270　　**12.** 350　　**13.** $\frac{1}{12}$　　**14.** $\frac{13}{18}$

15. $\frac{3}{32}$　　**16.** $\frac{31}{32}$　　**17.** $\frac{5}{6}$　　**18.** $\frac{1}{7}$　　**19.** $\frac{2}{5}$

20. $\frac{5}{12}$　　**21.** $\frac{9}{14}$　　**22.** $\frac{3}{4}$　　**23.** $\frac{168}{361}$　　**24.** $\frac{2}{21}$

25. $\frac{19}{26}$

INDEX

area A
perimeter P
length l

width w
surface area S
altitude (height) h

base b
circumference C
radius r

volume V
area of base B
slant height s

Rectangle

$A = lw \qquad P = 2l + 2w$

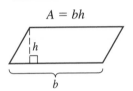

Parallelogram

$A = bh$

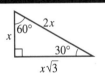

30°–60° Right Triangle

Right Circular Cylinder

$V = \pi r^2 h \qquad S = 2\pi r^2 + 2\pi rh$

Triangle

$A = \frac{1}{2}bh$

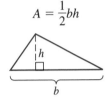

Trapezoid

$A = \frac{1}{2}h(b_1 + b_2)$

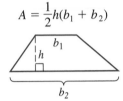

Right Triangle

$a^2 + b^2 = c^2$

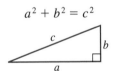

Sphere

$S = 4\pi r^2 \qquad V = \frac{4}{3}\pi r^3$

Square

$A = s^2 \qquad P = 4s$

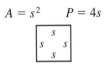

Circle

$A = \pi r^2 \qquad C = 2\pi r$

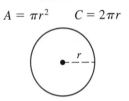

Isosceles Right Triangle

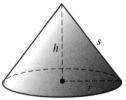

Right Circular Cone

$V = \frac{1}{3}\pi r^2 h \qquad S = \pi r^2 + \pi rs$

Pyramid

$V = \frac{1}{3}Bh$

Prism

$V = Bh$

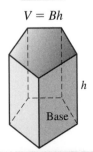